RADAR DATA PROCESSING WITH APPLICATIONS

RADAR DATA PROCESSING WITH APPLICATIONS

He You
Xiu Jianjuan
Guan Xin

電子工業出版社
PUBLISHING HOUSE OF ELECTRONICS INDUSTRY

WILEY

This edition first published 2016

Published by John Wiley & Sons Singapore Pte. Ltd., 1 Fusionopolis Walk, #07-01 Solaris South Tower, Singapore 138628, under exclusive license granted by [copub] for all media and languages excluding Simplified and Traditional Chinese and throughout the world excluding Mainland China, and with non-exclusive license for electronic versions in Mainland China.

For details of our global editorial offices, for customer services and for information about how to apply for permission to reuse the copyright material in this book please see our website at www.wiley.com.

Library of Congress Cataloging-in-Publication Data

Names: You, He, 1956– author. | Jianjuan, Xiu, 1971– author. | Xin, Guan, 1978– author.
Title: Radar data processing with applications / He You, Xiu Jianjuan, Guan Xin.
Description: Singapore : John Wiley & Sons, Inc., [2016] | Includes bibliographical references and index.
Identifiers: LCCN 2016010486 | ISBN 9781118956861 (cloth) | ISBN 9781118956885 (epub) | ISBN 9781118956892 (Adobe PDF)
Subjects: LCSH: Radar–Mathematics. | Radar–Data processing.
Classification: LCC TK6578 .H425 2016 | DDC 621.38480285–dc23
LC record available at https://lccn.loc.gov/2016010486

Set in 10/12pt Times by SPi Global, Pondicherry, India

1 2016

Contents

About the Authors

He You (Fellow of IET, Academician of Chinese Academy of Engineering) was born in Jilin Province, People's Republic of China, in 1956. He received a Ph.D. from the Department of Electronic Engineering of Tsinghua University, Beijing, People's Republic of China, in 1997. Then, he won a National Outstanding Doctoral Dissertation Award in 2000.

From October 1991 to November 1992, he was with the Institute of Communication at the Technical University of Braunschweig, Germany, as a senior visiting scholar researching automatic radar detection theory and constant false alarm rate processing. In December 1994, he joined the Naval Aeronautical and Astronautical University (NAAU) in Yantai, People's Republic of China, as a Professor in the Department of Electronic Engineering, and engaged in research on signal processing and information fusion with applications. Now, he is Director of the Ministerial-level Laboratory of Information Sensing and Fusion Technology, and Chief of the Information Fusion Research Institute, NAUU, as well. He has acquired over 20 authorized National Invention Patents, and is co-author of more than 200 peer-reviewed technical articles. Moreover, he is first author of the books *Radar Target Detection and CFAR Processing* (Tsinghua University Press, 1st edition 1999 and 2nd edition 2011), *Multi-sensor Information Fusion with Applications* (Publishing House of Electronics Industry, 1st edition 2000 and 2nd edition 2007), *Radar Data Processing with Application* (Publishing House of Electronics Industry, 1st edition 2006, 2nd edition 2009, 3rd edition 2013, and translated into English), *Information Fusion Theory with Application* (Publishing House of Electronics Industry, 2010). These published books and papers have been cited more than 7000 times by other scholars. He has provided leadership in many research projects and gained significant scientific research achievements including three 2nd Awards of National Science and Technology Progress, one 1st Award and one 2nd Award of National Teaching Achievements. He is leader of the Information Fusion Team, which has been ranked as Ministerial-level Innovation Team of Science and Technology, as well as Excellence Innovation Team of Shandong Province.

Dr. He has served as a Fellow of IET, Committee Member of National "863" Experts, Member of National Radar Detection Technology Group, Standing Director of Chinese Society of Aeronautics and Astronautics (CSAA), Founder and Chairman of Information Fusion Branch in CSAA, Standing Director of Chinese Society of Command and Control, Fellow of Chinese Institute of Electronics (CIE), Vice Chairman of Radio Positioning Technology Branch in CIE, and so on. He is a Vice Chairman on the Editorial Boards of *Ship Electronic Engineering*, *Radar Science and Technology*, and *Fire Control & Command Control*. He has also been appointed a Member on the Editorial Boards of *Chinese Journal of Aeronautics* (in English), *Acta Aeronautica et Astronautica Sinica*, *Signal Processing (China)*, *Journal of Radars*, *Journal of Data Acquisition & Processing*, and so on.

Xiu Jianjuan was born in Shandong Province, People's Republic of China, in April 1971. She received her master and Ph.D. degree from Naval Aeronautical and Astronautical University, Yantai, People's Republic of China, in 2000 and 2004. She is now a professor in Research Institute of Information Fusion of Naval Aeronautical and Astronautical University. His research interests include radar data processing and passive location.

Guan Xin was born in Jinzhou, Liaoning province, People's Republic of China, in 1978. She received her bachelor degree in communication engineering from Liaoning University, Liaoning, People's Republic of China, in 1999 and received master and PhD degree from Naval Aeronautical and Astronautical University in 2002 and 2006, respectively. She is now a professor and doctor tutor in Department of Electronics and Communication of Naval Aeronautical and Astronautical University. She also serves as a director of Chinese institute of command and control and a senior member of Chinese Aviation Society. She is major in target identification and evidence reasoning, and has published over 80 papers and 3 academic monographs.

Preface

Advances in radar technology and application demands have promoted the fast development of radar signal processing and data processing technology. In recent years, with the continual emergence of new types of radar, significant progress has been made in related hardware, algorithms, and computer performance, and the signal processing capacity has been constantly improved, which demands the application of new algorithms in related radar data processing equipment to implement the simultaneous processing of multiple targets in the cluttered environments and allow the data association and tracking of multiple targets and information fusion of multiple radars in complex environments. That is why we decided to publish *Radar Data Processing with Applications*.

This book begins with the basic linear and nonlinear filtering approaches, and introduces the development and latest research findings on radar data processing technology thoroughly and systematically. Its main contents are as follows.

1. The initial discussion deals with the static and dynamic parameter estimation for linear and nonlinear discrete-time systems, providing such classical filtering algorithms as the Kalman filter, the extended Kalman filter, the unscented Kalman filter, and the particle filter.
2. Measurement preprocessing techniques are discussed, including time and space registration, radar error correction, and data compression.
3. Such practical issues as multi-target track initiation, data association, and tracking are introduced, of which multi-target data association is divided into the maximum likelihood and Bayesian approach. Maneuvering target tracking, group target tracking, and track termination are also discussed.
4. The final analysis is the practical application of radar data processing, including passive radar data processing, pulse Doppler radar data processing, phased array radar data processing, radar network error registration, radar network data processing, radar data processing performance evaluation, and simulation techniques.

Acknowledgments

The authors gratefully acknowledge the contributions of those colleagues from the English Department who have been involved in the translation work: Associate Prof. Chen Li, Associate Prof.

Yang An-liang, Associate Prof. Liu Hong-ying, Lecturer Liu Hui, Lecturer Qu Lei, Lecturer Wang Xue-sheng, Lecturer Xu Xiao-juan, Lecturer Zhang Dong-li, Lecturer Zhu Zi-jian, and Lecturer Guan Hui-jie. The authors would like to express their appreciation to Dr. Dong Kai, Dr. Wang Hai-peng, Dr. Cui Ya-qi, and postgraduates Miao Xu-bin, Wang Wang-song, and Sun Shun for their participation in proofreading and revision. Special thanks go to the Electronic Industry Publishing House, especially to Editor Qu Xin, for support in the publication of this book.

It is expected that the publication of this book will not only provide a very readable reference for those engaged in information engineering, pattern recognition, military command, etc., but also lay a theoretical foundation for their work and further study.

Any advice and suggestions from readers of this book are most welcome.

1

Introduction

1.1 Aim and Significance of Radar Data Processing

Generally, a modern radar system consists of two important components: a signal processor and a data processor. The signal processor is used for target detection (i.e., the suppression of undesirable signals produced by ground or sea surface clutter, meteorological factors, radio frequency interference, noise sources, and man-made interference) [1–3]. When the video output signal, after signal processing and constant false alarm rate (CFAR) detection fusion, exceeds a certain detection threshold, it can be determined that a target has been discovered. Then, the discovered target signal will be transmitted to the data recording device, where the space position, amplitude value, radial velocity, and other characteristic parameters of the target are recorded, usually by computers. The measurement output from the data recording device needs to be processed in the data processor, which associates, tracks, filters, smooths, and predicts the obtained measurement data – such as the target position (radial distance, azimuth, and pitch angle) and the motion parameters [4–6] – for the effective suppression of random errors occurring during the measurement, estimation of the trajectory and related motion parameters (velocity and acceleration, etc.) of the target in the control area, prediction of the target's position at the next moment, and formation of a steady target track, so that highly accurate real-time tracking is realized [7–9].

In terms of the level at which radar echo signals are processed, radar signal processing is usually viewed as the primary processing of the information detected by the radar unit. It is done at each radar station, with information obtained from the same radar and the same scanning period and distance unit, with the aim of extracting useful target information from clutter, noise, and various active and passive jamming backgrounds. Radar data processing is usually viewed as secondary processing of the radar information [10–13]. Making use of information from the same radar, but with different scanning periods and distance units, it can be done both at each independent radar station and at the information processing center or system command center of the radar network. Data fusion of multiple radars can be viewed as a third or tertiary processing of the radar information, which is usually done at the information processing center. Specifically, the information the processing

Radar Data Processing with Applications, First Edition. He You, Xiu Jianjuan, and Guan Xin.

center receives is the measurement from the primary processing or the track from the secondary processing (usually called the local track) by multiple radars, and the track after fusion (called the global track or system track). The function of the secondary processing of radar information, based on the primary processing, is to filter and track several targets, and estimate the targets' motion parameters and characteristic parameters. Secondary processing is done strictly after primary processing, while there is no strict time limit between secondary and tertiary processing. The third level of processing is the expansion and extension of secondary processing, which is mainly reflected in space and dimension.

1.2 Basic Concepts in Radar Data Processing

The input to the radar data processing unit is the measurement from the front, which is the object of data processing, while the output is the track formed after data processing is conducted. Generally, functional modules of radar data processing include measurement pretreatment, track initiation and termination, and data association and tracking. A wave gate must be set up between the association and the tracking process, and their relationship is shown in the block diagram in Figure 1.1. The content and related concepts of the functional modules of radar data processing are briefly discussed as follows.

1.2.1 Measurements

Measurements, also called observations, refer to noise-corrupted observations related to the state of a target [14]. The measurements are not usually raw data points, but the output from the data recording device after signal processing. Measurements can be divided, according to whether they are associated with the known target track, into free measurements and correlated measurements. Free measurements are spots that are not correlated with the known target track, while correlated measurements are spots that are correlated with the known target track.

1.2.2 Measurement Preprocessing

Although modern radar adopts many signal processing technologies, there will always be a small proportion of clutter/interference signals left out. To relieve the computers doing the follow-up

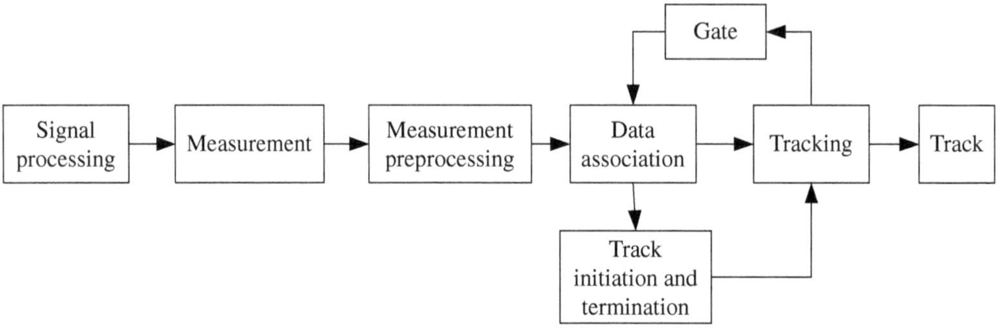

Figure 1.1 Radar data processing relation diagram

processing job from a heavy burden, prevent computers from saturation, and improve system performance, the measurement given by the primary processing needs to be preprocessed, which is called "measurement preprocessing": the preprocessing of secondary processing of radar information. The preprocessing is a precondition of correct processing of radar data, since an effective measurement data processing method can actually help yield twice the result with half the effort, with the target tracking accuracy improved while the computational complexity of the target tracking is reduced. The measurement preprocessing technology mainly involves system error registration, time synchronization, space alignment, outlier rejection, and saturation prevention.

1.2.2.1 System Error Registration

The measurement data from radars contains two types of error. One is random error, resulting from the interior noise of the measurement system. Random error may vary with each measurement, and may be eliminated to some extent by increasing the frequency of measurement and minimizing its variance in the statistical sense by means of methods like filtering. The other is system error, resulting from measurement environments, antennas, servo systems, and such non-calibration factors in the data correction process as the position error of radar stations and the zero deviation of altimeters. System error is complex, slowly varying, and non-random, and can be viewed as an unknown variable in a relatively long period of time. As indicated by the findings in Ref. [15], when the ratio of system errors to random errors is greater than or equal to 1, the effect of distributed track fusion and centralized measurement fusion deteriorates markedly, and at this point system errors must be corrected.

1.2.2.2 Time Synchronization

Owing to the possible difference in each radar's power-on time and sampling rate, the target measurement data recorded by data recording devices may be asynchronous. Therefore, these observation data must be synchronized in multiple-radar data processing. Usually, the sampling moment of a radar is set as the benchmark for the time of other radars.

1.2.2.3 Space Alignment

Space alignment is the process of unifying the coordinate origin, coordinate axis direction, etc. of the data from the radar stations in different places, so as to bring the measurement data from several radars into a unified reference framework, paving the way for the follow-up radar data processing.

1.2.2.4 Outlier Rejection

Outlier rejection is the process of removing the obviously abnormal values from radar measurement data.

1.2.2.5 Saturation Prevention

Saturation prevention mainly deals with saturation in the following two cases.

1. In the design of a data processing system, there is a limit to the number of target data. However, in a real system, saturation occurs when the data to be processed exceed the processing capacity.

2. The time used to process data is limited. Saturation occurs when the number of measurements, or batches of targets, reaches a certain extent. In this case, the processing of the data from one observation has to be interrupted before the processor starts to deal with the next batch of data.

1.2.3 Data Association

In the single-target, clutter-free environment, where there is only one measurement in the target-related wave gate, only tracking is involved. Under multi-target circumstances, where a single measurement falls in the intersection area of several wave gates or several measurements fall in the related wave gate of a single target, data association is involved. For instance, suppose two target tracks have been established before the radar's *n*th scanning, and two echoes are detected in the *n*th scanning, are the echoes from two new targets or from the two established tracks at that time? If they are from the two established tracks at that time, then in what way can the echoes resulting from the two scans and the two tracks be correctly paired? The answer involves data association, the establishment of the relationship between the radar measurements at a given moment and the measurements (or tracks) at other moments, to check whether these measurements originate from the processing of the same target (or to ensure a correct process of measurement-and-track pairing).

Data association, also called "data correlation" or "measurement correlation," is a crucial issue in radar data processing. False data association could pair the target with a false velocity, which could result in the collision of aircraft with air traffic control radars, or the loss of target interception with military radars. Data association is realized through related wave gates, which exclude the true measurements of other targets and the false measurements of noise and interference.

Generally, data association can be categorized, according to what is being associated with what, into the following classes [16]:

1. measurement-to-measurement (track initiation);
2. measurement-to-track (track maintenance or track updating);
3. track-to-track, also called track correlation (track fusion).

1.2.4 Wave Gate

In the process of target track initiation and tracking, a wave gate is often used to solve data association problems. What then is a wave gate? How many categories is it divided into? A brief discussion of these questions follows.

An initial wave gate is a domain centering on free measurements, used to determine the region where the target's observations may occur. At the track initiation stage, the initial wave gate is normally bigger for better target acquisition.

A correlation wave gate (or tracking wave gate, validation gate) is a domain centering on the predicted position of the tracked target, used to determine the region where the target's observations may occur [17].

The size of the wave gate is related to the magnitude of radar measurement error, the probability of correct echo reception, etc. That is to say, when deciding the wave gate's shape and size, one should make it highly probable that the true measurement falls in the wave gate, while making sure that there are not many unrelated measurements in the correlation wave gate. The echo falling in the correlation wave gate is called a candidate echo. The size of the tracking gate reflects the error in the predicted target position and velocity, which is related to the tracking method, radar measurement error, and required correct correlation rate. The size of the correlation wave gate is not fixed in the tracking process, but adaptive adjustment should be made among small, medium, and large wave gates in accordance with the tracking conditions.

1. For a target in uniform rectilinear motion (e.g., a civil airliner flying smoothly at high altitude), a small wave gate should be set up, with its minimum size no less than three times the mean square root value of the measurement error.
2. When the target maneuver is relatively small (e.g., when the aircraft is taking off, landing, or making a slow turn), a medium wave gate should be set up, by adding one or two times the mean square root value of the measurement error to the small wave gate.
3. When the target maneuver is relatively big (e.g., when the aircraft is making a fast turn, or when the target is lost and recaptured), a large wave gate should be set up. Besides, at the track initiation stage, a large wave gate should be adopted to effectively capture the target's initial wave gate.

1.2.5 Track Initiation and Termination

Track initiation refers to the process from the entrance (and detection) of a target into the radar coverage area to the establishment of the target track. Target initiation is important in radar data processing. If the track initiation is incorrect, target tracking is impossible.

Since the target being tracked may escape the surveillance zone at any time, once it goes beyond the radar detection range, the tracker must make relevant decisions to eliminate the unwanted track files for track termination.

1.2.6 Tracking

Tracking is one of the two primary issues in radar data processing. It refers to the processing of the target's measurements for the constant estimation of the target's current state [16]. The multiple-radar and multi-target tracking system is a highly complex large-scale system, whose complexity is mainly due to the uncertainty in radar data processing.

1. From the perspective of measurement data, the received radar measurements form a random sequence, which may be obtained by non-equal interval sampling, and the observation noises are non-Gaussian. This should be considered in real measurement data processing.
2. From the perspective of multi-target tracking, the complexity of the tracking problem lies mainly in:
   a. the uncertainty of measurement origin – since there are multiple targets and false alarms, many measurements may be produced in radar environments, which will lead to the uncertainty of the measurements used for filtering;
   b. the uncertainty of the target model parameter – since targets could be on maneuvers at any time, the model parameter initially set could be incorrect. Therefore, adjustments must be made to the model parameter in accordance with the tracking conditions; hence maneuvering target tracking.
3. From the perspective of the system, the tracking system could be nonlinear, with a complex construction. On the one hand, the system tracking performance under complex circumstances depends chiefly on the filtering algorithm's capability to deal with the uncertainty of measurement origins and target model parameters, or its capability to effectively solve the problem of measurement correlation and adaptive target tracking. On the other hand, the nonlinear characteristics of the system itself should also be taken into consideration.

For the effective tracking of the target under these complex circumstances, the following two problems need to be solved.

First, the establishment of the target motion model and the observation model. Estimation theory, which provides a foundation for radar data processing, requires the establishment of a system model describing the dynamic characteristics of target and radar measurement processes. A valuable method of describing the system model, the state variable method, is based on the system state equation and the observation equation. According to this method, the state variable, system state equation, system observation equation, system noise and observation noise, system input and output (i.e., the estimated value of the state variable) are the five essential elements of the target tracking system modeling. The five elements above reflect the basic characteristics of a system, and can be viewed as a complete expression of a dynamic system. The introduction of the state variable is the core of creating an optimum control and estimation theory, because in the state space, the state variable defined should be a batch of variables with minimum dimensions that can fully reflect the system dynamic characteristics. The state variable at any given time is expressed as a function of the state variable prior to that time, and the input/output relationship of the system is described by the state transition model and the output observation model in the time domain. The state reflects the system's "interior condition." The input can be described by the state equation, which is composed of the decided time function and the random process representing the unpredictable variable or noise. The output is a function of the state vector, usually disturbed by the random observation error, and can be described by measurement equations. In the system modeling process, the use of the system state equation and the observation equation in the description of the dynamic characteristics of the target is therefore the most successful method in common use. The relation between the state equation and the measurement equation is shown in Figure 1.2.

Second, the tracking algorithm. The tracking filtering algorithm in the state space is actually a matter of optimum estimation based on state space. The following two points are of major concern.

1. Multiple maneuvering target tracking. Maneuvers are both the basic attribute of the target and the forms of motion commonly used in attacks or escapes. Therefore, maneuvering multi-target tracking is the focus of target tracking, dealing with the problem of a maneuvering target model, testing and tracking algorithm.
2. The optimality, robustness, and rapidity of tracking algorithms. That is to say, an overall consideration is needed of the tracking timeliness, tracking accuracy, and robustness of the algorithm.

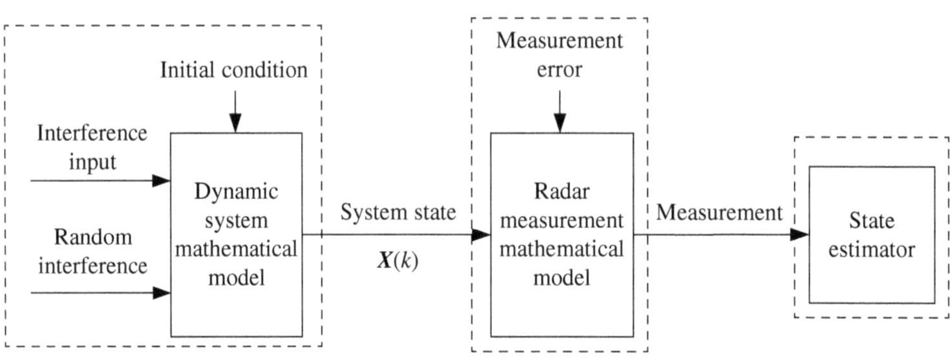

Figure 1.2 Filtering diagram

1.2.7 Track

A track is a trajectory which is formed with the states of a target estimated from a set of measurements of the same target (i.e., tracking trajectory). The radar, when conducting multi-target data processing, designates an identity (ID) for each tracking trajectory, namely the track ID, which serves as a point of reference for all the parameters related to a given track. The measurement of the track's reliability can be described by the track quality which, if properly controlled, can help both promptly and accurately initiate a track so that a new target file is set up, and cancel a track so that the redundant target files are cleared up. Tracks are the ultimate result of data processing, as shown in Figure 1.3.

The concepts related to tracks also include the following.

1. *Possible track.* The possible track is a track composed of a single measurement point.
2. *Tentative track.* Tentative tracks are tracks composed of two or more measurement points with low track quality. They could be target tracks, or random interference, namely false tracks. After initial correlation is complete, a possible track is turned into a tentative track or a canceled track. The tentative track is also called a temporary track.
3. *Confirmed track.* A confirmed track, also called a reliable track or a stable track, is a track with stable output or a track whose track quality exceeds a given value. It is the formal track set up by the data processor, and is generally considered as a true target track.
4. *Fixed track.* A fixed track is a track composed of clutter measurements, whose position does not change much with the scans of a radar set.

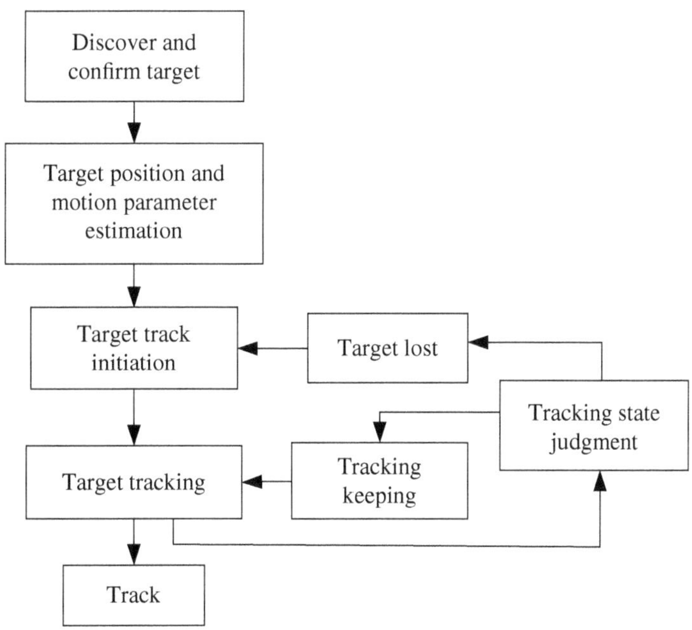

Figure 1.3 Data processing flowchart

The following sequence can be determined in the correlation process of measurements and tracks: fixed tracks first, then reliable tracks, and finally tentative tracks. That is to say, after a batch of observation measurements is obtained, the correlation of these measurements and the fixed track is done first. The measurements that can be correlated with the fixed track are deleted from the measurement file and are used to update the fixed track (i.e., to replace the old clutter points with the measurements that are correlated). If these measurements cannot be correlated with the fixed track, they should be correlated with the existing confirmed track. The successfully correlated measurements are used to update the confirmed track. The measurements that cannot be correlated with the confirmed track should be correlated with the tentative track, which finally either disappears or is turned into a confirmed track or a fixed track. The confirmed track has priority over the tentative track, which excludes the possibility that the tentative track obtains measurements from the reliable track.

5. *Canceled track.* When its quality is lower than a given value or is composed of isolated random interference points, the track is called a canceled track, and the process is called track cancellation or track termination. Track cancellation is the process of erasing the track when it does not conform to a certain rule, which means the track is not a track of a true target, or that the corresponding target has moved out of the radar coverage range. Specifically, when a certain track cannot be correlated with any measurement in a certain scan, an extrapolation should be done according to the latest velocity. Any track that does not receive a measurement in a certain number of successive scans should be canceled. The primary task of track cancellation is to promptly cancel a false track with the true one being retained.

   There are three possible instances of track cancellation.

   i. Possible tracks (with only track heads) to be canceled as long as there is no measurement in the first scanning period that follows them.
   ii. Tentative tracks (such as a newly initiated track) to be erased from the database as long as there is no measurement in the three successive scanning periods that follow them.
   iii. Confirmed tracks, whose cancellation should be done with caution. If no measurement falls in the relevant wave gates in four to six successive scanning periods, cancellation of the track can be considered. It is worth noting that extrapolation must be used several times to expand the wave gates to recapture the lost target. Of course, track quality management can also be used to cancel a track.

6. *Redundant tracks.* Two or more tracks being allocated to the same true target is called track redundancy. The unnecessary track is called a redundant track.
7. *Track interruption.* If a certain track is allocated to a true target at time t, but no track is allocated to the target at time $t+m$, then track interruption happens at time t, where m is a parameter set by the tester, usually $m=1$.
8. *Track switch.* If a certain track is allocated to a true target at time t, while another track is allocated to the target at time $t+m$, then track switch happens at time t, where m is a parameter set by the tester, usually $m=1$.
9. *Track life* (the length of a track; the times the track is successively correlated). Based on whether the terminated track is false or true, it can be divided into [18, 19]:
   a. *False track life.* The average times of radar scanning from the initiation of a false track to its deletion is called false track life. False track can sometimes last for a long time when false measurements are highly dense.

b. *True track life*. The average times of radar scanning of a true track mistaken for a false one and deleted after it is initiated.

True track maintenance time is restricted by two factors:

1. The measurement track correlation error (the true measurement is measured but is correlated with other tracks, which commonly occurs in dense target environments or crossed target environments) could lower the quality of a true track, or even result in the deletion of a true track mistaken for a false one.
2. The times that measurements are successively lost reach a given threshold, so the track is deleted as a lost target, which commonly happens when the signal-to-noise ratio is low or there is strong interference.

1.3 Design Requirements and Main Technical Indexes of Radar Data Processors

1.3.1 Basic Tasks of Data Processors

As can be seen from the discussion and elaboration of the relevant basic concepts in radar data processing, the basic tasks of data processors include:

a. measurement pretreatment;
b. determination of the correlation area and correlation principle, and the distinction between true and false measurements;
c. the establishment of new tracks;
d. the correlation of measurements and existing tracks, track maintenance;
e. the correlation between and fusion of tracks;
f. track termination and track management, including quality grade determination and track quality management;
g. situation display, including the display of tracks and measurements.

1.3.2 The Engineering Design of Data Processors

The engineering design of data processors is a comprehensive design. Generally, the following three issues need to be considered.

First, the balanced relationship between tracking accuracy, robustness, and real-time performance. Target tracking algorithms are mostly obtained when the probability distribution function of the system noise and measurement noise is subject to certain assumptions, and usually the assumed system noise and observation noise are both Gaussian white noise. However, in real systems it is hardly possible to find a matrix that accords completely with Gaussian distribution because the mutation of the electromagnetic environment, the immaturity and failure of the observation equipment, etc. can result in the deviation of observations from the Gaussian distribution. When the system's actual noise distribution deviates from the assumed noise distribution, tracking algorithms can effectively exclude the interference of the uncertainty factors and abnormal values in the system, and consequently ensure that there is not much change in the estimation effect and the estimation accuracy. Simply put, the tracking algorithms can ensure the robustness of estimation algorithms in this case, so that the system can operate normally. This is robust tracking (estimation). In other words, a relatively "loose" assumption of the noise distribution mode is allowed, which may not be the optimum one for a certain specific distribution mode, but can exclude the interference of the abnormal values and help improve the anti-interference ability of the system.

Basically, research on the robust estimation theory aims to find estimation algorithms that can both exclude or resist the influence of the abnormal value (cases) and basically possess the good characteristics of traditional estimation algorithms (i.e., algorithms that incorporate considerations of optimality and robustness of estimation in a balanced manner). What optimality emphasizes is an algorithm that makes the system index function reach its minimum (or maximum), while what robustness focuses on is an algorithm that sacrifices some indices of the system to improve its anti-interference performance. Therefore, an optimal balance between robustness and optimality is what needs to be taken into consideration in the whole process of robust tracking system design. Some efficiency has to be sacrificed to robustness [10].

Common problems in the balance between tracking accuracy, robustness, and real-time performance are:

1. Excessive emphasis is put on the tracking accuracy index, while the robustness index is neglected. As a result, the accuracy of the target tracking result is high at the simulation stage, but declines markedly at the actual engineering test stage, which reduces the algorithm's engineering value.
2. Too idealized an index design results in complexity of the algorithm structure, which badly affects its real-time performance.

As for engineering algorithms, the index of robustness is the first priority, followed by the tracking accuracy and the real-time index. However, in an engineered index design, the three indexes mentioned above are the basic technical indexes on which compromises must be made.

The second issue is one of reliability. An algorithm that is simple in structure, highly reliable, easy to realize, and mature in engineering should be used in the engineering design of radar data processing. Otherwise, the system cannot operate normally and continuously. Meanwhile, the design of the software system data processor needs to be modularized, visible, and revisable.

The third issue is that of intelligence information processing. Although the function modules contained in data processors are basically the same, different radars have different requirements for the data processor design. For example, the core of the skywave over-the-horizon radar is the ionosphere mathematical model. Specifically, the echo multipath resulting from the multipath structure of the ionosphere, and the severe attenuation of the echo signal resulting from the severe shortwave environment noise and ionosphere transmission characteristic can result in a higher probability of false alarms and missed alarms in radar measurements, leading to discontinuity of the track. However, the striking problem with the groundwave over-the-horizon radar is the rejection of false tracks and the maintenance of stable tracks. Therefore, in the design of data processors, an analysis of the data processor's characteristics should be made first according to the system's index requirements for data processors, including observation characteristics such as the measurements' temporal and spatial distribution characteristics, noise distribution and statistical characteristics, the variation of the signal-to-noise ratio, the intensiveness of the targets, etc. Besides, the system's resolution, probability of detector false alarms and discovery, accumulated time and coordinate system, etc. are also included in the analysis, to provide a basis for the assignment of data processor indexes and the emphasis of the design.

1.3.3 The Main Technical Indexes of Data Processors

The main technical indexes of data processors are as follows.

1. *Immediacy.* If the adopted tracking algorithm is too complex and takes too long a time to process the data, it is possible that the second batch of data will come before the processing of the first batch of data is complete, resulting in saturation of data processing. As a consequence, the processing effect and the immediacy of the situation display may be affected, so that the situation display cannot reflect the current target position information accurately.
2. *Tracking capacity.* The tracking capacity is the largest number of targets that the data processor can track simultaneously. The index becomes increasingly demanding with increasing intensiveness of targets, the complexity of the environment in which the sensors work, and the processing speed of the hardware system. Meanwhile, due to factors like undetected data, there could be discontinuous target tracks, so that one target track could be mistaken for several target tracks and assigned different target numbers, which increases the system tracking capacity.
3. *Probability of true target loss and false targets.* These are two mutually restricted crucial indexes. To ensure the initial probability of the true track, a large correlated wave gate must be built. This, on the one hand, makes it more probable that the true target will fall in the wave gate but, on the other hand, increases the number of other unrelated measurements falling in the wave gate, which is bad for the reduction in false track probability because the initiation of true targets is ensured at the expense of initiating a large number of false targets. Conversely, if the probability of false tracks is to be lowered, a small wave gate should be built; as a consequence, true targets may not fall in the wave gate, which could result in a loss of true targets. This requires a reasonable wave gate design, employing different principles according to the different emphases on the two indexes in engineering, or different detecting areas. In a specific system the test of this index is closely related to that of the detector index, requiring an overall consideration of the detector and data processor index [10].
4. *Tracking accuracy.* Tracking accuracy is a key index of the data processor. It depends mainly on the measurement accuracy of the detector, the data correlation, and the filtering algorithm adopted.

1.3.4 The Evaluation of Data Processors

The performance evaluation of the data processor mainly includes the following four aspects.

1. *Data association.* This is a comparatively complicated evaluation index. Data correlations are normally evaluated using the data under various circumstances – such as the existence of outliers, dense target environments, cross-target environments (see Chapter 17, Figure 17.1), target approaching and leaving (see Chapter 17, Figure 17.2), maneuvering multi-target environments, etc. – and by calculating indexes like the target's correct correlation probability, false correlation probability, missed correlation probability, etc.
2. *Tracking batches.* This reflects directly the tracking capacity and the processing capability of the system.
3. *The accuracy of the tracking filter.* The balance between indexes – including tracking accuracy, immediacy, and robustness (anti-interference ability) – should be considered comprehensively.
4. *Immediacy.* Actual measurement data should be used to test the processing speed of the data processor.

The evaluation of data processing is crucial to a radar system, because the test of many of its indexes – like coverage range, system resolution, tracking batches, tracking accuracy, target classification, and the estimation of threat – is ultimately determined by the evaluation of data processors. Related information will be discussed especially in Chapter 17.

1.4 History and Present Situation of Research in Radar Data Processing Technology

The earliest radar data processing method was the least-squares algorithm put forward by Gauss in 1795. Gauss used this method to predict Kamiya's orbit for the first time, and opened up the scientific field in which mathematical methods were used to process observation and experimental data. Despite its faults, such as neglect of the statistical characteristics of the observation data, the algorithm has its merits in that it is comparatively simple in calculation. Therefore, it is still a widely used estimation method, from which some forms suitable for real-time operation have developed through the generations. This method is used when accurate system dynamic errors and statistical characteristics of the observation data cannot be acquired.

The maximum likelihood method, proposed by R. A. Fisher in 1912, deals with the estimation problem from the perspective of probability density, and has made an important contribution to estimation theory. The estimation of a random process was not developed until the 1930s, while modern filtering theory was based on probability theory and random process theory. In 1940, American scholar N. Wiener, one of the originators of control theory, put forward a method to design statistical filters in frequency domains according to the requirement for fire control – the famous Wiener filtering. Since its proposal, the method has been used in the fields of communication, radar, and control, with great success. During the same period, former Soviet Union scholar Kolmogorov proposed and for the first time solved the problem of the prediction and extrapolation of the discrete stationary random sequence. The Wiener filtering, together with the Kolmogorov filtering method, opened up a new field in which the statistical method was used to deal with the random control problem, and established a foothold for the research and development of modern filter theories.

The Wiener filter, which adopts the frequency domain design method, is difficult in analysis and solution and complicated in operation [20]. What's more, the batch processing method it adopts demands large storage space. Consequently, its application is quite limited and it is only applicable to one-dimensional stationary random signal filtering. This defect in the Wiener filter forced people to seek other optimal filter design methods. An important contribution was made in this field by American scholar R. E. Kalman, who proposed the discrete-time system Kalman filter in 1960. In 1961, he worked with S. S. Bucy in extending this filtering theory to continuous-time systems [21], and formulated a complete theory of Kalman filter estimation.

The Kalman filter introduces the method for analysis of state variables to filtering theory, and obtains the time-domain solution of the minimum mean square error estimation problem. Moreover, the Kalman filter theory, which has broken through the limitations of the Wiener filter, can be used in non-stationary and multi-variable linear time-varying systems. With a recursive structure, the Kalman filter is more suitable for computer computation, requires lower computational complexity and smaller data memory, and has stronger real-time performance. It is because of its advantages over the other filtering methods mentioned above that the Kalman filter found practical engineering applications once proposed [22, 23]. The Apollo lunar landing program and the design of the C-5A aircraft navigation system were the most successful examples of its early engineering applications. Because of the Kalman filter's wide application and simple design method, steady-state gain filtering was proposed on its basis to further lower the computational complexity [24, 25]. At present, the Kalman filter theory, as one of the most important optimum estimation theories, is widely used in various fields, such as target tracking, inertial guidance, GPS, air traffic control, fault diagnosis, etc. In the over 200-year history of filtering theory, Gauss, Wiener, and Kalman have made important contributions, laying theoretical foundations for radar data processing.

Since the filtering theory initiated by Kalman is only applicable to linear systems and requires that the observation equation should be linear, in the following 10 years Bucy, Sunahara, and coworkers were committed to research on the extension of the Kalman filtering theory to nonlinear systems and observations, and proposed a filtering method applicable to nonlinear systems – the extended Kalman filter [16, 25]. Then, successively, in the early 1970s Singer *et al.* proposed a series of maneuvering target tracking methods [26], and in the mid-1970s Pearson, Shibata, and coworkers successfully applied Kalman filtering technology to the airborne radar tracking system [27]. The traditional Kalman filtering theory is based on the precondition that the model is accurate and the statistical characteristics of random interference signals are known. However, in an actual system, sometimes the model is inaccurate, and/or the statistical characteristics of interference signals are not completely known, which could greatly lower the traditional Kalman filter's estimation accuracy and result in filtering divergence in severe cases. That is why some scholars introduced the idea of robust control to filtering theory, producing the robust filtering theory [28].

The increasingly complex application environment in recent years requires that radars be capable of tracking several targets simultaneously. The concept of multi-target tracking was advanced by Wax in an article published in *Applied Physics* in 1955 [29]. Then, in 1964, the article "The association of optimum data in monitoring theory," published by Sittler in *IEEE Transactions on Military Electronics*, became the pioneering work of multi-target tracking [30]. However, since the Kalman filter was not widely used at that time, he adopted the track splitting algorithm [16]. In the early 1970s, the Kalman filtering method began to be used systematically for multi-target processing in the case of false alarms [31]. The nearest-neighbor algorithm proposed by Singer in 1971 is the simplest method of solving data association problems [32], but this method has a low association rate in clutter environments. In this period, Y. Bar-Shalom played an important role, proposing in 1975 the probabilistic data association algorithm, which is especially applicable to single-target tracking in clutter environments [33]; on its basis, Fortmann, Bar-Shalom, and coworkers put forward the joint probabilistic data association algorithm (JPDA) to effectively solve the problem of multi-target tracking in clutter environments [34]. Based on Bar-Shalom's poly concept, in 1979, Reid proposed using the multiple hypothesis method to solve the problem of multi-target tracking [35].

With the development of science and technology in recent years, targets have to make maneuvers to avoid being tracked and attacked. Therefore, since 1970, Singer, Bar-Shalom, Birmiwal, and coworkers have successively proposed tracking maneuvering targets with the Singer algorithm, variable dimension filtering algorithm, interacting multiple model algorithm, etc. [32, 36–39]. In 1986, S. S. Blackman *et al.* started to do research on the group target tracking issue. In 1988, Carlson put forward the federated filter [40], aimed at providing a theoretical basis for the design of the fault tolerance combined navigation system [41]. In order to effectively solve the filter problem in nonlinear systems, Julier *et al.* put forward unscented KF (UKF) [42], which takes samples of the estimated vector's probability density function (PDF) so as to decide its mean value and covariance, and acquires an estimation accuracy which is better than the first-order EKF algorithm and has the same magnitude as the second-order EKF algorithm. In Ref. [43] the particle filter (PF) algorithm is proposed, which is close to the UKF algorithm in performance except that it has a higher computational complexity. The PF has also been used in research tracking before detecting, etc. in recent years.

With further study being carried out on various aspects of the radar data processing technology, large numbers of treatises [5, 44–51], academic papers [52–56], and research reports [18, 23, 57] have emerged. Now data processing technology has been transformed from initial single-radar to multiple-radar, and from multiple-radar to multiple-sensor, with the emergence of a large number of treatises and papers on multiple-sensor information fusion [15, 58–64].

In the radar data processing field, many scholars and outstanding experts have made rewarding contributions, including Professor Bar-Shalom of the University of Connecticut, USA, who, since the end of the 1980s, has successively published many highly theoretical and systematical treatises on multi-target tracking with his students, originating many new theories and methods, especially in aspects of data association and multi-target/multi-sensor tracking data fusion. Their research features clear concepts, rigorous deduction, and strong theoretical dimensions. Another example is S. S. Blackman, an expert with Air America, whose research is characterized by its higher practicability, or stronger relevance, to actual engineering applications. Still another is Professor Farina of Naples University, one of the earliest scholars in radar data processing research.

1.5 Scope and Outline of the Book

Whether in modern defense or air and marine traffic control systems, multi-target tracking is an indispensable technology. Especially with developments in the "informatization" and networking of modern warfare, multi-target tracking technology is coming to the fore in all countries, as an active research field. For example, for air traffic control centers, the management of the aircraft in air and terminal areas, approach management, collision warning, and collision avoidance, etc. cannot be realized without a target tracking system, which requires the system to detect and track the aircraft, and accurately determine position, heading, and speed parameters, thus improving the safety of air traffic and the utilization of resources.

This book absorbs the latest developments in the field of radar data processing in recent years, aimed at providing people of the same profession with a foundation for further theoretical research and practical application. The main content and chapters are as follows.

Chapter 1: Introduction

This chapter discusses many basic concepts in radar data processing. Some of the practical issues addressed include engineering design requirements, principal technical indicators, and assessment of radar data processors.

Chapter 2: Parameter Estimation

Starting with the basic concept of time-constant parameter estimation, the chapter discusses some estimator properties like unbiasedness, variance of estimators, consistency and efficiency of estimators, etc. on the basis of the introduction of several frequently used time-constant parameter estimation techniques, such as maximum a posteriori (MAP), maximum likelihood (ML), minimum mean squared error (MMSE), and least squares (LS) estimators. Finally, the chapter analyzes the estimation of non-time-varying vectors, and discusses the LS, MMSE, and LMMSE estimators under vector circumstances.

Chapter 3: Linear Filtering Approaches

On the basis of the introduction of the measurement state and measurement equations for a Kalman filter, including the constant velocity model, constant acceleration model, and coordinate turn model, this chapter discusses the relevant filter models and the initiation of the Kalman filter. Finally, the chapter studies the steady-state Kalman filter, including a mathematical definition and judgment of stable filters, controllability and observability of random linear systems, etc.

Chapter 4: Nonlinear Filtering Approaches

This chapter discusses the nonlinear filtering approaches in radar data processing, including extended Kalman filter (EKF), unscented Kalman filter (UKF), and particle filter (PF), giving the filtering model of each approach. After a simulation analysis of two linear filter algorithms (Kalman filter and unbiased converted measurement Kalman filter), as well as two nonlinear filter algorithms (extended Kalman filter and unscented Kalman filter) in the same simulation environment on the condition that the posterior PDF of the system state is a Gauss hypothesis, a comparison of the tracking accuracy and computational complexity of these methods is made, and relevant conclusions drawn. The chapter also makes a simulation analysis using three nonlinear filter algorithms (extended Kalman filter, unscented Kalman filter, and particle filter), tracking the same target in the same simulation environment, compares the tracking accuracy and computational complexity of these approaches, and makes a comprehensive evaluation of the advantages and disadvantages of each approach.

Chapter 5: Measurement Preprocessing Techniques

This chapter deals with measurement processing. In a process where several sensors are used to track targets, in order to improve the tracking accuracy, it is necessary to fuse the information of several targets, while the primary problem to be solved in the fusion of multiple-sensor information is the synchronization of different sensors in time and space. The chapter first analyzes and discusses two issues: the time registration method; selection and transformation of the coordinate system. Since the selection of the coordinate system is closely related to practical application, and can directly influence the tracking effect of the whole system, the chapter starts with a discussion of some commonly used coordinate systems, and then studies some coordinate transformation techniques, to ensure that all the data information formats can be united in the same coordinate system. Finally, the chapter analyzes the problem of data compression to minimize the computational load and improve the track effect.

Chapter 6: Track Initiation in Multi-target Tracking

On the basis of the analysis of the shape, dimensions, and varieties of initial wave gates and correlation wave gates in track initiation, this chapter studies track initiation techniques in multi-target tracking, including target-oriented sequential processing techniques and batch processing techniques. Usually the sequential processing technique applies to target track initiation in clutter-free environments, and the target track initiates more quickly, while the batch processing technique is quite effective when applied to the initiation of the target track in strong clutter environments, which, however, is at the expense of increased computational complexity, and needs multiple scans to effectively initiate a track. Finally, a comparative analysis is made of the effects of several commonly used track initiation algorithms in the same simulation environment, including a logic-based method, modified logic-based method, Hough transformation method, and modified Hough transformation method, and relevant conclusions are drawn.

Chapter 7: Maximum Likelihood Class Multi-target Data Association Methods

This chapter mainly discusses the maximum likelihood class association methods, including the track splitting method, united maximum likelihood algorithm, 0–1 integer programming algorithm, and generalized correlation algorithm. The main feature of the maximum likelihood class filter algorithm is that it makes judgments on the basis of the likelihood ratio of the observation sequence, and does not create a probability that the sequence is correct. Specifically, the track splitting method makes use of likelihood functions to conduct pruning, excluding the measurement sequences that are unlikely to come from the target. The united maximum likelihood algorithm calculates the likelihood functions of different feasible partitions of all measurement sequences, and when the likelihood function reaches its maximum, the measurement sequence with feasible partition is considered the correct sequence from different targets. The principle of the 0–1 integer programming algorithm is similar to that of the united maximum likelihood algorithm, and it is further deduced from the united maximum likelihood algorithm. The generalized correlation algorithm defines a score function, which is used to initiate, confirm, and cancel tracks.

Chapter 8: Bayesian Multi-target Data Association Approach

This chapter mainly discusses the Bayesian association approach, which is concerned with studies on the latest determined measurement sets, including the nearest-neighbor algorithm, probabilistic data association algorithm (PDA), integrated probabilistic data association algorithm (IPDA), joint probabilistic data association algorithm (JPDA), etc. In the JPDA section of this chapter, a very simple and practical method of determining matrix separation is introduced, another merit of which is that errors are not likely to occur. Finally, the chapter compares and analyzes the track performance, consumed time, error tracking rate, etc. of various algorithms through simulation experiments.

Chapter 9: Tracking Maneuvering Targets

This chapter mainly discusses the tracking method of maneuvering targets. Generally, maneuvering target tracking methods can be divided into two classes: tracking algorithms with maneuvering detection capability (including the white noise model with adjustable level, variable dimension filtering algorithm, etc.) and adaptive tracking algorithms (including the modified input estimation algorithm, Singer model algorithm, the current model and its modified algorithm, Jerk model algorithm, multi-model algorithm, and interactive multi-model algorithm, etc.). The chapter discusses two typical maneuvering target tracking algorithms, makes a simulation analysis and comparison of the above two classes of methods through simulation examples, and draws conclusions.

Chapter 10: Group Target Tracking

This chapter mainly discusses the issue of group tracking. Because of problems typical of group tracking itself, the development and research in this area falls behind other techniques. The chapter starts with a discussion of the initiation of a group, and discusses several typical group initiation algorithms, including the definition, separation, and correlation of the group, and the estimation of the speed of the group. On this basis, it discusses refined track initiation of the targets in a group in a cluttered environment, proposes a group target refined track initiation algorithm based on the gray theory, and makes simulation verification and analysis. Besides, the chapter investigates centered group tracking, and analyzes and discusses such aspects as the tracking updating, merging, and splitting of a group. In order to further solve the tracking problems of targets in a group, the chapter also studies the formation group tracking algorithm. Finally, an overall simulation analysis and summary is made of the group tracking algorithm.

Chapter 11: Multi-target Track Termination Theory and Track Management

This chapter starts with research on a multi-target tracking termination technique, discussing the relevant algorithms based on "nearest neighbor," including the sequence probabilistic ratio test (SPRT) algorithm, tracking gate method, cost function method, Bayesian algorithm, and all-neighbors Bayesian algorithm, followed by a comparative analysis and relevant conclusions of the termination moment and false termination rate of the above-mentioned algorithms in the same simulation environment. The second part of the chapter relates to track ID management in the track management technique, including the management of single-track IDs, storage of track data, and management of double-track IDs. This part also discusses track quality management and track file management in information fusion systems, and analyzes the selection of initiation principles and the cancellation of tracks using track quality, as well as the management of track quality under single-station and multi-station circumstances.

Chapter 12: Passive Radar Data Processing

This chapter first discusses the space correlation of passive radar measurements, including passive location and tracking using the phase changing rate algorithm and Doppler shift changing rate multiple-model algorithm. The chapter also analyzes and discusses optimal deployment based on the area of the minimum concentration ellipse principle for passive sensors, as well as passive location using time difference of arrival, etc.

Chapter 13: Pulse Doppler Radar Data Processing

On the basis of the introduction of the basic characteristics of pulse Doppler (PD) radars, this chapter discusses the retrieval of radar data in single-target tracking and multi-target tracking systems. On this basis, a study is done on several typical tracking algorithms of PD radars, including optimal distance–velocity coupled tracking, radar target tracking with Doppler measurements, etc. The radar target tracking algorithm with Doppler measurements focuses on the unbiased sequential extended Kalman filter algorithm, unbiased sequential unscented Kalman filter algorithm, unscented Kalman filter algorithm with Doppler measurements, and unscented Kalman filter algorithm of maneuvering targets. A comparative analysis is made of several algorithms with Doppler measurements respectively in two simulation environments, and relevant conclusions are drawn.

Chapter 14: Phased Array Radar Data Processing

This chapter starts with an analysis and discussion of the phased array radar's main indexes and features, and on this basis investigates the system structure and work process of phased radars, and provides relevant system structure block diagrams and phased radar flowcharts. In the phased radar data processing part, research is carried out on multi-target processing, variable sampling interval filtering, and resource scheduling strategies on the basis of the discussion of tracking filtering methods. With regard to variable sampling interval filtering, the chapter analyzes and discusses adaptive sampling with steady-state gain filters, adaptive sampling based on the interactive multiple model, adaptive sampling based on the forecast error covariance threshold, and adaptive sampling with sampling intervals defined in advance. Finally, the chapter presents a simulation analysis and comparison of the performance of phased array radar track algorithms.

Chapter 15: Radar Network Error Registration Algorithm

This chapter starts with a discussion of the make-up of system errors and their influence, and in particular analyzes the influence of large range-finding system errors on tracks. Large range-finding system errors can result in the affine transformation of target tracks, as well as their shift and rotation, distorting the whole track, while azimuth-finding system errors can enlarge the target track shift only slightly, having a very small influence on the target track. On this basis, the chapter studies fixed radar error registration algorithms, including the RTQC error registration algorithm, LS error registration algorithm, GLS error registration algorithm, accurate maximum likelihood registration algorithm, and ECEF error registration algorithm. The chapter also deals with research into maneuvering radar error registration algorithms, particularly the maneuvering radar system modeling method, maneuvering radar registration algorithm with target locations known, MLRM algorithm, and ASR algorithm. The chapter ends with a simulation analysis and discussion of the performance of the above-mentioned algorithms.

Chapter 16: Radar Network Data Processing

On the basis of the introduction of performance evaluation indexes of radar networks, this chapter investigates data processing of a single-base radar, double-base radar, and multi-base radar network. Finally, the chapter studies the track correlation technique in radar network data processing, and focuses on the sequential track association algorithm in the case of multiple local nodes based on statistics.

Chapter 17: Evaluation of Radar Data Processing Performance

Radar data processing performance depends on various factors, which means that many factors are involved in the evaluation of radar data processing performance. This chapter mainly discusses the indexes of the evaluation of radar data processing performance in terms of average track initiation time, accumulative number of track interruptions, track ambiguity, accumulative number of track switches, track accuracy, maneuvering target tracking capability, false track ratio, divergence, track capacity, radar network detection probability, response time, etc. Finally, the chapter studies some evaluation methods of radar data processing performance, such as the Monte Carlo, semi-physical simulative evaluation, and testing methods.

Chapter 18: Radar Data Processing Simulation Technology

On the basis of the introduction of basic knowledge on system simulation, this chapter analyzes the methods of creating a uniform distribution, normal distribution, and arbitrarily distributed random numbers, as well as the simulation of the target motion model and the simulation of the observation process under different target motion circumstances. Finally, the chapter gives simulation examples of radar data processing algorithms, to help readers better understand the system simulation technology and radar data processing technology, and combine the two technologies in analyzing and solving practical problems in radar data processing.

Chapter 19: Practical Application of Radar Data Processing

This chapter discusses some typical uses of radar data processing technology in practical applications, including air traffic control systems, shipboard navigation radar, clutter suppression of shipboard radar, ground laser radar, marine surveillance systems, fleet aerial defense systems, airborne early warning radar, aircraft warning radar networks, phased array radar, etc. In practical applications, the use of radar data processing technology to estimate the track of a target and predict its future location is not the ultimate purpose of a radar system. Users should make use of the information to make judgments and take actions which meet the specific requirements.

Chapter 20: Review, Suggestions, and Outlook

This chapter provides a review of the main theoretical research achievements in the book, and some suggestions on key problems in radar data processing technology. Finally, prospects are given for research directions and development trends of radar data processing technology.

2

Parameter Estimation

2.1 Introduction

This chapter mainly discusses the estimation of time-invariant parameters. Specifically, it covers the basic concepts of time-invariant parameter estimation; four time-invariant parameter estimation techniques – maximum a posteriori (MAP), maximum likelihood (ML), minimum mean square error (MMSE), and least squares (LS) estimators; properties of estimators; estimation of static (time-invariant) vectors; and the linear minimum mean square error (LMMSE) estimator. Topics concerning the time-varying parameter estimation will be discussed in Chapter 3.

2.2 The Concept of Parameter Estimation

Estimation theory is concerned with estimating the value of unknown parameters from a set of observations related to the unknown parameters [14]. The concept of parameter estimation can be illustrated by the following example of the estimation of a parameter x.

Given the measurements of the parameter x

$$z(j) = h[j, x, w(j)], \quad j = 1, 2, \ldots, k \tag{2.1}$$

made in the presence of random noise $w(j)$, where j represents a discrete time, find a function of k observations

$$\hat{x}(k) = \hat{x}\left[k, Z^k\right] \tag{2.2}$$

that is an estimation of the parameter x, where Z^k is the set of cumulative measurements up to the time k.

Radar Data Processing with Applications, First Edition. He You, Xiu Jianjuan, and Guan Xin.

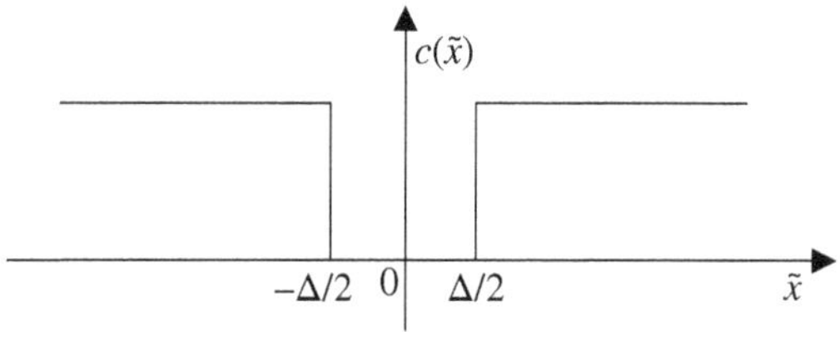

Figure 2.1 Uniform cost function

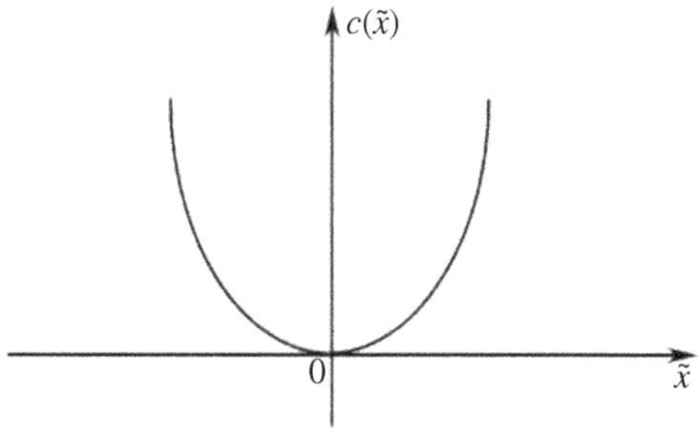

Figure 2.2 Squared error cost function

When estimating a value, the effect of noise may lead to errors, and a cost will be paid in estimating errors. The cost can be described with a cost function, $c(x,\hat{x})$. The cost function, also named "risk function," is a function of true values and estimated values. For a single-parameter estimation, the cost function is usually supposed to be a function of the estimation error $\tilde{x}=x-\hat{x}(z)$, that is, $c(x,\hat{x})=c(x-\hat{x})$. When the estimated parameter is a scalar value, three typical functions are given.

1. The uniform cost function, that is,

$$c(x,\hat{x})=\begin{cases}1, & |x-\hat{x}|\geq\dfrac{\Delta}{2}\\ 0, & |x-\hat{x}|<\dfrac{\Delta}{2}\end{cases}\tag{2.3}$$

   When $\Delta\to 0$ the estimate is very close to the true value and the cost is 0, otherwise the cost is 1 (see Figure 2.1). The MAP estimate is based on a uniform cost function.

2. The squared error cost function, that is,

$$c(x,\hat{x})=(x-\hat{x})^2\tag{2.4}$$

   Here the cost function grows rapidly with increasing error, as shown in Figure 2.2. The squared error cost function is most widely used because of its convenient mathematical processing. For example, it is the basis of the MMSE estimate.

3. The absolute value of error cost function, that is,

$$c(x,\hat{x})=|x-\hat{x}|\tag{2.5}$$

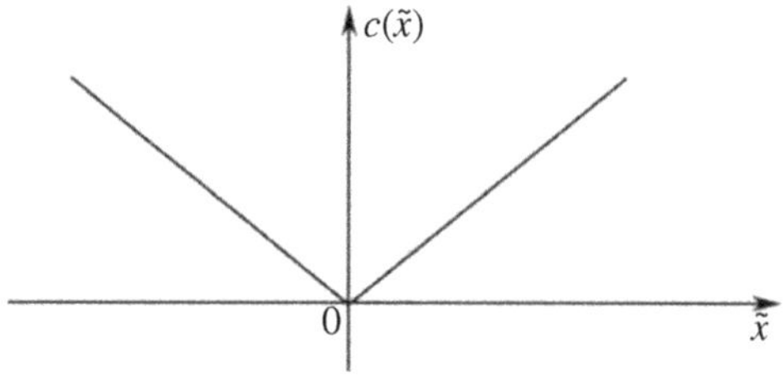

Figure 2.3 Absolute value of error cost function

Here a linear variation of the cost is found with the absolute value of errors, as shown in Figure 2.3. The conditional median estimate can be obtained from this cost function.

When the estimated parameter is an N-dimensional vector, three typical cost functions are given.

1. The uniform cost function, that is,

$$c(\boldsymbol{x},\hat{\boldsymbol{x}}) = \begin{cases} 1, & \|\tilde{\boldsymbol{x}}\|_S = (\tilde{\boldsymbol{x}}'\boldsymbol{S}\tilde{\boldsymbol{x}})^{1/2} \geq \dfrac{\Delta}{2} \\ 0, & \|\tilde{\boldsymbol{x}}\|_S = (\tilde{\boldsymbol{x}}'\boldsymbol{S}\tilde{\boldsymbol{x}})^{1/2} < \dfrac{\Delta}{2} \end{cases} \tag{2.6}$$

where $\|\tilde{\boldsymbol{x}}\|$ is the bound norm of the error vector, and $\boldsymbol{S}$ is the weighted non-negative definite matrix.

2. The quadratic cost function, or

$$c(\boldsymbol{x},\hat{\boldsymbol{x}}) = \|\tilde{\boldsymbol{x}}\|_S^2 = \tilde{\boldsymbol{x}}'\boldsymbol{S}\tilde{\boldsymbol{x}} \tag{2.7}$$

3. The bound norm cost function, or

$$c(\boldsymbol{x},\hat{\boldsymbol{x}}) = \|\tilde{\boldsymbol{x}}\|_S = (\tilde{\boldsymbol{x}}'\boldsymbol{S}\tilde{\boldsymbol{x}})^{1/2} \tag{2.8}$$

When the cost function is determined, an expression of average cost (average risk) can be obtained from the given cost function and the prior distribution function:

$$\bar{c} = \int_{-\infty}^{+\infty}\int_{-\infty}^{+\infty} c(x,\hat{x})p(x,z)\,\mathrm{d}x\mathrm{d}z \tag{2.9}$$

Bayesian estimation minimizes the average cost, that is, chooses $\hat{x}$ to obtain the minimum average cost. From the conditional probability density function (PDF), one can obtain

$$\bar{c} = \int_{-\infty}^{+\infty}\left[\int_{-\infty}^{+\infty} c(x,\hat{x})p(x|z)\,\mathrm{d}x\right]p(z)\,\mathrm{d}z \tag{2.10}$$

Define

$$\bar{c}(\hat{x}|z) = \int_{-\infty}^{+\infty} c(x,\hat{x})p(x|z)\,\mathrm{d}x \tag{2.11}$$

Obviously, the inner integral and $p(z)$ are not negative, so if the inner integral for $\hat{x}$ becomes minimum, the average cost will be minimum. $\bar{c}(\hat{x}|z)$ is called the conditional average cost or the conditional average risk, and thus the estimate $\hat{x}$ making the average cost minimum is equivalent to the estimation making the conditional average cost minimum.

2.3 Four Basic Parameter Estimation Techniques

According to the different principles, parameter estimation involves the following four basic techniques [16]: MAP, ML, MMSE, and LS estimators, which are introduced as follows.

2.3.1 Maximum A Posteriori Estimator

When the uniform average cost function is substituted into the expression of the conditional average cost function, we obtain the following:

$$\begin{aligned}\bar{c}(\hat{x}|Z^k) &= \int_{-\infty}^{+\infty} c(x,\hat{x})p(x|Z^k)\mathrm{d}x \\ &= \int_{-\infty}^{\hat{x}-\frac{\Delta}{2}} p(x|Z^k)\mathrm{d}x + \int_{\hat{x}+\frac{\Delta}{2}}^{+\infty} p(x|Z^k)\mathrm{d}x \\ &= \int_{-\infty}^{+\infty} p(x|Z^k)\mathrm{d}x - \int_{\hat{x}-\frac{\Delta}{2}}^{\hat{x}+\frac{\Delta}{2}} p(x|Z^k)\mathrm{d}x \\ &= 1 - \int_{\hat{x}-\frac{\Delta}{2}}^{\hat{x}+\frac{\Delta}{2}} p(x|Z^k)\mathrm{d}x\end{aligned} \tag{2.12}$$

If we want $\bar{c}(\hat{x}|z)$ to be minimum, the integral part on the right-hand side of the equation must be maximum. When $\Delta \to 0$, the maximization of the integral part is equivalent to choosing $\hat{x}$ to maximize the posterior PDF $p(x|z)$. So equivalently, the estimation rule of maximizing the posterior PDF $p(x|z)$ is called the MAP estimator.

For some random parameters, given the prior PDF $p(x)$, we can obtain its posterior PDF using Bayes' formula:

$$p(x|Z^k) = \frac{p(Z^k|x)p(x)}{p(Z^k)} \tag{2.13}$$

and the value of x that maximizes its posterior PDF is called the MAP estimate for the parameter x, that is,

$$\hat{x}^{\mathrm{MAP}}(k) = \arg\max_x p(x|Z^k) = \arg\max_x \left[p(Z^k|x)p(x)\right] \tag{2.14}$$

Significance: When the measurement Z^k is given, the probability of finding the parameter x in the neighborhood of the MAP estimate $\hat{x}^{\mathrm{MAP}}$ is higher than that in any other neighborhood of the same area, as shown in Figure 2.4.

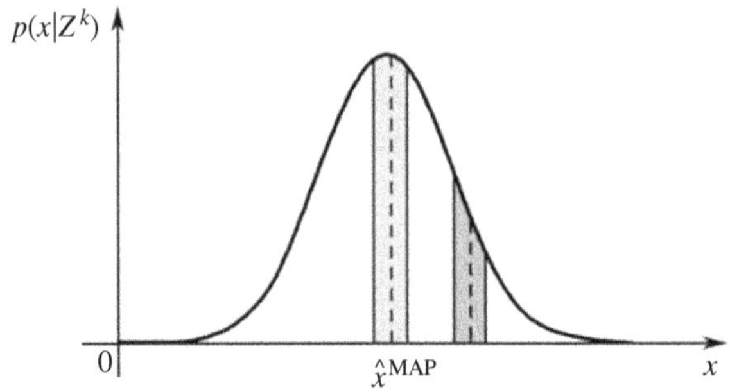

Figure 2.4 A posteriori probability density function

2.3.2 *Maximum Likelihood Estimator*

When the likelihood function $p(Z^k|x)$ reaches maximum, the value of x is called the maximum likelihood estimate for the parameter x, that is,

$$\hat{x}^{\mathrm{ML}}(k) = \arg\max_{x} p\left(Z^k|x\right) \tag{2.15}$$

Significance: When $x = \hat{x}^{\mathrm{ML}}$, the occurrence probability of the input cumulative measurement set Z^k reaches maximum. So when the input measurement set Z^k is observed, it can be decided that the measurements are caused by the parameter $\hat{x}^{\mathrm{ML}}$ which is most likely to result in its occurrence.

The MAP equation is

$$\left.\frac{\partial \ln p(x|z)}{\partial x}\right|_{x=\hat{x}^{\mathrm{MAP}}} = 0 \tag{2.16}$$

or

$$\left.\frac{\partial p(x|z)}{\partial x}\right|_{x=\hat{x}^{\mathrm{MAP}}} = 0 \tag{2.17}$$

The estimate determined by the maximum posterior equation is the MAP estimate.

Similarly, the likelihood equation can be deduced as follows:

$$\left.\frac{\partial \ln p(z|x)}{\partial x}\right|_{x=\hat{x}^{\mathrm{ML}}} = 0 \tag{2.18}$$

or

$$\left.\frac{\partial p(z|x)}{\partial x}\right|_{x=\hat{x}^{\mathrm{ML}}} = 0 \tag{2.19}$$

2.3.3 *Minimum Mean Square Error Estimator*

When the mean square error cost function is inserted into the conditional average cost function, we get

$$\begin{aligned}\bar{c}(\hat{x}|Z^k) &= \int_{-\infty}^{+\infty} c(x,\hat{x})p(x|Z^k)\mathrm{d}x \\ &= \int_{-\infty}^{+\infty} (x-\hat{x})^2 p(x|Z^k)\,\mathrm{d}x = \mathrm{E}\left[(\hat{x}-x)^2|Z^k\right]\end{aligned} \tag{2.20}$$

Choose $\hat{x}$ to minimize $\bar{c}(\hat{x}|Z^k)$, and the MMSE estimate can be obtained.

When the first-order derivative and the second-order derivative are taken for the conditional average cost function in (2.20), we obtain

$$\begin{aligned}&\frac{\mathrm{d}}{\mathrm{d}\hat{x}}\left(\int_{-\infty}^{+\infty} (x-\hat{x})^2 p(x|Z^k)\,\mathrm{d}x\right) \\ &= -2\left(\int_{-\infty}^{+\infty} (x-\hat{x})p(x|Z^k)\,\mathrm{d}x\right) \\ &= -2\int_{-\infty}^{+\infty} xp(x|Z^k)\,\mathrm{d}x + 2\hat{x}\int_{-\infty}^{+\infty} p(x|Z^k)\,\mathrm{d}x\end{aligned} \tag{2.21}$$

and

$$\frac{\mathrm{d}^2}{\mathrm{d}\hat{x}^2}\left(\int_{-\infty}^{+\infty} (x-\hat{x})^2 p(x|Z^k)\,\mathrm{d}x\right) = 2\int_{-\infty}^{+\infty} p(x|Z^k)\,\mathrm{d}x = 2 \tag{2.22}$$

Since the second-order derivative is greater than zero, the minimum value of the conditional average cost function exists. When the first-order derivative is equal to zero, we obtain

$$\hat{x} = \int_{-\infty}^{+\infty} xp(x|Z^k)\,\mathrm{d}x \tag{2.23}$$

To sum up, when the mean square error

$$\mathrm{E}\left[(\hat{x}-x)^2|Z^k\right] \tag{2.24}$$

reaches minimum, the estimation value of x is called the MMSE estimate, that is,

$$\hat{x}^{\mathrm{MMSE}}(k) = \arg\min_{x} \mathrm{E}\left[(\hat{x}-x)^2|Z^k\right] \tag{2.25}$$

Its solution is a conditional average value, and can be expressed by a conditional PDF as follows:

$$\hat{x}^{\mathrm{MMSE}}(k) = \mathrm{E}\left[x|Z^k\right] = \int xp(x|Z^k)\,\mathrm{d}x \tag{2.26}$$

Since the mean square error matrix in the MMSE estimator is smaller than or equal to that obtained by any other estimation rule, the MMSE estimator has the smallest error estimation variance matrix and is also called the "minimum variance estimator."

2.3.4 Least Squares Estimator

Among the estimation techniques mentioned above, the MAP and MMSE estimators require the given prior probability density of the values to be estimated, and the ML estimator requires the given likelihood function. If the probability density or likelihood functions are unknown, these techniques cannot be adopted. However, the LS estimator, which needs no assumption about statistic properties, is widely used.

For the measurements

$$z(j)=h(j,x)+w(j), \quad j=1,2,\ldots,k \tag{2.27}$$

the LS estimate of parameter x at time k refers to the value of x that minimizes the sum of error squares at this time, that is,

$$\hat{x}^{\mathrm{LS}}(k)=\arg\min_{x}\sum_{j=1}^{k}[z(j)-h(j,x)]^2 \tag{2.28}$$

The LS estimator treats signal parameter estimation as deterministic optimization, requiring no statistical knowledge of noises or parameters to be estimated.

2.4 Properties of Estimators

2.4.1 Unbiasedness

Unbiasedness is a basic and essential requirement for an estimator. For a non-random parameter x, the estimator of $\hat{x}$ is said to be unbiased if $\mathrm{E}[\hat{x}]=x_0$, where x_0 is the true value of the parameter. If in the limit case where $k\to\infty$, the above result still holds true, it can be called an asymptotic unbiased estimator; otherwise, it is a biased estimator [16, 25].

For a random variable x with a prior PDF $p(x)$, if $\mathrm{E}[\hat{x}]=\mathrm{E}[x]$, the estimator of $\hat{x}$ is unbiased, where $\mathrm{E}[\hat{x}]$ is the mathematical expectation of the joint PDF $p(Z^k, x)$ and $\mathrm{E}[x]$ is the mathematical expectation of the prior PDF. If in the limit case where $k\to\infty$, the above result still holds true, it can be called an asymptotic unbiased estimator; otherwise, it is a biased estimator.

2.4.2 The Variance of an Estimator

The quality of an estimator is often judged by its variance. The smaller the variance of an estimator, the narrower the range in which its value fluctuates around the parameter to be measured, and thus the more desirable the estimate.

For the unbiased estimator $\hat{x}$ of a non-random parameter x with true value x_0, its variance is $\mathrm{var}(\hat{x})=\mathrm{E}\left[(\hat{x}-x_0)^2\right]$. For the unbiased estimator $\hat{x}$ of a random parameter x, its variance is $\mathrm{var}(\hat{x})=\mathrm{E}\left[(\hat{x}-x)^2\right]$.

2.4.3 Consistent Estimators

An estimator is said to be a consistent estimator if the estimate converges to the true value with increasing number of observations available, that is, the probability of the estimate being different from the true value converges to zero.

The consistency of estimators can be judged by the convergence in mean square criterion, that is, if the non-random parameter satisfies the following:

$$\lim_{k\to\infty} \mathrm{E}\left\{[\hat{x}(k)-x_0]^2\right\}=0 \tag{2.29}$$

then the estimator is a consistent one.

If the random parameter satisfies the following:

$$\lim_{k\to\infty} \mathrm{E}\left\{[\hat{x}(k)-x]^2\right\}=0 \tag{2.30}$$

then the estimator is a consistent one. In other words, the estimator of a non-random or random parameter is consistent if the estimate converges to the true value in some stochastic sense.

Consistency is a basic requirement for estimators, however, its superiority is demonstrated only when the sampling size is quite large (e.g., the ML estimator), which is hard to achieve in practice. Thus, what is normally used in engineering is unbiasedness and efficiency (variance).

2.4.4 Efficient Estimators

In two unbiased estimators for the same parameter, the one with the smaller variance is better. However, how small can the variance of the estimator be? It can be proved that under some conditions a lower limit exists in any estimator, and that the estimator variance cannot be smaller than, but equal to or greater than the lower limit, which is called the Cramer–Rao lower bound (CRLB).

If the mean square error corresponding to the estimate of a parameter is not smaller than the CRLB, it can be regarded as an efficient estimator. To be specific, if the estimator $\hat{x}(k)$ of a non-random parameter x is unbiased and the variance is bounded, that is,

$$\mathrm{E}\left\{[\hat{x}(k)-x_0]^2\right\}\geq J^{-1} \tag{2.31}$$

then the estimate of a non-random parameter x which reaches the lower bound J^{-1} is an efficient estimator, where

$$J=-\mathrm{E}\left[\frac{\partial^2 \ln \Lambda_k(x)}{\partial x^2}\right]_{x=x_0}=\mathrm{E}\left[\frac{\partial \ln \Lambda_k(x)}{\partial x}\right]^2_{x=x_0} \tag{2.32}$$

is the Fisher information, $\Lambda_k(x)$ is the likelihood function, and x_0 is the true value of x.

If the estimator $\hat{x}(k)$ of a random parameter x is unbiased and its variance is bounded, that is,

$$\mathrm{E}\left\{[\hat{x}(k)-x]^2\right\}\geq J^{-1} \tag{2.33}$$

then the estimate of a random parameter x which reaches the lower bound J^{-1} is an efficient estimator,

$$J=-\mathrm{E}\left[\frac{\partial^2 \ln p(Z^k,x)}{\partial x^2}\right]=\mathrm{E}\left[\frac{\partial \ln p(Z^k,x)}{\partial x}\right]^2 \tag{2.34}$$

where J^{-1} is called the CRLB, which is a certain quantity related to the likelihood function.

2.5 Parameter Estimation of Static Vectors

Since in the ML and MAP estimators we need to know the likelihood function $p(Z^k|x)$ and the prior PDF $p(x)$, which is hard to achieve in practice, in this section we extend only the LS and MMSE estimators to static (time-invariant) vectors [15].

2.5.1 *Least Squares Estimator*

For vectors, the estimate $\hat{X}(k)$ that minimizes the quadratic error

$$J(k)=\sum_{i=1}^{k}[z(i)-\boldsymbol{H}(i)\boldsymbol{X}]'[z(i)-\boldsymbol{H}(i)\boldsymbol{X}]=\left[\boldsymbol{Z}^k-\boldsymbol{H}^k\boldsymbol{X}\right]'\left[\boldsymbol{Z}^k-\boldsymbol{H}^k\boldsymbol{X}\right] \tag{2.35}$$

is the LS estimate of the non-random vector $\boldsymbol{X}$, where the measurement at time i is

$$z(i)=\boldsymbol{H}(i)\boldsymbol{X}+\boldsymbol{W}(i) \tag{2.36}$$

Here, $\boldsymbol{H}(i)$ is the measurement matrix, $\boldsymbol{W}(i)$ is the measurement noise with covariance $\boldsymbol{R}(i)$, and

$$\boldsymbol{Z}^k=\begin{bmatrix} z(1) \\ \vdots \\ z(k) \end{bmatrix} \quad \boldsymbol{H}^k=\begin{bmatrix} \boldsymbol{H}(1) \\ \vdots \\ \boldsymbol{H}(k) \end{bmatrix} \tag{2.37}$$

$$\boldsymbol{W}^k=\begin{bmatrix} \boldsymbol{W}(1) \\ \vdots \\ \boldsymbol{W}(k) \end{bmatrix} \quad \boldsymbol{R}^k=\begin{bmatrix} \boldsymbol{R}(1) & \cdots & 0 \\ \vdots & \ddots & \vdots \\ 0 & \cdots & \boldsymbol{R}(k) \end{bmatrix} \tag{2.38}$$

The LS estimate of the non-random vector $\boldsymbol{X}$ can be obtained by setting its quadratic error gradient with respect to $\hat{X}(k)$ to zero, that is,

$$\nabla_X J(k)=2\left(\boldsymbol{H}^k\right)'\left[\boldsymbol{Z}^k-\boldsymbol{H}^k\boldsymbol{X}\right]=0 \tag{2.39}$$

which yields

$$\hat{X}(k)=\left[\left(\boldsymbol{H}^k\right)'\boldsymbol{H}^k\right]^{-1}\left(\boldsymbol{H}^k\right)'\boldsymbol{Z}^k \tag{2.40}$$

Generally, since the covariance matrix $\boldsymbol{R}(i)$ of the measurement noise $\boldsymbol{W}(i)$ is not identically distributed, the estimate $\hat{X}(k)$ that minimizes the weighted sum of squares of the error

$$J(k)=\sum_{i=1}^{k}[z(i)-\boldsymbol{H}(i)\boldsymbol{X}]'\boldsymbol{R}^{-1}(i)[z(i)-\boldsymbol{H}(i)\boldsymbol{X}]=\left[\boldsymbol{Z}^k-\boldsymbol{H}^k\boldsymbol{X}\right]'\left(\boldsymbol{R}^k\right)^{-1}\left[\boldsymbol{Z}^k-\boldsymbol{H}^k\boldsymbol{X}\right] \tag{2.41}$$

is more reasonable, and $\hat{X}(k)$ is called the weighted LS estimate of the non-random vector X, that is,

$$\hat{X}(k)=\left[(H^k)'(R^k)^{-1}H^k\right]^{-1}(H^k)'(R^k)^{-1}Z^k \tag{2.42}$$

From (2.40) and (2.42) it can be found that when the error covariance matrix R^k is the unit matrix, the weighted LS estimate is then the LS estimate. Therefore, we simply discuss the weighted LS estimate.

Since

$$\mathrm{E}\left[\hat{X}(k)\right]=\left[(H^k)'(R^k)^{-1}H^k\right]^{-1}(H^k)'(R^k)^{-1}\mathrm{E}\left[H^kX+W^k\right]=X \tag{2.43}$$

the weighted LS estimate of (2.42) for vectors is unbiased, whose estimation error is

$$\tilde{X}(k)=X-\hat{X}(k)=-\left[(H^k)'(R^k)^{-1}H^k\right]^{-1}(H^k)'(R^k)^{-1}W^k \tag{2.44}$$

Based on (2.44), for vectors, the error covariance weighted LS estimate is

$$\begin{aligned}
P(k)&=\mathrm{E}\left[\tilde{X}(k)\tilde{X}'(k)\right]\\
&=\left[(H^k)'(R^k)^{-1}H^k\right]^{-1}(H^k)'(R^k)^{-1}\mathrm{E}\left[W^k(W^k)'\right](R^k)^{-1}(H^k)\left[(H^k)'(R^k)^{-1}H^k\right]^{-1}\\
&=\left[(H^k)'(R^k)^{-1}H^k\right]^{-1}
\end{aligned} \tag{2.45}$$

For Gaussian disturbances, the LS estimate and the ML estimate of the non-random vector X are consistent. The LS estimates obtained from (2.40) and (2.42) are processing k data points simultaneously, that is, the batch processing mode, generally taking a lot of calculation. The following presents the recursive form of the LS estimate.

When the new measurement $z(k+1)$ is available, the measurements from time 1 to $k+1$ are used to construct the multi-vector matrix, the multi-measurement matrix, the measurement error multi-vector matrix, and the corresponding block diagonal positive definite matrix, expressed as

$$Z^{k+1}=\begin{bmatrix}Z^k\\ z(k+1)\end{bmatrix},\quad H^{k+1}=\begin{bmatrix}H^k\\ H(k+1)\end{bmatrix},\quad W^{k+1}=\begin{bmatrix}W^k\\ W(k+1)\end{bmatrix},\quad R^{k+1}=\begin{bmatrix}R^k & 0\\ 0 & R(k+1)\end{bmatrix}$$

From (2.45) we get the inverse of the error covariance matrix at time $k+1$,

$$P^{-1}(k+1)=(H^{k+1})'(R^{k+1})^{-1}H^{k+1}=(H^k)'(R^k)^{-1}H^k+H'(k+1)R^{-1}(k+1)H(k+1) \tag{2.46}$$

Hence, the information in the sense of Fisher (inverse covariance matrix) at $k+1$ is equal to the sum of the information at k and the new information about vector X from the measurements $z(k+1)$.

Using the matrix inversion lemma

$$(P^{-1}+H'R^{-1}H)^{-1}=P-PH'(HPH'+R)^{-1}HP \tag{2.47}$$

the recursion for the error covariance can be rewritten as

$$\boldsymbol{P}(k+1)=\boldsymbol{P}(k)-\boldsymbol{P}(k)\boldsymbol{H}'(k+1)[\boldsymbol{H}(k+1)\boldsymbol{P}(k)\boldsymbol{H}'(k+1)+\boldsymbol{R}(k+1)]^{-1}\boldsymbol{H}(k+1)\boldsymbol{P}(k) \tag{2.48}$$

Defining

$$\boldsymbol{S}(k+1)=\boldsymbol{H}(k+1)\boldsymbol{P}(k)\boldsymbol{H}'(k+1)+\boldsymbol{R}(k+1) \tag{2.49}$$

$$\boldsymbol{K}(k+1)=\boldsymbol{P}(k)\boldsymbol{H}'(k+1)\boldsymbol{S}^{-1}(k+1) \tag{2.50}$$

the recursion for the covariance can also be expressed as

$$\begin{aligned}\boldsymbol{P}(k+1)&=\boldsymbol{P}(k)-\boldsymbol{K}(k+1)\boldsymbol{H}(k+1)\boldsymbol{P}(k)\\&=[\boldsymbol{I}-\boldsymbol{K}(k+1)\boldsymbol{H}(k+1)]\boldsymbol{P}(k)=\boldsymbol{P}(k)-\boldsymbol{K}(k+1)\boldsymbol{S}(k+1)\boldsymbol{K}'(k+1)\end{aligned} \tag{2.51}$$

Using (2.48), one has

$$\begin{aligned}&\boldsymbol{P}(k+1)\boldsymbol{H}'(k+1)\boldsymbol{R}^{-1}(k+1)\\&=\{\boldsymbol{P}(k)\boldsymbol{H}'(k+1)-\boldsymbol{P}(k)\boldsymbol{H}'(k+1)\boldsymbol{S}^{-1}(k+1)\boldsymbol{H}(k+1)\boldsymbol{P}(k)\boldsymbol{H}'(k+1)\}\boldsymbol{R}^{-1}(k+1)\\&=\boldsymbol{P}(k)\boldsymbol{H}'(k+1)\boldsymbol{S}^{-1}(k+1)\{\boldsymbol{S}(k+1)-\boldsymbol{H}(k+1)\boldsymbol{P}(k)\boldsymbol{H}'(k+1)\}\boldsymbol{R}^{-1}(k+1)\\&=\boldsymbol{P}(k)\boldsymbol{H}'(k+1)\boldsymbol{S}^{-1}(k+1)=\boldsymbol{K}(k+1)\end{aligned} \tag{2.52}$$

which is the other expression for the gain $\boldsymbol{K}(k+1)$.

From (2.40) one can obtain the recursion of the estimate

$$\begin{aligned}\hat{\boldsymbol{X}}(k+1)&=\left[(\boldsymbol{H}^{k+1})'(\boldsymbol{R}^{k+1})^{-1}\boldsymbol{H}^{k+1}\right]^{-1}(\boldsymbol{H}^{k+1})'(\boldsymbol{R}^{k+1})^{-1}\boldsymbol{Z}^{k+1}\\&=\boldsymbol{P}(k+1)(\boldsymbol{H}^{k})'(\boldsymbol{R}^{k})^{-1}\boldsymbol{Z}^{k}+\boldsymbol{P}(k+1)\boldsymbol{H}'(k+1)\boldsymbol{R}^{-1}(k+1)z(k+1)\\&=[\boldsymbol{I}-\boldsymbol{K}(k+1)\boldsymbol{H}(k+1)]\boldsymbol{P}(k)(\boldsymbol{H}^{k})'(\boldsymbol{R}^{k})^{-1}\boldsymbol{Z}^{k}+\boldsymbol{K}(k-1)z(k+1)\\&=[\boldsymbol{I}-\boldsymbol{K}(k+1)\boldsymbol{H}(k+1)]\hat{\boldsymbol{X}}(k)+\boldsymbol{K}(k+1)z(k+1)\\&=\hat{\boldsymbol{X}}(k)+\boldsymbol{K}(k+1)\left[z(k+1)-\boldsymbol{H}(k+1)\hat{\boldsymbol{X}}(k)\right]\end{aligned} \tag{2.53}$$

The new estimate $\hat{\boldsymbol{X}}(k+1)$ is equal to the previous estimate $\hat{\boldsymbol{X}}(k)$ plus a correction term, which is composed of the gain $\boldsymbol{K}(k+1)$ and the new information.

2.5.2 Minimum Mean Square Error Estimator

Suppose that $\boldsymbol{x}$ is the vector to be estimated, z is the observation of the vector $\boldsymbol{x}$, and both the random vectors are of joint normal distribution, that is,

$$\boldsymbol{y}=\begin{bmatrix}\boldsymbol{x}\\ z\end{bmatrix}\sim \mathrm{N}\left[\bar{y},\boldsymbol{P}_{yy}\right] \tag{2.54}$$

where

$$\bar{y} = \begin{bmatrix} \bar{x} \\ \bar{z} \end{bmatrix}, \quad P_{yy} = \begin{bmatrix} P_{xx} & P_{xz} \\ P_{zx} & P_{zz} \end{bmatrix} \tag{2.55}$$

Here, $\bar{x}, P_{xx}$ and $\bar{z}, P_{xx}$ are the mean value and the autocovariance of the random vectors x and z, respectively, and P_{xz} is the cross covariance. Since

$$p(x,z) = p(y) = \mathrm{N}(y;\bar{y},P_{yy}) = |2\pi P_{yy}|^{-\frac{1}{2}} \exp\left\{-{}^1/_2(y-\bar{y})' P_{yy}^{-1}(y-\bar{y})\right\} \tag{2.56}$$

$$p(z) = \mathrm{N}(z;\bar{z},P_{zz}) = |2\pi P_{zz}|^{-\frac{1}{2}} \exp\{-{}^1/_2(z-\bar{z})' P_{zz}^{-1}(z-\bar{z})\} \tag{2.57}$$

one can obtain

$$p(x|z) = \frac{p(x,z)}{p(z)} = \frac{|2\pi P_{yy}|^{-\frac{1}{2}}}{|2\pi P_{zz}|^{-\frac{1}{2}}} \exp\left\{-{}^1/_2\left[(y-\bar{y})' P_{yy}^{-1}(y-\bar{y}) - (z-\bar{z})' P_{zz}^{-1}(z-\bar{z})\right]\right\} \tag{2.58}$$

Set

$$y - \bar{y} = \begin{bmatrix} x-\bar{x} \\ z-\bar{z} \end{bmatrix} = \begin{bmatrix} \xi \\ \eta \end{bmatrix} \tag{2.59}$$

$$P_{yy}{}^{-1} = \begin{bmatrix} T_{xx} & T_{xz} \\ T_{zx} & T_{zz} \end{bmatrix} \tag{2.60}$$

From (2.55) and (2.60) it then follows that

$$T_{xx}{}^{-1} = P_{xx} - P_{xz} P_{zz}{}^{-1} P_{zx}, \quad P_{zz}{}^{-1} = T_{zz} - T_{zx} T_{xx}{}^{-1} T_{xz} \quad T_{xx}{}^{-1} T_{xz} = -P_{xz} P_{zz}{}^{-1} \tag{2.61}$$

Now let

$$\begin{aligned} q &= (y-\bar{y})' P_{yy}^{-1}(y-\bar{y}) - (z-\bar{z})' P_{zz}^{-1}(z-\bar{z}) \\ &= \xi' T_{xx}\xi + \eta' T_{zx}\xi + \xi T_{xz}\eta + \eta' T_{zz}\eta - \eta' P_{zz}^{-1}\eta \\ &= \xi' T_{xx}\xi + \eta' T_{zx}\xi + \xi' T_{xz}\eta + \eta' T_{zx} T_{xx}{}^{-1} T_{xz}\eta + \eta' T_{zz}\eta - \eta' T_{zx} T_{xx}{}^{-1} T_{xz}\eta - \eta' P_{zz}^{-1}\eta \\ &= \left(\xi' + \eta' T_{zx} T_{xx}{}^{-1}\right) T_{xx}\xi + \left(\xi' + \eta' T_{zx} T_{xx}{}^{-1}\right) T_{xz}\eta + \eta'\left(T_{zz} - T_{zx} T_{xx}{}^{-1} T_{xz}\right)\eta - \eta' P_{zz}^{-1}\eta \\ &= \left(\xi + T_{xx}{}^{-1} T_{xz}\eta\right)' T_{xx}\left(\xi + T_{xx}{}^{-1} T_{xz}\eta\right) \end{aligned} \tag{2.62}$$

Since q is the quadratic form of x, the conditional PDF of x for given z is also Gaussian. Another reason is that

$$\xi + T_{xx}^{-1} T_{xz}\eta = x - \bar{x} - P_{xz} P_{zz}^{-1}(z-\bar{z}) \tag{2.63}$$

Thus we get the MMSE estimate of $\boldsymbol{x}$ in terms of $\boldsymbol{z}$ as

$$\hat{\boldsymbol{x}} = \mathrm{E}[\boldsymbol{x}|\boldsymbol{z}] = \bar{\boldsymbol{x}} + \boldsymbol{P}_{xz}\boldsymbol{P}_{zz}^{-1}(\boldsymbol{z}-\bar{\boldsymbol{z}}) \tag{2.64}$$

In Gaussian conditions, the MMSE estimate of $\boldsymbol{x}$ in terms of $\boldsymbol{z}$ is the conditional mean value of $\boldsymbol{x}$ for a given $\boldsymbol{z}$. The corresponding conditional error covariance matrix is

$$\boldsymbol{P}_{xx|z} = \mathrm{E}\left[(\boldsymbol{x}-\hat{\boldsymbol{x}})(\boldsymbol{x}-\hat{\boldsymbol{x}})'|\boldsymbol{z}\right] = \boldsymbol{T}_{xx}^{-1} = \boldsymbol{P}_{xx} - \boldsymbol{P}_{xz}\boldsymbol{P}_{zz}^{-1}\boldsymbol{P}_{zx} \tag{2.65}$$

2.5.3 Linear Minimum Mean Square Error Estimator

If the random vectors $\boldsymbol{x}$ and $\boldsymbol{z}$ are not jointly Gaussian, it is very difficult to obtain the conditional mean value in general. However, we can deduce the optimal linear estimate of $\boldsymbol{x}$ in terms of $\boldsymbol{z}$. The orthogonal principle is the necessary and sufficient condition for the linear estimator to become the optimal estimator [16]. According to the orthogonal principle, the estimation error $\tilde{\boldsymbol{x}}$ of the optimal linear estimate is unbiased, and is orthogonal to the observation $\boldsymbol{z}$.

Assume that

$$\hat{\boldsymbol{x}} = \boldsymbol{A}\boldsymbol{z} + \boldsymbol{b} \tag{2.66}$$

is the optimal linear estimate under non-Gaussian conditions. Because the estimation error $\tilde{\boldsymbol{x}}$ of the optimal linear estimate is unbiased, we can obtain

$$\mathrm{E}[\tilde{\boldsymbol{x}}] = \mathrm{E}[\boldsymbol{x}-\hat{\boldsymbol{x}}] = \bar{\boldsymbol{x}} - (\boldsymbol{A}\bar{\boldsymbol{z}} + \boldsymbol{b}) = 0 \tag{2.67}$$

It follows that

$$\boldsymbol{b} = \bar{\boldsymbol{x}} - \boldsymbol{A}\bar{\boldsymbol{z}} \tag{2.68}$$

Here the estimation error can be expressed as

$$\tilde{\boldsymbol{x}} = \boldsymbol{x} - \hat{\boldsymbol{x}} = \boldsymbol{x} - \boldsymbol{A}\boldsymbol{z} - \boldsymbol{b} = \boldsymbol{x} - \bar{\boldsymbol{x}} - \boldsymbol{A}(\boldsymbol{z}-\bar{\boldsymbol{z}}) \tag{2.69}$$

Because the optimal linear estimate must satisfy the condition that the estimation error $\tilde{\boldsymbol{x}}$ and the observation $\boldsymbol{z}$ are orthogonal, we find

$$\begin{aligned} \mathrm{E}[\tilde{\boldsymbol{x}}\boldsymbol{z}'] &= \mathrm{E}\left\{[(\boldsymbol{x}-\bar{\boldsymbol{x}}) - \boldsymbol{A}(\boldsymbol{z}-\bar{\boldsymbol{z}})](\boldsymbol{z}-\bar{\boldsymbol{z}}+\bar{\boldsymbol{z}})'\right\} \\ &= \mathrm{E}\left\{[(\boldsymbol{x}-\bar{\boldsymbol{x}}) - \boldsymbol{A}(\boldsymbol{z}-\bar{\boldsymbol{z}})](\boldsymbol{z}-\bar{\boldsymbol{z}})'\right\} = \boldsymbol{P}_{xz} - \boldsymbol{A}\boldsymbol{P}_{zz} = 0 \end{aligned} \tag{2.70}$$

From (2.70), the solution of $\boldsymbol{A}$ can be obtained as follows:

$$\boldsymbol{A} = \boldsymbol{P}_{xz}\boldsymbol{P}_{zz}^{-1} \tag{2.71}$$

From the simultaneous equations (2.68) and (2.71), we can obtain the expression of the LMMSE estimate

$$\hat{x} = Az + b = \bar{x} + A(z - \bar{z}) = \bar{x} + P_{xz}P_{zz}^{-1}(z - \bar{z}) \tag{2.72}$$

This estimate is the optimal linear estimate that minimizes the mean square error

$$J = \mathrm{E}\left[(x - \hat{x})'(x - \hat{x})\right] \tag{2.73}$$

under non-Gaussian conditions. Note that the expression for (2.72) is identical to that for the MMSE estimate (2.64) under Gaussian conditions, and that it is still the linear function of the measurement z, instead of the conditional mean value.

From (2.69) we get the mean square error corresponding to (2.72):

$$\begin{aligned}
\mathrm{E}\{\tilde{x}\tilde{x}'\} &= \mathrm{E}\{(x - \hat{x})(x - \hat{x})'\} \\
&= \mathrm{E}\left\{\left(x - \bar{x} - P_{xz}P_{zz}^{-1}(z - \bar{z})\right)\left(x - \bar{x} - P_{xz}P_{zz}^{-1}(z - \bar{z})\right)'\right\} \\
&= \mathrm{E}\left\{\left[(x - \bar{x}) - P_{xz}P_{zz}^{-1}(z - \bar{z})\right]\left[(x - \bar{x})' - (z - \bar{z})'P_{zz}^{-1\prime}P_{xz}'\right]\right\} \\
&= P_{xx} - P_{xz}P_{zz}^{-1}P_{zx} - P_{xz}P_{zz}^{-1}P_{zx} + P_{xz}P_{zz}^{-1}P_{zx} \\
&= P_{xx} - P_{xz}P_{zz}^{-1}P_{zx}
\end{aligned} \tag{2.74}$$

where

$$P_{xx} = \mathrm{E}\{(x - \bar{x})(x - \bar{x})'\} \quad P_{xz} = \mathrm{E}\{(x - \bar{x})(z - \bar{z})'\} \quad P_{zz} = \mathrm{E}\{(z - \bar{z})(z - \bar{z})'\}$$

It has the same expression as (2.65), but (2.72) is not the conditional mean value, so strictly speaking the above equation is not a covariance matrix.

2.6 Summary

This chapter discusses some basic parameter estimation techniques in radar data processing, including the ML, MAP, LS, and MMSE estimators. For the ML estimator, only the likelihood function is required; for the MAP estimator, we need to know the likelihood function and the prior PDF of the parameter to be estimated; for the MMSE estimator, we are only required to know the first-order and second-order statistical matrix of the related parameters, without any requirement for other probability assumptions. For the LS estimator, which excludes all probability assumptions, treating the problem of estimation as a problem of deterministic optimization can be regarded as the last step in constantly reducing statistical demands.

3

Linear Filtering Approaches

3.1 Introduction

Time-invariant parameter estimation techniques are analyzed and discussed in Chapter 2, the subject of which is static estimation. This chapter starts to study the estimation of time-varying parameters, that is, state estimation, which refers to smoothing the past motion state (including the location, velocity, acceleration, etc.) of a target, filtering the present motion state, and predicting the future motion state of the target [20, 65–72]. For instance, the positioning of the stations in a radar network or satellite orbits involves static estimation, while the tracking and interception of a moving target involves dynamic estimation. Both parameter estimation and state estimation calculate the values of unknown parameters in accordance with a set of measurements related to unknown parameters. But when processing observations we must consider the time evolution of the unknown parameters and the observed data, since the unknown parameters are time functions in the state estimation. This chapter mainly covers the Kalman filter (KF), including system model building, corresponding filtering models, filter initialization methods, the definition and judgment methods of filter stability, the controllability and observability of stochastic linear systems, steady-state Kalman filters, etc.

3.2 Kalman Filter

Kalman filters are the best not only among all linear filters, but also among all filters when the noise process is Gaussian. Kalman filtering requires no conditions, except that the system noise and the measurement noise be Gaussian white noise and that their secondary moments be given. Therefore, it is fully applicable to the estimation of non-stationary, multi-dimensional random sequences.

Radar Data Processing with Applications, First Edition. He You, Xiu Jianjuan, and Guan Xin.

3.2.1 System Model

The state variable method is a very valuable way of describing a dynamic system, where the relation of system input and output is described by the state transition model and the output observation model in a time domain. The input is expressed by the determined time function and the state equations composed of the unpredictable variables or the random processes of noises, while the output (often disturbed by random measurement errors) is a function of the state vectors, and thus can be described by measurement equations.

3.2.1.1 State Equation

Constant Velocity (CV) Model

The state equation is the assumption about the target motion laws, for example, assuming that the target moves in a straight line with a constant velocity in a two-dimensional plane. Then, in the discrete-time system, the target state (x_k, y_k) at time t_k can be denoted as

$$x_k = x_0 + v_x t_k = x_0 + v_x kT \tag{3.1}$$

$$y_k = y_0 + v_y t_k = y_0 + v_y kT \tag{3.2}$$

where (x_0, y_0) is the target location at the initial moment, v_x and v_y are the velocity along the x axis and the y axis, respectively, and T is the sampling interval.

Equations (3.1) and (3.2) can be denoted in the recursive form as

$$x_{k+1} = x_k + v_x T = x_k + \dot{x}_k T \tag{3.3}$$

$$y_{k+1} = y_k + v_y T = y_k + \dot{y}_k T \tag{3.4}$$

It is impossible to acquire an accurate model of targets, and there are plenty of unpredictable phenomena. In other words, targets are unlikely to be in an absolute constant-velocity motion, and their velocity must be experiencing some minor random fluctuations, for example, in the constant-velocity motion of the target, the pilot or environmental disturbance, etc. can result in some unpredictable changes in the velocity, such as the effects on a plane's speed of clouds and gusts of wind during flight. These minor changes in velocity can be seen as process noise when building a model. So, with process noise introduced, (3.3) and (3.4) should be expressed as

$$x_{k+1} = x_k + \dot{x}_k T + \frac{1}{2} v_x T^2 \tag{3.5}$$

$$y_{k+1} = y_k + \dot{y}_k T + \frac{1}{2} v_y T^2 \tag{3.6}$$

Here we emphasize that v_x and v_y refer to the random changes of x axis velocity and y axis velocity of the target. The velocity of the target can be described as

$$\dot{x}_{k+1} = \dot{x}_k + v_x T \tag{3.7}$$

$$\dot{y}_{k+1} = \dot{y}_k + v_y T \tag{3.8}$$

In constant-velocity models, the state vector $\boldsymbol{X}(k)$ used to describe the system's dynamic features is $\boldsymbol{X}(k) = [x_k \;\; y_k \;\; \dot{x}_k \;\; \dot{y}_k]'$, so (3.5)–(3.8) can be represented in matrix form as

$$\begin{bmatrix} x(k+1) \\ y(k+1) \\ \dot{x}(k+1) \\ \dot{y}(k+1) \end{bmatrix} = \begin{bmatrix} 1 & 0 & T & 0 \\ 0 & 1 & 0 & T \\ 0 & 0 & 1 & 0 \\ 0 & 0 & 0 & 1 \end{bmatrix} \begin{bmatrix} x(k) \\ y(k) \\ \dot{x}(k) \\ \dot{y}(k) \end{bmatrix} + \begin{bmatrix} 0.5T^2 & 0 \\ 0 & 0.5T^2 \\ T & 0 \\ 0 & T \end{bmatrix} \begin{bmatrix} v_x \\ v_y \end{bmatrix} \tag{3.9}$$

That is, the target state equation is

$$\boldsymbol{X}(k+1) = \boldsymbol{F}(k)\boldsymbol{X}(k) + \boldsymbol{\Gamma}(k)\boldsymbol{v}(k) \tag{3.10}$$

where $\boldsymbol{v}(k) = [v_x, v_y]'$ is the process noise vector,

$$\boldsymbol{F}(k) = \begin{bmatrix} 1 & 0 & T & 0 \\ 0 & 1 & 0 & T \\ 0 & 0 & 1 & 0 \\ 0 & 0 & 0 & 1 \end{bmatrix} \tag{3.11}$$

is the state transition matrix, and

$$\boldsymbol{\Gamma}(k) = \begin{bmatrix} 0.5T^2 & 0 \\ 0 & 0.5T^2 \\ T & 0 \\ 0 & T \end{bmatrix} \tag{3.12}$$

is the process noise distribution matrix.

If the target is three-dimensional with state vector $\boldsymbol{X}(k) = [x_k \;\; y_k \;\; z_k \;\; \dot{x}_k \;\; \dot{y}_k \;\; \dot{z}_k]'$, then its process noise vector $\boldsymbol{v}(k) = [v_x, v_y, v_z]'$, while the state transition matrix and the process noise distribution matrix of the system are

$$\boldsymbol{F}(k) = \begin{bmatrix} 1 & 0 & 0 & T & 0 & 0 \\ 0 & 1 & 0 & 0 & T & 0 \\ 0 & 0 & 1 & 0 & 0 & T \\ 0 & 0 & 0 & 1 & 0 & 0 \\ 0 & 0 & 0 & 0 & 1 & 0 \\ 0 & 0 & 0 & 0 & 0 & 1 \end{bmatrix} \tag{3.13}$$

$$\boldsymbol{\Gamma}(k) = \begin{bmatrix} 0.5T^2 & 0 & 0 \\ 0 & 0.5T^2 & 0 \\ 0 & 0 & 0.5T^2 \\ T & 0 & 0 \\ 0 & T & 0 \\ 0 & 0 & T \end{bmatrix} \tag{3.14}$$

Constant Acceleration (CA) Model

If the target moves with constant acceleration along a straight line in a two-dimensional plane, considering the random changes in velocity, the location and velocity can be expressed in recursive form as

$$x_{k+1} = x_k + \dot{x}_k T + \frac{1}{2}\ddot{x}_k T^2 + \frac{1}{2} v_x T^2 \tag{3.15}$$

$$y_{k+1} = y_k + \dot{y}_k T + \frac{1}{2}\ddot{y}_k T^2 + \frac{1}{2} v_y T^2 \tag{3.16}$$

$$\dot{x}_{k+1} = \dot{x}_k + \ddot{x}_k T + v_x T \tag{3.17}$$

$$\dot{y}_{k+1} = \dot{y}_k + \ddot{y}_k T + v_y T \tag{3.18}$$

$$\ddot{x}_{k+1} = \ddot{x}_k + v_x \tag{3.19}$$

$$\ddot{y}_{k+1} = \ddot{y}_k + v_y \tag{3.20}$$

So, the target state equation obtained from (3.15)–(3.20) has the same form as (3.10), but now the state vector is $\boldsymbol{X}(k) = [x_k \ \ \dot{x}_k \ \ \ddot{x}_k \ \ y_k \ \ \dot{y}_k \ \ \ddot{y}_k]'$, the process noise vector is $\boldsymbol{v}(k) = [v_x, \ v_y]'$, and the corresponding state transition matrix and process noise distribution matrix are, respectively,

$$F(k) = \begin{bmatrix} 1 & T & \frac{1}{2}T^2 & 0 & 0 & 0 \\ 0 & 1 & T & 0 & 0 & 0 \\ 0 & 0 & 1 & 0 & 0 & 0 \\ 0 & 0 & 0 & 1 & T & \frac{1}{2}T^2 \\ 0 & 0 & 0 & 0 & 1 & T \\ 0 & 0 & 0 & 0 & 0 & 1 \end{bmatrix} \quad \Gamma(k) = \begin{bmatrix} \frac{1}{2}T^2 & 0 \\ T & 0 \\ 1 & 0 \\ 0 & \frac{1}{2}T^2 \\ 0 & T \\ 0 & 1 \end{bmatrix} \tag{3.21}$$

Equally, when the target moves with constant velocity and constant acceleration in a three-dimensional space, the corresponding state vector is $\boldsymbol{X}(k) = [x_k \ \ \dot{x}_k \ \ \ddot{x}_k \ \ y_k \ \ \dot{y}_k \ \ \ddot{y}_k \ \ z_k \ \ \dot{z}_k \ \ \ddot{z}_k]'$ and the process noise vector is $\boldsymbol{v}(k) = [v_x, \ v_y, \ v_z]'$. Then, the system's state transition matrix and process noise distribution matrix are

$$F(k) = \begin{bmatrix} 1 & T & \frac{1}{2}T^2 & 0 & 0 & 0 & 0 & 0 & 0 \\ 0 & 1 & T & 0 & 0 & 0 & 0 & 0 & 0 \\ 0 & 0 & 1 & 0 & 0 & 0 & 0 & 0 & 0 \\ 0 & 0 & 0 & 1 & T & \frac{1}{2}T^2 & 0 & 0 & 0 \\ 0 & 0 & 0 & 0 & 1 & T & 0 & 0 & 0 \\ 0 & 0 & 0 & 0 & 0 & 1 & 0 & 0 & 0 \\ 0 & 0 & 0 & 0 & 0 & 0 & 1 & T & \frac{1}{2}T^2 \\ 0 & 0 & 0 & 0 & 0 & 0 & 0 & 1 & T \\ 0 & 0 & 0 & 0 & 0 & 0 & 0 & 0 & 1 \end{bmatrix} \quad \Gamma(k) = \begin{bmatrix} \frac{1}{2}T^2 & 0 & 0 \\ T & 0 & 0 \\ 1 & 0 & 0 \\ 0 & \frac{1}{2}T^2 & 0 \\ 0 & T & 0 \\ 0 & 1 & 0 \\ 0 & 0 & \frac{1}{2}T^2 \\ 0 & 0 & T \\ 0 & 0 & 1 \end{bmatrix} \tag{3.22}$$

Coordinate Turn (CT) Model

The principle of the CT model is shown in Figure 3.1.

We see that

$$v_x = v\cos\varphi \quad v_y = v\sin\varphi \tag{3.23}$$

owing to the fact that

$$\omega = \frac{d\varphi}{dt} \tag{3.24}$$

so

$$\frac{dv_x}{dt} = -v\sin\varphi\frac{d\varphi}{dt} = -v_y\omega \tag{3.25}$$

$$\frac{dv_y}{dt} = v\cos\varphi\frac{d\varphi}{dt} = v_x\omega \tag{3.26}$$

Then, we get

$$\begin{bmatrix} \dot{x}(t) \\ \ddot{x}(t) \\ \dot{y}(t) \\ \ddot{y}(t) \end{bmatrix} = \begin{bmatrix} 0 & 1 & 0 & 0 \\ 0 & 0 & 0 & -\omega(t) \\ 0 & 0 & 0 & 1 \\ 0 & \omega(t) & 0 & 0 \end{bmatrix} \begin{bmatrix} x(t) \\ \dot{x}(t) \\ y(t) \\ \dot{y}(t) \end{bmatrix} + V(t) \tag{3.27}$$

where $V(t)$ is the process noise, usually assumed as Gaussian white noise. After discrete treatment and Laplace transform of the above equation, we obtain the system state matrix of the turn model as follows [50, 51]:

$$\boldsymbol{F}(k) = \begin{bmatrix} 1 & \frac{\sin\omega T}{\omega} & 0 & \frac{\cos\omega T - 1}{\omega} \\ 0 & \cos\omega T & 0 & -\sin\omega T \\ 0 & \frac{1-\cos\omega T}{\omega} & 1 & \frac{\sin\omega T}{\omega} \\ 0 & \sin\omega T & 0 & \cos\omega T \end{bmatrix} \tag{3.28}$$

when the rate of turn (ω) is given.

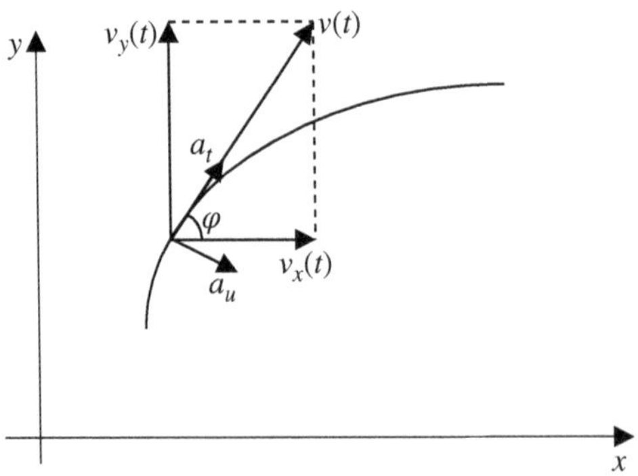

Figure 3.1 Diagram of coordinate turn model

When the rate of turn cannot be known accurately, the standard CT model should be extended to have an additional element ω in the state vector, that is,

$$F(k)=\begin{bmatrix} 1 & \dfrac{\sin\omega T}{\omega} & 0 & \dfrac{\cos\omega T-1}{\omega} & 0 \\ 0 & \cos\omega T & 0 & -\sin\omega T & 0 \\ 0 & \dfrac{1-\cos\omega T}{\omega} & 1 & \dfrac{\sin\omega T}{\omega} & 0 \\ 0 & \sin\omega T & 0 & \cos\omega T & 0 \\ 0 & 0 & 0 & 0 & 1 \end{bmatrix} \tag{3.29}$$

Now the corresponding process noise distribution matrix is

$$\Gamma(k)=\begin{bmatrix} T^2/2 & T & 0 & 0 & 0 \\ 0 & 0 & T^2/2 & T & 0 \\ 0 & 0 & 0 & 0 & 1 \end{bmatrix}' \tag{3.30}$$

In addition, it should be noted that the positions of the elements in the state vector $\boldsymbol{X}(k)$ can be interchanged randomly, and that those of the elements in the state transition matrix and process noise distribution matrix should be interchanged accordingly. As an increase in dimension of the state vector will lead to an increase in calculation complexity as well as an increase in accuracy of the estimation, simple mathematical models should be adopted as far as possible when the requirements for model accuracy and tracking performance are met.

Considering that there can be some control signals during the target motion process, the state equation is generally written as

$$\boldsymbol{X}(k+1)=\boldsymbol{F}(k)\boldsymbol{X}(k)+\boldsymbol{G}(k)\boldsymbol{u}(k)+\boldsymbol{V}(k) \tag{3.31}$$

where $\boldsymbol{G}(k)$ is the input control matrix, $\boldsymbol{u}(k)$ is the known input or control signal, and $\boldsymbol{V}(k)$ is a white Gaussian noise sequence with zero mean and covariance $\boldsymbol{Q}(k)$, that is, $\mathrm{E}[\boldsymbol{V}(k)\boldsymbol{V}'(j)]=\boldsymbol{Q}(k)\delta_{kj}$, where δ_{kj} is the Kronecker delta function. This property shows that the process noises are mutually independent at different times. If the process noise $\boldsymbol{V}(k)$ is replaced by $\boldsymbol{\Gamma}(k)\boldsymbol{v}(k)$, then $\boldsymbol{Q}(k)$ will change into $\boldsymbol{\Gamma}(k)\boldsymbol{q}(k)\boldsymbol{\Gamma}'(k)$.

3.2.1.2 Measurement Equation

The measurement equation is the hypothesis for a radar measurement process. For linear systems, the measurement equation can be denoted as

$$\boldsymbol{Z}(k+1)=\boldsymbol{H}(k+1)\boldsymbol{X}(k+1)+\boldsymbol{W}(k+1) \tag{3.32}$$

where $\boldsymbol{Z}(k+1)$ is the measurement vector, $\boldsymbol{H}(k+1)$ is the measurement matrix, and $\boldsymbol{W}(k+1)$ is a white Gaussian measurement noise sequence with zero mean and covariance $\boldsymbol{R}(k+1)$, that is, $\mathrm{E}[\boldsymbol{W}(k)\boldsymbol{W}'(j)]=\boldsymbol{R}(k)\delta_{kj}$. This property shows that measurement noises are divergent at different times, and it assumes that the process noise sequence, the measurement noise sequence, and the target initial state are independent.

When modeling a target moving with constant velocity or constant acceleration in a two-dimensional plane, the corresponding vectors are $X(k) = [x_k \ y_k \ \dot{x}_k \ \dot{y}_k]'$ and $X(k) = [x_k \ \dot{x}_k \ \ddot{x}_k \ y_k \ \dot{y}_k \ \ddot{y}_k]'$, respectively. In these two conditions the measurement vectors are both $Z(k) = [x_k \ y_k]'$, and the measurement matrixes are, respectively,

$$H(k) = \begin{bmatrix} 1 & 0 & 0 & 0 \\ 0 & 1 & 0 & 0 \end{bmatrix} \tag{3.33}$$

$$H(k) = \begin{bmatrix} 1 & 0 & 0 & 0 & 0 & 0 \\ 0 & 0 & 0 & 1 & 0 & 0 \end{bmatrix} \tag{3.34}$$

When modeling a target moving with constant velocity or constant acceleration in a three-dimensional space, the corresponding state vectors are, respectively, $X(k) = [x_k \ y_k \ z_k \ \dot{x}_k \ \dot{y}_k \ \dot{z}_k]'$ and $X(k) = [x_k \ \dot{x}_k \ \ddot{x}_k \ y_k \ \dot{y}_k \ \ddot{y}_k \ z_k \ \dot{z}_k \ \ddot{z}_k]'$. In these two conditions the measurement vectors are both $Z(k) = [x_k \ y_k]'$, and the measurement matrixes are, respectively,

$$H(k) = \begin{bmatrix} 1 & 0 & 0 & 0 & 0 & 0 \\ 0 & 1 & 0 & 0 & 0 & 0 \\ 0 & 0 & 1 & 0 & 0 & 0 \end{bmatrix} \tag{3.35}$$

$$H(k) = \begin{bmatrix} 1 & 0 & 0 & 0 & 0 & 0 & 0 & 0 & 0 \\ 0 & 0 & 0 & 1 & 0 & 0 & 0 & 0 & 0 \\ 0 & 0 & 0 & 0 & 0 & 0 & 1 & 0 & 0 \end{bmatrix} \tag{3.36}$$

When the CT model is applied, the state vectors are $X(k) = [x_k \ \dot{x}_k \ y_k \ \dot{y}_k]'$ and $X(k) = [x_k \ \dot{x}_k \ y_k \ \dot{y}_k \ \omega]'$, respectively. In these cases, the measurement vectors are both $Z(k) = [x_k \ y_k]'$, and the corresponding measurement matrixes are, respectively,

$$H(k) = \begin{bmatrix} 1 & 0 & 0 & 0 \\ 0 & 0 & 1 & 0 \end{bmatrix} \tag{3.37}$$

$$H(k) = \begin{bmatrix} 1 & 0 & 0 & 0 & 0 \\ 0 & 0 & 1 & 0 & 0 \end{bmatrix} \tag{3.38}$$

The above discrete-time linear system can also be denoted by the block diagram in Figure 3.2, and the system contains the following prior information [16, 25]:

1. The initial state $X(0)$ is Gaussian, with mean value $\hat{X}(0|0)$ and covariance $P(0|0)$.
2. The initial state is unrelated to the process and measurement noise sequences.
3. The process and measurement noise sequences are not mutually related.

In the above hypothetical conditions, the linear property of the state equation [see (3.31)] and the measurement equation [see (3.32)] can maintain the Gaussian property of states and measurements.

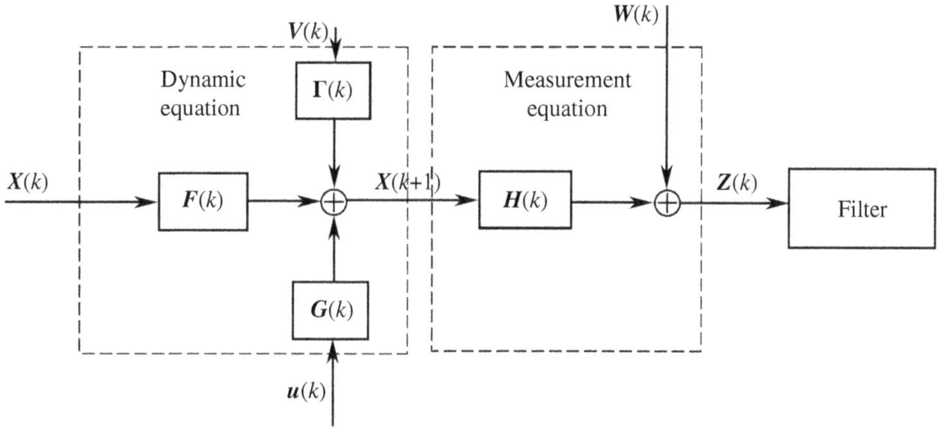

Figure 3.2 Discrete-time linear system

If the estimation of the state $\boldsymbol{X}(k)$ made at time k is marked as $\hat{\boldsymbol{X}}(k|j)$ according to the known measurements at time j and before time j, then the estimation can be summed up, in terms of the time indicated by the state estimation, in the following three types:

1. Filtering if $k=j$, with $\hat{\boldsymbol{X}}(k|j)$ as the filtering value of the state $\boldsymbol{X}(k)$ at time k.
2. Prediction if $k>j$, with $\hat{\boldsymbol{X}}(k|j)$ as the predicted value of the state $\boldsymbol{X}(k)$ at time k.
3. Smoothing if $k<j$, with $\hat{\boldsymbol{X}}(k|j)$ as the smoothed value of the state $\boldsymbol{X}(k)$ at time k.

From now on, we discuss prediction and filtering rather than smoothing.

3.2.2 Filtering Model

In all linear filters, the linear mean square estimation filter is optimal [25]. Filters under the linear mean square error criterion include the Wiener filter and the Kalman filter. These two filters are consistent in static conditions, but the Kalman filter, applicable to the non-stationarity of finite observation intervals, is a recursive algorithm for computation.

In Section 2.5.2 the MMSE estimate of the random vector $\boldsymbol{x}$ in the static (time-invariant) condition is

$$\hat{\boldsymbol{x}} = \mathrm{E}[\boldsymbol{x}|z] = \bar{\boldsymbol{x}} + \boldsymbol{P}_{xz}\boldsymbol{P}_{zz}^{-1}(z-\bar{z}) \tag{3.39}$$

with corresponding conditional error covariance matrix

$$\boldsymbol{P}_{xx|z} = \mathrm{E}\left[(\boldsymbol{x}-\hat{\boldsymbol{x}})(\boldsymbol{x}-\hat{\boldsymbol{x}})'|z\right] = \boldsymbol{P}_{xx} - \boldsymbol{P}_{xz}\boldsymbol{P}_{zz}^{-1}\boldsymbol{P}_{zx} \tag{3.40}$$

Similarly, the MMSE estimate in the dynamic condition can be defined as

$$\hat{\boldsymbol{x}} \to \hat{\boldsymbol{X}}(k|k) = \mathrm{E}\left[\boldsymbol{X}(k)|\boldsymbol{Z}^k\right] \tag{3.41}$$

where

$$\boldsymbol{Z}^k = \{\boldsymbol{Z}(j),\ \ j = 1, 2, \ldots, k\} \tag{3.42}$$

The state error covariance matrix accompanying (3.41) can be defined as

$$\boldsymbol{P}(k|k) = \mathrm{E}\left\{\left[\boldsymbol{X}(k) - \hat{\boldsymbol{X}}(k|k)\right]\left[\boldsymbol{X}(k) - \hat{\boldsymbol{X}}(k|k)\right]' |\boldsymbol{Z}^k\right\} = \mathrm{E}\left\{\tilde{\boldsymbol{X}}(k|k)\tilde{\boldsymbol{X}}'(k|k)|\boldsymbol{Z}^k\right\} \tag{3.43}$$

Applying the expectation operator with $\mathbf{Z}^k$ as the condition in (3.31), we obtain the one-step prediction of the state

$$\begin{aligned}\bar{\boldsymbol{x}} \rightarrow \hat{\boldsymbol{X}}(k+1|k) = \mathrm{E}\left[\boldsymbol{X}(k+1)\boldsymbol{Z}^k\right] &= \mathrm{E}\left[\boldsymbol{F}(k)\boldsymbol{X}(k) + \boldsymbol{G}(k)\boldsymbol{u}(k) + \boldsymbol{V}(k)|\boldsymbol{Z}^k\right] \\ &= \boldsymbol{F}(k)\hat{\boldsymbol{X}}(k|k) + \boldsymbol{G}(k)\boldsymbol{u}(k)\end{aligned} \tag{3.44}$$

The error of the predicted value is

$$\tilde{\boldsymbol{X}}(k+1|k) = \boldsymbol{X}(k+1) - \hat{\boldsymbol{X}}(k+1|k) = \boldsymbol{F}(k)\tilde{\boldsymbol{X}}(k|k) + \boldsymbol{V}(k) \tag{3.45}$$

The one-step predicted covariance is

$$\begin{aligned}\boldsymbol{P}_{xx} \rightarrow \boldsymbol{P}(k+1|k) &= \mathrm{E}\left[\tilde{\boldsymbol{X}}(k+1|k)\tilde{\boldsymbol{X}}'(k+1|k)|\boldsymbol{Z}^k\right] \\ &= \mathrm{E}\left\{\left[\boldsymbol{F}(k)\tilde{\boldsymbol{X}}(k|k) + \boldsymbol{V}(k)\right]\left[\tilde{\boldsymbol{X}}'(k|k)\boldsymbol{F}'(k) + \boldsymbol{V}'(k)\right]|\boldsymbol{Z}^k\right\} \\ &= \boldsymbol{F}(k)\boldsymbol{P}(k|k)\boldsymbol{F}'(k) + \boldsymbol{Q}(k)\end{aligned} \tag{3.46}$$

Note: The one-step predicted covariance $\boldsymbol{P}(k+1|k)$ is a symmetric matrix, which can be used to judge the prediction uncertainty, so the smaller $\boldsymbol{P}(k+1|k)$ is, the more accurate the prediction will be.

By taking the expectation values of (3.32) with Z^k as the condition at time $k+1$, we can similarly get the measurement prediction

$$\begin{aligned}\bar{\boldsymbol{Z}} \rightarrow \hat{\boldsymbol{Z}}(k+1|k) = \mathrm{E}\left[\boldsymbol{Z}(k+1)|\boldsymbol{Z}^k\right] &= \mathrm{E}\left[(\boldsymbol{H}(k+1)\boldsymbol{X}(k+1) + \boldsymbol{W}(k+1))|\boldsymbol{Z}^k\right] \\ &= \boldsymbol{H}(k+1)\hat{\boldsymbol{X}}(k+1|k)\end{aligned} \tag{3.47}$$

It follows that we can acquire the difference between the predictions and the measurements, as

$$\tilde{\boldsymbol{Z}}(k+1|k) = \boldsymbol{Z}(k+1) - \hat{\boldsymbol{Z}}(k+1|k) = \boldsymbol{H}(k+1)\tilde{\boldsymbol{X}}(k+1|k) + \boldsymbol{W}(k+1) \tag{3.48}$$

The prediction covariance (or innovation covariance) of the measurement is

$$\begin{aligned}\boldsymbol{P}_{zz} \rightarrow \boldsymbol{S}(k+1) &= \mathrm{E}\left[\tilde{\boldsymbol{Z}}(k+1|k)\tilde{\boldsymbol{Z}}'(k+1|k)|\boldsymbol{Z}^k\right] \\ &= \mathrm{E}\left\{\left[\boldsymbol{H}(k+1)\tilde{\boldsymbol{X}}(k+1|k) + \boldsymbol{W}(k+1)\right]\left[\tilde{\boldsymbol{X}}'(k+1|k)\boldsymbol{H}'(k+1) + \boldsymbol{W}'(k+1)\right]|\boldsymbol{Z}^k\right\} \\ &= \boldsymbol{H}(k+1)\boldsymbol{P}(k+1|k)\boldsymbol{H}'(k+1) + \boldsymbol{R}(k+1)\end{aligned} \tag{3.49}$$

Note: The innovation covariance $\boldsymbol{S}(k+1)$ is also a symmetric matrix, used to judge the uncertainty of innovations, so the smaller the innovation covariance is, the more accurate the measurements will be.

The covariance between the state and the measurement is

$$\boldsymbol{P}_{xz} \to \mathrm{E}\left[\widetilde{\boldsymbol{X}}(k+1|k)\widetilde{\boldsymbol{Z}}'(k+1|k)|\boldsymbol{Z}^k\right] = \mathrm{E}\left\{\widetilde{\boldsymbol{X}}(k+1|k)\left[\boldsymbol{H}(k+1)\widetilde{\boldsymbol{X}}(k+1|k)+\boldsymbol{W}(k+1)\right]'|\boldsymbol{Z}^k\right\} \\ = \boldsymbol{P}(k+1|k)\boldsymbol{H}'(k+1) \tag{3.50}$$

The gain is

$$\boldsymbol{P}_{xz}\boldsymbol{P}_{zz}^{-1} \to \boldsymbol{K}(k+1) = \boldsymbol{P}(k+1|k)\boldsymbol{H}'(k+1)\boldsymbol{S}^{-1}(k+1) \tag{3.51}$$

The amount of gain reflects the contribution of the recent measurement to the state estimate. It follows that we can get the state updating equation at time $k+1$:

$$\hat{\boldsymbol{X}}(k+1|k+1) = \hat{\boldsymbol{X}}(k+1|k) + \boldsymbol{K}(k+1)\boldsymbol{v}(k+1) \tag{3.52}$$

where $\boldsymbol{v}(k+1)$ is the innovation or the measurement residual, that is,

$$\boldsymbol{v}(k+1) = \widetilde{\boldsymbol{Z}}(k+1|k) = \boldsymbol{Z}(k+1) - \hat{\boldsymbol{Z}}(k+1|k) \tag{3.53}$$

Equation (3.52) shows that the estimation $\hat{\boldsymbol{X}}(k+1|k+1)$ at time $k+1$ is equal to the state prediction value $\hat{\boldsymbol{X}}(k+1|k)$ at this time plus a correction, which is related to the gain $\boldsymbol{K}(k+1)$ and the innovation.

The covariance updating equation is

$$\boldsymbol{P}(k+1|k+1) = \boldsymbol{P}(k+1|k) - \boldsymbol{P}(k+1|k)\boldsymbol{H}'(k+1)\boldsymbol{S}^{-1}(k+1)\boldsymbol{H}(k+1)\boldsymbol{P}(k+1|k) \tag{3.54}$$

$$= [\boldsymbol{I} - K(k+1)\boldsymbol{H}(k+1)]\boldsymbol{P}(k+1|k) \tag{3.55}$$

$$= \boldsymbol{P}(k+1|k) - \boldsymbol{K}(k+1)\boldsymbol{S}(k+1)\boldsymbol{K}'(k+1) \tag{3.56}$$

$$= [\boldsymbol{I} - \boldsymbol{K}(k+1)\boldsymbol{H}(k+1)]P(k+1|k)[\boldsymbol{I} + \boldsymbol{K}(k+1)\boldsymbol{H}(k+1)]' \\ - \boldsymbol{K}(k+1)\boldsymbol{R}(k+1)\boldsymbol{K}'(k+1) \tag{3.57}$$

where $\boldsymbol{I}$ is the unit matrix with the same dimensions as the covariance. Equation (3.57) can guarantee the symmetry and positive definiteness of the covariance matrix $\boldsymbol{P}$.

Another form of the filter gain is

$$P(k+1|k+1)H'(k+1)R^{-1}(k+1) \\ = [P(k+1|k)H'(k+1) - P(k+1|k)H'(k+1)S^{-1}(k+1)H(k+1)P(k+1|k)H'(k+1)]R^{-1}(k+1) \\ = P(k+1|k)H'(k+1)S^{-1}(k+1)[S(k+1) - H(k+1)P(k+1|k)H'(k+1)]R^{-1}(k+1) \\ = K(k+1) \tag{3.58}$$

Kalman filters are the best not only among all linear filters, but also among all filters when the process noise is Gaussian. Kalman filtering requires no conditions, except that the system noise and

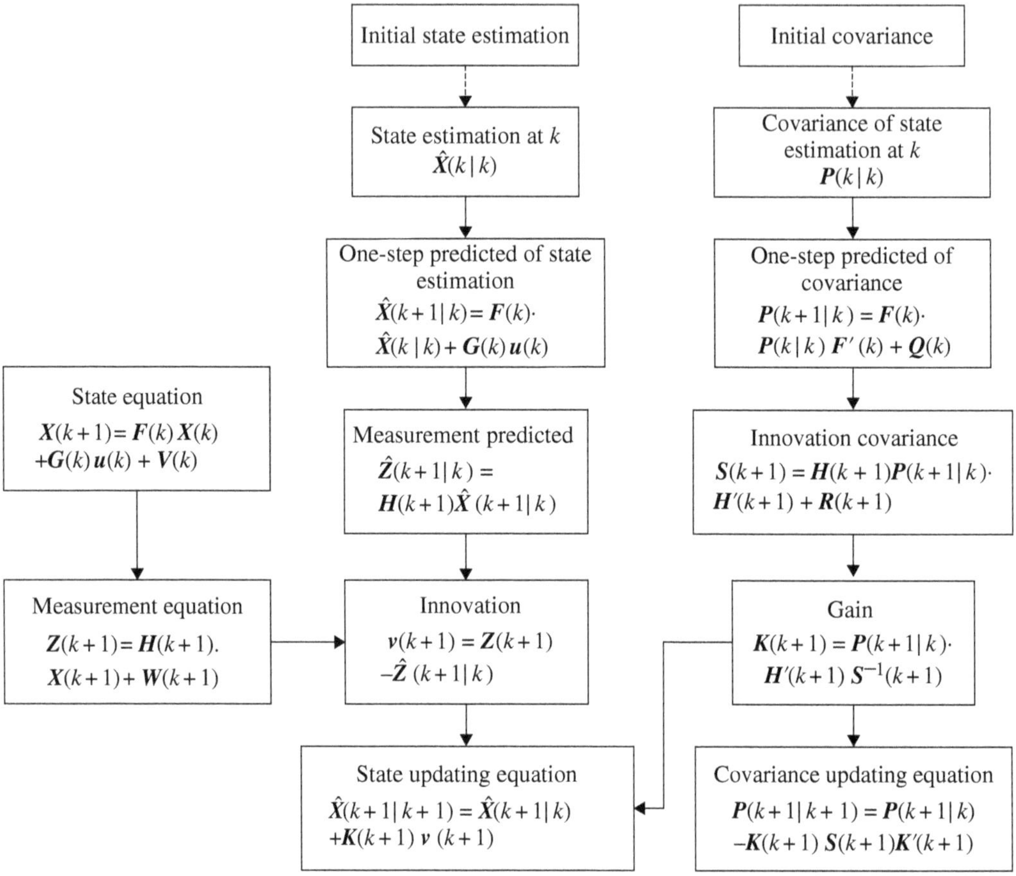

Figure 3.3 Kalman filter algorithm

the measurement noise be Gaussian white noise and that their secondary moments be known, so it is fully applicable to the estimation for non-stationary, multi-dimensional random sequences. Figure 3.3 shows the equations and filtering flow contained in the Kalman filter. Figure 3.4 illustrates one cycle of the Kalman filter; the rest of the cycles can be produced in the same way.

3.2.3 Initialization of Kalman Filters

The initialization of the state estimation discussed in this section is an important prerequisite for the use of the Kalman filter. Only when initialization is done can the Kalman filter be used to track targets.

3.2.3.1 Initialization of Two-Dimensional State Vector Estimation

The system's state equation and measurement equation are the same as (3.31) and (3.32), where the state vector is expressed as $\boldsymbol{X} = [x, \dot{x}]'$, the measurement noise as $W(k) \sim \mathrm{N}(0, r)$, and they are independent of the process noise. In this case the initialization of state estimation can be achieved by the

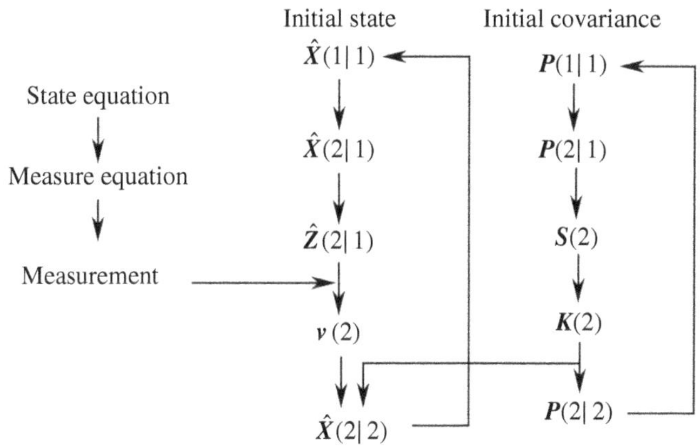

Figure 3.4 Single-cycle flow of Kalman filter algorithm

difference method of two points, which only uses two measurements $Z(0)$ and $Z(1)$ at the first and second time to make initialization, so the initial state is

$$\hat{X}(1|1)=\begin{bmatrix}\hat{x}(1|1)\\ \hat{\dot{x}}(1|1)\end{bmatrix}=\begin{bmatrix}Z(1)\\ \dfrac{Z(1)-Z(0)}{T}\end{bmatrix} \tag{3.59}$$

where T is the sampling interval. The initial covariance is

$$\boldsymbol{P}(1|1)=\begin{bmatrix}r & r/T\\ r/T & 2r/T^2\end{bmatrix} \tag{3.60}$$

Hence, the state estimation and filtering start from time $k=2$.

When the algorithm goes through the Monte Carlo test many times, there must be new noises in each test, and then the same method should be used for initialization. In Monte Carlo tests, the repeated use of the same initial conditions will lead to biased estimation, so for each test the initial state estimation should be randomly selected again. The two-dimensional Kalman filter is usually used in conditions of x, y, and z axis decoupling filtering.

3.2.3.2 Initialization of Four-Dimensional State Vector Estimation

The initialization in this case is the processing of two-coordinate radar data. If the system's state vector is denoted as

$$X(k)=[x\ \dot{x}\ y\ \dot{y}]' \tag{3.61}$$

and the measurement $\boldsymbol{Z}(k)$ in the Cartesian coordinate system is

$$\boldsymbol{Z}(k)=\begin{bmatrix}Z_1(k)\\ Z_2(k)\end{bmatrix}=\begin{bmatrix}x(k)\\ y(k)\end{bmatrix}=\begin{bmatrix}\rho\cos\theta\\ \rho\sin\theta\end{bmatrix} \tag{3.62}$$

where ρ and θ are, respectively, the target's radial distance and the azimuth measurement data in the polar coordinate system, then the system's initial state can be determined by the measurements **Z**(0) and **Z**(1) at the last two times, that is,

$$\hat{\boldsymbol{X}}(1|1)=\begin{bmatrix}Z_1(1) & \dfrac{Z_1(1)-Z_1(0)}{T} & Z_2(1) & \dfrac{Z_2(1)-Z_2(0)}{T}\end{bmatrix}' \tag{3.63}$$

The measurement noise covariance at time k in the Cartesian coordinate system is

$$R(k)=\begin{bmatrix}r_{11} & r_{12}\\ r_{12} & r_{22}\end{bmatrix}=A\begin{bmatrix}\sigma_\rho^2 & 0\\ 0 & \sigma_\theta^2\end{bmatrix}A' \tag{3.64}$$

where σ_ρ^2 and σ_θ^2 denote the covariance of the distance and azimuth measurement error, respectively, and

$$\boldsymbol{A}=\begin{bmatrix}\cos\theta & -\rho\sin\theta\\ \sin\theta & \rho\cos\theta\end{bmatrix} \tag{3.65}$$

From the elements in the measurement noise covariance, we get the initial covariance matrix in four-dimensional state vector conditions as

$$\boldsymbol{P}(1|1)=\begin{bmatrix}r_{11}(1) & r_{11}(1)/T & r_{12}(1) & r_{12}(1)/T\\ r_{11}(1)/T & 2r_{11}(1)/T^2 & r_{12}(1)/T & 2r_{12}(1)/T^2\\ r_{12}(1) & r_{12}(1)/T & r_{22}(1) & r_{22}(1)/T\\ r_{12}(1)/T & 2r_{12}(1)/T^2 & r_{22}(1)/T & 2r_{12}(1)/T^2\end{bmatrix} \tag{3.66}$$

3.2.3.3 Initialization of Six-Dimensional State Vector Estimation

This case describes the processing of three-coordinate radar data. If the system's state vector is denoted as

$$\boldsymbol{X}(k)=[x\ \ \dot{x}\ \ y\ \ \dot{y}\ \ z\ \ \dot{z}]' \tag{3.67}$$

then the measurement **Z**(k) in the Cartesian coordinate system is

$$\boldsymbol{Z}(k)=\begin{bmatrix}Z_1(k)\\ Z_2(k)\\ Z_3(k)\end{bmatrix}=\begin{bmatrix}x(k)\\ y(k)\\ z(k)\end{bmatrix}=\begin{bmatrix}\rho\cos\theta\cos\varepsilon\\ \rho\sin\theta\cos\varepsilon\\ \rho\sin\varepsilon\end{bmatrix} \tag{3.68}$$

where ρ and θ are defined as in four-dimensional vector conditions, and ε is the measurement data of target pitching. So, the initial state of the system is determined by the two measurements **Z**(0) and **Z**(1) at the last two times, that is,

$$\hat{X}(1|1) = \left[Z_1(1), \frac{Z_1(1)-Z_1(0)}{T}, Z_2(1), \frac{Z_2(1)-Z_2(0)}{T}, Z_3(1), \frac{Z_3(1)-Z_3(0)}{T}\right]' \tag{3.69}$$

In this case, the measurement noise covariance at time k in the Cartesian coordinate system is

$$\boldsymbol{R}(k) = \begin{bmatrix} r_{11} & r_{12} & r_{13} \\ r_{12} & r_{22} & r_{23} \\ r_{13} & r_{23} & r_{33} \end{bmatrix} = \boldsymbol{A}\begin{bmatrix} \sigma_\rho^2 & 0 & 0 \\ 0 & \sigma_\theta^2 & 0 \\ 0 & 0 & \sigma_\varepsilon^2 \end{bmatrix}\boldsymbol{A}' \tag{3.70}$$

where σ_ρ^2 and σ_θ^2 are defined as in four-dimensional vector conditions, σ_ε^2 is the variance of pitching measurement error, and

$$\boldsymbol{A} = \begin{bmatrix} \cos\theta\cos\varepsilon & -\rho\sin\theta\cos\varepsilon & -\rho\cos\theta\sin\varepsilon \\ \sin\theta\cos\varepsilon & \rho\cos\theta\cos\varepsilon & -\rho\sin\theta\sin\varepsilon \\ \sin\varepsilon & 0 & \rho\cos\varepsilon \end{bmatrix} \tag{3.71}$$

From the elements in the measurement noise covariance, we get the initial covariance matrix in the six-dimensional vector condition

$$\boldsymbol{P}(1|1) = \begin{bmatrix} r_{11}(1) & r_{11}(1)/T & r_{12}(1) & r_{11}(1)/T & r_{13}(1) & r_{13}(1)/T \\ r_{11}(1)/T & 2r_{11}(1)/T^2 & r_{12}(1)/T & 2r_{12}(1)/T^2 & r_{13}(1)/T & 2r_{13}(1)/T^2 \\ r_{12}(1) & r_{12}(1)/T & r_{22}(1) & r_{22}(1)/T & r_{23}(1) & r_{23}(1)/T \\ r_{12}(1)/T & 2r_{12}(1)/T^2 & r_{22}(1)T^2 & 2r_{22}(1)/T^2 & r_{23}(1)/T & 2r_{23}(1)/T^2 \\ r_{13}(1) & r_{13}(1)/T & r_{23}(1) & r_{23}(1)/T & r_{23}(1) & r_{33}(1)/T \\ r_{13}(1)/T & 2r_{13}(1)/T^2 & r_{13}(1)/T & 2r_{13}(1)/T^2 & r_{33}(1)/T & 2r_{13}(1)/T^2 \end{bmatrix} \tag{3.72}$$

3.2.3.4 Initialization of Nine-Dimensional State Vector Estimation

In this case, the system state vector is denoted as

$$\boldsymbol{X}(k) = [x \ \dot{x} \ \ddot{x} \ y \ \dot{y} \ \ddot{y} \ z \ \dot{z} \ \ddot{z}]' \tag{3.73}$$

Compared with the six-dimensional case, only the acceleration item is added, so the target measurement value $\boldsymbol{Z}(k)$ and the measurement noise covariance $\boldsymbol{R}(k)$ in the Cartesian coordinate system are the same as those in six-dimensional conditions.

Because the acceleration is contained, the initial state of the system should be determined by the measurement values $\boldsymbol{Z}(0)$, $\boldsymbol{Z}(1)$, and $\boldsymbol{Z}(2)$ at the last three times, that is,

$$\hat{X}(2|2)=\begin{bmatrix} Z_1(2) \\ (Z_1(2)-Z_1(1))/T \\ [(Z_1(2)-Z_1(1))/T-(Z_1(1)-Z_1(0))/T]/T \\ Z_2(2) \\ (Z_2(2)-Z_2(1))/T \\ [(Z_2(2)-Z_2(1))/T-(Z_2(1)-Z_2(0))/T]/T \\ Z_3(2) \\ (Z_3(2)-Z_3(1))/T \\ [(Z_3(2)-Z_3(1))/T-(Z_3(1)-Z_3(0))/T]/T \end{bmatrix} \tag{3.74}$$

The initial covariance matrix is

$$\boldsymbol{P}(2|2)=\begin{bmatrix} \boldsymbol{P}_{11} & \boldsymbol{P}_{12} & \boldsymbol{P}_{13} \\ \boldsymbol{P}_{12} & \boldsymbol{P}_{22} & \boldsymbol{P}_{23} \\ \boldsymbol{P}_{13} & \boldsymbol{P}_{23} & \boldsymbol{P}_{33} \end{bmatrix} \tag{3.75}$$

where $\boldsymbol{P}_{11}$, $\boldsymbol{P}_{12}$, $\boldsymbol{P}_{13}$, $\boldsymbol{P}_{22}$, $\boldsymbol{P}_{23}$, and $\boldsymbol{P}_{33}$ are block matrixes, and

$$\boldsymbol{P}_{ij}=\begin{bmatrix} r_{ij}(2) & \dfrac{r_{ij}(2)}{T} & \dfrac{r_{ij}(2)}{T^2} \\ \dfrac{r_{ij}(2)}{T} & \dfrac{r_{ij}(2)+r_{ij}(1)}{T^2} & \dfrac{r_{ij}(2)+2r_{ij}(1)}{T^3} \\ \dfrac{r_{ij}(2)}{T^2} & \dfrac{r_{ij}(2)+2r_{ij}(1)}{T^3} & \dfrac{r_{ij}(2)+4r_{ij}(1)+r_{ij}(0)}{T^4} \end{bmatrix}, \quad i=1,2,3, \ j=1,2,3 \tag{3.76}$$

3.3 Steady-State Kalman Filter

In Section 3.2 we presented in detail the basic equations of Kalman filtering for linear systems. Since the Kalman filter adopts a recursive calculation, the initial values of the state and the estimation error variance matrix must be given when the algorithm starts to calculate. When $\hat{\boldsymbol{X}}(0|0)=\mathrm{E}[\boldsymbol{X}(0)]$ and $\mathrm{E}\left[\widetilde{\boldsymbol{X}}(0|0)\widetilde{\boldsymbol{X}}'(0|0)\right]$, the filtering estimation is unbiased from the beginning, and the estimation error covariance matrix is the least. However, in practical applications, the initial state estimation and the initial covariance matrix of Kalman filters only make a hypothesis or estimation based on the measurement data, in other words, the values used for the initial state estimation and the initial covariance matrix are not the mean values and the corresponding estimation error covariance matrix. So how does the deviation of the initial state estimation from the hypothetical case of the initial covariance matrix affect the filtering results? Do these effects get stronger as the filtering time gets longer to cause filtering divergence, or wear off when the filtering goes on? This section will study under what conditions these effects weaken, that is, the effects of filtering initial values on Kalman filtering and the steady-state Kalman filter.

3.3.1 Mathematical Definition and Judgment Methods for Filter Stability

3.3.1.1 Mathematical Definition of Filter Stability

For any given positive number $\varepsilon > 0$, if another positive number $\delta > 0$ can be found and the initial states $\hat{\boldsymbol{X}}(0|0)$, $i = 1,2$ satisfy [20]

$$\left\| \hat{\boldsymbol{X}}^1(0|0) - \hat{\boldsymbol{X}}^2(0|0) \right\| < \delta \tag{3.77}$$

then there exists

$$\left\| \hat{\boldsymbol{X}}^1(k|k) - \hat{\boldsymbol{X}}^2(k|k) \right\| < \varepsilon \;\; \forall k \tag{3.78}$$

and the filter is steady.

Since filter stability means that the effect of the initial estimation value selected on the filtering value $\hat{\boldsymbol{X}}(k|k)$ lessens as the filtering time increases, in what conditions does the selection of the initial values of the filter gradually produce no effect on the filtering results? That is, in what conditions is the filtering steady? How should we make a judgment? That is what we will discuss next.

3.3.1.2 Stability Judgment

If the random linear system is consistently and completely controllable and consistently and completely observable, then the Kalman filter is consistently asymptotically steady [20], that is, when the filtering time is long enough, the Kalman filtering value will become asymptotically independent of the selection of the filtering initial value. The controllability of the random linear system is used to describe the ability of the system random noise to affect the system state, and observability means being able to obtain the target position information from observations by using a certain algorithm. For a random linear definite constant system, being consistently and completely controllable and consistently and completely observable can be thought of, respectively, as being completely controllable and completely observable. What is it to be completely controllable? And what is it to be completely observable? These questions will be discussed in the following section.

3.3.2 Controllability and Observability of Random Linear System

Definition (completely controllable): The necessary and sufficient conditions for the complete controllability of the random linear discrete system are that there exists a positive integer N, which makes the controllability matrix satisfy

$$\sum_{i=k-N+1}^{k} \boldsymbol{F}_{ik}\boldsymbol{\Gamma}_{i-1}\boldsymbol{Q}_{i-1}\boldsymbol{\Gamma}'_{i-1}\boldsymbol{F}'_{ik} > 0 \tag{3.79}$$

where $\boldsymbol{F}_{ik}$ is the state transition matrix from moment i to k, and $\boldsymbol{\Gamma}_{i-1}$ is the simplified form of the process noise distribution matrix $\boldsymbol{\Gamma}_{(i-1)i}$.

For a random linear definite constant system with identical sampling interval, the state transition matrix $\boldsymbol{F}_{ki}$ from time i to k can be expressed as $(k-i)$ one-step state transition matrixes multiplied, that is, $\boldsymbol{F}_{ki}=\boldsymbol{F}^{k-i}$, so from (3.79) we get

$$\sum_{i=k-N+1}^{k} \boldsymbol{F}^{k-i}\boldsymbol{\Gamma}\boldsymbol{Q}_{i-1}\boldsymbol{\Gamma}'(\boldsymbol{F}^{k-i})>0 \tag{3.80}$$

Definition (completely observable): The necessary and sufficient conditions for the random linear discrete system to be completely observable are that there exists a positive integer N, which makes the matrix satisfy

$$\sum_{i=k-N+1}^{k} \boldsymbol{F}'_{jk}\boldsymbol{H}'_j\boldsymbol{R}_j^{-1}\boldsymbol{H}_j\boldsymbol{F}_{jk}>0 \tag{3.81}$$

For a random linear definite constant system, the state transition matrix from time j to k can be expressed as $(k-j)$ one-step state transition matrixes multiplied, that is, $\boldsymbol{F}_{jk}=\boldsymbol{F}^{k-j}$, so from (3.81) we get

$$\sum_{j=k-N+1}^{k} (\boldsymbol{F}^{k-j})'\boldsymbol{H}'_j\boldsymbol{R}_j^{-1}\boldsymbol{H}_j\boldsymbol{F}^{k-j}>0 \tag{3.82}$$

Equations (3.80) and (3.82) are to obtain a sum, and both $(k-i)$ and $(k-j)$ vary within 0 to $(N-1)$, so they can be further streamlined by taking common factors, which will not be stated in detail.

If a random system is observable, it is perfectly possible to obtain the target position information from observations by using a certain algorithm.

For an ordinary system, if $\boldsymbol{Q}_{i-1}>0$ and $\boldsymbol{R}_j>0$, we can deduce that the necessary and sufficient conditions under which the random linear definite constant system is completely controllable and completely observable are

$$\sum_{l=0}^{n-1}\boldsymbol{F}^l\boldsymbol{\Gamma}\boldsymbol{\Gamma}'(\boldsymbol{F}^l)'>0 \tag{3.83}$$

$$\sum_{l=0}^{n-1}(\boldsymbol{F}^l)'\boldsymbol{H}'\boldsymbol{H}\boldsymbol{F}^l>0 \tag{3.84}$$

where n is the dimension number of state variables.

From the above analysis, we find that the controllability of a random linear definite constant system is related to its state transition matrix and process noise distribution matrix, while the observability is related to its state transition matrix and the observation matrix. When $\boldsymbol{P}(0|0)$, $\boldsymbol{Q}(k)$, and $\boldsymbol{R}(k)$ cannot be acquired accurately, if the possible range of their values is known, then we can use their possible and larger values, that is, the conservative values, which can prevent the actual estimation error variance matrix from diverging [20].

3.3.3 Steady-State Kalman Filter

When the observation time becomes longer and longer, steady-state Kalman filters can be used to describe the characteristics of the one-step prediction covariance and the state updating covariance.

In other words, when the observation time $k \to \infty$, the steady-state Kalman filter can be used to describe whether the covariance has a definite limit value, and under what conditions the definite limit value occurs. The one-step state prediction in (3.44) and the state updating equation in (3.52) can be combined to create a single recursive formula of the one-step prediction of the state:

$$\begin{aligned}\hat{\boldsymbol{X}}(k+1|k) &= \boldsymbol{F}(k)\hat{\boldsymbol{X}}(k|k)+\boldsymbol{G}(k)\boldsymbol{u}(k)\\ &= \boldsymbol{F}(k)\left[\hat{\boldsymbol{X}}(k|k-1)+\boldsymbol{K}(k)\boldsymbol{v}(k)\right]+\boldsymbol{G}(k)\boldsymbol{u}(k)\\ &= \boldsymbol{F}(k)\left[\hat{\boldsymbol{X}}(k|k-1)+\boldsymbol{F}(k)\boldsymbol{K}(k)\boldsymbol{Z}(k)-\boldsymbol{H}(k)\hat{\boldsymbol{X}}(k|k-1)\right]+\boldsymbol{G}(k)\boldsymbol{u}(k)\\ &= \boldsymbol{F}(k)[\boldsymbol{I}-\boldsymbol{K}(k)\boldsymbol{H}(k)]\hat{\boldsymbol{X}}(k|k-1)+\boldsymbol{F}(k)\boldsymbol{K}(k)\boldsymbol{Z}(k)+\boldsymbol{G}(k)\boldsymbol{u}(k)\end{aligned} \tag{3.85}$$

Similarly, we can obtain a single recursive formula of the one-step prediction covariance, that is, the discrete-time matrix Riccati equation

$$\begin{aligned}\boldsymbol{P}(k+1|k) &= \boldsymbol{F}(k)\boldsymbol{P}(k|k)\boldsymbol{F}'(k)+\boldsymbol{Q}(k)\\ &= \boldsymbol{F}(k)\left[\boldsymbol{P}(k|k-1)-\boldsymbol{P}(k|k-1)\boldsymbol{H}'(k)\boldsymbol{S}^{-1}(k)\boldsymbol{H}(k)\boldsymbol{P}(k|k-1)\boldsymbol{F}'(k)+\boldsymbol{Q}(k)\right.\\ &= \boldsymbol{F}(k)[\boldsymbol{P}(k|k-1)-\boldsymbol{P}(k|k-1)\boldsymbol{H}'(k)[\boldsymbol{H}(k)\boldsymbol{P}(k|k-1)\boldsymbol{H}'(k)\\ &\quad +\boldsymbol{R}(k)]^{-1}\boldsymbol{H}(k)\boldsymbol{P}(k|k-1)]\boldsymbol{F}'(k)+\boldsymbol{Q}(k)\end{aligned} \tag{3.86}$$

From (3.86) we find that the covariance $\boldsymbol{P}(k+1|k)$ at time $(k+1)$ predicted at time k is related only to the one-step prediction covariance $\boldsymbol{P}(k|k-1)$ at the last moment, the process noise covariance matrix $\boldsymbol{Q}(k)$, and the measurement noise covariance matrix $\boldsymbol{R}(k)$, and is unrelated directly to the measurement $\boldsymbol{Z}(k+1)$. So, in some particular conditions the one-step prediction covariance matrix can be calculated iteratively before the measurement.

If the system is time-invariant, that is, the state transition matrix $\boldsymbol{F}$ and the measurement matrix $\boldsymbol{H}$ are constant matrixes, and since the input is generally believed to be zero, all that is typically needed is that $\boldsymbol{F}$ and $\boldsymbol{H}$ are constant matrixes and the noise is steady, that is, $\boldsymbol{Q}$ and $\boldsymbol{R}$ are constant matrixes, and satisfy the following conditions:

1. $\boldsymbol{F}$ and $\boldsymbol{H}$ are completely observable;
2. $\boldsymbol{F}$ and $\boldsymbol{D}$ (the standard deviation of the process noise, that is, $\boldsymbol{Q}=\boldsymbol{D}\boldsymbol{D}'$) are completely controllable.

Then, with $k \to \infty$, the solution to the Riccati equation [see (3.86)] converges to a definite positive matrix $\bar{\boldsymbol{P}}$. To be specific, if the random linear system is consistently and completely controllable and observable, the Kalman filter will be consistently and asymptotically steady, and there exists only one definite positive matrix $\bar{\boldsymbol{P}}$, so that starting from any initial covariance matrix, $\boldsymbol{P}(0|0)\boldsymbol{P}(k+1|k) \to \bar{\boldsymbol{P}}$ when $k \to \infty$. At the same time, the Kalman filter steady-state gain produced by the constant covariance matrix $\bar{\boldsymbol{P}}$ is

$$\bar{\boldsymbol{K}} = \bar{\boldsymbol{P}}\boldsymbol{H}'\boldsymbol{S}^{-1}$$

The solution to $\boldsymbol{P}(k+1|k)$ determines the gain matrix of the discrete Kalman filter. So, when a completely observable and controllable random linear definite constant system reaches steady state,

$P(k|k-1) \to \bar{P}$, $P(k+1|k) \to \bar{P}$, $K(k+1) \to K$. From (3.86) we find that the Riccati difference equation degenerates to the Riccati algebraic equation

$$\bar{P} = F\left[\bar{P} - \bar{P}H'[H\bar{P}H' + R]^{-1}H\bar{P}\right]F' + Q \tag{3.87}$$

Regardless of the value of $P(k|k)$, the system process noise variance matrix $Q(k)$ always guarantees that a value is available for $P(k+1|k)$, and the measurement noise variance matrix $R(k)$ always guarantees that a value is available for $S(k+1)$, thus ensuring that a value is available for the gain $K(k+1)$ and so the calculation at each step is able to correct the estimation of the prior step with the observed update information and get a new real-time estimation. In addition, once the system reaches steady state, it will be controlled by the time constant type of (3.85):

$$\hat{X}(k+1|k) = F[I - \bar{K}H]\hat{X}(k|k-1) + F\bar{K}Z(k) + Gu(k) \tag{3.88}$$

Ideally, Kalman filtering is a linear unbiased estimation with minimum variance. According to the filtering stability theory, for consistently and completely controllable and observable systems, the steady filtering effect is irrelevant to the selection of the filtering initial values with the passage of time and the increase in number of measurements. The filtering estimation precision becomes higher, and the filtering error variance matrix tends to approach a steady-state value, or be bounded. That is to say, the filter is steady. These conclusions are based on the precondition of accurate system mathematical models. However, in practice, the state estimation obtained from filtering can be biased, and the estimation error variance could be massive, considerably beyond the variance scope given by the algorithm formula. Even worse, both the mean value and variance of the filtering error can approach infinity, leading to a divergence in filtering. Apparently, the filtering loses its effect when divergence occurs. Therefore, this must be restrained in practice.

3.4 Summary

This chapter focuses on the introduction of the Kalman filter in linear system conditions, including system model building, related filtering models, filter initialization, definition and judgment methods for filter stability, controllability and observability of random linear systems, steady-state Kalman filtering, etc.

The Kalman filtering algorithm takes precise mathematical modeling as a precondition, which requires the establishment of system equations and observation equations in the state space. The most difficult part of target tracking system modeling is building the system noise model, because the system noise directly reflects the maneuver features of the target in the system equation. Therefore, in practical applications, mismatches often occur between the model and the system. This type of difference between the theoretical model and the practical model is called "model error." However, it is not easy to set up precise mathematical models. Adaptive filtering theory has thus been created in order to further approach precise matching between the model and the system. How can we judge if the filter matches the system with observations? When the actual state changes, how can we modify the system model and filtering gain? This is what we need to solve by adaptive filtering theory.

4

Nonlinear Filtering Approaches

4.1 Introduction

In Chapter 3 we discussed and analyzed the filtering methods in linear systems, but many modern sensors such as infrared devices, electronic support measures (ESM), and passive sonar are passive detection systems, whose models cannot be established with linear systems [73–75]. The methods discussed in Chapter 3 cannot be used for nonlinear systems, so this chapter will introduce nonlinear filtering techniques. We will focus on some commonly used nonlinear filtering methods, including the extended Kalman filter (EKF), unscented Kalman filter (UKF), and particle filter (PF), and compare and analyze the three nonlinear filtering methods mentioned above through some simulation experiments [76–78].

4.2 Extended Kalman Filter

Kalman filters obtain dynamic estimation of targets under the linear Gaussian assumption by using the MMSE criterion, but in many actual cases the relations between the observed data and the dynamic parameters of targets are nonlinear. So far, perfect solutions remain to be found for nonlinear systems. The usual approach is to turn nonlinear filtering into approximate linear filtering using linearization techniques, and apply linear filtering theory to the suboptimal filtering algorithms for the original nonlinear filtering problems. The most commonly used linearization method is the Taylor series expansion, by which the filtering method of EKF [79–84] is achieved.

Radar Data Processing with Applications, First Edition. He You, Xiu Jianjuan, and Guan Xin.

4.2.1 Filter Model

The state equation of the nonlinear system is

$$X(k+1)=f(k,X(k))+V(k) \tag{4.1}$$

For simplicity, assume that there is no control input item, and that the process noise is the additive white noise with zero mean and variance

$$\mathrm{E}[V(k)V'(j)]=Q(k)\delta_{kj} \tag{4.2}$$

The measurement equation is

$$Z(k)=h[k,X(k)]+W(k) \tag{4.3}$$

where the measurement noise is also assumed to be the additive white noise with zero mean and variance

$$\mathrm{E}[W(k)W'(j)]=R(k)\delta_{kj} \tag{4.4}$$

Assume that the process and measurement noise sequences are not correlated, and possess the initial state estimation $\hat{X}(0|0)$ and the covariance matrix $P(0|0)$. Just as under linear conditions, we assume that the estimation at time k is

$$\hat{X}(k|k)\approx\mathrm{E}\left[X(k)|Z^k\right] \tag{4.5}$$

This is an approximate conditional mean value, and its companion covariance matrix is $P(k|k)$. Because $\hat{X}(k|k)$ is not a precise conditional mean value, strictly speaking, $P(k|k)$ is an approximate mean square error instead of a covariance. But it is customarily regarded as a covariance.

In order to obtain the predicted state $\hat{X}(k+1|k)$ for the nonlinear function in (4.1), we conduct a Taylor series expansion around $\hat{X}(k|k)$ to get the first-order or second-order terms with the aim of yielding a first-order or second-order EKF. The Taylor series expansion with second-order terms is

$$\begin{aligned}X(k+1)=&f\left(k,\hat{X}(k|k)\right)+f_X(k)\left[X(k)-\hat{X}(k|k)\right]\\&+\frac{1}{2}\sum_{i=1}^{n_x}e_i\left[X(k)-\hat{X}(k|k)\right]'f_{XX}^i(k)\left[X(k)-\hat{X}(k|k)\right]+(\text{higher-order terms})+V(k)\end{aligned} \tag{4.6}$$

where n_x is the number of dimensions of the state vector $X(k)$ and e_i is the ith Cartesian basic vector. For example, in the four-dimensional condition, there are four Cartesian basic vectors as follows:

$$e_1=\begin{bmatrix}1\\0\\0\\0\end{bmatrix}\quad e_2=\begin{bmatrix}0\\1\\0\\0\end{bmatrix}\quad e_3=\begin{bmatrix}0\\0\\1\\0\end{bmatrix}\quad e_4=\begin{bmatrix}0\\0\\0\\1\end{bmatrix} \tag{4.7}$$

In addition,

$$
f_X(k) = [\nabla_X f'(k,X)]'_{X=\hat{X}(k|k)} = \left[\begin{bmatrix} \partial/\partial x_1 \\ \vdots \\ \partial/\partial x_n \end{bmatrix} [f_1(X) \ \cdots \ f_n(X)] \right]'_{X=\hat{X}(k|k)}
$$

$$
= \begin{bmatrix} \frac{\partial f_1(X)}{\partial x_1} & \cdots & \frac{\partial f_n(X)}{\partial x_1} \\ \vdots & \cdots & \vdots \\ \frac{\partial f_1(X)}{\partial x_n} & \cdots & \frac{\partial f_n(X)}{\partial x_n} \end{bmatrix}'_{X=\hat{X}(k|k)} \tag{4.8}
$$

is the Jacobian matrix of the vector f, taking values from the latest state estimation, where $x_1, x_2, \ldots, x_{n_x}$ are the elements in the n_x-dimensional vector $X(k)$. Similarly, we get the Hessian matrix of the ith component of the vector f as follows:

$$
f^i_{XX}(k) = \left[\nabla_X \nabla'_X f^i(k,X)\right]_{X=\hat{X}(k|k)} = \begin{bmatrix} \frac{\partial^2 f^i(X)}{\partial x_1 \partial x_1} & \cdots & \frac{\partial^2 f^i(X)}{\partial x_1 \partial x_n} \\ \vdots & \cdots & \vdots \\ \frac{\partial^2 f^i(X)}{\partial x_n \partial x_1} & \cdots & \frac{\partial^2 f^i(X)}{\partial x_n \partial x_n} \end{bmatrix}_{X=\hat{X}(k|k)} \tag{4.9}
$$

The prediction of the state from time k to $(k+1)$ is obtained by taking the expectation of (4.6) conditional on Z^k. Neglecting higher-order terms, we get

$$
\hat{X}(k+1|k) = \mathrm{E}\left[X(k+1)|Z^k\right] = f\left(k, \hat{X}(k|k)\right) + \frac{1}{2}\sum_{i=1}^{n_x} e_i \mathrm{tr}\left[f^i_{XX}(k)P(k|k)\right] \tag{4.10}
$$

where we applied the identical equation

$$
\mathrm{E}[X'AX] = \mathrm{E}[\mathrm{tr}(AXX')] = \mathrm{tr}(AP) \tag{4.11}
$$

From (4.6) and (4.10) we obtain the estimation error of the prediction of the state with higher-order terms neglected, that is,

$$
\begin{aligned}
\tilde{X}(k+1|k) &= X(k+1) - \hat{X}(k+1|k) \\
&= f_X(k)\tilde{X}(k|k) + \frac{1}{2}\sum_{i=1}^{n_x} e_i \left[\tilde{X}'(k|k) f^i_{XX}(k)\tilde{X}(k|k) - \mathrm{tr}\left[f^i_{XX}(k)P(k|k)\right]\right] + V(k)
\end{aligned} \tag{4.12}
$$

Based on (4.12), we get the covariance accompanying (4.10), represented as

$$\begin{aligned}\boldsymbol{P}(k+1|k) &= \mathrm{E}\left[\tilde{\boldsymbol{X}}(k+1|k)\tilde{\boldsymbol{X}}'(k+1|k)|\boldsymbol{Z}^k\right] \\ &= \boldsymbol{f}_X(k)\boldsymbol{P}(k|k)\boldsymbol{f}'_X(k) + \frac{1}{2}\sum_{i=1}^{n_x}\sum_{j=1}^{n_x}\boldsymbol{e}_i\boldsymbol{e}'_j \ \mathrm{tr}\left[\boldsymbol{f}^i_{XX}(k)\boldsymbol{P}(k|k)\boldsymbol{f}^j_{XX}(k)\boldsymbol{P}(k|k)\right] + \boldsymbol{Q}(k)\end{aligned} \tag{4.13}$$

Here we applied the identical equation

$$\mathrm{E}\{[\boldsymbol{X}'A X - \mathrm{E}(\boldsymbol{X}'AX)][\boldsymbol{X}'\boldsymbol{B}\boldsymbol{X} - \mathrm{E}(\boldsymbol{X}'\boldsymbol{B}\boldsymbol{X})]\} = 2\,\mathrm{tr}(APBP) \tag{4.14}$$

For the second-order EKF, the prediction of measurement is

$$\hat{\boldsymbol{Z}}(k+1|k) = \boldsymbol{h}\left(k+1, \hat{\boldsymbol{X}}(k+1|k)\right) + \frac{1}{2}\sum_{i=1}^{n_z}\boldsymbol{e}_i\,\mathrm{tr}\left[\boldsymbol{h}^i_{XX}(k+1)\boldsymbol{P}(k+1|k)\right] \tag{4.15}$$

and the companion covariance (approximate mean square error) is

$$\begin{aligned}\boldsymbol{S}(k+1) &= \boldsymbol{h}_X(k+1)\boldsymbol{P}(k+1|k)\boldsymbol{h}'_X(k+1) \\ &+ \frac{1}{2}\sum_{i=1}^{n_z}\sum_{j=1}^{n_z}\boldsymbol{e}_i\boldsymbol{e}'_j\,\mathrm{tr}\left[\boldsymbol{h}^i_{XX}(k+1)\boldsymbol{P}(k+1|k)\boldsymbol{h}^j_{XX}(k+1)\boldsymbol{P}(k+1|k)\right] + \boldsymbol{R}(k+1)\end{aligned} \tag{4.16}$$

where $\boldsymbol{h}_X(k+1)$ is the Jacobian matrix, that is,

$$\boldsymbol{h}_X(k+1) = [\nabla_X \boldsymbol{h}'(k+1, \boldsymbol{X})]'_{X=\hat{X}(k+1|k)} \tag{4.17}$$

The Hessian matrix of the ith component is

$$\boldsymbol{h}^i_{XX}(k+1) = \left[\nabla_X \nabla'_X \boldsymbol{h}^i(k+1, \boldsymbol{X})\right]'_{X=\hat{X}(k+1|k)} \tag{4.18}$$

The gain is

$$\boldsymbol{K}(k+1) = \boldsymbol{P}(k+1|k)\boldsymbol{h}'_X(k+1)\boldsymbol{S}^{-1}(k+1) \tag{4.19}$$

The state update equation is

$$\hat{\boldsymbol{X}}(k+1|k+1) = \hat{\boldsymbol{X}}(k+1|k) + \boldsymbol{K}(k+1)\left\{\boldsymbol{Z}(k+1) - \boldsymbol{h}\left[k+1, \hat{\boldsymbol{X}}(k+1|k)\right]\right\} \tag{4.20}$$

The covariance update equation is

$$\begin{aligned}\boldsymbol{P}(k+1|k+1) &= [\boldsymbol{I} - \boldsymbol{K}(k+1)\boldsymbol{h}_X(k+1)]\boldsymbol{P}(k+1|k)[\boldsymbol{I} + \boldsymbol{K}(k+1)\boldsymbol{h}_X(k+1)]' \\ &- \boldsymbol{K}(k+1)\boldsymbol{R}(k+1)\boldsymbol{K}'(k+1)\end{aligned} \tag{4.21}$$

where $\boldsymbol{I}$ is the unit matrix.

Equations (4.10), (4.13), (4.16), and (4.19)–(4.21) constitute the formula system for the second-order EKF. From (4.19)–(4.21) we can see that in nonlinear conditions the gain, the state update equation, and the covariance update equation are similar to those in linear conditions, except that the measurement matrix $\boldsymbol{H}(k+1)$ is replaced by the Jacobian $\boldsymbol{h}_X(k+1)$ in this case. The way to obtain the first-order EKF formulas is similar to that for the second-order case, except that here Taylor series expansion is reserved only for first-order terms, that is,

$$\boldsymbol{X}(k+1)=\boldsymbol{f}\left[k,\hat{\boldsymbol{X}}(k|k)\right]+\boldsymbol{f}_X(k)\left[\boldsymbol{X}(k)-\hat{\boldsymbol{X}}(k|k)\right]+(\text{higher-order terms})+\boldsymbol{V}(k) \tag{4.22}$$

So, the first-order EKF formulas include:

- The one-step prediction of state, given by

$$\hat{\boldsymbol{X}}(k+1|k)=\boldsymbol{f}\left[k,\hat{\boldsymbol{X}}(k|k)\right] \tag{4.23}$$

- The one-step prediction of covariance, given by

$$\boldsymbol{P}(k+1|k)=\boldsymbol{f}_X(k)\boldsymbol{P}(k|k)\boldsymbol{f}'_X(k)+\boldsymbol{Q}(k) \tag{4.24}$$

- The prediction of measurements, given by

$$\hat{\boldsymbol{Z}}(k+1|k)=\boldsymbol{h}\left[k+1,\hat{\boldsymbol{X}}(k+1|k)\right] \tag{4.25}$$

- The companion covariance, given by

$$\boldsymbol{S}(k+1)=\boldsymbol{h}_X(k+1)\boldsymbol{P}(k+1|k)\boldsymbol{h}'_X(k+1)+\boldsymbol{R}(k+1) \tag{4.26}$$

- The gain, given by

$$\boldsymbol{K}(k+1)=\boldsymbol{P}(k+1|k)\boldsymbol{h}'_X(k+1)\boldsymbol{S}^{-1}(k+1) \tag{4.27}$$

- The state update equation, given by

$$\hat{\boldsymbol{X}}(k+1|k+1)=\hat{\boldsymbol{X}}(k+1|k)+\boldsymbol{K}(k+1)\left\{\boldsymbol{Z}(k+1)-\boldsymbol{h}\left[k+1,\hat{\boldsymbol{X}}(k+1|k)\right]\right\} \tag{4.28}$$

- The covariance update equation, given by

$$\begin{aligned}\boldsymbol{P}(k+1|k+1)=&[\boldsymbol{I}-\boldsymbol{K}(k+1)\boldsymbol{h}_X(k+1)]\boldsymbol{P}(k+1|k)[\boldsymbol{I}+\boldsymbol{K}(k+1)\boldsymbol{h}_X(k+1)]'\\&-\boldsymbol{K}(k+1)\boldsymbol{R}(k+1)\boldsymbol{K}'(k+1)\end{aligned} \tag{4.29}$$

where $\boldsymbol{I}$ is the unit matrix with the same number of dimensions as the covariance.

The covariance prediction formula for the first-order EKF is similar to that in linear filtering, but the Jacobian $f_x(k)$ is similar to the system state transition matrix $\boldsymbol{F}(k)$ [85, 86]. If the Taylor series expansion is reserved for third and fourth-order terms, we have the third and fourth-order EKFs. R. J. Phanenf carried out a simulation analysis on the performance of EKFs with different orders [87]. The simulation results showed that the performance of the second-order EKF is much better than that of the first-order one, but the performance of EKFs with higher orders is not obviously better than that of the second-order one, so EKFs with higher orders are not generally used. Although the second-order EKF is superior to the first-order one in performance, the amount of calculation required is huge, so we usually use the first-order EKF algorithm.

4.2.2 Some Problems in the Application of Extended Kalman Filters

Extended Kalman filtering is a widely used nonlinear filtering method, in which the existence of some nonlinear factors has a powerful influence on the filter stability and state estimation accuracy, and the filtering results have a lot to do with the statistical features of the process noise and the measurement noise [83]. Because in EKFs the pre-estimated process noise covariance $\boldsymbol{Q}(k)$ and the measurement noise covariance $\boldsymbol{R}(k)$ remain unchanged during filtering, the inaccurate estimation of these two noise covariance matrixes can lead to error accumulation and divergence in the filtering process. Additionally, for nonlinear systems with many dimensions, anomalies are likely to arise in the estimated process noise covariance matrix and measurement noise covariance matrix, that is, $\boldsymbol{Q}(k)$ loses half positive definiteness and $\boldsymbol{R}(k)$ loses positive definiteness, which can also cause filtering divergence. As for the use of EKFs in tracking a target, only when both the dynamic model and the observation model of the system are close to being linear (i.e., the error in the linearized model is very small) can EKF results be close to the true value. Another disadvantage of the EKF is that it is not easy to determine the initial values of the state, so if the errors in the hypothetical state initial value and the initial covariance are quite large, the filter will end up with divergence.

4.3 Unscented Kalman Filter

The basic idea behind the EKF algorithm is that through the first-order linearization truncation of the Taylor series expansion of a nonlinear function, nonlinear problems are transformed into linear ones, and then various linear estimation methods are adopted to achieve a suboptimal filtering algorithm of the original nonlinear filtering problem. Although widely used, EKFs still have some shortcomings: when the higher-order terms of a Taylor series expansion of the nonlinear function cannot be ignored, the model linearization error caused by linearization in the system tends to affect the final filtering precision, or even give rise to filter divergence. Additionally, the model linearization process is very complicated, and hard to achieve in practical applications. For this reason, this section will discuss the UKF [88], whose principle is displayed in Figure 4.1.

The UKF uses limited parameters to approximate the statistical characteristics of random variables, that is, a group of accurately selected δ points are mapped by nonlinear models to transmit the statistical characteristics of random variables, and these δ sampling points fully represent the actual mean value and covariance of Gaussian density. Then, the weighted statistical linear regression method is used to estimate the mean value and covariance of random variables, so the UKF does not need to calculate the Jacobian matrix. When these δ points are transmitted by any nonlinear system, the posterior mean value and covariance can be accurate to second-order level (i.e., insensitive to the system nonlinear intensity). Since there is no need to linearize the nonlinear system, and it can easily be applied to the state estimation of nonlinear systems [89–91], the UKF method is used widely in many fields, such as model parameter estimation [92], azimuth tracking of a man's head or hand, the state or parameter estimation of an aircraft [93], azimuth tracking of a target, etc.

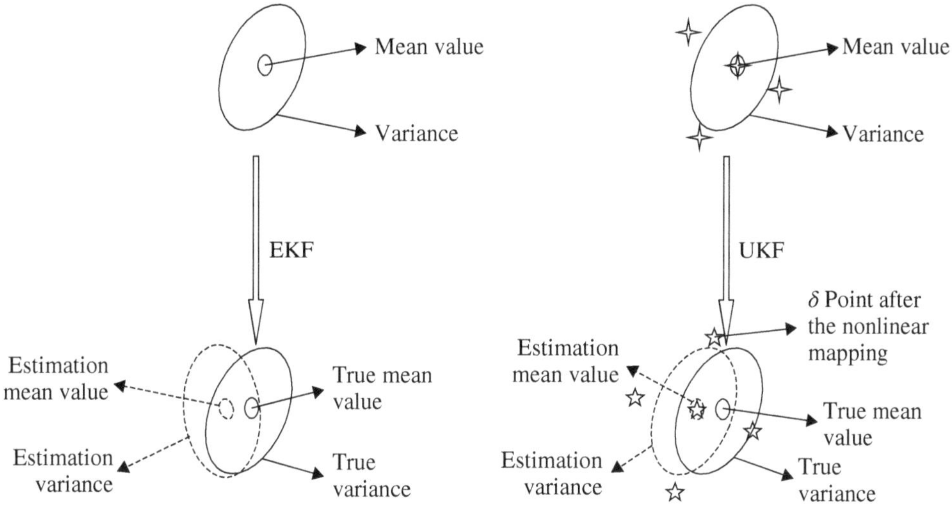

Figure 4.1 Principles of extended and unscented Kalman filters

4.3.1 Unscented Transformation

The UKF was developed on the basis of unscented transformation (UT). The basic idea of UT was first put forward by Julier and Uhlmann [42, 88], and it is a new method used in working out the statistical characteristics of random variables with nonlinear transformations. With no need to linearize the nonlinear state and the measurement model, UT approximates the PDF of the state vector. The approximated PDF is still Gaussian, but is represented as a series of selected δ sampling points.

Suppose that $\boldsymbol{X}$ is an n_x-dimensional random vector, $g:\boldsymbol{R}^{n_x}\rightarrow\boldsymbol{R}^{n_y}$ is a nonlinear function, and $y=g(x)$. The mean value and the covariance of $\boldsymbol{X}$ are, respectively, $\overline{\boldsymbol{X}}$ and $\boldsymbol{P}_x$. The steps to work out the UT are described as follows [94, 95].

1. First, calculate $(2n_x+1)$ δ sampling points $\boldsymbol{\xi}_i$ and the corresponding weight W_i:

$$\begin{cases} \boldsymbol{\xi}_0=\overline{\boldsymbol{X}} & i=0 \\ \boldsymbol{\xi}_i=\overline{\boldsymbol{X}}+\left(\sqrt{(n_x+\kappa)\boldsymbol{P}_x}\right)_i & i=1,\ldots,n_x \\ \boldsymbol{\xi}_{i+n_x}=\overline{\boldsymbol{X}}-\left(\sqrt{(n_x+\kappa)\boldsymbol{P}_x}\right)_i & i=1,\ldots,n_x \end{cases} \tag{4.30}$$

In this form, the number of δ sampling point sets should be $2n_x$ and they should be distributed symmetrically around the mean value of x, so it can possess a higher precision when dealing with random variables that satisfy various unimodal symmetric distributions which are mostly Gaussian.

$$\begin{cases} W_0=\dfrac{\kappa}{(n_x+\kappa)}, & i=0 \\ W_i=\dfrac{1}{[2(n_x+\kappa)]}, & i=1,\ldots,n_x \\ W_{i+n_x}=\dfrac{1}{[2(n_x+\kappa)]}, & i=1,\ldots,n_x \end{cases} \tag{4.31}$$

where κ is a scalar parameter and could be any value as long as $(n_x+\kappa)\neq 0$. $\left(\sqrt{(n_x+\kappa)\boldsymbol{P}_x}\right)_i$ is the ith row or ith column in the mean root matrix $(n_x+\kappa)\boldsymbol{P}_x$, and n_x is the number of dimensions of the state vector.

2. Each δ sampling point transmits through a nonlinear function, so we obtain

$$\boldsymbol{y}_i = g(\boldsymbol{\xi}_i),\quad i=0,\ldots,2n_x \tag{4.32}$$

3. The estimation mean value and covariance estimation of $\boldsymbol{y}$ are as follows:

$$\bar{\boldsymbol{y}} = \sum_{i=0}^{2n_x} W_i \boldsymbol{y}_i \tag{4.33}$$

$$\boldsymbol{P}_y = \sum_{i=0}^{2n_x} W_i(\boldsymbol{y}_i-\bar{\boldsymbol{y}})(\boldsymbol{y}_i-\bar{\boldsymbol{y}})' \tag{4.34}$$

4.3.2 Filtering Model

Assume that at time k the state estimation vector and state estimation covariance of the tracking system are, respectively, $\hat{\boldsymbol{X}}(k|k)$ and $\boldsymbol{P}(k|k)$, then we can make use of (4.30) and (4.31) to work out the relative δ point $\boldsymbol{\xi}_i(k|k)$ and its corresponding weight W_i. From the state equation (4.1), we can obtain the one-step prediction of the δ point:

$$\boldsymbol{\xi}_i(k+1|k) = \boldsymbol{f}(k,\boldsymbol{\xi}_i(k|k)) \tag{4.35}$$

Using the one-step prediction δ point $\boldsymbol{\xi}_i(k+1|k)$ and the weight W_i according to (4.32) and (4.33), we obtain the state prediction estimation and the state prediction covariance:

$$\hat{\boldsymbol{X}}(k+1|k) = \sum_{i=0}^{2n_x} W_i \boldsymbol{\xi}_i(k+1|k) \tag{4.36}$$

$$\boldsymbol{P}(k+1|k) = \sum_{i=0}^{2n_x} W_i \Delta\boldsymbol{X}_i(k+1|k)\Delta\boldsymbol{X}_i'(k+1|k) + \boldsymbol{Q}(k) \tag{4.37}$$

where

$$\Delta\boldsymbol{X}_i(k+1|k) = \boldsymbol{\xi}_i(k+1|k) - \hat{\boldsymbol{X}}(k+1|k) \tag{4.38}$$

Based on the measurement equation (4.3), we obtain the predicted measurement δ point

$$\boldsymbol{\varsigma}_i(k+1|k) = \boldsymbol{h}(k+1,\boldsymbol{\xi}_i(k+1|k)) \tag{4.39}$$

So, the predicted measurement and the corresponding covariance are

$$\hat{\boldsymbol{Z}}(k+1|k) = \sum_{i=0}^{2n_x} W_i \boldsymbol{\varsigma}_i(k+1|k) \tag{4.40}$$

$$P_{zz} = R(k+1) + \sum_{i=0}^{2n_x} W_i \Delta Z_i(k+1|k) \Delta Z_i'(k+1|k) \tag{4.41}$$

where

$$\Delta Z_i = \varsigma_i(k+1|k) - \hat{Z}(k+1|k) \tag{4.42}$$

Equally, we obtain the interaction covariance of the measurement and the state vector

$$P_{xz} = \sum_{i=0}^{2n_x} W_i \Delta X_i(k+1|k) \Delta Z_i' \tag{4.43}$$

If the measurement provided by the sensor at time $(k+1)$ is $Z(k+1)$, then the state update and the state update covariance can be denoted as

$$\hat{X}(k+1|k+1) = \hat{X}(k+1|k) + K(k+1)\left[Z(k+1) - \hat{Z}(k+1|k)\right] \tag{4.44}$$

$$P(k+1|k+1) = P(k+1|k) - K(k+1)S(k+1)K'(k+1) \tag{4.45}$$

$$K(k+1) = P_{xz}P_{zz}^{-1} = \sum_{i=0}^{2n_x} W_i \Delta X_i(k+1|k) \Delta Z_i' [R(k+1) + \sum_{i=0}^{2n_x} W_i \Delta Z_i(k+1|k) \Delta Z_i'(k+1|k)]^{-1} \tag{4.46}$$

4.3.3 Simulation Analysis

The simulation and analysis are carried out on target tracking by three-dimensional (3D) radars in the hypothetical Gaussian noise condition. Assume that the target is a plane with a velocity of 360 m/s and an altitude of 8 km, and that the initial distance between the target and the 3D radar is about 305 km. The target is flying at the same altitude, the angle between its horizontal direction and the x axis is $-120°$, the standard deviation of the distance measurement errors is 60 m, and the standard deviations of the azimuth and pitching angle measurement errors are both 1°. Here we compare the tracking errors of the KF, EKF, UKF, and unbiased converted measurements Kalman filter (UCMKF) when tracking the same target in the same environment. The initial state and the initial covariance of the KF, EKF, UKF, and UCMKF are given by (3.69) and (3.72) in Chapter 3. The KF, which has been discussed in detail in Chapter 3, will not be discussed in this section. The principle of the UCMKF is the same as that of the KF, except that two unbiased coefficients are added in the transformation of the measurement data from polar coordinates to Cartesian coordinates. Details of the UCMKF are given in Section 5.3.4 of Chapter 5. Here, we use the first-order EKF model. All four filtering models select constant velocity models, so the state equation of the target is

$$X(k+1) = F(k)X(k) + \Gamma(k)v(k) \tag{4.47}$$

where the state vector $\mathbf{X}(k) = [x \ \ \dot{x} \ \ y \ \ \dot{y} \ \ z \ \ \dot{z}]'$, and

$$\boldsymbol{F}(k) = \begin{bmatrix} 1 & T & 0 & 0 & 0 & 0 \\ 0 & 1 & 0 & 0 & 0 & 0 \\ 0 & 0 & 1 & T & 0 & 0 \\ 0 & 0 & 0 & 1 & 0 & 0 \\ 0 & 0 & 0 & 0 & 1 & T \\ 0 & 0 & 0 & 0 & 0 & 1 \end{bmatrix} \tag{4.48}$$

$$\boldsymbol{\Gamma}(k) = \begin{bmatrix} 0.5T^2 & 0 & 0 \\ T & 0 & 0 \\ 0 & 0.5T^2 & 0 \\ 0 & T & 0 \\ 0 & 0 & 0.5T^2 \\ 0 & 0 & T \end{bmatrix} \tag{4.49}$$

where $\boldsymbol{v}(k)$ is the Gaussian process noise with zero mean, whose covariance is $\sigma_v{}^2$.

The measurement equation is

$$\boldsymbol{Z}(k) = \boldsymbol{h}[\boldsymbol{X}(k)] + \boldsymbol{W}(k) \tag{4.50}$$

where $\boldsymbol{Z}(k) = [\rho(k) \ \ \theta(k) \ \ \gamma(k)]'$ and $\boldsymbol{W}(k)$ is assumed to be zero mean Gaussian noise, mutually independent of $\boldsymbol{V}(k)$, and its covariance matrix is $\boldsymbol{R} = \text{diag}\left(\sigma_\rho^2, \sigma_\theta^2, \sigma_\gamma^2\right)$, where σ_ρ^2, σ_θ^2, σ_γ^2 are, respectively, the measurement error covariances of distance, azimuth, and pitching, and the one-step prediction of the state

$$\boldsymbol{h}[\boldsymbol{X}(k)] = \begin{bmatrix} \sqrt{x^2(k) + y^2(k) + z^2(k)} \\ \arctan[y(k)/x(k)] \\ \arctan\left[z(k)/\sqrt{x^2(k) + y^2(k)}\right] \end{bmatrix} \tag{4.51}$$

is

$$\hat{\boldsymbol{X}}(k+1|k) = \boldsymbol{F}(k)\hat{\boldsymbol{X}}(k|k) \tag{4.52}$$

The one-step prediction of covariance is

$$\boldsymbol{P}(k+1|k) = \boldsymbol{F}(k)\boldsymbol{P}(k|k)\boldsymbol{F}'(k) + \boldsymbol{\Gamma}(k)\sigma_v^2\boldsymbol{\Gamma}'(k) \tag{4.53}$$

The innovation covariance is

$$\boldsymbol{S}(k+1) = \boldsymbol{h}_X(k+1)\boldsymbol{P}(k+1|k)\boldsymbol{h}_X'(k+1) + \boldsymbol{R}(k+1) \tag{4.54}$$

where the Jacobian matrix is

$$\boldsymbol{h}_X(k+1) = [\nabla_X \boldsymbol{h}'(k+1,\boldsymbol{X})]'_{\boldsymbol{X}=\hat{\boldsymbol{X}}(k+1|k)}$$
$$= \begin{bmatrix} \dfrac{\hat{x}(k+1|k)}{\hat{r}} & 0 & \dfrac{\hat{y}(k+1|k)}{\hat{r}} & 0 & \dfrac{\hat{z}(k+1|k)}{\hat{r}} & 0 \\ -\dfrac{\hat{y}(k+1|k)}{\hat{r}_{xy}^2} & 0 & \dfrac{\hat{x}(k+1|k)}{\hat{r}_{xy}^2} & 0 & 0 & 0 \\ -\dfrac{\hat{x}(k+1|k)\hat{z}(k+1|k)}{\hat{r}_{xy}\hat{r}^2} & 0 & -\dfrac{\hat{y}(k+1|k)\hat{z}(k+1|k)}{\hat{r}_{xy}\hat{r}^2} & 0 & \dfrac{\hat{r}_{xy}}{\hat{r}^2} & 0 \end{bmatrix} \tag{4.55}$$

where

$$\hat{r}_{xy} = \sqrt{\hat{x}^2(k+1|k) + \hat{y}^2(k+1|k)}, \quad \hat{r} = \sqrt{\hat{r}_{xy}^2 + \hat{z}^2(k+1|k)} \tag{4.56}$$

The filter gain is

$$\boldsymbol{K}(k+1) = \boldsymbol{P}(k+1|k)\boldsymbol{h}'_X(k+1)\boldsymbol{S}^{-1}(k+1) \tag{4.57}$$

The state update equation is

$$\hat{\boldsymbol{X}}(k+1|k+1) = \hat{\boldsymbol{X}}(k+1|k) + \boldsymbol{K}(k+1)\{\boldsymbol{Z}(k+1) - \boldsymbol{h}[k+1,\hat{\boldsymbol{X}}(k+1|k)]\} \tag{4.58}$$

where

$$\boldsymbol{h}[k,\hat{\boldsymbol{X}}(k+1|k)] = \begin{bmatrix} \sqrt{x^2+y^2+z^2} \\ \arctan[y/x] \\ \arctan\left[z/\sqrt{x^2+y^2}\right] \end{bmatrix}_{\hat{\boldsymbol{X}}(k+1|k)} \tag{4.59}$$

The covariance update equation is

$$\begin{aligned} \boldsymbol{P}(k+1|k+1) = {} & [\boldsymbol{I} - \boldsymbol{K}(k+1)\boldsymbol{h}_X(k+1)]\boldsymbol{P}(k+1|k)[\boldsymbol{I} + \boldsymbol{K}(k+1)\boldsymbol{h}_X(k+1)]' \\ & - \boldsymbol{K}(k+1)\boldsymbol{R}(k+1)\boldsymbol{K}'(k+1) \end{aligned} \tag{4.60}$$

where $\boldsymbol{I}$ is the unit matrix.

For the selection of all the parameters in the UKF, see Ref. [96]. The sampling point $\boldsymbol{\xi}_i$ is

$$\begin{cases} \boldsymbol{\xi}_0 = \overline{\boldsymbol{X}} & i = 0 \\ \boldsymbol{\xi}_i = \overline{\boldsymbol{X}} + \left(\sqrt{(n_x+k)}\sqrt{\boldsymbol{P}_x}\right)_i & i = 1,\ldots,n_x \\ \boldsymbol{\xi}_{i+n_x} = \overline{\boldsymbol{X}} - \left(\sqrt{(n_x+k)}\sqrt{\boldsymbol{P}_x}\right)_i & i = 1,\ldots,n_x \end{cases} \tag{4.61}$$

where $\overline{\boldsymbol{X}}$ and $\boldsymbol{P}_x$ are, respectively, the initial state and the initial covariance matrix; n_x is the number of dimensions of the state vector, $n_x = 6$, $\kappa = n_x(\alpha^2 - 1)$, and the parameter α ranges from 0.0001 to 1, that is, $0.0001 \le \alpha \le 1$ [96]. Here, $\alpha = 0.01$.

The corresponding weight W_i is

$$W_0^{(m)} = \frac{\lambda}{(n_x+\lambda)}, \qquad i=0 \tag{4.62}$$

$$W_0^{(c)} = \frac{\lambda}{(n_x+\lambda)} + 1 - \alpha^2 + \beta, \qquad i=0 \tag{4.63}$$

$$W_i^{(m)} = W_i^{(c)} = \frac{1}{[2(n_x+\lambda)]}, \qquad i=1,\dots,2n_x \tag{4.64}$$

where the parameter β is optimal when it takes 2 in the Gaussian noise condition [96], so we take $\beta = 2$. Superscript m denotes the weight in the state updating, and superscript c denotes the weight in the covariance updating. Hence, we can use (4.35)–(4.46) to conduct unscented Kalman filtering. Figure 4.2 shows the target position mean root errors of the above four filtering algorithms after 50 Monte Carlo experiments, and Figure 4.3 displays the amount of calculation required for the above four filtering algorithms.

From Figure 4.2 we see that in the simulation environment, KF, EKF, UKF, and UCMKF algorithms can fulfill the job of tracking the targets. As shown in Figure 4.3, KF requires the smallest amount of computation, followed by EKF and UCMKF, between which there is no distinct difference in amount, while UKF needs far more computation than the other three filtering algorithms. The reason for this is that EKF performs nonlinear filtering estimation by linearization, while UKF approximates the PDF of the state by samples. In terms of calculation speed, EKF has obvious advantages, but its performance will drop greatly as the nonlinear intensity increases. This problem can be solved satisfactorily by UKF, which does not use linearization. Whether by EKF or UKF, the Gaussian distribution is eventually adopted to approximate the state posterior possibility density. If the posterior possibility density function of the system state is non-Gaussian, both will generate substantial errors. Related to this problem, we discuss the particle filter next.

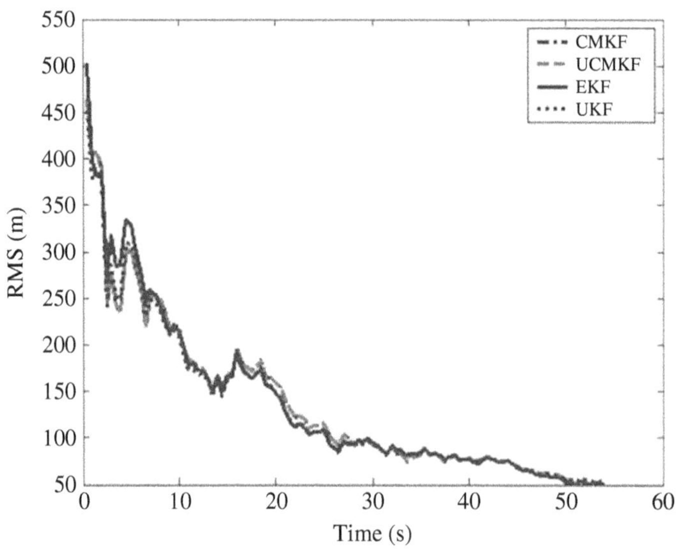

Figure 4.2 Root mean square error of target position

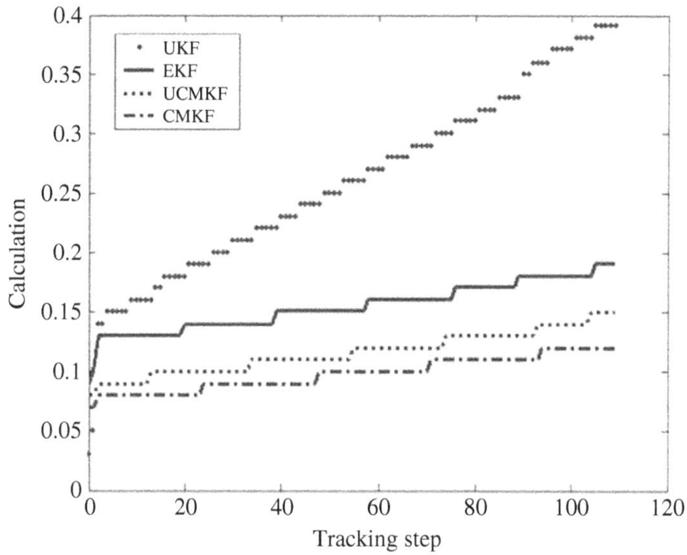

Figure 4.3 Comparison of calculations

4.4 Particle Filter

The particle filter (PF) is a type of nonlinear filtering algorithm [60, 97–122] which has emerged in recent years. It is an optimal recursive Bayesian filtering algorithm based on Monte Carlo simulation. The state vectors concerning this algorithm are denoted as a group of random samples with relative weights, and the state estimation can be worked out according to these samples and weights. Compared with other nonlinear filtering algorithms, such as EKF and UKF, this method is not confined by linearization errors or Gaussian noise hypothesis, so it can be applied to any state transformation or measurement model in any environment.

The system state equation and sensor measurement model are the same as (4.1) and (4.3), but the process noise vector $\boldsymbol{V}(k)$ and the measurement noise vector $\boldsymbol{W}(k)$ in (4.1) and (4.3), respectively, belong to non-Gaussian, independent, identically distributed noise sequences.

4.4.1 Filtering Model

Assume that at time k a group of random samples $\left\{\boldsymbol{X}_{0:k}^i, q_k^i\right\}_{i=1}^{N_s}$ are the samples obtained on the basis of posterior probability density $p(\boldsymbol{X}_{0:k}|\boldsymbol{Z}_{1:k})$, where $\boldsymbol{X}_{0:k}^i$ is the ith sample set from time 0 to k, that is, the particle set; q_k^i is the relative weight, which satisfies $\sum_{i=1}^{N_s} q_k^i = 1$; N_s is the sampling number, that is, the number of particles; $\mathbf{Z}_{1:k}$ represents the measurement set at time k; $\boldsymbol{X}_{0:k} = \left\{\boldsymbol{X}_j, j=0,\ldots,k\right\}$ represents all the state vector sets from time 0 to k. So at time k the posterior probability density can be denoted approximately as

$$p(\boldsymbol{X}_{0:k}|\boldsymbol{Z}_{1:k}) \approx \sum_{i=1}^{N_s} q_k^i \delta\left(\boldsymbol{X}_{0:k} - \boldsymbol{X}_{0:k}^i\right) \tag{4.65}$$

Since it is hard to draw samples directly from $p(\boldsymbol{X}_{0:k}|\boldsymbol{Z}_{1:k})$, we usually use an importance probability density $\pi(\boldsymbol{X}|\boldsymbol{Z})$ to obtain the sampling value [107]. So, the weight q_k^i can be obtained by sequential importance sampling. If $\boldsymbol{X}_{0:k}^i$ is the sample from $\pi(\boldsymbol{X}|\boldsymbol{Z})$, then according to Ref. [107], the non-normalized weight $\tilde{q}_k^i$ can be defined as

$$\tilde{q}_k^i = \frac{p\left(\boldsymbol{Z}_{1:k}|\boldsymbol{X}_{0:k}^i\right)p\left(\boldsymbol{X}_{0:k}^i\right)}{\pi\left(\boldsymbol{X}_{0:k}^i|\boldsymbol{Z}_{1:k}\right)} \tag{4.66}$$

If the selected importance probability density satisfies

$$\pi\left(\boldsymbol{X}_{0:k}^i|\boldsymbol{Z}_{1:k}\right) = \pi\left(\boldsymbol{X}_k^i|\boldsymbol{X}_{0:k-1}^i,\boldsymbol{Z}_{1:k}\right)\cdot\pi\left(\boldsymbol{X}_{0:k-1}^i|\boldsymbol{Z}_{1:k-1}\right) \tag{4.67}$$

then when (4.67) is substituted into (4.66), we obtain

$$\begin{aligned}\tilde{q}_k^i &= \frac{p\left(\boldsymbol{Z}_{1:k}|\boldsymbol{X}_{0:k}^i\right)p\left(\boldsymbol{X}_{0:k}^i\right)}{\pi\left(\boldsymbol{X}_k^i|\boldsymbol{X}_{0:k-1}^i,\boldsymbol{Z}_{1:k}\right)}\cdot\frac{1}{\pi\left(\boldsymbol{X}_{0:k-1}^i|\boldsymbol{Z}_{1:k-1}\right)}\\ &= \frac{p\left(\boldsymbol{Z}_k|\boldsymbol{X}_k^i\right)p\left(\boldsymbol{X}_k^i|\boldsymbol{X}_{k-1}^i\right)}{\pi\left(\boldsymbol{X}_k^i|\boldsymbol{X}_{0:k-1}^i,\boldsymbol{Z}_{1:k}\right)}\cdot\frac{p\left(\boldsymbol{Z}_{1:k-1}|\boldsymbol{X}_{0:k-1}^i\right)p\left(\boldsymbol{X}_{0:k-1}^i\right)}{\pi\left(\boldsymbol{X}_{0:k-1}^i|\boldsymbol{Z}_{1:k-1}\right)}\\ &= \frac{p\left(\boldsymbol{Z}_k|\boldsymbol{X}_k^i\right)p\left(\boldsymbol{X}_k^i|\boldsymbol{X}_{k-1}^i\right)}{\pi\left(\boldsymbol{X}_k^i|\boldsymbol{X}_{0:k-1}^i,\boldsymbol{Z}_{1:k}\right)}\tilde{q}_{k-1}^i\end{aligned} \tag{4.68}$$

In order to conveniently use the recursive Bayesian filtering algorithm, we hope that the importance probability density is related only to the measurement and state at the last time, that is,

$$\pi\left(\boldsymbol{X}_k^i|\boldsymbol{X}_{0:k-1}^i,\boldsymbol{Z}_{1:k}\right) = \pi\left(\boldsymbol{X}_k^i|\boldsymbol{X}_{k-1}^i,\boldsymbol{Z}_k\right) \tag{4.69}$$

By combining (4.68) and (4.69), the non-normalized weight $\tilde{q}_k^i$ can be denoted as

$$\tilde{q}_k^i = \frac{p\left(\boldsymbol{Z}_k|\boldsymbol{X}_k^i\right)p\left(\boldsymbol{X}_k^i|\boldsymbol{X}_{k-1}^i\right)}{\pi\left(\boldsymbol{X}_k^i|\boldsymbol{X}_{k-1}^i,\boldsymbol{Z}_k\right)}\cdot\tilde{q}_{k-1}^i \tag{4.70}$$

In the particle filtering algorithm, the weight of most particles will approach zero after several iterative cycles, that is, particle attenuation occurs [107]. This phenomenon is inevitable because the particle weight covariance becomes larger with increased time. The most direct way of reducing this effect is to use a large number of particles. Certainly, this is usually unrealistic. So at present, two methods are used: (1) choose optimal importance probability density; (2) conduct resampling.

According to Ref. [108], the optimal importance probability density is

$$\pi\left(\boldsymbol{X}_k^i|\boldsymbol{X}_{k-1}^i,\boldsymbol{Z}_k\right) = p\left(\boldsymbol{X}_k^i|\boldsymbol{X}_{k-1}^i,\boldsymbol{Z}_k\right) \tag{4.71}$$

The optimal importance probability density can minimize the covariance of the sampling weight. Nowadays it is adopted commonly in the case where $\boldsymbol{X}_k$ is a finite set, for example, the hopping Markov system used for tracking maneuvering targets in Ref. [103]. Also, in the case of nonlinear state equations and linear measurement equations [109]. For most systems, it is impossible to realize the optimal importance probability density, so the linearization technique is usually used to make a suboptimal approximation to the optimal importance probability density. For example, in Ref. [101], UT is used to conduct Gaussian approximation to the optimal importance probability density.

The basic idea of resampling is to reduce the small-weight particles and concentrate the larger-weight particles. Frequently used resampling methods include stratified sampling, residual sampling [123], and system resampling [124]. Resampling decreases the effect of attenuation on the system; meanwhile, it creates other problems. For one thing, all the particles must be rearranged, which hinders the system from parallel calculation. For another, the larger-weight particles could be selected multiple times, so that the existing particles cannot represent the current probability density. Hence, when the system process noise is quite weak, all the particles can possibly turn into a single point, which is often called "sampling depletion."

4.4.2 Examples of the Application of EKF, UKF, and PF

Assume that in a station there is a passive sensor tracking a target in uniform motion and that the dynamic equation of the target is

$$\boldsymbol{X}(k+1) = \boldsymbol{F}(k)\boldsymbol{X}(k) + \Gamma(k)\boldsymbol{v}(k) \tag{4.72}$$

where

$$\mathbf{X}(k) = [x \ \ y \ \ \dot{x} \ \ \dot{y}]' \tag{4.73}$$

$$\boldsymbol{F}(k) = \begin{bmatrix} 1 & 0 & T & 0 \\ 0 & 1 & 0 & T \\ 0 & 0 & 1 & 0 \\ 0 & 0 & 0 & 1 \end{bmatrix} \tag{4.74}$$

$$\boldsymbol{\Gamma}(k) = \begin{bmatrix} 0.5T^2 & 0 \\ 0 & 0.5T^2 \\ T & 0 \\ 0 & T \end{bmatrix} \tag{4.75}$$

Assume that the process noise is a two-dimensional Gaussian noise vector with zero mean, whose covariance is $\boldsymbol{Q} = q\boldsymbol{I}_{2\times 2}$, where $\boldsymbol{I}_{2\times 2}$ is a 2×2 unit matrix. There are two cases for the initial states of the target:

1. $\boldsymbol{X}(0) = [-10\,000 \text{ m}, 100 \text{ m/s}, 20\,000 \text{ m}, -200 \text{ m/s}]'$.
2. $\boldsymbol{X}(0) = [30\,000 \text{ m}, -200 \text{ m/s}, 20\,000 \text{ m}, 150 \text{ m/s}]'$.

Suppose that the measurement equation of the infrared sensor is

$$\boldsymbol{Z}(k) = \boldsymbol{h}(\boldsymbol{X}(k)) + \boldsymbol{W}(k) \tag{4.76}$$

where $\boldsymbol{W}(k)$ is assumed to be Gaussian white noise with zero mean and independent of $\boldsymbol{V}(k)$, and its covariance is $\boldsymbol{R}=\sigma_{\alpha}^2$. What is detected by the passive sensor is the azimuth of the target, that is,

$$\boldsymbol{Z}(k)=\alpha(k)=\arctan\left(\frac{x(k)}{y(k)}\right) \tag{4.77}$$

Then we obtain

$$\boldsymbol{H}(k)=\frac{\partial \boldsymbol{h}}{\partial \boldsymbol{X}}\bigg|_{X=\hat{X}(k|k-1)}=\begin{bmatrix}\frac{\hat{y}(k|k-1)}{\hat{r}_{xy}^2}, & -\frac{\hat{x}(k|k-1)}{\hat{r}_{xy}^2}, & 0, & 0\end{bmatrix} \tag{4.78}$$

where

$$\hat{r}_{xy}=\sqrt{[\hat{x}(k|k-1)]^2+[\hat{y}(k|k-1)]^2} \tag{4.79}$$

Suppose that the sampling interval of the sensor is 2 s, and the standard deviation of the azimuth measurement error is 0.00175 rad. Here, in the same environment, we track the same target with EKF, UKF, and PF, and compare the tracking results. In addition, notice that EKF takes the first-order model, whose details can be seen in Sections 4.2.1 and 4.3.3. 50 Monte Carlo simulations are conducted, and in each simulation the scanning time is 60 s.

In cases 1 and 2, the initial state estimations of the target are as follows.

1. $\hat{\boldsymbol{X}}(0|0)=[-10\,200\text{ m, }100\text{ m/s, }20\,300\text{ m, }-200\text{ m/s}]'$.
2. $\hat{\boldsymbol{X}}(0|0)=[30\,000\text{ m, }-200\text{ m/s, }20\,000\text{ m, }150\text{ m/s}]'$.

Assume that the corresponding state estimation covariance in both cases is

$$\boldsymbol{P}(0|0)=\begin{bmatrix}50\,000 & 0 & 0 & 0\\ 0 & 800 & 0 & 0\\ 0 & 0 & 50\,000 & 0\\ 0 & 0 & 0 & 800\end{bmatrix} \tag{4.80}$$

Figure 4.4(a) shows the target trajectory in case 1, and Figure 4.4(b) illustrates the x-axis root mean square (RMS) comparison of all the algorithms in case 1. Figure 4.5(a) shows the target trajectory in case 2, and Figure 4.5(b) shows the x-axis RMS comparison of all the algorithms in case 2.

In case 1, when the target trajectory passes the proximity of the detection station (the 50th step), the tangential change in measurement angle increases, and the non-linearization of the measurement equation has already intensified, so the linearization error of the model increases gradually, thus causing drop and divergence in the estimation precision of the first-order EKF. For UKF and PF, the system measurement equation does not need to be linearized, so the estimation precision is not affected by the linearization error. The precisions of UKF and PF near the measurement point drop to some degree, mainly because the measurement point is easy to hop here.

In case 2, when the target trajectory is far away from the measurement point, the tangential of the measurement angle does not change greatly, the non-linearization degree of the measurement equation is low, and so the linearization error of the model is quite small. Now the first-order EKF, UKF,

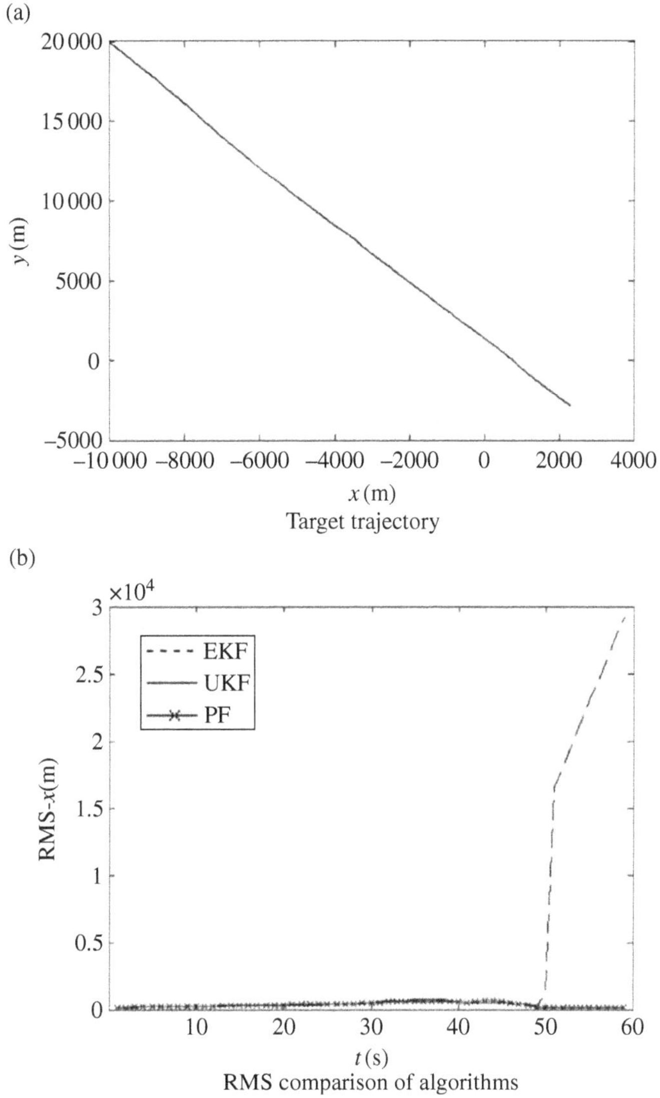

Figure 4.4 Simulation results in case 1

and PF present nearly identical precision, which shows that the estimation precision of the three algorithms is close when the non-linearization degree of the system is low (see Figure 4.5).

A comprehensive comparison of calculation speed, storage amount, environmental adaptability, etc. of EKF, UKF, and PF algorithms is given in Table 4.1, where the calculation speed refers to the calculation time used in each of the algorithms to calculate 60 time steps. The time does not concern the production, motion, and measurement of the target, but represents the time that each of the three algorithms spends on calculation.

From the results in Table 4.1, we see that EKF is the fastest of the three estimation algorithms, followed by UKF, and PF is the slowest (the number of particles is 5000). The storage requirements in Table 4.1 are just an approximate estimation in terms of the calculation process and complexity in

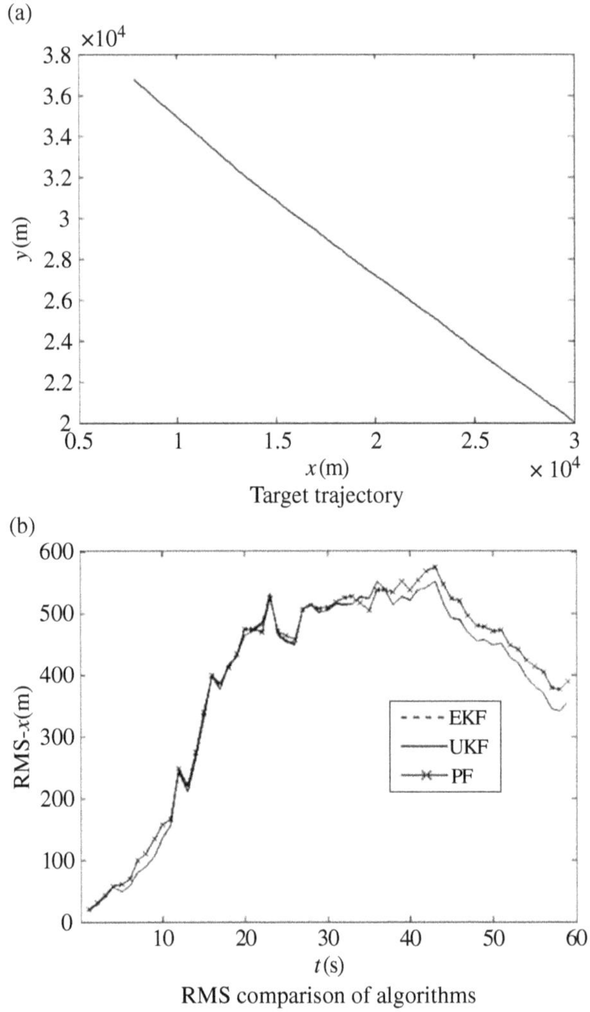

Figure 4.5 Simulation results in case 2

Table 4.1 Comprehensive comparison of the three algorithms

Algorithm	Calculation speed (case 1)	Calculation speed (case 2)	Storage amount	Environmental adaptability	Nonlinear intensity
EKF	0.039 s	0.038 s	low	Gaussian	weak
UKF	1.040 s	1.040 s	medium	Gaussian	not limited
PF	140.5164 s	40.5141 s	high	not limited	not limited

the algorithms. According to the analysis of the algorithms, the storage amount for PF is quite large, and increases with increasing number of particles; the storage amount for UKF stays in between; and that of EKF is quite small. In Table 4.1, environmental adaptability refers to the environmental noise (including measurement noise and state process noise) requirement. Nonlinear intensity means the requirement for nonlinear intensity in the system.

How do we make an overall assessment of the algorithms? First, EKF possesses apparent superiority in terms of calculation speed, and should be of top priority for practical engineering applications if other factors (system nonlinear intensity, environmental requirements, etc.) are not considered, However, from the simulation results in case 1, we see that when the nonlinear intensity increases and causes the linearization error to increase, the estimation precision of EKF drops significantly and divergence occurs. So, for such systems, we need to consider other filtering methods. UKF and PF have similar performance in terms of estimation precision, but the latter involves a much greater amount of calculation. Thus, UKF should be used in some ordinary nonlinear Gaussian environments. However, UKF is suitable only for Gaussian white noise environments and not more complicated non-Gaussian environments. PF, therefore, has good prospects in broad applications as the capability of computers improves.

4.5 Summary

In this chapter we have discussed nonlinear filtering techniques in radar data processing, including EKF, UKF, PF, etc. Finally, we have reached relative conclusions. We have also compared and analyzed the tracking precision and calculation amount of the same target with nonlinear algorithms (i.e., EKF, UKF, and PF) in the same simulation environment, and made a comprehensive assessment of their advantages and disadvantages. In 2002 Farina and Ristic [108] used CRLB methods to analyze EKF, the covariance analysis describing function technique (CADET), UKF, and PF. The simulation results show that EKF exhibits a better effect, because it possesses the advantages of small calculation amount and statistical effectiveness. We should notice that all these conclusions are reached in the simulation condition (quite ideal).

5

Measurement Preprocessing Techniques

5.1 Introduction

Target parameter estimation methods in linear and nonlinear systems have been analyzed and discussed in previous chapters. In modern complicated electronic environments, the precondition for processing radar data correctly is the preprocessing of measurements. Effective measurement preprocessing techniques can reduce computational complexity and improve tracking accuracy in radar data processing. This chapter mainly discusses time registration, space registration, outlier elimination [125–129], radar error calibration, and data compression in measurement preprocessing techniques.

In a multi-target tracking system, any observation model is established according to state-space models, therefore, proper selection of coordinate systems is very important, which directly affects tracking accuracy and computational complexity. In many radar tracking systems, the coordinate system of target measurements is inconsistent with that of data processing. In this case coordinate transformation techniques [17, 39, 40, 46, 130–139] are used to integrate all data formats into the same coordinate system.

In radar data processing systems, techniques like radar error calibration and data compression are closely associated with practical engineering [19, 39, 140–142]. Effective radar error calibration and data compression techniques can increase target tracking accuracy and reduce the computational complexity of the system.

5.2 Time Registration

Time registration, also called time synchronization, refers to the process by which the remaining time deviation is kept within the allowed range after the time alignment of many measurement units, including synchronization with astronomical time (absolute registration) and that with a high-precision master clock (relative registration). Time registration mainly solves the time

Radar Data Processing with Applications, First Edition. He You, Xiu Jianjuan, and Guan Xin.

synchronization problem in multi-sensor data fusion. For example, when radar and data links are used in target tracking fusion, due to the difference in data links between the two devices, time registration is necessary before fusion. During the registration process, radar measurements can be adopted as the benchmark against which the measurements of the radar link are calibrated. Registration methods include interpolation (extrapolation) methods, least-squares algorithms, and Lagrange methods, among which the interpolation (extrapolation) method uses the velocity.

5.2.1 Interpolation/Extrapolation Method Using Velocity

If the data rates of two different sensors (such as a data link and a radar) are inconsistent, time registration will be the first step. Here, radar measurements are viewed as the benchmark against which the time registration of data link tracks is implemented by the interpolation/extrapolation method. The specific steps are described below.

1. Assume that the first measurement of the data link and that of the radar are obtained at the same time, that is, $T_{a1} = T_{b1}$, as shown in Figure 5.1.
2. Estimate the times corresponding to the subsequent interpolation points of the data link. There are three cases.
   i. If the time difference between the corresponding time of the interpolation point and its previous sampling time is less than the sampling interval of the data link, then extrapolation should be done based on the data of the data link at the previous time. For example, the measurement of the data link corresponding to the measurement of the radar at time T_{b2} should be

$$X_{ab2} = X_{a1} + V_{a1}(T_{b2} - T_{a1}) \tag{5.1}$$

   where V_{a1} is the velocity.
   ii. If the time difference between the corresponding time of the interpolation point and its previous sampling time used for making a judgment is greater than the sampling interval of the data link, then extrapolation should be done from the data of the data link at the corresponding time which is nearest to the interpolation point. For example, the measurement of the data link corresponding to that of the radar at T_{b3} should be

$$X_{ab3} = X_{a2} + V_{a2}(T_{b3} - T_{a2}) \tag{5.2}$$

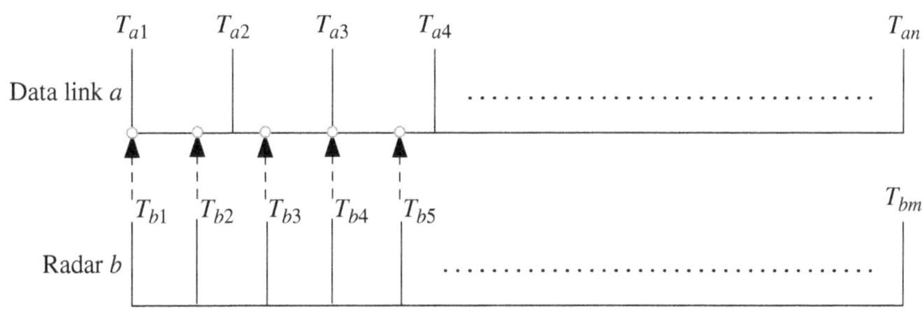

Figure 5.1 Time registration

iii. If the time difference between the corresponding time of the interpolation point and its previous sampling time used for making a judgment is equal to the sampling interval of the data link, then the data at that time should be kept unchanged (e.g., $X_{ab4} = X_{a3}$).
3. By analogy, conduct time synchronization of the data from the two different sensors so as to form target observations corresponding to the same time.

Additionally, when different sensors are used to track moving targets, time registration can be conducted of the state and corresponding covariance for tracks corresponding to the same target by the following method: take a certain sensor's state and corresponding covariance at different times as the benchmark, extrapolate the other sensors' states and corresponding covariances, and adopt the predicted state value and predicted covariance at the extrapolation time as the corresponding sensor's information at this time, that is,

$$\begin{cases} \hat{\boldsymbol{X}}(k+1|k) = \boldsymbol{F}(k)\hat{\boldsymbol{X}}(k|k) \\ \boldsymbol{P}(k+1|k) = \boldsymbol{F}(k)\boldsymbol{P}(k|k)\boldsymbol{F}'(k) + \boldsymbol{Q}(k) \end{cases} \tag{5.3}$$

where $\hat{\boldsymbol{X}}(k|k)$ is the target state filtering value at different times, similar to X_{ak}, $\mathbf{P}(k|k)$ is the estimation error covariance matrix corresponding to $\hat{\boldsymbol{X}}(k|k)$, and $\boldsymbol{Q}(k)$ is the process noise covariance matrix.

5.2.2 The Lagrange Interpolation Algorithm

The principle of the Lagrange interpolation algorithm is described as follows.

Suppose that the measurements at time t_{k-1}, t_k, t_{k+1} are, respectively, X_{k-1}, X_k, X_{k+1}, then the measurement value at time t_i $(t_{k-1} < t_i < t_{k+1})$ is

$$X_i = \frac{(t_i - t_k)(t_i - t_{k+1})}{(t_{k-1} - t_k)(t_{k-1} - t_{k+1})} \times X_{k-1} + \frac{(t_i - t_{k-1})(t_i - t_{k+1})}{(t_k - t_{k-1})(t_k - t_{k+1})} \times X_k + \frac{(t_i - t_{k-1})(t_i - t_k)}{(t_{k+1} - t_{k-1})(t_{k+1} - t_k)} \times X_{k+1} \tag{5.4}$$

5.2.3 Least-Squares Curve-Fitting Algorithm

In the case of curve fitting of given measurements (t_k, X_k) $(k = 1, 2, \ldots, n)$, the principle of the least-squares algorithm is to minimize the quadratic sum of deviations between measurements and fitted curves, so that the fitted curve is closer to the real function. The specific steps are described as follows.

Suppose that the unknown function is close to a linear function, and select

$$X(t) = a \cdot t + b \tag{5.5}$$

as its fitted curve. Assume that the obtained measurements are (t_k, X_k) $(k = 1, 2, \ldots, n)$, then the deviation between each observation data point and the fitted curve is

$$X(t_k) - X_k = a \cdot t_k + b - X_k \qquad k = 1, 2, \ldots, n \tag{5.6}$$

The quadratic sum of deviations is

$$F(a,b)=\sum_{k=0}^{n}(a\cdot t_k+b-X_k)^2 \tag{5.7}$$

According to the principle of least squares, a and b should minimize $F(a, b)$, that is, a and b should satisfy

$$\begin{cases} \dfrac{\partial F(a,b)}{\partial a}=2\sum_{k=0}^{n}(a\cdot t_k+b-X_k)\cdot t_k=0 \\ \dfrac{\partial F(a,b)}{\partial b}=2\sum_{k=0}^{n}(a\cdot t_k+b-X_k)=0 \end{cases} \tag{5.8}$$

That is,

$$\begin{cases} a\sum_{k=0}^{n}t_k^2+b\sum_{k=0}^{n}t_k=\sum_{k=0}^{n}t_kX_k \\ a\sum_{k=0}^{n}t_k+bn=\sum_{k=0}^{n}X_k \end{cases} \tag{5.9}$$

Solve the equation group above, to get the value of a and b.

5.3 Space Registration

5.3.1 Coordinates

For radar, target measurement is usually implemented in space polar coordinate systems, while the subsequent target measurement processing is conducted in rectangular coordinate systems. Furthermore, when radars are installed on different carriers (aircraft, ships, etc.), the coordinate systems used by different radar systems can be divided, according to their definitions, into the following categories: the north east down (NED) coordinate system, the carrier coordinate system, the radar antenna coordinate system, and the sight of target coordinate system, etc. This section will mainly introduce some frequently used coordinate systems related to radar measurement or data processing.

5.3.1.1 Descartes Rectangular Coordinate System

Choose three axes (that is, straight lines with positive direction) in space, which intersect at the same point and are perpendicular to each other. Customarily, one is the front–rear axis, called the horizontal axis (i.e., *OX* axis – *X* axis for short), whose positive direction is from rear to front. Another is the left–right axis, called the longitudinal axis (i.e., *OY* axis – *Y* axis for short), whose positive direction is from left to right. Still another is the up–down axis, called the vertical axis (i.e., *OZ* axis – *Z* axis for short), whose positive direction is from down to up. The *X*, *Y*, and *Z* axes are known collectively as the coordinate axes, whose point of intersection is called the origin, often

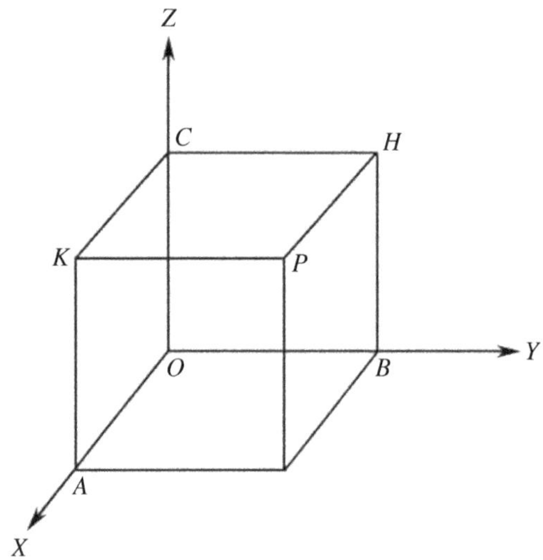

Figure 5.2 Right-hand space rectangular coordinate system

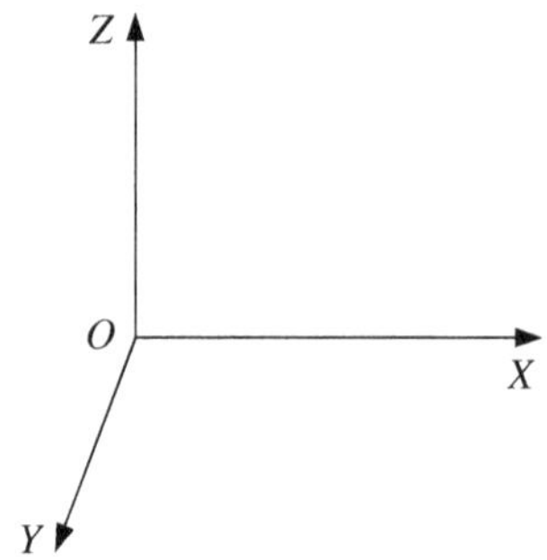

Figure 5.3 Left-hand space rectangular coordinate system

represented by the letter O. Planes YOZ, ZOX, and XOY are known collectively as the coordinate planes – YZ, ZX, and XY for short.

A supplementary explanation is given below of the rule of directions of coordinate systems. If the thumb and forefinger of the right hand, respectively, point in the directions of axes X and Y, and the direction of the middle finger is the same as that of the Z axis, then the coordinate system is called a right-hand coordinate system, as shown in Figure 5.2; otherwise, it is called a left-hand coordinate system, as shown in Figure 5.3. In order to be consistent with practical engineering applications, this book adopts right-hand coordinate systems. What's more, the axes in space coordinate systems are not necessarily required to be perpendicular to each other, they can cross obliquely, and this kind of coordinate system is called the Descartes oblique coordinate system. In practice, it is rather inconvenient and difficult to use oblique coordinate systems in equation deduction and application, so we usually choose right-hand rectangular coordinate systems.

As shown in Figure 5.2, suppose that P is a known point in space. Construct, through P, planes parallel to YZ, ZX, and XY, respectively, which cross X, Y, and Z at points A, B, and C.

The coordinates of A, B, C are a, b, c. The array (a, b, c) consisting of the three numbers is called the rectangular coordinates of the point P. a, b, c are the X, Y, Z coordinates of the point P, respectively.

5.3.1.2 Space Polar Coordinate System

In radar data processing, the advantage of rectangular coordinate systems lies in the fact that the filtering, interpolation, and extrapolation processes can be conducted in linear models. Normally, the measurement value of such sensors as radar is obtained in a space polar coordinate system, which is also called a spherical coordinate system. Definitions of coordinate axes and planes in space polar coordinate systems are the same as those in Descartes rectangular coordinate systems. The polar coordinate system is different from the rectangular coordinate system in that the coordinate definitions of spatial points are different in their respective coordinate system.

Assume that P is a known point in space, a vertical line is made from P to plane XY, and the perpendicular foot is L, r represents the radial distance, θ (elevation) and φ (azimuth) represent $\angle POB$ and $\angle XOB$, respectively. The three numbers r, θ, and φ determine the position of P. Therefore, array (r, φ, θ) is the spherical coordinates or space polar coordinates of the point P, as shown in Figure 5.4.

5.3.1.3 Earth Coordinate System

The earth coordinate system is an inertial coordinate system [132], whose origin lies at the earth's center. Axis X_g is the earth's spin axis, with direction from the earth's center to the North Pole. Axis Y_g is defined as the axis in the equatorial plane, with direction from the earth's center to the meridians. Axis Z_g is the orthogonal result of X_g and Y_g, as shown in Figure 5.5. It should be noticed that definitions of each coordinate axis in the earth coordinate system may be inconsistent in different practical implementations.

Usually, the target coordinates in this coordinate system are expressed by longitude, latitude, and altitude, therefore it is also called a geographical coordinate system. When the radar reports

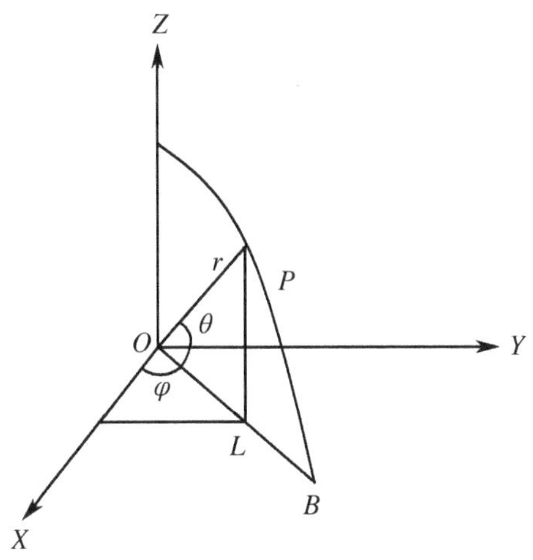

Figure 5.4 Space polar coordinate system

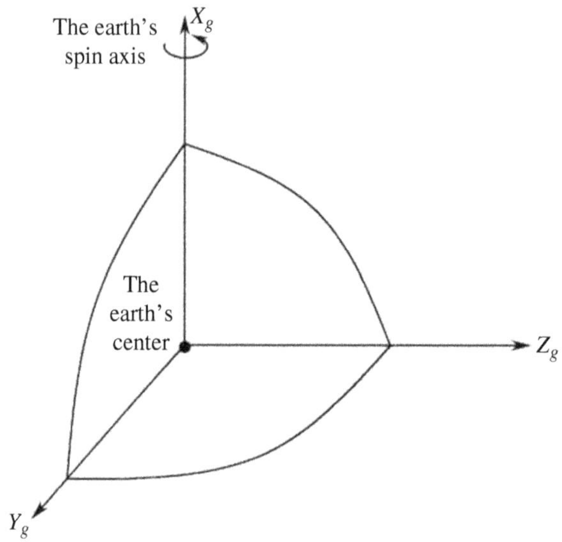

Figure 5.5 Earth coordinate system

the detected target information, the position information of the target is usually shown in geographical coordinates.

5.3.1.4 NED Coordinate System

The NED coordinate system, whose origin is at the carrier's mass center, is a local coordinate system [17, 39, 40, 46, 128]. As shown in Figure 5.6, N is the compass's direction, E is the tangential direction of the earth's rotation, and D is the direction from the carrier's mass center to the earth's center. For shipborne or ground sensors, the direction of axis D in the NED coordinate system should be from the earth's center to the carrier's mass center (O).

The NED coordinate system is a locally stable coordinate system. Technically, it is not an inertial coordinate system, because axis D will slowly change its pointing direction in space when the motion platform passes the surface of the earth. However, the influence of such movement can be ignored except when it moves near the North Pole. So, NED is an approximate inertial coordinate system for a motion platform. This coordinate system is suitable not only for airborne systems, but also for ground or shipborne tracking systems.

5.3.1.5 The Carrier Coordinate System

The origin of the carrier coordinate system [132] is the carrier's mass center. For shipborne sensors, axis X_d is the positive direction of the bow, axis Z_d is the plumb line of the deck plane, pointing to the sky, and axis Y_d is the positive direction of the starboard, as shown in Figure 5.7. For airborne sensors, X_d is the carrier's longitudinal axis, whose positive direction is head, axis Y_d is the positive direction of the right wing, and axis Z_d is determined by the right-hand spiral rule, pointing to the bottom of the fuselage.

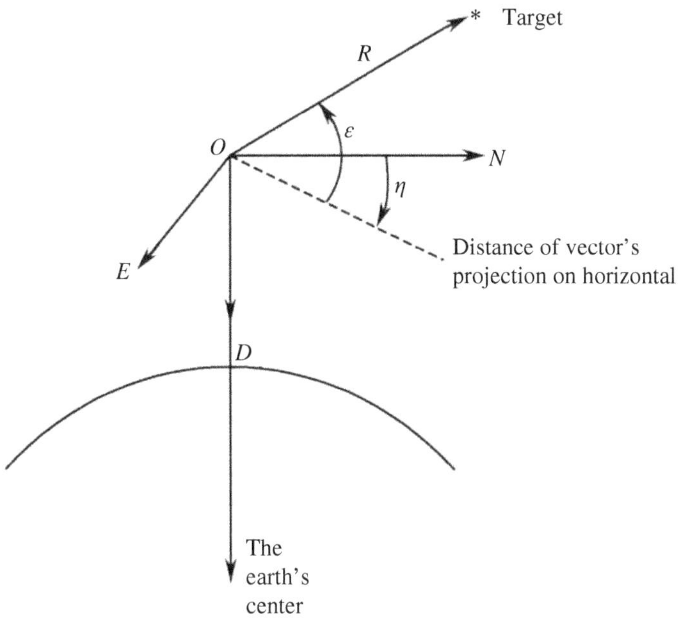

Figure 5.6 NED coordinate system

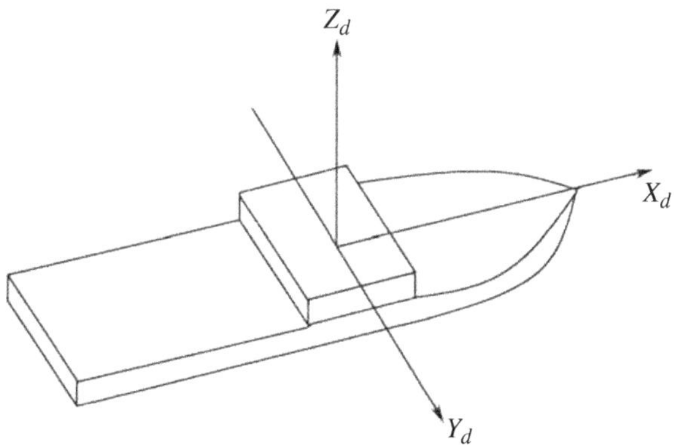

Figure 5.7 Shipborne carrier coordinate system

The carrier coordinate system is usually used in shipborne or airborne radars to measure space positions of targets.

5.3.1.6 Radar Antenna Coordinate System

The origin of the radar antenna coordinate system is the intersection of the pitching and beam axes of the radar. Axis R is the aiming direction of the beam axis, while E and D are a pair of orthogonal axes perpendicular to R. R, E, and D constitute a right-hand relationship in turn.

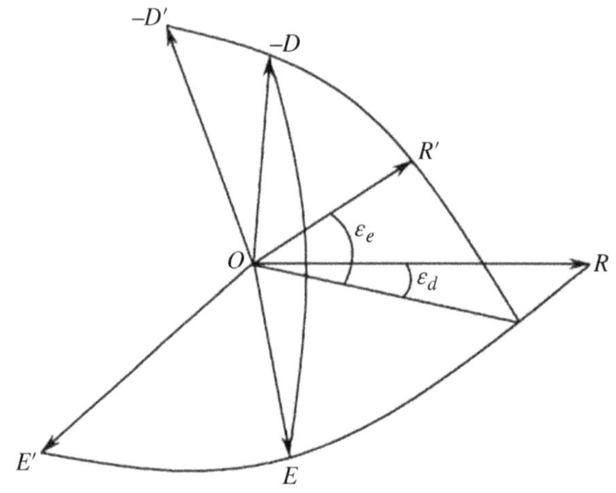

Figure 5.8 *ORED* and *O'R'E'D'*

5.3.1.7 The Sight of Target Coordinate System

The sight of target coordinate system has the same origin as the radar antenna coordinate system. Axis R' is the linking line direction from the focal point of the antenna to the target. E' and D' are a pair of orthogonal axes perpendicular to R'. R', E', and D' in turn constitute a right-hand relationship. The relationship between the radar antenna and the sight of target coordinate system is shown in Figure 5.8. These two coordinate systems are also often used to measure space positions of targets.

5.3.2 Coordinate Transformation

In radar tracking systems, coordinate transformation [130] refers to the following process: if two coordinate systems are known, the positional relation of two groups of coordinates of the same point can be determined based on their positional relation; according to this relation equation, the same target's space position can be expressed by different space coordinate systems, which is convenient for the target measurement and data processing of the whole radar tracking system.

Methods for coordinate transformation [130] fall into two classes: translation transformation and rotation transformation. The former changes the position of the origin with the axis direction kept unchanged, while the latter changes the axis direction instead of the origin position. The coordinate transformation of any system can be realized by both or either of these two methods.

5.3.2.1 Translation Transformation

As shown in Figure 5.9, the coordinate system is moved parallel from the first position *OX*, *OY*, *OZ* to the second position $O'X'$, $O'Y'$, $O'Z'$, that is, $O'X'$, $O'Y'$, $O'Z'$ are, respectively, parallel to *OX*, *OY*, *OZ*. This method is called "translation of coordinate systems."

Suppose that the coordinates of the new origin O' in the old coordinate system are (a, b, c), and that the coordinates of *P* in the old and new coordinate systems are, respectively, (x, y, z) and (x', y', z'). According to the spatial geometry relation in Figure 5.9, we obtain

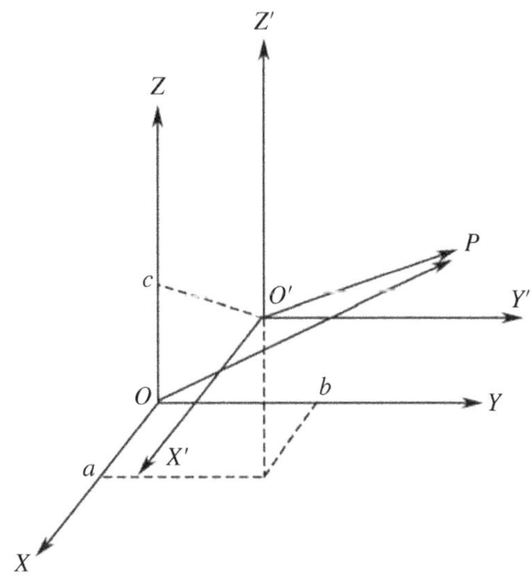

Figure 5.9 Coordinate translation transformation

$$\begin{cases} x = x' + a \\ y = y' + b \\ z = z' + c \end{cases} \tag{5.10}$$

or

$$\begin{cases} x' = x - a \\ y' = y - b \\ z' = z - c \end{cases} \tag{5.11}$$

Equations (5.10) and (5.11) are called coordinate transformation equations under the translation of coordinate axes, "translation equations" for short.

5.3.2.2 Rotation Transformation

The rotation of space coordinate systems is a motion that preserves the origin and unit segment while changing the direction of the coordinate axes. In order to elaborate on the equation deduction process of rotation transformation, we first study a rather simple case, in which one coordinate axis is kept fixed while the other two rotate around it, as shown in Figure 5.10. In this figure, OX and OY rotate around OZ in the same direction at angle θ, then we obtain OX' and OY', while OZ keeps fixed, which means that the coordinate system $OX'Y'Z'$ is obtained after $OXYZ$ is rotated counterclockwise.

If the coordinates of P in the old and new coordinate systems are, respectively, (x, y, z) and (x', y', z'), then obviously the Z coordinate is unchanged while the Y coordinate and X coordinate are changed.

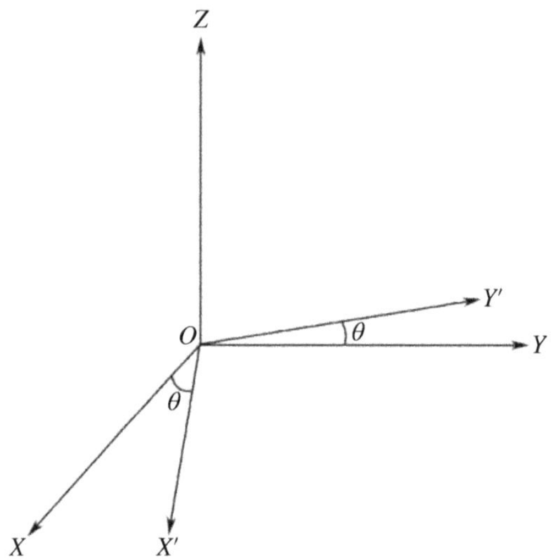

Figure 5.10 Rotation of single coordinate axes

According to the geometrical relationship between the points in Figure 5.10, the following can be obtained:

$$\begin{cases} x' = x\cos\theta + y\sin\theta \\ y' = -x\sin\theta + y\cos\theta \\ z' = z \end{cases} \tag{5.12}$$

Likewise, when the coordinates make a counterclockwise rotation around X or Y, similar equations can be obtained:

$$\begin{cases} x' = x \\ y' = y\cos\theta + z\sin\theta \\ z' = -y\sin\theta + z\cos\theta \end{cases} \tag{5.13}$$

$$\begin{cases} x' = x\cos\theta - z\sin\theta \\ y' = y \\ z' = x\sin\theta + z\cos\theta \end{cases} \tag{5.14}$$

In radar data processing systems, for convenience of expression, the vector is customarily used to indicate the position of a target in space. If a coordinate system in space is defined as $OX_aY_aZ_a$, then the coordinate X_a of any point P in this system can be expressed by the vector equation $\boldsymbol{X}_a = [X_{xa}\, X_{ya}\, X_{za}]'$, where X_{xa}, X_{ya}, X_{za}, respectively, represent the corresponding positions of P on the axes. Equations (5.12) and (5.14) can be expressed as

$$\boldsymbol{X}_b = \boldsymbol{L}_1 \boldsymbol{X}_a \tag{5.15}$$

$$\boldsymbol{X}_b = \boldsymbol{L}_2 \boldsymbol{X}_a \tag{5.16}$$

$$X_b = L_3 X_a \tag{5.17}$$

where

$$L_1(\theta) = \begin{bmatrix} \cos\theta & \sin\theta & 0 \\ -\sin\theta & \cos\theta & 0 \\ 0 & 0 & 1 \end{bmatrix} \tag{5.18}$$

$$L_2(\theta) = \begin{bmatrix} 1 & 0 & 0 \\ 0 & \cos\theta & \sin\theta \\ 0 & -\sin\theta & \cos\theta \end{bmatrix} \tag{5.19}$$

$$L_3(\theta) = \begin{bmatrix} \cos\theta & 0 & -\sin\theta \\ 0 & 1 & 0 \\ \sin\theta & 0 & \cos\theta \end{bmatrix} \tag{5.20}$$

which are called basic rotation matrixes around axes Z, X, and Y. The transformation relation between any two coordinate systems can be realized by the combination of basic rotation matrixes.

If coordinate system $OX_bY_bZ_b$ is obtained after $OX_aY_aZ_a$ rotates counterclockwise around axes X, Y, and Z, respectively, at angle φ_1, φ_2, and φ_3, and the coordinate vector of point P in the new coordinate system is expressed as $X_b = [X_{xb}\, X_{yb}\, X_{zb}]'$, then the rotation transformation relation between X_a and X_b is

$$X_b = L_{ba} X_a \tag{5.21}$$

where

$$L_{ba} = L_1(\varphi_1) \cdot L_2(\varphi_2) \cdot L_3(\varphi_3) \tag{5.22}$$

which is called the transformation matrix from coordinate system a to b.

It is not difficult to demonstrate that the coordinate transformation matrix L_{ba} meets the following reversible and orthogonal condition:

$$L'_{ba} = L_{ba}^{-1} = L_{ab} \tag{5.23}$$

5.3.3 *Transformation of Several Common Coordinate Systems*

5.3.3.1 Rectangular Coordinate System and Polar Coordinate System

The geometrical relation of space point P in the two coordinate systems is shown in Figure 5.4.

If the target position of point P is denoted as (r, φ, θ) in the space polar coordinate system and the position of its coordinates as (x, y, z) in the rectangular coordinate system, then the transformation relation between the polar and the rectangular coordinate system of the sensor is

$$\begin{cases} x = r\cos\varphi\cos\theta \\ y = r\sin\varphi\cos\theta \\ z = r\sin\theta \end{cases} \tag{5.24}$$

or

$$\begin{cases} r = \sqrt{x^2 + y^2 + z^2} \\ \varphi = \tan^{-1}\left(\frac{y}{x}\right) \\ \theta = \sin^{-1}\left(\frac{z}{r}\right) \end{cases} \tag{5.25}$$

5.3.3.2 NED Coordinate System and Shipborne Coordinate System

Suppose that the rolling angle of a ship is R (starboard rolling is the positive direction), the pitching angle is P (the pitch-up direction of the bow is positive), the angle between the course and due north is a_n (when the course makes starboard deviation from due north, a_n is positive). The rotation relation between the axes of the two coordinate systems is shown in Figure 5.11.

Hence, three basic rotation matrixes can be determined:

$$\boldsymbol{L}_1(R) = \begin{bmatrix} 1 & 0 & 0 \\ 0 & \cos R & \sin R \\ 0 & -\sin R & \cos R \end{bmatrix} \tag{5.26}$$

$$\boldsymbol{L}_2(P) = \begin{bmatrix} \cos P & 0 & -\sin P \\ 0 & 1 & 0 \\ \sin P & 0 & \cos P \end{bmatrix} \tag{5.27}$$

$$\boldsymbol{L}_3(a_n) = \begin{bmatrix} \cos a_n & \sin a_n & 0 \\ -\sin a_n & \cos a_n & 0 \\ 0 & 0 & 1 \end{bmatrix} \tag{5.28}$$

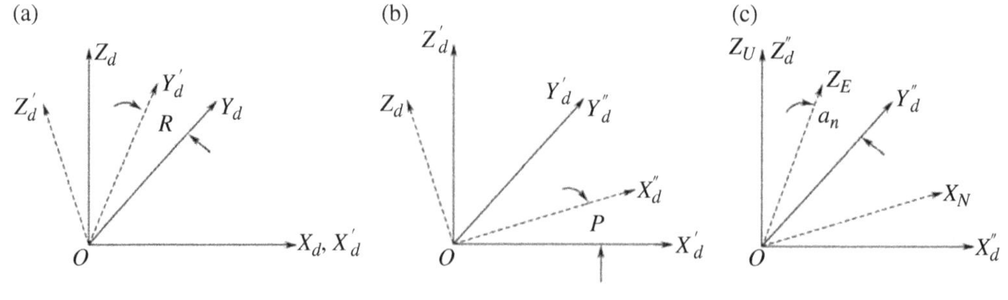

Figure 5.11 Transformation of NED coordinate and shipborne coordinate system

Eventually we get the coordinate transformation matrix $\boldsymbol{L}$:

$$\boldsymbol{L}=\boldsymbol{L}_1(R)\boldsymbol{L}_2(P)\boldsymbol{L}_3(a_n)=\begin{bmatrix} T_{11} & T_{12} & T_{13} \\ T_{21} & T_{22} & T_{23} \\ T_{31} & T_{32} & T_{33} \end{bmatrix} \tag{5.29}$$

where

$$T_{11}=\cos a_n \cos P$$
$$T_{12}=\sin a_n \cos P$$
$$T_{13}=-\sin P$$
$$T_{21}=\cos a_n \sin P \sin R-\sin a_n \cos R$$
$$T_{22}=\sin a_n \sin P \sin R+\cos a_n \cos R$$
$$T_{23}=\cos P \sin R$$
$$T_{31}=\cos a_n \sin P \cos R+\sin R \sin a_n$$
$$T_{32}=\sin a_n \sin P \cos R-\sin R \cos a_n$$
$$T_{33}=\cos P \cos R$$

The shipborne coordinate system is one type of carrier coordinate system. The transformation between the other types and NED coordinate systems can also be implemented by the methods mentioned above.

5.3.3.3 NED Coordinate System and Earth Rectangular Coordinate System

As indicated in Figure 5.12, assume that the longitude, latitude, altitude, and geodetic azimuth of the radar are, respectively, L, B, H, and A, then its coordinates in the earth rectangular coordinate system [130] are

$$\begin{cases} x_o=\left[N_R\left(1-e_1^2\right)+H\right]\sin B \\ y_o=(N_R+H)\cos B \cos L \\ z_o=(N_R+H)\cos B \sin L \end{cases} \tag{5.30}$$

where $e_1^2=\dfrac{(a^2-b^2)}{a^2}$ is the first eccentricity, $N_R=\dfrac{a}{\sqrt{1-e_1^2\sin^2 B}}$, a is the major semi-axis, and b is the short semi-axis. If coordinate system WGS-84 is adopted, then $a=6\,378\,137$ m, $b=6\,356\,752$ m.

Assume that the coordinate parameters of a target point in the NED and the earth rectangular coordinate system are, respectively, $\boldsymbol{X}_l=(x_l,y_l,z_l)$ and $\boldsymbol{X}_g=\left(x_g,y_g,z_g\right)$. According to Figure 5.12, the transformation relation between the two systems can be written as

$$\boldsymbol{X}_g=\boldsymbol{T}\boldsymbol{X}_l+\boldsymbol{X}_o \tag{5.31}$$

where

$$\boldsymbol{T}=\boldsymbol{R}_x(-L)\boldsymbol{R}_z(B)\boldsymbol{R}_y(A)$$

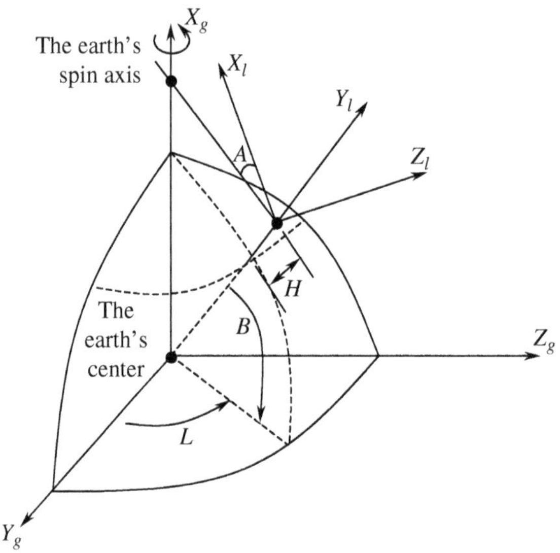

Figure 5.12 Transformation of NED coordinate and earth rectangular coordinate system

with

$$\boldsymbol{R}_x(\theta)=\begin{bmatrix}1 & 0 & 0\\ 0 & \cos\theta & \sin\theta\\ 0 & -\sin\theta & \cos\theta\end{bmatrix},\quad \boldsymbol{R}_y(\theta)=\begin{bmatrix}\cos\theta & 0 & -\sin\theta\\ 0 & 1 & 0\\ \sin\theta & 0 & \cos\theta\end{bmatrix},\quad \boldsymbol{R}_z(\theta)=\begin{bmatrix}\cos\theta & \sin\theta & 0\\ -\sin\theta & \cos\theta & 0\\ 0 & 0 & 1\end{bmatrix}$$

5.3.3.4 Antenna Coordinate System and the Sight of Target Coordinate System

When the target is tracked correctly, the antenna and the sight of target coordinate system coincide completely. When the target is not tracked correctly, the elevation angle error on plane $R'OD'$ is ε_e and the azimuth error on plane $R'OE'$ is ε_d, and the angle is positive, as shown in Figure 5.8. Assume that the target's coordinate parameters in the antenna and the sight of target coordinate system are, respectively, $\boldsymbol{X}_R=(x_r,e_r,d_r)$, $\boldsymbol{X}_{R'}=(x_{r'},e_{r'},d_{r'})$, then the transformation relation between them is

$$\boldsymbol{X}_{R'}=\boldsymbol{T}_{rr'}\boldsymbol{X}_R \tag{5.32}$$

where $\boldsymbol{T}_{rr'}=\boldsymbol{R}_E(\varepsilon_e)\boldsymbol{R}_D(\varepsilon_d)$ and

$$\boldsymbol{R}_E(\varepsilon_e)=\begin{bmatrix}\cos\varepsilon_e & 0 & -\sin\varepsilon_e\\ 0 & 1 & 0\\ \sin\varepsilon_e & 0 & \cos\varepsilon_e\end{bmatrix},\quad \boldsymbol{R}_D(\varepsilon_d)=\begin{bmatrix}\cos\varepsilon_d & \sin\varepsilon_d & 0\\ -\sin\varepsilon_d & \cos\varepsilon_d & 0\\ 0 & 0 & 1\end{bmatrix}$$

5.3.3.5 The Transformation of NED Coordinate Systems

For a radar network system, the data detected by all the radars should be transformed to one coordinate system in order to make full use of them. This process is the so-called "space registration" in radar network systems. The space registration method most frequently used is first to transform the measurement coordinate systems of all the radars to their respective NED coordinate systems, and then use the NED systems in the conversion between the different radar units. The transformation between the measurement coordinate system of different radars and the NED systems can be implemented by the methods introduced in Section 5.3.4. We focus on the NED coordinate transformation between different radar units here.

Assume that the longitude, latitude, altitude, and geodetic azimuth of the origins of the measurement coordinate systems of radars i and j are, respectively, L_i, B_i, H_i, A_i and L_j, B_j, H_j, A_j. Assume also that the position parameters of the target point in the two measurement coordinate systems are, respectively, $\boldsymbol{X}_{li} = (x_{li}, y_{li}, z_{li})$ and $\boldsymbol{X}_{lj} = (x_{lj}, y_{lj}, z_{lj})$. From (5.30), we obtain the earth rectangular coordinates of radars i and j, represented respectively as $\boldsymbol{X}_{oi}$ and $\boldsymbol{X}_{oj}$. From (5.31), we get the transformation relations between NED and earth rectangular coordinate systems of radars i and j, which are respectively

$$\boldsymbol{X}_g = \boldsymbol{T}_i\boldsymbol{X}_{li} + \boldsymbol{X}_{oi}$$

$$\boldsymbol{X}_g = \boldsymbol{T}_j\boldsymbol{X}_{lj} + \boldsymbol{X}_{oj}$$

According to the above equation, the transformation relation of the NED coordinate systems between radar i and radar j is

$$\boldsymbol{X}_{lj} = \boldsymbol{T}_j^{-1}\boldsymbol{T}_i\boldsymbol{X}_{li} + \boldsymbol{T}_j^{-1}\left(\boldsymbol{X}_{oi} - \boldsymbol{X}_{oj}\right) \tag{5.33}$$

In many engineering applications, some simple transformation methods can be used (as the situation dictates). For example, when the distance between radars is rather short, their NED coordinate systems are deemed to be approximately parallel with each other, so the coordinate transformation between them can be made with the translation transformation method. In the case of fixed ground radars, coordinate transformation relations of different radars can be calculated before the operation of the systems so as to reduce the computational amount required.

5.3.4 Selection of Tracking Coordinate Systems and Filtering State Variables

The design of tracking filters is greatly influenced by the mathematical models of the following [39]:

1. measurements (observations) provided by detectors;
2. movement of tracked targets.

Both models rely on the coordinate system adopted. Generally, we can use any type of tracking coordinate system. However, in terms of operating environments and convenience, rectangular, NED, or polar coordinate systems are usually used in the tracking of single targets in clutter-free environments. Mixed coordinate systems are more convenient in tracking multiple targets in multi-echo environments and in tracking single ones with multiple platforms. In any case, the coordinate system adopted should be:

- easy to use in describing the movement of the target;
- able to meet the bandwidth requirement of filters;
- convenient in state coupling and decoupling;

- small in dynamic and static deviation;
- able to reduce the computational complexity but not at the expense of tracking accuracy.

The general rule for state variables is to select a set of variables which have minimum dimensions but can fully describe the dynamic characteristics of targets in case the computational complexity increases with the number of state variables. It should be noted that state variables are directly related to the selection of tracking coordinate systems. Johson [136] proved that the choice of a proper coordinate system would reduce the calculating cost of state estimation considerably. Moreover, the introduction of velocity measurements in PD radars is an effective way to improve the tracking accuracy. This method can not only increase the bandwidth of the system, but also decrease its dynamic errors efficiently.

5.4 Radar Error Calibration Techniques

In the multi-radar fusion tracking process, even the best fusion method has its limitations when employed in directly fusing the information with errors collected from each radar, as proved in their military applications. For example, when a shipborne airplane takes off, three different routes are provided by three radars at a station, and the maximum error is greater than 10 km, in which case the fusion result of such data is certainly misleading. It is necessary, therefore, to make calibrations before the fusion so as to eliminate the errors in the ranging and angle measurement of the radar units.

Radar calibration methods widely adopted include static active and passive cooperative calibration, and non-cooperative calibration. These methods, however, require laser ranging and optical angle measuring sensors to acquire true values for calibrations. Active and passive calibrations are suitable for two-dimensional radars, while the use of optical measurement is limited to three-dimensional radars, which are relatively sophisticated and whose calibration structure (which is hard to put up) is subject to space restrictions.

The radar calibration unit incorporating the ADS-B system developed by Naval University of Aeronautics and Astronautics receives in real time the position data broadcast by civil aircraft, including information such as longitude, latitude, velocity, altitude, and type. Then, it compares those data with the three-dimensional information of the same batch of targets recorded by the radars. According to the comparison and analysis results, the unit testifies and evaluates whether the target detection performance of the radars reaches the required standard.

The ADS-B system has three main characteristics. Firstly, it provides real-time data. ADS-B obtains the information about its own position through the GNSS positioning system, and then transmits the real-time data to the ground receiving equipment and other neighboring civilian aircraft in the airspace. Secondly, the data which it sends are reliable. To be more exact, these data are highly accurate, because what the system provides is the GNSS navigation and positioning information, and the parameters acquired from other sophisticated avionic equipment. Thirdly, its data are convenient for reception. Since ADS-B transmits information by means of broadcasting, the ground receiving equipment can be simplified, that is to say, it is unnecessary for the ground equipment to send inquiry information to the target plane.

Radar calibration systems incorporating ADS-B are of modular design, and thus able to emendate the tactical performance of the radars by combination and extension of the modules. They are composed of seven software and hardware modules: ADS-B data receiving modules, GPS receiving modules, radar data real-time feed modules, ADS-B and radar target monitoring platforms, ADS-B and radar data comparing and processing modules, error analysis and calibration advice modules, and calibration report generating modules of radar performance. On the premise of not influencing the radars' performance, radar data are fed, through the data reporting access, into the ADS-B tactical performance calibration system, which provides the tactical performance assessment report of the radars by the data comparing and processing software. These ADS-B incorporating systems also have the function of tracking and directing radars, radar range assessment, radar target aided identification, and monitoring the intelligence quality.

5.5 Data Compression Techniques

In current data processing techniques, there are two types of data compression: one refers to compressing data at different times into the data at one time in the data processing system of a monostatic radar [39]; the other refers to compressing multiple-radar data into monostatic radar data [18, 143, 144] in the multi-radar (radar network) data processing system.

5.5.1 Data Compression in Monostatic Radar

Currently, the data rate of sensors of various types is becoming higher and higher, more and more motion information on targets can be obtained, so the tracking accuracy naturally becomes higher. The increase in filtering sampling rate, however, sets higher requirements for the computational speed and increases the tracking cost. Therefore, data compression techniques are frequently used in practical engineering to solve the contradictions between the filtering precision and the quantity of data [39]. The data compression techniques for monostatic radars can be divided into two groups: equal-weighted average preprocessing of measurements and variable-weighted average preprocessing of measurements.

5.5.1.1 Equal-Weighted Average Measurement Preprocessing

Suppose that the discrete state and the measurement equation of the target are, respectively,

$$\boldsymbol{X}(k+1)=\boldsymbol{F}(k+1,k)\boldsymbol{X}(k)+\boldsymbol{V}(k) \tag{5.34}$$

$$\boldsymbol{Z}(k+1)=\boldsymbol{H}(k+1)\boldsymbol{X}(k+1)+\boldsymbol{W}(k+1) \tag{5.35}$$

where $\boldsymbol{F}(k+1,k)$ and $\boldsymbol{H}(k+1)$ are, respectively, the state transformation and the observation matrix, $\boldsymbol{V}(k)$ and $\boldsymbol{W}(k+1)$ are mutually independent Gaussian white process noise and measurement noise vector.

Suppose that the filtering rate is k (1/s). Measure the target M times in each sampling period, then the measurement sequence is

$$\left\{\boldsymbol{Z}\left(k+\frac{1}{M}\right),\ldots,\boldsymbol{Z}\left(k+\frac{i}{M}\right),\ldots,\boldsymbol{Z}(k+1)\right\}$$

Define the equal-weighted average residual of the M measurements as $\boldsymbol{v}_{\text{pm}}(k+1)$, then from the standard Kalman filter equations in Section 3.2, we get the following relation:

$$\begin{aligned}
\boldsymbol{v}_{\text{pm}}(k+1) &= \frac{1}{M}\sum_{i=1}^{M}\boldsymbol{v}\left(k+\frac{i}{M}\right) \\
&= \frac{1}{M}\sum_{i=1}^{M}\left[\boldsymbol{Z}\left(k+\frac{i}{M}\right)-\boldsymbol{H}\left(k+\frac{i}{M}\right)\cdot\hat{\boldsymbol{X}}\left(k+\frac{i}{M}|k\right)\right] \\
&= \frac{1}{M}\sum_{i=1}^{M}\left[\boldsymbol{H}\left(k+\frac{i}{M}\right)\boldsymbol{X}\left(k+\frac{i}{M}\right)+\boldsymbol{W}\left(k+\frac{i}{M}\right)-\boldsymbol{H}\left(k+\frac{i}{M}\right)\boldsymbol{F}\left(k+\frac{i}{M},k\right)\hat{\boldsymbol{X}}(k|k)\right] \\
&= \frac{1}{M}\sum_{i=1}^{M}\boldsymbol{H}\left(k+\frac{i}{M}\right)\left[\boldsymbol{X}\left(k+\frac{i}{M}\right)-\boldsymbol{F}\left(k+\frac{i}{M},k\right)\hat{\boldsymbol{X}}(k|k)\right]+\frac{1}{M}\sum_{i=1}^{M}\boldsymbol{W}\left(k+\frac{i}{M}\right)
\end{aligned} \tag{5.36}$$

where

$$\boldsymbol{W}_{\text{pm}}(k+1) = \frac{1}{M}\sum_{i=1}^{M}\boldsymbol{W}\left(k+\frac{i}{M}\right)$$

is the equal-weighted average measurement noise, and its covariance matrix is

$$\begin{aligned}
\boldsymbol{R}_{\text{pm}}(k+1) &= \mathrm{E}\left[\boldsymbol{W}_{\text{pm}}(k+1)\boldsymbol{W}'_{\text{pm}}(k+1)\right] \\
&= \mathrm{E}\left[\frac{1}{M^2}\sum_{i=1}^{M}\sum_{j=1}^{M}\boldsymbol{W}\left(k+\frac{i}{M}\right)\boldsymbol{W}'\left(k+\frac{j}{M}\right)\right]
\end{aligned} \tag{5.37}$$

where $\boldsymbol{R}(k+1)$ is the covariance matrix of measurement noise $\boldsymbol{W}(k+1)$. Obviously, the influence of random measurement noise in the equal-weighted average residual has greatly reduced.

The basic idea of equal-weighted average measurement preprocessing is to replace the one-step measurement residual $\boldsymbol{v}(k+1)$ with the equal-weighted average residual $\boldsymbol{v}_{\text{pm}}(k+1)$ containing more target information but with less influence of measurement noise to calculate the target state estimation, which will undoubtedly increase the tracker's estimation precision substantially.

5.5.1.2 Variable-Weighted Average Measurement Preprocessing

Likewise, the core of variable-weighted average measurement preprocessing is to replace the one-step measurement residual $\boldsymbol{v}(k+1)$ with the variable-weighted average residual $\boldsymbol{v}_{\text{pm}}(k+1)$ containing more target information but with less influence of measurement noise in target state estimation, so as to enhance the effect of the latest measurement data on filtering.

The variable-weighted average residual of M measurements is defined as

$$\begin{aligned}\boldsymbol{v}_{\mathrm{vm}}(k+1) &= \frac{\sum_{i=1}^{M} i \cdot \boldsymbol{v}\left(k+\frac{i}{M}\right)}{\sum_{i=1}^{M} i} \\ &= \frac{1}{\sum_{i=1}^{M} i}\sum_{i=1}^{M} i \cdot \boldsymbol{H}\left(k+\frac{i}{M}\right)\left[\boldsymbol{X}\left(k+\frac{i}{M}\right)-\boldsymbol{F}\left(k+\frac{i}{M},k\right)\hat{\boldsymbol{X}}(k|k)\right]+\boldsymbol{W}_{\mathrm{vm}}(k+1)\end{aligned} \tag{5.38}$$

where $\boldsymbol{W}_{\mathrm{vm}}(k+1)$ is the variable-weighted average measurement noise

$$\boldsymbol{W}_{\mathrm{vm}}(k+1) = \frac{\sum_{i=1}^{M} i \cdot \boldsymbol{W}\left(k+\frac{i}{M}\right)}{\sum_{i=1}^{M} i} \tag{5.39}$$

and its covariance matrix is

$$\begin{aligned}\boldsymbol{R}_{\mathrm{vm}}(k+1) &= \mathrm{E}\left[\boldsymbol{W}_{\mathrm{vm}}(k+1)\boldsymbol{W}'_{\mathrm{vm}}(k+1)\right] \\ &= \frac{1}{\left(\sum_{i=1}^{M} i\right)^2}\mathrm{E}\left\{\sum_{i=1}^{M}\sum_{j=1}^{M} i \cdot j \cdot \boldsymbol{W}\left(k+\frac{i}{M}\right)\boldsymbol{W}'\left(k+\frac{j}{M}\right)\right\} = \frac{\sum_{i=1}^{M} i^2}{\left(\sum_{i=1}^{M} i\right)^2}\boldsymbol{R}(k+1)\end{aligned} \tag{5.40}$$

5.5.2 Data Compression in Multistatic Radar

The centralized structure is a data processing structure [18] commonly used in multistatic radar systems. This structure transmits the observations acquired by the sensors to the data processing center of the system for direct fusion. The computational complexity of this kind of system obviously increases with the number of targets. Therefore, data compression methods are frequently used to increase the real-time processing speed in many practical systems.

The data compression of radar network systems falls into two types [18]: measurement synthesis and serial combination. Measurement synthesis is to combine several radars' measurements of the same target at the same time and combine several detected data into one. Serial combination is to integrate multistatic radar data into detection measurements similar to those of monostatic radar instead of synthesizing many detection data into one.

5.5.2.1 Measurement Synthesis

Take the dual-radar system as an example. Assume that each radar can provide information about the range and angle of the target, so the process of data compression is as follows.

Assume that $\mathbf{Z}_k^1=\left\{z_{1,k}^1,\ldots,z_{N_{1,k},k}^1\right\}$ and $\mathbf{Z}_k^2=\left\{z_{1,k}^2,\ldots,z_{N_{2,k},k}^2\right\}$ are, respectively, the measurement set of radar 1 and radar 2 at time k, where $N_{1,k}$ and $N_{2,k}$ are the number of measurements obtained by radar 1 and radar 2 at time k, and $z_{i,k}^1=\left(\rho_{i,k}^1,\theta_{i,k}^1\right)^T$ $(i=1,\ldots,N_{1,k})$ and $z_{j,k}^2=\left(\rho_{j,k}^2,\theta_{j,k}^2\right)^T$ $(j=1,\ldots,N_{2,k})$ represent the ith and jth measurement of radar 1 and radar 2 at time k. If

$$\begin{cases}\left|\rho_{i,k}^1-\rho_{j,k}^2\right|\leq\rho_T\\ \left|\theta_{i,k}^1-\theta_{j,k}^2\right|\leq\theta_T\end{cases},\quad i=1,\ldots,N_{1,k},j=1,\ldots,N_{2,k}\tag{5.41}$$

then the measurement $z_{i,k}^1$ of radar 1 and the measurement $z_{j,k}^2$ of radar 2 are believed to be associated; ρ_T and θ_T are thresholds of correlated wave gates, whose values are related to sensors' measurement errors. If a certain measurement of radar 1 is associated with several measurements of radar 2, then the nearest measurement point is chosen to make an association by the nearest-neighborhood algorithm.

For any pair of associated measurements $z_{i,k}^1$ and $z_{j,k}^2$, make data compression according to the following equation:

$$\begin{cases}\hat{\rho}=\dfrac{1}{\sigma_{1,\rho}^2+\sigma_{2,\rho}^2}\left(\sigma_{2,\rho}^2\cdot\rho_{i,k}^1+\sigma_{1,\rho}^2\cdot\rho_{j,k}^2\right)\\ \hat{\theta}=\dfrac{1}{\sigma_{1,\theta}^2+\sigma_{2,\theta}^2}\left(\sigma_{2,\theta}^2\cdot\theta_{i,k}^1+\sigma_{1,\theta}^2\cdot\theta_{j,k}^2\right)\end{cases}\tag{5.42}$$

Then we can obtain an equivalent measurement and an equivalent measurement error:

$$\begin{cases}\hat{\sigma}_\rho^2=\left(\dfrac{1}{\sigma_{1,\rho}^2}+\dfrac{1}{\sigma_{2,\rho}^2}\right)^{-1}\\ \hat{\sigma}_\theta^2=\left(\dfrac{1}{\sigma_{1,\theta}^2}+\dfrac{1}{\sigma_{2,\theta}^2}\right)^{-1}\end{cases}\tag{5.43}$$

where $\sigma_{i,\rho}^2$ and $\sigma_{i,\theta}^2$ are, respectively, the ranging and angle measuring variance of the ith $(i=1,2)$ radar and $\hat{\sigma}_\rho^2$ and $\hat{\sigma}_\theta^2$ are the range and angle variance after compression.

From (5.42) and (5.43) we see that the nature of data compression is to conduct measurement weighting according to the precision, and the equivalent measurement after compression improves the precision, that is,

$$\begin{cases}\hat{\sigma}_\rho^2\leq\min\left\{\sigma_{1,\rho}^2,\sigma_{2,\rho}^2\right\}\\ \hat{\sigma}_\theta^2\leq\min\left\{\sigma_{1,\theta}^2,\sigma_{2,\theta}^2\right\}\end{cases}\tag{5.44}$$

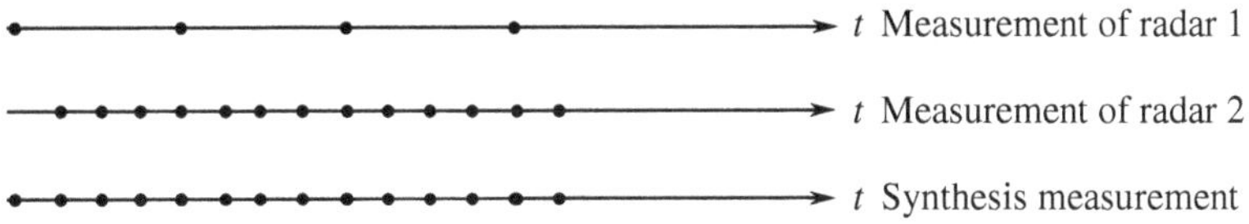

Figure 5.13 Synthesis measurement of centralized radars

The above-mentioned algorithm can also be applied to multistatic radar systems consisting of more than three radars. Assume that these radars' measurement vectors about the same target at the same time are, respectively, $\boldsymbol{Z}_1$, $\boldsymbol{Z}_2$, ..., $\boldsymbol{Z}_N$, and that the corresponding measurement error covariances are, respectively, $\boldsymbol{R}_1, \boldsymbol{R}_2, \ldots, \boldsymbol{R}_N$, then we can make data compression with the following equation:

$$\boldsymbol{Z} = \boldsymbol{R} \sum_{i=1}^{N} \boldsymbol{R}_i^{-1} \boldsymbol{Z}_i \tag{5.45}$$

$$R = \left[\sum_{i=1}^{N} R_i^{-1} \right]^{-1} \tag{5.46}$$

From (5.45) and (5.46) it can be found that the estimation result is weighting the measurements of each radar according to the precision [18].

5.5.2.2 Serial Combination

Serial combination, also known as measurement track synthesis, is widely used in practice. The principle of measurement data flow synthesis is shown in Figure 5.13 (taking the single target as an example), where the horizontal axis represents the time, and the point represents the detected measurement.

As can easily be seen from Figure 5.13, an outstanding characteristic of serial combination is that the rate of data flow increases after synthesis, which means an improvement in tracking accuracy, especially in the case of maneuvering targets. Moreover, the overall increase in data rate brings about an increase in the initial speed of tracks, which is especially important to low-altitude anti-penetration and low-altitude anti-missile operations. Meanwhile, it should be noted that in a radar system with a high data rate, the serial combination method makes no sense and measurement synthesis is the best choice.

5.6 Summary

Measurement preprocessing is a vital technical link in the chain of radar data processing. Efficient techniques for this stage can reduce computational complexity and improve target tracking precision, which will be a great help in improving the overall system performance. So, measurement preprocessing techniques are the main topic of this chapter, which starts with time registration and space registration problems and their corresponding solutions. Space registration involves the selection of coordinate systems, which directly influences the tracking result of the whole system. This chapter then presents some common coordinate systems, and the transformation relation of different coordinate systems. Finally, it analyzes and discusses the radar error calibration and data compression techniques. These techniques work well in reducing computational complexity and improving tracking efficiency.

6

Track Initiation in Multi-target Tracking

6.1 Introduction

Previous chapters have focused on filtering in radar data processing, whose essence is to solve the problem of the effect of noise on useful signals. In modern complex electromagnetic environments, however, this entails not only distinguishing wanted signals from noise (picking out desired signals from noise), but also correctly matching the measurements at adjacent moments. In other words, it involves solving the problem of the effects of clutter, false targets, or other targets on a certain target. The following three chapters (Chapters 6, 7, and 8) will discuss the problem of measurements matching. This chapter will mainly study track initiation, that is, the correct measurement-to-measurement association. To be more specific, it will focus on single-sensor multi-target tracking or single-sensor single-target tracking in a cluttered environment.

Track initiation is the first issue of multi-target tracking. Correct track initiation will effectively reduce the computational burden caused by the combinatorial explosion inherent in multi-target tracking. If track initiation is not correct, target tracking cannot be carried out at all, thus leading to the loss of targets. The old Chinese saying "A minimal error or deviation may result in wide divergence" best demonstrates the importance of track initiation. Additionally, track initiation is difficult to deal with due to factors like the greater range of targets, lower detecting resolution of sensors, lower accuracy of measurement, as well as the absence of real statistical rules on the occurrence of true and false targets. Multi-target track initiation in noisy environments is the most complex to handle, chiefly due to the complexity of track processing itself and the status of track initiation [145]. The formation of the correlated wave gate and confirmation area is the first issue in multi-target tracking. The correlated wave gate refers to a domain [17, 19, 145, 146] with the predicted position of the target being tracked as the center, used to determine the possible range of the observed value of the target. The size of this area is determined by the probability of correctly receiving the echo. That is to say, when determining the shape and size of the wave gate, true measurements should fall within the wave gate with high probability while there should be few unrelated measurements in this gate. The correlated wave gate is the decision threshold used to judge whether

Radar Data Processing with Applications, First Edition. He You, Xiu Jianjuan, and Guan Xin.

the measurements originate from the target. The echoes falling within the correlated wave gate are called "candidate echoes." Once the shape and size of the gate are determined, the detection probability (i.e., the probability of true target measurements being correctly detected) and the false alarm probability (i.e., the probability of false target measurements being falsely detected) are also determined. Owing to the fact that the detection probability and false alarm probability are often contradictory, choosing an appropriate correlated wave gate is very important. This chapter will discuss the formation of correlated wave gate and track initiation algorithms.

6.2 The Shape and Size of Track Initiation Gates

The correlated wave gate is the decision threshold used to judge whether measurements originate from the target. It is an area with the predicted position of the target being tracked as the center, used to determine the likely range of the observed values of the target at the next moment. The size of the gate is determined by:

- prediction errors (track extrapolation errors);
- probability of correctly receiving echoes (threshold probability);
- radar measurement errors;
- target moving (maneuvering) features;
- the choice of coordinate systems;
- antenna scanning periods.

The echoes which fall within the correlated wave gates are called "candidate echoes." Here we mainly discuss such frequently used correlated wave gates as the annular gate, the elliptic/ellipsoidal gate, the rectangular gate, and the sector gate in the polar coordinate system [14, 39, 147] (for other gates, see Ref. [39]). For the sake of convenience, the measurement equation, innovation (measurement residual), and innovation covariance discussed in Chapter 3 are presented again below.

The measurement equation is

$$\boldsymbol{Z}(k+1)=\boldsymbol{H}(k+1)\boldsymbol{X}(k+1)+\boldsymbol{W}(k+1) \tag{6.1}$$

where $\boldsymbol{H}(k+1)$ is a measurement matrix, $\boldsymbol{X}(k+1)$ is a state vector, and $\boldsymbol{W}(k+1)$ is a zero-mean Gaussian white noise sequence with covariance $\boldsymbol{R}(k+1)$.

The innovation is

$$\boldsymbol{v}(k+1)=\boldsymbol{Z}(k+1)-\hat{\boldsymbol{Z}}(k+1|k) \tag{6.2}$$

The innovation covariance is

$$\boldsymbol{S}(k+1)=\boldsymbol{H}(k+1)\boldsymbol{P}(k+1|k)\boldsymbol{H}'(k+1)+\boldsymbol{R}(k+1) \tag{6.3}$$

where $\boldsymbol{P}(k+1|k)$ is the one-step prediction of the covariance.

6.2.1 The Annular Gate

The annular gate usually serves as the initial gate in track initiation. It is a 360° large annular gate centering the track head, whose size is determined by the maximum and minimum movement

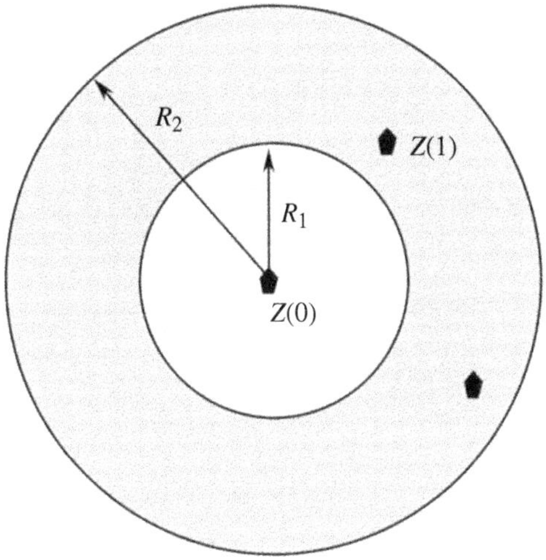

Figure 6.1 The annular gate

velocity of the target as well as the sampling interval. The initial gate needs to be large due to the greater range of the target, the lower detecting resolution of the sensor, and the poorer measurement accuracy. The inner and outer diameter of the gate should satisfy $R_1 = V_{\min}T$ and $R_2 = V_{\max}T$, as illustrated in Figure 6.1, where $V_{\min}$ and $V_{\max}$ are the maximum and minimum velocity, respectively, and T is the sampling interval.

6.2.2 *The Elliptic/Ellipsoidal Gate*

If the converted measurement $\mathbf{Z}_c(k+1)$ of a target in the rectangular coordinate system measured by the sensor satisfies

$$\begin{aligned}\widetilde{V}_{k+1}(\gamma) &\triangleq \left[\mathbf{Z}_c(k+1)-\hat{\mathbf{Z}}_c(k+1|k)\right]'\mathbf{S}^{-1}(k+1)\left[\mathbf{Z}_c(k+1)-\hat{\mathbf{Z}}_c(k+1|k)\right] \\ &= \mathbf{v}_c'(k+1)\mathbf{S}^{-1}(k+1)\mathbf{v}_c(k+1) \le \gamma\end{aligned} \tag{6.4}$$

then, the converted measurement is called the candidate echo. Formula (6.4) is called the "rule of the ellipsoidal gate," where the parameter γ can be obtained from the distribution table of χ^2. If the measurement $\mathbf{Z}_c(k+1)$ is n_z-dimensional, then $\widetilde{V}_{k+1}(\gamma)$ is the χ^2 distributed random variable with n_z degrees of freedom. The square root of the parameter ($g = \sqrt{\gamma}$) is called the "σ number" of the gate. When $n_z = 2$, the shape of the correlated ellipsoidal gate is as depicted in Figure 6.2.

The probability P_G that converted measurements fall within the gate varies with the value of γ and the measurement dimension n_z. Thus, P_G is defined as follows:

$$P_G = \Pr\left\{\mathbf{Z}_c(k+1) \in \widetilde{V}_{k+1}(\gamma)\right\} \tag{6.5}$$

The relationship between P_G and the measurement dimension n_z, as well as the parameter γ, can be found in Ref. [39]. Table 6.1 presents the probabilities corresponding to different γ when the measurement dimension n_z ranges from 1 to 3.

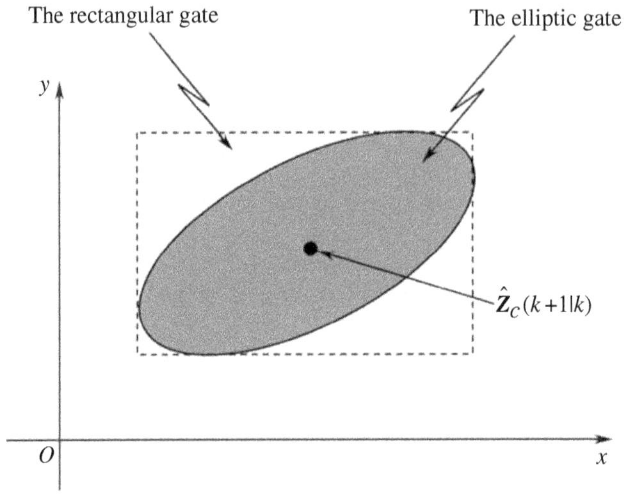

Figure 6.2 Correlated wave gate in the rectangular coordinate system

Table 6.1 The probability P_G that n_z-dimensional measurements fall within the gate

γ	1	4	9	16	25
$g=\sqrt{\gamma}$	1	2	3	4	5
$n_z=1$	0.683	0.954	0.997	0.99994	1.0
$n_z=2$	0.393	0.865	0.989	0.9997	1.0
$n_z=3$	0.199	0.739	0.971	0.9989	0.99998

The area/volume [39] of the n_z-dimensional elliptic/ellipsoidal gate is

$$V_{\text{elliptic}}(n_z)=c_{n_z}\gamma^{\frac{n_z}{2}}|\mathbf{S}(k+1)|^{\frac{1}{2}} \tag{6.6}$$

where

$$c_{n_z}=\begin{cases}\dfrac{\pi^{\frac{n_z}{2}}}{(n_z/2)!}, & n_z \text{ is even}\\[2ex] \dfrac{2^{n_z+1}\left(\dfrac{n_z+1}{2}\right)!\pi^{\frac{n_z-1}{2}}}{(n_z+1)!}, & n_z \text{ is odd}\end{cases} \tag{6.7}$$

When $n_z=1, 2, 3$, $c_{n_z}=2, \pi, 4\pi/3$, respectively.

After normalizing the n_z-dimensional elliptic/ellipsoidal gate using the standard deviation of innovation covariance, we get the volume of the gate:

$$V_{\text{elliptic}}(n_z)=c_{n_z}\gamma^{\frac{n_z}{2}} \tag{6.8}$$

6.2.3 The Rectangular Gate

The simplest method of setting up a correlated wave gate is to define a rectangular area in the tracking region, namely the rectangular gate [14, 39]. Assume that $v_{ci}(k+1)$, $z_{ci}(k+1)$, and $\hat{z}_{ci}(k+1|k)$ represent the *i*th component of the innovation $\boldsymbol{v}_c(k+1)$, the converted measurement $\mathbf{Z}_c(k+1)$, and the predicted value of the measurement $\hat{\mathbf{Z}}_c(k+1|k)$, respectively, and that S_{ij} represents the element in the *i*th row and *j*th column of innovation covariance $\boldsymbol{S}(k+1)$. When all components of the measurement $\mathbf{Z}_c(k+1)$ satisfy

$$|v_{ci}(k+1)| = \left|Z_{ci}(k+1) - \hat{Z}_{ci}(k+1|k)\right| \leq K_{\mathrm{G}}\sqrt{S_{ii}}, \qquad i = 1, 2, \ldots, n_z \tag{6.9}$$

then the converted measurement $\mathbf{Z}_c(k+1)$ is said to fall within the rectangular gate, and the measurement is a candidate echo, where K_{G} is a gate constant, often taking a bigger K_{G} value ($K_{\mathrm{G}} \geq 3.5$) in practical implementations.

The area/volume of the n_z-dimensional rectangular gate is

$$V_{\text{rectangular}}(n_z) = (2K_{\mathrm{G}})^{n_z} \prod_{i=1}^{n_z} \sqrt{S_{ii}} \tag{6.10}$$

After normalizing the n_z-dimensional rectangular gate using the standard deviation of innovation covariance, we get the area/volume of the gate:

$$V^u_{\text{rectangular}}(n_z) = (2K_{\mathrm{G}})^{n_z} \tag{6.11}$$

If the corresponding gate constant K_{G} varies with different components, then (6.10) and (6.11) are, respectively, changed as follows:

$$V_{\text{rectangular}}(n_z) = 2^{n_z} \prod_{i=1}^{n_z} K_{\mathrm{G}i} \sqrt{S_{ii}} \tag{6.12}$$

$$V^u_{\text{rectangular}}(n_z) = 2^{n_z} \prod_{i=1}^{n_z} K_{\mathrm{G}i} \tag{6.13}$$

From (6.8) and (6.11), it follows that when K_{G} is the same, the ratio of the area/volume of the elliptic/ellipsoidal gate to that of the rectangular gate is

$$\text{ratio}(n_z) = \frac{V^u_{\text{elliptic}}(n_z)}{V^u_{\text{rectangular}}(n_z)} = \frac{c_{n_z}\gamma^{n_z/2}}{(2K_{\mathrm{G}})^{n_z}} \tag{6.14}$$

With the gate constant K_{G} and parameters γ and n_z given, the ratio between the area/volume of the elliptic/ellipsoidal gate and that of the rectangular gate obtained from (6.14) is shown in Table 6.2.

6.2.4 The Sector Gate

The target measurements ρ and θ detected by the sensor can be said to fall within a sector gate, and the measurements are called candidate echoes, if the data is associated in the measurement coordinate system (polar coordinates) and ρ and θ satisfy

Table 6.2 The ratio between the area/volume of the elliptic/ellipsoidal gate and that of the rectangular gate

K_G	2.8			3.0			3.5		
γ	9	16	25	9	16	25	9	16	25
$n_z = 1$	0.9333	0.7000	0.5600	1.0000	0.7500	0.6000	1.1667	0.8750	0.7000
$n_z = 2$	1.1091	0.6239	0.3993	1.2732	0.7162	0.4584	1.7330	0.9748	0.6239
$n_z = 3$	1.5528	0.6551	0.3354	1.9099	0.8057	0.4125	3.0328	1.2795	0.6551
$n_z = 4$	2.4604	0.7785	0.3189	3.2423	1.0259	0.4202	6.0067	1.9006	0.7785
$n_z = 5$	6.3056	1.0217	0.3348	6.0793	1.4426	0.4727	13.1397	3.1181	1.0217

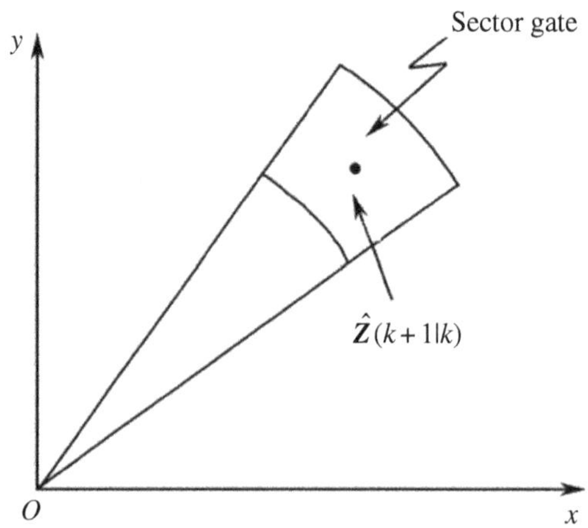

Figure 6.3 Shape of the sector gate

$$|\rho(k+1)-\hat{\rho}(k+1|k)| \le K_\rho \sqrt{\sigma_\rho^2 + \sigma_{\hat{\rho}(k+1|k)}^2} \tag{6.15}$$

$$|\theta(k+1)-\hat{\theta}(k+1|k)| \le K_\theta \sqrt{\sigma_\theta^2 + \sigma_{\hat{\theta}(k+1|k)}^2} \tag{6.16}$$

where K_ρ and K_θ refer to the square roots of parameters obtained from the χ^2 distribution table, σ_ρ^2 and σ_θ^2 denote the variances of measurement errors of ρ and θ in polar coordinates, respectively, and $\sigma_{\hat{\rho}(k+1|k)}^2$ and $\sigma_{\hat{\theta}(k+1|k)}^2$ are the variances of the corresponding predicted values, respectively. The shape of the sector gate is depicted in Figure 6.3, and its size is connected to the parameters obtained from the χ^2 distribution table, σ_ρ^2 and σ_θ^2 as well as $\sigma_{\hat{\rho}(k+1|k)}^2$ and $\sigma_{\hat{\theta}(k+1|k)}^2$.

6.3 Track Initiation Algorithms

The existing track initiation algorithms can be categorized into two types: sequential processing techniques and batch processing techniques. Sequential processing techniques generally apply to target track initiation in the environment of comparatively weak clutter, while batch processing techniques prove highly effective in initiating tracks in a strongly cluttered environment, but at the cost of increasing the computational burden. This section covers several frequently used track initiation algorithms including the logic-based method, modified logic-based method, Hough transform-based method, modified Hough transform-based method, Hough transform and logic-based method, and formation target method based on clustering and Hough transform.

6.3.1 Logic-Based Method

The logic-based method [16, 17, 145, 148] applies to the whole process of track processing, and thus certainly to track initiation. Logic-based methods and direct-vision methods are both concerned with the processing of sequential observations received from consecutive radar scans. The observation sequence represents the input of the sliding window containing N scans. When the detection number of the sliding window reaches a specified threshold, a track is initiated successfully. Otherwise, the sliding window will be moved forward (time increase direction) one. The difference between the two methods is that the direct-vision method reduces a possible track by two simple rules, of velocity and acceleration, while the logic-based method predicts and identifies a possible track by multiple hypotheses and correlated wave gates. A detailed discussion of logic-based methods follows.

Let $z_i^l(k)$ be the lth component of measurement i at time k, where $l = 1, \ldots, p, i = 1, \ldots, m_k$. Then, the lth component of the distance vector $\boldsymbol{d}_{ij}$ between observations $\boldsymbol{Z}_i(k)$ and $\boldsymbol{Z}_j(k+1)$ can be defined as

$$d_{ij}^l(t) = \max\left[0, z_j^l(k+1) - z_i^l(k) - v_{\max}^l t\right] + \max\left[0,\ -z_j^l(k+1) + z_i^l(k) + v_{\min}^l t\right] \tag{6.17}$$

where t denotes the time interval between two scans. Suppose that the observation error is independent Gaussian distributed with zero mean, and that the covariance is $\boldsymbol{R}_i(k)$, then the square of the normalized distance is

$$D_{ij}(k) \triangleq \boldsymbol{d}_{ij}'\left[\boldsymbol{R}_i(k) + \boldsymbol{R}_j(k+1)\right]^{-1}\boldsymbol{d}_{ij} \tag{6.18}$$

where $D_{ij}(k)$ is a random variable following a χ^2 distribution with p degrees of freedom. The threshold γ can be obtained from the χ^2 distribution table with p degrees of freedom by the given threshold probability. If $D_{ij}(k)$ ¡ Üγ, then it is determined that the two measurements $\boldsymbol{Z}_i(k)$ and $\boldsymbol{Z}_j(k+1)$ are associated.

The search program is executed in the following steps.

1. Set up a threshold with the measurements from the first scan as the track head, and an initial correlated wave gate with the velocity method, creating possible tracks for the second scan measurements falling within this gate.
2. Extrapolate every possible track. The subsequent correlated wave gate is centered with the extrapolated point, and its size is determined by the track extrapolation error covariance. Associate the third scan measurements falling within the subsequent correlated wave gate and closest to the extrapolated point.
3. If there is no measurement in the subsequent correlated wave gate, then cancel this possible track, or use the expanded gate limited by acceleration to check whether the measurements from the third scan fall within it.
4. Continue the steps above until a stable track is created, at which time the track initiation is completed.
5. In each scan, use the measurements that do not fall within the correlated wave gate to engage in data association discrimination as new track heads and turn to step 1.

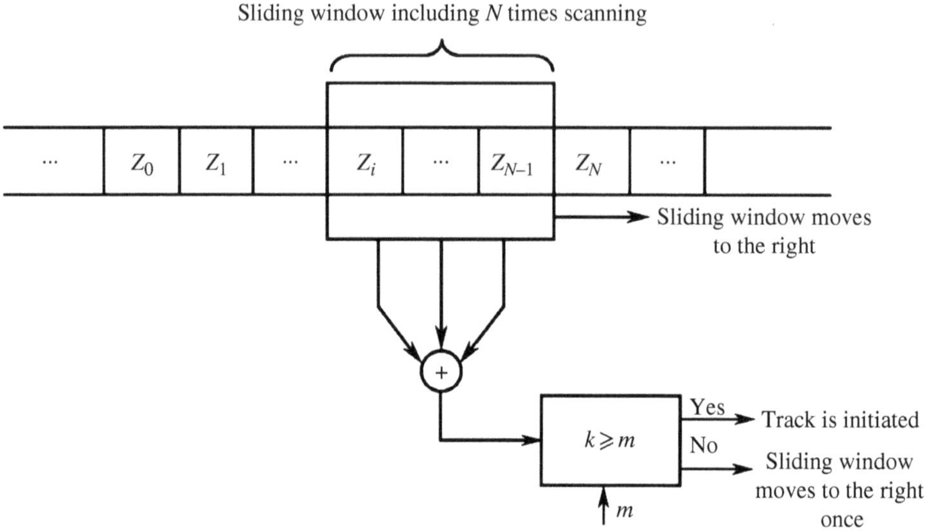

Figure 6.4 The *m/n* logic of sliding window method

When will a stable track be created by means of the logic-based method? It depends on the compromise between the complexity analysis of track initiation and its performance; to be more exact, it is determined by the performance of true and false targets, the density and distribution of these targets, the resolution and measurement error of searching sensors, etc. The commonly used method is the *m/n* logic of sliding window [17, 145], as shown in Figure 6.4.

The sequence $(z_1, z_2, \dots, z_i, \dots, z_n)$ denotes the input into the time window including n radar scans. If there are measurements in the correlated wave gate at the ith scan, the element z_i equals 1, otherwise it equals 0. When the number of detections in the time window reaches a certain value m, then track initiation is deemed successful, or else the sliding window moves to the right once, which means expanding the window time. The number of detections in track initiation and that of successive events in the sliding window together constitute the track initiation logic.

According to Ref. [145], it is most suitable to use 3/4 logic in the background simulation of military aircraft formation flying, for if $n = 5$, it will not produce a remarkable improving effect. To balance the effectiveness and computational complexity, $1/2 < m/n < 1$ is appropriate in several time scans, for $m/n > 1/2$ indicates that the number of associated measurements is more than half, or else it is not dependable as a possible track; if $m/n = 1$, it means that there are measurements associated with each scan, that is, overconfidence in a quiet environment. Therefore, only the following two values are chosen in engineering:

1. 2/3 ratio, used in fast initiation;
2. 3/4 ratio, used in normal track initiation.

6.3.2 *Modified Logic-Based Method*

In practical applications, the logic-based method can effectively initiate a target track only in the case of low false alarm probability. In order to fast initiate a track under circumstances of high false alarm probability, we can employ the modified logic-based method [149, 150]. This method has great practical value in engineering applications, for it has the same order of magnitude in terms of computational load as the logic-based method and is able to effectively initiate a target track.

Fundamentally, this algorithm is about adding a limiting condition to the measurements falling within the correlated wave gate in the track initiation period, so as to eliminate the measurements which, to some degree, are in a V-shape with the track. The algorithm search is carried out in five steps.

1. Let the measurement set obtained from the first scan be $\mathbf{Z}(1)=\{\mathbf{Z}_1(1),\ldots,\mathbf{Z}_{m_1}(1)\}$, and that from the second scan be $\mathbf{Z}(2)=\{\mathbf{Z}_1(2),\ldots,\mathbf{Z}_{m_2}(2)\}$. $\forall \mathbf{Z}_i(1)\in\mathbf{Z}(1)$ $(i=1,2,\ldots,m_1)$, $\forall \mathbf{Z}_j(2)\in\mathbf{Z}(2)$ $(j=1,2,\ldots,m_2)$, find $\boldsymbol{d}_{ij}$ and $D_{ij}(1)$ by (6.20) and (6.21), respectively, and if $D_{ij}(1)\leq\gamma$, then create a possible track $o_{s1}, s1=1,\ldots,q_1$.
2. Extrapolate every possible track o_{s1} with the linear extrapolation method, and build a subsequent correlated wave gate $\Omega_j(2)$ with the extrapolated point as the center, the gate size being determined by the track extrapolation error covariance. We can judge whether the measurement $\mathbf{Z}_j(3)$ falling within $\Omega_j(2)$ is associated with the track in the following way: let the included angle between this track and the line connecting $\mathbf{Z}_j(3)$ and the second point of track o_{s1} be α, if $\alpha\leq\sigma$ (where σ is determined by measurement error – a bigger σ can be selected to ensure a high probability of target track initiation), then it is believed that $z_j(3)$ is associated with this track.
3. If there are no measurements in the subsequent correlated wave gate $\Omega_j(2)$, then continue extrapolating the possible track $o_{s1}, s1=1,\ldots,q_1$ mentioned above with the linear extrapolation method. Build the subsequent correlated wave gate $\Omega_h(3)$ with the extrapolated point as the center, and the gate size determined by the track extrapolation error covariance. As for the measurement $\mathbf{Z}_h(4)$ falling within the subsequent correlated wave gate $\Omega_h(3)$ during the fourth scan, if the included angle β between this track and the line connecting $\mathbf{Z}_h(4)$ and the first point of track o_{s1} is less than σ, then it is believed that the measurement is associated with the track.
4. If there is no measurement in the subsequent correlated wave gate $\Omega_h(3)$ during the fourth scan, then terminate this possible track.
5. Use measurements not associated with any track in every period to initiate a new possible track, and turn to step 1.

When σ is chosen to be 360°, the modified logic-based method is simplified to the logic-based method. Generally speaking, when the target moves in a straight line, a smaller σ can be chosen to reduce the computational load and initiate target tracks effectively. When the target is maneuvering, σ should be properly amplified so as not to miss the target in track initiation. If the target's type of motion is not clear, then σ should be bigger.

6.3.3 *Hough Transform-Based Method*

Originally used in image processing, the Hough transform is a basic method of detecting image characteristics in image space [151, 152]. It mainly applies to the detection of straight lines in image space. Since the data obtained from several scans of a radar can be seen as an image, the method can be used to detect the target track. Now the Hough transform has been widely used in radar data processing, and has become an important method of multiple-sensor track initiation and low observable target detection [153]. In 1995, Carlson *et al.* [154–156] applied this method to search radars in the detection of low observable targets in linear motion or approximate linear motion. Reference [157] applies the method to track initiation, but initiating a track with it is slow, so Chen *et al.* [158] proposed the modified Hough transform to fast initiate a track.

The Hough transform method is to transform observed data (x, y) in the Cartesian coordinate system to (ρ, θ) in the parameter space, that is,

$$\rho = x\cos\theta + y\sin\theta \tag{6.19}$$

where $\theta \in [0, 180°]$. For a point (x_i, y_i) on a line, there must be only two parameters ρ_0 and θ_0 satisfying

$$\rho_0 = x_i\cos\theta_0 + y_i\sin\theta_0 \tag{6.20}$$

As illustrated in Figure 6.5, a straight line in Cartesian space can be defined by the distance ρ_0 from the origin to this straight line, and the included angle θ_0 between ρ_0 and the x axis.

Convert the points on the straight line in Figure 6.5 into curved lines in the parameter space, as illustrated in Figure 6.6.

Figure 6.6 clearly shows that the curved lines in the parameter space converted from the points on the straight line in Figure 6.5 intersect at a common point, which means that the coordinates in the Cartesian coordinate system corresponding to the curved lines that intersect at a common point must be in the same straight line.

In order to detect the target from the received radar data, plane $\rho-\theta$ needs to be discretely separated into several small checks, and the common intersecting point can be judged by detecting the peak value of the 3D histogram. The center of every check in the histogram is

$$\theta_n = \left(n - \frac{1}{2}\right)\Delta\theta, \quad n = 1, 2, \ldots, N_\theta \tag{6.21}$$

$$\rho_n = \left(n - \frac{1}{2}\right)\Delta\rho, \quad n = 1, 2, \ldots, N_\rho \tag{6.22}$$

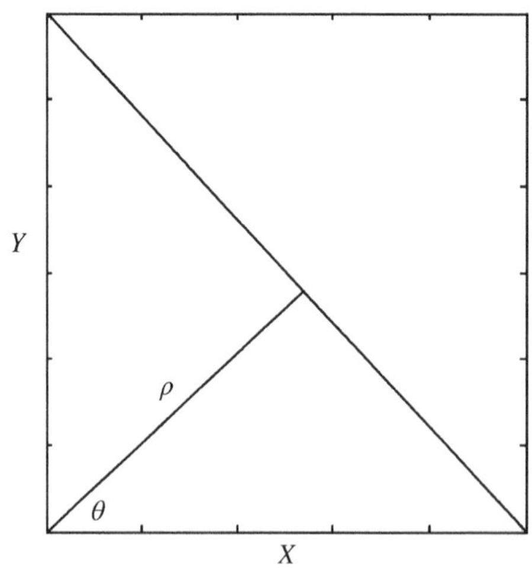

Figure 6.5 Straight line in the Cartesian coordinate system

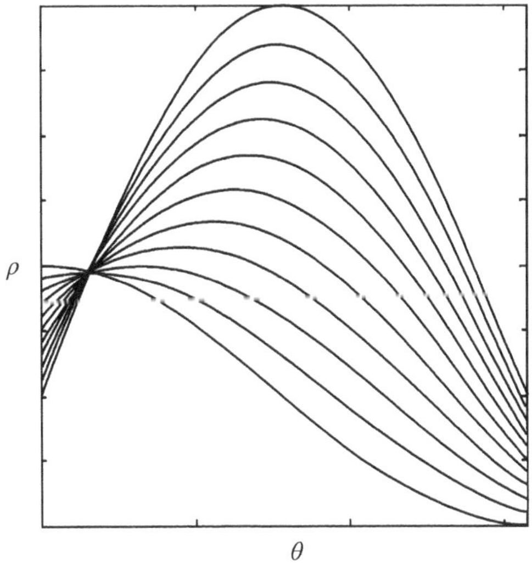

Figure 6.6 Hough transform

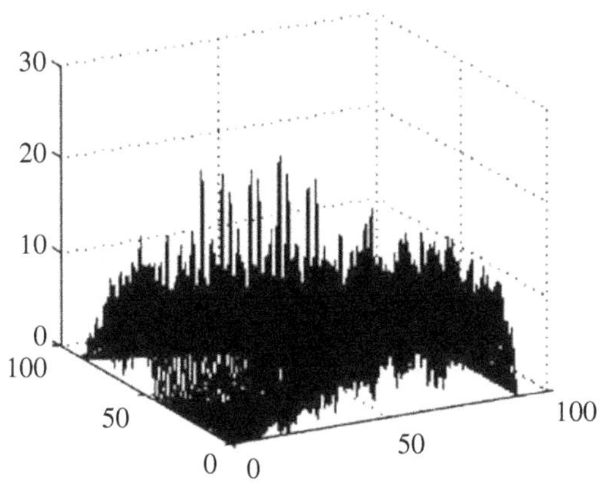

Figure 6.7 Histogram in the parameter space

where $\Delta\theta = \pi/N_\theta$, N_θ denotes the number of segments of parameter θ, and $\Delta\rho = L/N_\rho$. N_ρ denotes the number of segments of parameter ρ, and L is twice as much as the radar measurement range.

When there are a number of points that can be connected with a straight line, they will cluster in the corresponding checks of the $\rho-\theta$ plane. After several scans, the number of points in a certain unit will be accumulated for a target in linear motion. For example, in the histogram of the parameter space illustrated in Figure 6.7, the peak values suggest tentative tracks, but some of them are not yielded by the target track, but by clutter.

There are many ways to define data space. For example, a data image plane can be a 2D *RT* plane constituted of slope distance *R* and scanning time *T*, or a 2D *XY* plane constituted of the target coordinates (x, y) worked out with slope distance *R* and azimuth angle β. The 2D *RT* plane has the following characteristics: the track of a stationary or slow-speed target displays as a straight line perpendicular to the *R* axis; for a target with infinite velocity in the data image space, the slope of the track approximates to zero, but the track of a target moving with acceleration is a curved line. The 2D *XY* plane has the following characteristics: the track of a target moving with acceleration is still a straight line, but for a stationary target, it is a fixed point. Therefore, the ways to define a data space can be chosen according to practical need.

Carlson *et al.* [154] proposed a method of converting points in the Cartesian coordinate system to curved lines in the parameter space by using a simple multi-dimensional matrix.

The Hough transform applies to track initiation of targets in linear motion in a cluttered environment. The quality of initiating a track with this technique depends on the time of track initiation and the parameters $\Delta\theta$ and $\Delta\rho$. The longer it takes to initiate a track, the better the initiated track will be. Similarly, the quality of the initiated track improves with decreasing value of the parameters $\Delta\theta$ and $\Delta\rho$, but this tends to cause false dismissal of alarms. The values of $\Delta\theta$ and $\Delta\rho$ should be chosen according to the actual radar measurement error. In the case of larger measurement errors, larger values should be assigned to the parameters $\Delta\theta$ and $\Delta\rho$ lest alarms be falsely dismissed. The track of a maneuvering target is difficult to initiate with the Hough transform owing to the limitations inherent in the characteristics of the technique. It can be initiated with the generalized Hough transform, which, due to the heavy computational loads involved, is less applicable in reality and will not be elaborated on here.

6.3.4 *Modified Hough Transform-Based Method*

The classical Hough transform works effectively in track initiation only after many scans, which is not practical in engineering [158]. To tackle the slow track initiation and heavy computational loads involved in this method, Ref. [158] comes up with a modified Hough transform method, which can significantly increase the speed of track initiation. This modified technique, for example, can be employed in the fast track initiation for ballistic missiles, which proves very difficult otherwise, chiefly because of their massive acceleration in the initiation stage. The difficulty is further multiplied by their fast speed and high maneuverability.

Suppose that a radar receives three groups of data, $\boldsymbol{r}_n$, $\boldsymbol{r}_{n+1}$, and $\boldsymbol{r}_{n+2}$, at the nth, $(n+1)$th, and $(n+2)$th scan times, respectively. These data can be converted by (6.19) into three groups of curves ρ_n, ρ_{n+1}, and ρ_{n+2} in the parameter space, whereby the difference function is obtained as follows:

$$\Delta\rho_n = \rho_n - \rho_{n+1} \tag{6.23}$$

Mark the zero crossing point $\Delta\rho_n$ as $\Delta\rho_n(0)$. $\Delta\rho_n(0)$ provides the θ coordinates corresponding to the crossing points ρ_n and ρ_{n+1}, marked as $\theta_{\Delta\rho_n(0)}$. If considering the points in Cartesian coordinates, the sign of $\theta_{\Delta\rho_n(0)}$ is determined by the direction of the vector $(\boldsymbol{r}_n - \boldsymbol{r}_{n+1})$. Based on the two pieces of information above, two criteria can be obtained.

1. Zero crossing points $\theta_{\Delta\rho_n(0)}$ and $\theta_{\Delta\rho_{n+1}(0)}$ must be very close, that is,

$$\left|\theta_{\Delta\rho_n(0)} - \theta_{\Delta\rho_{n+1}(0)}\right| \leq \sigma_0 \tag{6.24}$$

   where $\Delta\theta \leq \sigma_0 \leq m\Delta\theta$ is the permissible error and m is any positive integer.
2. The signs of slopes at zero crossing points $\theta_{\Delta\rho_n(0)}$ and $\theta_{\Delta\rho_{n+1}(0)}$ have to be the same.

Criterion 1 can be used to judge whether data points are collinear. If the data of three consecutive scans received by a radar are collinear, then there should be a common intersection point in the parameter space. In practice, however, discrete intervals in the parameter space must be adjusted according to the margin of measurement error due to the presence of measurement noise, so that the intersecting points of most curved lines are in the same check. Criterion 2 can be used to determine the target moving direction to avoid the formation of an impractical track like a V-shaped one.

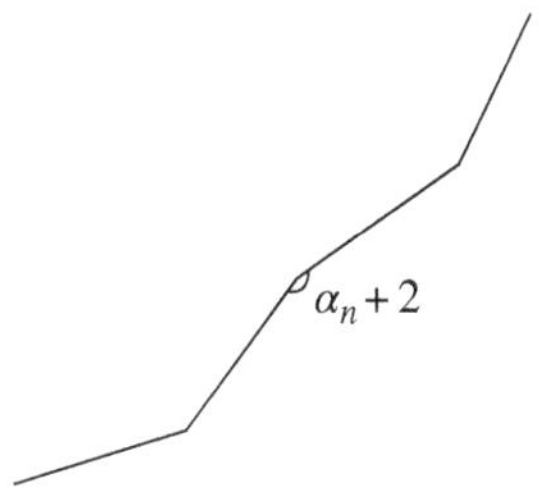

Figure 6.8 Angle restriction in track initiation

When both criteria are satisfied, we should also judge whether the tracks formed at the nth, $(n+1)$th, and $(n+2)$th times are collinear with those at the $(n+2)$th, $(n+3)$th, and $(n+4)$th times. Define the distance between $\boldsymbol{r}_{n+1}$ and $\boldsymbol{r}_{n+2}$ as $\boldsymbol{d}_{n+1,n+2}$, the included angle between vectors $(\boldsymbol{r}_{n+1}, \boldsymbol{r}_{n+2})$ and $(\boldsymbol{r}_{n+2}, \boldsymbol{r}_{n+3})$ as α_{n+2}.

Since a target's acceleration is restricted by its maximal acceleration, then

$$|d_{n+1,n+2}| \le c \times d_{n+2,n+3} \tag{6.25}$$

where c is determined by the maximal acceleration.

As illustrated in Figure 6.8, the included angle α_{n+2} between tracks has to satisfy

$$\beta_1 \le \alpha_{n+2} \le \beta_2 \tag{6.26}$$

Here, β_1 and β_2 should be properly chosen to avoid initiating a V-shaped track.

If the assumed track satisfies (6.25) and (6.26), then $\boldsymbol{r}_n$, $\boldsymbol{r}_{n+1}$, $\boldsymbol{r}_{n+2}$, $\boldsymbol{r}_{n+3}$, and $\boldsymbol{r}_{n+4}$ can form a track.

To enable it to initiate a track faster, Ref. [159] adds a condition to the modified Hough transform: it can be used to convert measurements to the parameter space only if they satisfy the condition of velocity gating below:

$$v_{\min} \le \left| \frac{x_i - x_{i-1}}{t_i - t_{i-1}} \right| \le v_{\max} \tag{6.27}$$

By using the condition of velocity gating, the number of measurements being transformed with the modified Hough transform can be reduced by a large margin, so that a fast track initiation can be realized.

6.3.5 Hough Transform and Logic-Based Method

The Hough transform-based track initiation algorithm can effectively initiate a target track in a densely cluttered environment, but it takes a long time and the parameters $\Delta\theta$ and $\Delta\rho$ are hard to select. The logic-based track initiation algorithm can initiate a target track in a short time, but it is hard to do that effectively in a densely cluttered environment. These problems can be handled effectively with the track initiation algorithm based on the Hough transform and logic, which combines the two algorithms. Track initiation with this algorithm is divided primarily into two steps: preliminary measurement association and association fuzziness elimination.

Basically, the first step is about eliminating, with the Hough transform, as many false clutter measurements as possible according to the difference in kinetic characteristics between the clutter and the target. In the initiation of tracks with the Hough transform, the selection of the value of $\Delta\theta$ and that of the threshold, due to the effect of the clutter, has a direct influence on the properties of the initiated track. At present, no general standard for selecting the two parameters is known. The principle of selecting the two parameters is to choose a bigger $\Delta\theta$ to ensure a higher probability of detecting all the true tracks. In this way, there are still some clutter measurements exceeding the threshold to yield fuzzy track initiation, but the clutter density has decreased substantially.

After a large number of clutter measurements have been eliminated at the first stage, the fuzzy associations between measurements can be eliminated with the *m/n* logic-based method.

6.3.6 Formation Target Method Based on Clustering and Hough Transform

This subsection combines the Hough transform mentioned above and the *k*-means clustering method to deal with formation target track initiation. The detailed procedure is as follows [146].

1. Let the sensor echoes at time k be separated into m formations, with $N_{ik} \geq 3$, $i=1,\ldots, m; k=1,\ldots,n$ members in each formation. If the number of formation i's members is $N_{ik} < 3$ (i.e., the number of target members in this formation is less than 3), then this formation is made up of points by default, which will be dismissed. This process eliminates most clutters or false measurements, and the m remaining formations serve as initialized particle points of the cluster. First, randomly select r $\left(r < \sum_{i=1}^{m} N_{ik}\right)$ particles $x_1, x_2, \ldots, x_r$ from all formation members $G_k(i),\ \ i=1,\ldots,m$ formed by the cycle threshold segmentation at time k as the initial points of k-means clusters $C_l\,(l=1,2,\ldots,r)$. The center of every cluster C_l might as well be marked as $o_l = x_l\,(l=1,2,\ldots,k)$.
2. For any particle $x_t \in G_k(i),\ \ i=1,\ldots,m$, calculate $d(x_t, o_l) = \|x_t - o_l\|_p\ \ (l=1,2,\ldots,r)$. If

$$d(x_t, o_l) = \min_{l \in \{1,2,\ldots,r\}} d(x_t, o_l) \tag{6.28}$$

   then distribute individual x_t to the lth cluster C_l. Follow this rule until all the particles in $x_t \in G_k(i),\ \ i=1,\ldots,m$ are distributed to different clusters $C_l\,(l=1,2,\ldots,r)$.
3. Recalculate to get the new center of every cluster $C_l\,(i=1,2,\ldots,r)$:

$$\bar{o}_l = \frac{1}{N_{lk}} \sum_{x_t \in C_l} x_t\,(t=1,2,\ldots,N_{lk}) \tag{6.29}$$

   where N_{lk} denotes the number of particles in C_l.
4. If $\bar{o}_i = o_i$, $i=1,2,\ldots,r$, output r clusters $C_i\,(i=1,2,\ldots,r)$ or let $o_i = \bar{o}_i$ and turn to step 2. The cluster centroid of the formation at each time is obtained by the procedures above, including the center of the true target and the centroid of the false target.
5. In the background of sparse formation, the formation structure can be judged from the measurements, so formations at adjacent times can be associated by judging the similarity in structure between them, requiring no linear judgment by the initial associated center with the Hough transform. Meanwhile, against the background of dense formation, the formation structure cannot be judged by the measurements, so it is necessary, after judging the distance of the formation clustering center and forming the initial association formation, to accomplish the formation track initiation by using the linear identification function of the Hough formation.

Substitute the coordinates of the initial association formation centers into (6.19) one by one. Convert the formation center points at the first three times into curved lines in the parameter space. If the curve lines in the parameter space intersect at a common point, the formation target is initiated successfully.

6.4 Comparison and Analysis of Track Initiation Algorithms

In order to demonstrate the effectiveness of the track initiation algorithms described in Section 6.3 more directly, this section makes a simulation comparison of the track initiation algorithms above in the same environment. The simulation is based on the assumption that five targets moving with constant velocity in a straight line are being tracked with a 2D radar, whose sampling period T, direction finding error σ_θ, and range error σ_r are 5 s, 0.3°, and 40 m, respectively (i.e., $T = 5$ s, $\sigma_\theta = 0.3°$, and $\sigma_r = 40$ m), and which identifies the initial positions of the targets, respectively, as (55 000 m, 55 000 m), (45 000 m, 45 000 m), (35 000 m, 35 000 m), (45 000 m, 25 000 m), and (55 000 m, 15 000 m), and their velocities as $v_x = 500$ m/s, $v_y = 0$ m/s.

The number of clutters in every period is determined by the Poisson distribution according to the method described in Ref. [160]. That is, with a given parameter λ, first yield the random number r uniformly distributed in section (0,1), then determine J from

$$e^{-\lambda}\sum_{j=0}^{J-1}\frac{\lambda^j}{j!} < r \le e^{-\lambda}\sum_{j=0}^{J}\frac{\lambda^j}{j!}, \quad J = 1, 2, \ldots \tag{6.30}$$

where J is the number of clutters to be produced. After J is determined, J clutters of every period are randomly distributed in the radar field of vision according to the uniform distribution.

When $\lambda = 50$, the situation of clutter measurements and true measurements of the target produced in four consecutive scan periods is illustrated in Figure 6.9, where ○ denotes the true track measurements, $*$ denotes clutter measurements in the first scan, □ in the second scan, + in the

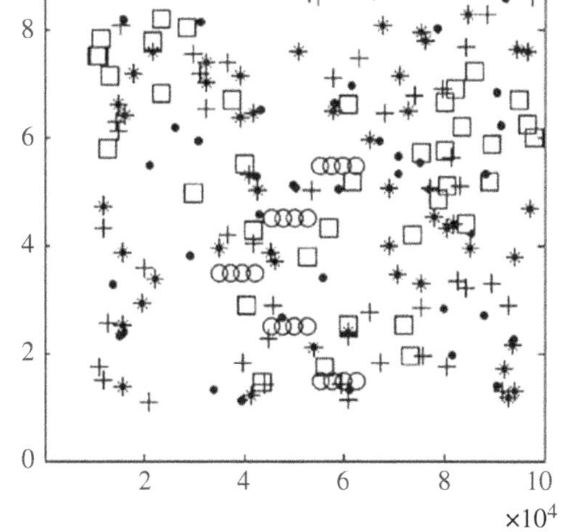

Figure 6.9 Situation map of clutter measurements and true measurements

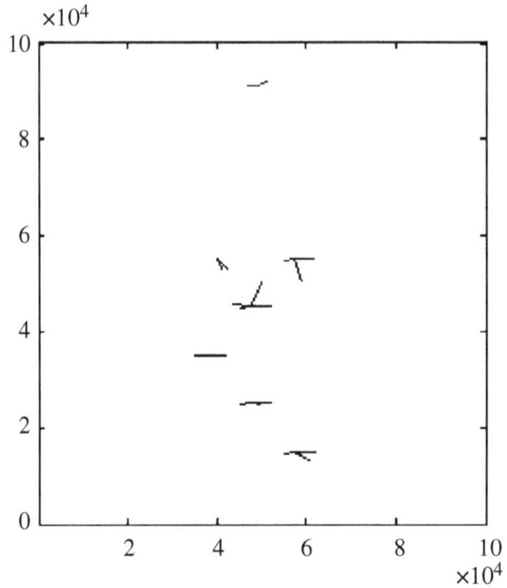

Figure 6.10 Chart of the track initiated with direct-vision method

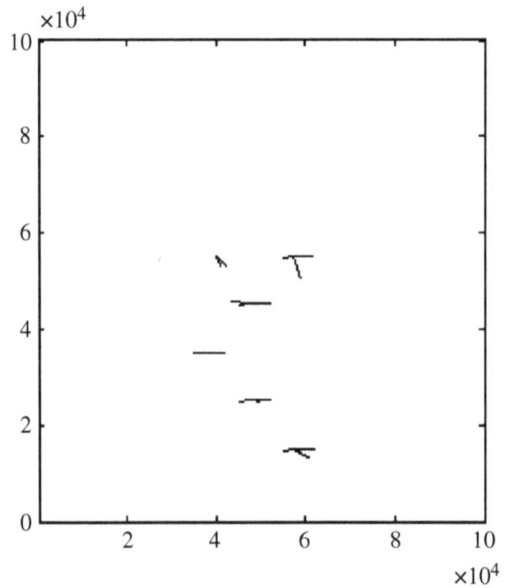

Figure 6.11 Chart of the track initiated with 3/4 logic-based method

third scan, and • in the fourth scan. Four scan periods are used for the situation map shown in Figure 6.9. The track initiated with the direct-vision method is illustrated in Figure 6.10; the track initiated with the 3/4 logic-based method in Figure 6.11; the track initiated with the modified 3/4 logic-based method in Figure 6.12; the track initiated with the Hough transform-based method in Figure 6.13, where $N_\theta = 90$, $N_\rho = 90$, and the threshold value of the parameter space is 4; the track

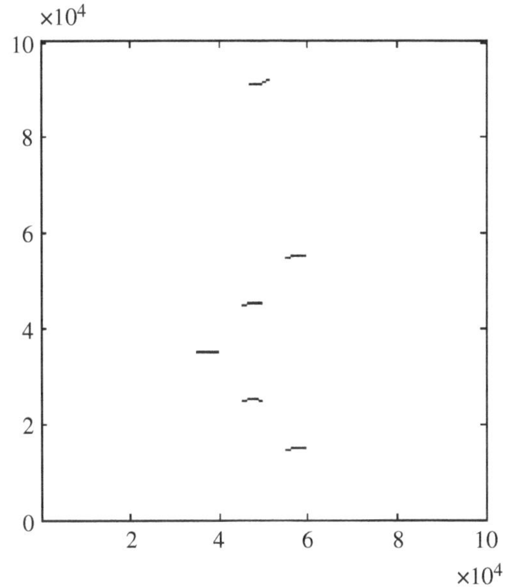

Figure 6.12 Chart of the track initiated with modified logic-based method

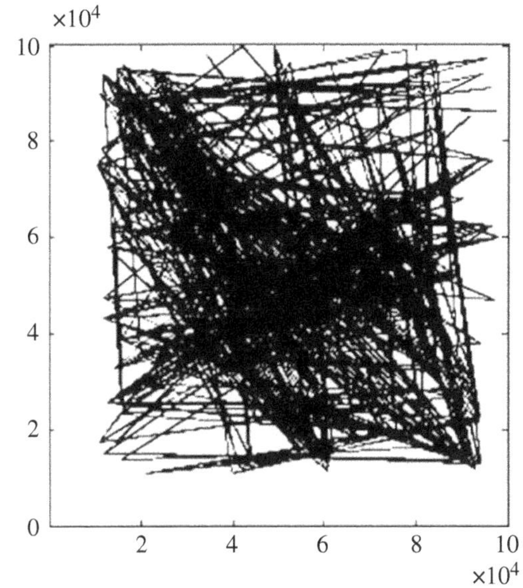

Figure 6.13 Chart of the track initiated with Hough transform-based method

initiated with the modified Hough transform method in Figure 6.14, where $N_\theta = 90$, $N_\rho = 90$, and the threshold value of the parameter space is 4; and the track initiated with the Hough transform and logic-based method in Figure 6.15, where $N_\theta = 90$, $N_\rho = 90$, and the threshold value of the Hough transform method is also 4 and the 3/4 logic-based method is used to initiate the track.

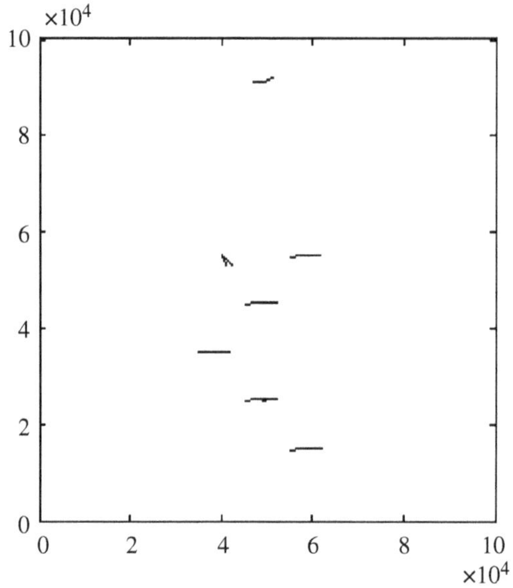

Figure 6.14 Chart of the track initiated with modified Hough transform method

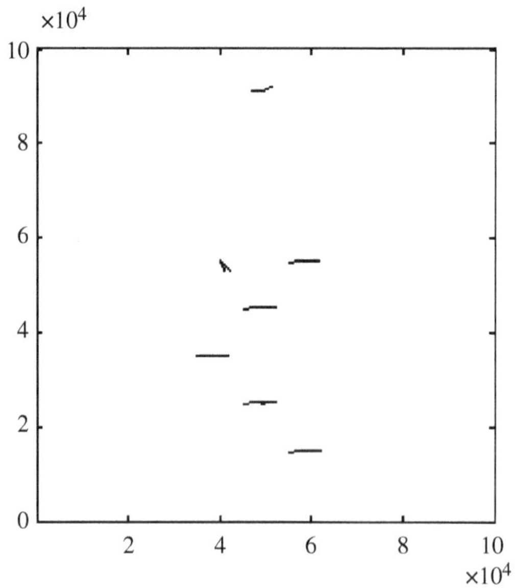

Figure 6.15 Chart of the track initiated with Hough transform and logic-based method

Comparing Figures 6.10–6.15, we find that it is the least effective to initiate a track with the Hough transform-based method, which cannot initiate a target track correctly at all, while in sparsely cluttered environments the performances of the other five track initiation algorithms are almost the same.

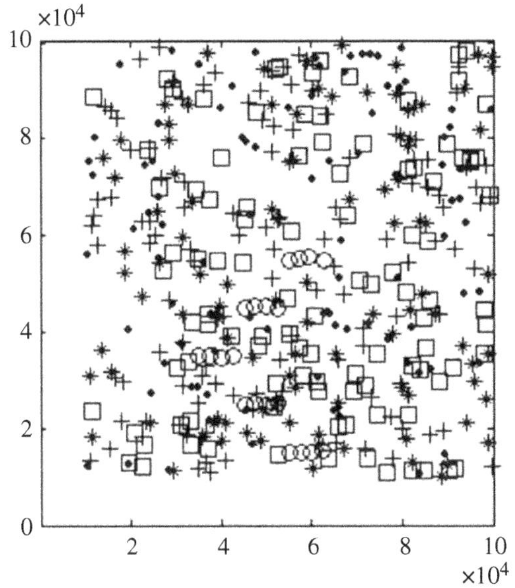

Figure 6.16 Situation map of clutter measurements and true measurements

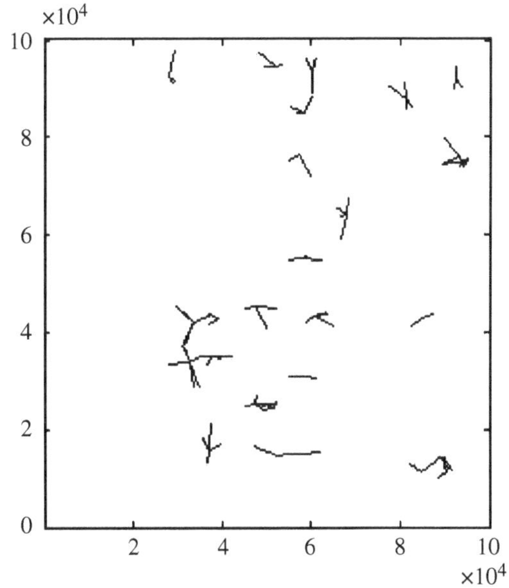

Figure 6.17 Chart of the tracks initiated with direct-vision method

To further demonstrate the effectiveness of track initiation with the algorithms above, let $\lambda = 100$, with simulation conditions the same as those mentioned above. The situation map under such conditions is illustrated in Figure 6.16. The track initiated with the direct-vision method is illustrated in Figure 6.17; the track initiated with the 3/4 logic-based method in Figure 6.18; the track initiated with the modified 3/4 logic-based method in Figure 6.19; the track initiated with the Hough

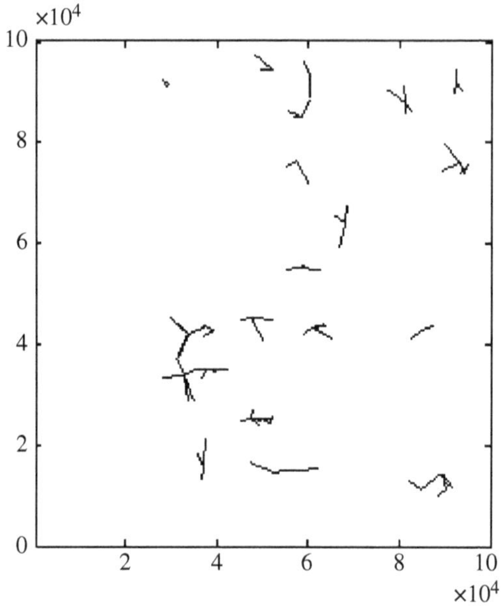

Figure 6.18 Chart of the tracks initiated with 3/4 logic-based method

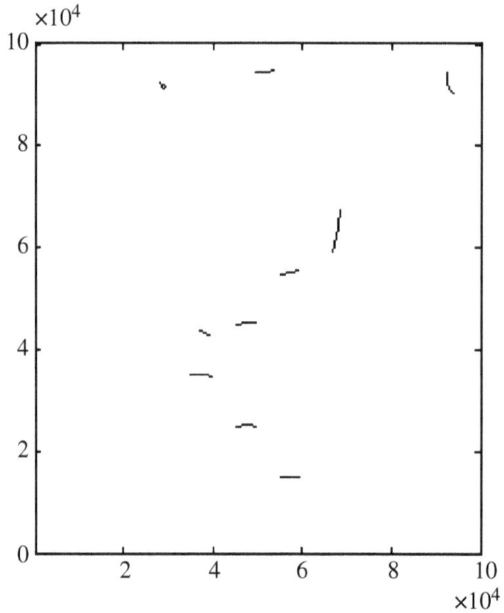

Figure 6.19 Chart of the tracks initiated with modified 3/4 logic-based method

transform-based method in Figure 6.20, where $N_\theta = 90$, $N_\rho = 90$, and the threshold value of the parameter space is 4; the track initiated with the modified Hough transform method in Figure 6.21, where $N_\theta = 90$, $N_\rho = 90$, and the threshold value of the Hough transform method is also 4 and the 3/4 logic-based method is used to initiate the track.

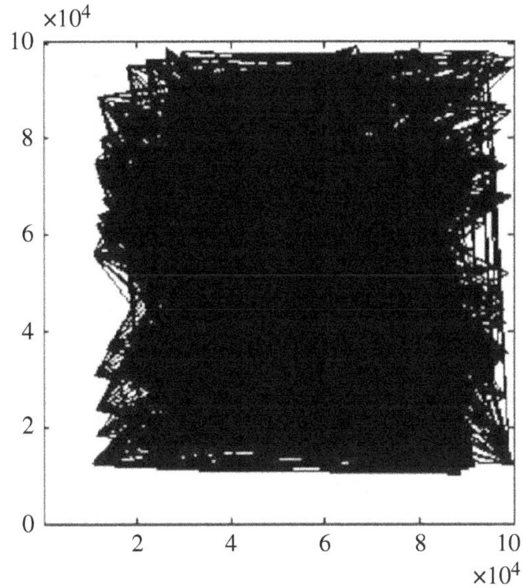

Figure 6.20 Chart of the tracks initiated with Hough transform-based method

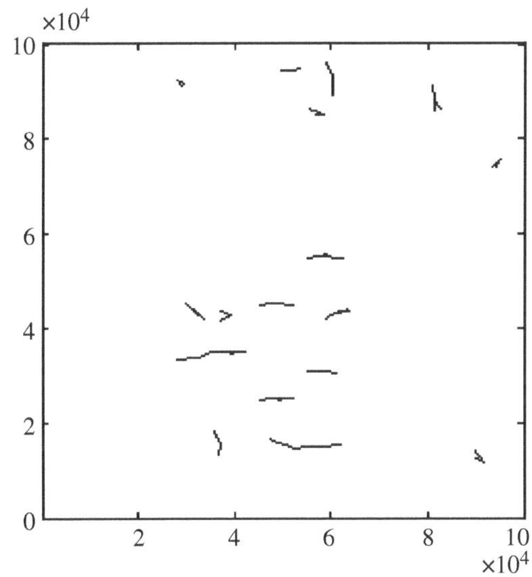

Figure 6.21 Chart of the tracks initiated with modified Hough transform-based method

Comparing Figures 6.17–6.21, we find that in a densely cluttered environment false alarms occur in all tracks initiated with the algorithms discussed above. Among them, the Hough transform-based algorithm involves the largest number of false tracks, the modified logic-based method is the most effective, the modified Hough transform method and the method based on the Hough transform and logic rank second in terms of effectiveness, the logic-based and the direct-vision method are the least effective.

6.5 Discussion of Some Issues in Track Initiation

As the first step in target tracking, track initiation is a decision-making technique to create a new target file, which principally involves temporary track formation and trajectory determination. The two basic requirements of track initiation are: (1) initiating true target tracks in every possible way; (2) initiating as few false target tracks as possible. There are usually two criteria of track initiation: track initiation probability and track false alarm probability. The former refers to the ratio of the number of correctly initiated tracks to the number of target tracks in existence, while the latter refers to the probability of forming false alarms. Track initiation involves an optimal compromise between fast track initiation and track initiation accumulation, which refers to the number of scans required to create a true track. Obviously the two considerations are contradictory: the longer the initiation takes, the less likely the false tracks will be initiated, and the greater the time lag of the true track initiation. Thus, in the case of a target moving with high velocity, it is required to initiate its tracks as quickly as possible.

Any automatic track initiation method aims to create the true tracks of targets as soon as they enter the radar power range, while avoiding the creation of false tracks due to the existence of inevitable false measurements. False tracks will reduce the confidence level of track data to a large extent, so track logic confirmation definitely takes some time. The fast speed and high probability of success required for track initiation are contradictory, because highly reliable detection of the true target track demands adequate information, which inevitably leads to a lag in reaction time of track initiation. A good track initiation method thus should strike the best balance between the capacity for fast track initiation and avoiding false tracks [145].

6.5.1 Main Indicators of Track Initiation Performance

Track reaction time. This mainly refers to the interval of time between the target's entrance into the radar power range and the creation of its track. As a random variable, it is often represented by the average scan times and the average rate of false association, particularly with the number of radar scan times as the unit.

Track quality. A number signifying track quality. It can be measured by the scoring method, or by the error of location speed, or represented by the accuracy of target indication. Additionally, it can also be represented by the average track purity. Chapter 11 will present a detailed discussion of track quality management methods.

The amount of computation and computation time. Here, "computation time" refers to the amount of time taken by all the programs to execute a period, a different notion from reaction time.

6.5.2 Demonstration of Track Initiation Scan Times

Track processing is often artificially divided into three stages such as track initiation, track maintenance, and track termination. In fact, the three stages have much in common. Track confirmation depends on high-quality track initiation, while accurate track initiation depends on correct choice of track heads. To fast initiate a track with high quality, the appropriate number of track initiation scan periods should be four according to theoretical analysis and practical engineering experience.

In actual environments, whether four scans can create stable tracks depends on the number of targets and their relative positions, the detection probability, measurement resolution, and false alarm probability. If four scans during one period cannot initiate a track, the data can still be processed for track initiation during the next period with many techniques.

6.6 Summary

Track initiation is of paramount importance in multi-target track processing. Its quality has a direct effect on the processing of subsequent tracks. This chapter has discussed techniques of multi-target track initiation, divided into two types: target-oriented sequential processing techniques and measurement-oriented batch processing techniques. The former, demanding a small amount of computation, is easy to carry out in engineering, but is suitable only in environments with sparse echoes and bad at identifying targets and false alarms, while the latter can yield satisfactory results, reducing false alarm probability effectively, but it is difficult to implement for the computational burden in the dense-echo environment. The track initiation algorithms introduced in this chapter (except the Hough transform algorithm) are all very effective at initiating tracks of targets moving in straight lines. They are also effective at initiating targets moving in non-straight lines, the tracks of which can be initiated within a matter of scanning periods and thus approximated as straight lines.

7

Maximum Likelihood Class Multi-target Data Association Methods

7.1 Introduction

Track initiation in the multi-target case of Chapter 6, which mainly attacks associations between measurements, is to be followed by associations between measurements and tracks. For one thing, initiated tracks may be false ones, which need to be affirmed or negated constantly by the subsequent measurements for confirmation or cancellation. For another, many measurements will be generated in an environment where a radar is operating in the presence of multiple targets and false alarms. Thus it is essential to decide which measurement should be used to update the state of a certain target at a given time, that is, to tackle the uncertainty of the measurements to be used for filtering. This results in association between measurements and tracks, the primary subject of this chapter and the following one.

The basic approaches to data association between measurements and tracks in multi-target tracking fall broadly into two categories [14, 17, 29, 40, 158, 161, 162]: maximum likelihood and Bayesian data association algorithms. The former, based on the likelihood ratio of the observation sequence, mainly include: manual plotting methods, track-splitting algorithms [16, 25, 40, 163], joint maximum likelihood algorithms [25, 163], integer programming algorithms [164–170], and generalized correlation algorithms [45, 171]. This chapter deals mostly with the last four algorithms. The latter category, which is based on the Bayesian principle [172–174], mainly includes: nearest-neighborhood algorithms, probability data association algorithms, united probability data association algorithms, optimal Bayesian algorithms, and multiple hypothesis tracking algorithms. These algorithms will be addressed in Chapter 8.

7.2 Track-Splitting Algorithm

The track-splitting algorithm, which was first proposed by Sitter in the early 1960s and then developed by Smith and Buechler in 1975, is a data association method based on likelihood function testing. The basic principles are described as follows.

Radar Data Processing with Applications, First Edition. He You, Xiu Jianjuan, and Guan Xin.

1. Set up a correlated wave gate centering the predicted position of the target at a certain moment, when the track has been initiated.
2. Choose from the measurement data by means of the correlated wave gate and identify the candidate measurements within the correlated wave gate as the effective measurements of targets at the moment.
3. Update the original target state and covariance by using each effective measurement value. Split the original target tracks into new tracks whose number corresponds to that of the original ones.
4. Calculate the likelihood function of each track.
5. The track with likelihood function lower than a certain threshold should be deleted, while the rest should be kept.

For the track-splitting algorithm, if many candidate echoes fall within the correlated wave gate, then an exponential increase may arise in the number of split tracks or a calculation explosion may even be caused. Therefore, pruning is needed in the process. A brief introduction to the algorithm will be given below.

7.2.1 Calculation of Likelihood Functions

In case of a constant choice of candidate echoes of the target by means of the correlated wave gate, denote the lth sequence of measurements beginning at time k as

$$\mathbf{Z}^{k,l} \triangleq \{z_{i_1 l}(1), \ldots, z_{i_k l}(k)\} \tag{7.1}$$

where $z_{i_j l}(j)$ is the i_jth measurement of the lth sequence of measurements at time j.

Denote the event that the measurements of the sequence are from the same target by

$$\theta^{k,l} \triangleq \{\mathbf{Z}^{k,l} \text{represents a true track}\} \tag{7.2}$$

Then, the likelihood functions can be expressed as

$$\begin{aligned}\Lambda(\theta^{k,l}) &= p[\mathbf{Z}^{k,l}|\theta^{k,l}] = p[z_{i_1 l}(1), \ldots z_{i_k l}(k)|\theta^{k,l}] = p[z_{i_k l}(k), \mathbf{Z}^{k-1,l}|\theta^{k,l}] \\ &= p[z_{i_k l}(k)|\mathbf{Z}^{k-1,l}, \theta^{k,l}]\, p[\mathbf{Z}^{k-1,l}|\theta^{k,l}]\end{aligned} \tag{7.3}$$

Likewise, $p[\mathbf{Z}^{k-1,l}|\theta^{k,l}]$ can be expressed as the multiplication of two conditional probability density functions, and then the following equation can be deduced by analogy:

$$\Lambda(\theta^{k,l}) = \prod_{j=1}^{k} p[z_{i_j l}(j)|\mathbf{Z}^{j-1,l}, \theta^{k,l}] \tag{7.4}$$

Note: The likelihood function contains the underlying assumption that the detection probability of the target is $P_D = 1$. If $P_D \neq 1$, then the equation of the above likelihood function is written

$$\Lambda(\theta^{k,l}) = \prod_{j=1}^{k} \{P_D p[z_{i_j l}(j)|\mathbf{Z}^{j-1,l}, \theta^{k,l}]\}^{1-\delta_{0i_j}} [1 - P_D]^{\delta_{0i_j}} \tag{7.5}$$

where

$$\delta_{0i_j} = \begin{cases} 1, & \text{omit detect} \\ 0, & \text{other} \end{cases} \tag{7.6}$$

Under the linear Gaussian or approximate Gaussian assumptions, the conditional probability density function of the measurement $z(j)$ at time j can be expressed as

$$\begin{aligned} p\left[z(j)\middle|Z^{j-1},\theta^{k,l}\right] &= \mathrm{N}[z(j);\hat{z}(j|j-1),S(j)] = \mathrm{N}[v(j);0,S(j)] \\ &= \frac{1}{(2\pi)^{n_z/2}\sqrt{|S(j)|}}\exp\left\{-\frac{1}{2}v'(j)S^{-1}(j)v(j)\right\} \end{aligned} \tag{7.7}$$

where n_z is the number of dimensions of the measurements, $\hat{z}(j|j-1)$ is the predicted measurement at time j, $v(j)$ is the innovation, and $S(j)$ is the innovation covariance.

From (7.7), the likelihood functions of (7.4) and (7.5) become

$$\begin{aligned} \Lambda\left(\theta^{k,l}\right) &= \prod_{j=1}^{k}\frac{1}{(2\pi)^{n_z/2}\sqrt{|S(j)|}}\exp\left[-\frac{1}{2}v'_{i_j l}(j)S_l^{-1}(j)v_{i_j l}(j)\right] \\ &= \left[\prod_{j=1}^{k}\frac{1}{(2\pi)^{n_z/2}\sqrt{|S(j)|}}\right]\exp\left[-\frac{1}{2}\sum_{j=1}^{k}v'_{i_j l}(j)S_l^{-1}(j)v_{i_j l}(j)\right] \end{aligned} \tag{7.8}$$

$$\Lambda\left(\theta^{k,l}\right) = \prod_{j=1}^{k}\left\{P_{\mathrm{D}}\frac{1}{(2\pi)^{n_z/2}\sqrt{|S(j)|}}\exp\left\{-\frac{1}{2}v'(j)S^{-1}(j)v(j)\right\}\right\}^{1-\delta_{0i_j}}[1-P_{\mathrm{D}}]^{\delta_{0i_j}} \tag{7.9}$$

where $v_{i_j l}(j)$ is the innovation of the i_j measurements corresponding to the lth sequence of measurements at time j, and $S(j)$ is the innovation covariance corresponding to the lth sequence of measurements at time j.

For simplicity, assuming that $P_D = 1$ (where P_D is the detection probability of the target), a modified logarithm likelihood function is defined as

$$\lambda(k) = -2\ln\frac{\Lambda\left(\theta^{k,l}\right)}{\prod_{j=1}^{k}\frac{1}{2\pi^{n_z/2}\sqrt{S(j)}}} = \sum_{j=1}^{k}v'_{i_j l}(j)S^{-1}(j)v_{i_j l}(j) \tag{7.10}$$

which can also be computed recursively as follows:

$$\lambda(k) = \lambda(k-1) + v'_{i_j l}(k)S^{-1}(k)v_{i_j l}(k) \tag{7.11}$$

7.2.2 *Threshold Setting*

Since the innovation $v_{i_j l}(j)$ follows the Gaussian distribution with zero mean and variance $\mathbf{S}(j)$, and when $n_z > 1$ (where n_z is the number of dimensions of the innovations), the innovation is denoted by vectors, $v'_{ijl}(j)S^{-1}(j)v_{ijl}(j)$ is a random variable subject to the χ^2 distribution with n_z degrees of freedom. Because the innovation is independent, the test statistic $\lambda(k)$ is a random variable subject to the χ^2 distribution with kn_z degrees of freedom. For a given confidence, its threshold value a can be obtained from the χ^2 distribution table, expressed as follows:

$$\Pr\left\{\chi^2_{kn_z} > a\right\} = \alpha \tag{7.12}$$

where α is the probability that a true track will be rejected. If the modified logarithm likelihood function $\lambda(k)$ satisfies

$$\lambda(k) \leq a \tag{7.13}$$

then the above sequences of measurements belong to the same track, and the latest measurements in the sequence can be used to update the state of the corresponding target, otherwise the above sequences of measurements may be regarded as false tracks, which will be discarded. In the presence of dense multiple echoes, this method demands a large amount of computation and memory, and may give rise to misjudgments.

7.2.3 Modified Likelihood Function

The track-splitting algorithm cannot be used to test a sequence of measurements with long time histories, because the likelihood function test equation given by (7.13) then becomes dominated by old measurements and responds very slowly to the latest ones. The usual solution to this problem is to replace the modified logarithm likelihood function $\lambda(k)$ with the fading-memory or finite-memory likelihood function to test the sequence, so that the new measurements containing much current information are strengthened, and the useless old measurements are weakened or abandoned.

The fading-memory likelihood function is

$$\rho(k) = \mu\rho(k-1) + v'(k)S^{-1}(k)v(k) = \sum_{i=1}^{k} \mu^{k-i}v'(i)S^{-1}(i)v(i) \tag{7.14}$$

where μ is a discount factor and $\mu < 1$.

The effective memory or window length of the fading-memory likelihood function $\rho(k)$ is $(1-\mu)^{-1}$. In the steady-state situation, the fading-memory likelihood function $\rho(k)$ is approximately considered as a random variable subject to the χ^2 distribution with $n_z(1+\mu)/(1-\mu)$ degrees of freedom, the mean value and variance of which are $n_z/(1-\mu)$ and $2n_z/(1-\mu^2)$, respectively. If a confidence level is given, its threshold a_ρ can be obtained from the χ^2 distribution table as follows:

$$\Pr\left\{\chi^2_{n_z(1+\mu)/(1-\mu)} > a_\rho\right\} = \alpha \tag{7.15}$$

If

$$\rho(k) \leq a_\rho \tag{7.16}$$

then the above sequences of measurements belong to the same track, otherwise the above sequences of measurements may be false tracks, which should be deleted.

The finite-memory likelihood function is

$$\xi(k)=\xi(k-1)+v'(k)S^{-1}(k)v(k)-v'(k-l)S^{-1}(k-l)v(k-l)=\sum_{i=k-l+1}^{k} v'(i)S^{-1}(i)v(i) \tag{7.17}$$

where l is the sliding-window length of the finite memory, which is generally 4 to 8.

Likewise, the finite-memory likelihood function $\xi(k)$ is a χ^2-distributed random variable with ln_z degrees of freedom, the mean value and variance of which are ln_z and $2ln_z$, respectively. If a confidence level is given, its threshold a_ξ can be obtained from the χ^2 distribution table as follows:

$$\Pr\left\{\chi^2_{ln_z}>a_\xi\right\}=\alpha \tag{7.18}$$

Likewise, if

$$\xi(k)\le a_\xi \tag{7.19}$$

then the above sequences of measurements belong to the same track, otherwise the above sequences of measurements may be false tracks, which should be deleted.

7.2.4 Characteristics of Track-Splitting Algorithm

The track-splitting algorithm chooses from the candidate echoes by means of the correlated wave gate. The number of new tracks into which the previous target track will be split is the same as that of the candidate echoes within the correlated wave gate of each target at the current moment. The principle followed by the target data association using the track-splitting algorithm is to maximize the likelihood function. This algorithm is a non-Bayesian method, which does not yield the probability that a sequence is correct. Therefore, it must be pruned by independently computing the likelihood function of each track. The track with likelihood function lower than a certain threshold should be deleted, while that with a higher one should be kept.

Strictly speaking, the track-splitting algorithm is not a perfect data association method. It is suitable in single-target environments, and in multi-target ones where no wave gates intersect or where no echo is available in the intersection of the wave gates, but not in multi-target environments where echoes are found within that of different targets. Besides, even in the environments where it applies, this method has some inherent weaknesses. Conditioned by the correlated wave gates and the logarithm likelihood function threshold, it associates one track with more than one measurement point in order to maximally associate the measurements of the real target at each time. As a result, one track is split into many possible tracks (only one of which is the correctly associated track, i.e., the real target track). And, at the next time when there are common echoes in the overlaps of some of the split tracks' correlated wave gates, these tracks will most probably be associated with the same measurement point, which will inevitably lead to competition between many split tracks for the same measurement.

Therefore, in practice, it is necessary to improve the track-splitting algorithm and expand its scope of application so as to make it more appropriate to single-target or multi-target environments. The handiest approach to solving the problems of maximal branching in the single-target environment and of competition for the same measurement between many of the real tracks in the multi-target one (i.e., the problem of assignment of the same measurement to different branches) is to calculate the likelihood function of the measurement corresponding to each possible branch when competition occurs, and to select that which minimizes the modified logarithm likelihood function as the effective association measurement. Additionally, feasible alternative schemes for measurement assignment and multi-target data association can be worked out by utilizing the joint maximum likelihood and the 0–1 integer programming algorithm as well, which will be discussed in Sections 7.3 and 7.4.

7.3 Joint Maximum Likelihood Algorithm

As mentioned in Section 7.2, the means of pruning entailed in the track-splitting method is applicable only in the single-target environment or the multi-target one where no intersecting wave gate is present or where no echo is found in the intersection of wave gates. It is inappropriate where echoes are detected in the overlaps of wave gates of different targets, causing competition for the same measurement between several branches. In this case, an alternative to this classical means of track splitting – the joint maximum likelihood algorithm [25, 163] – can be used to assign multiple measurements to multiple tracks effectively.

7.3.1 Establishment of Feasible Partitions

Assume that there are multiple targets in the surveillance region, each target's state is given by the state equation (3.31), the corresponding measurement of a target is given by the measurement equation (3.32), and the set of candidate echoes which fall into all the correlated wave gates at time k is

$$\mathbf{Z}(k) = \{z_i(k)\}_{i=1}^{m_k} \tag{7.20}$$

where m_k is the number of candidate echoes within the correlated wave gates.

Hence, the cumulative set of confirmed measurements up to time k is expressed as

$$\mathbf{Z}^k = \{\mathbf{Z}(j)\}_{j=1}^{k} \tag{7.21}$$

The total number of measurements in the set Z^k is

$$N^k = \sum_{j=1}^{k} m_j \tag{7.22}$$

Similar to (7.1), the lth sequence of measurements up to time k is denoted

$$\mathbf{Z}^{k,l} = \{z_{i_1 l}(1), \ldots, z_{i_k l}(k)\} \tag{7.23}$$

where $z_{i_j l}(j)$ is the i_jth measurement which belongs to the lth sequence of measurements at time j.

Denote the event that the measurements of this sequence are from the same target by

$$\theta^{k,l} = \{\mathbf{Z}^{k,l} \text{ represents a true track}\} \tag{7.24}$$

then use the modified logarithm, fading-memory, or finite-memory likelihood function discussed in Section 7.2 to test the measurement sequence $\mathbf{Z}^{k,l}$. Only sequences below the testing threshold are retained, while those above the testing threshold are deleted. Denote the retained sequences by γ_i $(i = 1, 2, \ldots, m)$ and divide them into multiple feasible partitions. Assume that

$$\tau = \{\gamma_l\}_{l=0}^{n} \tag{7.25}$$

is one of those feasible partitions. The measurement sequence γ_0 is the set of false measurements which are not associated with any track in the feasible partitions under consideration.

Denote the event that the measurements in a measurement sequence γ_i are false by

$$\theta^{k,0} = \{\gamma_i \text{ represents a track formed by false measurements}\} \tag{7.26}$$

Assume that the measurements originating from no target are subject to a uniform distribution in the radar surveillance region and different measurements are independent of each other. Then the PDF of the measurement sequence γ_0 constituted of false measurements can be expressed as

$$p\left(\gamma_0 \middle| \theta^{k,0}\right) = \left(\frac{1}{V}\right)^{N_0} \tag{7.27}$$

where V represents the size of the surveillance region and N_0 represents the number of false measurements in the false measurement sequence γ_0. The number of false measurements in γ_0 varies, since these measurements correspond to different feasible partitions. Therefore, N_0 varies with different feasible partitions.

The measurement sequences in the feasible partition τ should satisfy the following requirements:

$$\mathbf{Z}^k = \bigcup_{l=0}^{n} \gamma_l \tag{7.28}$$

and

$$\gamma_l \cap \gamma_j = \phi \quad l \neq j \tag{7.29}$$

where ϕ represents an empty set (i.e., (7.29) guarantees that a measurement originates only from one possible track, which means that track splitting will not arise).

Corresponding to the feasible partition τ, we have the event

$$\theta(\tau) = \{\text{the partition } \tau \text{ is true}\} \tag{7.30}$$

and define the set of all feasible partitions as

$$\Gamma = \{\tau\} \tag{7.31}$$

The most likely partition is obtained by calculating the joint likelihood function of each feasible partition in the set Γ of all feasible partitions and getting the maximum value.

From (7.29), we see that the measurements of each measurement sequence in the feasible partition τ are independent.

Hence,

$$p\left[\boldsymbol{Z}^k|\theta(\tau)\right]=\prod_{l=0}^{n}p\left(\gamma_l|\theta^{k,l}\right) \tag{7.32}$$

and the feasible partition satisfying

$$\max_{\tau\in\Gamma}p\left(\boldsymbol{Z}^k|\theta(\tau)\right)=\max_{\tau\in\Gamma}\prod_{l=0}^{n}p\left(\gamma_l|\theta^{k,l}\right) \tag{7.33}$$

is the correct partition, in which each measurement sequence corresponds to one target. In this case, they can be filtered using the Kalman filtering technique.

The likelihood function of the sequence γ_i can be expressed as

$$p\left(\gamma_i|\theta^{k,i}\right)=\left[\prod_{l=1}^{N_i}|2\pi\boldsymbol{S}_i(l)|^{-\frac{1}{2}}\right]\exp\left[-\frac{1}{2}\sum_{l=1}^{N_i}\boldsymbol{v}_i'(l)\boldsymbol{S}_i^{-1}(l)\boldsymbol{v}_i(l)\right] \tag{7.34}$$

where N_i is the number of measurements in the sequence γ_i, $\boldsymbol{v}_i(k)$ is the innovation corresponding to the possible track γ_i, $\boldsymbol{S}_i(k)$ is the corresponding innovation covariance, and

$$\boldsymbol{v}_i(k)=z(k)-\boldsymbol{H}(k)\hat{\boldsymbol{X}}_i(k|k-1) \tag{7.35}$$

where $z(k)$ is the measurement in the sequence γ_i at the latest moment, $\boldsymbol{H}(k)$ is the measurement matrix, and $\hat{\boldsymbol{X}}_i(k|k-1)$ is the one-step prediction of the state of the sequence γ_i. We have

$$\boldsymbol{S}_i(k)=\boldsymbol{H}(k)\boldsymbol{P}(k|k-1)\boldsymbol{H}'(k)+\boldsymbol{R}(k) \tag{7.36}$$

where $\boldsymbol{P}(k|k-1)$ is the one-step prediction covariance of the state, and $\boldsymbol{R}(k)$ is the measurement covariance.

7.3.2 Recursive Joint Maximum Likelihood Algorithm

The joint maximum likelihood algorithm discussed above implements the multi-target data association by using the joint likelihood functions of all the effective measurement sequences up to each moment, considering the feasible combination and partition of all the measurements, calculating the joint likelihood functions corresponding to all the feasible partitions, and taking the feasible partition which maximizes the joint likelihood function as the effective partition. Obviously this is a batch processing algorithm, but in practice, we need to consider the recursive implementation of the algorithm and construct the recursive joint likelihood algorithm to improve the real-time capability of the algorithm.

Assume that the correct partition τ_{k-1} of all the effective measurements has been formed at time $k-1$ (i.e., the partition satisfies (7.33)), which gives rise to

$$p\left[\mathbf{Z}^{k-1}|\theta(\tau_{k-1})\right] = \max_{\tau\in\Gamma} p\left[\mathbf{Z}^{k-1}|\theta(\tau)\right] = \max_{\tau\in\Gamma}\prod_{l=0}^{n} p\left(\gamma_l|\theta^{k-1,l}\right) \tag{7.37}$$

Therefore, the partition can form fixed data associations at time $k-1$ and constitute multiple target tracks. At time k, we only need to consider the assignment of all the effective measurements at this moment to the target tracks formed at time $k-1$. For the solution to the partition problem, we can refer to the structure and division of the confirmed matrix in the united probability data association algorithm, which will be discussed in Section 8.5. Similarly, by calculating the joint likelihood functions of all the feasible partitions, the feasible partition satisfying (7.33) can be taken as the correct partition of the target tracks by all the effective measurements at time k. In this case, the partition is that which maximizes the joint likelihood probability and thus is implemented recursively.

7.4 0–1 Integer Programming Algorithm

7.4.1 Calculation of the Logarithm Likelihood Ratio

In the mid-1970s, Morefield proved that multi-target tracking involves the decomposition and combination of the set in the 0–1 integer programming algorithm and deduced the 0–1 integer programming algorithm for multi-target tracking. The 0–1 linear integer programming algorithm is also a batch processing algorithm applicable for multi-target track initiation. The state and the covariance matrix of the target can be estimated using the filtering algorithm, according to each measurement sequence after the judgment is complete. This algorithm and the generalized correlation algorithm discussed below were proposed and widely used in the 1970s.

The negative logarithm likelihood function of the measurement sequence γ_i is obtained from (7.34). The function is expressed as

$$\lambda^{k,i} = -\ln\left[p\left(\gamma_i|\theta^{k,i}\right)\right] = \frac{N_i}{2}\ln(2\pi) + \frac{1}{2}\sum_{l=1}^{N_i}\ln(|\boldsymbol{S}_i(l)|) + \frac{1}{2}\sum_{l=1}^{N_i}\boldsymbol{v}_i'(l)\boldsymbol{S}_i^{-1}(l)\boldsymbol{v}_i(l) \tag{7.38}$$

and from (7.38) the following logarithm likelihood ratio can be obtained:

$$\tilde{\lambda}^{k,i} = -\ln\left\{\frac{\left[p\left(\gamma_i|\theta^{k,i}\right)\right]}{\left[p\left(\gamma_i|\theta^{k,0}\right)\right]}\right\} = \lambda^{k,i} - N_i\ln V \tag{7.39}$$

where N_i is the number of measurement data in the measurement sequence γ_i.

According to the discussion above, m measurement sequences in total are kept through threshold detection, and their logarithm likelihood ratios can be obtained from (7.39). These likelihood ratios form a column vector, written as

$$\boldsymbol{\lambda} = \left[\tilde{\lambda}^{k,1}, \tilde{\lambda}^{k,2}, \ldots, \tilde{\lambda}^{k,m}\right]' \tag{7.40}$$

Likewise, a logarithm likelihood ratio column vector corresponding to each feasible partition τ can be obtained in the way mentioned above.

Define an m-dimensional binary vector $\boldsymbol{\rho}$. If an element of the logarithm likelihood ratio column vector $\boldsymbol{\lambda}$ has a corresponding element in the feasible partition τ, then the relevant element in an m-dimensional binary vector $\boldsymbol{\rho}$ corresponding to the feasible partition τ is 1, otherwise 0, and each feasible partition τ corresponds to a binary vector $\boldsymbol{\rho}$. For example, assume that there are five elements in $\boldsymbol{\lambda}$, that is,

$$\boldsymbol{\lambda} = \left[\tilde{\lambda}^{k,1}, \tilde{\lambda}^{k,2}, \tilde{\lambda}^{k,3}, \tilde{\lambda}^{k,4}, \tilde{\lambda}^{k,5}\right]' \tag{7.41}$$

and the logarithm likelihood ratio in the feasible partition τ is $\left\{\tilde{\lambda}^{k,1}, \tilde{\lambda}^{k,3}, \tilde{\lambda}^{k,5}\right\}$. In this case, the binary vector $\boldsymbol{\rho}$ corresponding to τ can be expressed as

$$\boldsymbol{\rho} = \begin{bmatrix} 1 \\ 0 \\ 1 \\ 0 \\ 1 \end{bmatrix} \tag{7.42}$$

Assume that $\boldsymbol{\psi}^i$ $(i = 1, 2, \ldots, m)$ is an N-dimensional column vector and N is the total number of measurements in the cumulative set Z^k. The measurement in the sequence γ_i $(i = 1, 2, \ldots, m)$ corresponds to the element 1 in the column vector $\boldsymbol{\psi}^i$, otherwise to the element 0. These column vectors form a matrix with $N \times m$ dimensions, that is,

$$\boldsymbol{A} = \left[\boldsymbol{\psi}^1, \boldsymbol{\psi}^2, \ldots, \boldsymbol{\psi}^m\right] \tag{7.43}$$

According to the corresponding column vector $\boldsymbol{\rho}$, the conditional equations (7.28) and (7.29) of feasibility are expressed as

$$\boldsymbol{A\rho} \leq \boldsymbol{1} \tag{7.44}$$

where $\boldsymbol{1}$ is the N-dimensional column vector completely formed by 1.

For example, assume that Z^k consists of six measurements. If the measurement sequence γ_1 consists of the second, fourth, and fifth measurements, γ_2 of the first, third, and sixth measurements, γ_3 of the first, second, and fifth measurements, γ_4 of the third, fourth, and sixth measurements, then the assignment column vectors $\boldsymbol{\psi}^1$, $\boldsymbol{\psi}^2$, $\boldsymbol{\psi}^3$, and $\boldsymbol{\psi}^4$ of γ_1, γ_2, γ_3, and γ_4 are, respectively,

$$\boldsymbol{\psi}^1 = [0,1,0,1,1,0]', \ \boldsymbol{\psi}^2 = [1,0,1,0,0,1]', \ \boldsymbol{\psi}^3 = [1,1,0,0,1,0]', \ \boldsymbol{\psi}^4 = [0,0,1,1,0,1]' \tag{7.45}$$

At this time, the matrix $\boldsymbol{A}$ can be expressed as

$$\boldsymbol{A}=\begin{bmatrix}0&1&1&0\\1&0&1&0\\0&1&0&1\\1&0&0&1\\1&0&1&0\\0&1&0&1\end{bmatrix} \tag{7.46}$$

Hence, the set of column vectors $\boldsymbol{\rho}$ of all the feasible partitions at this time can be obtained from (7.44), the conditional equation of feasibility.

7.4.2 0–1 Linear Integer Programming Algorithm

From (7.32) we get

$$-\ln\{p[\mathbf{Z}^k|\theta(\tau)]\}=-\sum_{l=0}^{n}\ln[p(\gamma_l|\theta^{k,l})]=-\sum_{l=1}^{n}\ln[p(\gamma_l|\theta^{k,l})]-\ln[p(\gamma_0|\theta^{k,0})] \tag{7.47}$$

The binary vector $\boldsymbol{\rho}$ corresponds to a certain feasible partition τ, and therefore when the negative likelihood function $\lambda^{k,i}$ is a relevant element of the feasible partition τ, the relevant element of $\boldsymbol{\rho}$ is 1, otherwise it is 0. Hence, after using (7.38), (7.47) can be expressed as

$$-\ln[p(\mathbf{Z}^k|\theta(\tau))]=\sum_{i=1}^{m}\rho_i\lambda^{k,i}+\lambda^{k,0} \tag{7.48}$$

where ρ_i is the ith element of $\boldsymbol{\rho}$.

From (7.30) we find that the negative logarithm likelihood function of the measurement sequence γ_0 in a certain feasible partition τ is expressed as

$$\lambda^{k,0}=-\ln[p(\gamma_0|\theta^{k,0})]=N_0\ln V=\left(N-\sum_{i=1}^{m}\rho_iN_i\right)\ln V=N\ln V-\sum_{i=1}^{m}\rho_iN_i\ln V \tag{7.49}$$

where N_i is the number of dimensions of the measurement sequence γ_i.

Substituting (7.49) into (7.48), we obtain

$$-\ln\{p[\mathbf{Z}^k|\theta(\tau)]\}=\sum_{i=1}^{m}\rho_i(\lambda^{k,i}-N_i\ln V)+N\ln V \tag{7.50}$$

Substituting (7.39) into (7.50) gives

$$-\ln\{p[\mathbf{Z}^k|\theta(\tau)]\}=\sum_{i=1}^{m}\rho_i\tilde{\lambda}^{k,i}+N\ln V=\rho'\lambda+N\ln V \tag{7.51}$$

The last component $N\ln V$ in (7.51) is constant, and therefore minimizing $-\ln\{p[\mathbf{Z}^k|\theta(\tau)]\}$ is equivalent to minimizing $\boldsymbol{\rho}'\boldsymbol{\lambda}$, and minimizing $-\ln\{p[\mathbf{Z}^k|\theta(\tau)]\}$ is equivalent to maximizing $p[\mathbf{Z}^k|\theta(\tau)]$. Hence, the maximum likelihood function set of tracks can be obtained from

$$\min_{\rho}\{\boldsymbol{\rho}'\boldsymbol{\lambda}\} \tag{7.52}$$

The measurement sequence from the corresponding target can be determined if the m-dimensional binary vector $\boldsymbol{\rho}$ which can minimize $\boldsymbol{\rho}'\boldsymbol{\lambda}$ is obtained, so that the negative likelihood function can be expressed as a linear functional of the assumed $\boldsymbol{\rho}$, and the tracking problem will be transformed into that of integer programming.

Based on the results above, the 0–1 linear integer programming of multi-target tracking can be expressed as

$$\min_{\rho}\{\boldsymbol{\rho}'\boldsymbol{\lambda}\} \tag{7.53}$$

whose constraint can be expressed as

$$\boldsymbol{A}\boldsymbol{\rho} \leq \mathbf{1} \tag{7.54}$$

where $\mathbf{1}$ is the column vector completely formed of 1's and $\boldsymbol{\rho}$ is a binary vector. Equations (7.53) and (7.54) are defined as the combination problem of the sets. Once the matrix $\boldsymbol{A}$ is determined, the maximum possible set of tracks will be selected immediately and the calculation of the state estimation and the covariance matrix can be completed in the standard Kalman filter.

7.4.3 Recursive 0–1 Integer Programming Algorithm

Obviously, the core of the 0–1 integer programming algorithm is also (7.53), that is, the feasible tracks can be searched for through the overall optimization of various partitions of all the historic measurement sequences, which is considered as the embryo of the multi-hypothesis tracking algorithm. But two main defects exist in this algorithm: (1) using the exhaustive method to construct the set of feasible tracks; (2) the non-recursive form of the algorithm. Owing to the former, the algorithm can be used to solve the data association problem only where there are fewer measurements; owing to the latter, the historic information has to be processed repeatedly each time. These defects affect the application of the algorithm in engineering, and therefore we need its recursive form as well.

Assume that the correct partition of all the effective measurements has been formed at time $k-1$. The correct partition can form a fixed data association relationship and multiple target tracks. At this moment we can get the logarithm likelihood ratio column vector $\boldsymbol{\lambda}_{k-1}$ and the correct partition column vector of $\boldsymbol{\rho}_{k-1}$ at time $k-1$. Similarly, at time k, we only need to consider the feasible assignment of all the effective measurements at the moment to all the target tracks formed at time $k-1$. Assume that at this time the logarithm likelihood ratio column vector $\boldsymbol{\lambda}_k$ of the feasible tracks which satisfy the likelihood ratio threshold is expressed as

$$\boldsymbol{\lambda}_k = \left[\tilde{\lambda}^{k,1}, \tilde{\lambda}^{k,2}, \tilde{\lambda}^{k,3}, \ldots, \tilde{\lambda}^{k,m_k}\right]' \tag{7.55}$$

where m_k denotes the number of measurement sequences kept through the threshold testing at time k.

Because the correct partition column vector $\boldsymbol{\rho}_{k-1}$ has been obtained, a fixed set of measurement sequences has been formed at time $k-1$. Therefore, unlike in the case of the batch processing algorithm, what is needed is to update the possible assignment of the effective measurements at time k rather than to recalculate all the probabilities, since the correct measurement assignment of the matrix $\boldsymbol{A}_k$ at time $k-1$ is known.

At this time, the correct partition can be obtained from $\min_{\boldsymbol{\rho}_k}\left\{\boldsymbol{\rho}'_k\boldsymbol{\lambda}_k\right\}$ and its constraint $\boldsymbol{A}_k\boldsymbol{\rho}_k \leq \mathbf{1}$ through the 0–1 integer programming toolkit or the relevant solution algorithm. Thus, the state can be updated by applying the Kalman filter algorithm, on the basis of which recursion can be implemented.

7.4.4 Application of 0–1 Integer Programming Algorithm

Besides the data association between measurements and tracks in target tracking, the integer programming algorithm can be applied to the following problems:

1. Association between tracks [175].
2. Multi-person, multi-index group decision in the group decision scheme case. Specifically, in this case, two or more decision makers independently choose methods for decisions according to their own knowledge, experience, and preferences. After the set of decisions has been made, this algorithm can be used to form the optimal decision model and find the solution [176].
3. Investment decisions in information technology projects [177].

7.5 Generalized Correlation Algorithm

The generalized correlation algorithm [45, 171] was proposed by Stein and Blackman in 1975 based on the maximum likelihood algorithm. It is an optimal batch processing data association algorithm, which gives the general expression for calculating the optimal association and correlation (known as generalized correlation) between measurements and tracks. Its main characteristic is to define a score function, through which the initiation, confirmation, threshold logic, and cancellation of tracks can be implemented. Therefore, this section is concerned first with the process of setting the score functions, the key of the generalized correlation algorithm, and then with the applications of this algorithm.

7.5.1 Establishing the Score Function

Assume that r measurements are obtained after scanning a detection area with volume V, K times. These measurements include the updated values, false alarms, and echo values of some special targets ("special targets" refer to those targets which are detected only once in the K scans). Therefore, multiple probabilities exist in the association between the measurements and the tracks, and multiple hypotheses about the association between the measurements and the tracks will be obtained if a hypothesis is made about each possible situation. In order to accomplish the target tracking, it is necessary to determine which hypothesis is most likely to appear and therefore to calculate the probability of each hypothesis being correct. The generalized correlation algorithm gives the expression of finding this probability and defines its logarithmic expression as the score function.

The correct relative likelihood ratio Q_k of each hypothesis about the measurement-to-track association is defined as

$$Q_k = P_0(n_{FK}, n_K)\prod_{i=1}^{n_K} P_{TL}(D_i)P_{DT}(NU_i|D_i)\prod_{l=1}^{NU_i} P_{ER}(\tilde{y}_{il}) \tag{7.56}$$

where $P_0(n_{FK}, n_K)$ denotes the probability that n_K real targets and n_{FK} false targets occur in the detection area during the K scans, D_i denotes the number of scans during which the ith source is present in the detection area except for the initial detection (i.e., the track length of i), $P_{TL}(D_i)$ denotes the probability of the length being D_i, $P_{DT}(NU_i|D_i)$ denotes the probability of the track i having NU_i measurements when the track length is D_i, and $P_{ER}(\tilde{y}_{il})$ denotes the probability of the residual error of the lth measurement of track i being $\tilde{y}_{il}$. Equation (7.56) shows that each true track is determined by D_i and the number of measurements NU_i (with residual error $\tilde{y}_{il}$).

The establishment processes of all the expressions of (7.56) will be discussed briefly below. They are set based on standard models.

The probability of n_K real targets and n_{FK} false targets in the detection area during K scans is

$$P_0(n_{FK}, n_K) = (\beta_{FT}\Delta)^{n_{FK}}(\beta_{NT}\Delta)^{n_K}\mathrm{e}^{-\beta KV} \tag{7.57}$$

where β_{FT} is the density of false targets, β_{NT} is the density of real targets, V is the volume of the detection area, and Δ is the volume of each tiny unit in the detection area.

β (the density of the new sources) includes β_{NT} (the density of the real targets) and β_{FT} (the density of the false targets), that is, $\beta = \beta_{FT} + \beta_{NT}$. The density of the real targets can be determined by the detection probability P_D, the average track length D_E, and the given density of the total new targets β_{NT}, according to the following equation:

$$\beta_{NT} = \frac{P_\mathrm{D}\beta_T}{1-(1-P_\mathrm{D})\mathrm{e}^{-1/D_E}} \tag{7.58}$$

Assume that the PDF of the track length of the real targets is approximately subject to the following exponential distribution:

$$f(\tau) = \frac{\mathrm{e}^{-\tau/\tau_0}}{\tau_0}$$

where τ_0 is the expectation of the track length and $\tau_0 = D_E T$, where D_E is the average track length and T is the sampling interval.

The probability of the track length of track i being D_i is obtained from deduction as below:

$$P_{TL}(D_i) = \begin{cases} P_{TL1}(D_i), & \text{the source does not exist in the } k\text{th scan} \\ P_{TL2}(D_i), & \text{the source still exists in the } k\text{th scan} \end{cases} \tag{7.59}$$

$$P_{TL1}(D_i) = \frac{1}{\tau_0}\int_{D_iT}^{(D_i+1)T} \mathrm{e}^{-\tau/\tau_0}\mathrm{d}\tau = P_{TT}\mathrm{e}^{-D_i/D_E} \tag{7.60}$$

$$P_{TL2}(D_i) = \frac{1}{\tau_0}\int_{D_iT}^{\infty} \mathrm{e}^{-\tau/\tau_0}\mathrm{d}\tau = \mathrm{e}^{-D_i/D_E} \tag{7.61}$$

where P_{TT} is the parameter of track termination, which is defined as the following equation:

$$P_{TT} = 1 - e^{-1/D_E} \tag{7.62}$$

When the track length is D_i, the probability of track i containing NU_i measurements is as follows:

$$P_{DT_i} = P_D^{NU_i}(1 - P_D)^{(D_i - NU_i)} \tag{7.63}$$

where NU_i is the number of target points (i.e., measurements) of track i, except for the initial observation, and P_D is the detection probability.

Assume that, for NU_i measurements associated with the tracks, the PDF of the innovation vector is subject to the Gaussian distribution, that is,

$$f(\tilde{y}_{il}) = \frac{e^{-d_{il}^2/2}}{(2\pi)^{M/2}\sqrt{|S_{il}|}} \tag{7.64}$$

where M is the number of dimensions of the measurements, and $\tilde{y}_{il}(k)$ is the innovation, which is determined by the following equation:

$$\tilde{\boldsymbol{y}}_{il}(k) = \boldsymbol{y}_{il}(k) - H\hat{\boldsymbol{x}}_i(k|k-1) \tag{7.65}$$

where $|\mathbf{S}_{il}|$ is the determinant of the innovation covariance $\mathbf{S}_{il}$, and d_{il} is determined by the following equation:

$$d_{il}^2 = \tilde{\boldsymbol{y}}_{il}' \boldsymbol{S}_{il}^{-1} \tilde{\boldsymbol{y}}_{il} \tag{7.66}$$

The total innovation probability of all the innovations of all the tracks is derived as below:

$$P_{ER}(\tilde{\boldsymbol{y}}_{il}) = \Delta^{n_{uK}} \prod_{i=1}^{n_K} \prod_{l=1}^{NU_i} f(\tilde{\boldsymbol{y}}_{il}) \tag{7.67}$$

where n_{uK} is the total number of updating points of all the tracks, which satisfies the following equation:

$$n_{uK} = \sum_{i=1}^{n_K} NU_i = r - n_K - n_{FK} \tag{7.68}$$

Substituting (7.57), (7.59), (7.64), and (7.67) into (7.56), we obtain the probability function of the hypothesis that the measurements are associated with the tracks as below:

$$Q_k = C\beta_{FT}^{n_{FK}} \beta_{NT}^{n_K} \prod_{i=1}^{n_K} \left[P_{TL}(D_i) P_D^{NU_i} (1 - P_D)^{D_i - NU_i} \prod_{l=1}^{NU_i} f(\tilde{\boldsymbol{y}}_{il}) \right] \tag{7.69}$$

where $C = \Delta^r e^{-K\beta V}$. Δ (the volume of the tiny unit) and V (the volume of the detection area) are both included in the constant C, therefore they will not affect the ultimate probability.

To correctly track the targets, we must determine which measurement is most probably associated with the tracks. To calculate the maximum value of Q_k more easily, take logarithms on both sides of the equation for Q_k and substitute (7.64) into the resulting equation to obtain

$$\begin{aligned}L'_k &= n_{FK}\ln\beta_{FT} + n_K\ln\beta_{NT}\\&+\sum_{i=1}^{n_K}\left\{\ln[P_{TL}(D_i)] + (D_i - NU_i)\ln(1-P_{\mathrm{D}}) + \sum_{l=1}^{NU_i}\left\{\ln\left[\frac{P_{\mathrm{D}}}{(2\pi)^{M/2}\sqrt{|S_{il}|}}\right] - \frac{d_{il}^2}{2}\right\}\right\}\end{aligned} \tag{7.70}$$

where L'_K is defined as the score function.

To ensure that the values of the score function are positive when the effective tracks exist, minor modifications need to be made to (7.70). When there is no track formed, r measurements obtained after K scans are all false measurements, and therefore $n_{FK} = r$ and $n_K = 0$. Substituting these into (7.70), we have $L'_k = r\ln\beta_{FT}$, so the modified score function becomes

$$\begin{aligned}L_k &= L'_k - r\ln\beta_{FT}\\&= n_K\ln\frac{\beta_{NT}}{\beta_{FT}} + \sum_{i=1}^{n_K}\left\{\ln[P_{TL}(D_i)] + (D_i - NU_i)\ln(1-P_{\mathrm{D}}) + \sum_{l=1}^{NU_i}\left\{\ln\left[\frac{P_{\mathrm{D}}}{(2\pi)^{M/2}\sqrt{|S_{il}|}}\right] - \frac{d_{il}^2}{2}\right\}\right\}\end{aligned} \tag{7.71}$$

Now, the score function is successfully established. From the establishment process, we find that the score function is the multiplication of the probability of the new sources, the track length probability, the track update probability, and the track error probability. The value of the score function is related to the measurement-to-track association hypothesis. The association hypothesis which gets the maximum score can be used for track update and target tracking.

7.5.2 Application of the Generalized Correlation Algorithm

In Section 7.5.1 we treated the establishment and definition of the score function. This section will cover the application of this function, specifically the establishment of its suboptimal sequence-correlated recursive equation, and its application in track initiation, confirmation, maintenance, and cancellation.

7.5.2.1 Establishment of the Suboptimal Correlated Recursive Equation of Sequences of the Score Function

The score function defined by (7.71) is a common-sense batch processing method. It processes all the data obtained from the first scan to the last, which is generally infeasible in calculation. Therefore, it is necessary to construct the suboptimal sequence recursive equation of this function, that is, to explain it through sequence processing with its optimality and batch processing method kept maximally. On the premise of highlighting the effect of each track, we can evolve the recursive algorithm from (7.71) as

$$L_k = (n_K - n_{K-1})\ln\frac{\beta}{\beta'_{FT}} + \sum_{i=1}^{n_{K-1}} L_{i,k} \tag{7.72}$$

where $L_{i,k}$ is the score of the ith track at time k, that is,

$$L_{i,k} = L_{i,k-1} + \Delta L_{i,k} \tag{7.73}$$

$$\beta'_{FT} = \beta_{FT} + \beta_{NT}\left(1 - \mathrm{e}^{-1/D_E}\right) \tag{7.74}$$

In (7.73),

$$\begin{aligned} L_{i,k-1} = {} & \ln\frac{\beta}{\beta'_{FT}} + \ln\frac{\beta_{NT}}{\beta}U(D_{i,k-1}-1) + \left[D_{i,k-1} - m'_{i,k-1}\right]\ln(1-P_{\mathrm{D}}) \\ & - \frac{D_{i,k-1}}{D_E}\sum_{l=2}^{m_{i,k-1}}\left\{\ln\frac{P_{\mathrm{D}}}{\beta'_{FT}(2\pi)^{M/2}\sqrt{|S_{il}|}} - \frac{d_{il}^2}{2}\right\} + C_D \end{aligned} \tag{7.75}$$

where $m_{i,k-1}$ is the number of target detection points (measurements) of the ith track including the initial observation until time $k-1$, that is,

$$m'_{i,k-1} = m_{i,k-1} - 1 \tag{7.76}$$

$$C_D = \begin{cases} \ln\left(1 - e^{-1/D_E}\right) & \text{cancel track, } D_i \geq 0 \\ 0 & \text{maintain track} \\ \ln\dfrac{\beta'_{FT}}{\beta} & \text{cancel track, } D_i = 0 \end{cases} \tag{7.77}$$

$$U(D_{i,k-1}-1) = \begin{cases} 1, & D_{i,k-1} \geq 1 \\ 0, & D_{i,k-1} = 0 \end{cases} \tag{7.78}$$

In (7.73),

$$\Delta L_{i,k} = \begin{cases} \Delta L_{i,k}(1), \text{ have update} \\ \Delta L_{i,k}(2), \text{ no update} \end{cases} \tag{7.79}$$

where

$$\Delta L_{i,k}(1) = -\frac{1}{D_E} + \ln\frac{P_{\mathrm{D}}}{\beta'_{NT}(2\pi)^{M/2}\sqrt{|S_{il}|}} - \frac{d_{il}^2}{2} + \delta(D_{i,k}-1)\ln\frac{\beta_{NT}}{\beta} \tag{7.80}$$

$$\Delta L_{i,k}(2) = -\frac{1}{D_E} + \ln(1-P_{\mathrm{D}}) + \delta(D_{i,k}-1)\ln\frac{\beta_{NT}}{\beta} \tag{7.81}$$

$$\delta(D_{i,k}-1) = \begin{cases} 1, & D_{i,k} = 1 \\ 0, & \text{others} \end{cases} \tag{7.82}$$

To demonstrate the whole process of using the generalized correlated algorithm to track the targets more clearly, we establish the algorithm procedure using the suboptimal sequence correlated recursive algorithm of the score function, as shown in Figure 7.1.

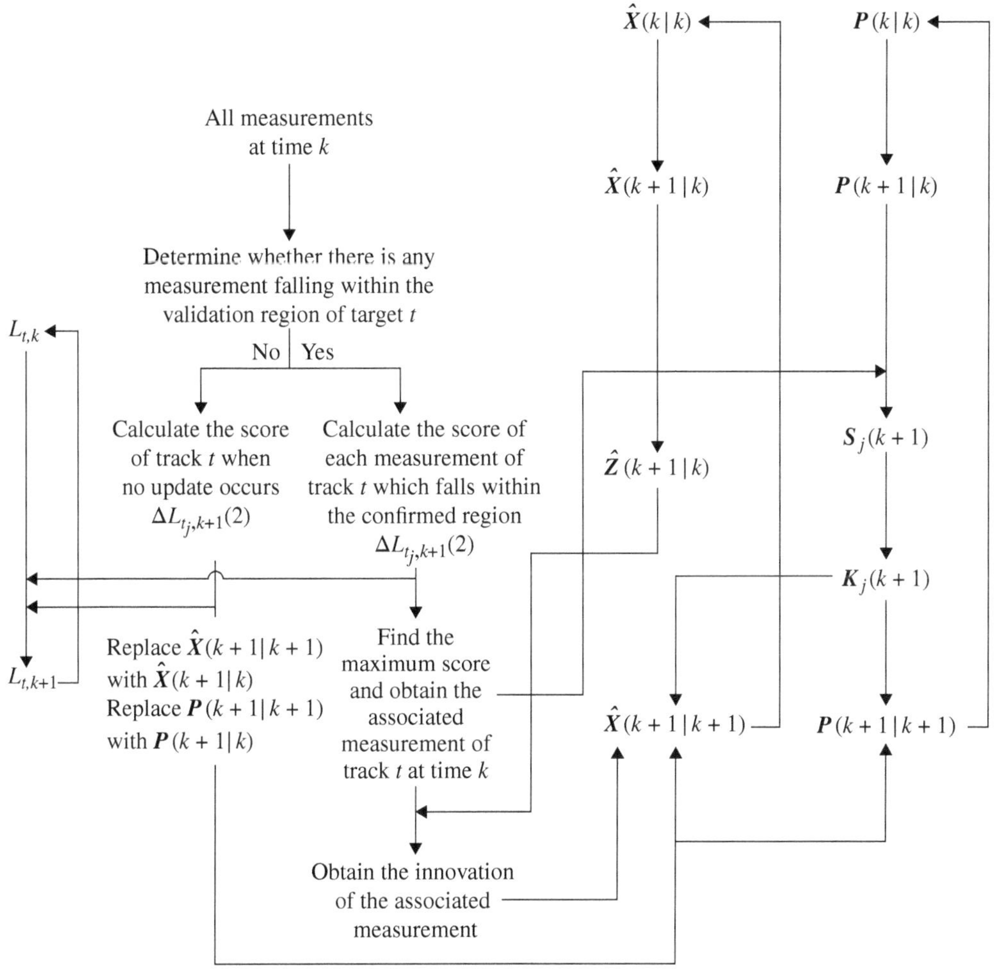

Figure 7.1 Single simulation cycle flowchart of the generalized correlation algorithm

7.5.2.2 Applications of the Score Function

Tracks can be initiated, confirmed, maintained, and cancelled by means of the recursive equation of the score function, which will be discussed briefly below.

1. *Track confirmation.* Considering the conversion of a non-confirmed track into a confirmed track, we decide that a non-confirmed track has turned into a confirmed one when its score is certainly positive, and that the score of the track is certainly positive when it is greater than the penalty score of the track cancellation. Therefore, deciding whether a track is confirmed is deciding whether its score is higher than the penalty score of the track cancellation. Assume that the score of the ith track at time k is $L_{i,k}$. The ith track is a confirmed track on condition that

$$L_{i,k} + \ln P_{TT} > 0 \tag{7.83}$$

where P_{TT} is the penalty score when the track is cancelled.

P_{TT} must be added to the score function when the track is cancelled, which is defined as follows:

$$P_{TT} = 1 - \mathrm{e}^{-1/D_E} \tag{7.84}$$

In most cases $D_E >> 1$, therefore

$$P_{TT} = 1 - \mathrm{e}^{-1/D_E} \sim 1 - \left(1 - \frac{1}{D_E}\right) \sim \frac{1}{D_E} \tag{7.85}$$

2. *Track initiation and track maintenance.* Considering the circumstances where the track is not cancelled, the following equation must be satisfied if measurement j is used to initiate or maintain track i:

$$d_{ij}^2 \le 2\min[G_i(1), G_i(2)] \tag{7.86}$$

where d_{ij}^2 is defined as in (7.66), and

$$G_i(1) = \ln \frac{P_{\mathrm{D}}}{(1 - P_{\mathrm{D}})\beta(2\pi)^{M/2}\sqrt{|\boldsymbol{S}_{ij}|}} \tag{7.87}$$

$$G_i(2) = G_i(1) + \ln(1 - P_{\mathrm{D}}) - \frac{1}{D_E} + \delta(D_{i,k-1})\ln\frac{\beta_{NT}}{\beta} - \Delta L_{\mathrm{G},i} + \Delta L_{L,i} \tag{7.88}$$

$$\Delta L_{\mathrm{G},i} = \begin{cases} L_{i,k_m} - L_{i,k-1} + \ln P_{TT}, & \text{confirmed track} \\ -L_{i,k-1}, & \text{unconfirmed track} \end{cases} \tag{7.89}$$

$$\Delta L_{L,i} = -\frac{1}{D_E} + \ln\frac{P_{\mathrm{D}}}{\beta_{FT}(2\pi)^{M/2}\sqrt{|\boldsymbol{S}_{ij,k+1}|}} - \ln\frac{\beta}{\beta'_{FT}} \tag{7.90}$$

where L_{i,k_m} is the score of the confirmed track with the highest score, which appears at time k_m.

3. *Track cancellation.* The principle of track cancellation can be expressed by the threshold. The following equation must be satisfied if the track i is cancelled:

$$G_i(2) \le G_i(1) + a_k \tag{7.91}$$

where

$$a_k = \frac{1}{2}\ln\frac{\boldsymbol{S}_{i,k+1}|\text{no update at time } k}{\boldsymbol{S}_{i,k+1}|\text{update at time } k} \tag{7.92}$$

where $\boldsymbol{S}_{i,k+1}$ is the innovation covariance of the ith track at time $k+1$.

Reference [178] extended the generalized correlation algorithm to distributed multi-sensor systems, presented the distributed multi-sensor generalized correlation algorithm, and made a simulation analysis using this algorithm.

7.6 Summary

This chapter mainly discusses several maximum likelihood-based data association algorithms, including track splitting, joint maximum likelihood, 0–1 integer programming, and generalized correlation. The basic estimation criterion of the four algorithms is to maximize the likelihood function, which denotes the probability of the vectors of target states conditioned on observations.

These algorithms are basically in the form of batch processing, hence incurring a large amount of calculation. The track-splitting algorithm requires the largest amount of computation, because it prunes out or eliminates, through the likelihood function, the measurement sequences which are unlikely to have originated from the targets.

The joint maximum likelihood algorithm calculates the likelihood functions of various feasible partitions of all the measurement sequences, and the measurement sequences under the feasible partitions which maximize the likelihood function are considered as the correct sequences from different targets.

The 0–1 integer programming algorithm, although evolving from the joint maximum likelihood algorithm, evaluates the binary vector $\boldsymbol{\rho}$ which minimizes the test statistic $\boldsymbol{\rho}'\boldsymbol{\lambda}$ instead of the feasible partition which maximizes the likelihood function. Compared with the joint maximum likelihood algorithm, it is similar in principle, but slightly different in the means of solving problems.

The generalized correlation algorithm defines a score function and uses it to initiate, confirm, and cancel tracks.

8

Bayesian Multi-target Data Association Approach

8.1 Introduction

With respect to multi-target data association, Chapter 7 explored maximum likelihood-based algorithms, while this chapter focuses on some Bayesian approaches, including:

- nearest neighbor, including the nearest-neighbor standard filter (NNSF) and the probabilistic nearest-neighbor filter (PNNF);
- probabilistic data association (PDA);
- integrated probabilistic data association (IPDA);
- joint probabilistic data association (JPDA).

The PDA, IPDA, and JPDA algorithms share a similar procedure of data association – they first compute the probabilities of being correct (i.e., having originated from the target) for each validated measurement at the current time, and then weight these probabilities to obtain the state estimate of the target. The difference between them is that the IPDA, which is based on the PDA and incorporates the concept of target existence, provides the existence probability estimations of potential targets while estimating the state of the target of interest so that the accuracy of track confirmation or termination can be improved, while the JPDA is mainly applied to the calculation of association probabilities in an environment of dense targets, where several tracks compete for the same measurement.

8.2 Nearest-Neighbor Algorithm

8.2.1 Nearest-Neighbor Standard Filter

In 1973, Singer and Sea proposed a filter which uses the characteristics of the prior statistics to estimate the correlated performances, that is, the NNSF [179]. Its working principle is as follows: set a tracking gate first, and then define the echoes obtained in the preliminary screening made by the

Radar Data Processing with Applications, First Edition. He You, Xiu Jianjuan, and Guan Xin.

tracking gate (correlation wave gate) as the candidate echoes so as to limit the number of targets [16, 17, 180–183] participating in the relevant differentiation.

As depicted in Chapter 6, the tracking gate is a subsection in the tracking space, centered at the predicted position of the target being tracked. Its size should be designed such that the correct echoes can be received with some probability. The measurements which fall within the gate are considered as the candidate echoes, that is, the identification of the candidate echoes depends on whether the measurement of the target $z(k+1)$ satisfies the following equation:

$$[z(k+1)-\hat{z}(k+1|k)]'S^{-1}(k+1)[z(k+1)-\hat{z}(k+1|k)] \leq \gamma \tag{8.1}$$

If only one measurement has fallen into the correlation wave gate of the target in track, it can be used to update the track directly, but if more than one track has fallen into the gate, the candidate echo with the shortest statistical distance should be taken as the target echo. That is, the measurement which minimizes the weighted norm of the innovation

$$d^2(z) = [z-\hat{z}(k+1|k)]'S^{-1}(k+1)[z-\hat{z}(k+1|k)] \tag{8.2}$$

is used in the NNSF to update the state of the target.

The NNSF algorithm has the advantage of being simple in computation. Its disadvantage is that in a multiple-echo environment the candidate echo nearest to the predicted position of a target of interest is not necessarily the real one (particularly so in cases where several targets are very close to each other or their trajectories intersect). In short, this algorithm is inadequate at resolving the problems of false tracking and track loss [184].

In this algorithm, state updating is done by using the measurement in the wave gate which is nearest to the predicted value. Therefore, it is applicable only for tracking non-maneuvering targets in the presence of sparse echoes, but inappropriate in the case of less observable maneuvering targets or when the probability of false alarms increases. A similar algorithm to this is the strongest neighbor filter (SNF) algorithm, which conducts state updating by using the strongest of the valid measurements that fall into the wave gate [185].

8.2.2 *Probabilistic Nearest-Neighbor Filter Algorithm*

The NNSF is widely applied because it uses the simplest method of data association and involves simple computation, as discussed in the previous subsection. To improve its tracking performance, Song *et al.* [186] proposed the PNNF algorithm and another nearest-neighbor association technique which updates the filter by using the nearest measurement in the wave gate. However, based on the NNSF, this algorithm accounts for the possibility that the nearest measurement might have come from false alarms and the situation where no echoes lie in the wave gate, thus modifying the corresponding update expression of the error covariance [187].

The PNNF algorithm defines three kinds of event, as follows:

1. there are no valid measurements within the wave gate (M_0);
2. the nearest measurement is target-originated (M_T);
3. the nearest measurement is false-alarm-originated (M_F).

The procedure of state prediction in the PNNF is essentially the same as in the standard Kalman filter of Section 3.2, and will not be repeated here, while that of state updating in this algorithm will be presented as follows.

When the event M_0 happens (i.e., when no echoes fall within the tracking gate with size $\sqrt{\gamma}$), the state prediction at time k is taken as the updated value at time $k-1$, given as

$$\hat{\boldsymbol{X}}(k|k)=\hat{\boldsymbol{X}}(k|k-1) \tag{8.3}$$

$$\boldsymbol{P}(k|k)=\boldsymbol{P}(k|k-1)_{M_0}=\boldsymbol{P}(k|k-1)+\frac{P_{\mathrm{D}}P_{\mathrm{G}}\left(1-C_{\tau g}\right)}{1-P_{\mathrm{D}}P_{\mathrm{G}}}\boldsymbol{K}(k)\boldsymbol{S}(k)\boldsymbol{K}'(k) \tag{8.4}$$

where P_{D} is the target detection probability, P_{G} is the gate probability, $C_{\tau g}=\dfrac{\int_0^{\gamma} q^{m/2}\mathrm{e}^{-q/2}\mathrm{d}q}{n\int_0^{\gamma} q^{m/2-1}\mathrm{e}^{-q/2}\mathrm{d}q}$, m is the dimension of the measurement vector, and when $m=2$, $C_{\tau g}=\left[1-\mathrm{e}^{-\gamma/2}(1+\gamma/2)\right]/\left(1-\mathrm{e}^{-\gamma/2}\right)$.

When the event $\bar{M}_0$ happens (i.e., when more than one echo falls within the tracking wave gate with size $\sqrt{\gamma}$), the nearest measurement within the gate is taken as the update

$$\hat{\boldsymbol{X}}(k|k)=\hat{\boldsymbol{X}}(k|k-1)+\boldsymbol{K}(k)\beta_1\boldsymbol{v}^*(k) \tag{8.5}$$

$$\bar{\boldsymbol{P}}_k^{M_F}(D)=\boldsymbol{P}(k|k-1)+\frac{P_{\mathrm{D}}P_R(D)[1-C_\tau(D)]}{1-P_{\mathrm{D}}P_R(D)}\boldsymbol{K}(k)\boldsymbol{S}(k)\boldsymbol{K}'(k) \tag{8.6}$$

$$D=\boldsymbol{v}^{*\prime}(k)\boldsymbol{S}^{-1}(k)\boldsymbol{v}^*(k) \tag{8.7}$$

$$\begin{aligned}\boldsymbol{P}(k|k)&=\beta_0\bar{\boldsymbol{P}}_k^{M_F}(D)+\beta_1[\boldsymbol{P}(k|k-1)-\boldsymbol{K}(k)\boldsymbol{S}(k)\boldsymbol{K}'(k)]+\beta_0\beta_1\boldsymbol{K}(k)\boldsymbol{v}^*(k)\boldsymbol{v}^{*\prime}(k)\boldsymbol{K}'(k)\\&=\boldsymbol{P}(k|k-1)+\left(\frac{\beta_0P_{\mathrm{D}}P_R(D)[1-C_\tau(D)]}{1-P_{\mathrm{D}}P_R(D)}-\beta_1\boldsymbol{K}(k)\boldsymbol{S}(k)\boldsymbol{K}'(k)\right)+\beta_0\beta_1\boldsymbol{K}(k)\boldsymbol{v}^*(k)\boldsymbol{v}^{*\prime}(k)\boldsymbol{K}'(k)\end{aligned} \tag{8.8}$$

where $\mathbf{v}^*(k)$ is the innovation corresponding to the nearest measurement within the wave gate, $\bar{\boldsymbol{P}}_k^{M_F}(D)$ is the state error covariance under the given condition D when the event M_F happens, $P_R(D)$ is the probability of the target existing within the wave gate of size $\sqrt{D}$, $P_R(D)=\dfrac{mC_m}{2^{m/2+1}\pi^{m/2}}\int_0^D q^{m/2-1}\mathrm{e}^{-q/2}\mathrm{d}q$, and when $m=2$, $P_R(D)=1-\mathrm{e}^{-D/2}$. $C_\tau(D)$ can be obtained by replacing γ in the equation for $C_{\tau g}$ with D. β_1 is the probability of this nearest measurement having originated from the target; $\beta_1=\dfrac{P_D\mathrm{e}^{-\lambda V_D}\mathrm{N}(v^*(k);0,\boldsymbol{S}(k))}{P_D\mathrm{e}^{-\lambda V_D}\mathrm{N}(v^*(k);0,\boldsymbol{S}(k))+(1-P_DP_R(D))\lambda\mathrm{e}^{-\lambda V_D}}\cdot\beta_0$ is the probability of this nearest measurement having originated from clutter; $\beta_0=1-\beta_1$. V_{D} is the volume of the wave gate with size $\sqrt{D}$, as shown in (6.6). This algorithm assumes that the amount of clutter within the wave gate is subject to the Poisson distribution with parameter λV_k.

8.3 Probabilistic Data Association Algorithm

The PDA algorithm is an all-neighbor approach which handles all the candidate echoes (validated measurements) falling in a correlation wave gate. It computes the probabilities of each candidate echo having originated from the target under discussion according to different associations. The PDA derives its name from this procedure, since it associates all the neighbors with the said target based on these probabilities. Then, these probabilistic data are used in the PDA filter (PDAF) to weight different echoes within the gate, and the sum of the weightings of the candidate echoes is taken as the equivalent echo which is employed finally to update the state of the target.

As a suboptimal filter approach in which only the latest measurements are decomposed, the PDAF algorithm is used chiefly to solve the single-radar, single-target tracking problem in cluttered environments. If more than one candidate echo is found in the correlation wave gate in a single-target environment, only one of them is assumed target-originated and the others due to false alarms or clutter. The advantages of this algorithm in tracking a single target in clutter are its smaller probability of tracking a false target and missing the real target and its smaller amount of calculation, only a little larger than that of the Kalman filter. The PDAF is one of the trends in the development of modern tracking technology [23, 188–194].

8.3.1 State Update and Covariance Update

Similarly to Chapter 7, $\mathbf{Z}(k)$ denotes the set of candidate echoes falling within the correlation wave gate of a certain target at time k and Z^k the accumulative set of validated measurements up to time k, that is,

$$\mathbf{Z}^k = \{\mathbf{Z}(j)\}_{j=1}^{k} \tag{8.9}$$

and

$$\mathbf{Z}(k) = \{z_i(k)\}_{i=1}^{m_k} \tag{8.10}$$

where m_k is the number of candidate echoes within the correlation wave gate.

Define the events

$\theta_i(k) \triangleq \{z_i(k)$ denotes the target-originated measurements$\}$, $i = 1, 2, \ldots, m_k$

$\theta_0(k) \triangleq \{$none of the measurements at time k is target-originated$\}$

with the conditional probability of the ith measurement $z_i(k)$ having originated from the target

$$\beta_i(k) \triangleq \Pr\{\theta_i(k) | \mathbf{Z}^k\} \tag{8.11}$$

conditioned on Z^k. These events are mutually exclusive and exhaustive, so $\sum_{i=0}^{m_k} \beta_i(k) = 1$, and the conditional mean of the state of the target at time k can be written as

$$\hat{\boldsymbol{X}}(k|k) = \mathrm{E}[\boldsymbol{X}(k)|\mathbf{Z}^k] = \sum_{i=0}^{m_k} \mathrm{E}[\boldsymbol{X}_i(k)|\theta_i(k), \mathbf{Z}^k] \Pr\{\theta_i(k)|\mathbf{Z}^k\} = \sum_{i=0}^{m_k} \beta_i(k) \hat{\boldsymbol{X}}_i(k|k) \tag{8.12}$$

where $\hat{X}_i(k|k)$ is the updated state estimate conditioned on the event $\theta_i(k)$, that is,

$$\hat{X}_i(k|k) = \hat{X}(k|k-1) + K(k)v_i(k) \tag{8.13}$$

where $v_i(k)$ is the innovation corresponding to the measurement of interest.

If none of the measurements is correct, or target-originated (i.e., for $i = 0$), then the state cannot be updated, and the state update at this moment should be approximately expressed by the predicted value, that is,

$$\hat{X}_0(k|k) = \hat{X}(k|k-1) \tag{8.14}$$

Substituting (8.13) and (8.14) into (8.12) yields the target state update equation, expressed as

$$\hat{X}(k|k) = \sum_{i=0}^{m_k} \beta_i(k)\hat{X}_i(k|k) = \hat{X}(k|k-1) + K(k)\sum_{i=1}^{m_k} \beta_i(k)v_i(k) = \hat{X}(k|k-1) + K(k)v(k) \tag{8.15}$$

where

$$v(k) = \sum_{i=1}^{m_k} \beta_i(k)v_i(k) \tag{8.16}$$

is called the "combined innovation."

The error covariance correlated with the updated state estimate is

$$P(k|k) = P(k|k-1)\beta_0(k) + [1-\beta_0(k)]P^c(k|k) + \widetilde{P}(k) \tag{8.17}$$

where

$$P^c(k|k) = [I - K(k)H(k)]P(k|k-1) \tag{8.18}$$

$$\widetilde{P}(k) = K(k)\left[\sum_{i=1}^{m_k} \beta_i(k)v_i(k)v_i'(k) - v(k)v'(k)\right]K'(k) \tag{8.19}$$

and $P(k|k-1)$, $K(k)$ are given respectively by (3.46), (3.51).

Example 8.1
To prove the validity of (8.17) by the definition of the state error covariance matrix.

Proof: From Chapter 3 it follows that the state error covariance matrix corresponding to the state update equation (8.15) is

$$\begin{aligned} P(k|k) &= \mathrm{E}\left\{[X(k)-\hat{X}(k|k)]\,[X(k)-\hat{X}(k|k)]'|Z^k\right\} \\ &= \sum_{i=0}^{m_k} \beta_i(k)\mathrm{E}\left\{\left[X(k)X'(k) - X(k)\hat{X}'(k|k) - X(k|k)\hat{X}'(k) + \hat{X}(k|k)\hat{X}'(k|k)\right]|\theta_i(k), Z^k\right\} \\ &\triangleq P^1 + P^2 + (P^2)' + P^3 \end{aligned} \tag{8.20}$$

Since

$$X(k) = \left[X(k) - \hat{X}_i(k|k)\right] + \hat{X}_i(k|k) = \widetilde{X}_i(k|k) + \hat{X}_i(k|k) \tag{8.21}$$

then

$$X(k)X'(k) = \widetilde{X}_i(k|k)\widetilde{X}_i'(k|k) + \hat{X}_i(k|k)\widetilde{X}_i'(k|k) + \widetilde{X}_i(k|k)\hat{X}_i'(k|k) + \hat{X}_i(k|k)\hat{X}_i'(k|k) \tag{8.22}$$

This gives

$$\begin{aligned} P^1 &= \sum_{i=0}^{m_k} \beta_i(k) \mathrm{E}\{X(k)X'(k)|\theta_i(k), Z^k\} = \sum_{i=0}^{m_k} \beta_i(k)\left[P_i(k|k) + \hat{X}_i(k|k)\hat{X}_i'(k|k)\right] \\ &= \beta_0(k)P_0(k|k) + [1-\beta_0(k)]P^c(k|k) + \sum_{i=0}^{m_k} \beta_i(k)\hat{X}_i(k|k)\hat{X}_i'(k|k) \end{aligned} \tag{8.23}$$

Likewise, we have

$$\begin{aligned} (P^2)' &= -\sum_{i=0}^{m_k} \beta_i(k) \mathrm{E}\{\hat{X}(k|k)X'(k)|\theta_i(k), Z^k\} = -\hat{X}(k|k)\sum_{i=0}^{m_k} \beta_i(k)\mathrm{E}\{X'(k)|\theta_i(k), Z^k\} \\ &= -\hat{X}(k|k)\sum_{i=0}^{m_k} \beta_i(k)\hat{X}'(k|k) = -\hat{X}(k|k)\hat{X}'(k|k) = P^2 \end{aligned} \tag{8.24}$$

$$\begin{aligned} P^3 &= \sum_{i=0}^{m_k} \beta_i(k) \mathrm{E}\left\{\hat{X}(k|k)\hat{X}'(k|k)|\theta_i(k), Z^k\right\} = \hat{X}(k|k)\hat{X}'(k|k)\sum_{i=0}^{m_k} \beta_i(k) \\ &= \hat{X}(k|k)\hat{X}'(k|k) \end{aligned} \tag{8.25}$$

Substituting P^1, P^2, and P^3 into (8.20) yields

$$\begin{aligned} P(k|k) &= \beta_0(k)P_0(k|k) + [1-\beta_0(k)]P^c(k|k) + \sum_{i=0}^{m_k} \beta_i(k)\hat{X}_i(k|k)\hat{X}_i'(k|k) - \hat{X}(k|k)\hat{X}'(k|k) \\ &= \beta_0(k)P(k|k-1) + [1-\beta_0(k)]P^c(k|k) + \widetilde{P}(k) \end{aligned} \tag{8.26}$$

where

$$\widetilde{P}(k) = \sum_{i=0}^{m_k} \beta_i(k)\hat{X}_i(k|k)\hat{X}_i'(k|k) - \hat{X}(k|k)\hat{X}'(k|k) \tag{8.27}$$

Then, (8.19) is proved equivalent to (8.13) below. Combining (8.13) and (8.16) into (8.27) gives

$$\begin{aligned}\widetilde{\boldsymbol{P}}(k) &= \sum_{i=0}^{m_k}\beta_i(k)\left[\hat{\boldsymbol{X}}(k|k-1)+\boldsymbol{K}(k)\boldsymbol{v}_i(k)\right]\left[\hat{\boldsymbol{X}}(k|k-1)+\boldsymbol{K}(k)\boldsymbol{v}_i(k)\right]' - \hat{\boldsymbol{X}}(k|k)\hat{\boldsymbol{X}}'(k|k)\\ &= \hat{\boldsymbol{X}}(k|k-1)\hat{\boldsymbol{X}}'(k|k-1)+\hat{\boldsymbol{X}}(k|k-1)\sum_{i=0}^{m_k}\beta_i(k)\boldsymbol{v}_i'(k)\boldsymbol{K}'(k)+\boldsymbol{K}(k)\sum_{i=0}^{m_k}\beta_i(k)\boldsymbol{v}_i(k)\hat{\boldsymbol{X}}'(k|k-1)\\ &\quad+\boldsymbol{K}(k)\sum_{i=0}^{m_k}\beta_i(k)\boldsymbol{v}_i(k)\boldsymbol{v}_i'(k)\boldsymbol{K}'(k)-\left[\hat{\boldsymbol{X}}(k|k-1)+\boldsymbol{K}(k)\boldsymbol{v}(k)\right]\left[\hat{\boldsymbol{X}}(k|k-1)+\boldsymbol{K}(k)\boldsymbol{v}(k)\right]'\\ &= \boldsymbol{K}(k)\left[\sum_{i=0}^{m_k}\beta_i(k)\boldsymbol{v}_i(k)\boldsymbol{v}_i'(k)-\boldsymbol{v}(k)\boldsymbol{v}'(k)\right]\boldsymbol{K}'(k)\end{aligned} \tag{8.28}$$

8.3.2 Calculation of the Association Probability

The association probability of (8.11) is calculated as follows. First, the measurement set $\boldsymbol{Z}^k$ is broken down into the past accumulative data $\boldsymbol{Z}^{k-1}$ and the latest data $\boldsymbol{Z}(k)$, then

$$\beta_i(k)=\Pr\{\theta_i(k)|\boldsymbol{Z}^k\}=\Pr\{\theta_i(k)|\boldsymbol{Z}(k),m_k,\boldsymbol{Z}^{k-1}\} \tag{8.29}$$

Use Bayes' rule,

$$\Pr(B_i|x)=\frac{p(x|B_i)\Pr(B_i)}{\sum_{j=1}^{n}p(x|B_j)\Pr(B_j)} \tag{8.30}$$

Equation (8.29) can be rewritten as

$$\beta_i(k)=\Pr\{\theta_i(k)|\boldsymbol{Z}(k),m_k,\boldsymbol{Z}^{k-1}\}=\frac{p\left[\boldsymbol{Z}(k)|\theta_i(k),m_k,\boldsymbol{Z}^{k-1}\right]\Pr\{\theta_i(k)|m_k,\boldsymbol{Z}^{k-1}\}}{\sum_{j=0}^{m_k}p\left[\boldsymbol{Z}(k)|\theta_j(k),m_k,\boldsymbol{Z}^{k-1}\right]\Pr\{\theta_j(k)|m_k,\boldsymbol{Z}^{k-1}\}} \tag{8.31}$$

If $z_i(k)$ denotes the target-originated measurement, then its PDF is

$$p\left[z_i(k)|\theta_i(k),m_k,\boldsymbol{Z}^k\right]=P_{\mathrm{G}}^{-1}\mathrm{N}[z_i(k);\hat{z}(k|k-1),\boldsymbol{S}(k)]=P_{\mathrm{G}}^{-1}\mathrm{N}[\boldsymbol{v}_i(k);0,\boldsymbol{S}(k)] \tag{8.32}$$

where P_{G} is the gate probability. If incorrect measurements are modeled as independently and uniformly distributed random variables in the correlation wave gate (validation region), then

$$p\left[\boldsymbol{Z}(k)|\theta_i(k),m_k,\boldsymbol{Z}^{k-1}\right]=\begin{cases}V_k^{-m_k+1}P_{\mathrm{G}}^{-1}\mathrm{N}[\boldsymbol{v}_i(k);0,\boldsymbol{S}(k)], & i=1,\ldots,m_k\\ V_k^{-m_k}, & i=0\end{cases} \tag{8.33}$$

where V_k is the volume of the correlation wave gate. From Ref. [2], it follows that the conditional probabilities of the events θ_i are

$$\gamma_i(m_k) = \Pr\{\theta_i(k)|m_k, \mathbf{Z}^{k-1}\} = \Pr\{\theta_i(k)|m_k\}$$

$$= \begin{cases} \dfrac{1}{m_k} P_{\mathrm{D}} P_{\mathrm{G}} \left[P_{\mathrm{D}} P_{\mathrm{G}} + (1 - P_{\mathrm{D}} P_{\mathrm{G}}) \dfrac{\mu_F(m_k)}{\mu_F(m_k - 1)} \right]^{-1}, & i = 1, 2, \ldots, m_k \\ (1 - P_{\mathrm{D}} P_{\mathrm{G}}) \dfrac{\mu_F(m_k)}{\mu_F(m_k - 1)} \left[P_{\mathrm{D}} P_{\mathrm{G}} + (1 - P_{\mathrm{D}} P_{\mathrm{G}}) \dfrac{\mu_F(m_k)}{\mu_F(m_k - 1)} \right]^{-1}, & i = 0 \end{cases} \tag{8.34}$$

where P_{D} is the target detection probability, that is, the probability of the correct measurement being detected at all, and $\mu_F(m_k)$ is the probability mass function (PMF) of the number of false measurements (clutter points).

There are two models for this PMF: parameter and non-parameter.

8.3.2.1 Parameter Models

The PMF of this model is a Poisson function with parameter λV_k:

$$\mu_F(m_k) = \Pr\{m_k^F = m_k\} = \mathrm{e}^{-\lambda V_k} \frac{(\lambda V_k)^{m_k}}{m_k!}, \quad m_k = 0, 1, 2, \ldots \tag{8.35}$$

where λ is the spatial density of false measurements (the number of false measurements per unit area), V_k is the volume of the validation region, so λV_k is the number of false measurements within the validation gate.

8.3.2.2 Non-parameter Models

The PMF of this model is a diffuse prior probability density function

$$\mu_F(m_k) = \frac{1}{N}, \quad m_k = 0, 1, \ldots, N - 1 \tag{8.36}$$

By using the Poisson parameter model (8.35) in (8.34), we get

$$\gamma_i(m_k) = \begin{cases} \dfrac{P_{\mathrm{D}} P_{\mathrm{G}}}{P_{\mathrm{D}} P_{\mathrm{G}} m_k + (1 - P_{\mathrm{D}} P_{\mathrm{G}}) \lambda V_k}, & i = 1, 2, \ldots, m_k \\ \dfrac{(1 - P_{\mathrm{D}} P_{\mathrm{G}}) \lambda V_k}{P_{\mathrm{D}} P_{\mathrm{G}} m_k + (1 - P_{\mathrm{D}} P_{\mathrm{G}}) \lambda V_k}, & i = 0 \end{cases} \tag{8.37}$$

Using the diffuse non-parameter prior model (8.36) in (8.34) gives

$$\gamma_i(m_k) = \begin{cases} P_{\mathrm{D}} P_{\mathrm{G}} / m_k, & i = 1, 2, \ldots, m_k \\ 1 - P_{\mathrm{D}} P_{\mathrm{G}}, & i = 0 \end{cases} \tag{8.38}$$

By replacing the Poisson parameter with the sample spatial density of the validated measurements, the non-parameter model can be derived directly from the Poisson model, that is, (8.38) can be obtained directly from (8.37) by letting $\lambda = m_k / V_k$.

Substituting (8.33) and (8.37) into (8.31) gives, after some cancellations, the probability with the Poisson clutter model

$$\begin{aligned} \beta_i(k) &= \frac{\mathrm{N}[\boldsymbol{v}_i(k);0,\boldsymbol{S}(k)]}{\lambda(1-P_\mathrm{D}P_\mathrm{G})/P_\mathrm{D} + \sum_{j=1}^{m_k}\mathrm{N}[\boldsymbol{v}_j(k);0,\boldsymbol{S}(k)]} \\ &= \frac{\exp\left\{-\frac{1}{2}\boldsymbol{v}_i'(k)\boldsymbol{S}^{-1}(k)\boldsymbol{v}_i(k)\right\}}{\lambda|2\pi\boldsymbol{S}(k)|^{\frac{1}{2}}(1-P_\mathrm{D}P_\mathrm{G})/P_\mathrm{D} + \sum_{j=1}^{m_k}\exp\left\{-\frac{1}{2}\boldsymbol{v}_j'(k)\boldsymbol{S}^{-1}(k)\boldsymbol{v}_j(k)\right\}}, \quad i=1,2,\ldots,m_k \end{aligned} \tag{8.39}$$

$$\beta_0(k) = \frac{\lambda|2\pi\boldsymbol{S}(k)|^{\frac{1}{2}}(1-P_\mathrm{D}P_\mathrm{G})/P_\mathrm{D}}{\lambda|2\pi\boldsymbol{S}(k)|^{\frac{1}{2}}(1-P_\mathrm{D}P_\mathrm{G})/P_\mathrm{D} + \sum_{j=1}^{m_k}\exp\left\{-\frac{1}{2}\boldsymbol{v}_j'(k)\boldsymbol{S}^{-1}(k)\boldsymbol{v}_j(k)\right\}} \tag{8.40}$$

Define

$$e_i \triangleq \exp\left\{-\frac{1}{2}\boldsymbol{v}_i'(k)\boldsymbol{S}^{-1}(k)\boldsymbol{v}_i(k)\right\} \tag{8.41}$$

$$b \triangleq \lambda|2\pi\boldsymbol{S}(k)|^{\frac{1}{2}}(1-P_\mathrm{D}P_\mathrm{G})/P_\mathrm{D} = |2\pi|^{\frac{1}{2}}\gamma^{-\frac{n_z}{2}}\frac{\lambda V_k}{c_{n_z}}(1-P_\mathrm{D}P_\mathrm{G})/P_\mathrm{D} \tag{8.42}$$

Then

$$\beta_0(k) = \frac{b}{b + \sum_{j=1}^{m_k} e_j} \tag{8.43}$$

$$\beta_i(k) = \frac{e_i}{b + \sum_{j=1}^{m_k} e_j}, \quad i=1,2,\ldots,m_k \tag{8.44}$$

If λV_k in (8.42) is replaced by m_k without any other alterations, then the non-parameter model's probabilities $\beta_i(k)$ and $\beta_0(k)$ can be derived. The probability $\beta_i(k)$ includes exponents, and thus (8.16) is highly nonlinear.

8.3.3 *Modified PDAF Algorithm*

A very important parameter in the PDAF algorithm is the clutter density (also called the expected number of false measurements within the wave gate), which directly affects the calculation of the association probability. Therefore, if this assumed parameter differs a lot from practical realities, the error in the obtained filtering result will be very large and the tracking accuracy of the algorithm will degrade. However, in many cases the parameter is hard to obtain in real time. This problem can be solved quite effectively by the modified PDAF algorithm proposed in Ref. [195], which, while making state estimations about the target, estimates the clutter density by using the sum of clutter obtained from previous scans as prior information.

In this algorithm, averaging the number of measurements which have fallen within the wave gate in all previous scans yields the mean clutter density

$$\lambda = \frac{1}{V_k k}\sum_{i=1}^{k} m_k \tag{8.45}$$

where λ is the clutter density, m_k is the number of measurements which fall within the wave gate at time k, and V_k is the volume of the validation region.

While the clutter density λ is estimated in real time, from (8.43) and (8.44) follow the association probabilities of all the valid measurements within the wave gate being false and the ith measurement being real (i.e., having originated from the target), and then this probabilistic information can be used to realize target tracking in clutter along the lines of filtering in the PDAF algorithm.

8.3.4 Performance Analysis

Here a simulation comparison between tracking effects for the PDA, NNSF, and PNNF algorithms is made in the cluttered single-target, uniform-motion simulation environment. Assume that the target being tracked is in constant planar motion, with the following numerical values: initial state $X(0) = (60\,000\ \text{m},\ 6\ \text{m/s},\ 60\,000\ \text{m},\ -14\ \text{m/s})$, process noise component $q_1 = 0.04$ and $q_2 = 0.03$, standard deviation of measurement errors 40 m for both the x and the y axis, sampling interval $T = 1$ s, and 100 simulation steps every time.

The discretized system equation is

$$\boldsymbol{X}(k+1) = \boldsymbol{F}(k)\boldsymbol{X}(k) + \boldsymbol{\Gamma}(k)v(k), \quad k = 0,\ldots,99 \tag{8.46}$$

where the state of the target is

$$\boldsymbol{X} = [x\ \ \dot{x}\ \ y\ \ \dot{y}]' \tag{8.47}$$

The state transition matrix is written as

$$\boldsymbol{F}(k) = \begin{bmatrix} 1 & T & 0 & 0 \\ 0 & 1 & 0 & 0 \\ 0 & 0 & 1 & T \\ 0 & 0 & 0 & 1 \end{bmatrix} \tag{8.48}$$

The process noise distribution matrix is given by

$$\Gamma(k) = \begin{bmatrix} \frac{1}{2}T^2 & 0 \\ T & 0 \\ 0 & \frac{1}{2}T^2 \\ 0 & T \end{bmatrix} \tag{8.49}$$

The process noise is zero-mean Gaussian white noise.

The measurement equation after converting the measurements is

$$z(k)=\boldsymbol{H}(k)\boldsymbol{X}(k)+\boldsymbol{W}(k) \tag{8.50}$$

where the measurement matrix is given as

$$\boldsymbol{H}(k)=\begin{bmatrix}1 & 0 & 0 & 0\\ 0 & 0 & 1 & 0\end{bmatrix} \tag{8.51}$$

$$z(k)=\begin{bmatrix}z_1(k)\\ z_2(k)\end{bmatrix} \tag{8.52}$$

The area of the validation region of two-dimensional measurements is

$$A_V=\pi\gamma|\boldsymbol{S}(k)|^{\frac{1}{2}} \tag{8.53}$$

where $S(k)$ is the innovation covariance.

Assuming that the parameter $\gamma=16$, we can obtain the gate probability mass $P_G=0.9997$ by referring to Table 6.2 according to γ and the measurement dimension n_z. False measurements were produced uniformly in the square centered at the correct measurements. The area of the square was $A=n_c/\lambda\approx 10A_V$, where λ is the number of false measurements per unit area. Let $\lambda=0.00004$ and n_c be the total number of false measurements, that is, $n_c=\text{INT}[10A_V\lambda+1]$, where INT[$x$] indicates that the largest integer taken is no greater than x. Then, the position of the ith false measurement is

$$x_i=a+(b-a)\text{RND},\ y_i=c+(d-c)\text{RND},\ \ i=1,2,\ldots,n_c \tag{8.54}$$

where RND denotes the uniformly distributed random numbers and

$$\begin{cases}a=x_k-q, & b=x_k+q\\ c=y_k-q, & d=y_k+q\end{cases} \tag{8.55}$$

where $(x_k,\ y_k)$ is the location of the correct measurement, that is,

$$q=\sqrt{10A_V}/2 \tag{8.56}$$

A large number of false measurements were produced in the area $A\approx 10A_V$, and the number of false measurements λA_V in the validation region A_V was approximately Poisson. The above procedure to produce false measurements simulates quite accurately what really happens in the case of random clutter or a high rate of false alarms.

Requirements. (1) Write out the target tracking steps using the PDAF. (2) Draw up the single-time real track of the target, and the filtering track and the RMS diagram of each algorithm.

Solution.

(1) Real tracks and measurement tracks. The real states $\boldsymbol{X}(1)$, $\boldsymbol{X}(2)$, …, $\boldsymbol{X}(99)$ of the target at other times can be obtained according to the discrete system equation and the given initial state $\boldsymbol{X}(0)$. Substituting these values into the measurement equation yields measurement conversions; the converted measurements of the target positions $z(1)$, $z(2)$, …, $z(99)$.

(2) Filtering tracks. Initializing the obtained measurements of the target position with the two-point difference method mentioned in Chapter 3, we get the initial state and the initial covariance of the target

$$\hat{X}(1|1)=\begin{bmatrix} z_1(1) & \frac{z_1(1)-z_1(0)}{T} & z_2(1) & \frac{z_2(1)-z_2(0)}{T} \end{bmatrix}' \tag{8.57}$$

$$P(1|1)=\begin{bmatrix} R_{11} & R_{11}/T & R_{12} & R_{12}/T \\ R_{11}/T & 2R_{11}/T^2 & R_{12}/T & 2R_{12}/T^2 \\ R_{21} & R_{21}/T & R_{22} & R_{22}/T \\ R_{21}/T & 2R_{21}/T^2 & R_{22}/T & 2R_{22}/T^2 \end{bmatrix} \tag{8.58}$$

The Kalman filtering equation mentioned in Chapter 3 is used before introducing clutter. After clutter is introduced, it is advisable to select the candidate echoes through setting up correlation wave gates, that is, to determine whether the following equation holds true:

$$v_i'(k+1)S^{-1}(k+1)v_i(k+1)\le\gamma \tag{8.59}$$

where $v_i(k+1)$ is the innovation associated with the ith measurement and $S(k+1)$ the innovation covariance. If the measurement satisfies (8.59), it will be retained as a candidate echo, otherwise it will be discarded as clutter. The combined innovation and association probabilities $\beta_0(k)$ and $\beta_i(k)$ of the candidate echoes can be obtained from (8.16), (8.43), and (8.44). Then, the updated values of the state and covariance of the target in clutter can be obtained from (8.15) and (8.17).

Figure 8.1 depicts the real and filtering tracks of the target, where the horizontal ordinate denotes the x-axis position of the target and the ordinate the y-axis position. Figure 8.2 is an enlarged version

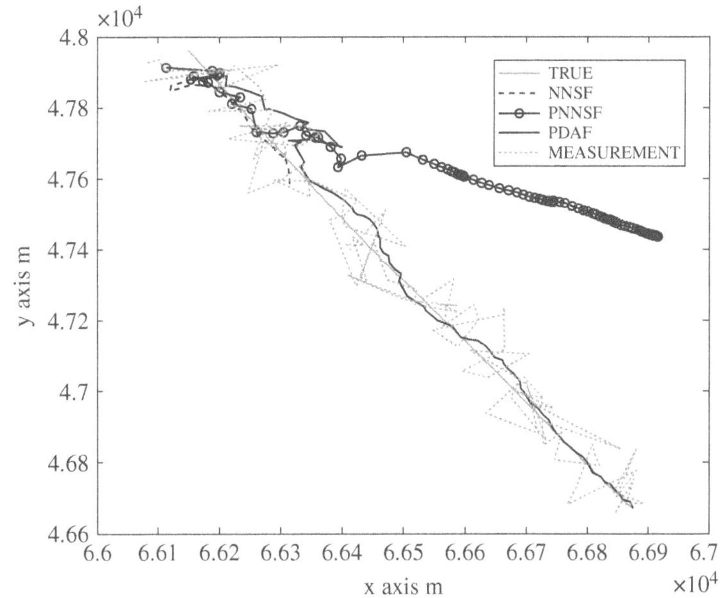

Figure 8.1 Real and filtering trajectory of the target

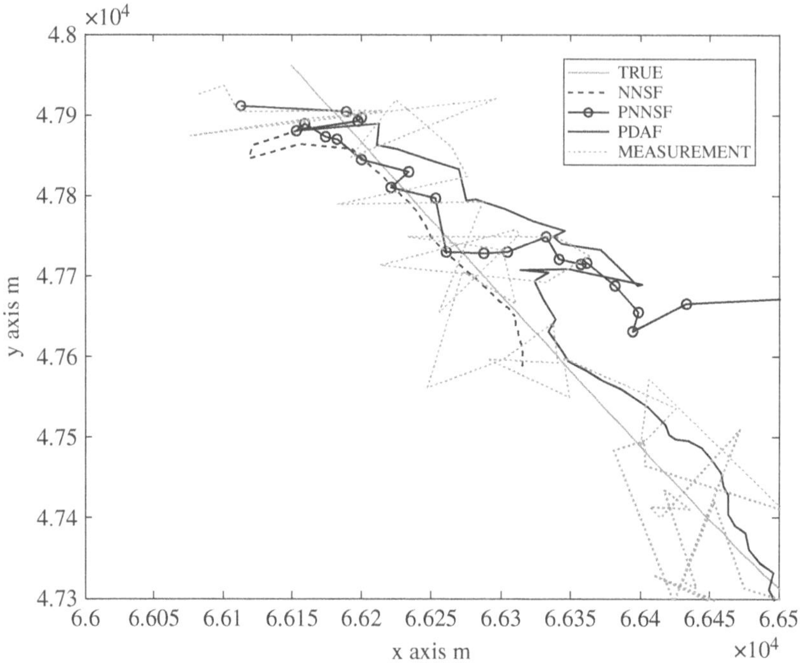

Figure 8.2 Enlarged version of Figure 8.1

of Figure 8.1. Figures 8.3 and 8.4 are the RMS of the x and y-axis target position obtained from the PDA algorithm through 50 Monte Carlo experiments. To be specific, assume that, for a target in the rectangular coordinate system, the noiseless real measurement is (x, y) and the filtering value is $(\hat{x}, \hat{y})$. Then, the RMS errors of the x and y-axis position are

$$\sigma_x = \sqrt{\frac{1}{N}\sum_{i=1}^{N}(\hat{x}-x)^2} \tag{8.60}$$

$$\sigma_y = \sqrt{\frac{1}{N}\sum_{i=1}^{N}(\hat{y}-y)^2} \tag{8.61}$$

respectively, where N denotes the number of Monte Carlo experiments.

As can be seen from Figures 8.1 and 8.2, the PDAF performs better in tracking a target in clutter than the NNSF and PNNF. Note that the target tracking performances of all three algorithms are affected by the density level of clutter, while the effect on that of the PDAF is weaker. The NNSF and PNNF prove effective at tracking targets in sparse clutter; even in the region of high clutter density, they may yield good results in some Monte Carlo simulation or other.

Since filtering divergence occurred in the NNSF and PNNF in this simulation, we present the simulation results only from the PDAF in multiple Monte Carlo simulations. Figures 8.3 and 8.4 indicate that the RMS errors of the x and y-axis positions obtained from the PDAF filtering converge quite quickly as the duration of the tracking extends. The following results from the three algorithms were also found in the simulation:

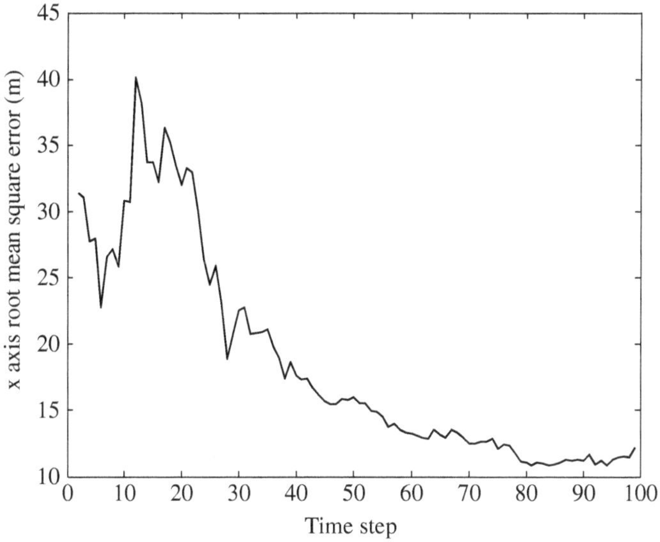

Figure 8.3 RMS of the x axis (PDAF filtering)

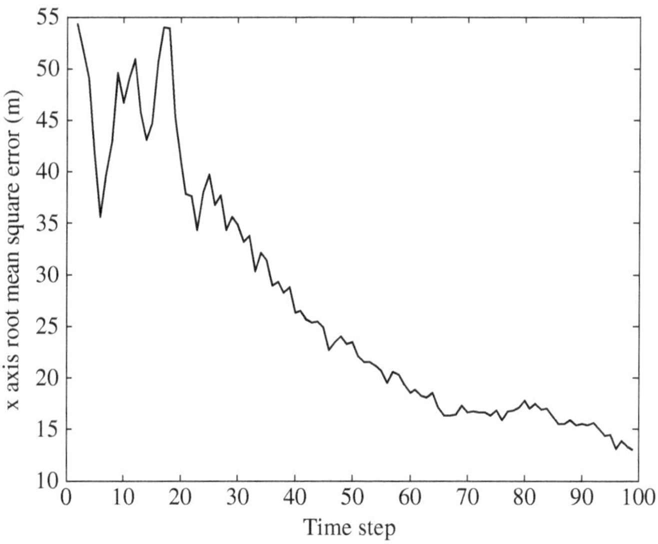

Figure 8.4 RMS of the y axis (PDAF filtering)

- The NNSF requires least computation, the PNNF a little more, and the PDAF most.
- The PNNF yields fewer occurrences of false tracking than the NNSF and thus has higher tracking accuracy.
- The PDAF has the strongest tracking capability in the presence of clutter, thanks to its fewest occurrences of false tracking and highest tracking accuracy, but it is the most time-consuming among the three algorithms.

8.4 Integrated Probabilistic Data Association Algorithm

Multi-target positioning and tracking in clutter by sensors involves the validation of measurements, their association with a real target, and the difficulty of determining the number of targets. To resolve these problems (especially that of determining the number of targets), Musicki and Evans [196, 197] introduced the concept of target existence into the PDA and, based on the assumption that no track exists, derived the IPDA algorithm. This algorithm is capable of estimating the probabilities of track existence of targets while tracking them, and has computational requirements equivalent to that of the PDA.

8.4.1 Judgment of Track Existence

In order to solve the problem of effectively estimating the number of targets in multi-target passive tracking when the number of targets in clutter is unknown and varying, track existence can be modeled by the Markov process and the events correlated with track existence can be defined as:

1. The targets exist and are visible (the targets exist with detection probability P_D).
2. The targets exist but are invisible.
3. The targets do not exist.

That is to say, the track existence is viewed as a three-state Markov chain:

1. $X_k^{t,o}$ indicates that target t exists at time k and is detected with probability P_D.
2. $X_k^{t,n}$ indicates that target t exists at time k but is not detected.
3. $\bar{X}_k^t$ indicates that target t does not exist at time k.

The probability matrix of transition between the three states is

$$\boldsymbol{P} = \begin{bmatrix} p_{11} & p_{12} & p_{13} \\ p_{21} & p_{22} & p_{23} \\ p_{31} & p_{32} & p_{33} \end{bmatrix} \tag{8.62}$$

where $0 \le p_{ij} \le 1$ and $\sum_{j=1}^{3} p_{ij} = 1, \quad i = 1,2,3.$

Assume that the probabilities of the three states (i.e., target t exists and is visible, target t exists but is invisible, target t does not exist) at time $k-1$ are known, that is, $\Pr\{X_{k-1}^{t,o} | Z^{k-1}\}$, $\Pr\{X_{k-1}^{t,n} | Z^{k-1}\}$, and $\Pr\{\overline{X_{k-1}^t} | Z^{k-1}\}$ are known, where Z^{k-1} denotes the set of accumulated measurements within the tracking gate up to time $k-1$. Then, by using the probabilities of the first state at time $k-1$, and those of the transition between the three states obtained from (8.62), the predicted value of $X_k^{t,o}$ can be obtained from the total probability theorem as follows:

$$\Pr\{X_k^{t,o} | Z^{k-1}\} = \Pr\{X_{k-1}^{t,o} | Z^{k-1}\} p_{11} + \Pr\{X_{k-1}^{t,n} | Z^{k-1}\} p_{21} + \Pr\{\overline{X_{k-1}^t} | Z^{k-1}\} p_{31} \tag{8.63}$$

That is, the sum of probabilities of transition from the three states at time $k-1$ to the first state at time k. Likewise, we have the predicted probabilities of the other states:

$$\Pr\{X_k^{t,n} | Z^{k-1}\} = \Pr\{X_{k-1}^{t,o} | Z^{k-1}\} p_{12} + \Pr\{X_{k-1}^{t,n} | Z^{k-1}\} p_{22} + \Pr\{\overline{X_{k-1}^t} | Z^{k-1}\} p_{32} \tag{8.64}$$

$$\Pr\{\overline{X_k^t}|Z^{k-1}\} = \Pr\{X_{k-1}^{t,o}|Z^{k-1}\}p_{13} + \Pr\{X_{k-1}^{t,n}|Z^{k-1}\}p_{23} + \Pr\{\overline{X_{k-1}^t}|Z^{k-1}\}p_{33} \tag{8.65}$$

From (8.64) and (8.65) it follows that the probability of the existence of the target t (including the states of being visible and invisible) is

$$\Pr\{X_k^t|Z^{k-1}\} = \Pr\{X_k^{t,o}|Z^{k-1}\} + \Pr\{X_k^{t,n}|Z^{k-1}\} \tag{8.66}$$

and that the probability that t does not exist is

$$\Pr\{\overline{X_k^t}|Z^{k-1}\} = 1 - \Pr\{X_k^t|Z^{k-1}\} \tag{8.67}$$

Let m_k^t denote the number of measurements within the tracking gate of the target t at time k, V_k^t the area (volume) of the gate at time k, and

$$V_k^t = c_{n_z}\gamma^{\frac{n_z}{2}}|S^t(k)|^{\frac{1}{2}} \tag{8.68}$$

where n_z is the dimension of the measurements. When $n_z = 1, 2, 3$, $Cn_z = 2$, π, and $4\pi/3$, respectively, and $\mathbf{S}^t(k)$ is the innovation covariance correlated with target t.

The number of false measurements within the tracking gate is denoted by $\hat{m}_k^t$ in the following situations.

1. If the number of false measurements is Poisson distributed and the parameter λ is known, then $\hat{m}_k^t = \lambda V_k^t$.
2. If the number of false measurements is Poisson distributed but the parameter λ is unknown, then

$$\hat{m}_k^t = \begin{cases} 0 & m_k^t = 0 \\ m_k^t - P_D P_G \Pr\{X_k^{t,o}|Z^{k-1}\} & m_k^t > 0 \end{cases} \tag{8.69}$$

   where P_G denotes the gate probability.
3. If the prior distribution of the number of false measurements is unknown, then $\hat{m}_k^t = m_k^t$.

This gives the updated probability correlated with the existence of target t at time k [196, 197]:

$$\Pr\{X_k^{t,o}|Z^k\} = \frac{(1-\delta_k^t)\Pr\{X_k^{t,o}|Z^{k-1}\}}{1-\delta_k^t\Pr\{X_k^{t,o}|Z^{k-1}\}} \tag{8.70}$$

$$\Pr\{X_k^{t,n}|Z^k\} = \frac{\Pr\{X_k^{t,n}|Z^{k-1}\}}{1-\delta_k^t\Pr\{X_k^{t,o}|Z^{k-1}\}} \tag{8.71}$$

$$\Pr\{\overline{X_k^t}|Z^k\} = 1 - \Pr\{X_k^{t,o}|Z^{k-1}\} - \Pr\{X_k^{t,n}|Z^{k-1}\} \tag{8.72}$$

where

$$\delta_k^t = \begin{cases} P_D P_G & m_k^t = 0 \\ P_D P_G - P_D P_G \dfrac{V_k^t}{\hat{m}_k^t}\sum_{i=1}^{m_k^t}\Lambda_i^t(k) & m_k^t > 0 \end{cases} \tag{8.73}$$

with $\Lambda_i^t(k)$ denoting the likelihood function of the ith measurement $z_i(k)$ within the tracking gate having originated from the target t at time k, that is,

$$\Lambda_k^i(k) = P_G^{-1}\mathrm{N}[\boldsymbol{v}_i(k);0,\boldsymbol{S}^t(k)] \tag{8.74}$$

where $v_i(k)$ is the innovation correlated with the measurement $z_i(k)$, and $\boldsymbol{S}^t(k)$ the innovation covariance associated with target t.

8.4.2 Data Association

The association probability that the ith measurement $z_i(k)$ within the tracking gate originated from target t at time k is [196, 197]:

$$\beta_i(k) = \frac{P_D P_G \frac{V_k^t}{\hat{m}_k^t}\Lambda_i^t(k)\Pr\{X_k^{t,o}|Z^{k-1}\}}{(1-\delta_k^t)\Pr\{X_k^{t,o}|Z^{k-1}\} + \Pr\{X_k^{t,n}|Z^{k-1}\}} \tag{8.75}$$

$$\beta_0(k) = \frac{(1-P_D P_G)\Pr\{X_k^{t,o}|Z^{k-1}\} + \Pr\{X_k^{t,n}|Z^{k-1}\}}{(1-\delta_k^t)\Pr\{X_k^{t,o}|Z^{k-1}\} + \Pr\{X_k^{t,n}|Z^{k-1}\}} \tag{8.76}$$

The state update equation of target t is

$$\hat{\boldsymbol{X}}(k|k) = \sum_{i=0}^{m_k^t}\beta_i(k)\hat{\boldsymbol{X}}_i(k|k) \tag{8.77}$$

where

$$\hat{\boldsymbol{X}}_0(k|k) = \hat{\boldsymbol{X}}(k|k-1) \tag{8.78}$$

The covariance update equation of target t is

$$\boldsymbol{P}(k|k) = \sum_{i=0}^{m_k^t}\beta_i(k)\{\boldsymbol{P}_i(k|k) + [\hat{\boldsymbol{X}}_i(k|k) - \hat{\boldsymbol{X}}(k|k)] - \hat{\boldsymbol{X}}(k|k)]'\} \tag{8.79}$$

where $\boldsymbol{P}_0(k|k) = \boldsymbol{P}(k|k-1)$.

Other filtering equations used in the IPDA are the same as in the PDA.

8.5 Joint Probabilistic Data Association Algorithm

The JPDA algorithm was proposed by Bar-Shalom and his students based on the PDA algorithm, which is appropriate only in single-target tracking. The JPDA is a good algorithm applicable to multi-target data association in the cluttered environment. Similarly to the PDA, it computes a weighted residual for track updating based on all the measurements within the validation gate. In this algorithm, unlike in the PDA, the target origins of all the measurements have to be considered when echoes are present in the overlap between correlation wave gates of the various targets. That is, the competition of several tracks for the same measurement has to be accounted for in the computation of association probabilities, and the weighted value of the target having participated in the competition should be somewhat decreased to reflect competition for the measurement under consideration.

The multi-target data association technique in cluttered environments is the most important and difficult problem encountered in multi-target tracking. If the wave gates of the two or more targets being tracked do not intersect, or no echoes lie in the intersection of the wave gates, then the problem of multi-target data association can be simplified to a problem of data association of several individual targets, which can be solved by the PDA algorithm discussed in Section 8.3. However, if echoes are present in the intersection, then the data association problem will be significantly more complex, as discussed here.

8.5.1 Basic Models of JPDA

8.5.1.1 Validation Matrix

When echoes are found in the intersection of correlation wave gates of various targets, the origins of the measurements have to be considered in an all-round way. To indicate the complex relationship between the valid echoes and the target tracking gates, Bar-Shalom introduced the concept of a "validation matrix."

The validation matrix is defined as

$$\Omega = \left[\omega_{jt}\right] = \begin{bmatrix} \omega_{10} & \cdots & \omega_{1T} \\ \vdots & \cdots & \vdots \\ \omega_{m_k 0} & \cdots & \omega_{m_k T} \end{bmatrix} \tag{8.80}$$

where ω_{jt} is a binary variable, $\omega_{jt}=1$ indicates that measurement j ($j=1,2,\ldots,m_k$) lies in the validation gate of target t ($t=0,1,\ldots,T$), while $\omega_{jt}=0$ shows that no measurement j is in this gate. The index $t=0$ stands for "no target" and all the elements ω_{j0} in the corresponding column of $\mathbf{\Omega}$ are 1, for each measurement is likely to have originated from clutter or a false alarm.

8.5.1.2 Association Matrix (Joint Events)

For a multi-target tracking problem, given the validation matrix (or association cluster matrix) $\mathbf{\Omega}$ reflecting the associations of the valid echoes with the targets or clutter or false alarms, the separation of this matrix yields all the association matrixes indicating the association events. This must be done under two basic assumptions [198]:

- A measurement can have only one source (i.e., a measurement must have originated either from the target or from clutter or a false alarm). In other words, indistinguishable detections are not considered here.
- No more than one measurement can originate from a target. If several measurements seem to match a target, one of them is assumed real and the others false.

In other words, the separation of the validation matrix must be done according to two criteria [19]:

- In each row of the validation matrix, one and only one “1” can be selected as the only non-zero element of the association matrix in this row, such that the feasible joint events indicated by the feasible matrix satisfy the first assumption (i.e., that a measurement can have only one origin).
- In each column, except for the first, of the feasible matrix, there can be only one non-zero element such that the feasible events indicated by the association matrix satisfy the second assumption (i.e., that no more than one measurement can originate from a target).

8.5.1.3 Calculation of Association Probability

The purpose of the JPDA is to calculate the probability that every measurement is associated with its various possible source targets. When echoes are found in the overlap of the correlation wave gates of these various targets, the target sources of all the measurements have to be considered at the same time.

Denote by $\theta_{jt}(k)$ the event of measurement j having originated from target t ($0 \le t \le T$), and by $\theta_{j0}(k)$ that of j having originated from clutter or false alarms. Then the definition of the conditional probability in the single-target PDAF gives

$$\beta_{jt}(k) = \Pr\{\theta_{jt}(k)|Z^k\}, \quad j=0,1,\ldots,m_k,\ t=0,1,\ldots,T \tag{8.81}$$

which denotes the probability of the jth measurement being associated with target t, and

$$\sum_{j=0}^{m_k} \beta_{jt}(k) = 1 \tag{8.82}$$

Then the estimate of the state of target t at time k is

$$\hat{\boldsymbol{X}}^t(k|k) = \mathrm{E}\left[\boldsymbol{X}^t(k)|\boldsymbol{Z}^k\right] = \sum_{j=0}^{m_k} \mathrm{E}\left[\boldsymbol{X}^t(k)|\theta_{jt}(k),\boldsymbol{Z}^k\right]\Pr\left\{\theta_{jt}(k)|\boldsymbol{Z}^k\right\} = \sum_{j=0}^{m_k} \beta_{jt}(k)\hat{\boldsymbol{X}}_j^t(k|k) \tag{8.83}$$

where

$$\hat{\boldsymbol{X}}_j^t(k|k) = \mathrm{E}\left[\boldsymbol{X}^t(k)|\theta_{jt}(k),\boldsymbol{Z}^k\right], \quad j=0,1,\ldots,m_k \tag{8.84}$$

indicates the estimate of the state obtained through the Kalman filtering of target t by the jth measurement at time k, while $\hat{\boldsymbol{X}}_0^t(k|k)$ indicates that no measurements are target-originated at time k, in which case the predicted value $\hat{\boldsymbol{X}}^t(k|k-1)$ should be used as a substitute.

The probability of the jth measurement being associated with the target can be obtained through the following equation:

$$\beta_{jt}(k) = \Pr\{\theta_{jt}(k)|\boldsymbol{Z}^k\} = \Pr\left\{\bigcup_{i=1}^{n_k} \theta_{jt}^i(k)|\boldsymbol{Z}^k\right\} = \sum_{i=1}^{n_k} \hat{\omega}_{jt}^i[\theta_i(k)]\Pr\{\theta_i(k)|\boldsymbol{Z}^k\} \tag{8.85}$$

where $\theta_{jt}^{i}(k)$ denotes the event of measurement j having originated from target t $(0 \le t \le T)$ in the ith joint event, $\theta_i(k)$ the ith joint event, and n_k the number of joint events, while

$$\hat{\omega}_{jt}^{i}(\theta_i(k)) = \begin{cases} 1, & \text{if } \theta_{jt}^{i}(k) \subset \theta_i(k) \\ 0, & \text{otherwise} \end{cases} \tag{8.86}$$

indicates whether measurement j originated from target t in the ith joint event; it is 1 if j originated from t, and 0 otherwise.

Define the ith joint event under normal circumstances as

$$\theta_i(k) = \bigcap_{j=1}^{m_k} \theta_{jt}^{i}(k) \tag{8.87}$$

which indicates a probability that m_k measurements match different targets.

Define the association matrix corresponding to the joint event as

$$\hat{\Omega}(\theta_i(k)) = \left[\hat{\omega}_{jt}^{i}(\theta_i(k))\right] = \begin{bmatrix} \hat{\omega}^{i}{}_{10} & \cdots & \hat{\omega}^{i}{}_{1T} \\ \vdots & \cdots & \vdots \\ \hat{\omega}^{i}{}_{m_k 0} & \cdots & \hat{\omega}^{i}{}_{m_k T} \end{bmatrix} \quad j=1,2,\ldots,m_k;\ i=1,2,\ldots,n_k;\ t=0,1,\ldots,T \tag{8.88}$$

From the two basic assumptions above, it is easy to deduce that the association matrix satisfies

$$\begin{aligned} &\sum_{t=0}^{T} \hat{\omega}_{jt}^{i}[\theta_i(k)] = 1, \qquad j=1,2,\ldots,m_k \\ &\sum_{j=1}^{m_k} \hat{\omega}_{jt}^{i}[\theta_i(k)] \le 1, \qquad t=1,2,\ldots,T \end{aligned} \tag{8.89}$$

8.5.1.4 Applications

Consider two target tracks, around whose measurement predictions two wave gates are set up. Assume that the scan at the next moment yields three echoes, the relationship between whose positions and those of the correlation wave gates are shown in Figure 8.5.

Write out their validation matrix and association matrixes, and find the probability $\beta_{jt}(k)$ of the measurements being associated with different targets.

Solution: Using the method for constructing a validation matrix defined in (8.80) yields the validation matrix

$$\mathbf{\Omega} = [\omega_{jt}] = \overbrace{\begin{bmatrix} 1 & 1 & 0 \\ 1 & 1 & 1 \\ 1 & 0 & 1 \end{bmatrix}}^{t} \left.\begin{matrix} 1 \\ 2 \\ 3 \end{matrix}\right\} j \tag{8.90}$$

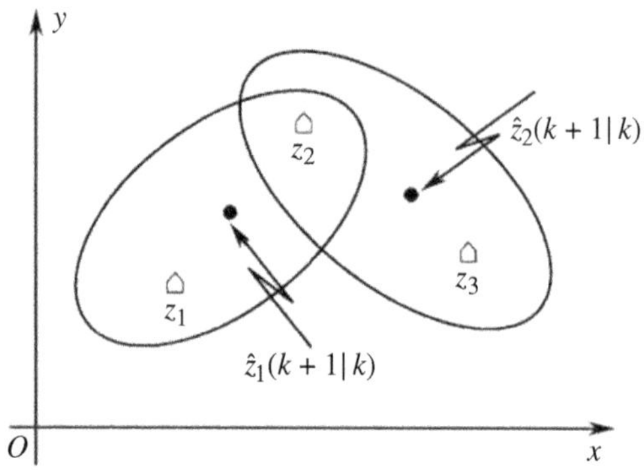

Figure 8.5 Example of the validation matrix and the formation of association events

Based on this matrix and the two criteria above, we have, through an exhaustive search, eight association matrixes and their corresponding joint events (feasible association events), as given below:

$$\hat{\Omega}[\theta_1(k)] = \begin{bmatrix} 1 & 0 & 0 \\ 1 & 0 & 0 \\ 1 & 0 & 0 \end{bmatrix}, \quad \theta_1(k) = \theta_{10}^1(k) \cap \theta_{20}^1(k) \cap \theta_{30}^1(k) \tag{8.91}$$

$$\hat{\Omega}[\theta_2(k)] = \begin{bmatrix} 0 & 1 & 0 \\ 1 & 0 & 0 \\ 1 & 0 & 0 \end{bmatrix}, \quad \theta_2(k) = \theta_{11}^2(k) \cap \theta_{20}^2(k) \cap \theta_{30}^2(k) \tag{8.92}$$

$$\hat{\Omega}[\theta_3(k)] = \begin{bmatrix} 0 & 1 & 0 \\ 0 & 0 & 1 \\ 1 & 0 & 0 \end{bmatrix}, \quad \theta_3(k) = \theta_{11}^3(k) \cap \theta_{22}^3(k) \cap \theta_{30}^3(k) \tag{8.93}$$

$$\hat{\Omega}[\theta_4(k)] = \begin{bmatrix} 0 & 1 & 0 \\ 1 & 0 & 0 \\ 0 & 0 & 1 \end{bmatrix}, \quad \theta_4(k) = \theta_{11}^4(k) \cap \theta_{20}^4(k) \cap \theta_{32}^4(k) \tag{8.94}$$

$$\hat{\Omega}[\theta_5(k)] = \begin{bmatrix} 1 & 0 & 0 \\ 0 & 1 & 0 \\ 1 & 0 & 0 \end{bmatrix}, \quad \theta_5(k) = \theta_{10}^5(k) \cap \theta_{21}^5(k) \cap \theta_{30}^5(k) \tag{8.95}$$

$$\hat{\Omega}[\theta_6(k)] = \begin{bmatrix} 1 & 0 & 0 \\ 0 & 1 & 0 \\ 0 & 0 & 1 \end{bmatrix}, \quad \theta_6(k) = \theta_{10}^6(k) \cap \theta_{21}^6(k) \cap \theta_{32}^6(k) \tag{8.96}$$

$$\hat{\Omega}[\theta_7(k)] = \begin{bmatrix} 1 & 0 & 0 \\ 0 & 0 & 1 \\ 1 & 0 & 0 \end{bmatrix}, \quad \theta_7(k) = \theta_{10}^7(k) \cap \theta_{22}^7(k) \cap \theta_{30}^7(k) \tag{8.97}$$

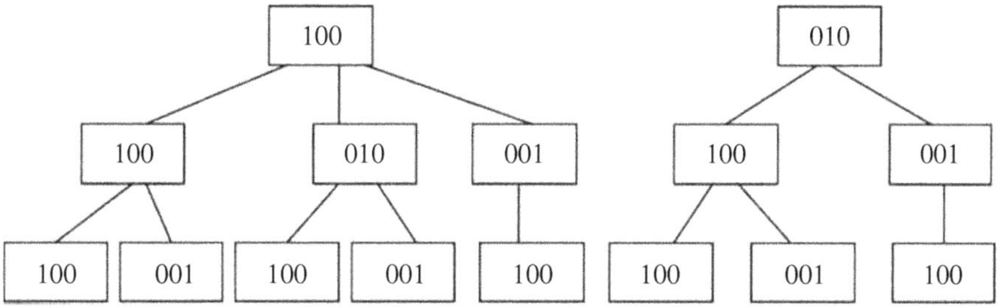

Figure 8.6 Association matrix block diagram

$$\hat{\Omega}[\theta_8(k)] = \begin{bmatrix} 1 & 0 & 0 \\ 1 & 0 & 0 \\ 0 & 0 & 1 \end{bmatrix}, \quad \theta_8(k) = \theta_{10}^8(k) \cap \theta_{20}^8(k) \cap \theta_{32}^8(k) \tag{8.98}$$

The procedure of deriving association matrixes from the validation can also be illustrated by the block diagram in Figure 8.6. Row 1 of the validation matrix derived from (8.90) can be separated, according to the separation criteria for validation matrixes, into [1 0 0] and [0 1 0]. Specifically, the first measurement originated from a false target or target 1; when the origin of the first measurement is a false target, the second one (row 2 of the validation matrix) can be further split into [1 0 0], [0 1 0], and [0 0 1]. That is, the second measurement belonged to a false target, target 1 or target 2; when the source of the first is target 1, the second can be divided onwards into [1 0 0] and [0 0 1], that is, the second measurement derived from a false target or target 2.

Notice that in this case, since the first measurement came from target 1, according to separation criterion 2 for validation matrixes, the second cannot have originated from target 1, hence the situation [0 1 0] does not exist. Likewise, if the first two measurements belonged to different targets, we can separate the source of the third one (row 3 of the validation matrix) and get a block diagram as shown in Figure 8.6.

The probability $\beta_{jt}(k)$ of different measurements being associated with different targets is given below:

$$\begin{aligned}\beta_{11}(k) &= \Pr\{\theta_{11}(k)|\mathbf{Z}^k\} = \Pr\left\{\bigcup_{i=1}^{8}\theta_{11}^i(k)|\mathbf{Z}^k\right\} = \sum_{i=1}^{8}\hat{\omega}_{11}^i[\theta_i(k)]\Pr\{\theta_i(k)|\mathbf{Z}^k\} \\ &= \sum_{i=2}^{4}\Pr\{\theta_i(k)|\mathbf{Z}^k\}\end{aligned} \tag{8.99}$$

$$\begin{aligned}\beta_{21}(k) &= \Pr\{\theta_{21}(k)|\mathbf{Z}^k\} = \Pr\left\{\bigcup_{i=1}^{8}\theta_{21}^i(k)|\mathbf{Z}^k\right\} = \sum_{i=1}^{8}\hat{\omega}_{21}^i[\theta_i(k)]\Pr\{\theta_i(k)|\mathbf{Z}^k\} \\ &= \sum_{i=5}^{6}\Pr\{\theta_i(k)|\mathbf{Z}^k\}\end{aligned} \tag{8.100}$$

$$\beta_{31}(k) = \Pr\{\theta_{31}(k)|\mathbf{Z}^k\} = \Pr\left\{\bigcup_{i=1}^{8}\theta_{31}^i(k)|\mathbf{Z}^k\right\} = \sum_{i=1}^{8}\hat{\omega}_{31}^i[\theta_i(k)]\Pr\{\theta_i(k)|\mathbf{Z}^k\} = 0 \tag{8.101}$$

$$\beta_{12}(k) = \Pr\{\theta_{12}(k)|\mathbf{Z}^k\} = \Pr\left\{\bigcup_{i=1}^{8}\theta_{12}^i(k)|\mathbf{Z}^k\right\} = \sum_{i=1}^{8}\hat{\omega}_{12}^i[\theta_i(k)]\Pr\{\theta_i(k)|\mathbf{Z}^k\} = 0 \tag{8.102}$$

$$\beta_{22}(k)=\Pr\{\theta_{22}(k)|\mathbf{Z}^k\}=\Pr\left\{\bigcup_{i=1}^{8}\theta_{22}^i(k)|\mathbf{Z}^k\right\}=\sum_{i=1}^{8}\hat{\omega}_{22}^i[\theta_i(k)]\Pr\{\theta_i(k)|\mathbf{Z}^k\}$$
$$=\Pr\{\theta_3(k)|\mathbf{Z}^k\}+\Pr\{\theta_7(k)|\mathbf{Z}^k\} \tag{8.103}$$

$$\beta_{32}(k)=\Pr\{\theta_{32}(k)|\mathbf{Z}^k\}=\Pr\left\{\bigcup_{i=1}^{8}\theta_{32}^i(k)|\mathbf{Z}^k\right\}=\sum_{i=1}^{8}\hat{\omega}_{32}^i[\theta_i(k)]\Pr\{\theta_i(k)|\mathbf{Z}^k\}$$
$$=\Pr\{\theta_4(k)|\mathbf{Z}^k\}+\Pr\{\theta_6(k)|\mathbf{Z}^k\}+\Pr\{\theta_8(k)|\mathbf{Z}^k\} \tag{8.104}$$

Likewise, the probability β_{0j} $(j=1,2)$ that all the valid echoes of targets 1 and 2 are false measurements can be represented as

$$\beta_{01}=\Pr\{\theta_1(k)|\mathbf{Z}^k\}+\sum_{i=7}^{8}\Pr\{\theta_i(k)|\mathbf{Z}^k\},\ \beta_{02}=\sum_{i=1}^{2}\Pr\{\theta_1(k)|\mathbf{Z}^k\}+\Pr\{\theta_5(k)|\mathbf{Z}^k\} \tag{8.105}$$

As can be seen from the example above, there is a one-to-one match between each association matrix and each feasible association event. In practical use, feasible association events are determined typically with the association matrixes derived from validation matrixes. According to the separation criteria, one validation matrix can be separated into many feasible association matrixes. The number of association matrixes grows, usually exponentially with the number of targets and that of valid echoes, and with the size of the intersection of wave gates as well. The example also indicates that the key to the evaluation of the probabilities of the jth measurement being associated with the target is that of the association event $\theta_i(k)$ $(i=1,2,\dots,n_k)$.

8.5.2 Calculation of the Probability of Joint Events

For the sake of discussion, two binary variables are introduced.

1. The measurement association indicator

$$\tau_j[\theta_i(k)]=\sum_{t=1}^{T}\hat{\omega}^{jti}(\theta_i(k))=\begin{cases}1\\0\end{cases} \tag{8.106}$$

indicates whether measurement j is associated with a real target in the joint event.

2. The target detection indicator

$$\delta_t[\theta_i(k)]=\sum_{j=1}^{m_k}\hat{\omega}^{jti}[\theta_i(k)]=\begin{cases}1\\0\end{cases} \tag{8.107}$$

indicates whether any measurement is associated with target t in the joint event $\theta_i(k)$, that is, whether it has been detected.

Denoting by $\phi[\theta_i(k)]$ the number of false measurements in the joint event $\theta_i(k)$ yields

$$\phi[\theta_i(k)]=\sum_{j=1}^{m_k}\{1-\tau_j[\theta_i(k)]\} \tag{8.108}$$

Using Bayes' rule gives the conditional probability of the joint event $\theta_i(k)$ at time k,

$$\Pr\{\theta_i(k)|\mathbf{Z}^k\} = \Pr\{\theta_i(k)|\mathbf{Z}(k),\mathbf{Z}^{k-1}\} = \frac{1}{c}p[\mathbf{Z}(k)|\theta_i(k),\mathbf{Z}^{k-1}]\Pr\{\theta_i(k)|\mathbf{Z}^{k-1}\} \tag{8.109}$$

where c is the normalization constant

$$c = \sum_{j=0}^{n_k} p[\mathbf{Z}(k)|\theta_j(k),\mathbf{Z}^{k-1}]\Pr\{\theta_j(k)\} \tag{8.110}$$

Assume that the false measurements (i.e., those unassociated with any target) are uniformly distributed in the validation region of volume V, and that the target-associated measurements have the Gaussian PDF $\mathrm{N}_t[z_j(k)] = \mathrm{N}[z_j(k); \hat{z}_t(k|k-1), \boldsymbol{S}_t(k)]$. The only difference from the single-target case is that all the validation regions are assumed to coincide with the entire surveillance region; this suggests that the gate probability $P_{\mathrm{G}} = 1$. By simulating the solution of the single-target PDA approach, we get

$$p\left[z_j(k)|\theta_{jt}^i(k),\mathbf{Z}^{k-1}\right] = \begin{cases} \mathrm{N}_t[z_j(k)], & \text{if } \tau_j[\theta_i(k)] = 1 \\ V^{-1}, & \text{if } \tau_j[\theta_i(k)] = 0 \end{cases} \tag{8.111}$$

Hence

$$p[\mathbf{Z}(k)|\theta_i(k),\mathbf{Z}^{k-1}] = \prod_{j=1}^{m_k} p\left[z_j(k)|\theta_{jt}^i(k),\mathbf{Z}^{k-1}\right] = V^{-\phi[\theta_i(k)]}\prod_{j=1}^{m_k}\mathrm{N}_t[z_j(k)]^{\tau_j[\theta_i(k)]} \tag{8.112}$$

As we know, given $\theta_i(k)$, the target detection indicator $\delta_t(\theta_i(k))$ and the number of false measurements $\phi(\theta_i(k))$ are completely determined. Thus,

$$\Pr\{\theta_i(k)\} = \Pr\{\theta_i(k), \delta_t[\theta_i(k)], \phi[\theta_i(k)]\} \tag{8.113}$$

Using the multiplication theorem, the above equation becomes

$$\Pr\{\theta_i(k)\} = \Pr\{\theta_i(k)|\delta_t(\theta_i(k)), \phi(\theta_i(k))\}\Pr\{\delta_t(\theta_i(k)), \phi(\theta_i(k))\} \tag{8.114}$$

In fact, given the number of false measurements, the joint event $\theta_i(k)$ will be defined only by the target detection indicator function $\delta_t[\theta_i(k)]$. But there are $C_{m_k}^{\phi[\theta_i(k)]}$ events in total, which include $\phi[\theta_i(k)]$ false measurements. These such events also contain $m_k - \phi[\theta_i(k)]$ real measurements, and there are $\{m_k - \phi[\theta_i(k)]\}!$ possible associations between these such real measurements and the target. Therefore,

$$\Pr\{\theta_i(k)|\delta_t(\theta_i(k)), \phi(\theta_i(k))\} = \frac{1}{(m_k - \phi(\theta_i(k))!C_{m_k}^{\phi(\theta_i(k))}} = \frac{\phi(\theta_i(k))!}{m_k!} \tag{8.115}$$

The last factor in (8.114) is

$$\Pr\{\delta_t(\theta_i(k)), \phi(\theta_i(k))\} = \prod_{t=1}^{T}(P_{\mathrm{D}}^t)^{\delta_t(\theta_i(k))}(1-P_{\mathrm{D}}^t)^{1-\delta_t(\theta_i(k))}\mu_F(\phi(\theta_i(k)) \tag{8.116}$$

where P_{D}^{t} denotes the detection probability of target t and $\mu_F\{\phi[\theta_i(k)]\}$ the prior PMF of the number of false measurements. Substituting (8.115) and (8.116) into (8.114) gives the prior probability of the joint event $\theta_i(k)$:

$$\Pr\{\theta_i(k)\} = \frac{\phi[\theta_i(k)]!}{m_k!}\mu_F\{\phi[\theta_i(k)]\}\prod_{t=1}^{T}\left(P_{\mathrm{D}}^{t}\right)^{\delta_t[\theta_i(k)]}\left(1-P_{\mathrm{D}}^{t}\right)^{1-\delta_t[\theta_i(k)]} \tag{8.117}$$

Similarly, combining (8.112) and (8.117) into (8.109), we get the posterior probability of $\theta_i(k)$ as

$$\Pr\{\theta_i(k)|\mathbf{Z}^k\} = \frac{1}{c}\frac{\phi[\theta_i(k)]!}{m_k!}\mu_F\{\phi[\theta_i(k)]\}V^{-\phi[\theta_i(k)]}\prod_{j=1}^{m_k}\mathrm{N}_{t_j}\left[\mathbf{Z}_j(k)\right]^{\tau_j[\theta_i(k)]}\prod_{t=1}^{T}\left(P_{\mathrm{D}}^{t}\right)^{\delta_t[\theta_i(k)]}\left(1-P_{\mathrm{D}}^{t}\right)^{1-\delta_t[\theta_i(k)]} \tag{8.118}$$

The JPDA filter has two forms according to the model for the PMF $\mu_F(\phi(\theta_i(k))$ of the number of false measurements [16]. The parametric JPDA filter uses the Poisson distribution, that is,

$$\mu_F(\varphi(\theta_i(k))) = \mathrm{e}^{-\lambda V}\frac{(\lambda V)^{\varphi(\theta_i(k))}}{\varphi(\theta_i(k))!} \tag{8.119}$$

where λ is the space density of false measurements, and λV is the expected number of false measurements in the gate. Inserting (8.119) into (8.118) yields

$$\Pr\{\theta_i(k)|Z^k\} = \frac{\lambda^{\phi[\theta_i(k)]}}{c'}\prod_{j=1}^{m_k}\mathrm{N}_{t_j}\left[\mathbf{Z}_j(k)\right]^{\tau_j[\theta_i(k)]}\prod_{t=1}^{T}\left(P_{D}^{t}\right)^{\delta_t[\theta_i(k)]}\left(1-P_{D}^{t}\right)^{1-\delta_t[\theta_i(k)]} \tag{8.120}$$

where c' is the new normalization constant.

The non-parametric JPDA uses the uniformly distributed $\mu_F(\phi(\theta_i(k))$, that is, $\mu_F(\phi(\theta_i(k))=\varepsilon$. Substitute $\mu_F(\phi(\theta_i(k))=\varepsilon$ into (8.118) and cancel the constants ε and $m_k!$ in every expression, then (8.118) becomes

$$\Pr\{\theta_i(k)|Z^k\} = \frac{1}{c''}\frac{\phi[\theta_i(k)]!}{V^{\phi[\theta_i(k)]}}\prod_{j=1}^{m_k}N_{t_j}\left[\mathbf{Z}_j(k)\right]^{\tau_j[\theta_i(k)]}\prod_{t=1}^{T}\left(P_{\mathrm{D}}^{t}\right)^{\delta_t[\theta_i(k)]}\left(1-P_{\mathrm{D}}^{t}\right)^{1-\delta_t[\theta_i(k)]} \tag{8.121}$$

where c'' is the new normalization constant.

8.5.3 Calculation of the State Estimation Covariance

From the Kalman filtering equation, we can obtain the covariance of the state estimate $\hat{\boldsymbol{X}}_j^t(k|k)$ of target t based on the jth measurement

$$\boldsymbol{P}_j^t(k|k) = \mathrm{E}\left\{\left[\boldsymbol{X}^t(k)-\hat{\boldsymbol{X}}_j^t(k|k)\right]\left[\boldsymbol{X}^t(k)-\hat{\boldsymbol{X}}_j^t(k|k)\right]'|\theta_{jt}(k),\mathbf{Z}^k\right\} \tag{8.122}$$

$$= \mathbf{P}^t(k|k-1)-\mathbf{K}^t(k)\mathbf{S}^t(k)\mathbf{K}^{t'}(k) \tag{8.123}$$

where $\boldsymbol{K}^t(k)$ denotes the gain matrix of the target at time k, and $\boldsymbol{S}^t(k)$ the corresponding innovation covariance.

As we know, when no targets come from t (i.e., when no valid echoes are used to update the state of the target), the state estimate of the target is equal to its predicted value. Therefore,

$$\begin{aligned}\boldsymbol{P}_0^t(k|k) &= \mathrm{E}\left\{\left[\boldsymbol{X}^t(k)-\hat{\boldsymbol{X}}_0^t(k|k)\right]\left[\boldsymbol{X}^t(k)-\hat{\boldsymbol{X}}_0^t(k|k)\right]'|\theta_{0t}(k),\boldsymbol{Z}^k\right\}\\ &= \mathrm{E}\left\{[\boldsymbol{X}^t(k)-\hat{\boldsymbol{X}}^t(k|k-1)][\boldsymbol{X}^t(k)-\hat{\boldsymbol{X}}^t(k|k-1)]'\theta_{0t}(k),\boldsymbol{Z}^k\right\} = \boldsymbol{P}^t(k|k-1)\end{aligned} \tag{8.124}$$

The covariance of the state estimate $\hat{\boldsymbol{X}}^t(k|k)$ is

$$\begin{aligned}\boldsymbol{P}^t(k|k) &= \mathrm{E}\left\{[\boldsymbol{X}^t(k)-\hat{\boldsymbol{X}}^t(k|k)][\boldsymbol{X}^t(k)-\hat{\boldsymbol{X}}^t(k|k)]'|\boldsymbol{Z}^k\right\}\\ &= \sum_{j=0}^{m_k}\beta_{jt}(k)\mathrm{E}\left\{[\boldsymbol{X}^t(k)-\hat{\boldsymbol{X}}^t(k|k)][\boldsymbol{X}^t(k)-\hat{\boldsymbol{X}}^t(k|k)]'|\theta_{jt}(k),\boldsymbol{Z}^k\right\}\\ &= \sum_{j=0}^{m_k}\beta_{jt}(k)\mathrm{E}\left\{\left[\left(\boldsymbol{X}^t(k)-\hat{\boldsymbol{X}}_j^t(k|k)\right)+\left(\hat{\boldsymbol{X}}_j^t(k|k)-\hat{\boldsymbol{X}}^t(k|k)\right)\right]\right.\\ &\quad \left.\left[\left(\boldsymbol{X}^t(k)-\hat{\boldsymbol{X}}_j^t(k|k)\right)+\left(\hat{\boldsymbol{X}}_j^t(k|k)-\hat{\boldsymbol{X}}^t(k|k)\right)\right]'|\theta_{jt}(k),\boldsymbol{Z}^k\right\}\\ &= \sum_{j=0}^{m_k}\beta_{jt}(k)\mathrm{E}\left\{\left[\boldsymbol{X}^t(k)-\hat{\boldsymbol{X}}_j^t(k|k)\right]\left[\boldsymbol{X}^t(k)-\hat{\boldsymbol{X}}_j^t(k|k)\right]'|\theta_{jt}(k),\boldsymbol{Z}^k\right\}\\ &\quad + \sum_{j=0}^{m_k}\beta_{jt}(k)\mathrm{E}\left\{\left[\boldsymbol{X}^t(k)-\hat{\boldsymbol{X}}_j^t(k|k)\right]\left[\hat{\boldsymbol{X}}_j^t(k|k)-\hat{\boldsymbol{X}}^t(k|k)\right]'|\theta_{jt}(k),\boldsymbol{Z}^k\right\}\\ &\quad + \sum_{j=0}^{m_k}\beta_{jt}(k)\mathrm{E}\left\{\left[\hat{\boldsymbol{X}}_j^t(k|k)-\hat{\boldsymbol{X}}^t(k|k)\right]\left[\boldsymbol{X}^t(k)-\hat{\boldsymbol{X}}_j^t(k|k)\right]'|\theta_{jt}(k),\boldsymbol{Z}^k\right\}\\ &\quad + \sum_{j=0}^{m_k}\beta_{jt}(k)\mathrm{E}\left\{\left[\hat{\boldsymbol{X}}_j^t(k|k)-\hat{\boldsymbol{X}}^t(k|k)\right]\left[\hat{\boldsymbol{X}}_j^t(k|k)-\hat{\boldsymbol{X}}^t(k|k)\right]'|\theta_{jt}(k),\boldsymbol{Z}^k\right\}\end{aligned} \tag{8.125}$$

It follows from (8.122)–(8.124) that

$$\begin{aligned}&\sum_{j=0}^{m_k}\beta_{jt}(k)\mathrm{E}\left\{\left[\boldsymbol{X}^t(k)-\hat{\boldsymbol{X}}_j^t(k|k)\right]\left[\boldsymbol{X}^t(k)-\hat{\boldsymbol{X}}_j^t(k|k)\right]'|\theta_{jt}(k),\boldsymbol{Z}^k\right\} = \sum_{j=0}^{m_k}\beta_{jt}(k)\boldsymbol{P}_j^t(k|k)\\ &= \beta_{0t}(k)\boldsymbol{P}_0^t(k|k)+\sum_{j=1}^{m_k}\beta_{jt}(k)\left[\boldsymbol{P}^t(k|k-1)-\boldsymbol{K}^t(k)\boldsymbol{S}^t(k)\boldsymbol{K}^{t'}(k)\right]\\ &= \boldsymbol{P}^t(k|k-1)-[1-\beta_{0t}(k)]\boldsymbol{K}^t(k)\boldsymbol{S}^t(k)\boldsymbol{K}^{t'}(k)\end{aligned} \tag{8.126}$$

From (8.125) we get

$$\begin{aligned}&\sum_{j=0}^{m_k}\beta_{jt}(k)\mathrm{E}\left\{\left[\boldsymbol{X}^t(k)-\hat{\boldsymbol{X}}_j^t(k|k)\right]\left[\hat{\boldsymbol{X}}_j^t(k|k)-\hat{\boldsymbol{X}}^t(k|k)\right]'|\theta_{jt}(k),\boldsymbol{Z}^k\right\}\\ &= \sum_{j=0}^{m_k}\beta_{jt}(k)\pm\left[\mathrm{E}\left[\boldsymbol{X}^t(k)|\theta_{jt}(k),\boldsymbol{Z}^k\right]-\hat{\boldsymbol{X}}_j^t(k|k)\right]\left[\hat{\boldsymbol{X}}_j^t(k|k)-\hat{\boldsymbol{X}}^t(k|k)\right]' = 0\end{aligned} \tag{8.127}$$

Likewise, we have

$$\sum_{j=0}^{m_k}\beta_{jt}(k)\mathrm{E}\left\{\left[\hat{\boldsymbol{X}}_j^t(k|k)-\hat{\boldsymbol{X}}^t(k|k)\right]\left[\boldsymbol{X}^t(k)-\hat{\boldsymbol{X}}_j^t(k|k)\right]'|\theta_{jt}(k),\boldsymbol{Z}^k\right\}=0 \tag{8.128}$$

$$\begin{aligned}&\sum_{j=0}^{m_k}\beta_{jt}(k)\mathrm{E}\left\{\left[\hat{\boldsymbol{X}}_j^t(k|k)-\hat{\boldsymbol{X}}^t(k|k)\right]\left[\hat{\boldsymbol{X}}_j^t(k|k)-\hat{\boldsymbol{X}}^t(k|k)\right]'|\theta_{jt}(k),\boldsymbol{Z}^k\right\}\\&=\sum_{j=0}^{m_k}\beta_{jt}(k)\left[\hat{\boldsymbol{X}}_j^t(k|k)\hat{\boldsymbol{X}}_j^{t'}(k|k)-\hat{\boldsymbol{X}}^t(k|k)\hat{\boldsymbol{X}}_j^{t'}(k|k)-\boldsymbol{X}_j^t(k|k)\hat{\boldsymbol{X}}^{t'}(k|k)+\hat{\boldsymbol{X}}^t(k|k)\hat{\boldsymbol{X}}^{t'}(k|k)\right]\\&=\sum_{j=0}^{m_k}\beta_{jt}(k)\hat{\boldsymbol{X}}_j^t(k|k)\hat{\boldsymbol{X}}_j^{t'}(k|k)-\hat{\boldsymbol{X}}^t(k|k)\hat{\boldsymbol{X}}^{t'}(k|k)\end{aligned} \tag{8.129}$$

Thus, combining (8.126)–(8.129) into (8.124) yields the covariance of $\hat{\boldsymbol{X}}^t(k|k)$:

$$\begin{aligned}\boldsymbol{P}^t(k|k)=&\boldsymbol{P}^t(k|k-1)-(1-\beta_{0t}(k))\boldsymbol{K}^t(k)\boldsymbol{S}^t(k)\boldsymbol{K}^{t'}(k)\\&+\sum_{j=0}^{m_k}\beta_{jt}(k)\hat{\boldsymbol{X}}_j^t(k|k)\hat{\boldsymbol{X}}_j^{t'}(k|k)-\hat{\boldsymbol{X}}^t(k|k)\hat{\boldsymbol{X}}^{t'}(k|k)\end{aligned} \tag{8.130}$$

8.5.4 Simplified JPDA Model

The JPDA algorithm, through the permutation and combination of all targets and measurements, selects reasonable joint events to calculate the joint probability. Therefore, it accounts for the fact that the various measurements from other targets are likely to lie in the same target association region. In this case, the algorithm has good performance in solving the problem of the measurements of several targets appearing in the same association region in a cluttered environment. But this algorithm is quite complex, and it has excessive computing requirements: with the increase in number of targets, the computational load for separating the validation matrix grows exponentially, even generating combination explosion. In view of these problems, the JPDA is quite difficult to implement in engineering.

For the sake of engineering implementations, many simplified algorithms are advanced based on the JPDA. An empirical equation for probability calculation was first proposed by Robert Fitzgerald (see Ref. [199]). This was followed by the suboptimal version presented by Roecher and Phillis [200]. The depth-first search (DFS) method and its condensed form initiated by B. Zhou can be found in Ref. [201]. These reduced versions perform differently with respect to tracking accuracy and real-time features in various cases. For convenience, three simplified models of the JPDA are given below.

The equation to calculate the empirical probability in the cheap JPDA algorithm [15, 199, 202] has the characteristics of the JPDA. To be specific, only the measurements in the association region of one track are heavily weighted, while those which lie in the overlap of the association regions of several tracks and contradict each other are lightly weighted. This algorithm assigns a higher weighting to the approximation of predicted positions and the association with fewest tracks. Generally (except in the presence of dense clutter), $B=0$ gives satisfactory results. In the cheap JPDA, only two to three measurements with the highest probabilities are used for updating the state of the target. This constraint accounts for the computational load. Another limit on the number of measurements is that the empirical probability may overweight false measurements, so that the covariance matrix increases unrestrainedly in the presence of dense targets.

The suboptimal JPDA [15, 200, 202, 203] adopts the concept of partial joint events. Assume that the associations between these partial joint events are non-contradictory. To be more precise, assume that measurement j_1 is associated with track t_1, and that measurement j_2 is associated with track t_2. In this case, if tracks t_1 and t_2 share a common measurement in the overlap (if any) of their association regions, this event is called a partial joint event. Events of this type are known as sub-optimal joint events, since they are all subsets of optimal joint events.

Data association is deemed equivalent to combination in terms of multi-target tracking, while in terms of JPDA implementations it primarily concerns how to generate hypothesis matrixes and calculate association probabilities effectively and quickly. A well-known model for combination is the constrained limited exhaustive search (CLES) method, by combining which the DFS algorithm [15, 201, 202, 204] attacks the problem of data association, or that of the production and separation of association matrixes for the JPDA.

8.5.5 *Performance Analysis*

The JPDA algorithm and its several simplified versions discussed above are capable of tracking dense targets in clutter. This section presents a simulation comparison between them.

Assume that a filter is tracking two targets under discussion which are in cross motion, with the following initial positions.

Target 1: $X(0) = (-29\,500\,\text{m}, 400\,\text{m/s}, 34\,500\,\text{m}, -400\,\text{m/s})$
Target 2: $X(0) = (-26\,250\,\text{m}, 296\,\text{m/s}, 34\,500\,\text{m}, -400\,\text{m/s})$
Process noise quantity $q_1 = q_2 = 0.01$
Range error of radar $\sigma_r = 100\,\text{m}$
Angle error of radar $\sigma_\theta = 0.02$ rad
Detection probability $P_\text{D} = 0.98$
Gate probability $P_\text{G} = 0.9997$
$\gamma = 16$
$m_k = 2$
Sampling interval $T = 1$ s

There are 70 simulation steps every run, and 50 runs in total. The state and measurement equations of the system are the same as in Section 8.3.4.

False measurements were produced uniformly in a square centered at the correct measurements, with an area of $A = n_c/\lambda \approx 10A_V$, where λ denotes the number of false measurements per unit area, and $\lambda = 0.0004$; n_c is the total number of false measurements, that is, $n_c = \text{INT}[10A_V\lambda + 1]$, $A_V = \pi\gamma|S(k)|^{\frac{1}{2}}$, and parameter $\gamma = 16$.

The real movement trajectories of the two targets and the curves of their RMS position errors calculated by the JPDA, cheap JPDA, suboptimal JPDA, and depth-first JPDA algorithms are illustrated in Figures 8.7, 8.8, and 8.9. The time consumed by these algorithms and their false tracking rates are shown in Table 8.1.

As can be seen from the simulation results, all the algorithms are capable of tracking two crossing targets in clutter. Compared with its condensed versions, the JPDA had higher tracking accuracy and lower false tracking rates; but its real-time performance was worse.

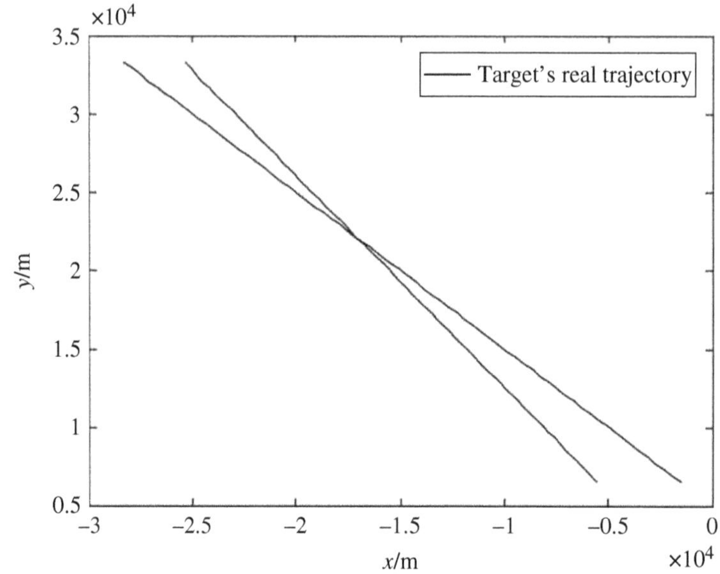

Figure 8.7 Real trajectories of the targets

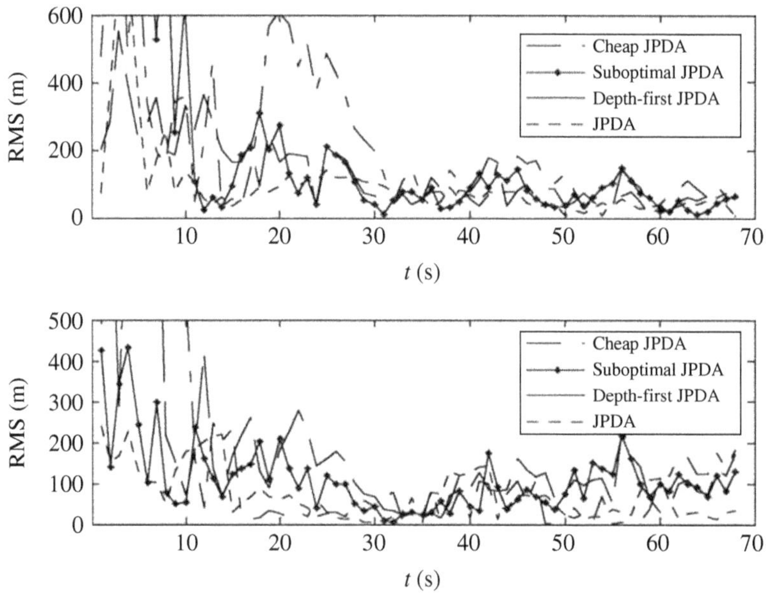

Figure 8.8 RMS position errors of target 1 on axis x (top) and y (bottom)

All three simplified algorithms have proved appropriate for engineering implementations. Among them, the depth-first JPDA performed best in terms of tracking accuracy and false tracking rate, and its real-time behavior was desirable. The suboptimal JPDA ranked high in tracking accuracy and low in error tracking rate, but poor in real-time performance. The cheap JPDA came out top in real-time performance, but at the expense of its highest false tracking rate.

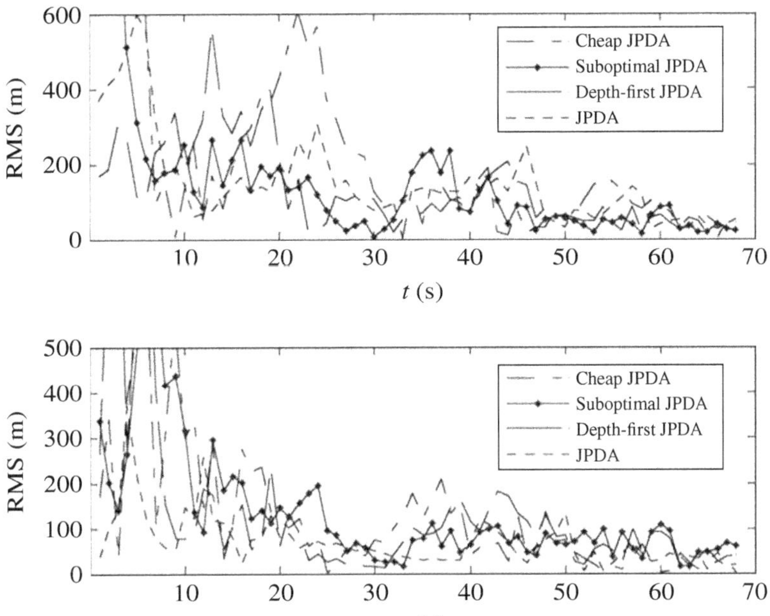

Figure 8.9 RMS position errors of target 2 on axis x (top) and y (bottom)

Table 8.1 Time consumed and false tracking rate

Algorithm	Time consumed per step (s)	False tracking rate (%)
JPDA	0.2206	3
Cheap JPDA	0.1406	10
Suboptimal JPDA	0.1540	6
Depth-first JPDA	0.1427	6

The various performances of the three reduced forms in this simulation can be attributed to the following factors.

- The cheap JPDA is direct and simple because it incorporates the essential properties of the JPDA in association probability computation, but the sum of these probabilities calculated is not 1 for any track, as a result of which empirical probabilities tend to overweight incorrect echoes.
- The suboptimal JPDA, by accounting for partial joint events, yields more reasonable association probabilities and higher tracking accuracy, which only compounds its complexity.
- The simplified form of depth-first JPDA evaluates association probabilities directly, and thus produces similar tracking results to those of the JPDA in medium-target-density cases, and with high speed, but its complex equations incur, in the presence of dense targets, extra computational load on the system.

8.6 Summary

This chapter focuses on Bayesian data association algorithms. Compared with batch processing approaches such as maximum-likelihood-based ones, they find wider applications in engineering and serve as a major source of in-depth research. Bayesian algorithms fall into two broad categories.

One is the optimal approach which processes only the latest set of validated measurements, mainly including the NNSF, PNNF, PDA, IPDA, and JPDA algorithms. These algorithms have proved convenient for engineering applications, with their reasonably low computing and memory requirements.

Second is the optimal Bayesian approach, which deals with all combinations of measurements from the initial right through to the current time and evaluates probabilities for each sequence of measurements. Major members in this category are the optimal Bayesian filter (OBF) and multiple hypothesis filter (MHT). These approaches demand more calculation and memory but can be somewhat simplified for practical uses. The optimal approaches are not covered in this chapter.

Based on the analysis of the suboptimal Bayesian algorithms under discussion, this chapter makes a comparative analysis of the performance of the JPDA and its three simplified varieties in a simulation experiment before concluding with a summary.

9

Tracking Maneuvering Targets

9.1 Introduction

In Chapters 3 and 4 we discussed some basic filtering methods of radar data processing, usually based on the assumption that the target of interest is in uniform or uniformly accelerated motion. If the radar moves faster than the target, say a ground or sea-surface target being tracked by an airborne radar, then the target can be viewed as being in approximately uniform or uniformly accelerated motion, or even static. However, the target tends to take turns, evasions, or other special attack postures due to its increasing maneuverability and the operator's manipulations at any time in the process of moving, so it is unlikely to always move with constant speed or uniform acceleration. To put it differently, maneuvers are likely to occur for a target in motion. Therefore, this chapter focuses on tracking maneuvering targets, or rather, the problem of uncertainty in establishing target model parameters in the filtering process. The uncertainty of target model parameters and measurement origins discussed in previous chapters constitute the two fundamental problems of target tracking.

The approaches to tracking maneuvering targets fall into two broad categories: tracking algorithms with maneuver detection and adaptive tracking algorithms without maneuver detection. The former category can be further divided into two types according to the parameters adjusted after the detection of maneuvers.

1. *Adjusting the gain of the filter*. The specific procedure is as follows: restart the gain sequence of the filter, enlarge the variance of the input noise, and augment the covariance matrix for the target state estimate. The adjustable white noise algorithm [16, 39, 205–208] in Section 9.2.1 falls into this type. This approach adjusts the gain of a filter by adjusting the variance of the noise input.
2. *Adjusting the structure of the filter*. The specific procedure is as follows: switch among different tracking filters and increase the number of dimensions of the target state. The variable-dimension filtering (VDF) approach [16, 37, 39, 209] in Section 9.2.2 is to increase the current dimensionality of the state of a target on confirming that the target is maneuvering, and to restore it to its original model on confirming that the maneuver is finished.

Radar Data Processing with Applications, First Edition. He You, Xiu Jianjuan, and Guan Xin.

The latter category need not detect maneuvers of a target, but corrects the gain of the filter while undertaking the estimation of the target.

Section 9.3.1 offers an adaptive tracking algorithm for maneuvering targets on the basis of the input estimation algorithm, a modified input estimation algorithm [210]. This algorithm combines Bayes' and Fisher's methods to realize the adaptive tracking for maneuvering targets.

The Singer model algorithm [16, 39, 211] in Section 9.3.2 holds that the noise process is colored, and that the target's acceleration should be modeled as the zero-mean random process with exponential autocorrelation.

The current model [40, 212, 213] of Section 9.3.3 estimates the mean value of the maneuvering acceleration while estimating the state of the target, corrects the distribution of acceleration through estimation in real time, and feeds back these values in the form of variance to the next-time gain of filters.

Section 9.3.4 presents the jerk model algorithm, which, similarly to the Singer algorithm, models the process noise as the color noise, and needs to find the derivative of the acceleration in real time, that is, to estimate the acceleration.

In the multiple model algorithm [16, 40, 214–219] of Section 9.3.5, a number of different noise levels can be assumed. The tracker may calculate the probability for each level, and then compute the weight sum of these probabilities. Or, the tracker may switch between levels according to a particular criterion (e.g., according to the probability of the noise level).

Section 9.3.6 discusses the interactive multiple model algorithm, a target tracking approach frequently used in recent years. Finally, a simulation analysis of the performance of the various algorithms in tracking is presented, pinpointing some common problems of these algorithms together with the solutions.

9.2 Tracking Algorithm with Maneuver Detection

The maneuver detection of the target is by nature a decision mechanism, in which detection is based on the measurements of the target and the theory of mathematical statistics. The basic idea in the tracking algorithm with maneuver detection is that maneuvers lead to the degradation of the original model, so that the target state estimate deviates from what it is in reality and the filtering residual properties change. Therefore, the estimation as to whether the maneuver has started or whether the maneuver is over can be made by observing the residual changes of the target's motion. Then, the tracking algorithm should be adjusted accordingly (i.e., noise variance adjustment or model transformation should be undertaken so that the target can be better tracked). As shown in

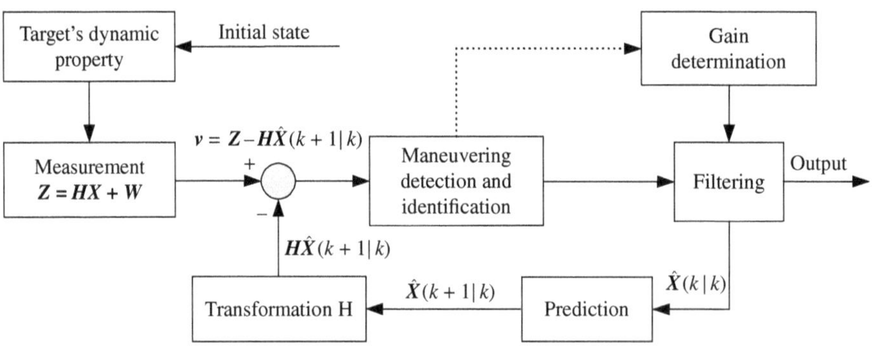

Figure 9.1 Schematic diagram of maneuvering target tracking

Figure 9.1 – a schematic diagram of the framework of this type of tracking algorithm for maneuvering targets – the measurement $\boldsymbol{Z}$ and the state prediction $\boldsymbol{H}\hat{\boldsymbol{X}}(k+1|k)$ first constitute the innovation vector $\boldsymbol{\nu}$. Then, the maneuvering detection is undertaken by observing the changes in $\boldsymbol{\nu}$ and, according to a particular criterion or logic, the filtering gain or structure of the filter is adjusted so that the maneuvering target can be tracked [220–222].

9.2.1 White Noise Model with Adjustable Level

The concept of the white noise model with adjustable level was propounded by Jazwinski in 1970 [223], and applied to the tracking of maneuvering targets in 1977 by Chang *et al.* [209]. This method detects the start and end of a maneuver by observing the changes in innovation of a target and adjusting the filter accordingly.

For convenience, the equation of motion for maneuvering targets will be repeated below:

$$\boldsymbol{X}(k+1)=\boldsymbol{F}(k)\boldsymbol{X}(k)+\boldsymbol{G}(k)\boldsymbol{u}(k)+\boldsymbol{V}(k) \tag{9.1}$$

where the process noise $\boldsymbol{V}(k)$ is a zero-mean, white, random sequence with covariance matrix $\boldsymbol{Q}(k)$, but the input $\boldsymbol{u}(k)$ is unknown. This input is the crux of the solution to the problem of tracking maneuvering targets.

Suppose that the dynamic equation for the target can be represented as in (9.1), then its corresponding measurement equation is

$$\boldsymbol{Z}(k+1)=\boldsymbol{H}(k+1)\boldsymbol{X}(k+1)+\boldsymbol{W}(k+1) \tag{9.2}$$

This method assumes that a maneuver manifests itself as a large innovation, and that the innovation will increase with the maneuver. A simple detection procedure for maneuvering targets is based on the normalized innovations squared,

$$\varepsilon_\nu(k)=\boldsymbol{\nu}'(k)S^{-1}(k)\boldsymbol{\nu}(k) \tag{9.3}$$

where the filtering residual (innovation) is

$$\boldsymbol{\nu}(k)=\boldsymbol{Z}(k)-\hat{\boldsymbol{Z}}(k|k-1) \tag{9.4}$$

with $\varepsilon_\nu(k)$ a χ^2-distributed random variable with n_z degrees of freedom, in which n_z represents the number of dimensions of the measurement. Supposing that $\varepsilon_{\max}$ is a threshold, and α the significance level, then a threshold value can be established such that, based on the non-maneuvering target model,

$$\Pr\{\varepsilon_\nu(k)\le\varepsilon_{\max}\}=1-\alpha \tag{9.5}$$

When the threshold is exceeded, the target is considered to begin to maneuver, and the covariance $\boldsymbol{Q}(k-1)$ of the process noise should be scaled up until $\varepsilon_\nu(k)$ is lessened to the threshold $\varepsilon_{\max}$. When $\varepsilon_\nu(k)>\varepsilon_{\max}$, the maneuver is considered over, and the original filtering model will be restored.

Alternatively, by multiplying the matrix of process noise $\boldsymbol{Q}(k-1)$ by the scale factor $\phi>1$, the innovation covariance becomes

$$\begin{aligned}\boldsymbol{S}(k)&=\boldsymbol{H}(k)\boldsymbol{P}(k|k-1)\boldsymbol{H}'(k)+\boldsymbol{R}(k)\\&=\boldsymbol{H}(k)[\boldsymbol{F}(k-1)\boldsymbol{P}(k-1|k-1)\boldsymbol{F}'(k-1)+\phi\boldsymbol{Q}(k-1)]\boldsymbol{H}'(k)+\boldsymbol{R}(k)\end{aligned}\tag{9.6}$$

In order to adjust the square of the normalized innovation, the noise can be adjusted by employing a sliding-window time average or a fading-memory likelihood function in place of the single-time test statistic.

9.2.2 *Variable-Dimension Filtering Approach*

The VDF approach was proposed by Bar-Shalom and Birmiwal in 1982 [37]. It does not rely on an a priori assumption about target maneuvers. In this approach, the maneuver of a target is regarded as a change inherent in its dynamics rather than modeled as noise. In the absence of maneuvers, the tracking filter operates in its original model. Once a maneuver is detected, the filter will use a different, higher-dimension state measurement; new state components are added.

Two models are employed here: a constant-velocity model in the absence of maneuvers and an approximately constant-acceleration model for maneuvering targets. In the former model, the state component for planar motion is

$$\boldsymbol{X}=[x\ \ \dot{x}\ \ y\ \ \dot{y}]'\tag{9.7}$$

In the latter model, the state component is

$$\boldsymbol{X}^{m}=[x\ \ \dot{x}\ \ y\ \ \dot{y}\ \ \ddot{x}\ \ \ddot{y}]'\tag{9.8}$$

In the case of the constant-velocity model, a maneuver occurs as follows. Denote by $\rho(k)$ the fading-memory average of the filtering innovation $\varepsilon_{\nu}(k)$,

$$\rho(k)=\mu\rho(k-1)+\varepsilon_{\nu}(k)\tag{9.9}$$

where μ is the discount factor $(0<\mu<1)$, $\mu=1-1/s$, s is the length of the sliding window (used to detect the existence of a maneuver), and $\varepsilon_{\nu}(k)$ is the normalized innovation squared in (9.3).

The hypothesis that a maneuver is happening is accepted when $\rho(k)$ exceeds the threshold set in (9.5), and at this moment the estimator switches from the non-maneuvering to the maneuvering model. Otherwise, the estimated accelerations are compared with their standard deviations. If they are not statistically significant, the maneuver hypothesis is rejected, at which point the estimator switches from the maneuvering to the non-maneuvering model.

The test statistic for the significance of the acceleration estimate is

$$\delta_a(k)=\hat{\boldsymbol{a}}'(k|k)\left[\boldsymbol{P}_a^{m}(k|k)\right]^{-1}\hat{\boldsymbol{a}}(k|k)\tag{9.10}$$

where $\hat{a}$ is the estimate of the acceleration component and $\boldsymbol{P}_a^{m}$ is the corresponding block from the covariance matrix of the maneuvering model. When the sum over a sliding window with length p,

$$\rho_a(k) = \sum_{j=k-p+1}^{k} \delta_a(j) \tag{9.11}$$

falls below the threshold, the acceleration is considered insignificant.

When the acceleration suddenly declines to 0 (i.e., when the maneuver ends abruptly), large innovations can be generated in the maneuvering model. This can be alleviated by allowing the estimator to switch to a lower-order level model when the innovations of the maneuvering model exceed 95% of the confidence region.

When a maneuver is detected at time k, the filter assumes that the target started to move with a constant acceleration at time $k-s-1$, where s is the length of the effective sliding window, and then corrects the state estimate appropriately for time $k-s$. First, the estimate at time $k-s$ for the acceleration is

$$\hat{X}^m_{4+i}(k-s|k-s) = \frac{2}{T^2}[z_i(k-s) - \hat{z}_i(k-s|k-s-1)], \quad i=1,2 \tag{9.12}$$

The position component of the estimate at time $k-s$ is taken as its corresponding measurement,

$$\hat{X}^m_{2i-1}(k-s|k-s) = z_i(k-s), \quad i=1,2 \tag{9.13}$$

while the velocity component is corrected with the acceleration estimate,

$$\hat{X}^m_{2i}(k-s|k-s) = \hat{X}_{2i}(k-s|k-s-1) + T\hat{X}^m_{4+i}(k-s|k-s), \quad i=1,2 \tag{9.14}$$

The covariance matrix corresponding to the corrected state estimate, whose derivation can be found in Ref. [38], is $\boldsymbol{P}^m(k-s|k-s)$ and is expressed as

$$\begin{cases} P^m_{11}(k-s|k-s) = R_{11},\ P^m_{12}(k-s|k-s) = 2R_{11}/T, P^m_{15}(k-s|k-s) = 2R_{11}/T^2 \\ P^m_{22}(k-s|k-s) = (4/T^2)*(R_{11}+P_{11}) + P_{22} + 4P_{12}/T \\ P^m_{25}(k-s|k-s) = (4/T^3)*(R_{11}+P_{11}) + (2/T)P_{22} + (6/T^2)P_{12} \\ P^m_{33}(k-s|k-s) = R_{22}, P^m_{34}(k-s|k-s) = 2R_{22}/T, P^m_{36}(k-s|k-s) = 2R_{22}/T^2 \\ P^m_{44}(k-s|k-s) = (4/T^2)*(R_{22}+P_{33}) + P_{44} + 4P_{34}/T \\ P^m_{46}(k-s|k-s) = (4/T^3)*(R_{22}+P_{33}) + (2/T)P_{44} + (6/T^2)P_{34} \\ P^m_{55}(k-s|k-s) = (4/T^4)*(R_{11}+P_{11}+2TP_{12}+T^2P_{22}) \\ P^m_{66}(k-s|k-s) = (4/T^4)*(R_{22}+P_{33}+2TP_{34}+T^2P_{44}) \\ P^m_{13} = P^m_{14} = P^m_{16} = P^m_{23} = P^m_{24} = P^m_{26} = P^m_{35} = P^m_{45} = P^m_{56} = 0 \\ P^m_{ij} = P^m_{ji}, \quad i,j=1,2,\ldots,6 \end{cases} \tag{9.15}$$

When a maneuver is detected, the state model of the target can be changed by bringing in an additional state component (i.e., the acceleration of the target). It can be seen from the above that the VDF detects maneuvers by a χ^2 test based on the innovation quantity of a fading memory. In the absence of maneuvers, it operates in the constant-velocity (CV) model. If the maneuver of a target is detected at time k, it assumes that the maneuver took place at time $k-s-1$, and starts up the constant-acceleration (CA) model at time $k-s$. Then this algorithm modifies the previous state estimates by using the succeeding measurements, and expands the target state. When the target is detected as having switched from the maneuvering to the non-maneuvering state, the algorithm does not remodify the preceding state estimates gained from the CA model. The reason is that the tracking performance of the algorithm decreases slightly when tracking non-maneuvering targets based on the CA model.

Another important approach to maneuvering target tracking and detection is the input estimation algorithm proposed in 1979 by Chan *et al.* [16, 40, 224]. This method does not rely on prior knowledge about the maneuverability of the target of interest. Instead, it treats the acceleration of a maneuver as an unknown deterministic input. The estimate of this acceleration is obtained by means of the least-squares method from the innovation, and is used to update the state of the target.

9.3 Adaptive Tracking Algorithm

9.3.1 Modified-Input Estimation Algorithm

Let the state equation of the discrete-time system be

$$\boldsymbol{X}(k+1)=\boldsymbol{F}(k)\boldsymbol{X}(k)+\boldsymbol{G}(k)\boldsymbol{u}(k)+\boldsymbol{\Gamma}(k)\boldsymbol{v}(k) \tag{9.16}$$

where $\boldsymbol{F}(k)$ is the state transition matrix, $\boldsymbol{X}(k)$ is the state vector, $\boldsymbol{G}(k)$ is the input control matrix, $\boldsymbol{u}(k)$ is the unknown input of the target maneuvering model, $\boldsymbol{\Gamma}(k)$ is the matrix of the process noise distribution, and $\boldsymbol{v}(k)$ is the process noise sequence, which is zero-mean, white, Gaussian and satisfies

$$E[\boldsymbol{v}(k)\boldsymbol{v}'(j)]=\boldsymbol{Q}(k)\delta_{kj} \tag{9.17}$$

In other words, the process noises at different moments are mutually independent statistically. Here, δ_{kj} is the Kronecker delta function. When the state vector is $\boldsymbol{X}(k)=[x\ \dot{x}\ y\ \dot{y}]'$ or $\boldsymbol{X}(k)=[x\ \dot{x}\ y\ \dot{y}\ z\ \dot{z}]'$, the unknown input of the corresponding target maneuvering model is $\boldsymbol{u}(k)=[a_x\ a_y]'$ or $\boldsymbol{u}(k)=[a_x\ a_y\ a_z]'$, while $\boldsymbol{G}(k)$ is, respectively,

$$\boldsymbol{G}(k)=\begin{bmatrix} T^2/2 & T & 0 & 0 \\ 0 & 0 & T^2/2 & T \end{bmatrix}' \tag{9.18}$$

or

$$\boldsymbol{G}(k)=\begin{bmatrix} T^2/2 & T & 0 & 0 & 0 & 0 \\ 0 & 0 & T^2/2 & T & 0 & 0 \\ 0 & 0 & 0 & 0 & T^2/2 & T \end{bmatrix}' \tag{9.19}$$

By adding the unknown input vector $\boldsymbol{u}(k)$ to the state vector $\boldsymbol{X}(k)$, the state equation of the maneuvering target given by (9.16) can be rewritten as the following non-maneuvering state equation [210]:

$$\begin{bmatrix} \boldsymbol{X}(k+1) \\ \boldsymbol{u}(k+1) \end{bmatrix} = \begin{bmatrix} \boldsymbol{F} & \boldsymbol{G} \\ \boldsymbol{0}_{2\times 4} & \boldsymbol{I} \end{bmatrix} \begin{bmatrix} \boldsymbol{X}(k) \\ \boldsymbol{u}(k) \end{bmatrix} + \begin{bmatrix} \boldsymbol{\Gamma} \\ \boldsymbol{0}_{2\times 2} \end{bmatrix} \boldsymbol{v}(k) \tag{9.20}$$

where $\boldsymbol{I}$ represents the 2×2 unit matrix, $\boldsymbol{0}_{2\times 4}$ the 2×4 all-zero matrix, and $\boldsymbol{0}_{2\times 2}$ the 2×2 all-zero matrix; $\boldsymbol{F}$, $\boldsymbol{G}$, $\boldsymbol{\Gamma}$ are the abbreviations, respectively, of $\boldsymbol{F}(k)$, $\boldsymbol{G}(k)$, $\boldsymbol{\Gamma}(k)$.

Define

$$\boldsymbol{X}_{Aug}(k+1) = \begin{bmatrix} \boldsymbol{X}(k+1) \\ \boldsymbol{u}(k+1) \end{bmatrix} \tag{9.21}$$

$$\boldsymbol{F}_{Aug}(k) = \begin{bmatrix} \boldsymbol{F} & \boldsymbol{G} \\ \boldsymbol{0}_{2\times 4} & \boldsymbol{I} \end{bmatrix} \tag{9.22}$$

$$\boldsymbol{\Gamma}_{Aug}(k) = \begin{bmatrix} \boldsymbol{\Gamma} \\ \boldsymbol{0}_{2\times 2} \end{bmatrix} \tag{9.23}$$

then the augmented state equation is

$$\boldsymbol{X}_{Aug}(k+1) = \boldsymbol{F}_{Aug}(k)\boldsymbol{X}_{Aug}(k) + \boldsymbol{\Gamma}_{Aug}(k)\boldsymbol{v}(k) \tag{9.24}$$

the measurement equation of the discrete-time system is

$$\begin{aligned} \boldsymbol{Z}(k+1) &= \boldsymbol{H}(k+1)\boldsymbol{X}(k+1) + \boldsymbol{W}(k+1) \\ &= \boldsymbol{H}(k+1)[\boldsymbol{F}(k)\boldsymbol{X}(k) + \boldsymbol{G}(k)\boldsymbol{u}(k) + \boldsymbol{\Gamma}(k)\boldsymbol{v}(k)] + \boldsymbol{W}(k+1) \end{aligned} \tag{9.25}$$

where $\boldsymbol{W}(k+1)$ is the zero-mean, white, Gaussian measurement noise sequence which satisfies $\mathrm{E}[\boldsymbol{W}(k)\boldsymbol{W}'(j)] = \boldsymbol{R}(k)\delta_{kj}$ and is uncorrelated with the process noise sequence.

It follows from (9.25) that

$$\boldsymbol{Z}(k+1) = [\boldsymbol{HF} \quad \boldsymbol{HG}] \begin{bmatrix} \boldsymbol{X}(k) \\ \boldsymbol{u}(k) \end{bmatrix} + \boldsymbol{H\Gamma v}(k) + \boldsymbol{W}(k+1) \tag{9.26}$$

For the sake of simplicity, $\boldsymbol{H}(k+1)$ is abbreviated as $\boldsymbol{H}$.

Define

$$\boldsymbol{Z}_{Aug}(k) = \boldsymbol{Z}(k+1) \tag{9.27}$$

$$\boldsymbol{H}_{Aug}(k) = [\boldsymbol{HF} \quad \boldsymbol{HG}] \tag{9.28}$$

$$\boldsymbol{W}_{Aug}(k) = \boldsymbol{H\Gamma v}(k) + \boldsymbol{W}(k+1) \tag{9.29}$$

Then (9.26) yields the augmented measurement equation

$$\boldsymbol{Z}_{Aug}(k) = \boldsymbol{H}_{Aug}(k)\boldsymbol{X}_{Aug}(k) + \boldsymbol{W}_{Aug}(k) \tag{9.30}$$

and combining (9.24) and (9.30) gives the modified-input estimation filtering algorithm.

The one-step prediction of the state is

$$\hat{\boldsymbol{X}}_{Aug}(k+1|k)=\boldsymbol{F}_{Aug}(k)\hat{\boldsymbol{X}}_{Aug}(k|k) \tag{9.31}$$

The one-step prediction of the covariance is

$$\boldsymbol{P}_{Aug}(k+1|k)=\boldsymbol{F}_{Aug}(k)\boldsymbol{P}_{Aug}(k|k)\boldsymbol{F}'_{Aug}(k)+\boldsymbol{\Gamma}_{Aug}(k)\boldsymbol{Q}(k)\boldsymbol{\Gamma}'_{Aug}(k) \tag{9.32}$$

The innovation covariance is

$$\boldsymbol{S}_{Aug}(k+1)=\boldsymbol{H}_{Aug}(k+1)\boldsymbol{P}_{Aug}(k+1|k)\boldsymbol{H}'_{Aug}(k+1)+\boldsymbol{R}_{Aug}(k+1) \tag{9.33}$$

where

$$\begin{aligned}\boldsymbol{R}_{Aug}(k+1)&=\mathrm{E}\left[\boldsymbol{W}_{Aug}(k+1)\boldsymbol{W}'_{Aug}(k+1)\right]\\&=\boldsymbol{H}(k+1)\boldsymbol{\Gamma}(k)\boldsymbol{Q}(k)\boldsymbol{H}'(k+1)\boldsymbol{\Gamma}'(k)+\boldsymbol{R}(k+1)\end{aligned} \tag{9.34}$$

The gain is

$$\boldsymbol{K}_{Aug}(k+1)=\boldsymbol{P}_{Aug}(k+1|k)\boldsymbol{H}'_{Aug}(k+1)\boldsymbol{S}^{-1}_{Aug}(k+1) \tag{9.35}$$

The measurement prediction is

$$\hat{\boldsymbol{Z}}_{Aug}(k+1|k)=\boldsymbol{H}_{Aug}(k+1)\hat{\boldsymbol{X}}_{Aug}(k+1|k) \tag{9.36}$$

The update equation of the state is

$$\hat{\boldsymbol{X}}_{Aug}(k+1|k+1)=\hat{\boldsymbol{X}}_{Aug}(k+1|k)+\boldsymbol{K}_{Aug}(k+1)\left[\boldsymbol{Z}_{Aug}(k+1)-\hat{\boldsymbol{Z}}_{Aug}(k+1|k)\right] \tag{9.37}$$

The update equation of the covariance is

$$\begin{aligned}\boldsymbol{P}_{Aug}(k+1|k+1)&=\boldsymbol{P}_{Aug}(k+1|k)-\boldsymbol{P}_{Aug}(k+1|k)\\&=\boldsymbol{H}(k+1)\boldsymbol{\Gamma}(k)\boldsymbol{Q}(k)\boldsymbol{H}'(k+1)\boldsymbol{\Gamma}'(k)+\boldsymbol{R}(k+1)\end{aligned} \tag{9.38}$$

The modified-input estimation algorithm combines the Bayesian method with Fisher's method, and regards the acceleration of a maneuvering target as an unknown input vector which is then added to the state equation, and tracks the target when the state vector is augmented. During the process of tracking, it estimates the acceleration of the target while estimating the original target state vector. This algorithm requires no maneuvering detection of the target and can be adapted to two working modes: maneuvering and quiescent, thereby realizing the adaptive tracking of the maneuvering target.

9.3.2 Singer Model Tracking Algorithm

The algorithm discussed in Section 9.2 models the maneuvering control as white noise. The white noise model is a relatively idealistic model. In fact, a more practical maneuvering model is one which models the maneuvering control as correlated (colored) noise. The Singer model approach regards the maneuvering model as a model of correlated noise, instead of a commonly hypothesized white noise model [16]. The acceleration $a(t)$ of the target is modeled as the zero-mean random process with exponential autocorrelation, that is,

$$R(\tau)=E[a(t)a(t+\tau)]=\sigma_m^2 e^{-\alpha|\tau|} \tag{9.39}$$

where σ_m^2 and α are indeterminate parameters determining the maneuverability of the target within $[t, t+\tau]$; σ_m^2 is the acceleration variance of the target; and α is the reciprocal of the constant of the maneuvering time (i.e., the maneuvering frequency).

The maneuvering acceleration variance σ_m^2 is calculated through the PDF of the maneuvering target. Usually, the distribution of the maneuvering acceleration is hypothesized as follows: (1) the probability of the maneuvering acceleration being a_{M} is p_{M}, while that for the maneuvering acceleration being $-a_{\mathrm{M}}$ is also p_{M}; (2) the probability of the acceleration being 0 is p_0 (the non-maneuvering probability); (3) the maneuvering acceleration is approximately subject to the uniform distribution within the interval $[-a_{\mathrm{M}}, a_{\mathrm{M}}]$. Based on the above hypotheses, the following PDF can be obtained:

$$p(a)=[\delta(a-a_{\mathrm{M}})+\delta(a+a_{\mathrm{M}})]p_{\mathrm{M}}+\delta(a)p_0+[1(a+a_{\mathrm{M}})-1(a-a_{\mathrm{M}})]\frac{1-p_0-2p_{\mathrm{M}}}{2a_{\mathrm{M}}} \tag{9.40}$$

where $1(\cdot)$ is the unit step function and $\delta(\cdot)$ is the Dirac pulse function.

The variance corresponding to (9.17) can be derived from the PDF mentioned above:

$$\sigma_m^2=\frac{a_{\mathrm{M}}^2}{3}(1+4p_{\mathrm{M}}-p_0) \tag{9.41}$$

The acceleration for maneuvering in the Singer model ($\alpha(t)$) is hypothesized as a relevant random process (colored noise), but is required to be an irrelevant white-noise process when applying Kalman filtering. Therefore, $\alpha(t)$ needs to be "whitened" before applying Kalman filtering. The transfer function of the whitening filter is $H(s)=s+\alpha$, and the Laplace transformation of the function correlated with the output signals from the whitening filter is

$$\Phi_w(s)=\Phi(s)H(s)H(-s)=2\alpha\sigma_m^2 \tag{9.42}$$

where

$$\Phi(s)=\sigma_m^2\frac{-2\alpha}{(s+\alpha)(s-\alpha)} \tag{9.43}$$

is the Laplace transformation corresponding to the time-correlation function $R(\tau)=\sigma_m^2 e^{-\alpha|\tau|}$.

Through a whitening filter, the maneuvering acceleration $\alpha(t)$ is output as a white-noise signal, the correlation function of which is $2\alpha\sigma_m^2\delta(\tau)$, that is,

$$\begin{aligned}H(s)=s+\alpha&=\frac{\widetilde{V}(s)}{A(s)}\\&\Rightarrow sA(s)+\alpha A(s)=\widetilde{V}(s)\\&\Rightarrow sA(s)=-\alpha A(s)+\widetilde{V}(s)\\&\Rightarrow \dot{a}(t)=-\alpha a(t)+\widetilde{v}(t)\end{aligned} \tag{9.44}$$

Equation (9.44) is the relational expression in which the maneuvering acceleration $\alpha(t)$ is expressed in the form of white noise. It is a first-order time-correlated model with white noise as its input (this dynamic model is the first-order Markov process). In the equation, $\tilde{v}(t)$ is Gaussian white noise with zero mean and variance $2\alpha\sigma_m^2$, that is,

$$E[\tilde{v}(t)\tilde{v}(\tau)] = 2\alpha\sigma_m^2\delta(t-\tau) \tag{9.45}$$

Let the state vector about the coordinate x be $\boldsymbol{X} = [x \;\; \dot{x} \;\; \ddot{x}]'$, where $\ddot{x} = a$. By using the state equation, the first-order time-correlated model above can be expressed as

$$\dot{\boldsymbol{X}}(t) = \boldsymbol{A}\boldsymbol{X}(t) + \widetilde{\boldsymbol{V}}(t) \tag{9.46}$$

This is the well-known Singer model, in which the system matrix is

$$\boldsymbol{A} = \begin{bmatrix} 0 & 1 & 0 \\ 0 & 0 & 1 \\ 0 & 0 & -\alpha \end{bmatrix} \tag{9.47}$$

and

$$\widetilde{\boldsymbol{V}} = [0 \;\; 0 \;\; \tilde{v}]' \tag{9.48}$$

is the process noise. Substituting **A** and $\widetilde{\boldsymbol{V}}$ into (9.46), we have

$$\dot{\boldsymbol{X}}(t) = \begin{bmatrix} \dot{x} \\ \ddot{x} \\ \dot{a} \end{bmatrix} = \boldsymbol{A}\boldsymbol{X}(t) + \widetilde{\boldsymbol{V}}(t) = \begin{bmatrix} 0 & 1 & 0 \\ 0 & 0 & 1 \\ 0 & 0 & -\alpha \end{bmatrix} \begin{bmatrix} x \\ \dot{x} \\ \ddot{x} \end{bmatrix} + \begin{bmatrix} 0 \\ 0 \\ \tilde{v} \end{bmatrix} \tag{9.49}$$

For the sample interval T, the discrete-time dynamic equation corresponding to (9.46) is

$$\boldsymbol{X}(k+1) = \boldsymbol{F}(k)\boldsymbol{X}(k) + \boldsymbol{V}(k) \tag{9.50}$$

where

$$\boldsymbol{F} = e^{AT} = \begin{bmatrix} 1 & T & (\alpha T - 1 + e^{-\alpha T})/\alpha^2 \\ 0 & 1 & (1 - e^{-\alpha T})/\alpha \\ 0 & 0 & e^{-\alpha T} \end{bmatrix} \tag{9.51}$$

whose discrete-time process noise $\boldsymbol{V}$ has covariance

$$\boldsymbol{Q} = 2\alpha\sigma_m^2 \begin{bmatrix} q_{11} & q_{12} & q_{13} \\ q_{21} & q_{22} & q_{23} \\ q_{31} & q_{32} & q_{33} \end{bmatrix} \tag{9.52}$$

which assumes $\alpha T << 1$, that is, the sampling interval T is assumed much less than the time constant of maneuvering autocorrelation $1/\alpha$. When the target tracking is done by a radar, this assumption ($\alpha T << 1$) is considered correct if the updating rate is high enough. Conversely, when the tracking is done by a long-distance sonar (active or passive), the opposite is correct (i.e., $\alpha T >> 1$). The measurement equation is similar to the Kalman filtering equation and the white-noise model, except for the measurement matrix $\boldsymbol{H}$. Here, $\boldsymbol{H} = [1 \quad 0 \quad 0]$ and the accurate expression of $\boldsymbol{Q}$ (a symmetrical matrix) is

$$\begin{cases} q_{11} = \dfrac{1}{2\alpha^5}\left[1 - e^{-2\alpha T} + 2\alpha T + \dfrac{2\alpha^3 T^3}{3} - 2\alpha^2 T^2 - 4\alpha T e^{-\alpha T}\right] \\ q_{12} = \dfrac{1}{2\alpha^4}\left[e^{-2\alpha T} + 1 - 2e^{-\alpha T} + 2\alpha T e^{-\alpha T} - 2\alpha T + \alpha^2 T^2\right] \\ q_{13} = \dfrac{1}{2\alpha^3}\left[1 - e^{-2\alpha T} - 2\alpha T e^{-\alpha T}\right] \\ q_{22} = \dfrac{1}{2\alpha^3}\left[4e^{-\alpha T} - 3 - e^{-2\alpha T} + 2\alpha T\right] \\ q_{23} = \dfrac{1}{2\alpha^2}\left[e^{-2\alpha T} + 1 - 2e^{-\alpha T}\right] \\ q_{33} = \dfrac{1}{2\alpha}\left[1 - e^{-2\alpha T}\right] \end{cases} \tag{9.53}$$

Note that in matrix $\boldsymbol{A}$ of (9.46), when $\alpha = 0$,

$$\boldsymbol{A} = \begin{bmatrix} 0 & 1 & 0 \\ 0 & 0 & 1 \\ 0 & 0 & 0 \end{bmatrix} \tag{9.54}$$

the state transition matrix is

$$\boldsymbol{F} = \begin{bmatrix} 1 & T & \frac{1}{2}T^2 \\ 0 & 1 & T \\ 0 & 0 & 1 \end{bmatrix} \tag{9.55}$$

and the covariance matrix of the process noise is

$$\boldsymbol{Q} = q \begin{bmatrix} T^5/20 & T^4/8 & T^3/6 \\ T^4/8 & T^3/3 & T^2/2 \\ T^3/6 & T^2/2 & T \end{bmatrix} \tag{9.56}$$

which is a model of uniformly accelerated rectilinear motion. In the equation, a smaller value can be taken for q (e.g., $q = 0.05$).

If $\boldsymbol{A}$ is assigned as

$$\boldsymbol{A} = \begin{bmatrix} 0 & 1 \\ 0 & 0 \end{bmatrix} \tag{9.57}$$

then the state transition matrix is

$$F = e^{AT} = \begin{bmatrix} 1 & T \\ 0 & 1 \end{bmatrix} \tag{9.58}$$

and the covariance matrix of the process noise is

$$Q = q \begin{bmatrix} T^3/3 & T^2/2 \\ T^2/2 & T \end{bmatrix} \tag{9.59}$$

which is a model of uniformly accelerated rectilinear motion. Hence, it can be found that the model of uniformly accelerated rectilinear motion and that of uniform rectilinear motion are two exceptions of the Singer model.

Actually, the Singer model algorithm laid a theoretical foundation for the subsequent maneuvering target model algorithms. It is clear that the Singer model algorithm is by nature an a priori model algorithm. But it is unrealistic to use an a priori maneuvering model to describe target maneuvers effectively. In addition, its hypothesis that target maneuvering acceleration in the interval $[-a_M, a_M]$ is approximately subject to a uniform distribution is inappropriate, for it keeps the mean value of the acceleration always at zero.

9.3.3 *Current Statistical Model Algorithm*

The current statistical model algorithm was proposed by Zhou *et al.* [39]. This algorithm is by nature a Singer model with adaptive, non-zero-mean acceleration. Unlike the Singer model algorithm, which has the hypothesis of an approximate uniform distribution, this algorithm describes the statistical features of maneuvering acceleration by use of the modified Rayleigh distribution. The merit of its hypothesized distribution is that the distribution changes with the mean, while the variance depends on the mean. Therefore, the algorithm, while estimating the target state, can identify the mean of the maneuvering acceleration, so as to modify the distribution of acceleration in real time, and feed back through the variance to the next-moment filtering gain. Thus, closed-loop adaptive tracking is realized.

Suppose that the state equation for the target motion is

$$\boldsymbol{X}(k+1) = \boldsymbol{F}(k)\boldsymbol{X}(k) + \boldsymbol{G}(k)\bar{a} + \boldsymbol{V}(k) \tag{9.60}$$

where $\boldsymbol{F}(k)$ is the same as described in (9.51) and $\boldsymbol{G}(k)$ is the input control matrix,

$$\boldsymbol{G}(k) = \begin{bmatrix} \frac{1}{\alpha}\left(-T + \frac{\alpha T^2}{2} + \frac{1-e^{-\alpha T}}{\alpha}\right) \\ T - \frac{1-e^{-\alpha T}}{\alpha} \\ 1-e^{-\alpha T} \end{bmatrix} \tag{9.61}$$

$V(k)$ is the discrete-time white noise sequence, and

$$Q(k) = \mathrm{E}[V(k)V'(k)] = 2\alpha\sigma_a^2 \begin{bmatrix} q_{11} & q_{12} & q_{13} \\ q_{12} & q_{22} & q_{23} \\ q_{13} & q_{23} & q_{33} \end{bmatrix} \tag{9.62}$$

where the expression of $Q(k)$ can be found in the Singer model (see Section 9.3.2), α is an autocorrelation time constant, σ_a^2 is the variance of maneuvering acceleration, and $\bar{a}(k)$ is the mean of maneuvering acceleration, that is,

$$\sigma_a^2 = \frac{4-\pi}{\pi}[a_{\max} - \bar{a}(k)]^2 \tag{9.63}$$

$$\bar{a}(k) = \hat{\ddot{x}}(k|k-1) \tag{9.64}$$

The one-step prediction equation of this algorithm is

$$\hat{X}(k|k-1) = F(k)\hat{X}(k-1|k-1) + G(k)\bar{a}(k) \tag{9.65}$$

where the measurement matrix is

$$H = [1 \quad 0 \quad 0] \tag{9.66}$$

The current statistical model algorithm is capable of adjusting the process noise adaptively according to the estimate of the acceleration for the last moment. It is concerned more with the current statistical characteristics of the maneuvering target. When the target is maneuvering at an acceleration, its acceleration at the next moment varies within a limited range, only in some neighborhood of the current acceleration. Therefore, compared with the Singer model algorithm, this algorithm can better reflect the changes in the maneuvering range and intensity of the target.

In view of the fact that it has difficulty selecting the autocorrelation time constant, the current model algorithm is modified with a multiple-model approach (see Ref. [220]), resulting in the modified current model algorithm.

The modified current model takes a comprehensive view of the impact of the autocorrelation time constants on the tracking effect in different maneuvering situations, sets models respectively for several typical autocorrelation time constants α which are likely to appear in real processes, and establishes a filter for each model. Then it gets, according to their likelihood functions, the probability that each model is correct. These probabilities are used to weight the updated state and covariance values derived from each model, the results of which are sent to each filter for circulation. Finally, the combination estimate $\hat{X}(k|k)$ and covariance $P(k|k)$ for r models are obtained, the expressions of which are, respectively,

$$\hat{X}(k|k) = \sum_{j=1}^{r} \mu_j(k)\hat{X}_j(k|k) \tag{9.67}$$

$$P(k|k) = \sum_{j=1}^{r} \mu_j(k)P_j(k|k) + \sum_{j=1}^{r} \mu_j(k)\left[\hat{X}_j(k|k) - \hat{X}(k|k)\right]\left[\hat{X}_j(k|k) - \hat{X}(k|k)\right]' \tag{9.68}$$

where $\mu_j(k)$ indicates the posterior probability that model j is correct at time k,

$$\mu_j(k) = \frac{\lambda_j(k)\mu_j(0)}{\sum_{l=1}^{r} \lambda_l(k)\mu_l(0)} \tag{9.69}$$

where $\mu_j(0)$ is the a priori probability that model j is correct at time k and $\lambda_j(k)$ is the likelihood function of the measurement.

9.3.4 Jerk Model Tracking Algorithm

The jerk model algorithm was proposed by Mehrotra and Mahapatra in 1997 [225], and holds that the main cause of the poor performance of maneuvering model algorithms in tracking complex maneuvering targets is that the state vector has insufficient derivative orders. Therefore, this algorithm adds one more dimension to the model of acceleration, in other words, it finds derivatives from the acceleration in real time (estimates the jerk). Hence, we can have a more accurate estimation of the acceleration so as to realize the tracking of maneuvering targets.

Similar to the Singer model algorithm, the exponential autocorrelation function of the target's jerk $j(t)$ is

$$R_j(\tau) = \mathrm{E}[j(t)j(t+\tau)] = \sigma_j^2 \mathrm{e}^{-\alpha|\tau|} \tag{9.70}$$

where σ_j^2 is the jerk variance of the target and α is an autocorrelation time constant. Conduct the Laplace transformation of $R_j(\tau)$,

$$R(s) = \mathrm{Laplace}\{R_j(\tau)\} = \frac{-2\alpha\sigma_j^2}{(s-\alpha)(s+\alpha)} = H(s)H(-s)V(s) \tag{9.71}$$

where

$$H(s) = 1/(s+\alpha) \tag{9.72}$$

$$V(s) = 2\alpha\sigma_j^2 \tag{9.73}$$

The differential equation of (9.72) is

$$\dot{j}(t) = -\alpha j(t) + v(t) \tag{9.74}$$

The inverse Laplace transform of the autocorrelation function of white noise $v(t)$ (i.e., that of (9.73)) is

$$r_v(\tau) = 2\alpha\sigma_j^2\delta(\tau) \tag{9.75}$$

Let the state vector about the coordinate x be

$$\boldsymbol{X} = [x \ \dot{x} \ \ddot{x} \ \dddot{x}]' \tag{9.76}$$

By use of the state equation, the first-order time-correlated model can be expressed as

$$\dot{\boldsymbol{X}}(t) = AX(t) + \boldsymbol{B}v(t) \tag{9.77}$$

where

$$\boldsymbol{A} = \begin{bmatrix} 0 & 1 & 0 & 0 \\ 0 & 0 & 1 & 0 \\ 0 & 0 & 0 & 1 \\ 0 & 0 & 0 & -\alpha \end{bmatrix} \tag{9.78}$$

is the system matrix.

$$\boldsymbol{B} = [0 \ \ 0 \ \ 0 \ \ 1]' \tag{9.79}$$

is the noise distribution matrix.

For sample interval T, the discrete-time dynamic equation corresponding to (9.77) is

$$\boldsymbol{X}(k+1) = \boldsymbol{F}(k)\boldsymbol{X}(k) + \boldsymbol{V}(k) \tag{9.80}$$

where

$$\boldsymbol{F}(k) = \begin{bmatrix} 1 & T & T^2/2 & p_1 \\ 0 & 1 & T & q_1 \\ 0 & 0 & 1 & r_1 \\ 0 & 0 & 0 & s_1 \end{bmatrix} \tag{9.81}$$

and

$$\begin{cases} p_1 = (2 - 2\alpha T + \alpha^2 T^2 - 2\mathrm{e}^{-\alpha T})/(2\alpha^3) \\ q_1 = (\mathrm{e}^{-\alpha T} - 1 + \alpha T)/\alpha^2 \\ r_1 = (1 - \mathrm{e}^{-\alpha T})/\alpha \\ s_1 = \mathrm{e}^{-\alpha T} \end{cases} \tag{9.82}$$

The covariance of the process noise is

$$\boldsymbol{Q}(k) = 2\alpha\sigma_j^2 \begin{bmatrix} q_{11} & q_{12} & q_{13} & q_{14} \\ q_{21} & q_{22} & q_{23} & q_{24} \\ q_{31} & q_{32} & q_{33} & q_{34} \\ q_{41} & q_{42} & q_{43} & q_{44} \end{bmatrix} \tag{9.83}$$

where the exact expression of the symmetrical matrix $\boldsymbol{Q}(k)$ is

$$
\begin{aligned}
q_{11} &= \frac{1}{2\alpha^7}\left(\frac{\alpha^5T^5}{10}-\frac{\alpha^4T^4}{2}+\frac{4\alpha^3T^3}{3}-2\alpha^2T^2+2\alpha T-3+4\mathrm{e}^{-\alpha T}+2\alpha^2T^2\mathrm{e}^{-\alpha T}-\mathrm{e}^{-2\alpha T}\right)\\
q_{12} &= \frac{1}{2\alpha^6}\left(1-2\alpha T+2\alpha^2T^2-\alpha^3T^3+\frac{\alpha^4T^4}{4}+\mathrm{e}^{-2\alpha T}+2\alpha T-2\mathrm{e}^{-\alpha T}-\alpha^2T^2\mathrm{e}^{-\alpha T}\right)\\
q_{13} &= \frac{1}{2\alpha^5}\left(2\alpha T-\alpha^2T^2-\frac{\alpha^3T^3}{3}-3-2\mathrm{e}^{-2\alpha T}+4\mathrm{e}^{-\alpha T}+\alpha^2T^2\mathrm{e}^{-\alpha T}\right)\\
q_{14} &= \frac{1}{2\alpha^4}\left(1+\mathrm{e}^{-2\alpha T}-2\mathrm{e}^{-\alpha T}-\alpha^2T^2\mathrm{e}^{-\alpha T}\right)\\
q_{22} &= \frac{1}{2\alpha^5}\left(1-\mathrm{e}^{-2\alpha T}+\frac{2\alpha^3T^3}{3}+2\alpha T-2\alpha^2T^2-4\alpha T\mathrm{e}^{-\alpha T}\right)\\
q_{23} &= \frac{1}{2\alpha^4}\left(1+\alpha^2T^2-2\alpha T+2\alpha T\mathrm{e}^{-\alpha T}+\mathrm{e}^{-2\alpha T}-2\mathrm{e}^{-\alpha T}\right)\\
q_{24} &= \frac{1}{2\alpha^3}\left(1-\mathrm{e}^{-2\alpha T}-2\alpha T\mathrm{e}^{-2\alpha T}\right) \quad q_{33} = \frac{1}{2\alpha^3}\left(4\mathrm{e}^{-\alpha T}-\mathrm{e}^{-2\alpha T}+2\alpha T-3\right)\\
q_{34} &= \frac{1}{2\alpha^2}\left(1-2\mathrm{e}^{-\alpha T}+\mathrm{e}^{-2\alpha T}\right) \quad q_{44} = \frac{1}{2\alpha}\left(1-\mathrm{e}^{-2\alpha T}\right)
\end{aligned}
\tag{9.84}
$$

9.3.5 *Multiple Model Algorithm*

When maneuvering target tracking is done by use of the adaptive filtering algorithm based on a single model, the tracking results tend to be poor because the model usually needs prior setting, and thus cannot match the target's maneuvering motion very well. Nowadays, as a result of targets' increasing maneuverability and great variations in the structures and parameters of targets' motion models, the single-model algorithm has difficulty in accurately identifying the maneuver parameters in time, which leads to inaccuracy of the model and degradation of the algorithm's performance.

In response to these limits of the single model, the multiple-model adaptive tracking algorithm is brought into being on the basis of the multiple-model adaptive control approach to automatic control. Hence, maneuvering target tracking is regarded as a question of mixed estimation. In the adjustable white-noise model of Section 9.2.1, there is only one level for the noise, so the noise covariance increases when a maneuver happens and the model reverts to its original form when the maneuver ends. In contrast, the multiple model algorithm is another modeling approach with unknown input as the white noise. It assumes two or more process noise levels and establishes a filter for each model. The filter switches between them according to a particular rule, or calculates the probability of being correct for each model based on its likelihood function, and then obtains the weighted sum of these probabilistic data.

Denote by M_j the event that model j with prior probability $\Pr\{M_j\}=\mu_j(0)\ (j=1,2,\ldots,r)$ is correct. The likelihood function of the measurements up to time k under the assumption of model j is

$$\lambda_j(k)=\Pr\left[\boldsymbol{Z}^k|\boldsymbol{M}_j\right]=\prod_{i=1}^{k}p\left[\boldsymbol{v}_j(i)\right] \tag{9.85}$$

in which the PDF of the innovation from filter j under the Gaussian hypothesis is

$$p[\boldsymbol{v}_j(k)] = |2\pi S_j(k)|^{-1/2} \exp\left[-{}^1/_2 \boldsymbol{v}_j'(k)\boldsymbol{S}_j^{-1}(i)\boldsymbol{v}_j(k)\right] \tag{9.86}$$

The use of Bayes' rule yields the posterior probability for model j being correct at time k as

$$\mu_j(k) \triangleq \Pr(\boldsymbol{M}_j|\boldsymbol{Z}^k) = \frac{\Pr(\boldsymbol{Z}^k|\boldsymbol{M}_j)\Pr(\boldsymbol{M}_j)}{\Pr(\boldsymbol{Z}^k)} = \frac{\Pr(\boldsymbol{Z}^k|\boldsymbol{M}_j)\Pr(\boldsymbol{M}_j)}{\sum_{l=1}^{r}\Pr(\boldsymbol{Z}^k|\boldsymbol{M}_l)\Pr(\boldsymbol{M}_l)} = \frac{\lambda_j(k)\mu_j(0)}{\sum_{l=1}^{r}\lambda_l(k)\mu_l(0)} \tag{9.87}$$

The state estimate of the target is a weighted average of the model-conditioned estimates with the above probabilistic data as weights,

$$\mathrm{E}\{\boldsymbol{X}(k)|\boldsymbol{Z}^k\} = \sum_{j=1}^{r}\mathrm{E}\{\boldsymbol{X}(k)|\boldsymbol{M}_j,\boldsymbol{Z}^k\}\Pr\{\boldsymbol{M}_j|\boldsymbol{Z}^k\} \tag{9.88}$$

At last we have the combined estimate $\hat{\boldsymbol{X}}(k|k)$ and the covariance $\boldsymbol{P}(k|k)$,

$$\hat{\boldsymbol{X}}(k|k) = \sum_{j=1}^{r}\mu_j(k)\hat{\boldsymbol{X}}_j(k|k) \tag{9.89}$$

$$\boldsymbol{P}(k|k) = \sum_{j=1}^{r}\mu_j(k)\boldsymbol{P}_j(k|k) + \sum_{j=1}^{r}\mu_j(k)\left[\hat{\boldsymbol{X}}_j(k|k)-\hat{\boldsymbol{X}}(k|k)\right]\left[\hat{\boldsymbol{X}}_j(k|k)-\hat{\boldsymbol{X}}(k|k)\right]' \tag{9.90}$$

The block diagram of this algorithm is shown in Figure 9.2. In this algorithm, the filters work in parallel, and the combined estimate is an MMSE estimate calculated probabilistically for all the models. But in practice, it might be necessary to use the probability estimated from

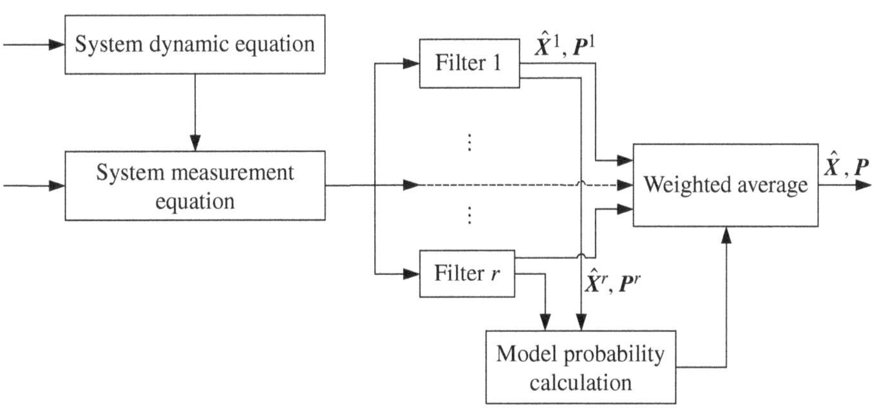

Figure 9.2 Structure chart of multiple model algorithm

the model with the highest $\mu_j(k)$ to eliminate the models with low probabilities, or to utilize some other scheme.

9.3.6 Interacting Multiple Model Algorithm

From 1984 to 1989, Blom and Bar-Shalom, on the basis of the generalized pseudo-Bayesian algorithm, proposed a structural adaptive algorithm with Markov transition probability, the interacting multiple model (IMM) algorithm [16, 39, 226–231]. This algorithm, on the basis of the multiple model algorithm, assumes that the transition between different models is subject to the finite Markov chain of the given transition probability, and obtains the state estimate of the target by accounting for the interaction between several models.

The IMM consists of several filters (with their corresponding models), a model probability estimator, an interacting actuator (at the input end of the filter), and an estimate mixer (at the output end of the filter). The models interact with one another to track the maneuver of a target. The IMM algorithm of N models is illustrated in Figure 9.3.

In Figure 9.3, $\hat{\boldsymbol{X}}(k|k)$ is the state estimate based on N models, and $\hat{\boldsymbol{X}}^j(k|k)$ $(j=1,2,\ldots,N)$ is the state estimate of model j. $\boldsymbol{\Lambda}(k)$ is a model possibility vector, and $\boldsymbol{u}(k)$ is a model probability vector. $\hat{\boldsymbol{X}}^j(k-1|k-1)$ $(j=1,2,\ldots,N)$ is the output of the jth filter at time $k-1$. $\hat{\boldsymbol{X}}^{oj}(k-1|k-1)$ $(j=1,2,\ldots,N)$ is the result of the interaction of $\hat{\boldsymbol{X}}^j(k-1|k-1)$ $(j=1,2,\ldots,N)$ and serves as the input of filter j at time k. $\mathbf{Z}(k)$ is the measurement at time k.

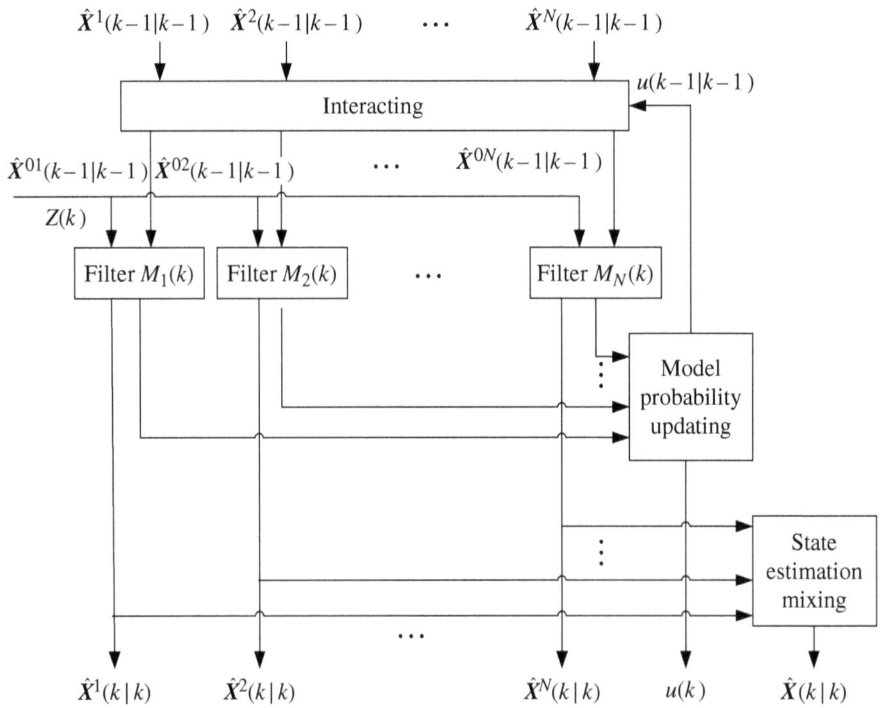

Figure 9.3 Diagram of interacting multiple model algorithm

Suppose that the model probability switches under the Markov chain; the interacting actuator employs the model probability and the model transition probability to calculate the interacting estimate for each filter. At the beginning and end of the filtering cycle, each filter makes use of the interacting estimate and measurement data to calculate a new estimate and the possibility of the model. Then the probability, possibility, and transition probability of the model at the last moment are used to calculate the new model probability. Hence, the general state can be calculated through the new state estimate and its corresponding model probability. The recursive process of the IMM algorithm with N models from time $k-1$ to k is described below.

9.3.6.1 Interacting State Estimates

Let the transition probability from model i to model j in Figure 9.3 be

$$\boldsymbol{P}_{t_{ij}} = \begin{bmatrix} \boldsymbol{P}_{t_{11}} & \boldsymbol{P}_{t_{12}} & \cdots & \boldsymbol{P}_{t_{1r}} \\ \boldsymbol{P}_{t_{21}} & \boldsymbol{P}_{t_{22}} & \cdots & \boldsymbol{P}_{t_{2r}} \\ \vdots & \vdots & \vdots & \vdots \\ \boldsymbol{P}_{t_{r1}} & \boldsymbol{P}_{t_{r2}} & \cdots & \boldsymbol{P}_{t_{NN}} \end{bmatrix} \tag{9.91}$$

Let $\hat{\boldsymbol{X}}^j(k-1|k-1)$ be the state estimate of filter j at time $k-1$, $\boldsymbol{P}^j(k-1|k-1)$ its corresponding state covariance matrix, and $u_{k-1}(j)$ the probability of model j at time $k-1$ ($i, j = 1, 2, \ldots, N$). Then, the input of r filters at time k after the interacting calculation is

$$\hat{\boldsymbol{X}}^{oj}(k-1|k-1) = \sum_{i=1}^{N} \hat{\boldsymbol{X}}^i(k-1|k-1)u_{k-1|k-1}(i|j) \tag{9.92}$$

where

$$\begin{cases} u_{k-1|k-1}(i|j) = \dfrac{1}{\bar{C}_j}\boldsymbol{P}_{t_{ij}}u_{k-1}(i) \\ \bar{C}_j = \displaystyle\sum_{i=1}^{N} \boldsymbol{P}_{t_{ij}}u_{k-1}(i) \end{cases} \tag{9.93}$$

$$\begin{aligned} \boldsymbol{P}^{oj}(k-1|k-1) = \sum_{i=1}^{N} \Big\{ & \boldsymbol{P}^i(k-1|k-1) + \big[\hat{\boldsymbol{X}}^i(k-1|k-1) \\ & -\hat{\boldsymbol{X}}^{oj}(k-1|k-1)\big]\big[\hat{\boldsymbol{X}}^i(k-1|k-1) - \hat{\boldsymbol{X}}^{oj}(k-1|k-1)\big]' \Big\} u_{k-1|k-1}(i|j) \end{aligned} \tag{9.94}$$

9.3.6.2 Model Modification

By using the state vector $\hat{\boldsymbol{X}}^{oj}(k-1|k-1)$, its variance $\boldsymbol{P}^{oj}(k-1|k-1)$, and observation $\boldsymbol{Z}(k)$ together as the inputs into the jth model at time k, a calculation of these inputs through a standard Kalman filter yields the output of each model, $\hat{\boldsymbol{X}}^j(k|k)$, $\boldsymbol{P}^j(k|k)$, $j = 1, 2, \ldots, N$.

9.3.6.3 Calculation of Model Possibility

If the filtering residual of model j is $\boldsymbol{v}_k^j$, and its corresponding covariance is $\boldsymbol{S}_k^j$, then the possibility of model j under Gaussian assumptions is

$$\Lambda_k^j = \frac{1}{\sqrt{|2\pi \boldsymbol{S}_k^j|}} \exp\left[-\frac{1}{2}(\boldsymbol{v}_k^j)'(\boldsymbol{S}_k^j)^{-1}\boldsymbol{v}_k^j\right] \tag{9.95}$$

where

$$\begin{cases} \boldsymbol{v}_k^j = \boldsymbol{Z}(k) - \boldsymbol{H}^j(k)\hat{\boldsymbol{X}}^j(k|k-1) \\ \boldsymbol{S}_k^j = \boldsymbol{H}^j(k)\boldsymbol{P}^j(k|k-1)\boldsymbol{H}^j(k)' + \boldsymbol{R}(k) \end{cases} \tag{9.96}$$

9.3.6.4 Update of Model Probability

The probability update of model j is

$$u_k(j) = \frac{1}{C}\Lambda_k^j \bar{C}_j \tag{9.97}$$

where

$$C = \sum_{i=1}^{N} \Lambda_k^i \bar{C}_i \tag{9.98}$$

9.3.6.5 Model Output

Let $\hat{\boldsymbol{X}}(k|k)$, $\boldsymbol{P}(k|k)$, respectively, be the interacting outputs at time k, then

$$\hat{\boldsymbol{X}}(k|k) = \sum_{i=1}^{N} \hat{\boldsymbol{X}}^i(k|k) u_k(i) \tag{9.99}$$

$$\boldsymbol{P}(k|k) = \sum_{i=1}^{N} u_k(i)\left\{\boldsymbol{P}^i(k|k) + \left[\hat{\boldsymbol{X}}^i(k|k) - \hat{\boldsymbol{X}}(k|k)\right]\left[\hat{\boldsymbol{X}}^i(k|k) - \hat{\boldsymbol{X}}(k|k)\right]'\right\} \tag{9.100}$$

The IMM algorithm is completed by means of this recursive process.

The features of the IMM algorithm include:

1. Its employment of measurement information is shown not only in the filtering, but also in the model probability, whose variation is used to adjust the model in an adaptive way.
2. It is modularized. Therefore, the filtering modules can adopt various linear or nonlinear filtering algorithms according to their different applications.
3. Its calculation efficiency is improved since its modules compute in parallel.

The IMM estimation is usually considered the most effective hybrid estimation scheme, possessing efficiency of estimation and advantages in calculation. It has been successfully applied to many tracking systems. However, as in the case of the multiple model algorithm, the performance of the IMM relies heavily on the model set it uses. This leads to a contradiction that is difficult to reconcile (i.e., more models are needed to match the target's motion in order to improve the performance of state estimation). These additional models increase the computational load of the system greatly, and even degrade the performance of the trackers in many cases. In addition, applying the IMM to less maneuverable or non-maneuvering targets leads to a waste of resources.

All the methods discussed above are adaptive tracking algorithms without maneuver detection. They are also called "tracking algorithms for maneuver identification." Their advantage is that no maneuver detection is required, so that there is no time delay for value estimation. Their disadvantage is that a practical a priori hypothesis needs to be established for the maneuverability of the target, but this hypothesis is hard to verify in practice.

9.4 Performance Comparison of Maneuvering Target Tracking Algorithms

The various maneuvering target tracking algorithms discussed in the previous sections perform differently in various environments since they are based on different principles. A few typical environments for target maneuvering have been chosen below. These algorithms are put in the same simulation environment, and a comprehensive comparison is made between them in terms of tracking accuracy, tracking lifetime, and instantaneity.

9.4.1 Simulation Environment and Parameter Setting

In this section, five typical environments for maneuvering targets are chosen.

Environment 1

Initial state of the target: $X(0) = [120\,000\,\text{m}, -426\,\text{m/s}, 2000\,\text{m}, 0\,\text{m/s}]'$. Duration of the motion: 90 s. The time when the maneuver happens and the magnitude of the acceleration are shown in Table 9.1.

Table 9.1 Target maneuvers in environment 1

Time when the maneuver happens	$t=31$	$t=38$	$t=49$	$t=61$	$t=65$	$t=66$	$t=81$
Acceleration in the direction of X (m/s^2)	5	−8	10	0	−10	−5	5
Acceleration in the direction of Y (m/s^2)	−10	18	−20	30	−8	0	−10

Environment 2

Initial state of the target: $X(0) = [47\,000\,\text{m}, -426\,\text{m/s}, 12\,000\,\text{m}, 0\,\text{m/s}]'$. Duration of the motion: 120 s. The time when the maneuver happens and the acceleration of the target are shown in Table 9.2.

Table 9.2 Target maneuvers in environment 2

Time when the maneuver happens	$t=31$	$t=38$	$t=61$	$t=71$	$t=91$
Acceleration in the direction of X (m/s^2)	10	0	−5	−10	50
Acceleration in the direction of Y (m/s^2)	−10	−10	30	0	−2

Environment 3

In this simulation environment, a motion model is established to simulate the diving motion of an aircraft with the kinematic equation retained and the kinetic equation simplified. The target's motion lasts for 100 s, its dipping height is 2500 m, dipping distance is 15 000 m, and the initial state of the target is $X(0) = [8000\,\text{m},\ 191\,\text{m/s},\ 6000\,\text{m},\ 0\,\text{m/s}]'$.

Environment 4

As in environment 3, a motion model is established to simulate the climbing movement of an aircraft. The target's motion lasts for 100 s, its climbing height is 400 m, climbing distance is 15 000 m, and the initial state of the target is $X(0) = [8000\,\text{m},\ 110\,\text{m/s},\ 6000\,\text{m},\ 0\,\text{m/s}]'$.

In the process of simulation, suppose that the radar sampling interval is $T=1$ s, the range measurement error is $\rho_r = 100$ m, and the angle measurement error is $\rho_\theta = 0.03$ rad. The number of simulations is $N=50$. The parameters for the algorithms are set out as follows.

1. Adjustable white-noise model: length of detection sliding window $l=8$, judgment threshold for maneuver happening $\varepsilon_{\max} = 5.07$, primary process noise $q_0 = 0.005$.
2. VDF algorithm: length of detection sliding window $l=8$, judgment threshold of maneuver happening $\varepsilon_{\max} = 20.1$, threshold of maneuver completion $\varepsilon_{\min} = 13.4$.
3. Input estimation algorithm: length of detection sliding window $l=8$, judgment threshold of maneuver happening $\varepsilon_{\max} = 5.07$.
4. IMM algorithm: there are three models adopted, each with quotient of process noise covariance $q_1 = 10$, $q_2 = 1$, $q_3 = 0.1$, a priori probability of the model $\mu_0 = [1/3, 1/3, 1/3]$, and transition probability of the Markov model

$$\boldsymbol{P}_{t_{ij}} = \begin{bmatrix} 0.8 & 0.15 & 0.05 \\ 0.3 & 0.4 & 0.3 \\ 0.05 & 0.15 & 0.8 \end{bmatrix} \tag{9.101}$$

5. Current model algorithm: autocorrelation time constant $\alpha = 1/20$, largest acceleration $a_{\max} = 100\text{m/s}^2$, $a_{-\max} = -100\text{m/s}^2$.
6. Singer model algorithm: autocorrelation time constant $\alpha = 1/20$, largest acceleration $a_{\max} = 100\text{m/s}^2$, largest probability $P_{\max} = 0.95$, $P_{\min} = 0.05$.

Environment 5

In the same simulation environment, conduct a simulation comparison of the tracking results, respectively, of the corrected input estimation, adjustable white noise, Singer, and jerk algorithms. Among them, the parameters of the Singer model algorithm are set as follows:

Table 9.3 Target maneuvers in environment 2

Time when the target maneuver happens	$t=400-600$	$t=610-660$	other times
Acceleration in the direction of X (m/s^2)	0.075	−0.3	0
Acceleration in the direction of Y (m/s^2)	0.075	−0.3	0

autocorrelation time constant $\alpha=1/20$, largest acceleration $a_{\max}=0.1\,m/s^2$, occurrence probability of the largest acceleration $P_M=0.2$, probability of non-occurrence of target maneuver $P_0=0.6$. The jerk variance of the jerk algorithm is 0.0009^2. In the process of simulation, suppose that the radar sampling interval is $T=2$ s, the range measurement error is $\sigma_r=100$ m, the angle measurement error is $\sigma_\theta=0.1°$, the number of simulations is $N=100$, the initial state of the target is $X(0)=[20\,000\,m, 4\,m/s, 80\,000\,m, -15\,m/s]'$, and the duration of target motion is 900 s. The time when the target maneuver happens and the acceleration of the maneuvering target are shown in Table 9.3.

9.4.2 Simulation Results and Analysis

Environment 1 (shown in Figures 9.4–9.7)

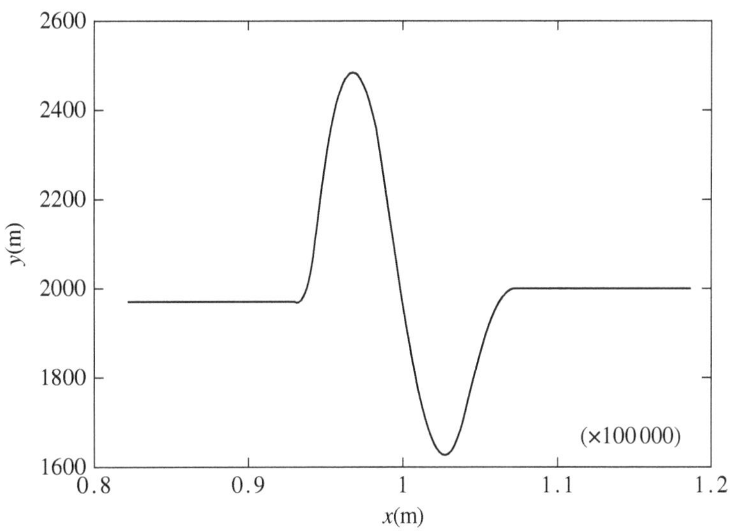

Figure 9.4 Target's trajectory in environment 1

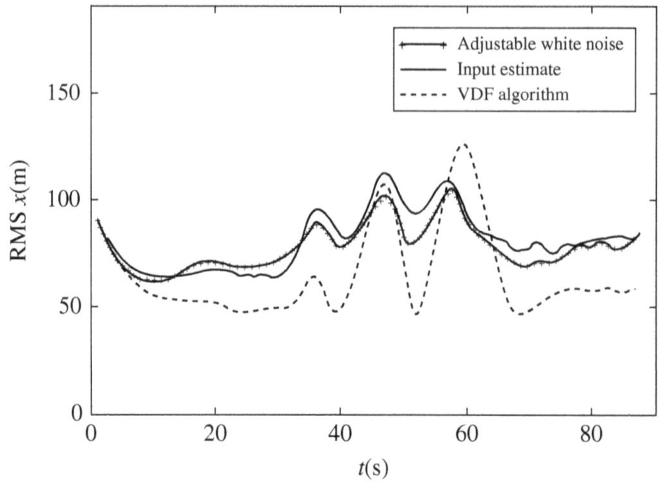

Figure 9.5 Comparison between Type 1 maneuvering target tracking algorithms

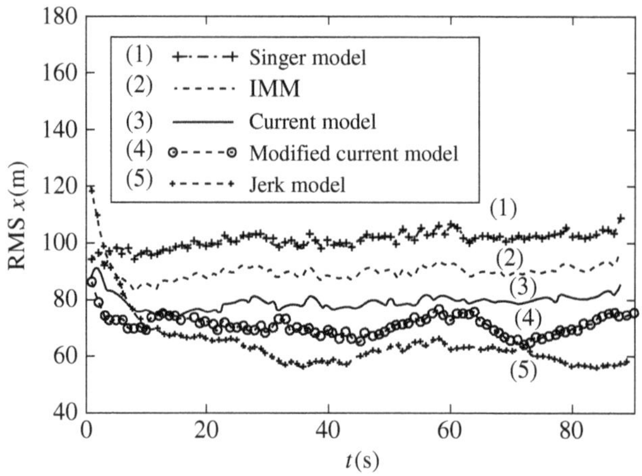

Figure 9.6 Comparison between Type 2 maneuvering target tracking algorithms

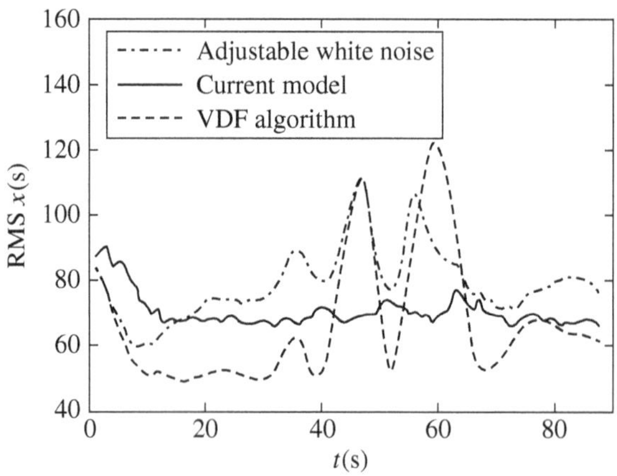

Figure 9.7 Comparison between maneuvering target tracking algorithms of Types 1 and 2

Environment 2 (shown in Figures 9.8–9.11)

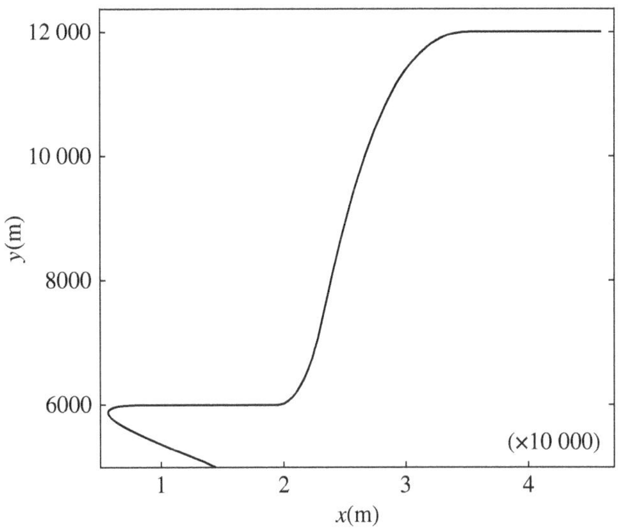

Figure 9.8 Target's trajectory in environment 2

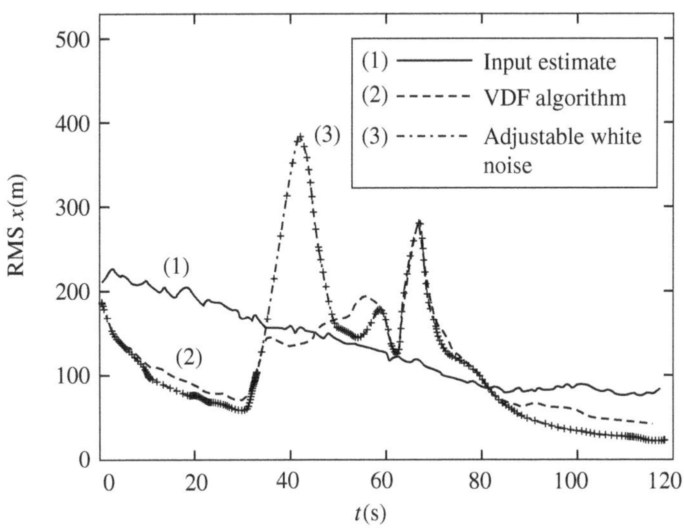

Figure 9.9 Comparison between Type 1 maneuvering target tracking algorithms

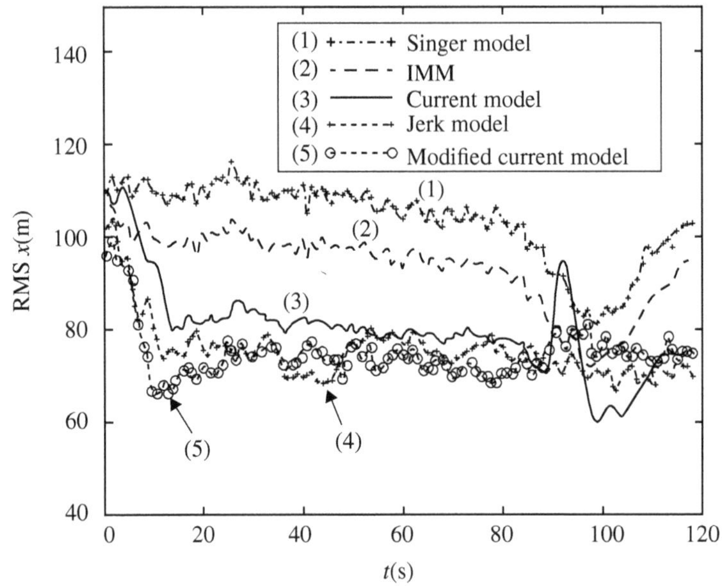

Figure 9.10 Comparison between Type 2 maneuvering target tracking algorithms

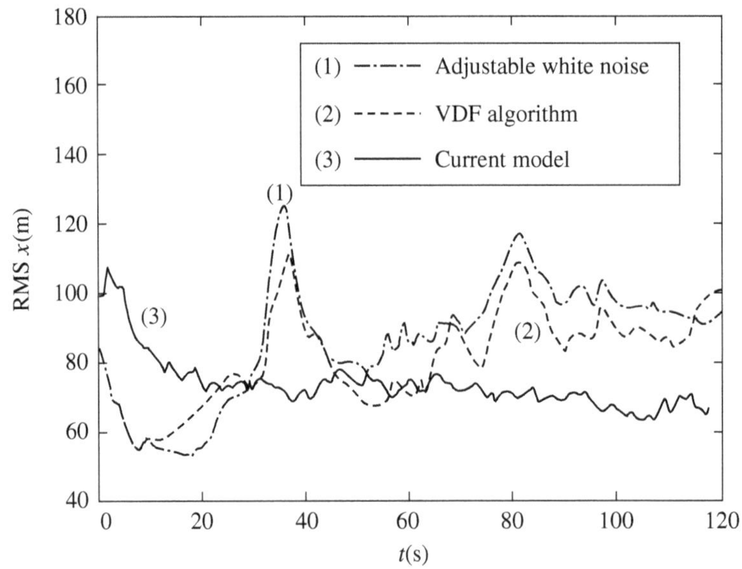

Figure 9.11 Comparison between maneuvering target tracking algorithms of Types 1 and 2

Environment 3 (shown in Figures 9.12–9.15)

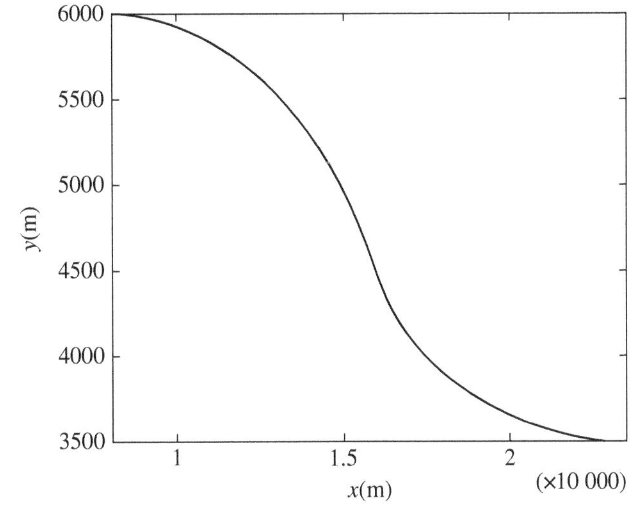

Figure 9.12 Target's trajectory in environment 3

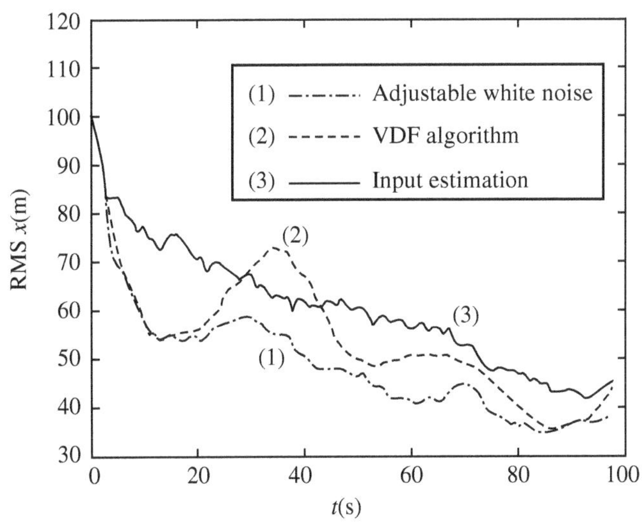

Figure 9.13 Comparison between Type 1 maneuvering target tracking algorithms

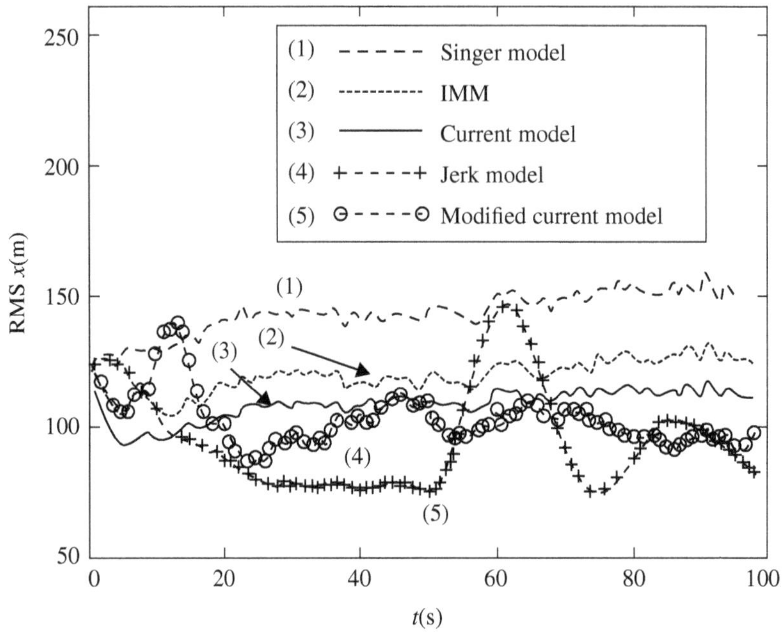

Figure 9.14 Comparison between Type 2 maneuvering target tracking algorithms

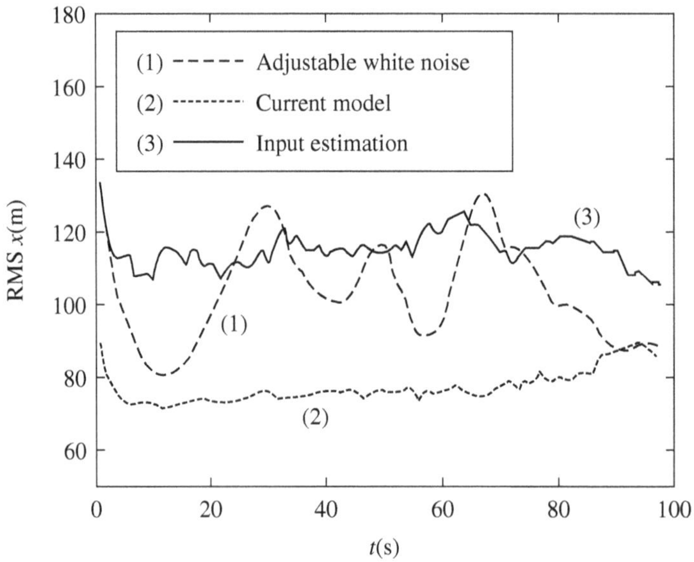

Figure 9.15 Comparison between maneuvering target tracking algorithms of Types 1 and 2

Environment 4 (shown in Figures 9.16–9.19)

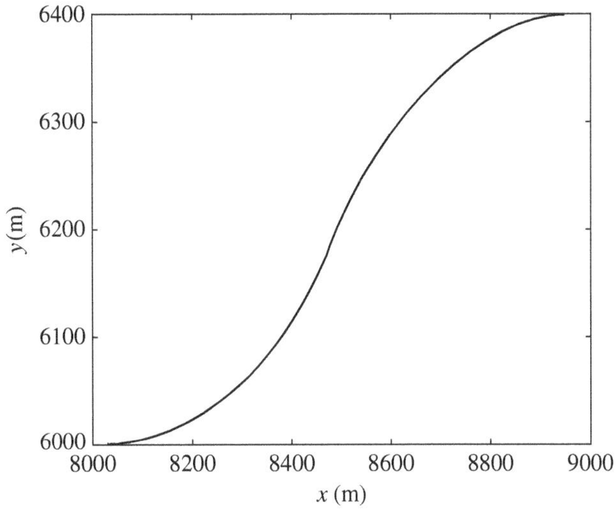

Figure 9.16 Target's trajectory in environment 4

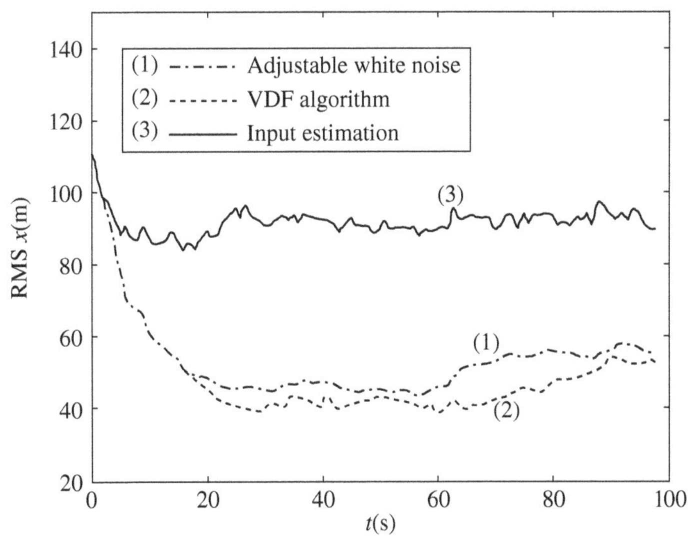

Figure 9.17 Comparison between Type 1 maneuvering target tracking algorithms

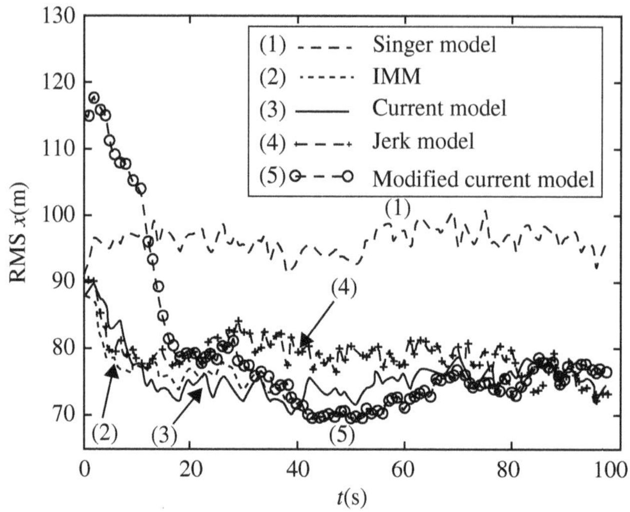

Figure 9.18 Comparison between Type 2 maneuvering target tracking algorithms

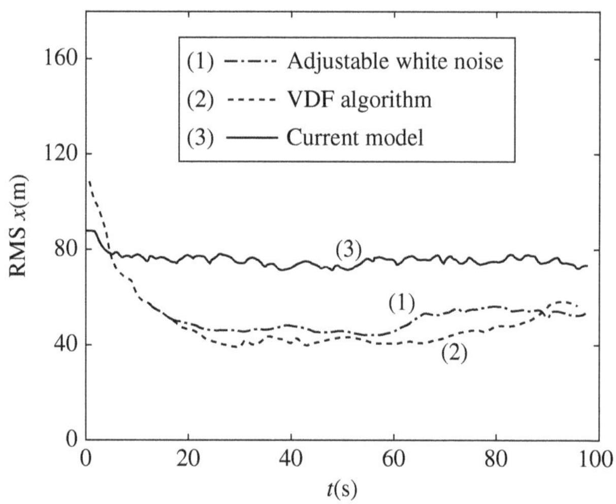

Figure 9.19 Comparison between maneuvering target tracking algorithms of Types 1 and 2

Table 9.4 Comparison of the track lifetime and the average elapsed time of a cycle of the algorithms from environment 1 to environment 4

Maneuvering targets tracking algorithms	Track lifetime (pace)				Average elapsed time of a cycle (ms)			
	1	2	3	4	1	2	3	4
Adjustable white noise	432	572	463	489	0.12	0.13	0.12	0.12
VDF	396	549	448	471	0.23	0.29	0.21	0.27
IMM	450	600	500	500	1.49	1.91	1.62	1.67
Singer model	428	577	492	500	0.15	0.16	0.15	0.15
Current statistical model	450	600	500	500	0.17	0.25	0.18	0.17
Jerk model	450	600	500	500	0.15	0.21	0.17	0.17
Modified current model	450	600	500	500	0.72	1.33	1.03	1.09

The track lifetime and the average elapsed time of a cycle of the algorithms from environment 1 to environment 4, respectively, are shown in Table 9.4.

Environment 5 (see Figures 9.20–9.22)
Figure 9.20 is the target's trajectory, Figures 9.21 and 9.22 show the x-axis and y-axis RMS error of 100 Monte Carlo experiments.

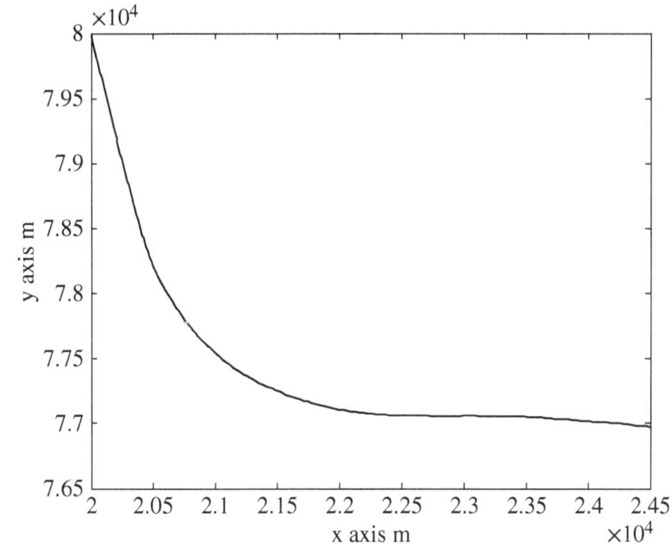

Figure 9.20 Target's trajectory in environment 5

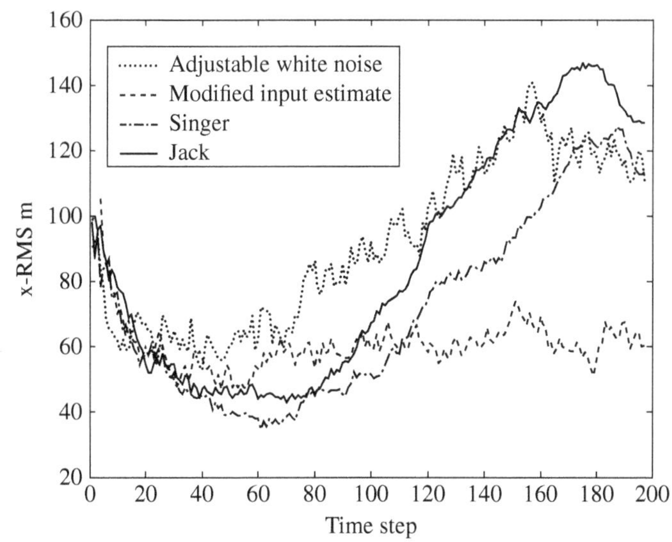

Figure 9.21 Target's x-axis RMS error

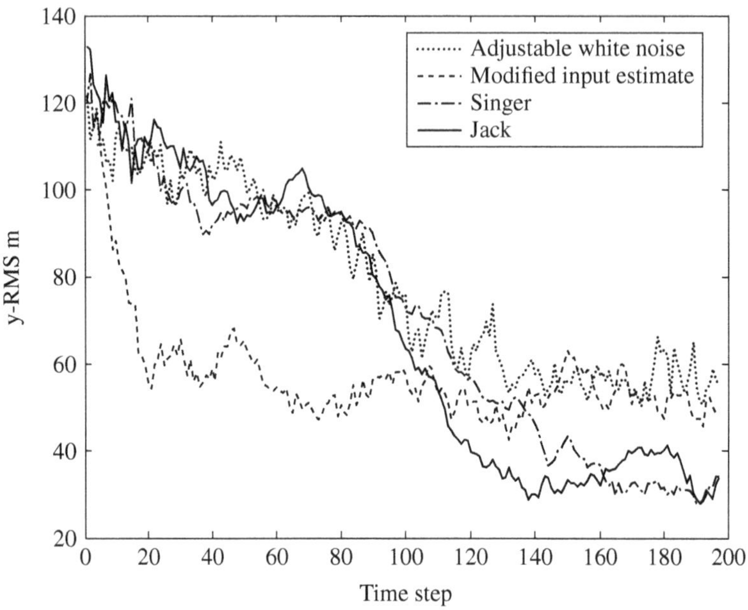

Figure 9.22 Target's y-axis RMS error

We can reach the following conclusions from the simulation results described above.

1. The VD filtering in the Type 1 algorithms may produce quite large errors when the maneuver happens, but generally the errors will decrease relatively fast, as shown in Figures 9.5, 9.9, and 9.17. Despite that, there will be very large errors sometimes, as shown in Figure 9.13.

   The tracking accuracy of the adjustable white noise algorithm is a little lower than that of the VDF algorithm, but its performance is quite stable in all the simulation experiments and very large errors have not occurred, as demonstrated in Figures 9.5, 9.9, 9.13, and 9.17.

   The input estimation algorithm has been relatively stable in the whole process of tracking, and its tracking error curve shows few drastic fluctuations. However, it is lower than either of the two algorithms in terms of tracking accuracy, as illustrated in Figures 9.5, 9.9, 9.13, and 9.17.
2. The current model in Type 2 algorithms is similar to the IMM algorithm in tracking accuracy. Sometimes the former is a little better than the latter, as shown in Figure 9.10, while sometimes it is the other way around, as illustrated in Figure 9.18. Sometimes the two algorithms are basically the same in tracking accuracy, as demonstrated in Figures 9.6 and 9.14. This is due to the a priori model set established by the IMM algorithm. If the motion model of the target is included in the model set, then the tracking accuracy of the algorithm will become relatively high.

   The tracking accuracy of the Singer model is lower than those of the two algorithms mentioned above according to all the simulation experiment results, as shown in Figures 9.6, 9.10, 9.14, and 9.18. The jerk model, featuring better tracking accuracy than the first three algorithms, is similar to the modified current model in terms of tracking accuracy in the whole tracking process, as shown in Figures 9.6, 9.10, and 9.14. But the RMS error of the jerk model showed a larger fluctuation occasionally, as demonstrated in Figure 9.14.
3. As shown in Figures 9.21 and 9.22, the modified input estimation, adjustable white noise, Singer, and jerk algorithms can all track the maneuvering target quite well in this environment.

In general, the modified input estimation algorithm is better at tracking than the other three algorithms mentioned above.

4. Comparing the two algorithm models with higher tracking accuracy of Type 1 and the current model algorithm of Type 2 in each environment, the current model is obviously higher in terms of tracking accuracy than the adjustable white noise model and the VDF filtering model in environments 2 and 3. In addition, in environment 1, the current model is obviously higher in terms of tracking accuracy than the adjustable white noise algorithm but lower than the VDF filtering model, while in environment 4, the current model is strikingly lower in terms of tracking accuracy than the adjustable white noise algorithm and the VDF filtering model. This is a result of the current model's a priori hypothesis of autocorrelation time constant of the target's maneuver. When the hypothesis is approximate to the target's real autocorrelation time constant for maneuvers, the tracking accuracy will become higher.
5. From Table 9.4 it can be seen that the current statistical model, IMM, jerk model, and modified current algorithm possess the longest track lifetime, and all are able to perform a complete tracking of the target's motion in the 50 simulations. The adjustable white noise algorithm, input estimation algorithm, and Singer model have a shorter track lifetime, and the VDF algorithm has the shortest track lifetime. In terms of instantaneity of algorithm, the adjustable white noise algorithm is the most instantaneous, while the VDF algorithm, Singer model, jerk model, and current statistical model are less instantaneous, and the modified current model and IMM algorithm are the least instantaneous. Therefore, from a comprehensive perspective, each algorithm has its own advantage in terms of tracking accuracy, track lifetime, and/or instantaneity.

9.5 Summary

The tracking algorithms for maneuvering targets discussed in this chapter generally fall into two categories: tracking algorithms with maneuver detection and adaptive tracking algorithms. The former category, based on the adjusted parameters after the maneuver is detected, is further divided into algorithms adjusting the filter gain and algorithms adjusting the filter structure. The latter can be subdivided into single-model algorithms and multiple-model algorithms. The single-model adaptive tracking algorithms for maneuvering targets include the modified input estimation, Singer model, current statistical model, and jerk model algorithms, while the multiple-model adaptive tracking algorithms include the multiple model and IMM algorithms.

Through a simulation analysis and comparison of the two categories of algorithms mentioned above, we can reach the following conclusion: the tracking algorithms with maneuver detection feature less calculation and greater instantaneity, but this category of models cause a larger error when the target maneuvers. In addition, because of maneuver detection, it is impossible for this category to avoid some estimated time delay, which will affect the tracking performance of the filter.

In contrast, the common advantages with adaptive tracking algorithms are the reduced time delay to maneuver tracking and the stable performance in tracking. Of these algorithms, the single-model tracking algorithms realize the adaptive tracking of the maneuvering targets mainly by the real-time estimation of the maneuvering target's states and model parameters. This group of algorithms does not cause a larger error when the target maneuvers, but usually needs a reasonable hypothesis about the maneuverability of the target, which means that when the hypothesis is incongruent with reality, its performance will degrade greatly. Additionally, the tracking accuracy of this group of algorithms will fall to some degree when the target is non-maneuvering [232].

The modified current statistical model algorithm among this group may not need to establish an a priori hypothesis of the target's maneuverability, and it has a good performance in tracking when the target is non-maneuvering. However, the instantaneity of this algorithm needs to be improved.

Through the effective combination of multi-target models or process noise levels, multiple-model algorithms among the adaptive tracking algorithms for maneuvering targets realize the adaptive estimation of the maneuvering target states, thus achieving a better effect in tracking maneuvering targets. Therefore, this is considered as the most important research achievement in maneuvering targets tracking in recent decades, and has become a mainstream tracking algorithm for maneuvering targets which is of practical value in an engineering sense.

10

Group Target Tracking

10.1 Introduction

In reality, a complicated group target will probably form within a small spatial distribution range because of uncontrollable or specific human factors, such as the splitting of space debris, a large amount of debris and decoy, along with the process of penetration of ballistic missiles, missiles and aircraft formations, etc. These targets occupy a smaller spatial distribution range, and their differences in the motion are obscure. They have a lower relative velocity of motion, with approximate characteristics. In the field of target tracking, this type of target is called a "group target."

Restrained by its measurement equipment's angular resolution, range resolution, power, and measurement accuracy, etc., a detection system in the process of tracking a group target will generally encounter three situations:

1. It is totally unable to identify the targets within the group. Then the group cannot be identified.
2. It is sometimes able to identify the targets within the group, but unable to gain a stable, constant, and effective measurement. This means that the group target is partly identifiable.
3. It is able to completely identify the targets within the group, which means that the group is identifiable.

In addition, when the target changes its range and angle to the sensor, the three situations will switch between each other. Therefore, the echo characteristics of the group are more complex than those of the traditional multi-targets.

However, conventional target tracking algorithms [41, 160, 233–236] underestimate the complexities of the echo of the group, thus being relatively simple in design, and defective in the solution of the problems of track initiation, track maintenance, maneuver processing, track cancellation, and so on. Hence, they are generally quite limited in tracking effect.

Radar Data Processing with Applications, First Edition. He You, Xiu Jianjuan, and Guan Xin.

In order to better solve the problems of group tracking, this chapter will first focus on the initiation techniques of the group in Section 10.2. Several typical models of group initiation will be discussed in terms of group segmentation, group association, and estimation of group velocity. Then, in Section 10.3, a gray fine track initiation algorithm of the group applicable to cluttered environments is proposed. Sections 10.4 and 10.5 will focus on the tracking algorithms, respectively, for center group targets and the formation of group targets. Finally, in Section 10.6, a few simulation environments close to reality are designed, and then a test and analysis of the comprehensive performance of the algorithms in this chapter are conducted.

10.2 Basic Methods for Track Initiation of the Group Target

The track initiation of the group target [41, 237, 238] is much more complex than that in the case of single or multiple-target tracking. Conventional algorithms for track initiation are unsatisfactory in this respect [236, 239–241]. First, because the range between the targets in a group is small, the initial gates for each target will seriously intersect if the direct-vision method or logical approach [242, 243] is adopted to establish the courses, respectively, for the group targets. Owing to the existence of measurement errors and extrapolation errors, the group targets tend to make mistaken cross-correlations with the measurements. Second, because the behavioral models of the group targets are similar to each other and the cross-correlations between targets are very strong at the time before and after the echoes, mistaken temporary tracks can find correlation values at follow-up moments. The rules for track confirmation in the direct-vision method, logical approach, or other conventional methods for target initiation cannot restrain the output of mistaken tracks so that the initiation rate of false track increases. Finally, if the track initiation algorithms [244, 245] based on the Hough transform are adopted to establish courses for the group targets, then the echoes of other targets are all clutter to the targets which need course establishment. The errors are most likely to cause local maximum values, cross-correlations of measurements between the group targets, and decreases in the initiation rate of correct tracks.

10.2.1 Group Definition

The typical initiation algorithm for the group target is usually composed of group division, group correlation, and group velocity estimation. But before group division is discussed, it is necessary to define the group. The group target, also known as the formation target, group target, etc. [17, 41, 45, 160, 233–238, 246–254] is by nature a group of targets which satisfy the following three conditions in terms of direction, range, and velocity [19, 236, 238]:

1. The same motion direction.
2. A much smaller distance between the members of a group than that between groups.
3. Basically the same velocity.

The tracking of group targets defined by the above conditions has three conceptual advantages over that of dense targets.

1. In terms of the typical radar tracking problems discussed by Taenzer [237], the group target's tracking saves radar resources. In the case of group target tracking it only needs to track the center of the group instead of every target in the group, and therefore the necessary radar operations will decrease. Since it is the single group rather than all the targets within the group that is

tracked, the number of track files that need processing will decrease. In the case of air-to-ground tracking, this advantage of group target tracking is crucial when the tracked target is a motorcade moving on the ground, since it is impossible, difficult, or unnecessary to track every single target in such environments.
2. When tracking dense targets in complex environments, mistaken correlations are unavoidable. The group target tracking in such environments will help to make target tracking more smooth and more stable.
3. When the sensor's target detection probability is low, it is easier for group target tracking to gain information from each target group, thereby achieving a better effect in tracking.

10.2.2 Group Segmentation

Based on the definition of the group in Section 10.2.1, this section discusses three typical methods of group segmentation, including distance division, threshold value circulation, and diagrammatical methods.

10.2.2.1 Distance Segmentation Method

In the definition of a group, it is required that the distance between the members of a group is far smaller than that between groups. Therefore, group segmentation can be completed by comparing the spatial distance between two measurements with a constant in terms of size [160, 236, 238].

Suppose that $\mathbf{Z}(k)$ is the measurement set gained by the sensor at time k, and

$$\mathbf{Z}(k) = \{z_i(k)\}_{i=1}^{m_k} \tag{10.1}$$

where m_k is the number of measurements at time k.

Define the distance between the ith measurement $z_i(k) = [x_{ik} \ \ y_{ik} \ \ z_{ik}]'$ and the jth measurement $z_j(k)$ in $\mathbf{Z}(k)$ at time as

$$d\left(z_i(k), z_j(k)\right) = \sqrt{\left(x_{ik} - x_{jk}\right)^2 + \left(y_{ik} - y_{jk}\right)^2 + \left(z_{ik} - z_{jk}\right)^2} \tag{10.2}$$

If

$$d\left(z_i(k), z_j(k)\right) < d_0 \tag{10.3}$$

then measurements $z_i(k)$ and $z_j(k)$ belong to the same group. Here, d_0 reflects the density of targets within the group. For mechanical scanning radars, the aim of adopting the group target tracking is to cope with the increase in covariance of filtering error as a result of mistaken correlations. Therefore, d_0 is the spatial range between two targets which cannot be distinguished by means of the nearest-neighbor approach. For phased array radars, the purpose of adopting the group target tracking is to save radar resources, and $d_0 = 3 \sim 5$ km can be taken.

By using (10.2) and (10.3) to calculate the range between any two measurements in $\mathbf{Z}(k)$, which is then compared with d_0, $\mathbf{Z}(k)$ can be segmented into different groups. Suppose that $\mathbf{Z}(k)$ can be segmented into m groups, denoted by $\{\boldsymbol{U}_1, \boldsymbol{U}_2, \ldots, \boldsymbol{U}_m\}$, and

$$U_i = \left\{\tilde{z}_j^i(k)\right\}_{j=1}^{\tilde{m}_i} \tag{10.4}$$

where $\tilde{z}_j^i(k)$ is the jth measurement of the ith group and the $\tilde{m}_j$ number of measurements of the ith group.

Define the range between two groups U_i and U_j,

$$d(U_i, U_j) = \min\left\{d\left(\tilde{z}_m^i(k), \tilde{z}_n^j(k)\right)\right\} \quad 1 \le m \le \tilde{m}_i, 1 \le n \le \tilde{m}_j \tag{10.5}$$

When $d(U_i, U_j) \ge d_0$, U_i and U_j are two separate groups. When $d(U_i, U_j) < d_0$, U_i and U_j are combined into one group. Therefore, each measurement in U_i must satisfy

$$\min\left\{d\left(\tilde{z}_m^i(k), \tilde{z}_n^i(k)\right)\right\} < d_0 \quad 1 \le m \le \tilde{m}_i, 1 \le n \le \tilde{m}_i, m \ne n \tag{10.6}$$

and the set $\left\{\tilde{z}_1^i(k), \tilde{z}_2^i(k), \ldots, \tilde{z}_{\tilde{m}_i}^i(k)\right\}$ cannot be segmented into two parts: U_1^* and U_2^*. Let $d(U_1^*, U_2^*) \ge d_0$.

10.2.2.2 Circulation Threshold Value Method

The circulation threshold value method includes the following four procedures:

1. Choose $z_i(k)$ as the center, and establish a wave gate with the threshold value of d_0.
2. Rebuild a gate with the threshold value of d_0 for each measurement falling in the gate, and look for the measurement falling in the latest gate.
3. Repeat step 2 until there is no measurement in the established gate. The measurements concerned in the process are defined as a group.
4. Choose any one measurement from the measurements which do not belong to the defined groups, then repeat the previous three steps until the last measurement so as to finish the group segmentation.

The value assignment of d_0 in this method is the same as that in the range segmentation method. As indicated by its procedure, this algorithm is based on the range segmentation method with the same results of segmentation, except for the considerable reduction in computing complexity.

10.2.2.3 Diagrammatical Method

The diagrammatical method is to segment the groups in the whole detection region. According to its definition, a group is any measurement set U which satisfies the following two conditions:

1. The distance between the measurements in U is less than the threshold value d_0.
2. The number of measurements in U is larger than the threshold value L.

The value assignment of d_0 and L relies on real tracking systems and their task requirements [246].

In view of the above rules, the diagrammatical method includes the following four steps.

1. As shown in Figure 10.1, the detection region is segmented into l^2 small regions, and l is determined by

$$l = \mathrm{INT}\left[\sqrt{\frac{V}{\pi d_0^2}} + 1\right] \tag{10.7}$$

where V is the area of the detection region and INT[x] indicates the largest integer not larger than x.

The segmentation process of the detection region is the key to the diagrammatical method. After segmentation, if the areas of small regions are too large, dense target regions might be ignored; if the small regions are too small, the calculation work will increase.

2. As shown in Figure 10.2, the number of measurements falling in each small region is counted and denoted as the number of measurements falling in the small region of the ith line and the jth row.

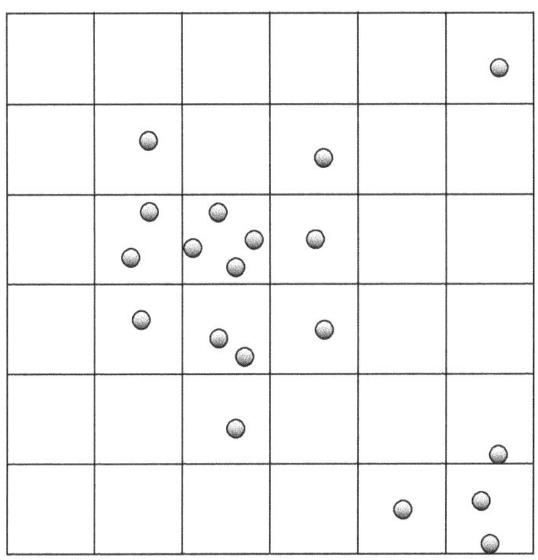

Figure 10.1 Diagram for segmentation of the detection region

0	0	0	0	0	1
0	1	0	1	0	0
0	2	4	1	0	0
0	1	2	1	0	0
0	0	1	0	0	1
0	0	0	0	1	2

Figure 10.2 Statistical diagram of the number of measurements

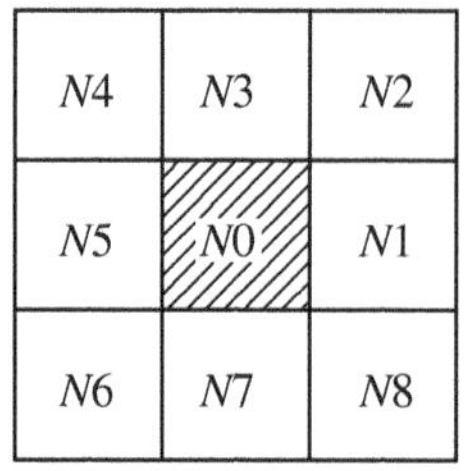

Figure 10.3 Diagram for value assignment and calculation in the small region

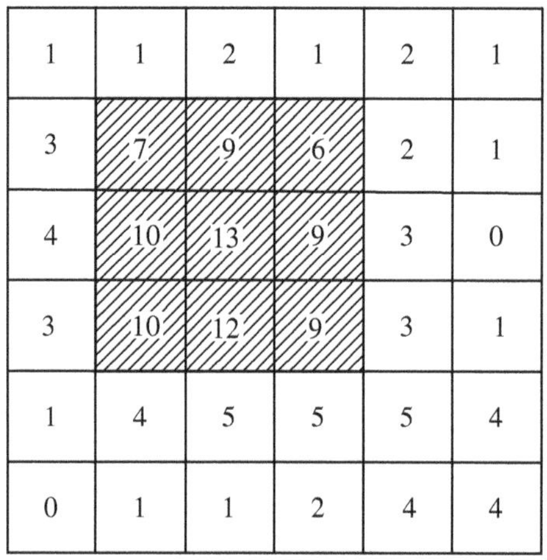

Figure 10.4 Diagram for determination of the dense region of measurements

3. As illustrated in Figure 10.3, choose any small region, denote by $N0$ the number of measurements falling in the small region, and then define the value M of the small region as

$$M = N0 + N1 + N2 + N3 + N4 + N5 + N6 + N7 + N8 \tag{10.8}$$

   where $N1, N2, \ldots, N8$ are the numbers of measurements, respectively, for the eight neighboring small regions of the one ($N0$) under discussion.
4. Compare the value of each small region in (10.8) with the threshold value L. The region with a value larger than L is defined as the dense region of measurements, illustrated by the shaded region in Figure 10.4. Then the measurements falling in the dense region of measurements form a group.

Straightforward and convenient, the diagrammatical method has a better segmentation effect and can be used to determine several groups for one time with less calculation work.

10.2.3 Group Correlation

Suppose that the measurement set $\mathbf{Z}(k)$ is finally segmented into m groups, denoted $\{\boldsymbol{U}_1, \boldsymbol{U}_2, \ldots, \boldsymbol{U}_m\}$. To correlate the groups [41, 160, 236, 237], the center of each group has to be calculated first. Define $\bar{\mathbf{Z}}_i(k)$ as the center of $\boldsymbol{U}_i$, the ith group, and

$$\bar{\boldsymbol{Z}}_i(k) = [\bar{X}_{ik} \;\; \bar{Y}_{ik} \;\; \bar{Z}_{ik}]' \tag{10.9}$$

$$\bar{X}_{ik} = \frac{1}{\tilde{m}_i}\sum_{l=1}^{\tilde{m}_i} x_{lk}, \quad \bar{Y}_{ik} = \frac{1}{\tilde{m}_i}\sum_{l=1}^{\tilde{m}_i} y_{lk}, \quad \bar{Z}_{ik} = \frac{1}{\tilde{m}_i}\sum_{l=1}^{\tilde{m}_i} z_{lk} \tag{10.10}$$

where $z_l^i(k) = [x_{lk}, y_{lk}, z_{lk}]'$ is the set of l measurements of the ith group and $\tilde{m}_i$ is the number of measurements in the ith group.

After gaining the centers of each group at the adjacent moments, define the range between the center of the ith group at time $k = m$ and that of the jth group at time $k = n$ as

$$d[\boldsymbol{Z}_i(m), \boldsymbol{Z}_j(n)] = \sqrt{(\bar{X}_{im} - \bar{X}_{jn})^2 + (\bar{Y}_{im} - \bar{Y}_{jn})^2 + (\bar{Z}_{im} - \bar{Z}_{jn})^2} \tag{10.11}$$

In order to be correlated, the two groups must satisfy

$$d[\boldsymbol{Z}_i(m), \boldsymbol{Z}_j(n)] < V_{\max} \cdot T \tag{10.12}$$

where T is the sampling interval, $n = m + 1$, and $V_{\max}$ is the largest velocity of the group, the value of which depends on the specific type of group target.

If the ith group at time $k = m$ correlates with the multiple groups at time $k = n$, then the nearest group is chosen as the group for correlation.

10.2.4 Group Velocity Estimation

After completing the division and correlation of the group, it is still necessary to conduct an estimation of the group velocity [160, 236, 238] before the completion of group initiation. In order to ensure a high estimation accuracy of group velocity without making the algorithm more complex, it is necessary to adopt different estimation algorithms of group velocity based on the number of targets in the group.

10.2.4.1 Direct Estimation Algorithm

If the number of targets in the group is $N > 6$, the group velocity can be calculated directly through the association groups at the previous three moments [238]. Suppose that $\boldsymbol{U}_i(1)$, the ith group at time $k = 1$, and $\boldsymbol{U}_j(2)$, the jth group at time $k = 2$, are associated, and that the jth group at time $k = 2$ associates with $\boldsymbol{U}_m(3)$, the mth group at time $k = 3$, then the group velocity estimate of $\boldsymbol{U}_m(3)$ is

$$\boldsymbol{V}_m(3) = \frac{\boldsymbol{V}_2 + \boldsymbol{V}_3}{2} \tag{10.13}$$

where

$$\boldsymbol{V}_2 = \frac{\bar{\boldsymbol{Z}}_j(2) - \bar{\boldsymbol{Z}}_i(1)}{T}, \quad \boldsymbol{V}_3 = \frac{\bar{\boldsymbol{Z}}_m(3) - \bar{\boldsymbol{Z}}_j(2)}{T} \tag{10.14}$$

and $\bar{\boldsymbol{Z}}_i(1)$, $\bar{\boldsymbol{Z}}_j(2)$, $\bar{\boldsymbol{Z}}_m(3)$ are the centers, respectively, of $\boldsymbol{U}_i(1)$, $\boldsymbol{U}_j(2)$, $\boldsymbol{U}_m(3)$.

10.2.4.2 Association and Distinction Algorithm

If the number of targets in the group is $N \leq 2$, then the estimation of the group velocity by the direct estimation algorithm will be inaccurate. In order to increase the estimation accuracy, another estimation algorithm of group velocity is needed, that is, the association and distinction algorithm. In this algorithm, estimation is done mainly according to the following three steps.

1. *Calculating the candidate value for the velocity.* Suppose that $\boldsymbol{U}_i(1)$ and $\boldsymbol{U}_j(2)$ are associated, and that $\boldsymbol{U}_j(2)$ and $\boldsymbol{U}_m(3)$ are associated, and establish validation gates with the size of $\boldsymbol{V}_{max} \cdot T$ around the measurements in $\boldsymbol{U}_i(1)$. For the measurements in $\boldsymbol{U}_j(2)$ falling in the validation gates, estimate the candidate value of the velocity by use of the difference value between the associated measurements at the adjacent moments. As illustrated in Figure 10.5, there are two measurements in $\boldsymbol{U}_i(1)$: z_{11} and z_{21}, and three measurements in $\boldsymbol{U}_j(2)$: z_{12}, z_{22}, and z_{32}, z_{12}, z_{22}, and z_{11} are associated; z_{32}, z_{22}, and z_{21} are associated. Therefore, we get four corresponding candidate values of velocity, that is,

$$\begin{cases} \boldsymbol{v}_{11} = \dfrac{z_{12} - z_{11}}{T}, \boldsymbol{v}_{21} = \dfrac{z_{22} - z_{11}}{T} \\ \boldsymbol{v}_{31} = \dfrac{z_{22} - z_{21}}{T}, \boldsymbol{v}_{41} = \dfrac{z_{32} - z_{21}}{T} \end{cases} \tag{10.15}$$

2. *Combining candidate values of the velocity.* Suppose that $\boldsymbol{v}_i = [v_{ix} \ \ v_{iy} \ \ v_{iz}]'$ and $\boldsymbol{v}_j = [v_{jx} \ \ v_{jy} \ \ v_{jz}]'$ are any two candidate values of velocity. If the two values satisfy (10.16), then make $\boldsymbol{v}_i$ roughly equal to $\boldsymbol{v}_j$:

$$|v_{ix} - v_{jx}| \leq \sigma_x, \quad |v_{iy} - v_{jy}| \leq \sigma_y, \quad |v_{iz} - v_{jz}| \leq \sigma_z \tag{10.16}$$

where σ_x, σ_y, and σ_z are, respectively, the standard deviations of measurement errors in the different directions. As illustrated in Figure 10.5, the candidate values of velocity $\boldsymbol{v}_{21}$ and $\boldsymbol{v}_{31}$ do not meet (10.16), while $\boldsymbol{v}_{11}$ and $\boldsymbol{v}_{41}$ satisfy (10.16). They can be combined into

$$\boldsymbol{v}_1 = \frac{\boldsymbol{v}_{11} + \boldsymbol{v}_{41}}{2} \tag{10.17}$$

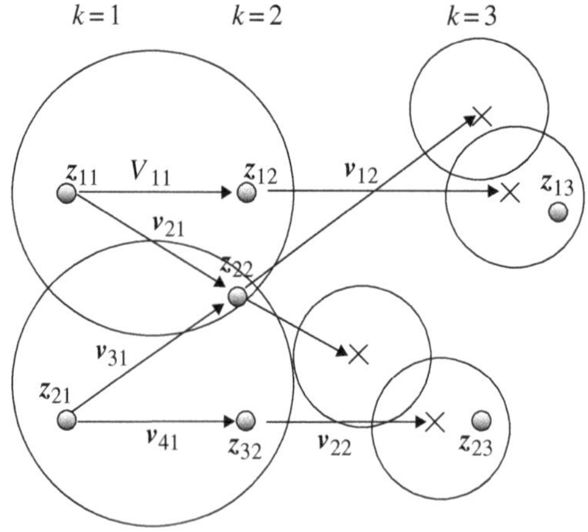

Figure 10.5 Diagram of association and distinction algorithm

3. *Estimating the group velocity.* If there are still several candidate values of velocity left after step 2 (as shown in Figure 10.5, $\boldsymbol{v}_1$, $\boldsymbol{v}_{21}$, and $\boldsymbol{v}_{31}$ still exist after combination), it is necessary to use these candidate values of velocity to extrapolate the measurement at time $k=2$ to time $k=3$, establish the validation gate around each extrapolation point, and associate the extrapolation point with the measurements in $U_m(3)$. If the measurement $z_{i3}=[x_{i3} \ \ y_{i3} \ \ z_{i3}]'$ at time $k=3$ falls in the gate of the extrapolation point $z'=[x', y', z']$, then the following condition must be met:

$$d(z_{i3}, z') = \sqrt{(x_{i3}-x')^2 + (y_{i3}-y')^2 + (z_{i3}-z')^2} \le c_1\sigma_x c_2\sigma_y c_3\sigma_z \tag{10.18}$$

where c_1, c_2, c_3 are constants.

If there is no measurement in the validation gate, then the candidate value of velocity used for extrapolation is proved to be ineffective, and cannot be applied to the estimation of group velocity. If there is any measurement in the validation gate, continue to calculate the candidate values of velocity and repeat the procedures mentioned above.

Since the number of targets in the group is small, it only takes a few moments to estimate the group velocity according to the rules.

As shown in Figure 10.5, z_{13} and z_{23} are two measurements in $\boldsymbol{U}_m(3)$, and z_{13} associates with the point obtained through the extrapolation of z_{12} by using the velocity $\boldsymbol{v}_1$. Then its candidate value of velocity is calculated as

$$\boldsymbol{v}_{12} = \frac{z_{13}-z_{12}}{T} \tag{10.19}$$

z_{23} associates with the point obtained through the extrapolation of z_{32} by using the velocity $\boldsymbol{v}_1$, and its candidate value of velocity is calculated as

$$\boldsymbol{v}_{22} = \frac{z_{23}-z_{32}}{T} \tag{10.20}$$

Using the rules defined by (10.16) to conduct verification, it can be found that $\boldsymbol{v}_{12}$ is roughly equal to $\mathbf{v}_{22}$. Thus, combining them yields

$$\boldsymbol{v}_2 = \frac{\boldsymbol{v}_{12}+\boldsymbol{v}_{22}}{2} \tag{10.21}$$

Since there is no measurement falling in the validation gates of the points obtained through the extrapolation of z_{22} by the velocity $\boldsymbol{v}_{21}$ and $\boldsymbol{v}_{31}$, $\boldsymbol{v}_2$ is the only candidate value of velocity at time $k=3$, and it is verified that $\boldsymbol{v}_2$ is roughly equal to $\boldsymbol{v}_1$. Combining them gives the final estimate of the group velocity as

$$\boldsymbol{v} = \frac{\boldsymbol{v}_1+\boldsymbol{v}_2}{2} \tag{10.22}$$

10.2.4.3 Center Extrapolation Algorithm

If the number of targets in the group is $2<N\le 6$ and the direct estimation algorithm is adopted to calculate the group velocity, the estimation error will become large. If the association and distinction

algorithm is adopted, there will be more calculation work to do. Therefore, it is necessary to adopt another algorithm for group velocity estimation, that is, the center extrapolation algorithm.

This algorithm conducts estimations under the following procedure. It first supposes that the relative position of each target in the group is basically fixed. Through several extrapolations, it deletes those measurements which do not belong to any group (e.g., false alarms). Then, from the measurements of the real target in the group, the center of the group is calculated, thereby estimating the group velocity finally.

1. *Establishing the basic set and candidate set.* To estimate the group velocity with this method, it is necessary to establish the basic set and candidate set first. Choose any two measurements $z_1 = [x_{w1} \; y_{w1} \; z_{w1}]'$ and $z_2 = [x_{w2} \; y_{w2} \; z_{w2}]'$ in $\boldsymbol{U}_i(1)$ as the basic set. If the line connecting $z_3 = [x_{w3} \; y_{w3} \; z_{w3}]'$ with $z_4 = [x_{w4} \; y_{w4} \; z_{w4}]'$ in $\boldsymbol{U}_j(2)$ parallels the line connecting z_1 with z_2 (i.e., if (10.23) is satisfied), then z_3 and z_4 can be chosen as the candidate set. For the same basic set, there can be many candidate sets. Suppose that m candidate sets can be found:

$$\begin{cases} |(x_{w1}-x_{w2})-(x_{w3}-x_{w4})| < \sigma_x \\ |(y_{w1}-y_{w2})-(y_{w3}-y_{w4})| < \sigma_y \\ |(z_{w1}-z_{w2})-(z_{w3}-z_{w4})| < \sigma_z \end{cases} \tag{10.23}$$

   If no measurement in $\boldsymbol{U}_j(2)$ can meet (10.23), it is necessary to reselect two measurements from $\boldsymbol{U}_i(1)$ as the basic set, and repeat the above procedure until the candidate set corresponding to the above basic set surely exists. If no basic set in $\boldsymbol{U}_i(1)$ can meet the above conditions, the initiation of the group track can start from time $k=2$ or $k=3$.

   As shown in Figure 10.6, if the measurements z_{11} and z_{21} in $\boldsymbol{U}_i(1)$ are selected as the basic set, then the measurements z_{12} and z_{22}, and z_{32} and z_{42} in $\boldsymbol{U}_j(2)$, can form two candidate sets. Then, $m=2$.
2. *Expansion of basic sets and reducing the number of candidate sets.* This can be done as follows. Select another measurement z^* from $\boldsymbol{U}_i(1)$, which had better not be in the same line with the members of the basic set. Then select a candidate set $\boldsymbol{C}$ from $\boldsymbol{U}_j(2)$, calculate the velocity $\boldsymbol{v}^*$ of $\boldsymbol{C}$'s center relative to the basic set's center, extrapolate the measurement z^* by use of the velocity $\boldsymbol{v}^*$, establish a validation gate around the extrapolation point, and then associate the extrapolation point with the measurements in $\boldsymbol{U}_j(2)$. If the association is successful, then the measurement z^* joins the basic set, the successfully associated measurements in $\boldsymbol{U}_j(2)$ join the candidate set $\boldsymbol{C}$, and update the center of the basic set and candidate set $\boldsymbol{C}$. If the association fails, delete the selected candidate set. If associations for all the candidate sets fail, select another measurement from $\boldsymbol{U}_i(1)$ and repeat the procedures above.

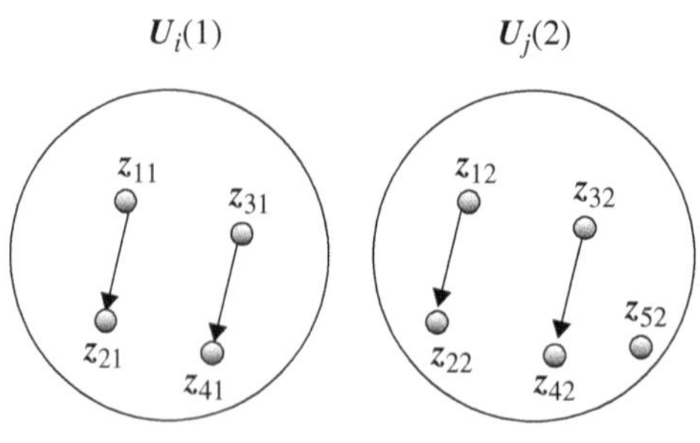

Figure 10.6 Diagram of establishment of basic set and candidate set

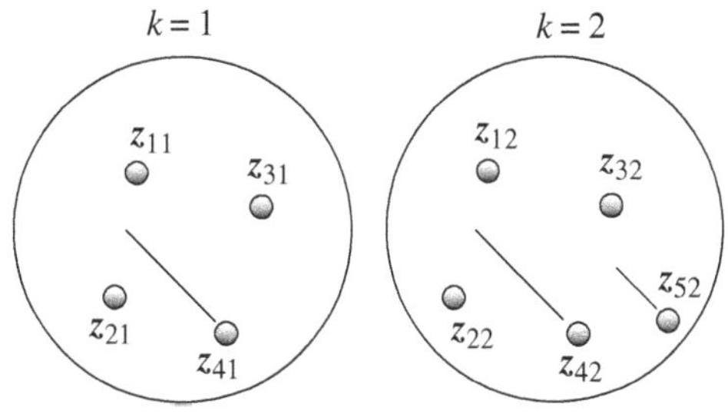

Figure 10.7 Diagram of expansion of basic sets

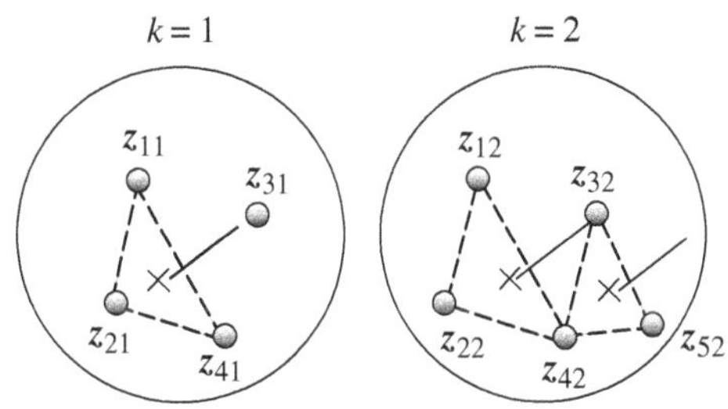

Figure 10.8 Diagram of reduction of the number of candidate sets

If the number of candidate sets is still larger than 1 after this process, repeat the procedures mentioned above until the number of candidate sets is 1 or there is no measurement to choose from in $\boldsymbol{U}_i(1)$.

As shown in Figure 10.7, select z_{41} and join it to the basic set, then the measurement z_{42} can join the first candidate set, and z_{52} can join the second candidate set. Hence, the number of candidate sets is $m = 2$, and thus it is necessary to further join measurement z_{31} to the basic set.

As shown in Figure 10.8, it is verified that measurement z_{32} can join the first candidate set and no measurement in $\boldsymbol{U}_j(2)$ can join the second one, so delete the second candidate set, then the number of candidate sets is $m = 1$.

3. *Estimating group velocity.* By use of the centers of the basic set and the candidate set, the candidate value of group velocity can be calculated. If this value is the only one, it is the group velocity. If there is more than one, extrapolate based on step 2 using each candidate value of the group velocity, establish the validation gates centered at the extrapolation points, and associate the extrapolation points with the measurements in $\boldsymbol{U}_m(3)$. The successfully associated one that has the largest number of measurements is the group velocity. If two or more candidate values of the group velocity have the same number of successfully associated measurements, it is necessary to continue to extrapolate the point to the next moment until the group velocity can be gained according to the rules. Since there are not many candidate values of velocity left after step 2, the computational load of step 3 will not be too heavy.

As illustrated in Figure 10.9, the number of candidate sets is $m = 1$, and the group velocity candidate value v is calculated as

$$v = \sqrt{v_x^2 + v_y^2 + v_z^2} \tag{10.24}$$

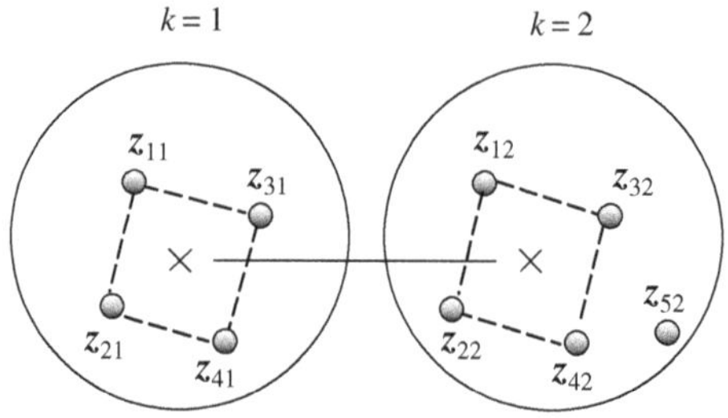

Figure 10.9 Final basic set and candidate set

$$v_x = \frac{z_{2x} - z_{1x}}{T}, \quad v_y = \frac{z_{2y} - z_{1y}}{T}, \quad v_z = \frac{z_{2z} - z_{1z}}{T} \tag{10.25}$$

where $[z_{1x}, z_{1y}, z_{1z}]'$ is the center of the basic set while $[z_{2x}, z_{2y}, z_{3z}]'$ is the center of the candidate set. Since v is the only one, it is the estimate of group velocity.

10.3 The Gray Fine Track Initiation Algorithm for Group Targets

As can be seen from Section 10.2, most of the existing algorithms [46, 146, 240] for track initiation of the group targets are based on the K method [236, 254], cluster seeding method [236, 254], diagramming method [246], etc. They first undertake group segmentation, make group association based on the equivalent measurements of the group, estimate the group velocity according to the number of targets within the group, and finally get the state values for the group's equivalent measurements. Their advantage is that the mistaken cross-association between the group targets is avoided to a large degree, and the calculation is reduced. The disadvantages mainly include:

1. Since it is likely that new members join the group at any time, or old members leave the group at any time, and that the measurement in the group might be lost when the radar resolution is low, it is inaccurate to conduct a direct group segmentation simply based on the spatial distance in the cluttered environment – the group association and the estimation of the group velocity will be unstable, and the precision of the initial track will be lower.
2. In practical implementations, like intercepting low-altitude-formation penetrating targets, or tracking group targets with special significance, etc., there is a real need to establish the course for individual targets within the group while tracking the whole group target. Hence, track initiation is needed for individual targets in the group, that is, fine track initiation of group targets is required. However, most of the existing track initiation algorithms for group targets are only able to gain the state of the group centers, and have not yet solved the problem of fine track initiation of group targets in cluttered environments.

In order to further solve these problems, this section begins with a complete framework for track initiation of group targets, specified as follows.

Suppose that $\mathbf{Z}(k)$ is the kth measurement set gained by the sensors, that is,

$$\mathbf{Z}(k) = \{z_i(k)\} \quad i = 1, 2, \ldots, m_k \tag{10.26}$$

where m_k indicates the number of measurements, $z_i(k) = [x, y, t]'$, and t is the actual time when the radar system outputs measurement $z_i(k)$. Because some of the existing radar systems output

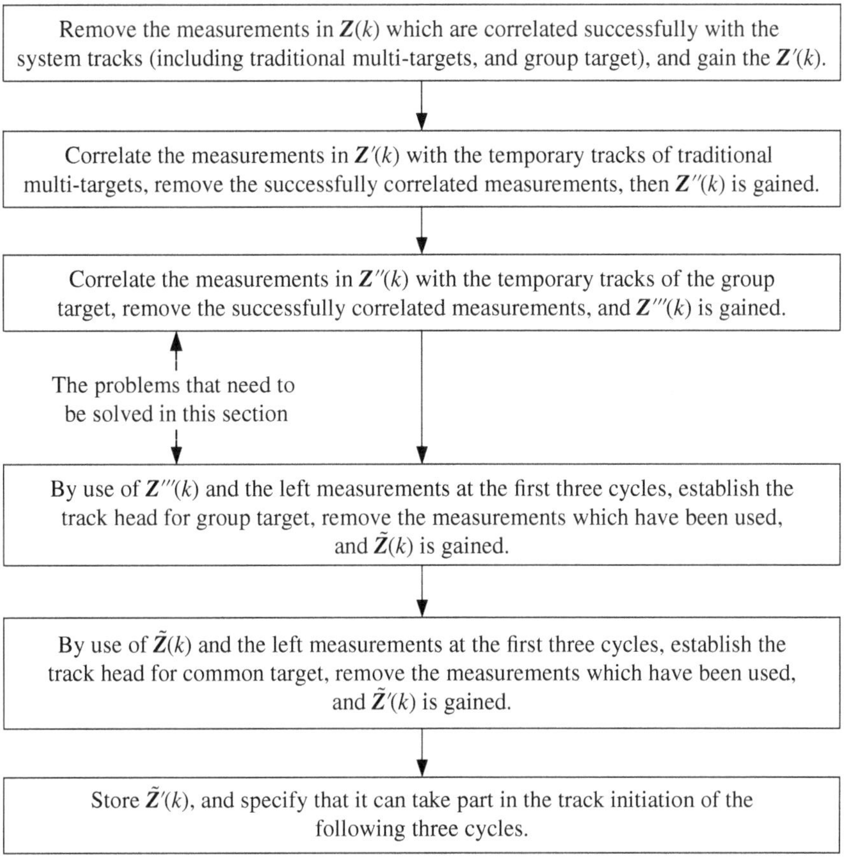

Figure 10.10 Framework for track initiation of group targets

measurements by section, even in the same detection cycle, the output times of each measurement might differ.

Suppose that at time k the sensor has already confirmed that the system tracks are composed of traditional multi-target tracks and group target tracks, and that track initiation includes dual track initiation of the conventional multi-target and the group targets. The complete initiation framework is shown in Figure 10.10.

In order to complete the third and fourth steps of the framework for track initiation of group targets, this section has proposed the gray fine track initiation algorithm for group targets based on the relative position vector. The detailed procedure is shown in Figure 10.11.

10.3.1 Gray Fine Association of Targets within the Group Based on the Relative Position Vector of the Measurement

As indicated from the definition of group targets, the relative position of each target in the group drifts slowly, the target echoes in the same group in neighboring cycles can form a whole with a relatively stable structure, and the extent to which affine transformation happens to the whole is small, which is mainly affected by measurement error. In the phase of track initiation, for group

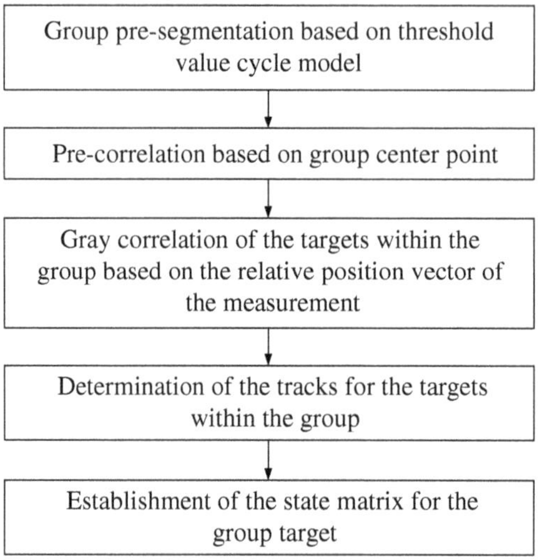

Figure 10.11 Flowchart of the algorithm

targets which have successfully pre-associated at neighboring moments, the relative positional relationship of the target echoes inside the group is basically unchanged, but it is simply translation and rotation as a whole. The appearance of clutter within the neighboring cycles is random rather than associated as a whole, as with the echoes of real targets. This is the theoretical basis for the gray fine association of targets within the group based on the relative position vector of the measurement.

10.3.1.1 Establishment of Relative Position Vector of Measurement

Suppose that $\mathbf{Z}_1$ and $\mathbf{Z}_2$ are two groups which pre-interconnect successfully with one another at neighboring cycles. $\mathbf{Z}_1$ is at the front while $\mathbf{Z}_2$ is at the back, and

$$\mathbf{Z}_1 = \left\{z_{l_1}^1\right\}_{l_1=1}^{\tilde{m}_1}, \quad \mathbf{Z}_2 = \left\{z_{l_2}^2\right\}_{l_2=1}^{\tilde{m}_2} \tag{10.27}$$

where $\tilde{m}_1$ and $\tilde{m}_2$ are the numbers of measurements, respectively, of the two groups. The relative position vector for each measurement in $\mathbf{Z}_1$ and $\mathbf{Z}_2$ can be established as follows.

Establish Corresponding Coordinate Systems

1. *Establish the basic coordinate system and the reference coordinate system.* The basic and the candidate set can be established through center extrapolation, as in Section 10.2.4. First select any two measurements $z_1^1 = \left[x_1^1, y_1^1\right]'$ and $z_2^1 = \left[x_2^1, y_2^1\right]'$ from $\mathbf{Z}_1$. If there exist two measurements $z_1^2 = \left[x_1^2, y_1^2\right]'$ and $z_2^2 = \left[x_2^2, y_2^2\right]'$ in $\mathbf{Z}_2$, and the length and direction of their connection line is basically the same as that of the connection line of z_1^1 and z_2^1, then (10.28) is satisfied. Set the midpoint of the connection line of z_1^1 and z_2^1 as the origin, and the basic coordinate system is established

according to the ground rectangular coordinate system. In a similar way, set up the reference coordinate system based on z_1^2 and z_2^2:

$$\begin{cases} |d_1-d_2|<a\sigma_\rho \\ |\theta_1-\theta_2|<b\sigma_\theta \\ d_1=\sqrt{\left(x_1^1-x_2^1\right)^2+\left(y_1^1-y_2^1\right)^2} \\ d_2=\sqrt{\left(x_1^2-x_2^2\right)^2+\left(y_1^2-y_2^2\right)^2} \\ \theta_1=c\pi+d\arcsin\dfrac{\left(y_1^1-y_2^1\right)}{\sqrt{\left(x_1^1-x_2^1\right)^2+\left(y_1^1-y_2^1\right)^2}} \\ \theta_2=c\pi+d\arcsin\dfrac{\left(y_1^2-y_2^2\right)}{\sqrt{\left(x_1^2-x_2^2\right)^2+\left(y_1^2-y_2^2\right)^2}} \end{cases} \tag{10.28}$$

where σ_ρ and σ_θ are the standard deviations of measurement error, respectively, of the directions of ρ and θ, a and b are the coefficients of the threshold, and c and d are related, respectively, to the quadrants in which the measurements $\left(z_1^2-z_1^1\right)$ and $\left(z_2^2-z_2^1\right)$ lie. If the measurement $\left(z_1^2-z_1^1\right)$ or $\left(z_2^2-z_2^1\right)$ lies in the first quadrant, then $c=0$ and $d=1$. If it lies in the second quadrant, then $c=1$ and $d=-1$. If it lies in the third quadrant, then $c=2$ and $d=-1$. If it lies in the fourth quadrant, then $c=1$ and $d=1$.

2. *Establish the comprehensive value of the coordinate origin.* For a basic coordinate system, there may be several reference coordinate systems meeting (10.28), but actually there is only one corresponding to it at most. As for the overall relationship between the coordinate origin and each measurement in the group, the corresponding coordinate system is the nearest one. Therefore, the comprehensive value of the coordinate origin can be established to describe the overall relationship between the coordinate origin and each measurement in the group, before the determination of the corresponding coordinate system is complete.

   Divide the basic and the reference coordinate systems clockwise from the polar axis into $\widehat{S}$ quadrants [255]. Connect all the measurements in the same quadrant of the basic and reference coordinate systems to their respective origins, and make judgments based on (10.28). Undertake summation of the Euclidean distances between the measurements meeting (10.28) in the quadrant and the coordinate origin, which will be the component of this quadrant. Take the reference coordinate system $\widehat{j}$, for example, and define the comprehensive value $C_{\widehat{j}}$ of the coordinate origin as

$$C_{\widehat{j}}=\left[\sum_{s=1}^{S_1}\rho^{0i_s^1},\ldots,\sum_{s=1}^{S_{\widehat{n}}}\rho^{0i_s^{\widehat{n}}},\ldots,\sum_{s=1}^{S_M}\rho^{0i_s^{\widehat{N}}}\right] \tag{10.29}$$

where $\rho^{0i_s^{\widehat{n}}}=\sqrt{\left(x_{\widehat{j}}^{20}-x_{\widehat{j}}^{i_s^{\widehat{n}}}\right)^2+\left(y_{\widehat{j}}^{20}-\hat{y}_{\widehat{j}}^{i_s^{\widehat{n}}}\right)^2}$ indicates the Euclidean distance between the coordinate origin $z_{\widehat{j}}^{20}=\left(z_{1\widehat{j}}^2+z_{2\widehat{j}}^2\right)/2=\left[x_{\widehat{j}}^{20},y_{\widehat{j}}^{20}\right]'$ of the reference coordinate system $\widehat{j}$ and the sth measurement in the group falling in the $\widehat{n}$th quadrant and satisfying (10.28). $S_{\widehat{n}}$ is the number of measurements meeting (10.28) in the quadrant $\widehat{n}$.

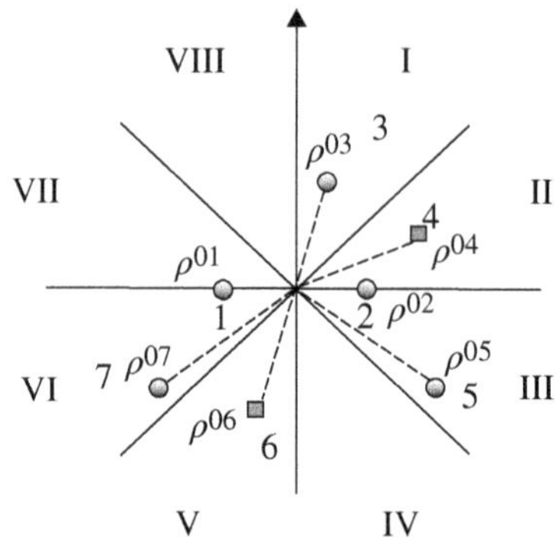

Figure 10.12 Relative position of measurements in reference coordinate system $\hat{j}$

Suppose that the quadrant number $\widehat{S}=8$, and that there are seven measurements in the group $\boldsymbol{Z}_2$, as shown in Figure 10.12. After detection, it is found that 5 measurements of group $\boldsymbol{Z}_2$ in the reference coordinate system $\hat{j}$ and the measurements of group $\boldsymbol{Z}_1$ in the basic coordinate system satisfy (10.28), while z_4^2 and z_6^2 do not. Then the comprehensive value of the coordinate origin of the reference coordinate system $\hat{j}$ is $C_{\hat{j}}=\left[\rho^{03},\rho^{02},\rho^{05},0,0,\rho^{01}+\rho^{07},0,0\right]$. The comprehensive values $\boldsymbol{B}_{\hat{i}}$ and $C_{\hat{j}}$, $\hat{j}=1,\ldots,\widehat{M}$, respectively, of the coordinate origins of the basic coordinate system $\hat{i}$ and $\widehat{M}$ reference coordinate systems can be obtained in a similar way.

3. *Determine corresponding coordinate systems.* To find the similarities between the reference and the basic coordinate system, establish the statistic $T_{\hat{i}\hat{j}}$ based on (10.30) and select the basic and reference coordinate systems which have the smallest $T_{\hat{i}\hat{j}}$ as the corresponding coordinate systems:

$$\frac{T_{\hat{i}\hat{j}}=1-B_{\hat{i}}C_{\hat{j}}^{T}}{\sqrt{|B_{\hat{i}}||C_{\hat{j}}|}};\quad \hat{i}=1,\ldots,\widehat{N};j=1,\ldots,\widehat{M} \tag{10.30}$$

where $\widehat{N}$ is the number of basic coordinate systems.

Establish Relative Position Vectors

After the corresponding coordinate systems are determined, establish the relative position vectors $\boldsymbol{W}_2=\left\{w_{l_2}^2\right\}$, $l_2=1,\ldots,n_2$ for the measurements in group $\boldsymbol{Z}_2$:

$$w_{l_2}^2=\mathrm{Pol}\left(\left[x_{l_2}^2-x_{j^*}^{20},y_{l_2}^2-y_{j^*}^{20}\right]'\right)=\left(\rho_{l_2}^2,\theta_{l_2}^2\right) \tag{10.31}$$

where $\mathrm{Pol}(\cdot)$ is the function which transforms rectangular coordinates into polar coordinates, $\left[x_{j^*}^{20},y_{j^*}^{20}\right]'$ is the coordinate in the ground rectangular coordinate system of the coordinate origin of the reference coordinate system j^*, and $\left(\rho_{l_2}^2,\theta_{l_2}^2\right)$ is the distance and bearing of the measurement

z_i^2 relative to the coordinate origin. The relative position vectors $\boldsymbol{W}_1 = \left\{ w_{l_1}^1 \right\}, l_1 = 1, \ldots, n_1$ of the measurements in $\boldsymbol{Z}_1$ can be gained in a similar way.

10.3.1.2 Establishment of the Gray Fine Association Model

For $\boldsymbol{Z}_1$ and $\boldsymbol{Z}_2$, the positions of the target echoes in their respective corresponding coordinate systems are basically the same. The relative position vectors describe the positions of the measurements in their corresponding coordinate systems. Therefore, based on the relative position vector of each measurement, similarity between the measurements in the pre-association group at different moments can be judged so as to remove the clutter and realize the fine association of the targets in the group. Here, the gray theory [175] is adopted to solve the problem.

Problem Description

For convenience, only the associated groups within two neighboring cycles are considered. By regarding the l_1 measurements from group $\boldsymbol{Z}_1$ as the l_1 known modes, and the measurement $z_{l_2}^2$ from $\boldsymbol{Z}_2$ as the mode to be identified, the fine association of the measurements of the targets in the pre-associated groups in different cycles has been transformed into a matter of typical mode identification.

Gray Association Degree between Relative Position Vectors of the Measurements

1. *Determination of data columns.* Select measurement $z_{l_2}^2$ of $\boldsymbol{Z}_2$ as the reference vector, denoted $w_0 = \left\{ w_{l_2}^2(g),\ g = 1, 2 \right\}$. Suppose that n_1 measurements in $\boldsymbol{Z}_1$ are vectors for comparison, denoted $w_{l_1} = \{ w_{l_1}(g),\ g = 1, 2,\ l_1 = 1, \ldots, n_1 \}$.
2. *Data standardization.* In order to make sure that the data is comparable, it is necessary to conduct a generative treatment of the data columns in the analysis of gray association. Here the interval value method is adopted to normalize the characteristic data of the relative position of the measurements:

$$w_{l_1}(g) = \frac{w_{l_1}(g) - \min\limits_{l_1} w_{l_1}(g)}{\max\limits_{l_1} w_j(g) - \min\limits_{l_1} w_j(g)} \quad l_1 = 1, \ldots, n_1 \tag{10.32}$$

$$w_0(g) = \frac{w_0(g) - \min\limits_{l_1} w_{l_1}(g)}{\max\limits_{l_1} w_j(g) - \min\limits_{l_1} w_j(g)} \quad l_1 = 1, \ldots, n_1 \tag{10.33}$$

3. *Calculating the coefficients of gray association.* It follows from the standard deviation of measurement error $\sigma = \left[\sigma_\rho, \sigma_\theta \right]'$ that the association coefficient of the reference vector w_0 and the comparison vector w_j is

$$\xi_{l_1}(g) = \frac{\sigma(g)}{\sigma(g) + |w_0(g) - w_{l_1}(g)| \bullet A(g)} \tag{10.34}$$

where $A(g) = \left(\max\limits_{l_1} w_{l_1}(g) - \min\limits_{l_1} w_{l_1}(g) \right)$. This gives the association coefficient of the reference vector w_0 and the comparison vector w_{l_1} as $\xi_{l_1} = \left\{ \xi_{l_1}(g), g = 1, 2 \right\}$.

4. *Calculating the gray association degree.* For the sake of comparison, it is necessary to incorporate each indicator of the association coefficients into one value, which is called the "gray association degree." Denote the gray association degree of the comparison vector w_{l_1} to the reference vector w_0 as $\gamma(\omega_0,\omega_{l_1})$ (γ_{l_1} for short). As can be seen from (10.31), the relative position vector of the measurements is composed of the distance and bearing of the measurements in the group relative to the origins of the corresponding coordinate systems. Without considering the system errors, the information on distance and bearing is affected by measurement noise. When the distance measurement noise is large, the detection distance of the target gained by the radar differs a lot from the real distance, and the contribution of the distance information to the relative positions of measurements is less reliable, at which point a smaller weight should be assigned to the distance information. The same occurs with the bearing information indicator. Define the gray association degree as

$$\gamma_{l_1}=\lambda_1\xi_{l_1}(\rho)+\lambda_2\xi_{l_1}(\theta)=\frac{\sigma(\theta)\sigma_{\max}(\rho)\xi_{l_1}(\rho)+\sigma(\theta)\sigma_{\max}(\rho)\xi_{l_1}(\theta)}{\sigma(\rho)\sigma_{\max}(\theta)+\sigma(\theta)\sigma_{\max}(\rho)} \tag{10.35}$$

where $\sigma_{\max}=[\sigma_{\max}(\rho),\sigma_{\max}(\theta)]'$ is the largest value of the standard deviation of radar measurement error. When it is impossible to determine $\sigma_{\max}$, make $\lambda_1=\lambda_2=0.5$, which can be used to meet common demands.

10.3.1.3 Rules of Fine Association for Gray Association Measurements

After getting the gray association degree describing the degree of closeness of two measurements at the relative positions in the corresponding coordinate system, it is necessary to judge whether or not these measurements associate with each other. In order to make the judgment on the association between the measurements $z_{l_2}^2$ and $z_{l_1}^1|l_1=1,\ldots,n_1$, it is necessary to rank the gray association degree from largest to smallest, thereby getting the gray association order. Here, the maximum association degree identification principle is adopted, that is,

$$\gamma_*=\max_{l_1}\gamma_{l_1} \tag{10.36}$$

and

$$\gamma_*>\varepsilon \tag{10.37}$$

Then, measurement $z_{l_2}^2$ is considered associated with $z_{l_1^*}^1$ and not be associated with any other measurements. Otherwise, $z_{l_2}^2$ is considered as clutter. ε is the threshold value parameter and $\varepsilon\leq 1$. Its actual value assignment is related to all the measurements constituting ω_0,ω_{l_1} and measurement error σ. The calculation formula is (10A.13) in Appendix 10A (for a detailed derivation, see Appendix 10A). Finally, the measurement set $\hat{\mathbf{Z}}=\{(z_c^1,z_c^2)\}_{c=1}^{C}$ composed of the corresponding and associated measurements in $\mathbf{Z}_1$ and $\mathbf{Z}_2$ is obtained, in which C is the number of association measurement pairs.

10.3.2 Confirmation of the Tracks within a Group

After the possible tracks are established based on the corresponding and associated measurements set, the output of the confirmed track for each target in the group is completed by means of the logic rule of 3/4, and the initiation rate of false tracks is further reduced.

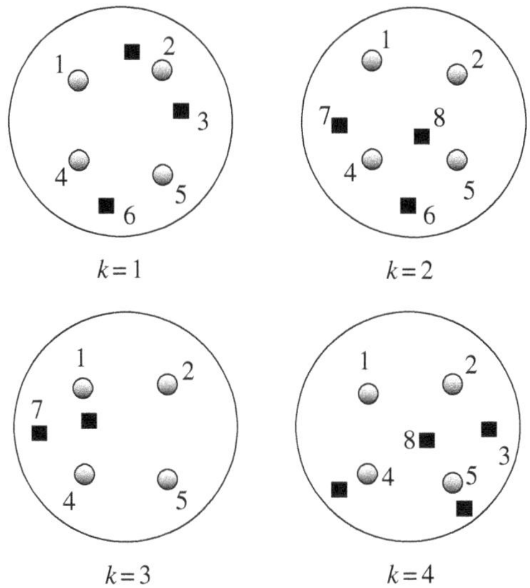

Figure 10.13 Measurement distribution diagram of pre-association group

In order to more clearly describe the confirmation of tracks, an example is cited here. Suppose that Figure 10.13 shows the measurement distribution in a pre-association group for four successive processing cycles. There are eight possible tracks formed on the basis of the gray fine association model. In the figure, measurements marked with the same sequence number belong to the same track, but the tracks with more than three association measurements are only $\{1, 2, 4, 5\}$. According to the logic rule of 3/4, only these four tracks are confirmed. Output $\{1, 2, 4, 5\}$ and cancel the other tracks.

10.3.3 Establishment of State Matrixes for Group Targets

In order to fully describe the state of group targets, establish a state matrix for them on the basis of (10.38). The first column indicates the state of the group center, while the other n columns are the states of the n confirmed tracks in the group. It needs to be clarified that the state of the group center and the state covariance are determined by the confirmed tracks in the group, and are unrelated to the results of pre-segmentation and pre-association.

$$\boldsymbol{X} = \begin{bmatrix} x_0 & x_1 & \cdots & x_n \\ \dot{x}_0 & \dot{x}_1 & \cdots & \dot{x}_n \\ y_0 & y_1 & \cdots & y_n \\ \dot{y}_0 & \dot{y}_1 & \cdots & \dot{y}_n \\ t_0 & t_1 & \cdots & t_n \end{bmatrix} \tag{10.38}$$

10.3.4 Simulation Verification and Analysis of the Algorithm

In order to verify the performance and effectiveness of the algorithm, a Monte Carlo simulation with $N = 100$ runs was conducted for the gray fine track initiation algorithm for group targets based on relative position vectors (group algorithm), modified logic method (logic algorithm), and track initiation algorithm of multi-formation based on cluster and Hough transformation in Ref. [146] (center algorithm). This section presents a comparison and analysis of the performances of these algorithms in this test in terms of track initiation.

10.3.4.1 Simulation Environment

Suppose that the sampling period of the radar $T = 1$ s, the radar's direction measurement error and distance measurement error are, respectively, $\sigma_\theta = 0.3°$ and $\sigma_r = 40$ m. In order to compare those algorithms in terms of performance in track initiation in different environments, the following three typical simulation scenarios are set.

Environment 1: Simulation of sparse group targets in clutter. Suppose that there are 10 targets in a two-dimensional plane, among which the eight targets divide into two groups, and the distance between the targets in the sparse group is usually in the range of (600 m, 1000 m). The first group is in uniform linear motion, and is composed of the first four targets; their initial positions are, respectively, (5000 m, 800 m), (5400 m, 1400 m), (5850 m, 1500 m), and (6100 m, 900 m); their initial velocity is uniform (0 m/s, 300 m/s).

The second group conducts maneuvers, and is composed of five to eight targets; their initial positions are, respectively, (−5000 m, 10 000 m), (−5200 m, 9400 m), (−4900 m, 8600 m), and (−5300 m, 8000 m); their initial velocity is uniform (−270 m/s, 270 m/s), and their initial accelerated velocity is uniform (5 m/s^2, −10 m/s^2). The other two targets are in uniform linear motion and their initial positions are, respectively, (10 000 m, −8000 m) and (−10 000 m, −8000 m); their initial velocities are, respectively, (−240 m/s, 200 m/s) and (200 m/s, 230 m/s).

The clutter generation in simulation is divided into two parts. For the common target T_0, establish a matrix with T_0 as the center and $[10\,\sigma_\rho, 10\,\sigma_\theta]$ as the side length in polar coordinates. In this matrix, λ_1 clutters are evenly produced. For the group targets $\boldsymbol{G}$, calculate the center point $\bar{G}$ of the group targets, and establish a matrix with $\bar{G}$ as the center and $[2\Delta G_\rho + 10\sigma_\rho, 2\Delta G_\theta + 10\sigma_\theta]$ as the side length in polar coordinates (ΔG_ρ and ΔG_θ are the maximum difference values of each measurement in $\boldsymbol{G}$ on the two coordinate axes of the polar coordinate system). In this matrix, λ_2 clutters are evenly produced. Here, $\lambda_1 = 2$ and $\lambda_2 = 4$.

Environment 2: Simulation of the dense group targets in clutter. The distance between the targets in dense group target cases is usually in the range of (100 m, 300 m). The initial positions of each target in the first group turn into (5000 m, 800 m), (5200 m, 850 m), (5350 m, 900 m), and (5550 m, 830 m). The initial positions of each target in the second group turn into (−5000 m, 10 000 m), (−5100 m, 9800 m), (−5000 m, 9650 m), and (−5050 m, 9500 m). The other parameters are the same as those in environment 1.

Environment 3. In order to verify the variation of the algorithms' comprehensive initiation capability with the variation in clutter and sensor measurement error, on the basis of environment 1, the value assignment of clutter (in numbers), the distance measurement error (in meters), and the angle measurement error (in degrees) are shown in Table 10.1.

10.3.4.2 Simulation Results and Analysis

Figure 10.14 is a partially enlarged drawing of the overall situation for the 10 targets, in which there are two group targets and two common targets. Figure 10.15 is the distribution diagram of sensor measurements at the first four moments, in which, compared with that of conventional targets,

Table 10.1 Value assignment for clutter and measurement error in environment 3

λ_1	1	2	3	4	5	6
λ_2	2	4	6	8	10	12
σ_ρ	20	40	60	70	80	100
σ_θ	0.1	0.3	0.5	0.7	0.9	1.2

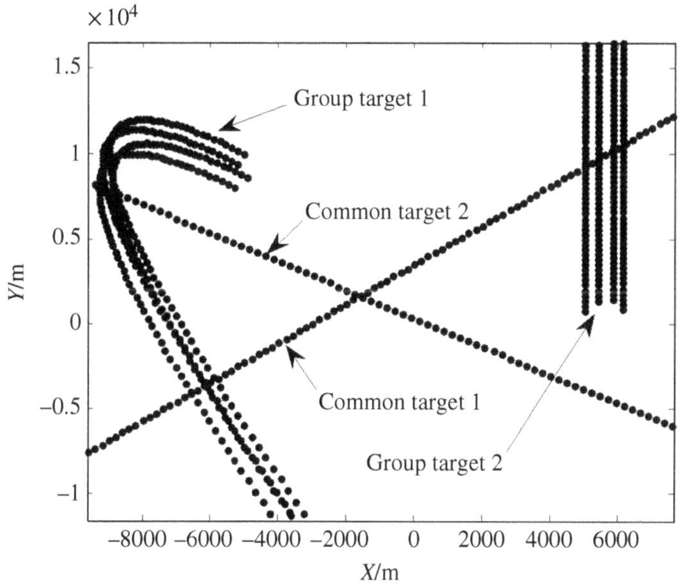

Figure 10.14 Targets' overall situation (environment 1)

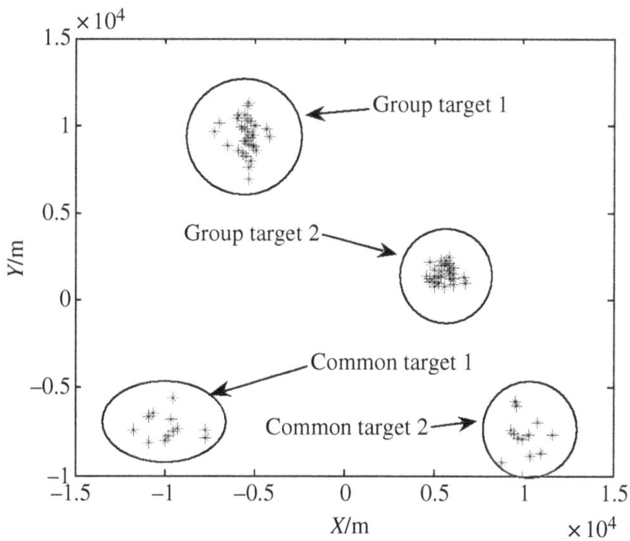

Figure 10.15 Measurements distribution at the first four moments (environment 1)

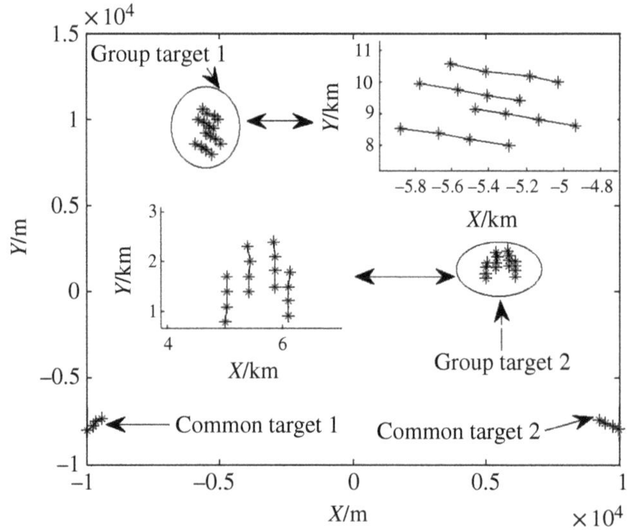

Figure 10.16 The real tracks of each target at the first four cycles (environment 1)

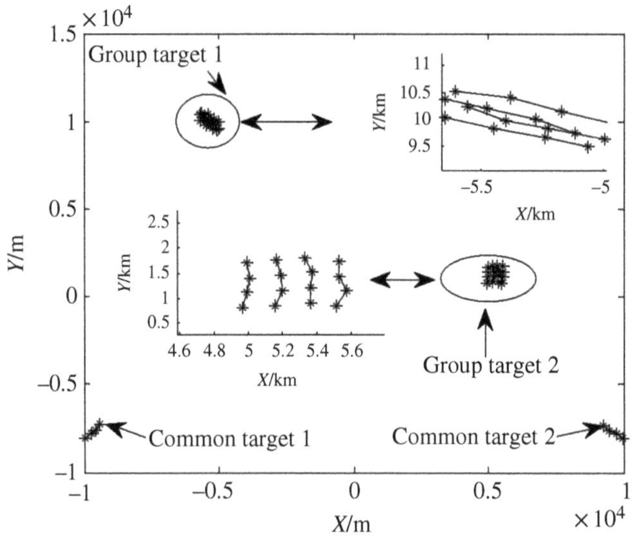

Figure 10.17 The real tracks of each target at the first four cycles (environment 2)

the group targets' measurement distribution is much denser. Figures 10.16 and 10.17 are, respectively, the targets' real track pictures in environments 1 and 2, and the group's motion states are similar in the two environments. But in terms of density of tracks in the group, the latter is greater than the former.

Figure 10.18 shows the comparison of track initiations in environment 1, between the logic, group, and center algorithms, respectively, for the first group targets (subgraphs (a)–(c)), the second group targets (subgraphs (d)–(f)), and the first common target (subgraphs (g)–(i)).

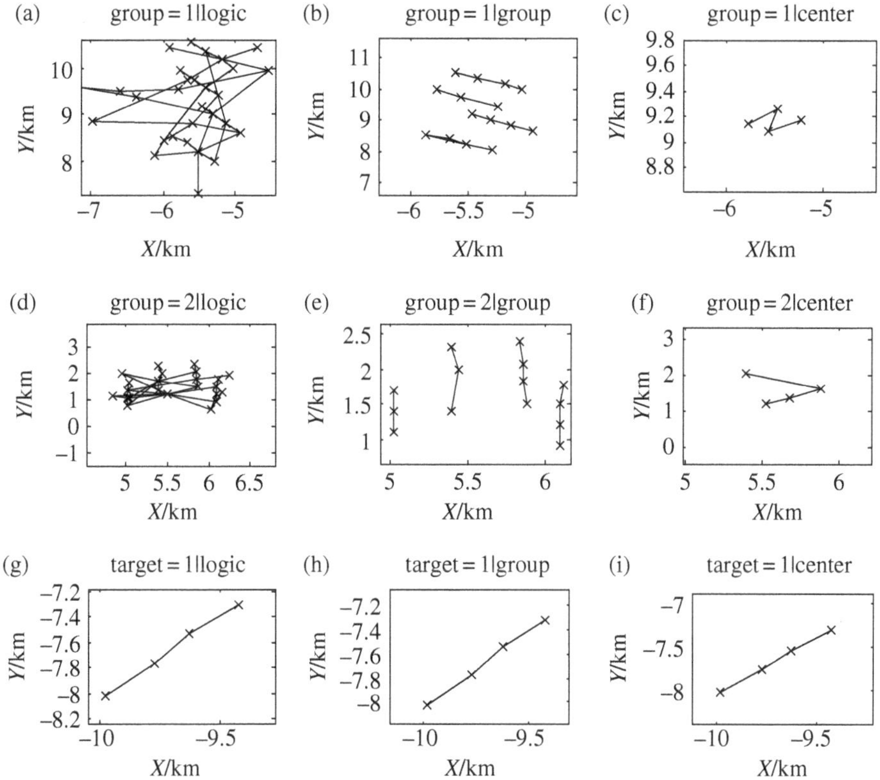

Figure 10.18 Comparison of track initiation of the three algorithms (environment 1)

Figure 10.19 shows the comparison of track initiations between the three algorithms in environment 2. Subgraphs (a)–(c) are for the first group targets, subgraphs (d)–(f) are for the second, and subgraphs (g)–(i) for the first common target.

As indicated by the comparison between Figures 10.18 and 10.19 with Figures 10.16 and 10.17, in terms of group measurements in the two environments, the logic algorithm initiated several false tracks, so that it is impossible to identify the real motion posture of the group. The center algorithm can establish only one track for each group, and has low track accuracy. The group algorithm can initiate each target in the group basically accurately, there is only one track crossing of the initiation for the first group in Figure 10.19, and its overall effect is obviously better than that of the logic and center algorithms. For the common target, since the initial logic adopted by each algorithm is the same, their respective initiation effects are the same.

The reasons for the results above are as follows. The logic algorithm is non-pre-emptive, that is, the measurement already used to establish tracks can still be used for other tracks so as to ensure a higher correct track initiation rate and track accuracy. But the initiation rate of false tracks will also be greatly increased. The center algorithm conducts initiation based on the group center point, and can initiate no more than one track. As a result, there will be a loss of situations. In addition, the clutter is likely to cause deviations of the group center point from the real value, and drops in the accuracy of the established track, so that even track establishment is impossible. The group algorithm is to conduct precise track establishment within the group for each association group based on relative vectors of measurements so that the impact of clutter is alleviated to the largest degree, and

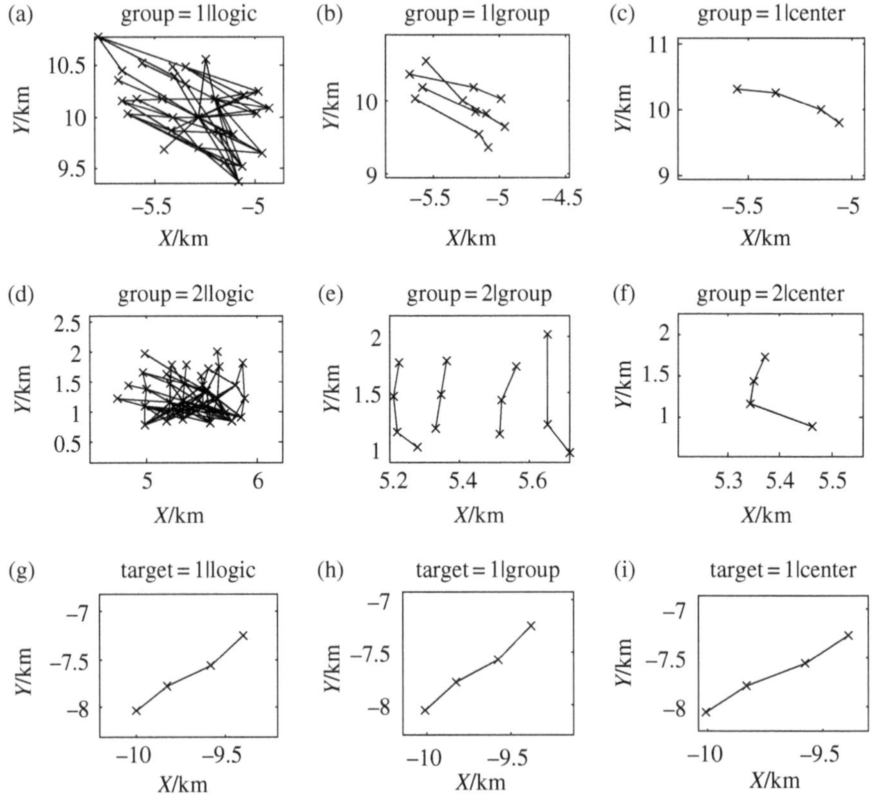

Figure 10.19 Comparison of track initiation of the three algorithms (environment 2)

the false tracks are removed on the basis of 3/4 logic to ensure higher correct track initiation rate and lower false track initiation rate.

In order to quantize the effectiveness of track initiation for each algorithm, we established two indicators – the overall initiation track quality indicator and the overall initiation track accuracy indicator – and offer a comparison diagram of the algorithms in terms of the two indicators in 50 simulations (each simulation includes 100 Monte Carlo simulations). The indicators are established in the following procedure.

1. *Verification of initiated tracks.* Suppose that the track initiation algorithms establish T_l tracks on the basis of the measurements of the four cycles, in which the ith track's state is $\hat{X}_i = \left[\hat{x}_i, \hat{v}_{ix}, \hat{y}_i, \hat{v}_{iy}\right]'$. To calculate the overall quality and accuracy of the tracks initiated by the algorithms, first it is necessary to determine the number of real tracks among the T_l tracks. Suppose that the real track of T targets at this moment is $\left\{X_j = \left[x_j, v_{jx}, y_j, v_{jy}\right]'\right\}_{j=1}^{T}$. If $\hat{X}_i$ and X_j satisfy (10.39), then $\hat{X}_i$ is the candidate track corresponding to the real track X_j. That is,

$$\begin{cases} |\hat{\rho}_i - \rho_j| < \xi_\rho \\ |\hat{\theta}_i - \theta_j| < \xi_\theta \\ \Delta d < \xi_d \end{cases} \tag{10.39}$$

where ξ_ρ, ξ_θ, ξ_d are the threshold values, respectively, for judging velocity magnitude, velocity direction, and position distance, and are related to measurement error. $(\hat{\rho}_i, \hat{\theta}_i) = \text{Pol}(\hat{v}_{ix}, \hat{v}_{iy})$, $(\rho_j, \theta_j) = \text{Pol}(v_{jx}, v_{jy})$, and $\text{Pol}(\cdot)$ is the function which transforms rectangular coordinates into polar coordinates. If the ith track includes four measurements, then

$$\Delta d = \sqrt{(\hat{x}_i - x_j)^2 + (\hat{y}_i - y_j)^2} \tag{10.40}$$

If the ith track includes only three tracks, then

$$\begin{aligned} \Delta d &= \min\left(\sqrt{(\hat{x}_i - x_j)^2 + (\hat{y}_i - y_j)^2}, d'\right) \\ d' &= \sqrt{(\hat{x}_i - x_{3j})^2 + (\hat{y}_i - y_{3j})^2} \end{aligned} \tag{10.41}$$

where (x_{3j}, y_{3j}) is the third measurement point of the jth real track.

In $\{\hat{X}_i\}_{i=1}^{T_l}$ there may be multiple candidate tracks corresponding to X_j. Define the comprehensive value D_{ij} to make judgments, that is,

$$\begin{cases} D_{ij} = |\hat{\rho}_i - \rho_j| + |\hat{\theta}_i - \theta_j| + \Delta d \\ i^* = \underset{i=1:T'}{\text{argmin}}\,(D_{ij}) \end{cases} \tag{10.42}$$

where T' is the number of candidate tracks corresponding to X_j. Set $\hat{X}_{i^*}$ and X_j as 0, and then they cannot be used to judge whether other tracks are true or false. Add 1 to l_{true} indicating the number of real tracks, and store $D_{l_{true}} = D_{i^*j}$.

2. *Establishment of overall initiation track quality.* In track initiation, it is necessary to initiate as many true tracks as possible, and as few false tracks as possible. Therefore, it is possible to comprehensively indicate whether an algorithm is good or bad by use of the correct track initiation rate, false track initiation rate, and omitted track initiation rate. Equation (10.43) is employed to define the overall initiation track quality P_{qu} of an algorithm, in which the numerator is the sum of the algorithm's false track initiation rate and omitted track initiation rate, while the denominator is the correct track initiation rate. Then, the smaller P_{qu} the better the track initiation.

$$P_{\text{qu}} = \frac{\left[\left(1 - \dfrac{l_{true}}{T}\right) + \dfrac{(T_l - l_{true})}{T}\right]}{\dfrac{l_{true}}{T}} \tag{10.43}$$

3. *Establishment of overall initiation track accuracy.* The track state required for track initiation should be maximally congruent with the true track, so we can judge whether an initiation algorithm is good or not by means of state accuracy parameters such as the position and velocity of initiated tracks. Equation (10.44) is used to define the overall initiation track quality P_{pr} for an algorithm. D_i fully contains the accuracy information of position, velocity magnitude, and velocity direction. Therefore, the smaller P_{pr} is, the better the track initiation is.

$$P_{pr} = \frac{\sum_{i=1}^{l_{true}} D_i}{l_{true}} \tag{10.44}$$

Figures 10.20 and 10.21 serve to compare the overall qualities of tracks initiated by the algorithms in the 50 time simulations, respectively, of environments 1 and 2. It can be seen that the overall track quality of the group algorithm is much better than that of the logic and the center algorithm. The rationale for this is that the logic algorithm is non-pre-emptive, so its initiation rate of real tracks is quite high while its initiation rate of omitted tracks is very low. However, its false track initiation rate is far higher than that of the other two algorithms, so its overall initiation track quality is low.

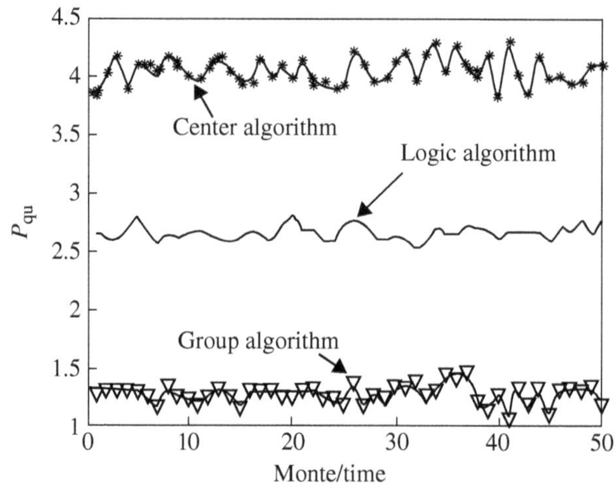

Figure 10.20 Comparison of overall initiation track qualities for the algorithms (environment 1)

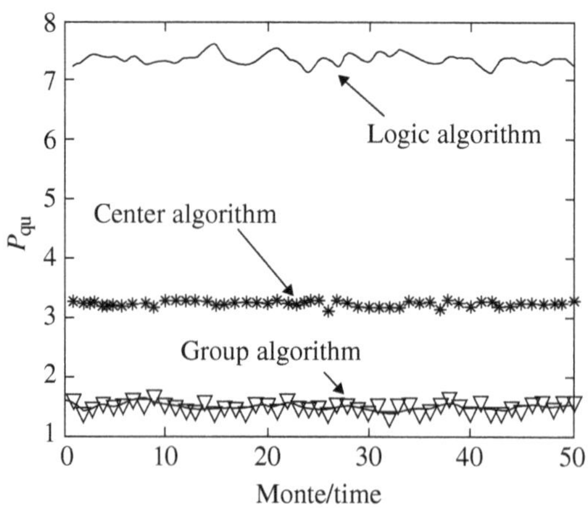

Figure 10.21 Comparison of overall initiation track qualities for the algorithms (environment 2)

The center algorithm conducts track establishment simply based on the center point of the group, and its false track initiation rate is relatively low. However, its real track initiation rate is low and the initiation rate of omitted tracks is high, so its overall quality of initiation track is low.

Compared with the logic algorithm, the group algorithm might have a slightly lower correct track initiation rate but a much lower false track initiation rate, so its overall track quality is higher. In comparison with the logic algorithm, its false track initiation rate might be the same, but its correct track initiation rate is far higher. Therefore, its overall track quality is far higher as well.

Figures 10.22 and 10.23 serve to compare the overall quality of initiation track for each algorithm in the 50 simulations, respectively, of environments 1 and 2. It can be seen that the logic algorithm has the highest overall track accuracy, the group algorithm takes second place, and the center algorithm is the worst. The reason is that each of the measurements in the logic algorithm is used to initiate many tracks, and the optimal association point can be found for each track. The

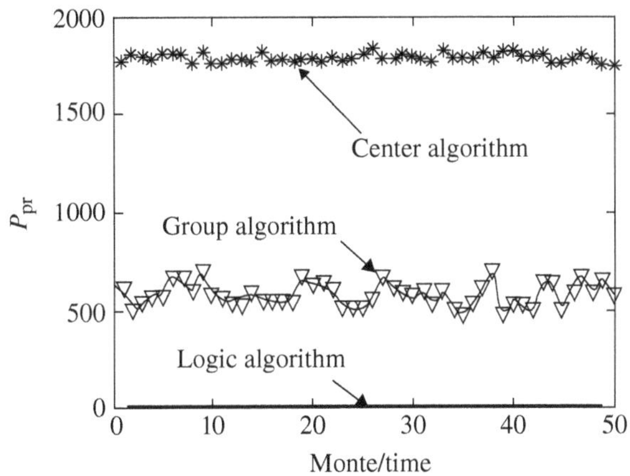

Figure 10.22 Comparison of overall initiation track accuracies for the algorithms (environment 1)

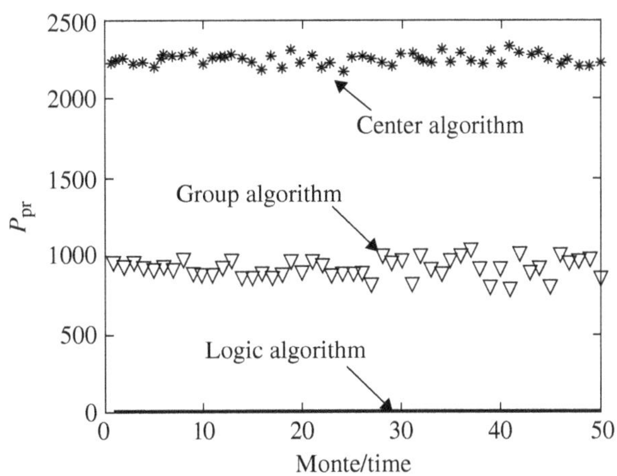

Figure 10.23 Comparison of overall initiation track accuracies for the algorithms (environment 2)

group algorithm is pre-emptive, and each measurement can associate with only one track. Measurement association errors of one track will cause disturbance to the other tracks in finding their true association points, so the overall quality of initiation track will be somewhat reduced. The center algorithm initiates tracks by use of the center point of the group, but in the presence of clutter, the clutter might be involved in the group which is being segmented, so that the center point deviates from the true value. Therefore, a group's accuracy of the track initiated by means of the center point is relatively low.

In order to verify the adaptability of each algorithm to clutter and sensor measurement error, two evaluation indicators are brought in.

i. The ratio of the number of correct tracks to the number of true tracks is defined as

$$P_{\text{Correct}} = \frac{T_{\text{initiation}}}{T_{\text{true}}} \tag{10.45}$$

where $T_{\text{initiation}}$ is the number of correct tracks initiated by algorithms, while T_{true} is the number of true tracks. The larger P_{correct} is, the more capable the algorithm is of initiating the track correctly.

ii. The ratio of the number of false tracks to the number of true tracks is defined as

$$P_{\text{Error}} = \frac{T_{\text{false}} + T_{\text{seep}}}{T_{\text{true}}} \tag{10.46}$$

where T_{false} is the number of false tracks initiated by algorithms, and $T_{\text{false}} = T_{\text{num}} - T_{\text{initiation}}$. T_{num} is the total number of tracks initiated by algorithms. T_{seep} is the number of true tracks that fail to be initiated successfully by the algorithms, and $T_{\text{seep}} = T_{\text{true}} - T_{\text{initiation}}$. It needs to be noted that since T_{false} may be larger than T_{true}, so P_{Error} may be larger than 1. The larger P_{Error} is, the less capable the algorithm is of initiating true tracks and suppressing false tracks.

In Tables 10.2 and 10.3, the variations of P_{Correct} and P_{Error} are compared for each algorithm in environment 3.

As shown in Table 10.2, with the increase in clutter, P_{Correct} of the logic algorithm remains the highest, and can nearly ensure the initiation of all the true tracks, because this algorithm conducts track establishment for all the measurements meeting association requirements without considering the repetitive use of measurements. Hence, the true tracks must be contained in the tracks established by it, and T_{seep} is nearly 0. However, in this way, the cost is that multiple false tracks have been established, and P_{Error} is much higher than those of the other two algorithms. When the

Table 10.2 Comparison of P_{Correct} and P_{Error} varying with the number of clutters between the algorithms

Clutter	λ_1	1	2	3	4	5	6
	λ_2	2	4	6	8	10	12
P_{Correct}	logic	1	1	0.9990	1	1	0.9990
	group	0.8850	0.7490	0.7050	0.6490	0.6090	0.5360
	center	0.3450	0.3570	0.3720	0.3730	0.2120	0.1970
P_{Error}	logic	1.3360	2.6470	4.4870	6.7910	9.4050	12.9870
	group	0.4280	0.6530	0.7490	0.9910	1.2703	1.5370
	center	0.7230	0.7980	0.9430	1.1000	1.0400	0.9680

Table 10.3 Comparison of P_{Correct} and P_{Error} varying with measurement errors between the algorithms

Measurement error	σ_ρ	20	40	60	70	80	100
	σ_θ	0.1	0.3	0.5	0.7	0.9	1.2
P_{Correct}	logic	1.0000	1.0000	0.9880	0.9180	0.8380	0.7360
	group	0.7950	0.7860	0.7380	0.6800	0.6190	0.5590
	center	0.3450	0.3430	0.3370	0.3070	0.2770	0.2340
P_{Error}	logic	1.3060	1.3070	1.2000	1.3250	1.5300	1.8240
	group	0.4120	0.4310	0.4560	0.5630	0.6270	0.6960
	center	0.7980	0.7210	0.7110	0.7590	0.8090	0.8680

clutter number is (6, 12), P_{Error} is as high as 12.9870, that is, the algorithm can initiate over 120 false tracks, which is more than 10 times that of the other two algorithms.

The center algorithm's P_{Correct} is reduced to some degree, and is always below 0.4, which cannot meet practical demands, because this algorithm can establish only one track for one group. Although T_{false} is lower, T_{seep} is higher and when the clutter is dense, its T_{false} will increase.

Since it undertakes special processing of the measurements within the group based on relative position vector, the group algorithm is less affected by clutter. Although P_{Correct} falls a little, and P_{Error} rises slightly, its overall initiation effectiveness keeps at a higher level, and its robustness towards clutter is better than that of the other two algorithms.

It can be seen in Table 10.3 that as the measurement error increases, P_{Correct} of the three algorithms all decreases to some degree, and P_{Error} all rises to some degree. The variation amplitude of the center algorithm is the smallest, because its track initiation is composed of the group track initiation and the track initiation of a common target, the measurement error impacts slightly on the group segmentation, and the number of group tracks formed by this algorithm is only related to the number of groups, thereby the formation of group tracks is less affected. Hence, the impact of measurement error on the algorithm results only from the impact on the track of a common target.

The variation amplitude of the logic algorithm is moderate, because this algorithm treats group targets as common targets. Hence, the measurement error also affects the tracks of the group targets.

The variation amplitude of the group algorithm is relatively large, because the premise for this algorithm to conduct fine initiation of the targets in a group is that the relative positions of each target in the group drift slowly. But larger measurement errors enlarge the amplitude of affine transformation of the overall shape of each target measurement in the same group at the phase of track initiation, and reduce the accuracy in judging the fine association relationship between the measurements in the group based on relative position vector, which leads to mistaken measurement association and increases the initiation rate of false tracks. Figures 10.24 and 10.25 show the measurement distribution of the group targets in the first four cycles, respectively, under the two measurement errors. It can be found that the amplitude of affine transformation for the same group targets in each cycle of the latter is obviously larger than that of the former.

10.3.5 Discussion

In order to solve the problems of fine association of the targets within a group, this section proposes a gray fine track initiation algorithm for the group targets based on relative position vector. The advantages of this algorithm are as follows.

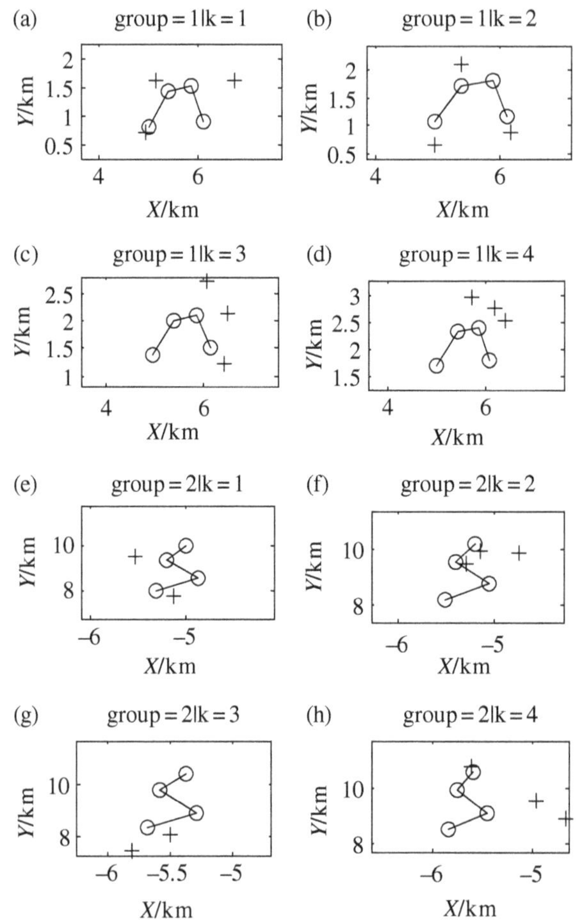

Figure 10.24 First four cycle measurements of group target ($\sigma_\theta = 0.3°$, $\sigma_r = 40\,\text{m}$)

1. It classifies targets into group targets and common targets, avoiding the high false track rate when traditional algorithms are used to initiate group targets.
2. It conducts a precise treatment of the successfully pre-associated group and establishes tracks, respectively, for the targets within the group based on relative position vector by means of gray theory, thus avoiding the loss of posture and decrease in track accuracy when the existing track initiation algorithms for group targets are used to establish tracks, simply based on the group's equivalent measurement.
3. It maximally removes the clutter in the group by means of the relative position vectors of the measurements within the group and has a greater adaptability to the clutter, thus avoiding the production of a large amount of false measurements, and ensuring the overall accuracy for the initiation tracks as well.

The disadvantages of this algorithm are that when the measurement error is very large, the affine transformation of the group shape might be large, and the algorithm is no longer applicable. And this algorithm, which accounts only for the case of single sensors, needs to be extended to multi-sensor

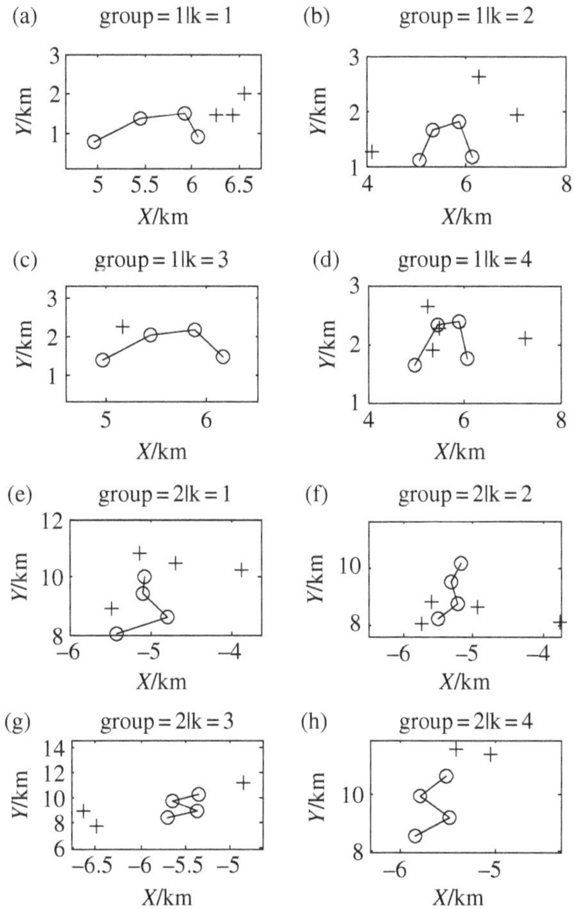

Figure 10.25 First four cycle measurements distribution of group targets ($\sigma_\theta = 1.2°$, $\sigma_r = 100\,\text{m}$)

systems. Therefore, it is necessary to further investigate the track features of the group targets, establish the unscented variables for the measurement errors, and solve the problems of fine track initiation for multi-sensor group targets when the measurement error is large.

10.4 Centroid Group Tracking

Centroid group tracking (CGT) was proposed by Frazier and Scott in the 1970s [248]. The CGT algorithm conducts track initiation by human assistance or automatically. It tracks the group center directly by use of the Kalman filter, and is the most direct tracking algorithm for group targets. This algorithm has good real-time performance, but yields poor results in cases of group measurement loss or in the presence of dense clutter. In the late 1990s, Blackman made a summary of the tracking algorithms for group targets in *Design and Analysis of Modern Tracking Systems*, focusing on the implementation procedures and merits/demerits of the CGT algorithm and formation tracking algorithm [45]. That summary is a milestone in the history of group target tracking.

The tracking coordinate system for the CGT algorithm is usually chosen between the Cartesian and the polar coordinate systems. References [248, 249] select the Cartesian coordinate system but given the distance variation rate, the polar coordinate system is preferable, and the distance and angle can be selected as the state variables.

State variables use standard state variables for tracking but in the case of ground-moving targets, their speed and azimuth angle should be included in the state variables [248]. Supposing that the amplitude of the speed's variation is less than that of the azimuth angle, the speed magnitude can be smoothed.

10.4.1 Initiation, Confirmation, and Cancellation of Group Tracks

The initiation of group tracks can be completed with the assistance of operators [248], and it can also be done automatically by the many measurements which have not been assigned to the existing group tracks. If the group initiation is automatic, the group segmentation can be done by use of the spatial distance segmentation or the circulatory threshold value method. And the estimation of group velocity can be done by means of the direct estimation method. These algorithms have been discussed fully in Section 10.2.

As with single-target or multi-target tracking algorithms, if a group track has been associated successfully at several successive moments, then it is considered determined. On the contrary, if a group track cannot be detected for several successive moments, it will be revoked.

10.4.2 Track Updating

Consider a sensor which is tracking T groups in clutter. The system's dynamic equation can be denoted by

$$\boldsymbol{X}^t(k+1)=\boldsymbol{F}(k)\boldsymbol{X}^t(k)+\boldsymbol{G}(k)\boldsymbol{V}^t(k), \quad k=1,2,\ldots, \quad t=1,2,\ldots,T \tag{10.47}$$

where $\boldsymbol{X}^t(k+1)$ is the overall state vector for the center of group t, and in the two-dimensional coordinate system $\boldsymbol{X}^t(k+1)=[x \ \dot{x} \ y \ \dot{y}]'$.

The measurement equation is denoted by

$$z^t(k)=\boldsymbol{H}(k)\boldsymbol{X}^t(k)+\boldsymbol{W}(k) \tag{10.48}$$

where measurement $z^t(k)$ is the measurement value of the center for the group targets t at time k, which can be gained by means of the measurements falling in group t. Suppose that $\boldsymbol{Z}(k)$ is the measurement set at time k, and

$$\boldsymbol{Z}(k)=\{z_i(k)\}_{i=1}^{m_k} \tag{10.49}$$

where m_k is the number of measurements at time k.

For the filter model on the basis of (10.47) and (10.48), the procedure for updating group tracks is composed of three steps.

1. *Establishing the tracking gate for the one-step predicted value of the group center.* For temporary tracks, either the rectangular gate or the ellipse gate can be considered as the tracking gate, the establishment of which is affected by many factors including the maneuvering of the targets in the group, the distribution of the targets in the group, measurement noise, etc.

Suppose that $\hat{X}^t(k|k-1)$ is the one-step predicted value of the state of the group track t at time k. The measurements falling into its validation gate must satisfy

$$d_i^2(k) = v_i^t(k)' S_G^{t\,-1}(k) v_i^t(k) < d_{\max}^2 \tag{10.50}$$

where $v_i^t(k)$ is the innovation of the one-step predicted value $\hat{z}^t(k|k-1)$ of the measurement $z_i(k)$ corresponding to the group track t. $d_i^2(k)$ is a normalized distance. $d_{\max}^2$ is the allowed maximum value of the normalized distance. $S_G^t(k)$ is a normalized variance matrix [248] defined by Franzier and Scott, the definition equation of which is

$$S_G^t = \hat{S}_D^t + R_G^t + HP^tH' \tag{10.51}$$

where the matrix $\hat{S}_D^t$ is related to the distribution of the measurements in group t, which is an estimate of the distribution of the measurements in the group. R_G^t is the covariance matrix of measurement error for group t. It is different from the covariance matrix defined in the case of single-target tracking. Usually, R_G^t can be derived by the following equation:

$$R_G^t = CR_m^tC' \tag{10.52}$$

where R_m^t is the error covariance matrix of group measurement t under the measurement coordinate system, and C is the transformation matrix from the measurement coordinate system (e.g., the antenna coordinate system) to the tracking coordinate system.

The matrix HP^tH' represents the uncertainty of the center of group t caused by the dynamic change or maneuver of the target. In this matrix, it includes the one-step prediction covariance matrix $P^t(k|k-1)$ for group t and the measurement matrix H, which is similar to the corresponding parameters in single-target tracking.

As a result of extrapolation, the ellipse gate will probably be increased too much. Hence, on the basis that (10.50) is satisfied, it is still necessary to check the actual (un-normalized) distance [248] from the measurement to the group center.

2. *Establishing new groups based on the measurements falling in the validation gate.* New groups are established as follows.
   i. Suppose that measurements meeting (10.50) exist. Select from them the measurement with the minimum normalized distance d^2 as the seed measurement for a group t. On the basis of the seed measurement, establish the group G_0.
   ii. Check the other measurements meeting (10.50). If the measurements meet an approximate standard set for the seed measurement, for example, if the measurement falls in an ellipse gate with the seed measurement as the center, then the measurement will be temporarily joined in the group G_0.
   iii. Set an additional logic process to determine which measurements can finally be retained in the group G_0.

Select the measurements which have not fallen into the group G_0 as the seed measurements. Repeat the above procedure until there is no measurement for choice. After the establishment of all the groups, calculate the group center and distribution matrix.

The establishment of a group is a process in which new measurements are constantly added to the measurements already in the group. The groups established must follow a particular rule, such as:

- Any measurement in the group must meet the two distance standards with respect to the group center and the seed measurement.
- The number of measurements for each group cannot surpass the upper limit set for the number of measurements in the group.

- Given the group's distance range in each sampling interval, the distance variation of all the measurements in the group must be confined to this range.

  If the above rules cannot be followed in establishing a new group around the seed measurement, a new group should be established surrounding the next nearest measurement until the group is successfully established or there is no measurement for choice. This procedure can be applied to all the existing group tracks.

3. *Associating group measurements with group tracks after the establishment of all the groups.* It is likely that we will encounter problems inherent in the association of measurements and tracks, that is, how the group measurements are assigned to the group tracks. The most direct method to solve this problem is to directly assign the group measurements to the group tracks producing seed measurements.

Suppose there are multiple group measurements in the validation gate of the group track t, and define an association algorithm in which the group track t can possibly associate with all measurements in the group. Just as the problem of association of tracks and measurements for a single target is solved, we can calculate the one-step predicted value $\hat{z}(k|k-1)$ for the center of the group track t and the normalized distance for each group measurement center. By use of the normalized distance, their association can finally be realized.

The group measurement set G_0 is used to update the group track t, and Kalman filtering is used to conduct state updating of the group track t. The first step is to determine the covariance matrix $\boldsymbol{R}_c^t$ of measurement noise used to calculate the Kalman gain, that is,

$$\boldsymbol{R}_c^t = \frac{\boldsymbol{R}_G^t}{N_0} + f\left(N_0, \hat{N}_t\right)\hat{\boldsymbol{S}}_D^t \tag{10.53}$$

where N_0 is the number of measurements in G_0; $\hat{N}_t$ is the predicted number of measurements in the gate of the group track t; $\hat{\boldsymbol{S}}_D^t$ is related to the measurements distribution in the group t, representing an estimate of the measurement distribution in t; and $f\left(N_0, \hat{N}_t\right)$ is a weight factor, which is a function of the number of measurements in the group and the expected number of measurements in the gate of the group track.

The covariance matrix $\boldsymbol{R}_c^t$ of measurement noise determined by (10.53) is a sum of two matrixes. The first matrix is $\boldsymbol{R}_G^t/N_0$, which indicates the error of the center of the group measurement resulting from the radar measurement error, and decreases with the number of measurements in the group. The second denotes the uncertainty of the group center. Since not all measurements in the group can be observed, indeterminacy exists. If the number of measurements in the group is known as N_t, and there is no false echo (so $N_0 \le N_t$), then the weight factor is defined as

$$f(N_0, N_t) = \frac{(N_t - N_0)}{(N_t - 1)N_0} \tag{10.54}$$

It follows from (10.54) that the weight factor is 0 when all the measurements in the group are detected, and 1 when there is only one measurement detected.

It follows from the $\boldsymbol{R}_c^t$ given by (10.53) that standard Kalman filtering can be used in state estimation. As for the number of measurements in the group track and the group distribution matrix, their recursive estimate can be gained by use of α trackers, that is,

$$\begin{cases} \hat{N}_t(k) = (1-\alpha)\hat{N}_t(k-1) + \alpha N_0 \\ \hat{\boldsymbol{S}}_D^t(k) = (1-\alpha)\hat{\boldsymbol{S}}_D^t(k-1) + \alpha \hat{\boldsymbol{S}}_{DG0}(k) \end{cases} \tag{10.55}$$

where k is the time mark, α is the common filtering gain of the position component of the target state, and $\hat{\boldsymbol{S}}_{DG0}$ is the estimate distribution matrix corresponding to G_0. Finally, it should be noted that (10.55) should be used to update $\hat{N}_i$ and $\hat{\boldsymbol{S}}_{Di}$ before calculating $\boldsymbol{R}_c^t$ with (10.53).

10.4.3 Other Questions

As can be found from the updating of group tracks, when the CGT approach is employed to conduct group tracking it is necessary to estimate the number of elements in the group for each time, which can be gained by modifying the number of measurements in group detection. In addition, Franzier and Scott [248] defined the maximum number of detected measurements as the estimate number of measurements in the group. This method features easy calculation, but when there is a very high clutter density, this method will not be precise.

When the tracking of a group is conducted by means of CGT, it is still necessary to estimate the matrix $\hat{\boldsymbol{S}}_D$ related to the measurement distribution in the group. As for the two-dimensional tracking (x, y), $\hat{\boldsymbol{S}}_D$ is defined as

$$\hat{\boldsymbol{S}}_D = \begin{bmatrix} s_x^2 & s_{xy}^2 \\ s_{xy}^2 & s_y^2 \end{bmatrix} \tag{10.56}$$

where s_x^2 and s_y^2 are the group's estimate covariances, respectively, in the direction of x and y, and s_{xy}^2 is the cross-covariance in the direction of x and y. For a given group, the standard method can be used to calculate the estimate of center variance $\left(s_x^2, s_y^2\right)$ and mean $(\bar{x}, \bar{y})$. The cross-covariance s_{xy}^2 is

$$s_{xy}^2 = \frac{1}{N}\sum_{i=1}^{N}(x_i - \bar{x})(y_i - \bar{y}) \tag{10.57}$$

where (x_i, y_i) is the ith measurement in the group, and N is the number of measurements in the group.

Another important question in group tracking is to deal with splitting and combining. This is crucial to ensuring the timely adjustment of the size of group targets and the stable tracking of group targets. This question focuses on the elements, rules, algorithms, etc. for splitting and combining the members in the group targets. If the size of the group tracking gate and the range of the measurements in the group are restrained, a split target group should automatically form a new group track. On the contrary, when two or more groups are combined into a new one, this group will contain the measurements of other groups. The centers of the groups will also be combined, and in order to retain only one group track in the same target set, it is necessary to conduct redundancy detection for each group track.

CGT using the group center to track the group directly features less calculation and is easier to understand. However, when there are false measurements in the environment for tracking or the loss of true measurements happens as a result of external causes, the tracking effectiveness of CGT might worsen, because the false measurements within the group will destroy the original distribution matrix of the group, thus leading to an over-large number of measurements in the group. The loss of true target measurements may seriously affect the estimation of group velocity when the target is covered. For example, suppose a motorcade being tracked as a group target enters a covered region, where the pace car cannot be discovered, then the estimate of the group velocity will deviate seriously from its true value.

10.5 Formation Group Tracking

10.5.1 Overview of Formation Group Tracking

In order to reduce the impact of clutter and measurement loss on the calculation of a group center, Flad [250] and Taenzer [237, 251] proposed the formation group tracking (FGT) algorithm. This algorithm is used to track the targets in the group while tracking the center of a group, but because of the necessity to maintain the position estimation of individual targets, the calculation of the result is hard to deal with. In the 1980s, in order to enhance the engineering utility of tracking group targets, Farina came up with the functional flowchart [17] for formation target tracking. Here it needs to be clarified that although the ideas are similar, the specific tracking methods proposed by Flad and Taenzer are to some degree different from each other. The algorithm proposed by Flad is used to track only individual targets in the group instead of the center of the formation, without considering the impact of measurement loss. The algorithm proposed by Taenzer, by use of the estimated center track and the relative position between the center and the formation members, establishes track files, so the center's stability can still be maintained when the measurement is lost.

10.5.2 Logic Description of Formation Group Tracking

The FGT algorithm calculates the group velocity also by means of the variation of the group center. This algorithm extrapolates the measurements in the group by using the group velocity and sampling interval, and the next moment, so that the maintenance of a single-target track is implemented. The advantages of this algorithm can be summarized as follows [45, 233]:

1. Saving the radar and computer resources.
2. Providing position estimation of a single target.
3. Weakening the bad influence of measurement loss and false measurements.
4. The same tracking logic is adopted for both group tracking and single-target tracking.

Both FGT and CGT have the first advantage. When group tracking is conducted, multiple targets (as a group) are tracked by the radar. It is unnecessary to track each target individually, so radar resources are saved. The other three advantages represent FGT's improvement over CGT.

The FGT algorithm makes an improvement on the CGT algorithm discussed in the previous section. Examples are given as follows. The process under consideration is that of a group composed of four targets entering a covered region [45]. Suppose that all four targets can be detected in the first scanning, the two front targets are covered in the second scanning, and the two front targets are detected when leaving the covered region while the two rear targets are covered in the third scanning. z_{ij} is the position of target i in the jth scanning. $\mathbf{Z}_j$ and $\mathbf{Z}_j^*$ are, respectively, the measurement center and the true center of the group in the jth scanning. Note that $\mathbf{Z}_1$ and $\mathbf{Z}_1^*$ are the same, since all four targets in the group can be detected in the first scanning.

In such a scenario, the procedure for track updating with FGT can be described as follows.

1. *Prediction of measurements*. The single target in the group should be extrapolated by use of the estimate of group velocity and the sampling interval to get the predicted position for each target. The single measurement z_{i1} gained in the first scanning is extrapolated by

$$\hat{z}_{i2} = z_{i1} + v_G T \tag{10.58}$$

where $\mathbf{v}_G$ is the estimate of group velocity and T is the sampling interval.

2. *Association of measurements*. Establish the gate with each extrapolation point as the center, and associate the extrapolation point with the measurement point gained in the next scanning. If the association is done successfully, the associated measurements are the measurements in the group. If there is one measurement which succeeds in association, the measurement will be used in estimation of the center and velocity of the group. If there are multiple measurements succeeding in association, select the measurement with the minimum normalized distance for the estimation of the group center and velocity. If the association fails, the measurement is lost. Then, the lost measurement should be replaced with the predicted position of the target so as to estimate the group center and velocity.

   The gate is established with each extrapolation point as the center, and the extrapolation point is associated with the measurement point gained in the second scanning. In the same way as the tracking of single targets or multi-targets, the tracking gate for the predicted position $\hat{z}_{i2}$ can be represented finally as $d^2 = \tilde{y}'_{ij} S_i^{-1} \tilde{y}_{ij}$ through the variance matrix. Therefore, if measurement z_{j2} is to be considered to have fallen into the tracking gate of the group center, it must satisfy

$$d^2 = \tilde{y}'_{ij} \boldsymbol{S}_i^{-1}(k) \tilde{y}'_{ij} < d^2_{\max} \tag{10.59}$$

   where $d^2_{\max}$ is the allowed maximum value of the normalized distance function; $\boldsymbol{S}_i$ is the covariance matrix of measurement error; $\tilde{\boldsymbol{y}}_{ij}$ is the error of the measurement z_{j2} from the predicted position $\hat{z}_{i2}$, which can be represented as

$$\tilde{\boldsymbol{y}}_{ij} = z_{j2} - \hat{z}_{i2} \tag{10.60}$$

   After testing, in the second scanning, the measurements z_{12} and z_{22} fell into the tracking gates of the extrapolation points $\hat{z}_{12}$ and $\hat{z}_{22}$. However, as targets 3 and 4 were covered, there is no measurement falling into the tracking gates of the extrapolation points $\hat{z}_{32}$ and $\hat{z}_{42}$. In the updating of group tracks, substitute the predicted values of $\hat{z}_{32}$ and $\hat{z}_{42}$ for the lost measurements. Finally, use z_{12}, z_{22}, $\hat{z}_{32}$, and $\hat{z}_{42}$ to calculate the measurement center of the group to update the estimate of the group velocity.
3. *Updating of group tracks*. By use of associated measurements and the predicted position of the target, FGT estimates the center and velocity of the group, and updates the group tracks.

In the group tracking environment with a covered region, the group center gained by means of CGT is not precise. The group velocity estimated from the group center is unstable. If CGT is used for tracking in such an environment, when the front targets emerge from the covered region, the group track will probably be lost. However, FGT undertakes position estimation of the single targets in the group, and substitutes the predicted values for the lost measurements. So, a stable group center can be gained. The group velocity estimated in this way is more precise than that estimated by CGT.

FGT also establishes a simple revocation rule to ensure that the target which has sent false alarms or has broken away from the group will not be extrapolated. For example, in the system described in Refs [237, 251], if the measurement of a target is lost two successive times, the track of this target will be revoked. Before the estimate of group velocity is updated, the group measurement which is determined to be revoked should not be involved in the current or previous group center calculation, so that the group velocity gained by this algorithm is more precise.

For FGT, there are some differences between Flad's and Taenzer's approaches in determining which measurements belong to the group track. Flad holds that the new measurements can be joined in the group only after having fallen into the tracking gate of the measurements in the group. In contrast, Taenzer takes all the measurements falling into the tracking gates of distance and angle surrounding the predicted value of the group center into consideration.

Finally, what needs to be emphasized is that, for FGT, although it is necessary to conduct a position estimation of a single target in the group, it is unnecessary to undertake position smoothing. The measurement detected can be regarded as the optimal current estimate of the state for each target in the group, and these estimates can be extrapolated forward in time by use of the estimate of group velocity. Hence, FGT describes the group by use of the position estimate of each target in the group, and directly employs these unsmoothed measurements to calculate the center and velocity of the group.

10.6 Performance Analysis of Tracking Algorithms for Group Targets

10.6.1 Simulation Environment

In order to analyze and compare the performances of the tracking algorithms for group targets in this chapter, two typical simulation environments are hypothesized.

Environment 1: Tracking 20 targets. The first 10 targets constitute the first group, in which the initial position of each target is randomly selected from (500 m, 1000 m) and (−5000 m, −3000 m). The initial velocity is (200 m/s, 400 m/s). The remaining 10 targets form the second group, in which the initial position of each target is randomly selected from (2000 m, 2500 m) and (2000 m, 3000 m), with initial velocity (400 m/s, 200 m/s).

Environment 2: On the basis of environment 1, a covered region is added. From the 10th sampling interval to the 25th sampling interval, the rear 5 targets of the first group are in the covered region. From the 26th sampling interval to the 40th sampling interval, the front 5 targets in the first group are in the covered region.

The process noise component is $q_1 = q_2 = 0.01$, the sampling time interval for 2D radar is 1 s, the sampling duration is 50 s, the distance measurement error is $\sigma_r = 30\,\text{m}$, the angle measurement error is $\sigma_\theta = 0.03$ rad, the detection probability is $P_D = 0.997$, and the gate probability is $P_G = 0.997$. The non-parametric Poisson distribution clutter model is adopted in simulation, and the expected number of false measurements within the gate is $m = 1.8$.

10.6.2 Simulation Results

The simulation results are shown in Figures 10.26–10.35.

10.6.3 Simulation Analysis

Figure 10.26 shows the tracks of the 20 batches of targets in environment 1. As shown in the figure, the 20 batches of targets are clearly divided into two groups when they are moving. Figure 10.27 serves to compare the true motion tracks of the two group centers with the filtering tracks for the two tracking algorithms for group targets. Figures 10.28–10.31, respectively, compare the RMS position and velocity errors when the sensor detection probability $P_D = 0.997$ for CGT and FGT is used to track the two groups. As shown in the figure, the tracking accuracy of FGT is more stable and generally better than that of CGT, whose fluctuation range of tracking accuracy is relatively large.

Figure 10.32 compares the algorithms' single-update durations varying with clutter numbers. As shown in the figure, with the increase in average number of clutters within the gate, the

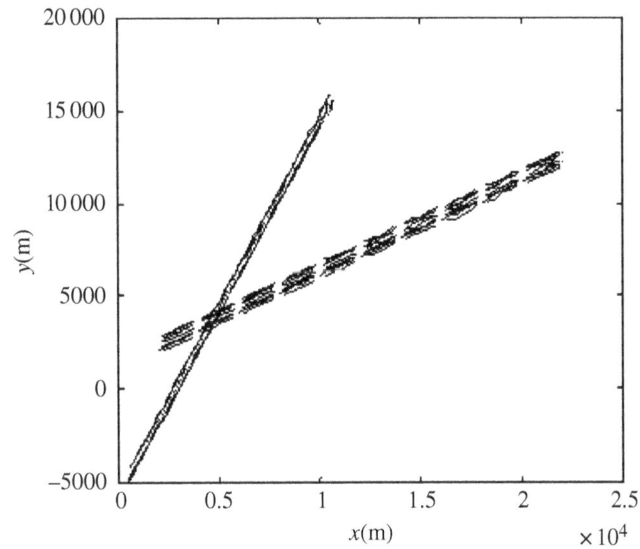

Figure 10.26 Tracks of the targets of 20 batches in environment 1

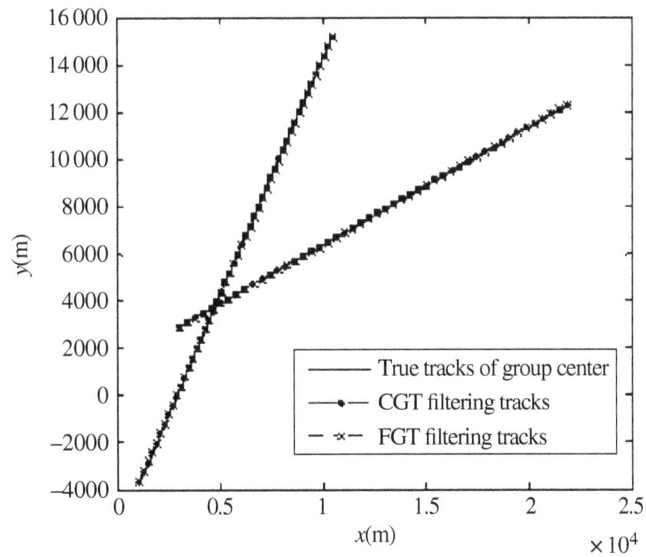

Figure 10.27 Comparison of filtering tracks of two group tracking algorithms

single-update duration for FGT is far larger than that for CGT, and the increase in amplitude of duration for the former is higher than that of the latter.

Figure 10.33 compares the algorithms' correct association rates varying with clutter numbers. As shown in the figure, for the same clutter number, the effective tracking rate of FGT is higher than that of CGT. With the increase in average number of clutters within the gate, the decrease in amplitude of the former is lower than that of the latter.

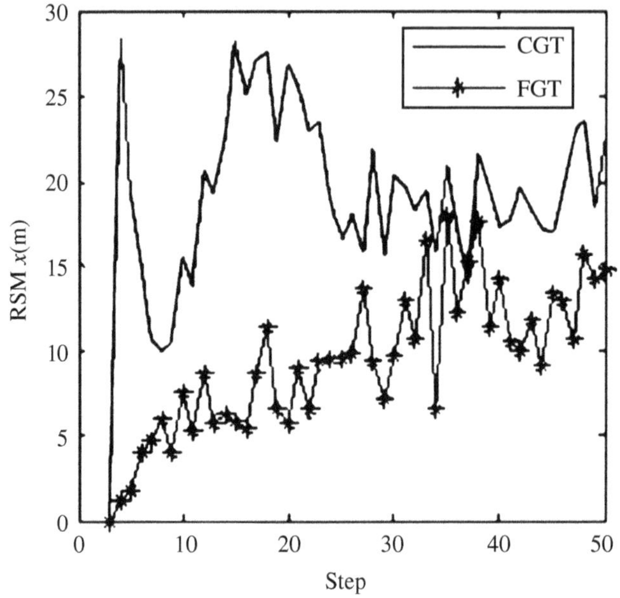

Figure 10.28 Comparison of RMS position errors in x axis for group 1

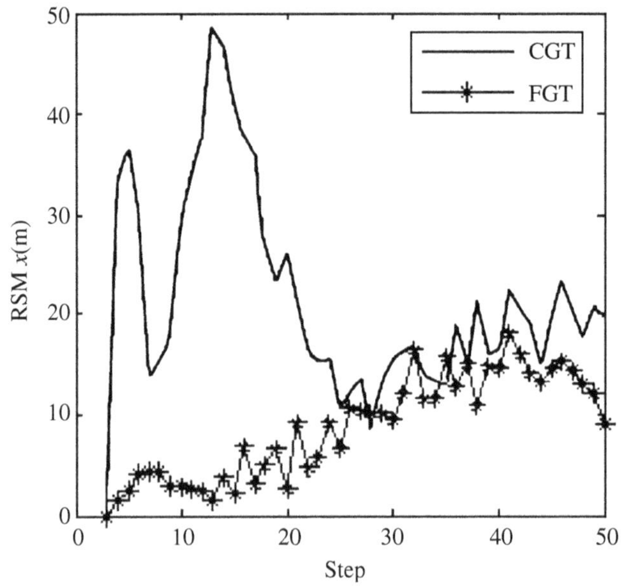

Figure 10.29 Comparison of RMS position errors in x axis for group 2

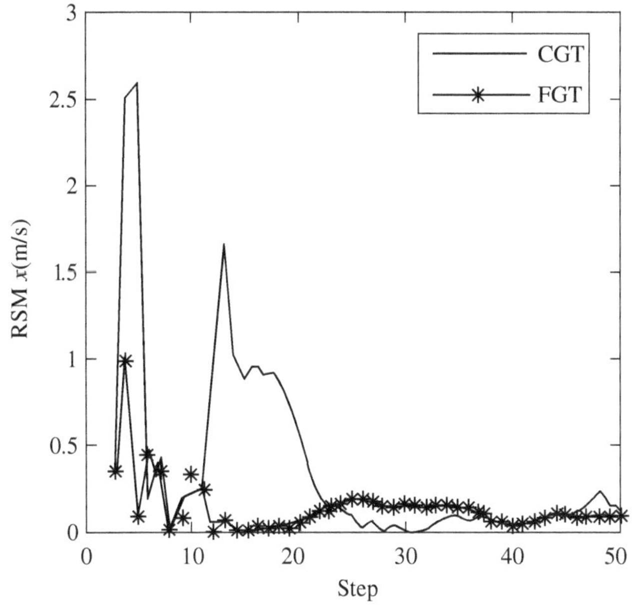

Figure 10.30 Comparison of RMS velocity errors in x axis for group 1

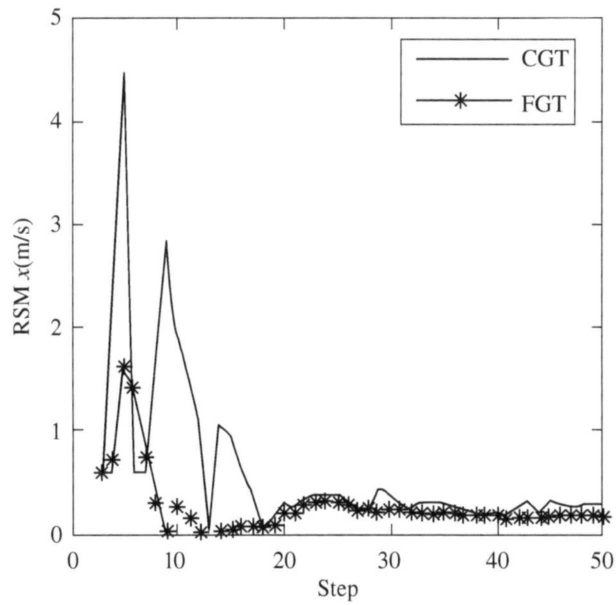

Figure 10.31 Comparison of RMS velocity errors in x axis for group 2

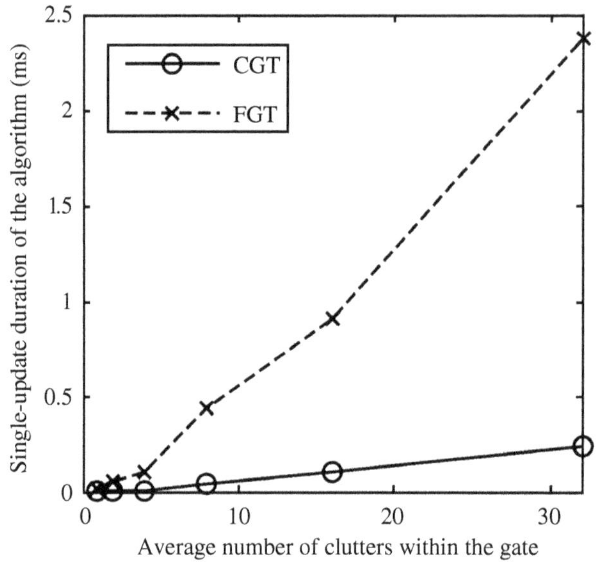

Figure 10.32 Comparison of algorithms' single-update durations varying with number of clutters

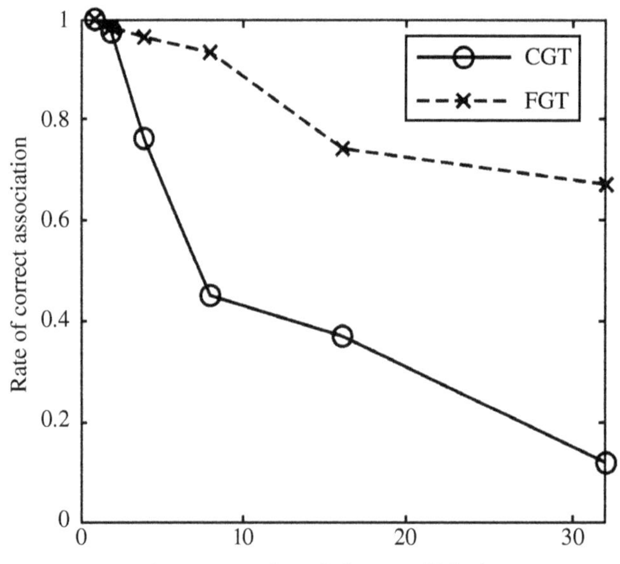

Figure 10.33 Comparison of algorithms' correct association rates varying with number of clutters

Figures 10.34 and 10.35 compare the RMS position and velocity errors, respectively, of CGT and FGT in environment 2. As shown in the figure, compared with CGT, FGT is more stable in tracking accuracy. A great fluctuation of the tracking performance of CGT occurs. Especially when some targets in the group are in the covered region, the RMS position error for CGT is greater than 100 m. Hence, the algorithm is unable to conduct effective tracking of the group. In contrast, the

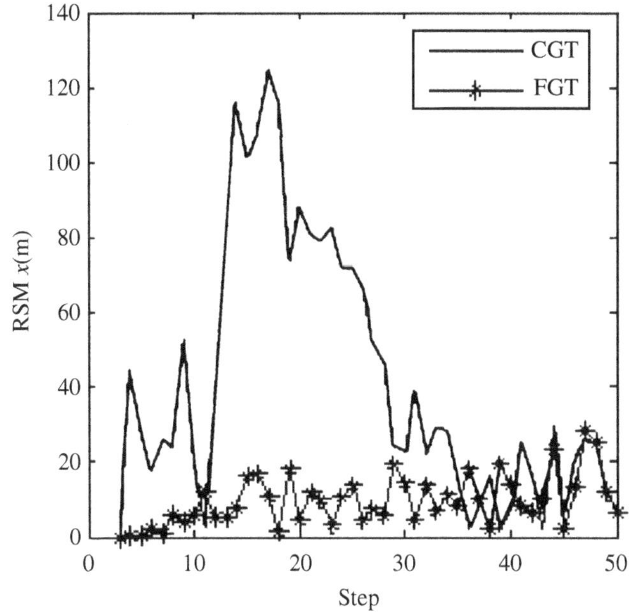

Figure 10.34 Comparison of RMS errors in x axis in environment 2

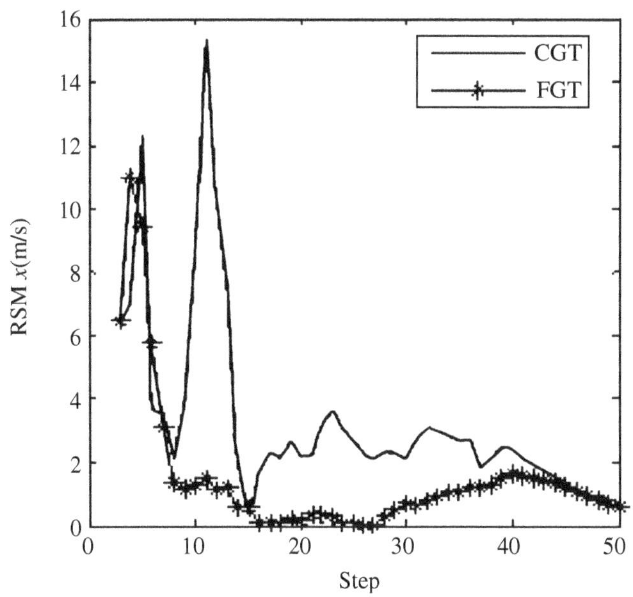

Figure 10.35 Comparison of RMS velocity errors in x axis in environment 2

RMS position error for FGT is less than 20 m, and it can still effectively track the group targets. Therefore, it has a better tracking effect than CGT.

It is known from the above comprehensive comparison that, compared with CGT, FGT has better tracking effectiveness, but it consumes more time than CGT, because the latter only conducts treatment on the group center, but the former needs to process the measurements falling in the group while processing the group center. That is why FGT is obviously better in terms of tracking effectiveness than CGT when a covered region exists.

10.7 Summary

This chapter begins in Section 10.2 with a discussion of the basic algorithms for initiation of the tracks of group targets, and points out that group initiation usually includes group segmentation, group association, and group velocity estimation. Group segmentation and group velocity estimation are two difficult points for group initiation algorithms, which are different from other algorithms for target tracking. A few algorithms have been discussed concerning these two aspects.

In Section 10.3, in order to solve the problem of fine track initiation of targets in a group in a cluttered environment, a complete framework for the track initiation of group targets is given, and a gray fine track initiation algorithm for the group targets based on the relative position vector is proposed. After being verified by the simulation data, this algorithm has a better comprehensive performance in terms of initiation of true tracks, suppressing false tracks, and clutter robustness compared with the modified logic algorithm or the initiation algorithm for multi-formation track based on cluster and Hough transformation.

Sections 10.4 and 10.5, respectively, discuss two typical group target tracking algorithms: CGT and FGT. CGT is the most direct algorithm for group tracking, since it undertakes prediction and estimation only of the group center without tracking the individual targets in the group. However, when a group element is lost or a false alarm is mistakenly taken for an element in the group, this algorithm is unreliable in calculation of the group center. Then, the group velocity cannot be estimated accurately so the track is lost.

The FGT algorithm is able to use the information contained in a separate measurement to realize the tracking of a single target in the group. The advantage of this is that the estimation of the group center is more stable, and the estimation of the group velocity is more stable as well. However, since it is necessary to make a position estimate of a single target, the amount of calculation will be greatly increased. FGT is most applicable to the tracking of air targets, as the number of air group targets is limited, and the single target is given more consideration. For the tracking of ground targets, when the covered region exists or the specific tactical significance of the single target needs to be identified, this algorithm can bring its own advantages into full play and achieve better tracking effectiveness.

Finally, Section 10.6 conducts a test and analysis of comprehensive tracking performance for this chapter's two algorithms in a simulation environment. It can be found from the simulation results that, compared with CGT, FGT has better tracking effectiveness. However, in terms of algorithm duration, CGT consumes less time than FGT.

In addition to the algorithms discussed in this chapter, scholars have also proposed other algorithms like group tracking algorithms based on conventional data association methods [8, 120, 180, 256–258] like JPDA [176], MHT [177], particle filter [178], and Bayesian recursion [259], etc., and group tracking algorithms based on the genetic algorithm [260], dynamic network [261], general Janossy measurement density equation [262], and probabilistic hypothesis density filter (PHDF) [240]. These algorithms have partly solved the problem of tracking group targets and targets in a group, but the premise for most of them is that the detection system is able to completely identify the targets in the group. In reality, because of the targets' coverage of each other and inadequate sensor resolution, etc., group targets are usually only partly identifiable.

Therefore, Ref. [240] investigates the group structure and state estimation when the measurement origin is obscure. Meanwhile, Refs. [263, 264], based on the random set, explore the data association and track maintenance of partly identifiable group targets and extended targets, and Ref. [265] proposes a tracking algorithm for partly identifiable group targets based on SMC-PHDF, which can directly gain the group number and the state/shape of the group mass center. However, these algorithms are all unable to gain precise tracks of targets in a partly identifiable group, and their derivation environment is relatively singular. Also, it is hard for them to be applied in complex situations involving multiple-sensor system error [266–269] and unequal dimensions of the sensor [270], etc.

In the region of maneuver tracking, current research focuses mostly on working out the logical relationship of groups splitting, combining, and crossing in terms of position, direction, and track history, etc. Then, the maneuver treatment for the group is completed based on the PDA [271], mode space [272], Markov chain Monte Carlo (MCMC) particle filter [273], sequential Monte Carlo probabilistic hypothesis density filter (SMC-PHDF) [265], etc. However, research still focuses on the whole group, less attention is paid to the track changes of the targets in the group in the maneuver situation, and especially the effectiveness in maneuver treatment needs improvement for the targets in the group when the group is partly identifiable. From a comprehensive perspective, the current algorithms underestimate the complexities of the echoes of the targets in the group, and are defective to some degree in track initiation, track maintenance, and maneuver tracking, etc. Details are specified as follows.

1. So far, track initiation for the group targets has not been solved effectively. Especially, track initiation for the targets in a partly identifiable group is not yet covered in the references.
2. The existing references are unable to solve the problems caused by the measurement loss, target glint, and other specific situations, which make it difficult for the tracks of the targets in a group to be maintained when the detection system can only partly identify the targets in the group.
3. The available references contain a lot of theoretical description about maneuver treatment of the targets in the group, but their practical tracking results are not as good as expected. In addition to the typical splitting, combining, and crossing, there are another two special maneuvers: the group's sudden dispersal (as in a flock of birds being scared or the tactical dispersal of aircraft) and the parallel maneuver of the group (as in the formation of aircraft firing missiles in salvoes). As yet, there is no research on such subjects.
4. In order to improve the precise tracking effectiveness for partly identifiable targets in a group, in practice, it is necessary to use different equipment to gain group target measurements in different measurement directions from the perspective of the measurement system to undertake data association and integration, etc. However, the available references are only concerned with group tracking with a single sensor, without considering the more complex multi-sensor cases.

Appendix 10A

Suppose that $\mathbf{Z}_1$ and $\mathbf{Z}_2$ are two pre-associated groups at neighboring moments. For the sake of our discussion, define that $\{z_{11}, z_{12}, z_{13}\}$ in $\mathbf{Z}_1$ and $\{z_{21}, z_{22}, z_{23}\}$ in $\mathbf{Z}_2$ correspond to the target $\{t_1, t_2, t_3\}$ at the two respective moments. z_{12}, z_{13} constitute the basic coordinate system, and z_{22}, z_{23} the reference coordinate system.

1. *Derive the impact of measurement error on measurement right-angle coordinates.* Suppose that the standard deviation of measurement error is $\sigma = [\sigma_\rho, \sigma_\theta]'$, $z_{11} = \mathrm{Pol}[x_{11}, y_{11}]' = [\rho_{11}, \theta_{11}]'$, and the corresponding true measurement for the target is $z'_{11} = \mathrm{Pol}[x'_{11}, y'_{11}]' = [\rho'_{11}, \theta'_{11}]'$. Then

$$\begin{aligned} x'_{11} &= \rho'_{11}\cos\theta'_{11} = (\rho_{11} + \Delta\rho_{11})\cos(\theta_{11} + \Delta\theta_{11}) \\ &= \rho_{11}\cos\theta_{11}\cos\Delta\theta_{11} - \rho_{11}\sin\theta_{11}\sin\Delta\theta_{11} + \Delta\rho_{11}\cos\theta_{11}\cos\Delta\theta_{11} + \Delta\rho_{11}\sin\theta_{11}\sin\Delta\theta_{11} \\ &\approx x_{11}\cos\Delta\theta_{11} - y_{11}\sin\Delta\theta_{11} + \Delta\rho_{11}\cos\theta_{11} \end{aligned} \tag{10A.1}$$

where $\Delta\rho_{11} \in [-\sigma_\rho, \sigma_\rho]$, $\Delta\theta_{11} \in [-\sigma_\theta, \sigma_\theta]$, then

$$x_{11} - x'_{11} = x_{11}(1 - \cos\Delta\theta_{11}) - y_{11}\sin\Delta\theta_{11} + \Delta\rho_{11}\cos\theta_{11} \tag{10A.2}$$

Likewise,

$$y_{11} - y'_{11} = y_{11}(1 - \cos\Delta\theta_{11}) + x_{11}\sin\Delta\theta_{11} + \Delta\rho_{11}\sin\theta_{11} \tag{10A.3}$$

Through derivation, we get

$$\begin{aligned} x_{11} - x'_{11} &\in [x_{11}(1-\cos\sigma_\theta) - |y_{11}\sin\sigma_\theta| - |\sigma_\rho\cos\theta_{11}|, x_{11}(1-\cos\sigma_\theta) + |y_{11}\sin\sigma_\theta| + |\sigma_\rho\cos\theta_{11}|] \\ y_{11} - y'_{11} &\in [y_{11}(1-\cos\sigma_\theta) - |x_{11}\sin\sigma_\theta| - |\sigma_\rho\sin\theta_{11}|, y_{11}(1-\cos\sigma_\theta) + |x_{11}\sin\sigma_\theta| + |\sigma_\rho\sin\theta_{11}|] \end{aligned} \tag{10A.4}$$

A similar procedure can be applied to z_{12}, z_{13}.

2. *Derive the impact of measurement error on relative position vector in the right-angle coordinate system.* Suppose that $z_{01} = z_{11} - \frac{z_{12} + z_{13}}{2} = [x_{01}, y_{01}]'$, with a true value $z'_{01} = z'_{11} - \frac{z'_{12} + z'_{13}}{2} = [x'_{01}, y'_{01}]'$, then

$$\begin{aligned} x_{01} - x'_{01} &= \frac{2(x_{11} - x'_{11}) - (x_{12} - x'_{12}) - (x_{13} - x'_{13})}{2} \\ &\geq \frac{[(2x_{11} - x_{12} - x_{13})(1 - \cos\sigma_\theta) - (|y_{11}| + |y_{12}| + |y_{13}|)\sin, \sigma_\theta, - (|\cos\theta_{11}| + |\cos\theta_{12}| + |\cos\theta_{13}|)\sigma_\rho}{2} \end{aligned} \tag{10A.5}$$

Since $\{z_{11}, z_{12}, z_{13}\}$ belong to the same pre-segmentation group,

$$|x_{11} - x_{12}| \le d_0, \quad |x_{11} - x_{13}| \le d_0 \tag{10A.6}$$

in which d_0 is the threshold value for group segmentation, so $\frac{|2x_{11} - x_{12} - x_{13}|}{2} \le d_0$. Because σ_θ is assigned a smaller value, $\frac{(2x_{11} - x_{12} - x_{13})(1 - \cos\sigma_\theta)}{2} \approx 0$, $\sin\Delta\theta \approx \Delta\theta$. Equation (10A.5) can be simplified as

$$x_{01}-x'_{01} \geq -\frac{(|y_{11}|+|y_{12}|+|y_{13}|)\sigma_\theta+(|\cos\theta_{11}|+|\cos\theta_{12}|+|\cos\theta_{13}|)\sigma_\rho}{2} = -\left(A\sigma_\theta+B\sigma_\rho\right) \quad (10A.7)$$

where $A_{01}=\frac{|y_{11}|+|y_{12}|+|y_{13}|}{2}$, $B_{01}=\frac{|\cos\theta_{11}|+|\cos\theta_{12}|+|\cos\theta_{13}|}{2}$. It follows that

$$x_{01}-x'_{01} \in \left[-\left(A\sigma_\theta+B\sigma_\rho\right), A\sigma_\theta+B\sigma_\rho\right] \quad (10A.8)$$

Likewise,

$$y_{01}-y'_{01} \in \left[-\left(C\sigma_\theta+D\sigma_\rho\right), C\sigma_\theta+D\sigma_\rho\right] \quad (10A.9)$$

where $C_{01}=\frac{|x_{11}|+|x_{12}|+|x_{13}|}{2}$, $D_{01}=\frac{|\sin\theta_{11}|+|\sin\theta_{12}|+|\sin\theta_{13}|}{2}$.

3. *Derive the impact of the measurement error on the relative position vector in the polar coordinate system.* Suppose that $w_{01}=\mathrm{Pol}(x_{01},y_{01})=(\rho_{01},\theta_{01})$, with a true value $w'_{01}=\mathrm{Pol}(x'_{01},y'_{01})=(\rho'_{01},\theta'_{01})$. Equations (10A.8) and (10A.9) define the range of deviation from the relative position vector to the true value in the right-angle coordinate system. Through derivation, we have

$$|\rho_{01}-\rho'_{01}| \in [-E_{01},E_{01}],\ |\theta_{01}-\theta'_{01}| \in [-F_{01},F_{01}] \quad (10A.10)$$

where

$$\begin{aligned} E_{01} &= \sqrt{\left(A_{01}{}^2+B_{01}{}^2\right)\sigma_\theta^2+\left(C_{01}{}^2+D_{01}{}^2\right)\sigma_\rho^2+2(A_{01}B_{01}+C_{01}D_{01})\sigma_\theta\sigma_\rho} \\ F_{01} &= \max\left(\left|\arctan\frac{A_{01}\sigma_\theta+B_{01}\sigma_\rho}{C_{01}\sigma_\theta+D_{01}\sigma_\rho}\right|, \left|\arctan\frac{C_{01}\sigma_\theta+D_{01}\sigma_\rho}{A_{01}\sigma_\theta+B_{01}\sigma_\rho}\right|\right) \end{aligned} \quad (10A.11)$$

Likewise, the range of deviation from the relative position vector w_{02} of z_{21} to its true value in the reference coordinate system is

$$|\rho_{02}-\rho'_{02}| \in [-E_{02},E_{02}],\ |\theta_{02}-\theta'_{02}| \in [-F_{02},F_{02}] \quad (10A.12)$$

4. From (10A.10) and (10A.11), and (10.34) and (10.35), it follows that $\gamma \in [\varepsilon,1]$, where

$$\varepsilon=\frac{\sigma(\theta)\sigma_{\max}(\rho)G(\rho)+\sigma(\theta)\sigma_{\max}(\rho)G(\theta)}{\sigma(\rho)\sigma_{\max}(\theta)+\sigma(\theta)\sigma_{\max}(\rho)} \quad (10A.13)$$

with $G(\rho)=\frac{\sigma(\rho)}{\sigma(\rho)+(E_{01}+E_{02})}$, $G(\theta)=\frac{\sigma(\theta)}{\sigma(\theta)+(F_{01}+F_{02})}$.

11

Multi-target Track Termination Theory and Track Management

11.1 Introduction

In addition to the filter estimation method, track initiation, data correlation, and maneuvering target tracking, track termination and management are also a concern in the field of multi-target tracking.

When the target being tracked escapes from the surveillance region at any time, the tracker must terminate the track and delete the redundant track files [40, 41]. Currently, the main multi-target track termination technologies are the sequential probability ratio test [40, 274–276], tracking gate [171, 277, 278], cost function [40, 279], Bayesian [280, 281], and all-neighbor Bayesian algorithms [282].

In view of the complicated relations between radar tracks in the increasingly complicated battlefield environment, the initiation, confirmation, maintenance, and cancellation criteria for target tracks have proven crucial in engineering applications. Consequently, battlefield track management [283–285], especially in terms of track batch [286, 287] and quality [17, 23, 49, 288], has become a vital link in the procedure for radar data processing. Thus, this chapter will address itself to track termination and management.

11.2 Multi-target Track Termination Theory

11.2.1 Sequential Probability Ratio Test Algorithm

The sequential probability ratio test (SPRT) algorithm proposed in Refs [40, 274–276] is mainly applied to track initiation as well as track termination. It adopts the hypothesis testing approach for track initiation or termination.

First, two hypotheses H_1 and H_0 should be established, among which H_1 refers to the track maintenance hypothesis and H_0 the track termination hypothesis.

Radar Data Processing with Applications, First Edition. He You, Xiu Jianjuan, and Guan Xin.

Second, the likelihood function of each hypothesis, P_{1k} and P_{0k}, should be calculated, respectively:

$$H_1 : P_{1k} = P_D^m (1 - P_D)^{k-m} \tag{11.1}$$

$$H_0 : P_{0k} = P_F^m (1 - P_F)^{k-m} \tag{11.2}$$

where P_D and P_F are the detection and the false probability, respectively, m the detection number, and k the scanned number. The likelihood ratio function corresponding to the two hypotheses is defined as

$$U_k = \frac{P_{1k}}{P_{0k}} \tag{11.3}$$

The corresponding thresholds are set as C_1 and C_2.

Third, the decision logic of the SPRT algorithm should be arranged as follows:

1. If $U_k \geq C_2$, hypothesis H_1 is accepted and the track maintained.
2. If $U_k \leq C_1$, hypothesis H_0 is accepted and the track maintained.
3. If $C_1 < U_k < C_2$, testing continues.

The decision thresholds C_1 and C_2 above satisfy

$$C_1 = \frac{\beta}{1 - \alpha} \tag{11.4}$$

$$C_2 = \frac{1 - \beta}{\alpha} \tag{11.5}$$

where α and β are the predefined permissible error probabilities. α is the probability that H_1 is accepted when hypothesis H_0 holds, or the probability of missed cancellations (the probability that no cancellation is decided for tracks which should be cancelled), while β is the probability that H_0 is accepted when hypothesis H_1 is true, or the probability of false cancellations (the probability that cancellation is decided in the presence of real tracks).

The logarithmic form of the likelihood ratio function can be obtained by evaluating the logarithm of (11.3) and capitalizing on (11.1) and (11.2). Thus, the decision logic equation becomes

$$\ln U_k = \ln(P_{1k}/P_{0k}) = m a_1 - k a_2 \tag{11.6}$$

where the parameters a_1 and a_2 are

$$a_1 = \ln \frac{P_D/(1 - P_D)}{P_F/(1 - P_F)} \tag{11.7}$$

$$a_2 = \ln \frac{1 - P_F}{1 - P_D} \tag{11.8}$$

Define the test statistical variables $ST(k)$:

$$ST(k) = m a_1 \tag{11.9}$$

$$\ln U_k = ST(k) - ka_2 \tag{11.10}$$

and define the decision threshold at time k as

$$T_U(k) = \ln C_2 + ka_2 \tag{11.11}$$

$$T_L(k) = \ln C_1 + ka_2 \tag{11.12}$$

Then, the decision logic of track termination can be expressed as follows:

1. If $ST(k) \geq T_U(k)$, hypothesis H_1 is accepted and the track maintained.
2. If $ST(k) \leq T_L(k)$, hypothesis H_0 is accepted and the track maintained.
3. If $T_L(k) < ST(k) < T_U(k)$, testing continues.

Specifically, if measurements fall into the wave gate of a track at time k, then the statistics $ST(k)$ increases by a_1; if there is no measurement in the wave gate of the track, $ST(k)$ remains unchanged, while thresholds $T_L(k)$ and $T_U(k)$ increase by a_2 at every moment. When the statistics $ST(k)$ is higher than the threshold $T_U(k)$, the algorithm decides track maintenance; when the statistics $ST(k)$ is lower than the threshold $T_L(k)$, it decides track termination and cancellation; otherwise, the test continues.

11.2.2 Tracking Gate Method

This method applies the optimal track threshold γ_0 [171, 277, 278] to the criteria for determining track termination, with γ_0 expressed as

$$\gamma_0 = 2\ln\frac{P_D}{(1-P_D)\beta_{new}(2\pi)^{M/2}\sqrt{|\mathbf{S}|}} \tag{11.13}$$

where P_D is the detection probability, β_{new} the new echo density, M the observation dimension, and $|\mathbf{S}|$ the determinant of the innovation covariance matrix.

The tracking gate rules show that the echo received by the detector most probably originated from the target being tracked when the innovation norm $\psi(k)$ generated by the tracking filter meets the relation equation

$$\psi(k) \leq \gamma_0 \tag{11.14}$$

Therefore, as long as $\gamma_0 > 0$, there exists a possibility of updating the track of the target. On the contrary, if $\gamma_0 < 0$, the echo received most probably came from the new target rather than the target being tracked:

$$\gamma_0 < \gamma_{min} \tag{11.15}$$

If the size of the elliptical tracking gate γ has been calculated, then a criterion follows naturally for track termination: a track is considered terminated when and only when

$$\gamma_0 < \gamma_{min} \tag{11.15}$$

holds valid. In the above, γ_{min} refers to a minimal threshold that could be obtained by the standard χ_M^2 distribution with M (observation dimension) degrees of freedom lest the track be terminated in the presence of a predefined track-update probability.

11.2.3 Cost Function Method

It is known that when the dynamic model of the target is accurate enough and the observation/track are correctly matched, the innovation norm of the target $\psi(k)$ follows a χ^2_M distribution with M degrees of freedom, where $\boldsymbol{v}(k)$ refers to the innovation vector of the target, $\boldsymbol{S}(k)$ refers to the innovation covariance matrix, and M is the observation dimension:

$$\psi(k) = \boldsymbol{v}'(k)\boldsymbol{S}^{-1}(k)\boldsymbol{v}(k) \tag{11.16}$$

Reference [279] defines a cumulative χ^2 cost function normalized by the renewal times N_i for track i:

$$C_i = \frac{1}{N_i}\sum_{k=1}^{N_i}\psi(k) \tag{11.17}$$

It follows from the definition above that N_iC_i follows a $\chi^2_{MN_i}$ distribution with MN_i degrees of freedom. Therefore, according to the $\chi^2_{MN_i}$ distribution or the Gauss approximation, the threshold can be set as

$$\eta_i = \mu_{c_i} + \alpha\sigma_{c_i}, \quad \forall\alpha \geq 3 \tag{11.18}$$

where μ_{c_i} and σ_{c_i} refer to the mean value and the standard deviation of C_i, respectively:

$$\mu_{c_i} = \mathrm{E}[C_i] \tag{11.19}$$

$$\sigma_{c_i} = \sqrt{\frac{2\mu_{c_i}}{N_i}} \tag{11.20}$$

Finally, when

$$C_i > \mu_{c_i} + \alpha\sigma_{c_i} \tag{11.21}$$

or

$$C_i < \mu_{c_i} - \alpha\sigma_{c_i} \tag{11.22}$$

is satisfied, the algorithm accepts the hypothesis of track termination.

In this algorithm, along with the increase in number of updates N_i, the previous data will be heavily weighted but new data lightly weighted by the cost function C_i defined in (11.17), which may result in false track termination. One of the solutions to this problem is to set the attenuation coefficient $\delta(k)$ in the cost function C_i, that is,

$$C_i^* = \frac{1}{N_i}\sum_{k=1}^{N_i}\delta(k)\psi(k) \tag{11.23}$$

where $\delta(N_i) = 1$ and $\delta(k+1) > \delta(k)$.

The modified cost function follows a $A\chi^2_{V_\Gamma}$ distribution with V_Γ degrees of freedom (refer to Ref. [41] for proof), where

$$A = \frac{\sum_{k=1}^{N_i} \delta^2(k)}{\sum_{k=1}^{N_i} \delta(k)} \tag{11.24}$$

$$V_\Gamma = M \frac{\left[\sum_{k=1}^{N_i} \delta(k)\right]^2}{\sum_{k=1}^{N_i} \delta^2(k)} \tag{11.25}$$

Another solution is to take a fixed number of updates N_i and conduct track termination testing with a sliding window of length N_i for the track such that the cost function C_i always includes the latest N_i innovation vectors.

11.2.4 Bayesian Algorithm

A Bayesian algorithm [280, 281] could be used for track initiation as well as track termination. First, the posterior probability $\Pr(T|\mathbf{Z})$ that a track is true, given the measurement set $\mathbf{Z}$, is calculated. According to the Bayesian algorithm,

$$\Pr(T|\mathbf{Z}) = \frac{\Pr(\mathbf{Z}|T)P_0(T)}{\Pr(\mathbf{Z})} \tag{11.26}$$

where

$$\Pr(\mathbf{Z}) = \Pr(\mathbf{Z}|T)P_0(T) + \Pr(\mathbf{Z}|F)P_0(F) \tag{11.27}$$

$$P_0(F) = 1 - P_0(T) \tag{11.28}$$

and $\Pr(\mathbf{Z}|T)$ and $\Pr(\mathbf{Z}|F)$ refer to the probabilities, respectively, of receiving the measurement set $\mathbf{Z}$ in the presence of true targets and false targets; $P_0(T)$ and $P_0(F)$ stand for the prior probability for the true target and the false target; $\Pr(\mathbf{Z})$ means the probability of receiving a measurement set.

Defining the likelihood ratio of statistics as $L(\mathbf{Z}) = \frac{\Pr(\mathbf{Z}|T)}{\Pr(\mathbf{Z}|F)}$ and combining (11.26)–(11.28) yields

$$\Pr(T|\mathbf{Z}) = \frac{L(\mathbf{Z})P_0(T)}{L(\mathbf{Z})P_0(T) + 1 - P_0(T)} \tag{11.29}$$

Assume that L_k is the statistics likelihood ratio of the kth scan and $\Pr(T|\mathbf{Z}_k)$ refers to the probability that the target is true by the end of the kth scan, then (11.29) becomes

$$\Pr(T|\mathbf{Z}_k) = \frac{L_k \Pr(T|\mathbf{Z}_{k-1})}{L_k \Pr(T|\mathbf{Z}_{k-1}) + 1 - \Pr(T|\mathbf{Z}_{k-1})} \tag{11.30}$$

where

$$L_k = \frac{\Pr(\mathbf{Z}_{k-1}|T)}{\Pr(\mathbf{Z}_{k-1}|F)} = \begin{cases} \dfrac{P_D V_j(k)\exp\left(-\psi_j^2(k)/2\right)}{P_F(2\pi)^{M/2}\sqrt{|\mathbf{S}_j(k)|}}, & \text{detection} \\ \dfrac{1-P_D}{1-P_F}, & \text{miss detection} \end{cases} \tag{11.31}$$

with $V_j(k)$ the volume of the correlation region of the jth target, $\mathbf{S}_j(k)$ the innovation covariance matrix, $\psi_j(k)$ the innovation norm, M the observation dimension, P_D the radar detection probability, P_F the false alarm probability, and

$$P_F = \beta_{FT} \cdot V_j \tag{11.32}$$

where β_{FT} is the false alarm density.

With the track termination threshold set as P_{el}, the track is considered terminated when and only when

$$\Pr(T|\mathbf{Z}_k) < P_{el} \tag{11.33}$$

Through this method, it is possible to carry out track confirmations and track terminations at the same time; after each track is confirmed, the track termination test will start with a larger initial target probability P_0.

11.2.5 All-Neighbor Bayesian Algorithm

The aforementioned Bayesian algorithms are nearest-neighbor algorithms, applicable to target track termination under the sparse echo environment. To conduct maneuvering multi-target track terminations (MMTT) in very dense multiple-echo environments, Bayesian algorithms can be modified such that modified probabilities are used for associations. To be more specific, the all-neighbor equivalent innovations in the secondary filter algorithm are used to replace the "nearest" innovations in the Bayesian algorithms so as to obtain a method for calculating a new data likelihood ratio.

The basic equation of the all-neighbor Bayesian algorithm is

$$\Pr(T|\mathbf{Z}_k) = \frac{L_k \Pr(T|\mathbf{Z}_{k-1})}{L_k \Pr(T|\mathbf{Z}_{k-1}) + 1 - \Pr(T|\mathbf{Z}_{k-1})} \tag{11.34}$$

where

$$L_k = \begin{cases} \dfrac{P_D V_j(k)\exp\left(-\varphi_j(k)/2\right)}{P_F(2\pi)^{M/2}\sqrt{|\mathbf{S}_j(k)|}}, & \text{detection} \\ \dfrac{1-P_D}{1-P_F}, & \text{miss detection} \end{cases} \tag{11.35}$$

with $\varphi_j(k)$ the equivalent innovation norm of the jth target, namely

$$\varphi_j(k) = [\boldsymbol{Z}_j(k) - \boldsymbol{H}(k)\hat{\boldsymbol{X}}_j(k|k-1)]' \boldsymbol{S}_j^{-1}(k) [\boldsymbol{Z}_j(k) - \boldsymbol{H}(k)\hat{\boldsymbol{X}}_j(k|k-1)], \quad j = 1, \ldots, n \tag{11.36}$$

where n refers to the number of targets.

Because extra echoes form a uniform distribution in the surveillance region, in order to meet the needs of the high-density echo environment, redefine

$$P_F = \frac{V_j}{V_T}, \qquad j = 1, \ldots, n \tag{11.37}$$

where V_T refers to the volume of the surveillance region.

Likewise, set the track termination threshold P_{TT} such that the hypothesis of track termination is accepted when and only when

$$\Pr(T|\boldsymbol{Z}_k) < P_{TT} \tag{11.38}$$

11.2.6 Performance Analysis of Several Algorithms

11.2.6.1 Simulation Environment and Parameter Setting

Two typical multi-maneuvering-target environments are chosen below.

Environment 1. Initiation state of targets: $X(0) = [60\,000\,\text{m}, 0\,\text{m/s}, 40\,000\,\text{m}, 600\,\text{m/s}]'$ at $t = 30\,\text{s}$, $a_x = 35\,\text{m/s}^2$, $a_y = 35\,\text{m/s}^2$. The termination time of the target motion is $t = 50\,\text{s}$.

Environment 2. Assume that the targets keep emerging in the surveillance region, with the number of targets increasing gradually from 5 to 60. Their initial positions are subject to a Gaussian distribution in the area shown in Figure 11.1. The initial velocities and headings are uniformly distributed between 4–1200 m/s and 0–2.

In Figure 11.1, r_1 is the observation radius, r_1' is the radius of the blind zone, and o' is the local coordinate origin; $r_1 = 110\,\text{km}$, $r_1' = 2\,\text{km}$, $a = b = 125\,\text{km}$, $x_1 = 380\,\text{km}$, $y_1 = 270\,\text{km}$.

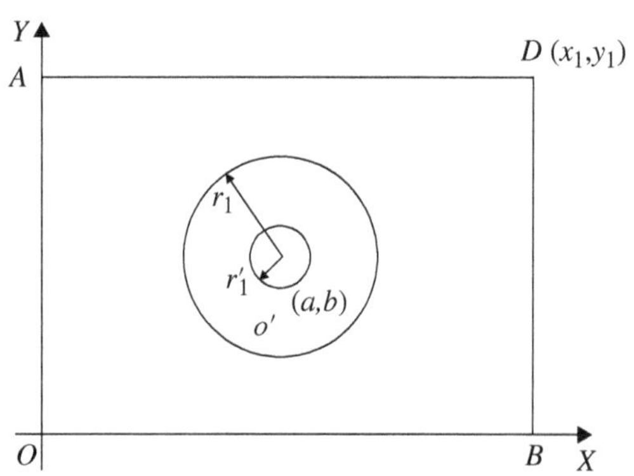

Figure 11.1 Schematic diagram for radar observation region

Other numerical values used in the simulation are: the interval for radar sampling, $T = 2$ s; range-finding errors, $\sigma_r = 100$ m; angle-finding errors, $\sigma_\theta = 0.03$ rad; simulation runs, $N = 50$. The parameters of each algorithm were set as follows.

1. *Sequential probability ratio test algorithm*: detection probability $P_D = 0.95$, false alarm probability $P_F = 0.1$, probability of missed cancellations $\alpha = 0.15$, probability of false cancellations $\beta = 0.1$.
2. *Tracking gate algorithm*: detection probability $P_D = 0.95$, new target echo density $\beta_{new} = 0$ (environment 1), $\beta_{new} = 0.2$ (environment 2), minimal threshold value $\gamma_{min} = 0.103$.
3. *Cost function algorithm*: sliding window length $N_i = 8$.
4. *Bayesian algorithm*: detection probability $P_D = 0.95$, false alarm probability $P_F = 0.1$, termination threshold $P_{el} = 0.7$, initiation probability $P_0 = 0.01$.
5. *All-neighbor Bayesian algorithm*: detection probability $P_D = 0.95$, false alarm probability $P_F = 0.1$, termination threshold $P_{TT} = 0.7$, initiation probability $P_0 = 0.01$.

11.2.6.2 Simulation Results and Analysis

Environment 1

Table 11.1 offers a comparison of the termination time required by the various track deletion algorithms. As shown by the simulation results, the all-neighbor Bayesian algorithm needs the shortest termination time, thus rapidly providing information about the deletion of the track; it is followed by the cost function algorithm. The Bayesian algorithm and the sequential probability ratio test algorithm need a longer termination time. The tracking gate algorithm consumes the longest time with weak real-time performance.

Environment 2

Figure 11.2 offers the comparison results of each algorithm's false termination rate varying with the target batch number. The false termination rate of the all-neighbor Bayesian algorithm and the sequential probability ratio test algorithm slowly rises with the target batch number, while that of the tracking gate algorithm and the Bayesian algorithm rapidly grows with the target batch number. However, compared with the tracking gate algorithm and the Bayesian algorithm, the false termination rate of the cost function algorithm is much lower when the target batch number is less than 40; with a target batch number over 40, its false termination rate will gradually exceed that of the aforementioned two algorithms.

Therefore, from the perspective of track termination time and the false termination rate of the algorithm: the all-neighbor Bayesian algorithm has better real-time performance and strong stability; the sequential probability ratio test algorithm requires a longer track termination time but

Table 11.1 Track termination time of each algorithm

Termination algorithm	Sequential probability ratio test algorithm	Tracking gate algorithm	Cost function algorithm	Bayesian algorithm	All-neighbor Bayesian algorithm
Termination time (s)	6	8	5	6	4

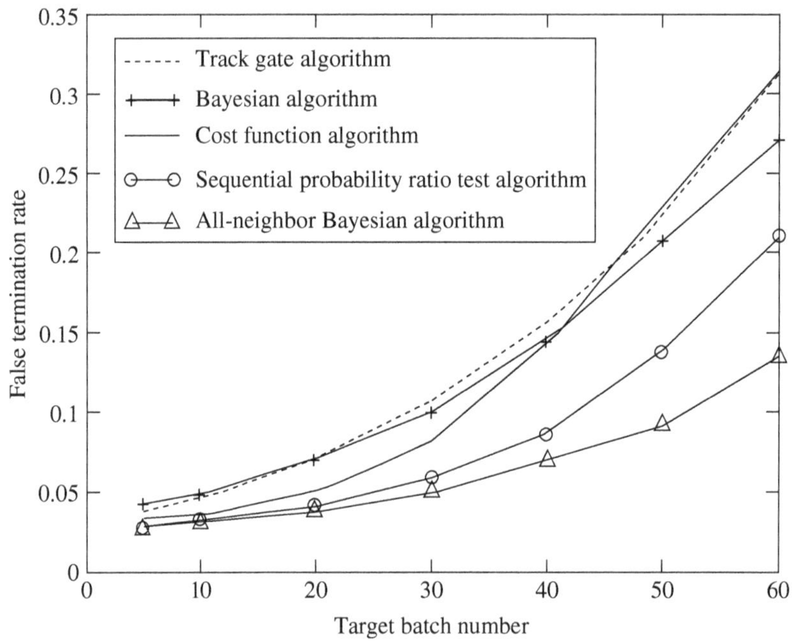

Figure 11.2 Comparison of each algorithm's false termination rate varying with the target batch number

acquires strong stability, with similar real-time performance to that of the all-neighbor Bayesian algorithm; the cost function algorithm has lower false termination rate when there is a smaller target batch number and higher false termination rate when there is a larger target batch number. The Bayesian algorithm and the tracking gate algorithm all consume a longer track termination time with poor stability. From the perspective of algorithm classification, the sequential probability ratio test algorithm, tracking gate algorithm, cost function algorithm, and Bayesian algorithm all belong to the nearest-neighbor algorithms, which determines that the application of these algorithms is limited only to the sparse target echo environment and non-maneuvering target environment. By utilizing all-neighbor information, the all-neighbor Bayesian algorithm is a track termination technology that could be applied in the very-dense multiple-echo environment and the multi-maneuvering-targets environment, because the algorithm provides reliable and rapid track termination decisions. For this reason, it possesses practical engineering values.

11.3 Track Management

11.3.1 Track Batch Management

Each radar tracking system must have its own track file management system. Track management is usually done by track numbering. The track batch, as it implies, is the numbering of the track, which serves as a reference for all the parameters related to the given track. For one thing, the track is marked in track management for the correlation processing. For another, it can be used for the statistical analysis of the processing effect afterwards. Furthermore, the management of the track batch could be utilized to describe the battlefield situation and be fed back to the information processing inside the track.

The application, cancellation, maintenance, as well as the calculation and operation of tracks in track batch management calls for the establishment of a track batch array and a track batch linked list. This is done as follows:

1. The track batch array DT with NN (where NN is an integer) dimensions is established and initialized to $DT(i)=i$, $i=1,2,3,\ldots,NN$.
2. Assume NU is the pointer entering into the radar surveillance region with initial value 0.
3. The application for track batch. $NU=NU+1$ when the new track appears in the surveillance region, with assigned track batch $NT=DT(NU)$.
4. The deletion of track batch. If the track $NT1$ is deleted, $DT(NU)=NT1$ and $NU=NU-1$.

Considering that the track batch serves as the first parameter for all operations of the track, it is necessary to establish the assignment track linked list for the radar surveillance region so as to facilitate the continuous and effective operation of all tracks in the region. The procedure of managing the track batch storage and conversion in the surveillance region is as below:

1. Set up the track batch storage array $IDT1$ with the initial value of $IDT1$ set to 0. Set a variable $TB1$ and store the first track batch $NT1$ in $TB1$.
2. The track batch is stored in $IDT1$ such that $IDT1(NT1)=NT2$ and $IDT1(NTm)=NTm+1$ accordingly.
3. With regard to the operation of the track in the surveillance region, the first track batch $NT1$ is obtained from $TB1$ and then in sequence $NTm=IDT1(NTm-1)$, till 0 appears.
4. The cancellation of the track.

Assume that NT is the current track batch under processing and NTL the last processed track batch, then the cancellation process is:

1. If $NTL=0$, $TB1=IDT1(NT)$ and $IDT1(NT)=0$.
2. If $NTL\neq 0$, $NT1(NTL)=IDT1(NT)$ and $IDT1(NT)=0$.

Besides, the track batch assignment serves as the key process in the track batch management. The following subsections offer an investigation and analysis of the single-track batch dispatch method in the radar tracking system, a multiple-track batch dispatch method on this basis, as well as a method for the expanded-track batch cubic figures. The introduction of track batch cubic figures, in particular, provides a new technique for track management.

11.3.1.1 Single-Track Batch Assignment Method

While executing the attack mission in a certain attacking area, the formations usually perform maneuvers, respectively, to enhance their combat capability. Under such circumstances, the formation tracking is similar to the case of a target entering into the highly cluttered area, with many measurements appearing in the shared wave gate. In general, the method of multiple hypothesis is adopted, according to which the track is split into two or more tracks in accordance with the measurement data in the wave gate; the originally accumulated track information is taken as the shared information after the track is split. After recursion for a few time steps, according to their own association points, the two split tracks are true if they can still be maintained, or one of them is terminated in the case of false alarms. In the process of treating splitting tracks, identifying the relationship

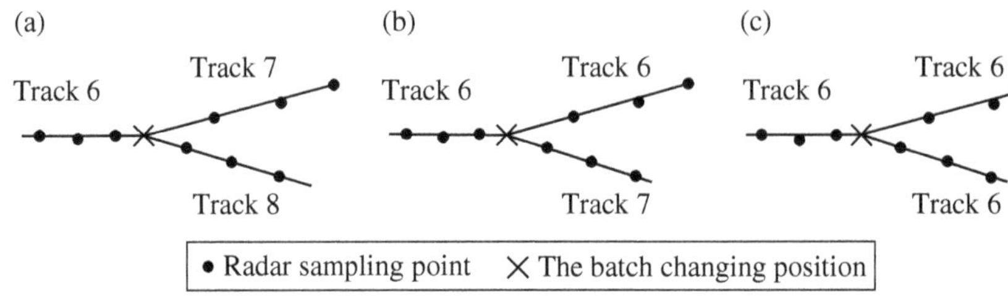

Figure 11.3 Description of splitting tracks with the single-track batch assignment method

between consecutive tracks can only be implemented through a change of track batch. How do we dispatch the track batch so as to identify the variation characteristics of the tracks?

There are three methods for splitting tracks.

Method 1. Assign according to Figure 11.3(a). Although the information on track 6 is reserved in the new tracks 7 and 8, the connections between tracks 7, 8, and 6 have already disappeared along with the innovation of the track batch at a higher level, which means that tracks 7 and 8 are similar to the reinitiation track, whose historical information suffers the greatest loss.

Method 2. Dispatch according to the method in Figure 11.3(b). The original track information is reserved in track 6 or 7. The original continuity of track 6 is maintained in the system, however, two situations emerge: (i) No information will be lost if the new track 6 is the real continuation of the original track 6; (ii) If the new track 6 is a false one and soon terminates, the loss of historical information is the same as in method 1. Here, track 7 could be regarded as the temporary track (unlike the temporary track at track initiation). After a few time steps, its existence is checked to determine whether to maintain it or not. If track 7 remains but track 6 terminates, then batch 7 is changed into batch 6 so as to retain this continual relationship. If tracks 6 and 7 are both reserved, then batch 7 is officially allocated to this initiated track.

Method 3. Dispatch according to the method in Figure 11.3(c). The application of one track batch to two tracks violates the common principle: there should be only one track batch identification. However, the advantage is the maintenance of the continual relationship of one track. These methods can be applied on occasions where the track batch can be repeated.

An in-depth investigation of method 2 reveals that the assignment of a temporary track batch and its maintenance for a certain period of time are a thorny problem. This method cannot identify from the track batch whether the track knows its state. Therefore, it is suggested to maintain two track batches on each track in areas where tracks are prone to intersect and split, such as within some specific track clusters. Among these two track batches, one serves as the former track batch and the other as the latter track batch. The former track refers to the assigned track batch before splitting tracks, which is free from intersection or mergence. The latter track refers to the assigned track batch after the intersection or mergence occur. In this way, the continual transformation relation is maintained, which makes the processing more flexible.

11.3.1.2 Double-Track Assignment Method

The double-track assignment method provides a series of assigning mechanisms. For convenience, the positive number x can be used to express the former track batch and the positive number y to refer to the latter track batch. Generally:

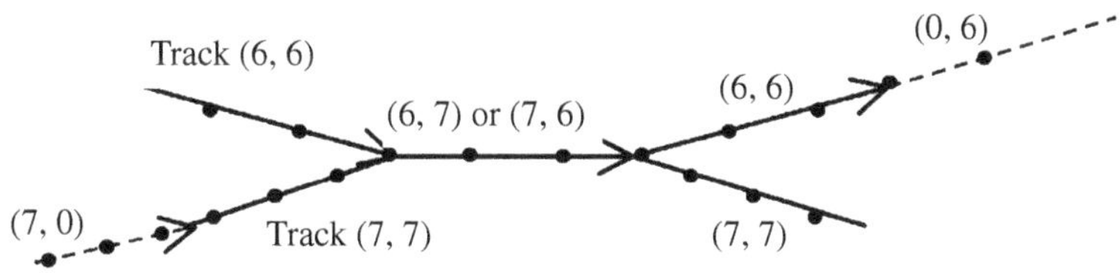

Figure 11.4 The typical process of the double-track batch descriptive method

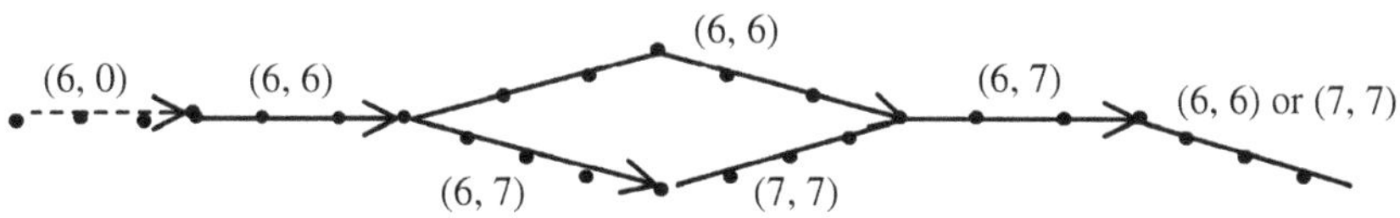

Figure 11.5 Track batch assignment of the whole process including track initiation, maintenance, splitting, and mergence

- at the track maintenance step, $x = y$;
- the track initiation step, $y = 0$, restricted time is τ_i;
- track cancellation step, $x = 0$, restricted time is τ_c;
- track transformation step (intersection, mergence, and splitting), $x \neq y$, restricted time is τ_t.

The process from track combination to track splitting and finally the deletion of one of the tracks is described in Figure 11.4.

The track at the initiation stage ($x = 7$, $y = 0$) will enter into the maintenance stage after a certain period of time ($x = y = 7$) and combines with another track ($x = y = 6$) that also exists at the maintenance stage. During the period of mergence, the track batch is ($x = 6$, $y = 7$), whose order remains obscure. If the track existence time is higher than τ_t, after mergence, the track batch is ($x = y = 6$) or ($x = y = 7$). If it splits during τ_t, and exists longer than τ_t, the track batch will revert to track ($x = y = 6$) and ($x = y = 7$). The corresponding relation of the track batch before and after intersection can be determined through the application of some rules and experience. For example, a common track usually has no "V" turn. If the track ($x = y = 6$) enters the cancellation stage at this moment, the track batch turns into ($x = 0$, $y = 6$).

The description of track initiation, maintenance, splitting, and mergence is presented in Figure 11.5.

11.3.1.3 Descriptive Diagram and Characteristics of the Double-Track Batch

According to the double-track dispatch method, the batch of a track includes two positive number values: x and y. If placed in the XY plane, it simply corresponds to a point in a plane (whose coordinate is positive). For a multi-track environment corresponding to a closely spaced target, the distribution of track batches forms an area, the size of which is susceptible to the limitation of the maximum (the maximum assignable track batch – e.g., if the batch limit is 120, then the track batch ranges from 1 to 120, and the maximum is 120) of tracking multiple targets.

As shown in Figure 11.6, axis x is the collection of the track batches at the initiation stage, axis y is the collection of the track batches at the cancellation stage, the straight line $x = y$ is the collection of the track batches at the maintenance stage, the area between x and $x = y$ is the collection of the track

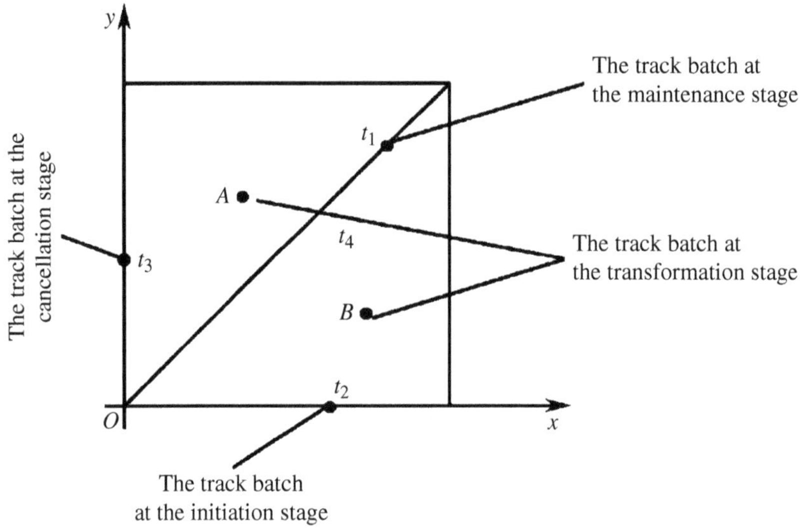

Figure 11.6 The plane graph of the track batch

batches at the combination stage, and the area between y and $x=y$ is the collection of the track batches at the split stage. Thus, the points within this area are characterized by the following features: the track batches at the maintenance stage are distributed on the straight line $x=y$, such as t_1; the track batches at the initiation stage are distributed on axis x $(y=0)$, such as t_2; the track batches at the cancellation stage are distributed on axis y $(x=0)$, such as t_3; the tracks at the transformation stage (intersection, mergence, and splitting process) are distributed in the two areas A and B without the boundaries in the first quadrant, such as t_4.

The general changing process of tracks includes: first, the track exists at the initiation point $(k, 0)$ on the x axis, then at the maintenance stage of the track, the point appears at the point of intersection (k, k) of the $x=k$, $x=y$ segment; if splitting occurs, the two split points (k, l_1) and (k, l_2) appear on the $x=k$ segment; if the track (k, l_1) is cancelled, it will appear on the point $(0, l_1)$ on the y coordinate axis with the track (k, l_2) returning to the state of (k, k) at the steady maintenance stage. The whole changing process is demonstrated in Figure 11.7. This method is used to identify the two track transformations mentioned above, namely, splitting and mergence, as shown in Figure 11.8.

The number of points on $x=k$ indicates the historical splitting of the track. The features of track mergence and intersection can be demonstrated as follows: there is a point (x, y) in the transformation area; if points exist on both (x, x) and (y, y) beforehand, then track (x, y) is formed after track mergence; if the two points (x, x) and (y, y) appear in the follow-up time, they are formed after track intersection; if only (x, x) or (y, y) appears, then it is obtained through track mergence and soon enters into the maintenance stage.

11.3.1.4 Solid Figure Description of the Double-Track Batch

Solid Figure

Previous studies focus on the transient dispatch characteristics of the track batch. If the existence time of each stage is taken into consideration, it is necessary to introduce the time axis to constitute the three-dimensional track batch structure, which contains more elements.

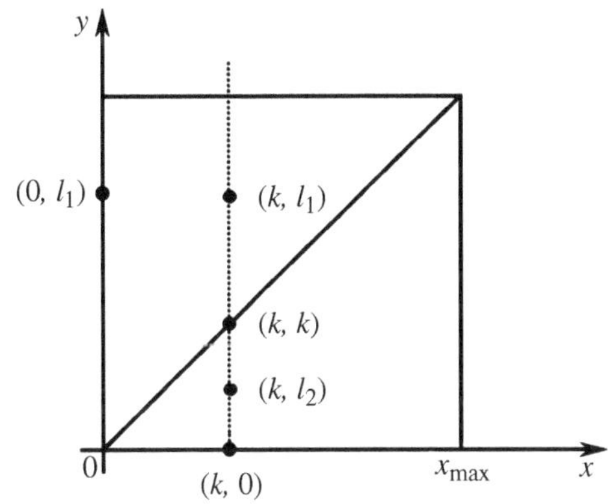

Figure 11.7 Description of typical track transformation process with track batch figure

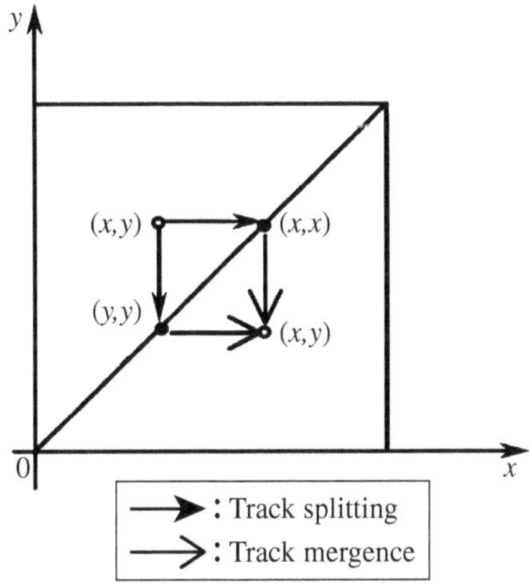

Figure 11.8 Description of track splitting and mergence with the track batch figure

Because radar records data at a certain sampling interval, the existence time of the track batch is discrete – just as the track batch itself. Its solid figure is presented in Figure 11.9(a).

The following subsection offers an analysis of the meaning of each plane: as shown in Figure 11.9(b), the points in the $x=y$ plane all refer to the tracks at the track maintenance stage, with the sustained time stretch standing for the track life in a general sense. In view of the fact that the track life is rated as an important factor reflecting the features of tracking, the number of tracks within this plane whose life exceeds a certain value could be calculated in a simulation so as to evaluate the track effect. The state estimate of the tracks that correspond to the points within this plane possesses the highest reliability and the best target identification capability.

The *XT* plane is shown in Figure 11.9(c). The whole process of simulation reflects the quantity of initiated tracks as well as the sampling points needed, which is attributable to the method adopted for track initiation. For example, with regard to the *n*/*m* rule of the sliding-window method, the number of points *k* and the quantity of track batches at the same location on the *x* coordinate demonstrate the adaptability of this method to the environment. This measure of adaptability, used to guide the selection of the track initiation method, is prone to two restraints and is a compromise between the

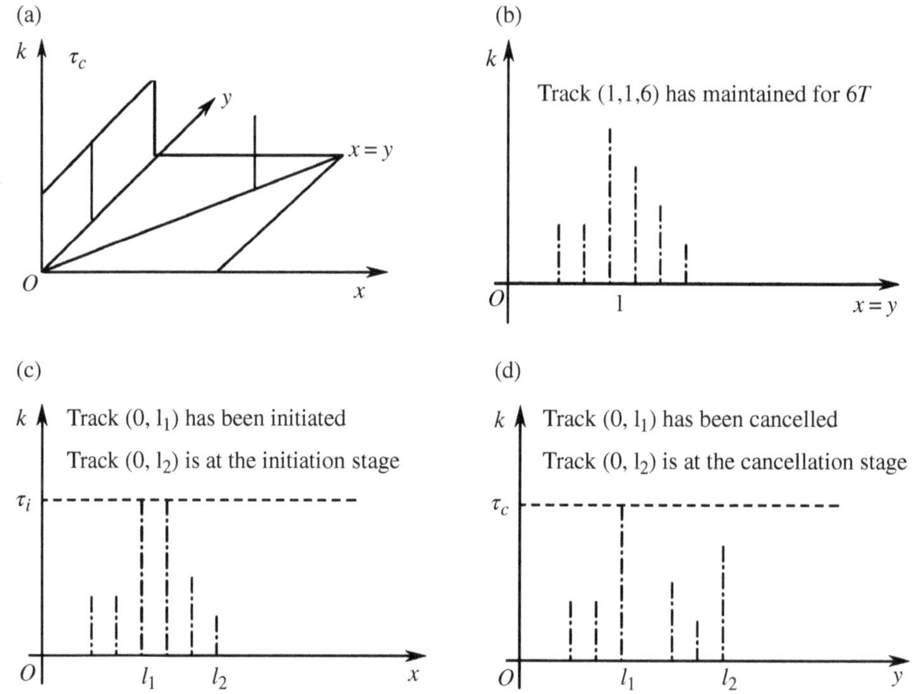

Figure 11.9 Solid figure

capacity of rapid initiation of track and the production of false tracks, for which the statistics of the *YT* planar points offers a fairly good basis. The restricted time for the track initiation is τ_i, with $k^*T < \tau_i$.

Figure 11.9(d) records the process of track cancellation. As the inverse process of track initiation, the statistics of its points is also of evaluation significance. The restricted time is τ_c, meaning that it is insignificant to reserve the information of the track whose cancellation time exceeds τ_c and can be cancelled. However, it is of prime importance to reserve the batch of the track that exists at the cancellation stage, because it can trace back the past and adjust the assignment of the correlative track batch. As shown in Figure 11.10, the batch of track (6, 6) at the cancellation stage is (0, 6), which could be used to transform the temporary track (6, 7) into the reliable track (6, 6).

The region between the four planes is the most complicated; the significance of points should be evaluated on the basis of their locations on the four planes. For example, (3, 6, 3) is interpreted as: track 3 splits into track 6 and has lasted for 3 *T*. The locations of the points describe the causal relationship of track transformation as well as the contemporary state, and provide a method for track batch description in processing tracks of multiple targets in clutter, which facilitates the realization of multi-target track algorithms such as the multiple-hypothesis filtering and JPDA. Meanwhile, as it reflects the dynamic combination and transformation features of the track in the overall situation, it is possible to roughly estimate the joint and cooperative operational characteristics of the target on the basis of combining other features of tracks, such as the spatial position and properties.

Initiation and Management

The description and realization of track batch management by the solid figure of the track batch necessitates the following initializations.

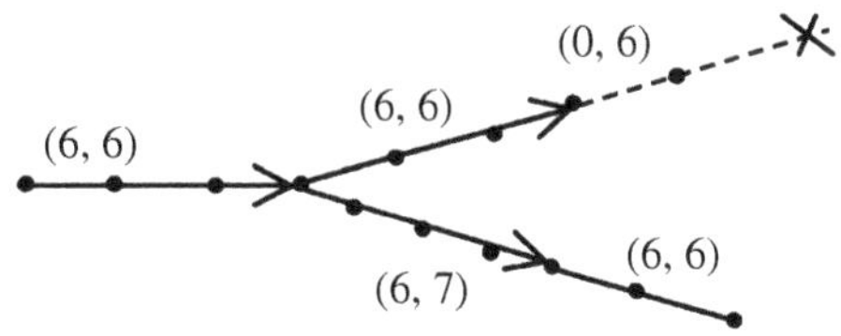

Figure 11.10 Significance of the track batch at the cancellation stage

1. The maximum track number that can be contained is $x_{max} = y_{max}$.
2. The track initiation time constant is τ_i.
3. The track cancellation time constant is τ_c.
4. The initialization of the transformation rules, concerning mainly the points that are not on the reference plane:
   i. the data correlation rules of the point (track) from the straight line $x = y$ to the transformation region, including those for measurement-to-track and track-to-track correlations;
   ii. the data association rule of the point (track) returning from the transformation region to the straight line $x = y$, mainly referred to as track-to-track correlations.

The region management algorithms of the solid figure are as follows.

1. Initiation face: track initiation algorithms, like the *n*/*m* rule.
2. Maintenance face: track maintenance algorithms, like the adaptive track algorithm.
3. Cancellation face: track cancellation algorithms, like the logic-based method.
4. A zone: track combination management algorithms.
5. B zone: track splitting management algorithms (see Figure 11.6).

The track batch solid figure can meet the urgent need of track management. Its main functions are as follows.

1. To evaluate the performance of methods used in track initiation, cancellation, and maintenance.
2. Dimension separation processing of the solid figure, which can be taken as a single index in self-adaptive control of a certain parameter in track management.
3. A solid figure with spatial position could be used to analyze attack attributes of the target, instruct the attribute identification, comprehend its tactical intention and threat level, and feed back to the state estimation for model prediction.

11.3.1.5 Storage of Track Data

Track data within a certain period of time has to be stored according to the requirement of data processing algorithms to facilitate subsequent data processing. But in real systems, various events such as crossing, bifurcation, combination, and disappearance may take place for the targets when they enter or leave the detection zone. Therefore, the storage of track data should be allocated and freed dynamically so as to manage data updating flexibly.

The track data of radars are stored in the form of a unidirectional linked list, as shown in Figure 11.11. When a new track turns up, a track batch *DT* (*NU*) is assigned according to the track batch array, and the memory space of a fixed number of storage units is dynamically opened to store the track data. When *DT* (*NU*) is removed, the memory space is released and the pointers in the two adjacent units are revised so that the two become neighboring units. At last, the track batch *DT* (*NU*) is revoked.

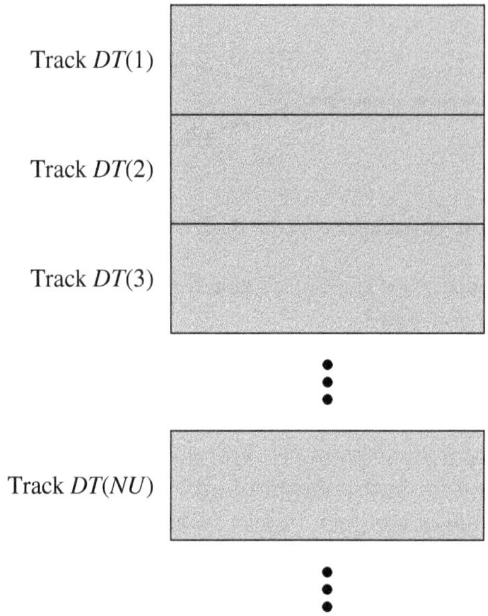

Figure 11.11 Sketch map of unidirectional chained list storage of radar tracks

The data of each track are stored in accordance with the sliding circulative time window, as shown in Figure 11.12. Assume that there are M storage units of track data, and the current moment is k. As illustrated in Figure 11.12(a), when $k < M$, the new measurement k of the track is stored at the corresponding storage point k, and is stored in turn afterwards. When $k = M$, the new measurement k is stored at the corresponding storage point M, as shown in Figure 11.12(b). When $k = M + 1$, the new measurement is stored in unit 1, as illustrated in Figure 11.12(c). The subsequent measurements are stored in the same manner. This guarantees that the track data stored in the circulative time window are always those at the latest M moments. M is set according to the specific system requirements.

11.3.2 Track Quality Management

As an integral part of track management, track quality management can help initiate tracks promptly and accurately to set up new target files, or revoke tracks in the same fashion to remove redundant target files. Track quality management has two main goals:

1. To initiate new tracks correctly and rapidly, and suppress false ones from initiating.
2. To delete the already set-up false tracks correctly and promptly, and keep true ones from being deleted.

11.3.2.1 Select Initiation Rule and Delete Tracks According to Track Quality

Initiation Response Time

To conduct a theoretical analysis of track quality management, track initiation and cancellation will be illustrated with the sliding-window detector. The sliding-window detector has been discussed in Chapter 6. As it requires a small amount of calculation and can be analyzed with the Monte Carlo method (or analytic method), it has been adopted in many practical tracking systems.

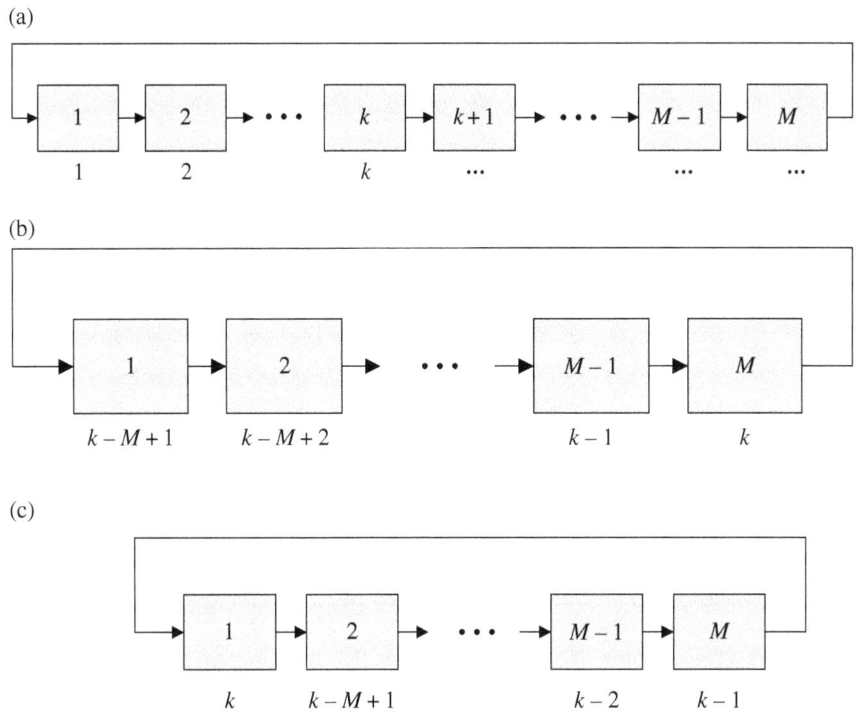

Figure 11.12 Sketch map of storage units in the sliding time window: (a) $k<M$; (b) $k=M$; (c) $k=M+1$

The general principle of sliding-window detectors is as follows. Assume the sequence (Z_1, Z_2, …, Z_N) represents N times of radar scanning. If there is measurement in the correlation gate of the ith scan, let $Z_i=1$; otherwise, let $Z_i=0$. If the detected number in a sliding window with size n reaches m, then track initiation is successful, otherwise the sliding window moves ahead.

At present, the common technical index for track initiation is the initiation response time. This refers to the duration from a target's entrance into a radar power coverage to the establishment of the track, and it usually takes the number of radar scanning cycles as its unit. A quick track initiation usually takes 3 or 4 scanning cycles, while a slow one usually takes 8 to 10 scanning cycles.

In sliding-window detection, the result of Hammers (see Table 11.2 [283, 284]) can be used directly to calculate the time that conventional criteria take to detect a target. In track initiation, the probability of a true track is equal to the probability of it being found by radar in target detection. Assume that p is the probability of $Z_i=1$, and P_c (N) is the probability state difference equation for succeeding at the Nth detection. It is shown in Table 11.2 that, under the circumstance of certain rule, the initiation response time is the uniform function of p. With a given initiation response time, a number of n/m rules with response time smaller than the rated response time can be singled out. Given different rules and p, the initiation response time and the probability of success at the Nth initiation can be obtained according to Table 11.2.

When the initiation response time is smaller than that required by the system index, the common criterion for n/m logic sliding detection is shown in Table 11.3.

As shown in the table, it is difficult to determine the required initiation rule only with a given initiation response time.

Table 11.2 State difference equation for the Nth initiation's probability of success

Rule	Initiation response time	State difference equation for the Nth initiation's probability of success
2/2	$\frac{(1+p)}{p^2}$	$P_c(N)=p^2\delta_{N2}+(2-p)P_c(N-1)-(1-p)^2P_c(N-2)-p(1-p)P_c(N-3)$ when $N<2$, $P_c(N)=0$
2/3	$\frac{2-q^2}{p(1-q^2)}$, in which $q=1-p$	$P_c(N)=p^2\delta_{N2}+p^2(1-p)\delta_{N3}+(2-p)P_c(N-1)+(p-1)P_c(N-2)+p(1-p)^2P_c(N-3)-p(1-p)^2P_c(N-4)$ when $N<2$, $P_c(N)=0$
3/3	$\frac{1+p+p^2}{p^3}$	$P_c(N)=p^3\delta_{N3}+(2-p)P_c(N-1)-(1-p)^2P_c(N-2)-p(1-p)^2P_c(N-3)-p^2(1-p)P_c(N-4)$ when $N<3$, $P_c(N)=0$

(*Note:* when $N=i$, $\delta_{Ni}=1$; otherwise, $\delta_{Ni}=0$).

Table 11.3 Criteria by which the initiation response time is smaller than the system index

Initiation time index criterion	$P_d=0.9$	$P_d=0.8$	$P_d=0.7$	$P_d=0.6$
Quick track initiation (response time ≤4)	2/2, 2/3, 3/3	2/2, 2/3	2/2, 2/3	
Slow track initiation (response time ≤8)	2/2, 2/3, 3/3	2/2, 2/3, 3/3	2/2, 2/3, 3/3	2/2, 2/3

False Track Initiation Probability

To reflect the inhibition of track quality management systems against false track initiation, define the probability of false track initiation as P_{FTI}. The characteristics of false track initiation have to be studied in order to calculate the probability of false tracks.

The probability of false targets turning up is

$$P_c=1-(1-P_F)^L \tag{11.39}$$

where P_F is the radar's probability of false alarms and L is the number of radar resolution units in the initiated wave gate. For 2D radars,

$$L=\frac{\pi(V_{max}T)^2}{\Delta\rho(\rho\Delta\theta)} \tag{11.40}$$

where $\Delta\rho$ and $\Delta\theta$ are the distance resolution and orientation resolution of the radar, respectively, ρ is the distance between the initiated wave gate and the radar, V_{max} is the maximum expected target velocity, and T is the radar scanning cycle.

Assume that $\rho=50$ km, $\Delta\theta=0.026$ rad, $\Delta\rho=300$ m, $\sigma_\theta=0.014$ rad, $\sigma_\rho=150$ m, $V_{max}=600$ m/s, $T=1$ s, and $P_F=10^5$, then $P_c=5.9\times10^{-4}$ can be obtained according to (11.39) and (11.40). Substituting P_c into the state difference equation in Table 11.2 as p (false targets' probability of detection) yields P_{FTI}, the probability of false track initiation under different rules, as listed in Table 11.4. Make it smaller than P_{FTIT}, the probability index of false track initiation of the system, that is, $P_{FTI}\le P_{FTIT}$. It is shown in Table 11.4 that P_{FTI} under the 3/3 rule is the minimum. Given $P_{FTIT}=5\times10^{-5}$, 2/2 can be chosen.

Table 11.4 Probability of false track initiation (conditioned on $P_c = 5.9 \times 10^4$)

Track initiation rule	P_{FTI}
2/2	3.4×10^5
2/3	6.8×10^3
3/3	2.0×10^8

False Track Life and True Track Life

The main task of track cancellation is to promptly delete false tracks and keep true tracks. For this purpose, false track life and true track life [288] are defined as follows.

Definition 11.1 False track life (L_{FT}) is the average number of scans of a false track from initiation to cancellation.

Definition 11.2 True track life is the average number of scans of a true track from initiation to cancellation because it is mistaken for a false track.

When a sliding-window detector is used, the rule for track cancellation is: If there is no measurement detected that is correlated with the track for m times out of n detections, then delete the track.

The number of resolution units in subsequent correlation gates is related to the size of the gate, which in turn is related to the accuracy of the filter. It is a function of the observation and prediction covariance matrix. Assume that prediction error equals observation error, then the number of resolution units in subsequent correlation gates is

$$L = \frac{2\chi^2 \sigma_\rho \sigma_\theta}{\Delta\rho \Delta\theta} \tag{11.41}$$

where χ^2 is the chi-squared distribution threshold with a given significance level. The measurement accuracy is generally positively correlated with resolution units, and the number of resolution units in subsequent gates is related to the SNR and accumulation method. The probability of no false target being in subsequent correlation gates is $\overline{P_c} = 1 - P_c$.

According to the Hammers method [284], assume that the parameters in (11.41) are the same as mentioned previously, and obtain $\overline{P_c} = 0.99996$, so the false track life under different rules is obtained, as listed in Table 11.5.

The following conditions have to be considered when studying true track life:

1. The echo from the target falls in the association region, and there is no false alarm.
2. The echo from the target falls in the association region, and there are false alarms.
3. There is no echo from the target, and there are false alarms.
4. There is no echo from the target, and there is no false alarm.

Under these conditions, it is assumed that the probability of measurements (echoes) falling in association regions, or the threshold probability, is $P_G = 1.0$. So, the probabilities of the four events are $P_1 = P_d \times (1 - P_c)$, $P_2 = P_d \times P_c$, $P_3 = (1 - P_d) \times P_c$, $P_4 = (1 - P_d) \times (1 - P_c)$. The measurement will get lost once under the last condition, so the probability that a true track loses measurements is

Table 11.5 False track life ($P_c = 4 \times 10^{-5}$)

Rules	False track life
2/2	2.00
2/3	2.00
3/3	3.00

Table 11.6 Measurement missing probability and true track life

Rule	$P_d = 0.9$	$P_d = 0.8$	$P_d = 0.7$	$P_d = 0.6$
2/2	0.1/110.0	0.2/30.4	0.3/14.3	0.4/8.7
2/3	0.1/62.4	0.2/18.9	0.3/9.7	0.4/6.3
3/3	0.1/1110.0	0.2/155	0.3/51.5	0.4/24.9

Note: "/" in the table means measurement missing probability/true track life.

$$\overline{P_{TL}} = 1 - P_d - P_c + P_d P_c \tag{11.42}$$

The probability of measurement loss and true track life under $P_c = 4.0 \times 10^5$ and different detection probability conditions is listed in Table 11.6.

It is shown in Tables 11.5 and 11.6 that, when $P_d = 0.9$, the 3/3 rule is capable of effectively deleting false tracks and keeping true ones.

11.3.2.2 Optimization of Track Quality Management under Mono-radar Circumstances

To sum up, the rule for track initiation and track cancellation depends mainly on the probability of detection P_d and the probability of a false target turning up P_c.When a signal detection system is given, both P_d and P_c are related to the SNR in the constant false-alarm radar.

According to the radar equation, the SNR of received signals is

$$(\text{SNR})_{\text{dB}} = (P_t)_{\text{dBw}} + 2(G)_{\text{dB}} + 2(\lambda)_{\text{dB cm}} + (\sigma)^2_{\text{dBm}} - 4(R)_{\text{dB sea mile}} - (B)_{\text{dBHz}} - \left(\overline{\text{NF}_0}\right)_{\text{dB}} - (L)_{\text{dB}} \tag{11.43}$$

where P_t is transmission power, G is antenna gain, λ is wavelength, σ is radar cross-section, R is distance from the target, B is system bandwidth, $\overline{\text{NF}_0}$ is effective noise coefficient, L is total system loss factor of the radar system. A decibel is defined as ten times the logarithm.

The detection probability of the optimum detection of the coherent radar system is

$$P_d = 1 - \Phi\left[\sqrt{\frac{1}{d}}\ \ln l_0 + \frac{1}{2}\sqrt{d} - d\right] \tag{11.44}$$

where l_0 is the threshold value, which depends on the judgment rule, $d = 2E_1/N_0$ is SNR, and $\Phi(x) = \int_{-\infty}^{x} \frac{1}{\sqrt{2\pi}} e^{-\frac{v^2}{2}} dv$ is the Gaussian distribution function.

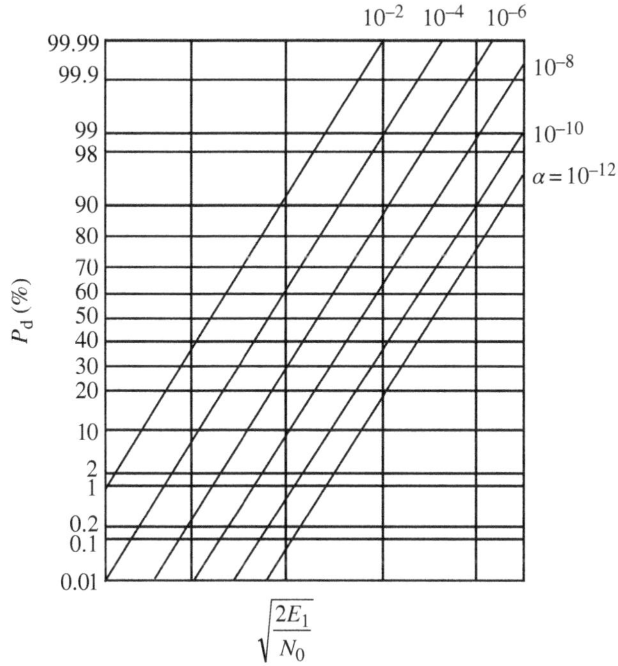

Figure 11.13 Detection curve

Given l_0, the relation curves (detection characteristic curves) of $P_d - \sqrt{d}$ under different false alarm rates are obtained, as drawn in Figure 11.13. When the rate is given, the relation between P_d and SNR (expressed in dB) is similar to a linear one. According to (11.43), SNR has a linear relation with distance. Therefore, the detection probability can be considered to have a linear relation with distance.

As the detection probability changes with distance, different rules for track initiation and cancellation are applied to different distances. To set up track management rules for different distances, define the optimum initiation and cancellation rules as below.

Assume that B is the set of available rules

$$B = \{B_i\} \tag{11.45}$$

The available rules are 2/2, 2/3, and 3/3; sometimes 3/4, 4/4, etc.

Assume that $S_i \subseteq B$, and that the track initiation response time T_I is smaller than the rated response time T_IT, that is,

$$S_i = \{B_i | T_\mathrm{I}(B_i) \le T_\mathrm{IT}\} \tag{11.46}$$

Assume that $P_\mathrm{FTI}(B_i)$ is the false track initiation probability in the set when rule B_i is adopted, then define the optimum initiation rule B_opt as

$$B_\mathrm{opt} = \left\{ B_i \middle| \min_{B_i \in S_i} P_\mathrm{FTI}(B_i) \right\} \tag{11.47}$$

Assume the set $S_P \subseteq B$, and the false track life L_{FT} is shorter than the rated false track life L_{FTT}. That is,

$$S_P = \{B_i | L_{FT} \le L_{FTT}\} \tag{11.48}$$

Assume that $L_{RT}(B_i)$ is the true track life when rule B_i is adopted. Define the optimum cancellation rule Q_{opt} as

$$Q_{opt} = \left\{ B_i | \min_{B_i \in S_p} L_{RT}(B_i) \right\} \tag{11.49}$$

If the system index specifies the rated track initiation response time T_{IT} and the rated probability of false track initiation P_{FTIT}, then define the pseudo-optimum initiation rule.

Assume S_q is the set of rules B_i when the probability of false track initiation $P_{FTI}(B_i)$ is smaller than the rated false track probability P_{FTIT}, that is,

$$S_q = \{B_i | P_{FTI}(B_i) < P_{FTIT}\} S_i \tag{11.50}$$

Define the pseudo-optimum track initiation rule as

$$B_{sopt} = \left\{ B_i | \min_{B_i \in S_q} T_I(B_i) \right\} \tag{11.51}$$

The pseudo-optimum track initiation rule is defined as such because the accuracy of track filtering is usually related to the number of filterings. Therefore, rules with a shorter track initiation response time are preferred.

11.3.2.3 Track Quality Management under Multiple Site Circumstances

Under multiple site circumstances, the detection probability, false alarm rate, size of resolution units, size of association regions, and data rate are different for each radar. So, introduce the concepts of average detection probability of true targets and false targets.

Assume that a certain region Ω_j is the common power coverage of N_{Rj} radars, and that the scanning cycle of each radar is $T_i \left(i = 1, \ldots, N_{Rj}\right)$, so the probability of the scanning beams of all radars pointing at one target is

$$P_i = \frac{1/T_i}{\sum_{i=1}^{N_{Rj}} 1/T_i} \tag{11.52}$$

Assume that the detection probability of the ith radar is P_{di}, so the average detect probability of the N_{Rj}th radar is

$$E\left[P_{dj}\right] = \sum_{i=1}^{N_{Rj}} P_{di} P_i \tag{11.53}$$

Assume that the probability of a false target turning up in the association region for the ith radar is P_{ci}, then the average probability of false targets turning up for the N_{Rj}th radar is

$$E[P_{cj}] = \sum_{i=1}^{N_{Rj}} P_{ci} P_i \tag{11.54}$$

As the power coverage for every radar is different, $E[P_{dj}]$ and $E[P_{cj}]$ are also not the same in different spaces Ω_j. Therefore, one track quality management system is not enough for a multiple-radar system. The optimum design method of the track quality management subsystem for multiple-radar systems is given below.

1. Determine the number N of different spaces Ω_j. The so-called space difference refers to the difference in model and number between the radars which detect the space of interest.
2. Each space Ω_j can be divided into a number of subspaces (usually based on distance), Ω_{jk} ($k = 1, \ldots, N_j$), such that a different optimum (or sub-optimum) track quality management rule is made for each subspace. The corresponding rule for each subspace is denoted R_{jk} ($k = 1, \ldots, N_j$).
3. Combine the $N_j \times N$ rules R_{jk} based on their variety. Assume there are N_B track quality management plans after combination. Establish rule allocation matrix A, which is ($N_B \times N_j \times N$)-dimensional. Its line numbers correspond to the numbers N_B of management rules, and its column numbers to those of $N_j \times N$ subspaces.

When evaluating the quality of a track, determine the subspace that the track belongs to and then find the corresponding track quality rule and use it to evaluate the track quality. In the other expression methods of track quality management, the compatibility of rules should be taken into account, that is, a corresponding relationship should be established between the state of a rule and that of another.

11.3.3 Track File Management in the Information Fusion System

Along with the rapid development of information fusion technology, the brand new data processing technology is applied to the operational command system by conducting multi-level and multi-side processing of observations from various sensors or multi-sources, so as to find the targets and identify their properties in real time, giving a comprehensive battlefield situation. The track file management serves as an indispensable and important part of the information fusion system. Unlike a single sensor, the fusion system needs to promptly carry out time and space registration, data correlation, track integration, and uniform batch numbering on the track data from all sensors and remove the redundant track and false track, thus forming a clear picture of what is happening on the battlefield.

To conduct efficient management on the local track from each sensor and the overall track from the fusion center, and facilitate the fusion center's multi-sensor data fusion algorithm, many original track tables are adopted in the system corresponding to the local tracks of the multi-sensors. Hence, these are fusion track tables coordinating with the fusion tracks generated by the fusion center. Generally, each fusion track corresponds to at least one original track. To easily find this correlation, there is a bi-directional linked list structure (correlation list) on each of the track tables that establishes the relationship between the fusion track and its related original track. The fusion track with an empty correlation list is called the "isolated track" [289].

Table 11.7 Example of the correlation list of the fusion track

Time	Track batch of sensor 1	Track batch of sensor 2	Track batch of sensor 3	Fusion track batch
k	1	1	—	1
	2	2	1	2
	3	—	2	3
	4	—	—	4
$k+1$	1	1	—	1
	2	2	1	2
	3	—	2	3

Examples pertaining to the correlation list of the fusion track are presented in Table 11.7. At time k, sensors 1, 2, and 3 obtain four, two, and two tracks, respectively, with numbering shown in the table. The number 1 track of sensor 1 is correlated with the number 1 track of sensor 2. Their fusion generates the fused track with batch 1. The number 2 track of sensor 1 is correlated with the number 2 track of sensor 2 and the number 1 track of sensor 3, the fusion of which generates the fusion track with batch 2. The number 3 track of sensor 1 is correlated with the number 2 track of sensor 3, which, after being fused, become the fusion track with batch 3. The number 4 track of sensor 1 has no correlated track, thus being called the isolated track and directly generating the fusion track with batch 4. At time $k+1$, the cancellation of number 4 track in sensor 1 is also accompanied by the cancellation of number 4 fusion track. Therefore, the example reveals that the correlation list is set up for each of the fusion tracks to record the source of the track. For one thing, in the process of the correlation judgment, it could avoid repeated judgment of the tracks and reduce the computational load; for another, the new fusion tracks may coincide with the batch of the deleted tracks, in which case the correlation list can indicate the different data sources of the fusion track.

The basic steps for fusion track management are as follows.

1. *Establishment of the fusion track.* The new original track obtained by the sensor is sent to the fusion center for correlation processing. Even if there is no correlated track, the batch is still rearranged to form a fusion track to reveal all the targets promptly. However, the problem is that the redundant track may emerge in the presence of false track correlations or missed correlations, which could be resolved by deleting the redundant track in the follow-up processing period.
2. *Maintenance of the fusion track.* The number of track correlations is recorded through the track correlation quality, in other words, adding 1 to the correlation quality of the successfully correlated track and adding 0 to that of the track which fails to be associated. As the 6/8 principle suggests, a track is confirmed as already correlated when the value of its correlation quality reaches 6 among the latest 8 times of correlation judgment, but it is considered as a tentative correlated track if the value is less than 6. The fused track corresponding to the confirmed correlated track is maintained and its corresponding correlation list is set, while the experimental correlated track will not be fused with its corresponding correlation list set as empty.
3. *Cancellation of the fusion track.* False tracks and redundant tracks may be generated in the initiation of original tracks and in processes of tracking and track correlation, and these tracks will be fused as well. However, correlation calculations are always conducted for tracks which are in the process of being fused, so that false and redundant ones will be cancelled as the fusion proceeds.

11.4 Summary

This chapter discussed the multi-target track termination and track management technologies. The multi-target track termination technology discussed is based mainly on the "nearest"-neighbor algorithms, and falls into two broad categories: target-oriented recursion methods and measurement-oriented batch processing methods. The former, demanding less calculation, are convenient for engineering applications but produce poor results in the presence of dense echoes, while the latter are difficult to apply in engineering because of their excessive computing requirements but yield significantly better results in the presence of dense echoes. The track management technology discussed in this chapter includes the track batch and quality management methods, as well as the track file management method for the information fusion system.

To be specific, this chapter discusses the single and double-track batch management methods and the track data storage method, which are capable of handling such complicated situations as the addition, cancellation, splitting, and combination of target tracks in the operation of real systems. It also analyzes the selection of initiation rules and track cancellations by means of track quality, as well as the track quality management, respectively, in the single and multiple-site cases, offering a theoretical basis for the optimal design of the track quality management system. The four technical indexes proposed in this chapter (namely, initiation response time, initiation probability of false tracks, life of true tracks, life of false tracks) are of vital importance to engineering applications.

For the information fusion system, the methods for managing track files using tables of original and fused tracks have been discussed, which can effectively manage local tracks from the individual sensors and global ones from the fusion center, and facilitate the fusion center's multi-sensor data fusion computation.

12

Passive Radar Data Processing

12.1 Introduction

"Passive radar" refers to the radar system that itself does not radiate electromagnetic waves, but receives by its antennas data carried by direct waves from the radiation sources of a target or by reflected or scattered waves from a target irradiated by external radiation sources. In contrast, "active radar" refers to the radar system that itself radiates electromagnetic waves, and receives reflected or scattered waves from a target, which carry information useful for target detection. These waves are sent to the receiver, which, through processing, extracts the useful information and eliminates the useless and interference. The resulting data, after being processed, are used to complete the location and tracking of the target.

Active radars are what are generally referred to as radars. For convenience, no distinction is made, however, between active and passive radars in this chapter, which focuses on the data processing of the latter. It starts with an analysis of the characteristics and advantages of passive radars, and proceeds to discuss the spatial association of passive radars' measurements, and the optimal deployment of passive sensors based on the principle of minimum area of positioning fuzzy ellipse and TDOA location, before concluding with a summary.

12.2 Advantages of Passive Radars

Compared with their active counterparts, passive detection systems are characterized by many advantages including their better ability to avoid detection and extract information on targets' attributes. These advantages can enhance their anti-reconnaissance, anti-jamming, and anti-hard/soft killing abilities in electronic warfare, resulting in their better survivability and performance. In recent years, the development and maturity of the passive detection technology has enabled it to be used not only for electronic reconnaissance, but also for surveillance, air defense, and precision positioning strikes. With current trends towards better covert attack and hard destruction

Radar Data Processing with Applications, First Edition. He You, Xiu Jianjuan, and Guan Xin.

performance in military electronic systems, the abilities of electronic countermeasures systems have proven crucial in deciding the process and outcome of wars. Generally speaking, passive radar systems have the following features and advantages [75, 78, 290–293].

1. *Good stealth ability and survivability*. Because passive detection systems themselves do not radiate high-powered electromagnetic signals and only receive electromagnetic waves radiated, reflected, or scattered from targets, they have better stealth ability, as a result of which the exposure probability of their carriers is reduced, and they are difficult to detect by enemy electronic reconnaissance equipment. Therefore, unlike active radars, which are prone to being attacked by anti-radiation missiles, cruise missiles, or other precision-guided weapons, passive radar systems have better survivability.
2. *Good anti-jamming capability*. Since passive detection systems are characterized by good stealth ability, enemies cannot implement any targeted electronic jamming. Moreover, adversary electronic jamming of conventional active radars also becomes a signal that can be used by passive radars to detect the sources of electronic jamming. Passive radars can use these signals to conduct passive detection, location, and attack.
3. *Enormous potential for detecting stealth aircraft*. Because they can detect and identify stealth targets by means of these targets' electromagnetic radiation signals, communications, electronic jamming, etc., passive radars have the potential to detect stealth aircraft [294].
4. *Good detecting ability for low-altitude targets*. Unlike active radars, passive radars cannot be confused by strong ground and sea-clutter jamming. Therefore, they perform better at detecting low and super-low-altitude targets [295].
5. *Further detection range*. Active radar systems receive reflection signals of targets, and the strength of the received signals is inversely proportional to the fourth power of the range. In contrast, passive detection systems receive radiation signals from their targets directly, which are one-way transmitted with the strength inversely proportional to the square of the range. Therefore, their operational range is much further than that of active ones.
6. *Good target identifying ability*. Passive detection is characterized by obtaining more target attribute information and measuring fine features of targets' electromagnetic radiation signals. Passive radars can conduct, by intercepting parameters of radiation sources and depending on their database, a fine-grained identification of the characteristics of targets, thereby determining their attributes, types, quantities, and working states.
7. *Broad airspace coverage*. Modern passive detection equipment usually has 360° azimuthal coverage, and 50–60° pitch coverage. Thus, it has broader instantaneous airspace coverage, higher interception probability, and faster response time [294, 295].
8. *Extensive applicability*. Any modern weapon system relies on military electronic equipment, such as radars, communications, and so on, and thus it is inevitable for them to produce electromagnetic radiation. Therefore, passive detection equipment is widely applied to detecting operation targets, whether in the air, on the land, or at sea, whether stationary or moving. As long as they have electromagnetic radiation signals, passive detection equipment can detect them.
9. *Extreme broad frequency coverage*. The bandwidth of modern passive detection equipment is up to tens of gigahertz, able to cover broad frequency domains, including meter, decimeter, centimeter, and millimeter waves, and the common working wave band of infrared and laser.
10. *Smaller volume, lighter weight, and lower costs*. Passive detection systems do not radiate high-powered detection signals. There is no need to build and maintain high-powered transmitters for them, which saves the TR switch and related electronic units, thus lowering their cost. Besides, they are small in size, light in weight, and easy to maneuver and camouflage.

12.3 Passive Radar Spatial Data Association

When tracking targets using bearings-only information of passive radars, operators must adopt angle measurements to identify the position and velocity of the targets of interest relative to the sensor's platform. This can be done by the acceleration of this platform. When the relative acceleration between this carrier and the target is zero, the target range state is unobservable, and thus its optimal estimation is not available [296, 297].

Quite a few approaches have been proposed in order to overcome this problem. The pseudo-linear filter algorithm put forward in Ref. [298] is characterized by the stability of algorithm, simplicity of calculations, and easy implementation. However, it involves biased estimates. If the extended Kalman filter (EKF) is adopted in tracking the target, then the angle measurements, as incomplete location observations, cannot be converted into rectangular coordinates for linear filtering. As has been proved by its application in recursive bearings-only tracking (state equations use rectangular coordinates while the measurements are a nonlinear function of the state), even without the unfavorable effects of false measurements or clutter, the EKF can only provide unstable estimation results and state [299, 300]. This is because in the rectangular coordinate system, the state equation is linear while the measurement equation is nonlinear, so a Jacobian matrix is needed for the filtering covariance calculation, which brings inaccurate valuations in the calculation of gain and covariance. Moreover, in the presence of large range and angle measuring errors, the EKF will introduce big truncation errors, which lead to filter divergence. In recent years, in addition to angle measurements, some other information is used for passive location and tracking, such as in the phase change rate method [301, 302], Doppler change rate method [303, 304], multi-model method [219, 305, 306], etc.

If there is more than one target in the detection area, then track initiation and data association in multi-target tracking must be realized on the basis of achieving multi-target passive location. Track initiation in passive location systems is similar to that in active location systems, except that it is more difficult to deal with, mainly due to the characteristics of passive location systems [307].

12.3.1 Phase Changing Rate Method

As shown in Figure 12.1, suppose that the phase difference of the coming waves received by the antenna array elements A_1 and A_2 on the moving platform is

$$\varphi(t) = \omega_0 \Delta t = \frac{2\pi D}{c} f_0 \sin\beta(t) \tag{12.1}$$

where ω_0 is the angular frequency of the coming waves, Δt is the time difference of the arrival of the coming waves to the antenna array elements A_1 and A_2, c is the propagation velocity of electromagnetic waves, D is the range between array elements (namely the length of the interferometer baseline), which is assumed far less than the range between the moving platform and the radiation source, f_0 is the frequency of the coming waves, and β is the azimuth angle of the coming waves.

The phase changing rate $\dot{\varphi}(t)$ can be derived from (12.1):

$$\dot{\varphi}(t) = \frac{2\pi D}{c} f_0 \cos\beta(t)\dot{\beta}(t) \tag{12.2}$$

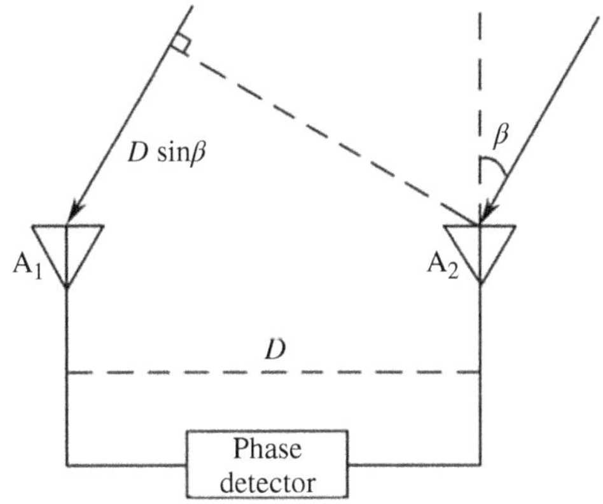

Figure 12.1 Difference of phase

where

$$\dot{\varphi}(t) = \frac{\mathrm{d}\varphi(t)}{\mathrm{d}t}, \quad \dot{\beta}(t) = \frac{\mathrm{d}\beta(t)}{\mathrm{d}t} \tag{12.3}$$

Then, we can obtain

$$\dot{\beta}(t) = \frac{\dot{\varphi}(t)}{\frac{2\pi D}{c} f_0 \cos\beta(t)} \tag{12.4}$$

In addition, as known from geometric knowledge, the target azimuth angle measured at time t satisfies

$$\beta(t) = \arctan\left[\frac{x_T(t) - x_o(t)}{y_T(t) - y_o(t)}\right] \tag{12.5}$$

where $[x_T(t), y_T(t)]$ and $[x_o(t), y_o(t)]$ are, respectively, the locations of radiation source and sensor platform at time t. For convenience, the moment t is dropped.

Derived from (12.5):

$$\dot{\beta}(t) = \frac{(\dot{x}_T - \dot{x}_o)(y_T - y_o) - (x_T - x_o)(\dot{y}_T - \dot{y}_o)}{(x_T - x_o)^2 + (y_T - y_o)^2} \tag{12.6}$$

If we denote $x = x_T - x_o$, $y = y_T - y_o$, and define the range from the sensor platform to the radiation source at time t as

$$r^2(t) = (x_T - x_o)^2 + (y_T - y_o)^2 \tag{12.7}$$

then (12.6) can be described as

$$\dot{\beta}(t) = \frac{\dot{x}\cos\beta(t) - \dot{y}\sin\beta(t)}{r(t)} \tag{12.8}$$

where

$$\dot{x} = \dot{x}_T - \dot{x}_o, \quad \dot{y} = \dot{y}_T - \dot{y}_o \tag{12.9}$$

When the radiation source is stationary, the following can be obtained from (12.9):

$$\dot{x} = -\dot{x}_o, \quad \dot{y} = -\dot{y}_o \tag{12.10}$$

Derived from (12.8), the relative range from the radiation source to the moving platform at time t is

$$r(t) = \frac{\dot{x}\cos\beta(t) - \dot{y}\sin\beta(t)}{\dot{\beta}(t)} \tag{12.11}$$

Substituting (12.4) into (12.11) yields

$$r(t) = \frac{2\pi D}{c} f_0 \cos\beta(t) \frac{-\dot{x}_o\cos\beta(t) + \dot{y}_o\sin\beta(t)}{\dot{\varphi}(t)} \tag{12.12}$$

Then, the location coordinates of the radiation source can be obtained:

$$x_T = x_o + r(t)\sin\beta(t) = x_o + \frac{2\pi D}{c} f_0 \cos\beta(t)\sin\beta(t) \frac{-\dot{x}_o\cos\beta(t) + \dot{y}_o\sin\beta(t)}{\dot{\varphi}(t)} \tag{12.13}$$

$$y_T = y_o + r(t)\cos\beta(t) = y_o + \frac{2\pi D}{c} f_0 \cos^2\beta(t) \frac{-\dot{x}_o\cos\beta(t) + \dot{y}_o\sin\beta(t)}{\dot{\varphi}(t)} \tag{12.14}$$

As indicated, when the radiation source is stationary, the location of the target can be identified by conducting calculations on the phase changing rate $\dot{\varphi}(t)$ and the azimuth angle $\beta(t)$ measured by sensors. The target can be tracked using the Kalman filter method mentioned in Chapter 2 after obtaining location coordinates of the target. As indicated in (12.11), an assumption implied here is $\dot{\beta}(t) \neq 0$. When both the radiation source and the sensor platform are stationary, $\dot{\beta}(t) = 0$, as shown in (12.8). When the sensor platform conducts radial motion relative to the target, then

$$\dot{x}_o(y_T - y_o) = (x_T - x_o)\dot{y}_o \tag{12.15}$$

and it follows from (12.6) that $\dot{\beta}(t) = 0$. When the target is located on the straight line overlapping the interferometer baseline, it follows from (12.1) that $|\varphi(t)|$ is maximum. From (12.2) it is known that in this case, $\dot{\varphi}(t) = 0$. Thus, $\dot{\beta}(t)$ cannot be determined by measurements. To sum up, in all three cases above, the location coordinates of the target cannot be obtained through (12.13) and (12.14). Hence, the target is unobservable at this time.

If the radiation source is moving, the situation is more complicated. Since in this case the motion velocity of the target is unknown, it is impossible to derive the relative motion velocity between the target and the sensor platform from (12.9). Therefore, the relative range between the radiation source and the motion platform cannot be derived from (12.11), nor can the target location be derived from (12.13) and (12.14). In such cases, the estimates of the target location can be obtained through filtering by using the nonlinear filter method if the observable condition is satisfied. As known from Refs [301, 302], when the observing platform is maneuvering, no matter what forms of maneuver it takes, the system is observable. If the system satisfies the observability condition, the EKF discussed in Chapter 4 can be applied for track processing.

The state equation of the system can be described as

$$\boldsymbol{X}(k+1)=\boldsymbol{F}(k)\boldsymbol{X}(k)-\boldsymbol{U}(k) \tag{12.16}$$

where

$$\boldsymbol{X}(k)=[x(k),y(k),\dot{x}(k),\dot{y}(k)]' \tag{12.17}$$

is the state vector at time k;

$$\boldsymbol{F}(k)=\begin{bmatrix}1&0&T&0\\0&1&0&T\\0&0&1&0\\0&0&0&1\end{bmatrix} \tag{12.18}$$

is the state transition matrix;

$$\boldsymbol{U}(k)=[u_1(k),u_2(k),u_3(k),u_4(k)]' \tag{12.19}$$

is the controlled variable, which denotes the maneuver of the sensor platform from time t_{k-1} to t_k, and

$$u_1(k)=\int_{t_{k-1}}^{t_k}(t_k-\tau)a_x(\tau)\mathrm{d}\tau \tag{12.20}$$

$$u_2(k)=\int_{t_{k-1}}^{t_k}(t_k-\tau)a_y(\tau)\mathrm{d}\tau \tag{12.21}$$

$$u_3(k)=\int_{t_{k-1}}^{t_k}a_x(\tau)\mathrm{d}\tau \tag{12.22}$$

$$u_4(k)=\int_{t_{k-1}}^{t_k}a_y(\tau)\mathrm{d}\tau \tag{12.23}$$

where $a_x(t)$ and $a_y(t)$ are components of motion acceleration of the sensor platform on axes x and y, respectively.

The measurement equation of the system can be described as

$$\boldsymbol{Z}(k)=\boldsymbol{h}[\boldsymbol{X}(k)]+\boldsymbol{W}(k) \tag{12.24}$$

where

$$\boldsymbol{h}[\boldsymbol{X}(k)]=\begin{bmatrix} h_1(\boldsymbol{X}(k)) \\ h_2(\boldsymbol{X}(k)) \end{bmatrix} \tag{12.25}$$

and

$$h_1[\boldsymbol{X}(k)]=\arctan\left(\frac{x_T-x_o}{y_T-y_o}\right) \tag{12.26}$$

$$h_2[\boldsymbol{X}(k)]=\dot{\varphi}(t)=\frac{2\pi D}{c}f_0\cos\beta(t)\dot{\beta}(t) \tag{12.27}$$

$\boldsymbol{W}(k)$ is the zero-mean, white, Gaussian measurement noise with variance $\boldsymbol{R}(k)$, and

$$R(k)=\begin{bmatrix} \sigma_\beta^2 & 0 \\ 0 & \sigma_{\dot{\varphi}}^2 \end{bmatrix} \tag{12.28}$$

where σ_β^2 and $\sigma_{\dot{\varphi}}^2$ are the variance for measurement noise of azimuth angle and phase changing rate, respectively.

The one-step prediction of the state is

$$\hat{\boldsymbol{X}}(k+1|k)=\boldsymbol{F}(k)\hat{\boldsymbol{X}}(k|k)-\boldsymbol{U}(k) \tag{12.29}$$

The one-step prediction of the covariance is

$$\boldsymbol{P}(k+1|k)=\boldsymbol{F}(k)\boldsymbol{P}(k|k)\boldsymbol{F}'(k) \tag{12.30}$$

The prediction covariance (innovation covariance) of measurement is

$$\boldsymbol{S}(k+1)=\boldsymbol{h}_X(k+1)\boldsymbol{P}(k+1|k)\boldsymbol{h}_X'(k+1)+\boldsymbol{R}(k+1) \tag{12.31}$$

where the Jacobian matrix is

$$\begin{aligned} \boldsymbol{h}_X(k+1) &= [\nabla_X \boldsymbol{h}'(k+1,\boldsymbol{X})]'_{X=\hat{X}(k+1|k)} \\ &= \left[\frac{\partial \boldsymbol{h}[\boldsymbol{X}(k)]}{\partial x}\ \frac{\partial \boldsymbol{h}[\boldsymbol{X}(k)]}{\partial y}\ \frac{\partial \boldsymbol{h}[\boldsymbol{X}(k)]}{\partial \dot{x}}\ \frac{\partial \boldsymbol{h}[\boldsymbol{X}(k)]}{\partial \dot{y}}\right]_{X=\hat{X}(k+1|k)} \end{aligned} \tag{12.32}$$

The filter gain is

$$\boldsymbol{K}(k+1)=\boldsymbol{P}(k+1|k)\boldsymbol{h}_X'(k+1)\boldsymbol{S}^{-1}(k+1) \tag{12.33}$$

The state update equation is

$$\hat{X}(k+1|k+1) = \hat{X}(k+1|k) + K(k+1)\{Z(k+1) - h_X[k+1, \hat{X}(k+1|k)]\} \tag{12.34}$$

The covariance update equation is

$$P(k+1|k+1) = [I - K(k+1)h_X(k+1)]P(k+1|k)[I + K(k+1)h_X(k+1)]' - K(k+1)R(k+1)K'(k+1) \tag{12.35}$$

Compared with the location methods using angle measuring information only, the location method, which adopts the phase changing rate on the basis of direction measuring and positioning technology, is less restricted; besides, it has faster speed and higher accuracy. This technology can be applied to counter incoherent radar, but the precision of the phase changing rate is critical to the realization of immediate and accurate location of the radiation source.

12.3.2 Doppler Changing Rate and Azimuth Joint Location

When there is a radial velocity between the target and the observing instrument, Doppler shifts occur on the observing instrument. Hence, passive location can be conducted using the Doppler changing rate of the target [304]. For the sake of simplicity, we might as well start with the location in a 2D plane. Assume that the relative velocity between the observing instrument and the radiation source is v. In the reference coordinate system which takes the observing instrument as origin, the relative velocity can be decomposed into the tangential velocity v_t and the radial velocity v_r, as shown in Figure 12.2.

Derived from (12.8),

$$\dot{\beta}(t) = \frac{v_x \cos\beta - v_y \sin\beta}{r(t)} \tag{12.36}$$

where v_x and v_y are components of relative velocity between the observing instrument and the radiation source on the x and y axes, respectively, that is,

$$v_x = \dot{x}_T - \dot{x}_o, \quad v_y = \dot{y}_T - \dot{y}_o \tag{12.37}$$

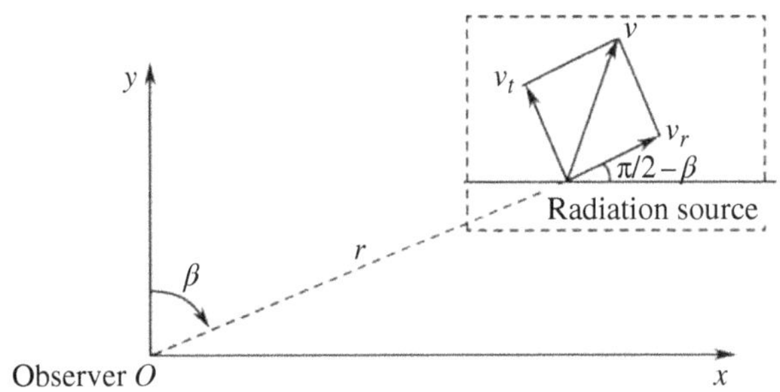

Figure 12.2 Location principle diagram

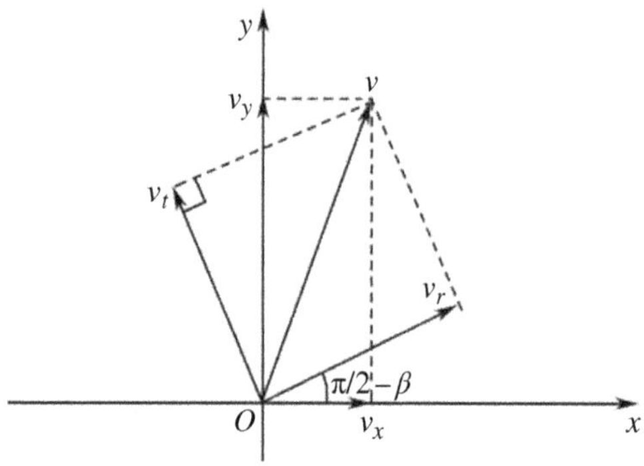

Figure 12.3 Amplified diagram of location principle

Figure 12.3 is an amplified diagram of this in the dotted-line square of Figure 12.2. From Figure 12.3, we obtain

$$\left(v_y - v_x \cot\beta\right)\sin\beta = v_t(t) \tag{12.38}$$

Equation (12.38) can be simplified as

$$v_y \sin\beta - v_x \cos\beta = v_t(t) \tag{12.39}$$

Substituting (12.39) into (12.36) gives

$$\dot{\beta}(t) = -\frac{v_t(t)}{r(t)} \tag{12.40}$$

Then $r(t)$, the range from the observing instrument to the radiation source, is

$$r(t) = -\frac{v_t(t)}{\dot{\beta}(t)} \tag{12.41}$$

where $\dot{\beta}(t)$ is the angle changing rate caused by relative motion, and t is the time. However, since the motion velocity of the radiation source (the target) is typically unknown, so is the motion velocity relative to the observing instrument, and $v_t(t)$ cannot be obtained. In this case, the range cannot be measured only using (12.41). In addition, according to the kinematic principle, there is another equation

$$\ddot{r}(t) = \frac{v_t^2(t)}{r(t)} \tag{12.42}$$

where the centrifugal acceleration $\ddot{r}$ is the second derivative of the scalar of the range r.

Combining (12.41) and (12.42) yields the relative equation

$$r(t) = \frac{\ddot{r}(t)}{\dot{\beta}^2(t)} \tag{12.43}$$

If the centrifugal acceleration $\ddot{r}(t)$ and the angular velocity $\dot{\beta}(t)$ at a certain moment can be obtained by measuring at this moment, instantaneous range measuring can be achieved. The observing instrument can generally receive the signals radiated by radiation sources. Hence, the centrifugal acceleration information can be obtained from the frequency domain of the signals. Its principle is as follows: according to Doppler effects, the relation between radial velocity v_r and Doppler frequency f_d can be described as

$$v_r(t)=\dot{r}(t)=-\lambda f_d(t) \tag{12.44}$$

The relation between $\ddot{r}(t)$ and Doppler frequency changing rate $\dot{f}_d(t)$ derived from the above equation is

$$\ddot{r}(t)=-\lambda\dot{f}_d(t) \tag{12.45}$$

Substituting (12.45) into (12.43) yields

$$r(t)=-\lambda\frac{\dot{f}_d(t)}{\dot{\beta}^2(t)} \tag{12.46}$$

which is the single-station range-measuring formula based on the Doppler frequency changing rate. Then, combining it with the target azimuth angle measured by the sensor gives the target coordinates $[x(t), y(t)]$ in the rectangular coordinate system as

$$\begin{cases} x(t)=r(t)\sin\beta(t) \\ y(t)=r(t)\cos\beta(t) \end{cases} \tag{12.47}$$

12.3.3 Doppler Changing Rate and Azimuth, Elevation Joint Location

When locating the spatial target using the azimuth angle, elevation angle, and Doppler frequency changing rate measured by sensors, suppose that the azimuth angle measured by the sensor at time t satisfies (12.5), the azimuth angle changing rate is shown in (12.36), and the elevation angle is

$$\varepsilon(t)=\mathrm{arctg}\left(\frac{z_T-z_o}{\sqrt{(x_T-x_o)^2+(y_T-y_o)^2}}\right) \tag{12.48}$$

where (x_T, y_T, z_T) and (x_o, y_o, z_o) are locations of the radiation source and sensor platform at time t.

The following can be derived from (12.48):

$$\begin{aligned}\dot{\varepsilon}(t)=&\frac{-(x_T-x_o)(z_T-z_o)(\dot{x}_T-\dot{x}_o)-(z_T-z_o)(y_T-y_o)(\dot{y}_T-\dot{y}_o)}{\left[(x_T-x_o)^2+(y_T-y_o)^2+(z_T-z_o)^2\right]\sqrt{(x_T-x_o)^2+(y_T-y_o)^2}}\\&+\frac{\left[(x_T-x_o)^2+(y_T-y_o)^2\right](\dot{z}_T-\dot{z}_o)}{\left[(x_T-x_o)^2+(y_T-y_o)^2+(z_T-z_o)^2\right]\sqrt{(x_T-x_o)^2+(y_T-y_o)^2}}\end{aligned} \tag{12.49}$$

Using the Doppler frequency changing rate, we have

$$r(t) = -\lambda \frac{\dot{f}_d(t)}{\left[\left(\dot{\beta}(t)\cos\varepsilon\right)^2 + \dot{\varepsilon}^2(t)\right]} \tag{12.50}$$

According to (12.50) and combining the azimuth angle and elevation angle measured by sensors, the target coordinates $(x(t), y(t), z(t))$ in the rectangular coordinate system can be derived as

$$\begin{cases} x(t) = r(t)\sin\beta(t)\cos\varepsilon(t) \\ y(t) = r(t)\cos\beta(t)\cos\varepsilon(t) \\ z(t) = r(t)\sin\varepsilon(t) \end{cases} \tag{12.51}$$

12.3.4 Multiple-Model Method

When using single passive sensors to measure the direction of the target, suppose that the target azimuth angle measured at time k is $\beta_m(k)$, and suppose that the possible furthest and nearest ranges of the target are $L_{\max}$ and $L_{\min}$, respectively. Divide this space into N sub-intervals [308, 309] with different range intervals. $2\sigma_{L(i)}$ and $L(i)$ denote the length and average radial range of the ith sub-interval, and they satisfy

$$L(i) = L(i-1) + \sigma_{L(i-1)} + \sigma_{L(i)} \tag{12.52}$$

$$\frac{2\sigma_{L(i)}}{L(i)} = \frac{2(\rho-1)}{\rho+1}, \quad i = 1,2,\ldots,N \tag{12.53}$$

where $L(0) = L_{\min}$, $\sigma_{L(0)} = 0$, $L_{\max} = L(N) + \sigma_{L(N)}$, $\rho = \left(\frac{L_{\max}}{L_{\min}}\right)^{\frac{1}{N}}$.

By simple mathematical calculations, $2\sigma_{L(i)}$ and $L(i)$ can be derived from (12.52) and (12.53):

$$2\sigma_{L(i)} = \rho^{i-1}(\rho-1)L_{\min} \tag{12.54}$$

$$L(i) = \frac{\rho^{i-1}(\rho+1)L_{\min}}{2} \tag{12.55}$$

Establish an EKF model on each sub-interval, then the initial state vector and covariance matrix of the ith sub-interval are, respectively,

$$\hat{X}^m(i,1|1) = \begin{bmatrix} L(i)\sin(\beta(1)) \\ L(i)\cos(\beta(1)) \\ v_s\sin(\beta(1)) \\ v_s\cos(\beta(1)) \end{bmatrix} \tag{12.56}$$

$$\boldsymbol{P}(i,1|1)=\boldsymbol{A}_{RP}=\begin{bmatrix} L^2(i)\sigma_\alpha^2 & 0 & 0 & 0 \\ 0 & \sigma_{L(i)}^2 & 0 & 0 \\ 0 & 0 & \hat{\sigma}_v^2 & 0 \\ 0 & 0 & 0 & \hat{\sigma}_v^2 \end{bmatrix} A'_{RP} \tag{12.57}$$

where v_s denotes the initial motion velocity of the baseline observing station, $\hat{\sigma}_v$ denotes the estimates of standard deviation of velocity measurement errors, and

$$A_{RP}=\begin{bmatrix} \cos[\beta_m(1)] & \sin[\beta_m(1)] & 0 & 0 \\ -\sin[\beta_m(1)] & \cos[\beta_m(1)] & 0 & 0 \\ 0 & 0 & 1 & 0 \\ 0 & 0 & 0 & 1 \end{bmatrix} \tag{12.58}$$

The initial probability of the target in the ith model is

$$\Pr(i,1)=\frac{2\sigma_{L(i)}}{L_{\max}-L_{\min}} \tag{12.59}$$

The state equation of the target is

$$\boldsymbol{X}(k+1)=\boldsymbol{F}(k)\boldsymbol{X}(k)+\boldsymbol{\Gamma}(k)\boldsymbol{v}(k) \tag{12.60}$$

where $\boldsymbol{X}(k)$ denotes the state vector of the target at time k, $\boldsymbol{F}(k)$ the state transition matrix, $\boldsymbol{v}(k)$ the zero-mean, white, Gaussian process noise with covariance matrix $\boldsymbol{Q}(k)$, and $\boldsymbol{\Gamma}(k)$ the process noise distribution matrix.

$$\boldsymbol{X}(k)=[x(k)y(k)\dot{x}(k)\dot{y}(k)]' \tag{12.61}$$

$$\boldsymbol{F}(k)=\begin{bmatrix} 1 & 0 & T & 0 \\ 0 & 1 & 0 & T \\ 0 & 0 & 1 & 0 \\ 0 & 0 & 0 & 1 \end{bmatrix} \tag{12.62}$$

$$\boldsymbol{\Gamma}(k)=\begin{bmatrix} \frac{T^2}{2} & 0 \\ 0 & \frac{T^2}{2} \\ T & 0 \\ 0 & T \end{bmatrix} \tag{12.63}$$

where T denotes the sampling interval.

The measurement equation of the target is

$$Z(k) = \boldsymbol{h}[\boldsymbol{X}(k)] + W(k) = \arctan\left(\frac{y-y_l}{x-x_l}\right) + W(k) \tag{12.64}$$

where (x_l, y_l) denotes the location of the observing station, (x, y) the location of the target, and $W(k)$ the measurement noise of the base observing station, which is independent of the process noise $v(k)$ and is the zero-mean, white, Gaussian noise with covariance matrix $\boldsymbol{R}(k)$.

The one-step prediction of the state equation in the ith model is

$$\hat{\boldsymbol{X}}(i,k+1|k) = \boldsymbol{F}(k)\hat{\boldsymbol{X}}(i,k|k) - \boldsymbol{U}(k+1) \tag{12.65}$$

where $\boldsymbol{U}(k+1)$ denotes changes in location of the observing station in the time interval from k to $k+1$. If the observing station is assumed to be in uniform rectilinear motion, then

$$\boldsymbol{U}(k+1) = \begin{bmatrix} v_{xs}T & v_{ys}T & 0 & 0 \end{bmatrix}' \tag{12.66}$$

where v_{xs}, v_{ys} denote the motion velocities of the observing station in the x and y axis directions, respectively, and T the sampling interval in the matrix.

The state prediction covariance matrix of the ith model is

$$\boldsymbol{P}(i,k+1|k) = \boldsymbol{F}(k)\boldsymbol{P}(i,k|k)\boldsymbol{F}(k)' + \boldsymbol{\Gamma}(k)\boldsymbol{Q}(k)\boldsymbol{\Gamma}(k)' \tag{12.67}$$

Then, the filter gain of the ith model can be obtained:

$$\boldsymbol{K}(i,k+1) = \boldsymbol{P}(i,k+1|k)(\boldsymbol{H}(i,k+1))'(\boldsymbol{S}(i,k+1))^{-1} \tag{12.68}$$

where

$$\boldsymbol{S}(i,k+1) = \boldsymbol{H}(i,k+1)\boldsymbol{P}(i,k+1|k)[\boldsymbol{H}(i,k+1)]' + \boldsymbol{R}(k+1) \tag{12.69}$$

is the innovation covariance, and

$$\boldsymbol{H}(i,k+1) = \frac{\partial \boldsymbol{h}}{\partial \hat{\boldsymbol{X}}^m(i,k+1|k)} = \left[\frac{-\hat{y}(i,k+1|k)}{(\hat{x}(i,k+1|k))^2 + (\hat{y}(i,k+1|k))^2}, \frac{\hat{x}(i,k+1|k)}{(\hat{x}(i,k+1|k))^2 + (\hat{y}(i,k+1|k))^2}, 0, 0\right] \tag{12.70}$$

is the measurement matrix.

From (12.65) and (12.68), the state update equation of the ith model can be derived as

$$\hat{\boldsymbol{X}}(i,k+1|k+1) = \hat{\boldsymbol{X}}(i,k+1|k) + \boldsymbol{K}(i,k+1)\left\{\boldsymbol{Z}(k+1) - h\left[\hat{\boldsymbol{X}}(i,k+1|k)\right]\right\} \tag{12.71}$$

The covariance update equation of the ith model can be derived from (12.67), (12.68), and (12.70) as

$$\begin{aligned}\boldsymbol{P}(i,k+1|k+1) = {} & [\boldsymbol{I} - \boldsymbol{K}(i,k+1)\boldsymbol{H}(i,k+1)]\boldsymbol{P}(i,k+1|k)[\boldsymbol{I} + \boldsymbol{K}(i,k+1)\boldsymbol{H}(i,k+1)]' \\ & - \boldsymbol{K}(i,k+1)\boldsymbol{R}(k+1)\boldsymbol{K}'(i,k+1)\end{aligned} \tag{12.72}$$

where $\boldsymbol{I}$ is the unit matrix.

The update probability of the target in the ith model can be derived according to Bayes' rule as

$$\Pr(i,k)=\frac{\Pr[\beta(k)|i]\Pr(i,k-1)}{\sum_{n=1}^{N}\Pr[\beta(k)|n]\Pr(n,k-1)} \tag{12.73}$$

where

$$\Pr[\beta(k)|i]=|2\pi S(i,k)|^{-\frac{1}{2}}\exp\left\{\frac{-[v(i,k)]'[S(i,k)]^{-1}v(i,k)}{2}\right\} \tag{12.74}$$

is the likelihood function of the ith model at time k, which correlates with the innovation $v(i, k)$ and the innovation covariance $S(i, k)$ of the ith model at time k.

When filtering, compare the update probability with a certain assumed detection threshold constantly, and only when the updating probability is higher than the detection threshold can the sub-interval be kept [310], and the state estimates and the covariance matrixes of each sub-interval at different moments can be derived with the constant iteration of filter equations set, including (12.65), (12.67), (12.68), (12.71), and (12.72). Take the updating probability derived from (12.73) as weight, and conduct weighted fusion on the filtering results of each sub-interval. The result after fusion is output as the state estimate and covariance corresponding to the target, that is,

$$\hat{X}(k|k)=\sum_{i=1}^{N}\Pr(i,k)\hat{X}(i,k|k) \tag{12.75}$$

$$P(k|k)=\sum_{i=1}^{N}\Pr(i,k)\{P(i,k|k)+\left[\hat{X}(i,k|k)-\hat{X}(k|k)\right]\left[\hat{X}(i,k|k)-\hat{X}(k|k)\right]'\} \tag{12.76}$$

12.4 Optimal Deployment of Direction-Finding Location

Research on direction-finding location focuses on the direction-finding location algorithm, optimal deployment of the sensor, and passive data correlation in multi-target environments [311–314]. This subsection deals chiefly with the optimal deployment of passive sensors under the criterion of minimum area of position concentration ellipse.

12.4.1 Area of the Position Concentration Ellipse

Suppose that the true location of the target at a certain moment is $X=(x,y)'$, passive sensor 1 is located at the origin of the coordinate system, and the coordinates of passive sensor 2 are (x_2, y_2), as shown in Figure 12.4. The azimuth angles measured by the two sensors are $\theta_i(i=1,2)$, $\theta_i\in[0,\pi]$, and $\theta_2>\theta_1$, that is, the target is located on the baseline or above the baseline of the two sensors. Also, suppose that the measurement noise at a certain time is independent and additive Gaussian white noise with zero mean, and variance $\sigma_{\theta_i}^2$ $(i=1,2)$, respectively. Suppose further that the angle-measuring errors of the two passive sensors are not the same; generally, suppose that the angle measurement error of passive sensor 1 is less than or equal to that of passive sensor 2,

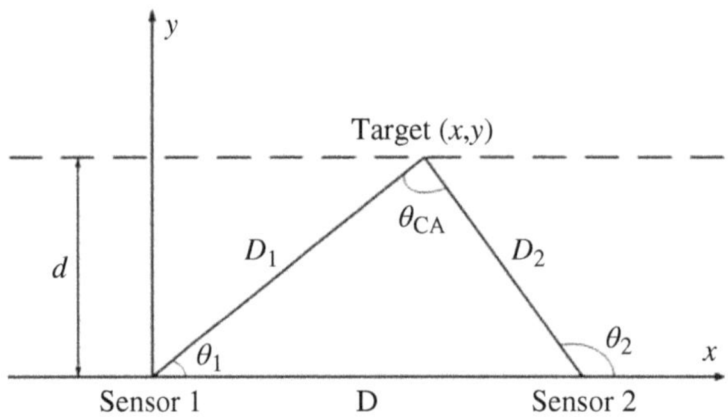

Figure 12.4 Diagram of direction-measuring cross locating under restricted conditions

that is, $\sigma_{\theta_1}^2 = k\sigma_{\theta_2}^2 = k\sigma^2$, $k \le 1$. What is worth mentioning is that the figure represents the geometric distribution between the target and the two sensors at a certain time, and subsequent conclusions can also be applied to the situation where the positions of the target and sensors are time-varying and to the situation where the target is located below the baseline.

As shown in Figure 12.4, the cut angle can be defined as $\theta_2 - \theta_1$, which can be denoted as θ_{CA}. The distances from the two sensors to the target are D_i $(i = 1,2)$, respectively. The length of the baseline of the two sensors is D. The vertical distance from the target to the baseline of the two sensors is d, and $d > 0$. Suppose $l = d/D$, that is, l denotes the ratio of the vertical distance to the length of the baseline. Then, according to the sine theorem, we have

$$D_1 = \frac{D\sin\theta_2}{\sin(\theta_2 - \theta_1)} \tag{12.77}$$

$$D_2 = \frac{D\sin\theta_1}{\sin(\theta_2 - \theta_1)} \tag{12.78}$$

The estimated location of the target is

$$\hat{x} = D_1 \cos\theta_1 \tag{12.79}$$

$$\hat{y} = D_1 \sin\theta_1 \tag{12.80}$$

Differentiating both sides of (12.79) and (12.80) yields

$$\mathrm{d}\hat{x} = \frac{D\sin\theta_2\cos\theta_2}{\sin^2(\theta_2 - \theta_1)}\mathrm{d}\theta_1 - \frac{D\sin\theta_1\cos\theta_1}{\sin^2(\theta_2 - \theta_1)}\mathrm{d}\theta_2 \tag{12.81}$$

$$\mathrm{d}\hat{y} = \frac{D\sin^2\theta_2}{\sin^2(\theta_2 - \theta_1)}\mathrm{d}\theta_1 - \frac{D\sin^2\theta_1}{\sin^2(\theta_2 - \theta_1)}\mathrm{d}\theta_2 \tag{12.82}$$

Then, the location error covariance matrix can be obtained as

$$\boldsymbol{P} = \begin{bmatrix} \sigma_x^2 & \sigma_{xy} \\ \sigma_{yx} & \sigma_y^2 \end{bmatrix} = \mathrm{E}\left\{ \begin{bmatrix} \mathrm{d}\hat{x} \\ \mathrm{d}\hat{y} \end{bmatrix} [\mathrm{d}\hat{x} \;\; \mathrm{d}\hat{y}] \right\} \tag{12.83}$$

where

$$\begin{cases} \sigma_x^2 = \dfrac{D^2}{\sin^4(\theta_2-\theta_1)}\left(\sin^2\theta_2\cos^2\theta_2\sigma_{\theta_1}^2 + \sin^2\theta_1\cos^2\theta_1\sigma_{\theta_2}^2\right) \\ \sigma_y^2 = \dfrac{D^2}{\sin^4(\theta_2-\theta_1)}\left(\sin^4\theta_2\sigma_{\theta_1}^2 + \sin^4\theta_1\sigma_{\theta_2}^2\right) \\ \sigma_{xy} = \sigma_{yx} = \dfrac{D^2}{\sin^4(\theta_2-\theta_1)}\left(\sin^3\theta_2\cos\theta_2\sigma_{\theta_1}^2 + \sin^3\theta_1\cos\theta_1\sigma_{\theta_2}^2\right) \end{cases} \tag{12.84}$$

As to the given vertical range d, it follows from Figure 12.4 that the equation

$$\frac{D\sin\theta_1\sin\theta_2}{\sin(\theta_2-\theta_1)} - d = 0 \tag{12.85}$$

is valid.

Under the previous assumptions, the PDF of the estimated location of the target $\hat{\boldsymbol{X}} = (\hat{x},\hat{y})'$ can be described approximately as [311]

$$p(\hat{x},\hat{y}) = (2\pi)^{-1}(|\boldsymbol{P}|)^{-1/2}\exp\left\{-\frac{1}{2}(\hat{\boldsymbol{X}}-\boldsymbol{X})'\boldsymbol{P}^{-1}(\hat{\boldsymbol{X}}-\boldsymbol{X})\right\} \tag{12.86}$$

where $\boldsymbol{P}$ denotes the location error covariance and $\boldsymbol{X} = (x,y)'$ the true location of the target. Then, the concentration area of the target location is an ellipse, called a "position concentration ellipse." The long and short semi-axes of the ellipse can be described as

$$a = \sqrt{\frac{m}{2}\left[\sigma_x^2 + \sigma_y^2 + \sqrt{\left(\sigma_x^2 - \sigma_y^2\right)^2 + 4\sigma_{xy}^2}\right]} \tag{12.87}$$

$$b = \sqrt{\frac{m}{2}\left[\sigma_x^2 + \sigma_y^2 - \sqrt{\left(\sigma_x^2 - \sigma_y^2\right)^2 + 4\sigma_{xy}^2}\right]} \tag{12.88}$$

where $m = -2\ \ln(1-P_e)$, and P_e is the probability that the estimated location $\hat{\boldsymbol{X}} = (\hat{x},\hat{y})'$ falls into the concentration ellipse.

The area of the position concentration ellipse is

$$S = \pi ab = \pi m\sqrt{\sigma_x^2\sigma_y^2 - \sigma_{xy}^2} \tag{12.89}$$

Substitute (12.84) into (12.9), to obtain

$$\begin{aligned} S &= \pi m\sqrt{\frac{D^4\sin^2\theta_1\sin^2\theta_2}{\sin^6(\theta_2-\theta_1)}\sigma_{\theta_1}^2\sigma_{\theta_2}^2} = \pi m D^2\sqrt{k}\sigma_{\theta_2}^2\sqrt{\frac{\sin^2\theta_1\sin^2\theta_2}{\sin^6(\theta_2-\theta_1)}} \\ &\hat{=}\pi m D^2\sqrt{k}\sigma_{\theta_2}^2\sqrt{g(\theta_1,\theta_2)} \end{aligned} \tag{12.90}$$

where D, σ_{θ_2}, and k are known constants.

12.4.2 Derivation of the Conditional Extremum Based on the Lagrange Multiplier Method

Minimizing the area of the concentration ellipse S is equivalent to minimizing the value of the dual function $g(\theta_1, \theta_2)$, that is,

$$\min g(\theta_1,\theta_2)=\frac{\sin^2\theta_1\sin^2\theta_2}{\sin^6(\theta_2-\theta_1)} \tag{12.91}$$

Take (12.85) as constraints, and obtain the conditional extremum by conducting joint optimization with (12.91). Here, the Lagrange multiplier method is applied [311, 312].

Let

$$G(\theta_1,\theta_2)=g(\theta_1,\theta_2)+\lambda\left(\frac{D\sin\theta_1\sin\theta_2}{\sin(\theta_2-\theta_1)}-d\right) \tag{12.92}$$

where λ denotes the Lagrange multiplier. Take partial derivatives of θ_1 and θ_2 on both sides of (12.92) and let the results be zero, to obtain

$$\begin{aligned}\frac{\partial G}{\partial\theta_1}&=\frac{2\sin\theta_1\sin^2\theta_2\cos\theta_1\sin(\theta_2-\theta_1)+6\sin^2\theta_1\sin^2\theta_2\cos(\theta_2-\theta_1)}{\sin^7(\theta_2-\theta_1)}\\&\quad+\lambda\left[\frac{D\sin\theta_1\sin\theta_2\cos(\theta_2-\theta_1)}{\sin^2(\theta_2-\theta_1)}+\frac{D\cos\theta_1\sin\theta_2}{\sin(\theta_2-\theta_1)}\right]=0\end{aligned} \tag{12.93a}$$

$$\begin{aligned}\frac{\partial G}{\partial\theta_2}&=\frac{2\sin^2\theta_1\sin\theta_2\cos\theta_2\sin(\theta_2-\theta_1)-6\sin^2\theta_1\sin^2\theta_2\cos(\theta_2-\theta_1)}{\sin^7(\theta_2-\theta_1)}\\&\quad+\lambda\left[-\frac{D\sin\theta_1\sin\theta_2\cos(\theta_2-\theta_1)}{\sin^2(\theta_2-\theta_1)}+\frac{D\sin\theta_1\cos\theta_2}{\sin(\theta_2-\theta_1)}\right]=0\end{aligned} \tag{12.93b}$$

It follows from (12.93) that

$$\lambda\sin(\theta_1+\theta_2)=-\frac{2\sin\theta_1\sin\theta_2}{D\sin^5(\theta_2-\theta_1)}\sin(\theta_1+\theta_2) \tag{12.94}$$

In order to obtain λ in (12.94), two situations ($\sin(\theta_1+\theta_2)\neq 0$ and $\sin(\theta_1+\theta_2)=0$) need to be considered, as follows.

1. When $\sin(\theta_1+\theta_2)\neq 0$,

$$\lambda=-\frac{2\sin\theta_1\sin\theta_2}{D\sin^5(\theta_2-\theta_1)} \tag{12.95}$$

Substituting (12.95) into (12.93a) yields

$$\sin^2\theta_1\sin^2\theta_2\cos(\theta_2-\theta_1)=0 \tag{12.96}$$

It follows from (12.96) and the previous assumptions $\theta_1\neq 0$ and $\theta_2\neq 0$ that

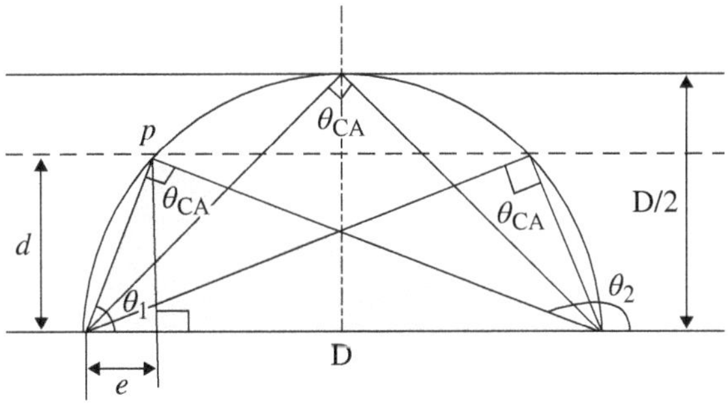

Figure 12.5 Optimal cut angle θ_{CA} when $l \le 0.5$

$$\theta_{CA} = \theta_2 - \theta_1 = \frac{\pi}{2} \tag{12.97}$$

Equation (12.97) indicates that at the moment the target is located on the circumference (except for the locations of the two passive sensors), the diameter is the baseline of the two sensors, as shown in Figure 12.5. In such cases $d \le D/2$, and (12.97) is valid when $l = d/D \le 0.5$.

From Figure 12.5 we have

$$\theta_1 = \arctan\left(\frac{d}{e}\right) \tag{12.98}$$

According to the shadowed right triangle in Figure 12.5, the following equation is valid:

$$e(D-e) = d^2 \tag{12.99}$$

Solving the above equation yields

$$e = \frac{D \pm \sqrt{D^2 - 4d^2}}{2} \tag{12.100}$$

As to (12.100), two situations will be discussed below:

- when $\theta_1 \in \left(0, \frac{\pi}{4}\right]$, $e \ge D/2$, such that

$$e = \frac{D + \sqrt{D^2 - 4d^2}}{2}, \ \theta_1 \in (0, \frac{\pi}{4}] \tag{12.101}$$

- when $\theta_1 \in \left[\frac{\pi}{4}, \frac{\pi}{2}\right)$, $e \le D/2 e \le \mathrm{D}/2$, such that

$$e = \frac{D - \sqrt{D^2 - 4d^2}}{2}, \ \theta_1 \in [\frac{\pi}{4}, \frac{\pi}{2}) \tag{12.102}$$

For convenience, the solutions to (12.97) and (12.98) can be described as (θ_1^0, θ_2^0). In order to analyze whether the area of the position concentration ellipse can reach the local minimum on the stable point $\left(\theta_1^0, \theta_2^0\right)$, the second-order partial derivatives of $g(\theta_1, \theta_2)$ can be obtained conditional on (12.85).

Using (12.85) gives

$$\theta_1 = \arccos\left(\frac{1 + lc}{\sqrt{1 + 2lc + l^2c^2 + l^2}}\right) \tag{12.103}$$

where

$$c = \frac{\cos\theta_2}{\sin\theta_2} \tag{12.104}$$

then

$$\cos\theta_1 = \frac{\sin\theta_2 + l\cos\theta_2}{\sqrt{(\sin\theta_2 + l\cos\theta_2)^2 + l^2\sin^2\theta_2}} \tag{12.105}$$

and

$$\sin\theta_1 = \frac{l\sin\theta_2}{\sqrt{(\sin\theta_2 + l\cos\theta_2)^2 + l^2\sin^2\theta_2}} \tag{12.106}$$

Combining (12.105) and (12.106) into $g(\theta_1, \theta_2)$ of (12.91) yields

$$g(\theta_1, \theta_2) \hat{=} \psi(\theta_2) = \frac{l^2(\cos^2\theta_2 - l\sin2\theta_2 - l^2 - 1)^2}{\sin^8\theta_2} \tag{12.107}$$

By using (12.107), the second-order derivatives of $\psi(\theta_2)$ with respect to θ_2 can be obtained as

$$\begin{aligned}\psi''(\theta_2) = {} & \frac{2l^2(-\sin2\theta_2 - 2l\cos2\theta_2)^2}{\sin^8\theta_2} - \frac{8l^2\cos2\theta_2(\cos^2\theta_2 - l\sin2\theta_2 - l^2 - 1)^2}{\sin^{10}\theta_2} \\ & - \frac{16l^2\sin2\theta_2(\cos^2\theta_2 - l\sin2\theta_2 - l^2 - 1)(-\sin2\theta_2 - 2l\cos2\theta_2)}{\sin^{10}\theta_2} \\ & + \frac{2l^2(\cos^2\theta_2 - l\sin2\theta_2 - l^2 - 1)(-2\cos2\theta_2 + 4l\sin2\theta_2)}{\sin^8\theta_2} \\ & + \frac{20l^2\sin^2 2\theta_2(\cos^2\theta_2 - l\sin2\theta_2 - l^2 - 1)^2}{\sin^{12}\theta_2}\end{aligned} \tag{12.108}$$

Based on the symbol $\psi''\left(\theta_2^0\right)$, we can determine whether the stable point is the local maximum, minimum, or saddle point [315]. From the representations of e in (12.101) and (12.102), two situations ($\theta_1 \in \left(0, \frac{\pi}{4}\right]$ and $\theta_1 \in \left[\frac{\pi}{4}, \frac{\pi}{2}\right)$) need to be considered, as follows.

- When $\theta_1 \in \left(0, \frac{\pi}{4}\right]$, substituting (12.101) into (12.98), after some cancellations, yields

$$\cos\theta_2 = -\sin\theta_1 = -\sqrt{\frac{1}{2}+\sqrt{\frac{1}{4}-l^2}}, \theta_1 \in \left(0, \frac{\pi}{4}\right] \tag{12.109}$$

$$\cos\theta_1 = \sin\theta_2 = \sqrt{\frac{1}{2}-\sqrt{\frac{1}{4}-l^2}}, \theta_1 \in \left(0, \frac{\pi}{4}\right] \tag{12.110}$$

Combining (12.109) and (12.110) into (12.108) yields

$$\psi''(\theta_2) = \frac{16l^2(1-4l^2)}{\left(\sqrt{1-4l^2}-1\right)^2}, \theta_1 \in \left(0, \frac{\pi}{4}\right] \tag{12.111}$$

- When $\theta_1 \in \left[\frac{\pi}{4}, \frac{\pi}{2}\right)$, inserting (12.102) into (12.98), after some cancellations, gives

$$\cos\theta_2 = -\sin\theta_1 = -\sqrt{\frac{1}{2}-\sqrt{\frac{1}{4}-l^2}}, \theta_1 \in \left[\frac{\pi}{4}, \frac{\pi}{2}\right) \tag{12.112}$$

$$\cos\theta_1 = \sin\theta_2 = \sqrt{\frac{1}{2}+\sqrt{\frac{1}{4}-l^2}}, \theta_1 \in \left[\frac{\pi}{4}, \frac{\pi}{2}\right) \tag{12.113}$$

Combining (12.112) and (12.113) into (12.108) gives

$$\psi''(\theta_2) = \frac{16l^2(1-4l^2)}{\left(\sqrt{1-4l^2}+1\right)^2}, \theta_1 \in \left[\frac{\pi}{4}, \frac{\pi}{2}\right) \tag{12.114}$$

As shown in (12.111) and (12.114), when $\sin(\theta_1+\theta_2) \neq 0$ and $l<0.5$, $\psi''(\theta_2)>0$, the area of the position concentration ellipse reaches the local minimum; when $l=0.5$, $\psi''(\theta_2)=0$, and it cannot be determined whether the area of the position concentration ellipse reaches the extremum.

As shown in Figure 12.5, when $l=0.5$, the stable point satisfies $(\theta_1^0, \theta_2^0) = (45°, 135°)$. Then

$$\psi'(135°) = \psi''(135°) = \psi'''(135°) = 0$$

and

$$\psi^{(4)}(135°) = 48 > 0$$

Thus, according to the extremum determination criterion in Ref. [315], the area of the position concentration ellipse reaches the local minimum.

2. When $\sin(\theta_1+\theta_2)=0$,

$$\theta_1 + \theta_2 = \pi \tag{12.115}$$

Here, the target and the two passive sensors form an isosceles triangle. As shown in Figure 12.4,

$$\theta_1 = \arctan\left(\frac{2d}{D}\right) \tag{12.116}$$

then

$$\sin\theta_1 = \sin\theta_2 = \frac{2l}{\sqrt{1+4l^2}} \tag{12.117}$$

$$\cos\theta_1 = -\cos\theta_2 = \frac{1}{\sqrt{1+4l^2}} \tag{12.118}$$

According to (12.94), the Lagrange multiplier λ is arbitrary. Now, it cannot be determined yet whether the solution of (12.115) is the stable point of function $\psi(\theta_2)$. Equation (12.107) yields

$$\begin{aligned}\psi'(\theta_2) = & \frac{2l^2(\cos^2\theta_2 - l\sin 2\theta_2 - l^2 - 1)(-\sin 2\theta_2 - 2l\cos 2\theta_2)}{\sin^8\theta_2} \\ & - \frac{4l^2\sin 2\theta_2(\cos^2\theta_2 - l\sin 2\theta_2 - l^2 - 1)^2}{\sin^{10}\theta_2}\end{aligned} \tag{12.119}$$

Substituting (12.117) and (12.118) into (12.119) gives

$$\psi'(\theta_2) = 0 \tag{12.120}$$

Then, it can be determined that the solution of (12.115) is the stable point for $\psi(\theta_2)$, and substituting (12.117) and (12.118) into (12.108) yields

$$\psi''(\theta_2) = -\frac{(1-4l^2)(1+4l^2)^4}{128l^4} \tag{12.121}$$

It follows from (12.121) that

$$\begin{cases} \psi''(\theta_2) > 0, & l > 0.5 \\ \psi''(\theta_2) = 0, & l = 0.5 \\ \psi''(\theta_2) < 0, & 0 < l < 0.5 \end{cases} \tag{12.122}$$

Therefore, when $l > 0.5$, the area of the position concentration ellipse reaches its local minimum; when $0 < l < 0.5$, it reaches its local maximum; when $l = 0.5$, it is uncertain whether it reaches its extremum. A procedure similar to that above can be used to prove that the area of the position concentration ellipse has reached the local minimum under this condition.

Under the criteria for position concentration ellipses, the cut angle has four characteristics.

1. When $l > 0.5$, the optimal cut angle can be obtained if the target and the two passive sensors form an isosceles triangle. The optimum mentioned here is in a relative sense, that is, the cut angle corresponding to the local minimum under constraints. The same is true below.

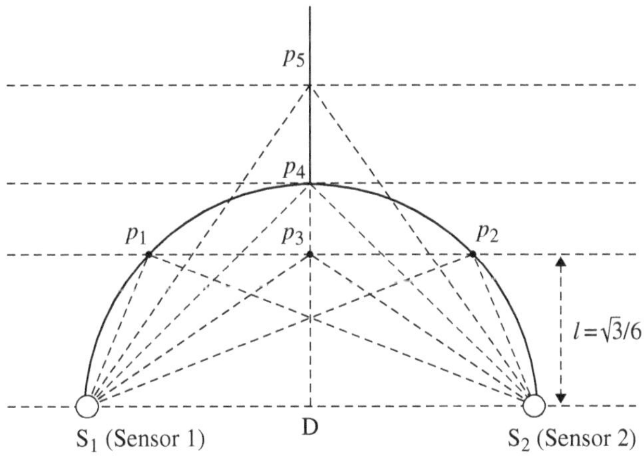

Figure 12.6 The target location distribution (on solid line and arc) when the area of the position concentration ellipse reaches the local minimum

2. When $l < 0.5$, the optimal cut angle can be obtained if the target is located on the circle (except for the locations of the two sensors), the diameter of which is the baseline of the two sensors.
3. When $l = 0.5$, the optimal cut angle can be obtained if the target and the two passive sensors form an isosceles right triangle.
4. The optimal cut angle is unrelated to the variance multiple k.

According to the above conclusions, under the criterion for the minimum position concentration ellipse, the optimum cut angle can be obtained if and only if the target is located on the solid line (or arc) in Figure 12.6.

12.4.3 Optimal Deployment by the Criterion that the Position Concentration Ellipse Area is Minimum

Discussions on how the change in l, which is the ratio of the vertical range to the length of the baseline, influences the local minimum of the position concentration ellipse are presented below.

- When $l \geq 0.5$, $\theta_1 \in \left[\frac{\pi}{4}, \frac{\pi}{2}\right)$ and $\theta_1 + \theta_2 = \pi$, as shown in Figure 12.6. Substituting (12.115) into $g(\theta_1, \theta_2)$ in (12.91), we obtain

$$g_1(\theta_1) = g(\theta_1, \theta_2)|_{\theta_2 = \pi - \theta_1} = g_1(\theta_1) = g(\theta_1, \theta_2)|_{\theta_2 = \pi - \theta_1} = \frac{1}{64 \sin^2\theta_1 \cos^6\theta_1} \tag{12.123}$$

The following can be derived from (12.123):

$$g_1'(\theta_1) = \frac{4\sin^2\theta_1 - 1}{32 \sin^3\theta_1 \cos^5\theta_1} > 0 \tag{12.124}$$

Then, in the interval $\left[\frac{\pi}{4}, \frac{\pi}{2}\right)$, $g_1(\theta_1)$ increases monotonically. As shown in Figure 12.6, θ_1 increases with l, that is, $g_1(\theta_1)$ increases with l.

- When $0<l<0.5$, $\theta_2=\theta_1+\pi/2$ and $\theta_1\in\left(0,\frac{\pi}{4}\right)$ or $\theta_1\in\left(\frac{\pi}{4},\frac{\pi}{2}\right)$, as shown in Figure 12.6. Substituting (12.97) into $g(\theta_1,\theta_2)$ in (12.91), we obtain

$$g_2(\theta_1)=g(\theta_1,\theta_2)|_{\theta_2=\frac{\pi}{2}+\theta_1}=\frac{1}{4}\sin^2 2\theta_1 \tag{12.125}$$

The following can be derived from (12.125):

$$g_2'(\theta_1)=\sin 2\theta_1\cos 2\theta_1 \tag{12.126}$$

$$\begin{cases} g_2'(\theta_1)>0, & \theta_1\in\left(0,\frac{\pi}{4}\right) \\ g_2'(\theta_1)<0, & \theta_1\in\left(\frac{\pi}{4},\frac{\pi}{2}\right) \end{cases} \tag{12.127}$$

Then $g_2(\theta_1)$ increases monotonically in the interval $\left(0,\frac{\pi}{4}\right)$, and decreases monotonically in the interval $\left(\frac{\pi}{4},\frac{\pi}{2}\right)$. As shown in Figure 12.6, θ_1 increases with l in the interval $\left(0,\frac{\pi}{4}\right)$, and θ_1 decreases as l increases in the interval $\left(\frac{\pi}{4},\frac{\pi}{2}\right)$. Thus, when $0<l<0.5$, $g_2(\theta_1)$ increases with l.

As can be seen from the two situations above, $g_2(\theta_1)$ increases monotonically with l in the interval $(0,+\infty)$, and the global minimum of $g_2(\theta_1)$ can be achieved gradually by decreasing l to zero. In this case, the target tends infinitely to the location of sensor 1 or sensor 2 along the solid line arc in Figure 12.6, at which moment the relevant stable point satisfies $(\theta_1^0,\theta_2^0)\to\left(\frac{\pi}{2},\pi\right)$ or $(\theta_1^0,\theta_2^0)\to\left(0,\frac{\pi}{2}\right)$.

Next a special situation will be discussed: conditional on $\theta_1+\theta_2=\pi$, by letting $g_1'(\theta_1)=0$, we can obtain from (12.123) or (12.124), $\theta_1=30°$ and $\theta_2=150°$ as

$$(30°,150°)=\arg\min_{\substack{\theta_1,\theta_2\\ \theta_1+\theta_2=\pi}} g(\theta_1,\theta_2) \tag{12.128}$$

The conclusions obtained here are consistent with those in Ref. [314]. Therefore, under the minimum position concentration ellipse principle, the stable point in Ref. [314] is optimal conditional on $\theta_1+\theta_2=\pi$.

Using (12.128) in (12.116) yields

$$l=\sqrt{3}/6<0.5 \tag{12.129}$$

At this moment, p_3 in Figure 12.6 is the target location. Based on the second characteristic of the optimal cut angle obtained previously, when

$$l=\sqrt{3}/6 \tag{12.130}$$

there are two target positions, denoted p_1 and p_2 in Figure 12.6, which have reached the minimum concentration ellipse.

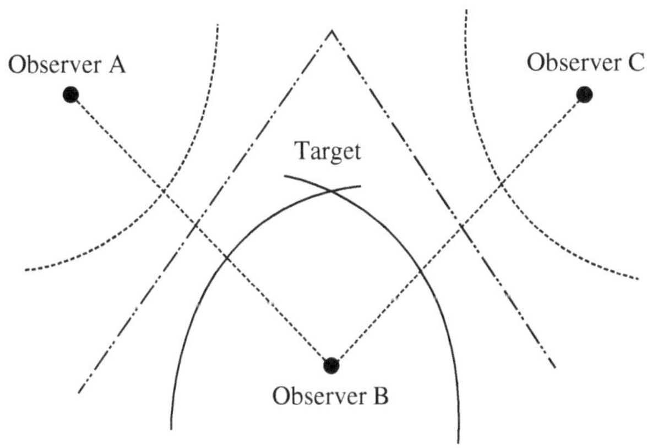

Figure 12.7 Principle of location based on TDOA measurements in planes

12.5 Passive Location Based on TDOA Measurements

12.5.1 Location Model

The time difference of arrival (TDOA) location, also called the "hyperbolic location," is the process of locating targets by processing the TDOA data of target signals received by three or more measurement stations [316]. At least three measurement stations are needed to locate 2D targets, and at least four for 3D targets. One of these stations is set as the main station, and the rest as auxiliary stations.

In a 2D, three measurement stations can form two pairs of hyperbolas, and the intersection point of two unilateral hyperbolic planes among these hyperbolas is the location of the target, as shown in Figure 12.7. In a 3D space, the time difference of the target signal arriving at the main station and auxiliary stations forms a pair of hyperbolic planes, which take the two stations as foci. Four measurement stations can form three pairs of hyperbolic planes. Among them, two of the unilateral hyperbolic planes get an intersecting line, which forms an intersection point with one of the third pair of hyperbolic planes. This intersection point is where the target is located.

Because the time difference location system can extend the baseline and increase the precision of time difference measurements, it has higher location precision than other methods, such as the direction-measuring cross location and single-station location.

12.5.2 Two-Dimensional Condition

Set the target location as (x, y), the ranges from the target location to the main station (x_0, y_0) and to the two assistant stations (x_i, y_i) as r_0 and r_i $(i=1,2)$, respectively, and their range difference as Δr_i:

$$\Delta r_i = r_i - r_0 = C \cdot \Delta t_{i0} \quad (i=1,2) \tag{12.131}$$

where Δt_{i0} denotes the time difference of the target signal arriving at the main and assistant stations on the observing platform, that is,

$$\Delta t_{10} = t_1 - t_0 \tag{12.132}$$

$$\Delta t_{20} = t_2 - t_0 \tag{12.133}$$

where t_0 and t_1, t_2 are the moments when the target signal arrives at the ground main and assistant stations, and

$$r_i^2 = (x - x_i)^2 + (y - y_i)^2 \tag{12.134}$$

$$r_0^2 = (x - x_0)^2 + (y - y_0)^2 \tag{12.135}$$

Move r_0 in (12.131) to the left side of the first equality mark, and square on both sides. Combining (12.134) and (12.135) into (12.131), after some cancellation, yields

$$(x_0 - x_i)x + (y_0 - y_i)y = k_i + r_0 \Delta r_i \tag{12.136}$$

where

$$k_i = \frac{1}{2}\left[\Delta r_i^2 + d_0^2 - d_i^2\right] \tag{12.137}$$

$$d_0^2 = x_0^2 + y_0^2 + z_0^2 \tag{12.138}$$

$$d_i^2 = x_i^2 + y_i^2 + z_i^2 \tag{12.139}$$

Write (12.136) in matrix form, as

$$\boldsymbol{AX} = \boldsymbol{B} \tag{12.140}$$

where

$$\boldsymbol{A} = \begin{bmatrix} x_0 - x_1 & y_0 - y_1 \\ x_0 - x_2 & y_0 - y_2 \end{bmatrix} \tag{12.141}$$

$$\boldsymbol{X} = \begin{bmatrix} x \\ y \end{bmatrix} \tag{12.142}$$

$$\boldsymbol{B} = \begin{bmatrix} k_1 + r_0 \Delta r_1 \\ k_2 + r_0 \Delta r_2 \end{bmatrix} \tag{12.143}$$

When the three sensors are not on the same line, matrix $\boldsymbol{A}$ is reversible. This gives

$$\hat{\boldsymbol{X}} = \boldsymbol{A}^{-1}\boldsymbol{B} \tag{12.144}$$

where

$$\boldsymbol{A}^{-1} = \frac{1}{(x_0 - x_1)(y_0 - y_2) - (y_0 - y_1)(x_0 - x_2)} \begin{bmatrix} y_0 - y_2 & y_1 - y_0 \\ x_2 - x_0 & x_0 - x_1 \end{bmatrix} \triangleq \begin{bmatrix} a_{11} & a_{12} \\ a_{21} & a_{22} \end{bmatrix} \tag{12.145}$$

Then, the solutions of x, y containing r_0 are

$$\hat{x} = m_1 r_0 + n_1 \tag{12.146}$$

$$\hat{y} = m_2 r_0 + n_2 \tag{12.147}$$

where

$$m_i = \sum_{j=1}^{2} a_{ij} \Delta r_j \quad i = 1,2 \tag{12.148}$$

$$n_i = \sum_{j=1}^{2} a_{ij} k_j \quad i = 1,2 \tag{12.149}$$

Substituting (12.146) and (12.147) into (12.135), after some cancellation, gives

$$(m_1 r_0 + n_1 - x_0)^2 + (m_2 r_0 + n_2 - y_0)^2 + z_0^2 = r_0^2 \tag{12.150}$$

Equation (12.150) can be further simplified as

$$a r_0^2 + 2 b r_0 + c = 0 \tag{12.151}$$

where

$$a = m_1^2 + m_2^2 - 1 \tag{12.152}$$

$$b = m_1 (n_1 - x_0) + m_2 (n_2 - y_0) \tag{12.153}$$

$$c = (n_1 - x_0)^2 + (n_2 - y_0)^2 + z_0^2 \tag{12.154}$$

The following can be derived from (12.151):

$$r_0 = \frac{-b \pm \sqrt{b^2 - ac}}{a} \tag{12.155}$$

Substituting r_0 derived from (12.155) into (12.146) and (12.147) yields estimates of the target location.

12.5.3 *Three-Dimensional Condition*

Set the target location as (x, y, z), and the range difference from the target location to the main station (x_0, y_0, z_0) and to the three assistant stations (x_i, y_i, z_i) as

$$\Delta r_i = r_i - r_0 = C \cdot \Delta t_{i0} \quad (i = 1,2,3) \tag{12.156}$$

where C denotes the velocity of light, Δt_{i0} the time difference of the target signal arriving at the main observing station and at the assistant observing stations, which can be derived from (12.132) and (12.133), and

$$r_0 = \sqrt{(x-x_0)^2+(y-y_0)^2+(z-z_0)^2} \tag{12.157}$$

$$r_i = \sqrt{(x-x_i)^2+(y-y_i)^2+(z-z_i)^2} \tag{12.158}$$

Move r_0 to the left side of the first equality mark in (12.156), and square on both sides. Substituting (12.157) and (12.158) into (12.156), with some cancellation, yields

$$(x_0-x_i)x+(y_0-y_i)y+(z_0-z_i)z=k_i+r_0\Delta r_i \quad i=1,2,3 \tag{12.159}$$

where

$$k_i = \frac{1}{2}\left[\Delta r_i^2 + d_0^2 - d_i^2\right] \tag{12.160}$$

$$d_0^2 = x_0^2 + y_0^2 + z_0^2 \tag{12.161}$$

$$d_i^2 = x_i^2 + y_i^2 + z_i^2 \tag{12.162}$$

Write (12.159) in matrix form, as

$$\boldsymbol{AX}=\boldsymbol{B} \tag{12.163}$$

where

$$\boldsymbol{A} = \begin{bmatrix} x_0-x_1 & y_0-y_1 & z_0-z_1 \\ x_0-x_2 & y_0-y_2 & z_0-z_2 \\ x_0-x_3 & y_0-y_3 & z_0-z_3 \end{bmatrix} \tag{12.164}$$

$$\boldsymbol{X} = \begin{bmatrix} x \\ y \\ z \end{bmatrix} \quad \boldsymbol{B} = \begin{bmatrix} k_1+r_0\Delta r_1 \\ k_2+r_0\Delta r_2 \\ k_3+r_0\Delta r_3 \end{bmatrix} \tag{12.165}$$

When the four sensors are not in the same plane, the rank of the coefficient matrix $\boldsymbol{A}$ is rank $(\boldsymbol{A})=3$, at which point matrix $\boldsymbol{A}$ is reversible. When they are in the same plane, dimension reduction has to be conducted, that is, conducting 2D location with the long-range target. When they are not only in the same plane but also on the same line, location is not accessible. If they are not in the same plane, solving (12.163) yields

$$\hat{\boldsymbol{X}}=\boldsymbol{A}^{-1}\boldsymbol{B} \tag{12.166}$$

where

$$\boldsymbol{A}^{-1} \triangleq \begin{bmatrix} a_{11} & a_{12} & a_{13} \\ a_{21} & a_{22} & a_{23} \\ a_{31} & a_{32} & a_{33} \end{bmatrix} \tag{12.167}$$

Then, the solutions of x, y, z containing r_0 are

$$\hat{x} = m_1 r_0 + n_1 \tag{12.168}$$

$$\hat{y} = m_2 r_0 + n_2 \tag{12.169}$$

$$\hat{z} = m_3 r_0 + n_3 \tag{12.170}$$

where

$$m_i = \sum_{j=1}^{3} a_{ij} \Delta r_j \quad i = 1,2,3 \tag{12.171}$$

$$n_i = \sum_{j=1}^{3} a_{ij} k_j \quad i = 1,2,3 \tag{12.172}$$

Combining (12.168), (12.169), and (12.170) into (12.157), with some cancellation, yields

$$a r_0^2 + 2 b r_0 + c = 0 \tag{12.173}$$

where

$$a = m_1^2 + m_2^2 + m_3^2 - 1 \tag{12.174}$$

$$b = m_1 (n_1 - x_0) + m_2 (n_2 - y_0) + m_3 (n_3 - z_0) \tag{12.175}$$

$$c = (n_1 - x_0)^2 + (n_2 - y_0)^2 + (n_3 - z_0)^2 \tag{12.176}$$

It can be derived from (12.173) that

$$r_0 = \frac{-b \pm \sqrt{b^2 - ac}}{a} \tag{12.177}$$

Substituting r_0 derived from (12.177) into (12.168), (12.169), and (12.170) gives estimates of the target location, which leads to the passive location of the target using TDOA.

12.6 Summary

Compared with the active location system, the passive location system has better stealth ability, which makes it difficult to detect by an enemy's reconnaissance system. Because it receives direct waves from radiation sources, it has advantages in terms of range, which results in longer early warning time, thus improving the survivability and combat ability of the system in electronic warfare.

Based on the analysis of the characteristics and advantages of passive radars, this chapter discusses the spatial correlation of the passive radar measurement data, including passive location and tracking using the phase changing rate method, Doppler changing rate, and multi-model method. Finally, this chapter addresses the optimal deployment under the principle of minimum area of the position concentration ellipse, and time difference measurements' passive location.

13

Pulse Doppler Radar Data Processing

13.1 Introduction

Pulse Doppler (PD) radar is a kind of pulse radar using the Doppler effect to detect target information. It is widely applied in meteorological detection, air traffic control, ground air-defense warning, low-altitude detection, missile guidance, and shipborne fire control, especially in airborne early warning and fire control. The most important characteristic of the PD radar is that it can obtain not only target range, azimuth, and elevation (which can all be obtained by conventional radars), but also radial velocity using Doppler frequency calculations. Therefore, the crux of the data processing of PD radar lies in how to make full use of this new information, enhancing radar's capability to evaluate the target state. The current research mainly involves assisting track initiation, multi-target data association, and track filtering using Doppler measurements.

In the first place, this chapter introduces the characteristics of PD radar and the composition of its tracking system. Then, it puts emphasis on the typical data processing algorithms of PD radar, namely, optimal distance–velocity coupled tracking, multi-target tracking, and target tracking algorithms with Doppler measurements. Since Doppler measurements are characterized by their strong nonlinearity, a detailed discussion is made of several nonlinear filtering methods in the target tracking category which deal with Doppler measurements. Finally, we present a test and analysis of the tracking capabilities of several PD radar tracking algorithms.

13.2 Overview of PD Radar Systems

13.2.1 Characteristics of PD Radar

The Doppler effect is the change in frequency of the received signal for a receiver in radial motion relative to its source. It is named after the Austrian physicist Christian Doppler, who proposed it in his acoustic studies in 1842. In the case of a radar moving relative to its target, the Doppler effect manifests itself as the inequality in frequencies between echoes and transmitted signals. When the

Radar Data Processing with Applications, First Edition. He You, Xiu Jianjuan, and Guan Xin.

electromagnetic wave signals transmitted by a radar come across a target which is approaching the radar, their back-and-forth propagations between the radar and the target due to the Doppler effect produce electromagnetic wave signals at a higher frequency.

The phase change of signals transmitted by the radar and echo signals is

$$\varphi(t) = \frac{2\pi}{\lambda} 2R(t) \tag{13.1}$$

where $R(t)$ is the time-varying one-way distance for a radar moving relative to its target, and λ is the radar signal wavelength. When a target is moving towards a static radar with speed v, then $R(t) = vt$. Thus, (13.1) can be written as

$$\varphi(t) = 2\pi \frac{2v}{\lambda} t = 2\pi f_{\text{d}} t \tag{13.2}$$

where $f_{\text{d}} = \frac{2v}{\lambda}$, which is the Doppler effect caused by the moving target. The Doppler frequency is proportional to the relative speed between the target and the radar when the wavelength of the radar signal is fixed.

Those radars that conduct target information extraction and processing by utilizing the Doppler effect caused by the relative motion between the radar and its target are called Doppler radars. Those which transmit a pulse modulated radio frequency are called PD radars.

PD radars should have:

1. A sufficiently high pulse repetition frequency (PRF), lest the clutter or the observed target suffer from velocity fuzziness.
2. An ability for Doppler filtering of the single spectral line of the pulse train frequency spectrum (i.e., frequency domain filtering ability).
3. Great possibilities to produce range fuzziness for the observed target due to its high PRF.

Practical PD radars usually vary greatly [317, 318] in their function and composition. As to their data processors, the main functions involve data association and filtering, range tracking, angle tracking, ambiguity-resolving calculations, antenna angle error modification, control of operating modes, formation of scanning graphs, choice of PRF, prediction of clutter frequency and interfaces of other systems, etc.

Because of their extraordinary performance in clutter suppression, PD radars have become the center of world attention. Improvements in the performance and navigation methods of modern air vehicles enable them to fly at low and extremely low altitudes, which means that defense against low-altitude invasions has become a key issue. Therefore, airborne radars, including early warning and fire-control radars, should have a downward-looking ability, the ability to find weak target signals in strong ground clutters. That's the reason why modern airborne early warning and fire-control radars both choose PD systems, which can suppress ground clutters efficiently, and have favorable resistance to passive jamming and active jamming as well [319]. A detailed discussion of the data extraction, typical filtering algorithms, and multi-target tracking of PD radars will be presented in Section 13.2.2.

13.2.2 PD Radar Tracking System

The raw data for data processing originates from the azimuth, elevation, range, and velocity tracking loop of the radar. There are two kinds of tracking systems for PD radars: single-target and multi-target.

13.2.2.1 Single-Target Velocity (Doppler Frequency) Tracking System

Frequency tracking loops can be classified into frequency-locked and phase-locked according to the difference in frequency-sensitive elements. The former type use frequency discriminators as their sensitive elements. Their functional block diagram is shown in Figure 13.1.

At the very beginning, the tracking loop operates in the search mode. A voltage with periodic variation is given at inputs to the voltage-controlled oscillator (VCO), which is then enabled to vary within the scope of the predicted Doppler frequency. Once the target is located, with target echo frequency $f_0 + f_d$, subtract the VCO frequency $f_0 - f_2 + f'_d$, then the resulting beat signals at frequency $f'_2 = f_2 + f_d - f'_d$ enter the frequency discriminator via the narrow band filter. At this point, the additional intercept circuit can be used to control the loop to switch from the search to the track mode. If $f'_d > f_2$, the central frequency of the beat signals' spectrum is $f'_2 < f_2$. Then, the frequency discriminator will output a positive voltage to reduce the frequency of the VCO, and f'_d is very close to f_d after closed-loop adjustment. After being switched through the frequency output circuit, the frequency deviation of the VCO f'_d can output the velocity data of the target. When there is any change in the Doppler frequency of the target echo, the frequency discriminator determines the dimensions and direction of these changes, and outputs a control voltage, thereby activating corresponding changes in the frequency of the VCO and accomplishing the tracking of automatic frequencies.

The functional block diagram of phase-locked frequency trackers is shown in Figure 13.2. As indicated in the diagram, they are essentially the same as frequency-locked frequency trackers except that the element sensitive to frequency changes is replaced with a phase discriminator.

In order that the phase-locked system remains in tracking mode, the phase of the VCO should synchronize with the changes in signal phase on the whole, in other words, the errors between them should stay within a certain fraction of the signal period, which sets a high standard for the stability of the radar facilities. Then, the band of the phase-locked system should be wide enough to ensure that the dynamic lag in the phase arising from the maneuvers of the target is maintained within tolerance. But the enlargement of the bandwidth can increase the tracking errors brought about by noise. The phase-locked system may be subject to the limits of the highest acceleration of the trackable target with its bandwidth fixed.

13.2.2.2 Single-Target Range Tracking System

The basic methods of range tracking are the same as those of conventional pulse radar; the difference lies in the fact that the velocity selection is added to the range loop of the PD radar. Since signals

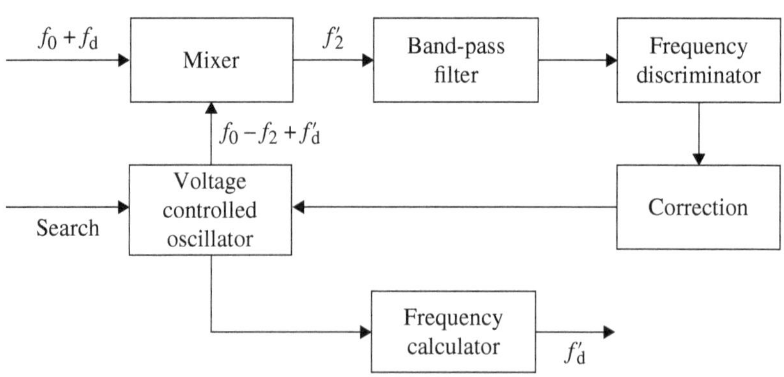

Figure 13.1 Frequency-locked frequency tracking loop

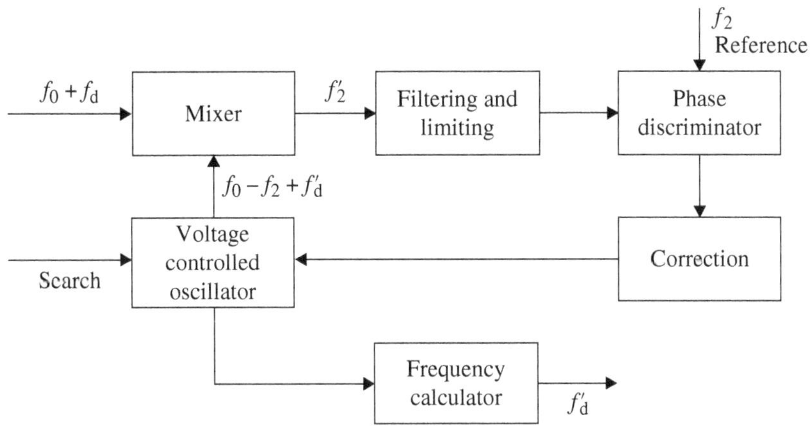

Figure 13.2 Phase-locked frequency tracker

approximate continuous waves without range messages after going through the single side band and the narrow band Doppler filter, the range gate must be added to the broad band intermediate frequency amplifier before velocity selection, as shown in Figure 13.3.

13.2.2.3 Multi-target Tracking System

Multi-target tracking can be done by multiplex reception channels. In this system, the range gate covers the whole pulse interval at different times, and there is a group of Doppler filters in every channel. When the antenna is scanning, all of the targets will be tested, and be tracked by discrete data. When the multiple pulse repetition frequency is used to measure the range, the time when the target is being scanned by the antenna must be divided into several periods in order to meet the needs of repeated observations required by ranging systems. Since the accumulating time is shortened, the bandwidth of the Doppler filter must be widened correspondingly, which will reduce the detection range [320–324]. Meanwhile, in order to avoid ranging fuzziness, several detections must be conducted, which will further weaken its ranging capability. Although the multi-target tracking system is not specific to pulse Doppler radars, it has unparalleled advantages (as in the single-target tracking mode) over conventional radars in many cases, especially under the interference of strong clutters.

13.3 Typical Algorithms of PD Radar Tracking

To make the various parameters of describing a target's moving state as accurate as possible, the data obtained by radar measurements need further processing because there are unavoidably measurement errors and noises. The algorithms commonly used include the LS algorithm, constant gain α–β filtering, constant gain α–β–γ filtering, Kalman filtering, extended Kalman filtering, etc.

Theoretically, Kalman filtering is linear unbiased minimum variance estimation with a time-varying structure, so it is applicable to estimation of the non-stationary process, the detailed description of whose model is given in Chapter 3. However, there are still many problems to deal with when this algorithm is applied in the practical data processing system of PD radars. These problems mainly involve the filtering mathematical model, real-time ability, and numerical divergence.

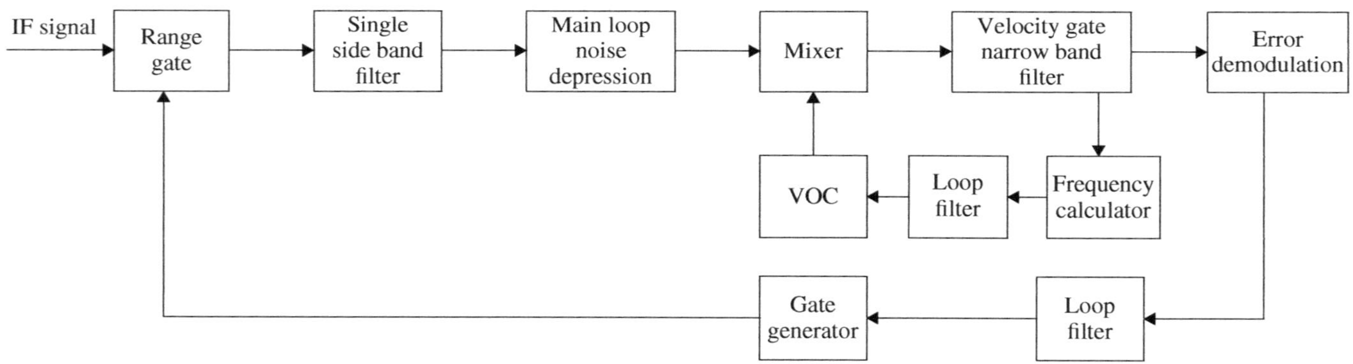

Figure 13.3 Range tracking loop

The Kalman filter performs best only when its mathematical model is certain. The steps to build a mathematical model include the establishment of state and measurement equations, and confirmation of the statistical characteristics of initial estimation vectors, system noise, and measurement noise. However it is quite difficult, sometimes even impossible, to establish an accurate filtering model in most practical implementations. If the filtering model cannot match the mathematical rules in practical processes, the precision of the filter will suffer. Worse still, divergence of the filter may occur.

Errors caused by an inaccurate filtering model are allowable under some conditions, and these errors may gradually disappear with time. However, if model errors go beyond the permitted limits or more precise filtering is needed, the filtering model should be modified, which requires the employment of model identification and adaptive filtering techniques. Actually, in the procedure of applying Kalman filtering to PD radar data processing, the main work lies in the establishment of a filtering mathematical model and the pursuit of an applicable adaptive filtering algorithm.

A typical example is how to use radar to track maneuvering targets. In most cases, the air tactical targets that radar has to track are manned and unmanned air vehicles in the atmosphere. Generally speaking, the target flies linearly at constant velocity. In that case, extrapolation along a straight line should be used to predict its location in the future. It is reasonable to consider that this way of flying is most frequently adopted by an air vehicle based on the fact that its effective velocity goes down and the time to fulfill its task and stay in dangerous zones extends if it takes lots of turns in the flight. However, the air vehicle will take such actions as accelerations, decelerations, climbs, dives, and turns under the condition that it encounters obstacles or radar illumination, or has to dodge enemy firepower, or perform maneuvers in air battles. Since both the overload that an air vehicle can bear and its power are limited, so is its maneuver ability, thereby necessitating the utilization of various adaptive filtering algorithms suitable for tracking maneuvering targets. The corresponding model description and comparative analysis of their performance have been discussed in detail in Chapter 9.

13.3.1 Optimal Range–Velocity Mutual Coupling Tracking

Although there is a strict correspondence between the range and velocity of the target, range and velocity tracking loops are independent of one another in conventional PD radars. Actually, range tracking error and velocity tracking error are two components of its innovation vector if a Kalman filter is inserted into the tracking loop. With filtering computing, a connection can naturally be established between the two tracking loops. This kind of tracking is known as "optimal range–velocity mutual coupling tracking." The procedure for getting a closed-form solution to optimal range–velocity mutual coupling tracking will be discussed below.

Assume that the target's system model on a certain coordinate axis is a second-order differential equation

$$\ddot{x}(t) = v(t) \tag{13.3}$$

where $x(t)$ is the real range of the target and $v(t)$ is the random acceleration whose statistical characteristic is $\mathrm{E}[v(t)] = 0$, $\mathrm{E}[v(t)v(\tau)] = \sigma^2\delta(t-\tau)$.

The state equation of the target is

$$\dot{\boldsymbol{X}}(t) = \boldsymbol{A}\boldsymbol{X}(t) + \boldsymbol{\Gamma}v(t) \tag{13.4}$$

and

$$X(t)=\begin{bmatrix}x(t)\\ \dot{x}(t)\end{bmatrix},\quad A=\begin{bmatrix}0&1\\0&0\end{bmatrix},\quad \Gamma=\begin{bmatrix}0\\1\end{bmatrix} \tag{13.5}$$

The measuring model of the PD radar is

$$Z(t)=HX(t)+W(t) \tag{13.6}$$

where

$$Z(t)=\begin{bmatrix}Z_x(t)\\ Z_{\dot{x}}(t)\end{bmatrix},\quad H=\begin{bmatrix}1&0\\0&1\end{bmatrix},\quad W(t)=\begin{bmatrix}W_x(t)\\ W\dot{x}(t)\end{bmatrix}$$

Generally speaking, $W_x(t)$ and $W_{\dot{x}}(t)$ are irrelevant zero-mean, white noise processes with statistical characteristics $\mathrm{E}[W(t)]=\mathbf{0}$, $\mathrm{E}[W(t)W'(\tau)]=R\delta(t-\tau)$, where R is a stationary diagonal matrix, $R=\mathrm{diag}\left[\sigma_r^2,\sigma_v^2\right]$, with σ_r^2 and σ_v^2 as the range and velocity measurement noise variance, respectively.

Based on the continuous Kalman filtering theory, the optimal estimate of the target state for the above-mentioned model can be given by the following differential equations group:

$$\hat{X}(t+1)=A\hat{X}(t)+K(t)\left[Z(t)-H\hat{X}(t)\right] \tag{13.7}$$

$$K(t)=P(t)H'R^{-1} \tag{13.8}$$

$$P(t+1)=AP(t)+P(t)A'+\Gamma\sigma^2\Gamma'-P(t)H'R^{-1}HP(t) \tag{13.9}$$

Equation (13.9) is the famous Riccati equation, in which $K(t)$ is the filter gain matrix, $P(t)$ is the estimation error covariance matrix, and $\left[Z(t)-H\hat{X}(t)\right]$ is the filtering innovation.

In view of the steady filtering problem, (13.9) is degenerated to the matrix Riccati algebra equation:

$$AP+PA'+\Gamma\sigma^2\Gamma'-PH'R^{-1}HP-P=0 \tag{13.10}$$

Solving (13.10) yields

$$P=\begin{bmatrix}\sigma_r\sigma_v\sin\varphi & \sigma_r\sigma\cos\varphi\\ \sigma_r\sigma\cos\varphi & \sigma_v\sigma\sin\varphi\end{bmatrix} \tag{13.11}$$

where

$$\cos\varphi=\frac{1}{1+\sigma_r\sigma\sigma_v^{-2}} \tag{13.12}$$

Let $\alpha=\sigma_v/\sigma_r$ and $\beta=\sigma/\sigma_v$, then the gain matrix is

$$K=\begin{bmatrix}\alpha\sin\varphi & \dfrac{\beta}{\alpha}\cos\varphi\\ \alpha\beta\cos\varphi & \beta\sin\varphi\end{bmatrix} \tag{13.13}$$

and

$$\cos\varphi = \frac{\alpha}{\alpha+\beta} \tag{13.14}$$

The steady mutual coupling Kalman filter is steady for any group α, β, and the pole of the system can be defined by the characteristic equation

$$s^2 + \sqrt{\beta^2 + 2\alpha\beta}s + \alpha\beta = 0 \tag{13.15}$$

The acceleration error constant of the system is

$$K_a = \frac{\alpha^2\beta}{\alpha+\beta} \tag{13.16}$$

Therefore, as long as α gets a limited value, the system can track the input process of steady acceleration with limited tracking error.

In order to estimate the performance of the optimal mutual coupling tracking system, its range error variance can be compared with the range tracking error of the tracking loop without mutual coupling. The measurement equation of the single ranging tracking loop is

$$Z(t) = [1 \quad 0]\boldsymbol{X}(t) + w(t) \tag{13.17}$$

Let $\sigma_v^{-2} = 0$ in (13.12), then the steady tracking error covariance matrix of the system can be obtained, which indicates that the ranging tracking loop is the degenerate form of the mutual coupling tracking loop when the velocity measurement noise covariance approaches infinity. From (13.12) and (13.14), we get

$$p_{11}^* = \sigma_r\sqrt{2\sigma_r\sigma} \tag{13.18}$$

Define the performance index J as the ratio of the range error covariance of the two systems, then

$$J = \frac{p_{11}^*}{p_{11}} = \frac{1+x}{\sqrt{1+x/2}} \tag{13.19}$$

where

$$x = \sigma_r\sigma\sigma_v^{-2} \tag{13.20}$$

It can be concluded that the performance index increases monotonously from 1 to infinity and J goes gradually close to $\sqrt{2x}$ when $x \gg 1$. For any kind of practical PD radar, there must be $\sigma_v^2 < \infty$, so the mutual coupling tracking system always performs better than the single ranging one.

The state model of the target is a three or higher-order differential equation when the random acceleration of the target is a correlation process. In this case, it is difficult to get the closed-form solution to the filters. Therefore, computers should serve as an aid in analyzing the performance of the filters.

Notice that the precondition of performing mutual coupling filtering is to make sure that velocity measuring ambiguity [325, 326] is eliminated.

The Kalman filtering model established in line-of-sight coordinate systems has comparatively high estimation accuracy and adaptability. The tracking system closed by this kind of filter has better stability, shown particularly in its angel tracking system, which needs only two-axis stabilization (by antenna stabilization loops). By doing so, the radar stabilized platform can be spared in the whole process and the radar structure is greatly simplified. Therefore, it is one of the most frequently used systems in modern fire control. Yet one of the disadvantages of this filtering model is that it cannot be used to track multiple targets.

13.3.2 Multi-target Tracking

With the increase in the number of targets that radars have to deal with, it becomes an essential system for many types of radar to track multiple targets simultaneously. For instance, while implementing long-distance interception, the airborne multi-function radar is required to track multiple targets at the same time and choose as its targets those which present the greatest threat or have the highest percentage of hits. Even when used to attack single targets, it still needs to monitor the air intelligence in the vicinity. Each kind of early warning radar must be provided with the capability to track multiple targets of different batches simultaneously. Besides, they should be able to identify and catalog the targets automatically.

Multi-target tracking algorithms mainly include the track splitting, joint maximum likelihood, 0–1 programming, generalized correlation, probabilistic data association, nearest-neighbor, joint probabilistic data association, optimal Bayesian, and multiple hypothesis algorithms. The differences between those algorithms lie in the number of targets, detection probability, correlation between multiple scans, and amount of calculation when they are put into practice, which have been discussed in detail in Chapters 7 and 8.

13.3.3 Target Tracking with Doppler Measurements

The radars used in practice, especially Doppler radars, can usually perform Doppler measurements. It has been proved by theoretical computation and practice that tracking accuracy can be greatly improved by making full use of Doppler measurements.

The EKF is the most frequently used method to solve the target tracking problem with Doppler measurements. However, since the relationship between radar measurements and the moving state of the target is completely nonlinear, the result of estimation is often unsatisfactory. For the current radar target tracking algorithms with Doppler measurements, the common method is to hypothesize that errors of range, angle, and Doppler measurements are statistically independent. But for some wave forms, range and Doppler measurements are statistically dependent in terms of errors based on the latest research [326].

13.3.3.1 Unbiased Sequential Extended Kalman Filtering (USEKF) Algorithm

In order to make the best of Doppler measurements, the following discussion will focus on how to extend the 2D debiased consistent converted measurement Kalman filtering algorithm, which only considers position measurements, to the case where Doppler measurements are also included, and range errors and Doppler measurement errors are statistically dependent.

Problem Description

In Cartesian coordinates, the moving model of the target can usually be expressed as

$$\boldsymbol{X}(k)=\boldsymbol{F}(k-1)\boldsymbol{X}(k-1)+\boldsymbol{G}(k-1)\boldsymbol{u}(k-1)+\boldsymbol{V}(k-1) \tag{13.21}$$

where $\boldsymbol{X}(k)=\left[x(k),y(k),\dot{x}(k),\dot{y}(k),s_{1\times(n-4)}\right]'$ is the moving state of the target, $x(k)$ and $y(k)$ are position components of the target in two directions x, y, respectively, $\dot{x}(k),\dot{y}(k)$ are their corresponding velocity components, $s_{1\times(n-4)}$ is the rest of the state components, $\boldsymbol{F}(k)\in\boldsymbol{R}^{n\times n}$ is the state transition matrix, $\boldsymbol{G}(k)$ is the coefficient matrix of proper dimensions, $\boldsymbol{u}(k)$ is the deterministic input vector, and $\boldsymbol{V}(k)$ is a Gaussian white noise sequence whose mean value is zero and variance $\boldsymbol{Q}_k$.

Assume that a 2D radar is located at the origin of the coordinate, then the radar's measurement equation in polar coordinates can be expressed as

$$\boldsymbol{z}^m(k)=[\rho^m(k),\theta^m(k),\dot{\rho}^m(k)]'=\boldsymbol{f}_k(\boldsymbol{X}(k))+\boldsymbol{v}^m(k)=[\rho(k),\theta(k),\dot{\rho}(k)]'+\boldsymbol{v}^m(k) \tag{13.22}$$

where

$$\begin{cases}\rho(k)=\sqrt{x^2(k)+y^2(k)}\\ \theta(k)=\arctan(y(k)/x(k))\\ \dot{\rho}(k)=(x(k)\dot{x}(k)+y(k)\dot{y}(k))/\sqrt{x^2(k)+y^2(k)}\\ \boldsymbol{v}^m(k)=\left[\widetilde{\rho}(k),\widetilde{\theta}(k),\widetilde{\dot{\rho}}(k)\right]'\end{cases} \tag{13.23}$$

with $\rho^m(k)$, $\theta^m(k)$, and $\dot{\rho}^m(k)$ the radar's measurement values of target range, azimuth angle, and Doppler, respectively. $\rho(k)$, $\theta(k)$, and $\dot{\rho}(k)$ are their corresponding true values; $\widetilde{\rho}(k)$, $\widetilde{\theta}(k)$, and $\widetilde{\dot{\rho}}(k)$ are their corresponding additive measurement errors. Suppose that all of them are zero-mean, Gaussian, white noise sequences with variances σ_ρ^2, σ_θ^2, and $\sigma_{\dot{\rho}}^2$, respectively, $\widetilde{\rho}(k)$ and $\widetilde{\theta}(k)$ are uncorrelated, so are $\widetilde{\theta}(k)$ and $\widetilde{\dot{\rho}}(k)$, and the correlative coefficient of $\widetilde{\rho}(k)$ and $\widetilde{\dot{\rho}}(k)$ is r.

Measurement Conversion

Converting the position (range and azimuth angle) measurements in polar coordinates to Cartesian coordinates gives

$$\begin{cases}x^c(k)=\rho^m(k)\cos\theta^m(k)=x(k)+\widetilde{x}(k)\\ y^c(k)=\rho^m(k)\sin\theta^m(k)=y(k)+\widetilde{y}(k)\end{cases} \tag{13.24}$$

where $\widetilde{x}(k)$ and $\widetilde{y}(k)$ are components of position-converted measurement errors in two directions x, y of Cartesian coordinates, respectively.

The strength of nonlinearity between Doppler measurements and the moving state of the target can be weakened by using the estimated measurement converted equation

$$\xi^c(k)=\rho^m(k)\dot{\rho}^m(k)=x(k)\dot{x}(k)+y(k)\dot{y}(k)+\widetilde{\xi}(k) \tag{13.25}$$

where $\widetilde{\xi}(k)$ is the converted error of estimated measurement $\xi(k)$ in Cartesian coordinates.

It follows from (13.24) and (13.25) that the radar measurements, after being converted from polar to Cartesian coordinates, can be expressed on the whole as

$$\begin{aligned} z^c(k) &= [x^c(k), y^c(k), \xi^c(k)]' = \boldsymbol{h}_k(\boldsymbol{X}(k)) + \boldsymbol{v}^c(k) \\ &= [x(k), y(k), x(k)\dot{x}(k) + y(k)\dot{y}(k)]' + \left[\widetilde{x}(k), \widetilde{y}(k), \widetilde{\xi}(k)\right]' \end{aligned} \tag{13.26}$$

True Bias and Covariance of Converted Measurement Errors

Given the real position of the target and Doppler, the true mean and covariance of converted measurement errors can be obtained based on the known condition of radar measurement errors from (13.22) as

$$\begin{cases} \boldsymbol{\mu}_t(k) = \mathrm{E}[\boldsymbol{v}^c(k)|\rho(k), \theta(k), \dot{\rho}(k)] = [\mu_t^x(k), \mu_t^y(k), \mu^\xi(k)]' \\ \boldsymbol{R}_t(k) = \mathrm{cov}[\boldsymbol{v}^c(k)|\rho(k), \theta(k), \dot{\rho}(k)] = \begin{bmatrix} R_t^{xx}(k) & R_t^{xy}(k) & R_t^{x\xi}(k) \\ R_t^{yx}(k) & R_t^{yy}(k) & R_t^{y\xi}(k) \\ R_t^{\xi x}(k) & R_t^{\xi y}(k) & R_t^{\xi\xi}(k) \end{bmatrix} \end{cases} \tag{13.27}$$

and

$$\begin{cases} \mu_t^x = \rho(k)\cos\theta(k)\left(\mathrm{e}^{-\frac{\sigma_\theta^2}{2}} - 1\right) \\ \mu_t^y = \rho(k)\sin\theta(k)\left(\mathrm{e}^{-\frac{\sigma_\theta^2}{2}} - 1\right) \\ \mu_t^\xi(k) = r\sigma_\rho\sigma_{\dot{\rho}} \\ R_t^{xx} = \mathrm{var}[\widetilde{x}|\rho(k), \theta(k)] = (\rho(k))^2 e^{-\sigma_\theta^2}\left[\cos^2\theta(k)\left(\cosh\left(\sigma_\theta^2\right) - 1\right) + \sin^2\theta(k)\left(\sinh\left(\sigma_\theta^2\right)\right)\right] \\ \qquad + \sigma_\rho^2 e^{-\sigma_\theta^2}\left[\cos^2\theta(k)\cosh\left(\sigma_\theta^2\right) + \sin^2\theta(k)\left(\sinh\left(\sigma_\theta^2\right)\right)\right] \\ R_t^{yy} = \mathrm{var}[\widetilde{y}|\rho(k), \theta(k)] = (\rho(k))^2 e^{-\sigma_\theta^2}\left[\sin^2\theta(k)\left(\cosh\left(\sigma_\theta^2\right) - 1\right) + \cos^2\theta(k)\left(\sinh\left(\sigma_\theta^2\right)\right)\right] \\ \qquad + \sigma_\rho^2 e^{-\sigma_\theta^2}\left[\sin^2\theta(k)\cosh\left(\sigma_\theta^2\right) + \cos^2\theta(k)\left(\sinh\left(\sigma_\theta^2\right)\right)\right] \\ R_t^{xy} = R_t^{yx} = \mathrm{var}[\widetilde{x}, \widetilde{y}|\rho(k), \theta(k)] = \sin\theta(k)\cos\theta(k)e^{-2\sigma_\theta^2}\left\{\sigma_\rho^2 + [\rho(k)]^2\left(1 - e^{\sigma_\theta^2}\right)\right\} \\ R_t^{x\xi}(k) = R_t^{\xi x}(k) = \left(\sigma_\rho^2\dot{\rho}(k) + \rho(k)r\sigma_\rho\sigma_{\dot{\rho}}\right)\cos[\theta(k)]\mathrm{e}^{-\sigma_\theta^2/2} \\ R_t^{y\xi}(k) = R_t^{\xi y}(k) = \left(\sigma_\rho^2\dot{\rho}(k) + \rho(k)r\sigma_\rho\sigma_{\dot{\rho}}\right)\sin[\theta(k)]\mathrm{e}^{-\sigma_\theta^2/2} \\ R_t^{\xi\xi}(k) = [\rho(k)]^2\sigma_{\dot{\rho}}^2 + \sigma_\rho^2[\dot{\rho}(k)]^2 + (1 + r^2)\sigma_\rho^2\sigma_{\dot{\rho}}^2 + 2\rho(k)\dot{\rho}(k)r\sigma_\rho\sigma_{\dot{\rho}}. \end{cases} \tag{13.28}$$

which are the true bias and covariance of converted measurement errors.

True Bias and Covariance of Average Converted Measurement Errors

Equation (13.27) cannot be applied directly in practice, since the real position of the target and Doppler are always unknown. In order to make it applicable, the mathematical expectation of the above-mentioned true mean and covariance can be solved, under the condition that measurements are known, as

$$\begin{cases} \boldsymbol{\mu}_a(k) = \mathrm{E}[\boldsymbol{\mu}_t(k)|\rho^m(k),\theta^m(k),\dot{\rho}^m(k)] = [\mu_a^x(k),\mu_a^y(k),\mu_a^\xi(k)]' \\ \boldsymbol{R}_a(k) = \mathrm{E}[\boldsymbol{R}_t(k)|\rho^m(k),\theta^m(k),\dot{\rho}^m(k)] = \begin{bmatrix} R_a^{xx}(k) & R_a^{xy}(k) & R_a^{x\xi}(k) \\ R_a^{yx}(k) & R_a^{yy}(k) & R_a^{y\xi}(k) \\ R_a^{\xi x}(k) & R_a^{\xi y}(k) & R_a^{\xi\xi}(k) \end{bmatrix} \end{cases} \tag{13.29}$$

and

$$\begin{cases} \mu_a^x = \rho^m(k)\cos[\theta^m(k)]\left(e^{-\sigma_\theta^2} - e^{-\sigma_\theta^2/2}\right) \\ \mu_a^y = \rho^m(k)\sin[\theta^m(k)]\left(e^{-\sigma_\theta^2} - e^{-\sigma_\theta^2/2}\right) \\ \mu_a{}^\xi(k) = r\sigma_\rho\sigma_{\dot{\rho}} \\ R_a^{xx} = [\rho^m(k)]^2 e^{-2\sigma_\theta^2}\{\cos^2[\theta^m(k)][\cosh(2\sigma_\theta^2) - \cosh(\sigma_\theta^2)] + \sin^2[\theta^m(k)][\sinh(2\sigma_\theta^2) - \sinh(\sigma_\theta^2)]\} \\ \qquad + \sigma_\rho^2 e^{-2\sigma_\theta^2}\{\cos^2[\theta^m(k)][2\cosh(2\sigma_\theta^2) - \cosh(\sigma_\theta^2)] + \sin^2[\theta^m(k)][2\sinh(2\sigma_\theta^2) - \sinh(\sigma_\theta^2)]\} \\ R_a^{yy} = (\rho^m(k))^2 e^{-2\sigma_\theta^2}\{\sin^2[\theta^m(k)][\cosh(2\sigma_\theta^2) - \cosh(\sigma_\theta^2)] + \cos^2[\theta^m(k)][\sinh(2\sigma_\theta^2) - \sinh(\sigma_\theta^2)]\} \\ \qquad + \sigma_\rho^2 e^{-2\sigma_\theta^2}\{\sin^2[\theta^m(k)][2\cosh(2\sigma_\theta^2) - \cosh(\sigma_\theta^2)] + \cos^2[\theta^m(k)][2\sinh(2\sigma_\theta^2) - \sinh(\sigma_\theta^2)]\} \\ R_a^{xy} = R_a^{yx} = \sin[\theta^m(k)]\cos[\theta^m(k)]e^{-4\sigma_\theta^2}\left[\sigma_\rho^2 + [\rho^m(k)]^2 + \sigma_\rho^2\right)\left(1 - e^{-\sigma_\theta^2}\right)] \\ R_a{}^{x\xi}(k) = R_a{}^{\xi x}(k) = \left[\sigma_\rho^2\dot{\rho}^m(k) + \rho^m(k)r\sigma_\rho\sigma_{\dot{\rho}}\right]\cos[\theta^m(k)]\mathrm{e}^{-\sigma_\theta^2/2} \\ R_a{}^{y\xi}(k) = R_a{}^{\xi y}(k) = \left[\sigma_\rho^2\dot{\rho}^m(k) + \rho^m(k)r\sigma_\rho\sigma_{\dot{\rho}}\right]\sin[\theta^m(k)]\mathrm{e}^{-\sigma_\theta^2/2} \\ R_a{}^{\xi\xi}(k) = [\rho^m(k)]^2\sigma_{\dot{\rho}}^2 + \sigma_\rho^2[\dot{\rho}^m(k)]^2 + 3(1+r^2)\sigma_\rho^2\sigma_{\dot{\rho}}^2 + 2\rho^m(k)\dot{\rho}^m(k)r\sigma_\rho\sigma_{\dot{\rho}}. \end{cases} \tag{13.30}$$

Generally speaking, although converted measurement errors are not Gaussian distributed, (13.29) can be proven as the consistent estimation of the first second-order moment of converted measurement errors, which, during the process of tracking filtering, can thus be approximately considered as Gaussian. In that case, the converted measurement equation (13.26) can replace the radar's practical measurement equation (13.22).

Tracking Filter

It follows from (13.26) that converted measurement is the nonlinear function of the moving state of targets. Therefore, linearization of $\boldsymbol{h}_k[\boldsymbol{X}(k)]$ must be done in order to finish tracking filtering. And one of the most commonly used methods is to extend $\boldsymbol{h}_k[\boldsymbol{X}(k)]$ by Taylor series around the one-step prediction value $\hat{\boldsymbol{X}}(k|k-1)$ of the state. However, since position-converted measurement is a linear function of the moving state of targets, the corresponding state filtering value $\hat{X}^p(k|k)$ should be obtained by sequential filtering estimation (i.e., dealing with converted position measurements in preference), after which the nonlinear function of estimated measurement can be extended by Taylor series according to $\hat{X}^p(k|k)$, which would certainly reduce the errors generated from linearization. It can be concluded from (13.29) that there is correlation between position and converted errors of estimated measurements. So, this correlation must be removed first to realize sequential filtering estimation. The partition of the covariance matrix $\boldsymbol{R}_a(k)$ based on position and estimated measurement can be described as follows:

$$\boldsymbol{R}_a(k)=\begin{bmatrix}\boldsymbol{R}_a^p(k) & \left(\boldsymbol{R}_a^{\xi p}(k)\right)' \\ \boldsymbol{R}_a^{\xi p}(k) & \boldsymbol{R}_a^{\xi}(k)\end{bmatrix} \tag{13.31}$$

Suppose

$$\begin{cases}\boldsymbol{L}(k)=-\boldsymbol{R}_a^{\xi p}(k)\left[\boldsymbol{R}_a^p(k)\right]^{-1}=\left[\boldsymbol{L}^1(k),\boldsymbol{L}^2(k)\right] \\ \boldsymbol{B}(k)=\begin{bmatrix}\boldsymbol{I}_2 & \boldsymbol{0} \\ \boldsymbol{L}(k) & 1\end{bmatrix}\end{cases} \tag{13.32}$$

Pre-multiply both sides of (13.26) and $\mathbf{B}(k)$, then the following can be obtained by Cholesky factorization of the matrix:

$$\begin{cases}\boldsymbol{z}^{c,p}(k)=\boldsymbol{H}^{c,p}(k)\boldsymbol{X}(k)+\boldsymbol{v}^{c,p}(k) \\ \varepsilon^c(k)=h_k{}^{\varepsilon}[\boldsymbol{X}(k)]+\widetilde{\varepsilon}(k)\end{cases} \tag{13.33}$$

where

$$\begin{cases}\boldsymbol{z}^{c,p}(k)=[x^c(k),y^c(k)]' \\ \boldsymbol{H}^{c,p}=\left[\boldsymbol{I}_2,\boldsymbol{0}_{2\times(n-2)}\right] \\ \boldsymbol{v}^{c,p}(k)=[\widetilde{x}(k),\widetilde{y}(k)]' \\ \mathrm{E}[\boldsymbol{v}^{c,p}(k)]=\boldsymbol{\mu}_a^p(k)=\left[\mu_a^x(k),\mu_a^y(k)\right]' \\ \mathrm{cov}\left[\boldsymbol{v}^{c,p}(k)\right]=\boldsymbol{R}_a^p(k) \\ \varepsilon^c(k)=L^1(k)x^c(k)+L^2(k)y^c(k)+\xi^c(k) \\ h_k^{\varepsilon}(\boldsymbol{X}(k))=L^1(k)x(k)+L^2(k)y(k)+x(k)\dot{x}(k)+y(k)\dot{y}(k) \\ \widetilde{\varepsilon}(k)=L^1(k)\widetilde{x}(k)+L^2(k)\widetilde{y}(k)+\widetilde{\xi}(k) \\ \mathrm{E}[\widetilde{\varepsilon}(k)]=\mu_a^{\varepsilon}(k)=L^1(k)\mu_a^x(k)+L^2(k)\mu_a^y(k)+\mu_a^{\xi}(k) \\ \mathrm{var}[\widetilde{\varepsilon}(k)]=\boldsymbol{R}_a^{\varepsilon}(k)=\boldsymbol{R}_a^{\xi}(k)-\boldsymbol{R}_a^{\xi p}(k)\left(\boldsymbol{R}_a^p(k)\right)^{-1}\left[\boldsymbol{R}_a^{\xi p}(k)\right]'\end{cases} \tag{13.34}$$

and $\widetilde{\varepsilon}(k)$ and $\mathbf{v}^{c,p}(k)$ are uncorrelated.

Thus the filtering estimation of the moving state of targets can be carried out sequentially with the targets' moving state equation (13.21) and the measurement equation (13.33) by following the four steps below.

- Step 1: update filtering estimation with time.

$$\begin{cases} \hat{\boldsymbol{X}}(k|k-1) = \boldsymbol{F}(k-1)\hat{\boldsymbol{X}}(k-1|k-1) + \boldsymbol{G}(k-1)\boldsymbol{u}(k-1) \\ \boldsymbol{P}(k|k-1) = \boldsymbol{F}(k-1)\boldsymbol{P}(k-1|k-1)[\boldsymbol{F}(k-1)]' + \boldsymbol{Q}(k-1) \end{cases} \tag{13.35}$$

- Step 2: update filtering estimation with position measurement.

$$\begin{cases} \boldsymbol{K}^p(k) = \boldsymbol{P}(k|k-1)[\boldsymbol{H}^{c,p}(k)]'\left[\boldsymbol{H}^{c,p}(k)\boldsymbol{P}(k|k-1)[\boldsymbol{H}^{c,p}(k)]' + \boldsymbol{R}_a^p(k)\right]^{-1} \\ \hat{\boldsymbol{X}}^p(k|k) = \hat{\boldsymbol{X}}(k|k-1) + \boldsymbol{K}^p(k)\left[\boldsymbol{z}^{c,p}(k) - \boldsymbol{\mu}_a^p(k) - \boldsymbol{H}^{c,p}(k)\hat{\boldsymbol{X}}(k|k-1)\right] \\ \boldsymbol{P}^p(k|k) = (\boldsymbol{I}_n - \boldsymbol{K}^p(k)\boldsymbol{H}^{c,p}(k))\boldsymbol{P}(k|k-1) \end{cases} \tag{13.36}$$

- Step 3: update filtering estimation with pseudo measurement.
  It follows from (13.33) that the pseudo-measurement is the quadratic function of the moving state of targets. Consequently, the nonlinear tracking filtering of the targets' moving state can best be achieved using second-order EKF only. That is,

$$\begin{cases} \boldsymbol{K}^\varepsilon(k) = \boldsymbol{P}^p(k|k)[\boldsymbol{H}^\varepsilon(k)]'\left[\boldsymbol{H}^\varepsilon(k)\boldsymbol{P}^p(k|k)[\boldsymbol{H}^\varepsilon(k)]' + R_a^\varepsilon(k) + A(k)\right]^{-1} \\ \hat{\boldsymbol{X}}^\varepsilon(k|k) = \hat{\boldsymbol{X}}^p(k|k) + \boldsymbol{K}^\varepsilon(k)\left[\varepsilon^c(k) - \mu_a^\varepsilon(k) - \boldsymbol{h}^\varepsilon{}_k(\hat{\boldsymbol{X}}^p(k|k)) - \frac{1}{2}\Delta^2(k)\right] \\ \boldsymbol{P}^\varepsilon(k|k) = (\boldsymbol{I}_n - \boldsymbol{K}^\varepsilon(k)\boldsymbol{H}^\varepsilon(k))\boldsymbol{P}^p(k|k) \end{cases} \tag{13.37}$$

where $\boldsymbol{H}^\varepsilon(k)$ is still the Jacobian matrix of $\boldsymbol{h}^\varepsilon{}_k[\boldsymbol{X}(k)]$ in the position of $\hat{\boldsymbol{X}}^p(k|k)$, that is,

$$\boldsymbol{H}^\varepsilon(k) = \left[L^1(k) + \hat{x}^p(k|k), \hat{\dot{x}}^p(k|k), L^2(k) + \hat{y}^p(k|k), \hat{\dot{y}}^p(k|k), \boldsymbol{0}_{1\times(n-4)}\right] \tag{13.38}$$

Meanwhile $\delta^2(k)$ is formed by the second-order derivative of $\boldsymbol{h}^\varepsilon{}_k[\boldsymbol{X}(k)]$, that is,

$$\delta^2(k) = 2P_k^p(1,3) + 2P_k^p(2,4) \tag{13.39}$$

and $A(k)$ is

$$\begin{aligned} A(k) = {} & P_k^p(1,1)P_k^p(2,2) + P_k^p(3,3)P_k^p(4,4) \\ & + 2P_k^p(1,3)P_k^p(2,4) + 2P_k^p(1,4)P_k^p(2,3) + \left[P_k^p(1,2)\right]^2 + \left[P_k^p(3,4)\right]^2 \end{aligned} \tag{13.40}$$

where $P_k^p(i,j)$ stands for the element which is in row i and column j of the position filtering error covariance matrix $\boldsymbol{P}^p(k|k)$.

- Step 4: final filtering estimation.

$$\begin{cases} \hat{\boldsymbol{X}}(k|k) = \hat{\boldsymbol{X}}^\varepsilon(k|k) \\ \boldsymbol{P}(k|k) = \boldsymbol{P}^\varepsilon(k|k) \end{cases} \tag{13.41}$$

13.3.3.2 Unbiased Sequential Unscented Kalman Filtering (USUKF) Algorithm

The way that unscented Kalman filters can be applied to deal with the estimation of strong nonlinear systems has been introduced in Section 4.3. Therefore, after the position measurement is filtered in the USEKF algorithm in Section 13.3.3.1, the UKF method can be used in place of the second-order EKF algorithm. The unbiased converted measurement filtering process of position measurement, and the decorrelation process of distance measurement and Doppler measurement, are the same as (13.21)–(13.34) in the USEKF algorithm. Then, the UKF algorithm can be used to deal with Doppler measurements [327, 328] under the following procedure.

- Step 1: update filtering estimation with time.

$$\begin{cases} \hat{\boldsymbol{X}}(k|k-1) = \boldsymbol{F}(k-1)\hat{\boldsymbol{X}}(k-1|k-1) + \boldsymbol{G}(k-1)\boldsymbol{u}(k-1) \\ \boldsymbol{P}(k|k-1) = \boldsymbol{F}(k-1)\boldsymbol{P}(k-1|k-1)[\boldsymbol{F}(k-1)]' + \boldsymbol{Q}(k-1) \end{cases} \tag{13.42}$$

- Step 2: update filtering estimation with position measurement.

$$\begin{cases} \boldsymbol{K}^p(k) = \boldsymbol{P}(k|k-1)[\boldsymbol{H}^{c,p}(k)]'\left[\boldsymbol{H}^{c,p}(k)\boldsymbol{P}(k|k-1)[\boldsymbol{H}^{c,p}(k)]' + \boldsymbol{R}_a^p(k)\right]^{-1} \\ \hat{\boldsymbol{X}}^p(k|k) = \hat{\boldsymbol{X}}(k|k-1) + \boldsymbol{K}^p(k)\left[z^{c,p}(k) - \boldsymbol{\mu}_a^p(k) - \boldsymbol{H}^{c,p}(k)\hat{\boldsymbol{X}}(k|k-1)\right] \\ \boldsymbol{P}^p(k|k) = (\boldsymbol{I}_n - \boldsymbol{K}^p(k)\boldsymbol{H}^{c,p}(k))\boldsymbol{P}(k|k-1) \end{cases} \tag{13.43}$$

- Step 3: update filtering estimation with pseudo measurement.
  i. Calculate $(2n_x+1)$ δ sampling points $\boldsymbol{\xi}_i$ and their corresponding weight values $\boldsymbol{W}_i$:

$$\begin{cases} \boldsymbol{\xi}_0(k|k) = \hat{\boldsymbol{X}}^p(k|k), \quad i=0 \\ \boldsymbol{\xi}_i(k|k) = \hat{\boldsymbol{X}}^p(k|k) + \left(\sqrt{(n_x+\lambda)}\sqrt{\boldsymbol{P}^p(k|k)}\right)_i, \quad i=1,\ldots,n_x \\ \boldsymbol{\xi}_{i+n_x}(k|k) = \hat{\boldsymbol{X}}^p(k|k) - \left(\sqrt{(n_x+\lambda)}\sqrt{\boldsymbol{P}^p(k|k)}\right)_i, \quad i=1,\ldots,n_x \end{cases} \tag{13.44}$$

where n_x is the dimension of the state vector, and here $n_x=4$; $\lambda = n_x(\alpha^2-1)$, and $\alpha=0.01$; $\left(\sqrt{(n_x+\lambda)\boldsymbol{P}^p(k|k)}\right)_i$ is column i in the mean square root matrix $(n_x+\lambda)\boldsymbol{P}^p(k|k)$.

The corresponding weight value $\boldsymbol{W}_i$ is

$$W_0^{(m)} = \frac{\lambda}{(n_x+\lambda)}, \qquad i=0 \tag{13.45}$$

$$W_0^{(c)} = \frac{\lambda}{(n_x+\lambda)} + 1 - \alpha^2 + \beta, \quad i=0 \tag{13.46}$$

$$W_i^{(m)} = W_i^{(c)} = \frac{1}{[2(n_x+\lambda)]}, \quad i=1,\ldots,2n_x \tag{13.47}$$

where $\beta=2$, superscript m stands for the weight value in the state update, and superscript c is the weight value in the covariance update.

ii. Measurement update

$$\varsigma_i(k|k) = \boldsymbol{h}_k^{\varepsilon}[k, \boldsymbol{\xi}_i(k|k)] \tag{13.48}$$

where $\boldsymbol{h}_k^{\varepsilon}$ is the measurement equation, then the predicted measurement and corresponding covariance are

$$\hat{\boldsymbol{Z}}^{\varepsilon}(k|k) = \sum_{i=0}^{2n_x} W_i \varsigma_i(k|k) \tag{13.49}$$

$$\boldsymbol{P}_{zz} = \boldsymbol{R}_a^{\varepsilon}(k) + \sum_{i=0}^{2n_x} W_i \Delta \boldsymbol{Z}_i^{\varepsilon}(k|k) \Delta \boldsymbol{Z}_i^{\varepsilon'}(k|k) \tag{13.50}$$

where $\Delta \boldsymbol{Z}_i^{\varepsilon}(k|k) = \varsigma_i(k|k) - \hat{\boldsymbol{Z}}^{\varepsilon}(k|k)$.

Likewise, the interactive covariance of measurements and state vectors can be obtained as

$$\boldsymbol{P}_{xz} = \sum_{i=0}^{2n_x} W_i \Delta \boldsymbol{X}_i^{\varepsilon}(k|k) \Delta \boldsymbol{Z}_i^{\varepsilon} i'(k|k) \tag{13.51}$$

where $\Delta \boldsymbol{X}_i^{\varepsilon}(k|k) = \boldsymbol{\xi}_i(k|k) - \hat{\boldsymbol{X}}^{p}(k|k)$.

iii. The state update and its covariance can be expressed as

$$\hat{\boldsymbol{X}}^{\varepsilon}(k|k) = \hat{\boldsymbol{X}}^{p}(k|k) + \boldsymbol{K}^{\varepsilon}(k)\left[\varepsilon^{c}(k) - \mu_a^{\varepsilon}(k) - \hat{\boldsymbol{Z}}^{\varepsilon}(k|k)\right] \tag{13.52}$$

$$\boldsymbol{P}^{\varepsilon}(k|k) = \boldsymbol{P}^{p}(k|k) - \boldsymbol{K}^{\varepsilon}(k)\boldsymbol{P}_{zz}\boldsymbol{K}^{\varepsilon'}(k) \tag{13.53}$$

$$\boldsymbol{K}^{\varepsilon}(k) = \boldsymbol{P}_{xz}\boldsymbol{P}_{zz}^{-1} \tag{13.54}$$

- Step 4: final filtering estimation.

$$\begin{cases} \hat{\boldsymbol{X}}(k|k) = \hat{\boldsymbol{X}}^{\varepsilon}(k|k) \\ \boldsymbol{P}(k|k) = \boldsymbol{P}^{\varepsilon}(k|k) \end{cases} \tag{13.55}$$

13.3.3.3 Doppler Measurement Unscented Kalman Filtering (DUKF) Algorithm

It can be concluded from the measurement equation of Doppler radars (13.23) that every measurement, especially Doppler measurement, of the measurement vector is nonlinear. As for the nonlinearity of range and position measurements, converted measurement Kalman filters (CMKFs) are usually adopted. But they cannot be used to deal with Doppler measurements. Therefore, the sequential filtering method should be used after the CMKF to deal with the filtering of Doppler measurements. Since the UKF is not sensitive to the strength of a system's nonlinearity, the posterior mean and covariance from filtering can both be accurate to second order. Thus, the UKF method can be applied directly to filter the measurement vectors composed of $\rho^m(k)$, $\theta^m(k)$, and

$\dot{\rho}^m(k)$, which avoids converted position measurement filtering and the decorrelation process between range measurements and Doppler measurements. The specific implementation procedures are as follows (see Section 4.3). First of all, calculate $(2n_x+1)$ δ sampling points and their weight values, and obtain the one-step prediction of point δ according to the state equation. Then, calculate the predicted measurement by the measurement equation. Finally, update states and state covariance with the measurements provided by sensors.

What should be emphasized is that the measurement vector discussed in this part includes range, position, as well as Doppler measurement. Assume that the state estimation vector and the state estimation covariance of the tracking system at time k are $\hat{\boldsymbol{X}}(k|k)$ and $\boldsymbol{P}(k|k)$, respectively, then the covariance matrix of measurement noise $\boldsymbol{v}^m(k)$ in polar coordinates is

$$\boldsymbol{R}(k)=\text{cov}[\boldsymbol{v}^m(k)]=\begin{bmatrix}\sigma_\rho^2 & 0 & r\sigma_\rho\sigma_{\dot{\rho}}\\ 0 & \sigma_\theta^2 & 0\\ r\sigma_\rho\sigma_{\dot{\rho}} & 0 & \sigma_{\dot{\rho}}^2\end{bmatrix} \tag{13.56}$$

13.3.3.4 Unscented Kalman Filtering Algorithm for Maneuvering Targets

In the maneuvering target tracking case, the interactive multiple model is a practical multiple-model estimation algorithm. The IMM-based UKF algorithm can be obtained by a combination of IMM and UKF on the basis of the establishment of UKF multi-filter models. Furthermore, the IMM-based DUKF algorithm can be obtained by applying the UKF, which has splendid nonlinearity, to the case of maneuvering targets with Doppler measurements.

The IMM-DUKF algorithm has some obvious advantages [329] over the IMM-UKF algorithm (without Doppler measurement).

- *Tracking performance:* the IMM-DUKF estimator can diminish the errors in position and velocity estimation during the maneuvering and non-maneuvering period of the target.
- *Maneuvering sensitivity:* the IMM-DUKF estimator has a faster speed of response when the target is taking maneuvering flight.
- *Data association:* the IMM-DUKF estimator has a smaller error covariance matrix, thus decreasing the number of false associations.

The basic procedures for the IMM-DUKF algorithm are:

1. Establish the UKF model sepatately for the constant velocity (CV), constant acceleration (CA), and constant turn rate (CT) moving modes.
2. Establish the IMM algorithm formed by the above-mentioned models. For details of the mathematical model of IMM, see Section 9.3.5.
3. Introduce Doppler measurements; extend measurement $z=[x,y]$ into $z=[x,y,\dot{\rho}]$ for filtering estimation. For details, refer to the previous two sections.

13.4 Performance Analysis on PD Radar Tracking Algorithms

13.4.1 Simulation Environments and Parameter Settings

Simulation environment 1. Suppose that the values of the initial state of the target $X(0)$, radar ranging error σ_ρ, angle measurement error σ_θ, and Doppler velocity measurement error $\sigma_{\dot{\rho}}$ are as shown in Table 13.1. The correlation coefficient of range measurement and Doppler velocity measurement $\gamma = 0.5$, and the scanning period $T = 5$ s.

Then, the CMKF, UCMKF, EKF, USEKF will be compared against each other. For every algorithm, the simulation will be conducted 500 times, 30 steps each.

Simulation environment 2. Suppose that the initial state of the target $X(0) = [10\,000, 30, 10\,000, 30]$ (m, m/s, m, m/s), ranging error $\sigma_\rho = 150$ m, angle measurement error $\sigma_\theta = 5$ mrad, and scanning period $T = 5$ s. The values of the correlation coefficient γ of Doppler velocity, measurement error $\sigma_{\dot{\rho}}$, and range and Doppler measurement are shown in Table 13.2.

The UKF, USEKF, USUKF, and DUKF will be compared below. For every algorithm, the simulation will be conducted 500 times, 30 steps each.

Simulation environment 3. Consider a target which is moving with initial state $X(0) = [10\,000, -160, 2000, 50]$ (m, m/s, m, m/s). It has moved for 100 s in a 2D surface. It was in constant motion within the periods 0–20 s and 40–60 s, respectively, in constant turning within the periods 20–40 s and 60–80 s, and has been in constant acceleration within the period 80–100 s. The angle velocities within the periods 20–40 s and 60–80 s are 10°/s and −10°/s, respectively, and the accelerations in two directions are, respectively, 5 m/s^2 and 5 m/s^2 within the period 80–100 s. The ranging error $\sigma_\rho = 150$ m, angle measurement error $\sigma_\theta = 5$ mrad, scanning period $T = 5$ s, and the correlation coefficient of Doppler velocity measurement error $\sigma_{\dot{\rho}}$ and range and Doppler measurement $\gamma = 0.1$.

Then the IMM-DUKF and IMM-UKF algorithms will be compared with each other. For every algorithm the simulation will be conducted 500 times, 30 steps each. The true movement track of the target is shown in Figure 13.13.

Table 13.1 Simulation environment 1 parameter settings

No.	$X(0)$ (m, m/s, m, m/s)	σ_ρ (m)	σ_θ (mrad)	$\sigma_{\dot{\rho}}$ (m/s)	Simulation figure
1	[10 000, 30, 10 000, 30]	150	5	1	Figure 13.4
2	[100 000, 30, 100 000, 30]	150	5	1	Figure 13.5
3	[10 000, 30, 10 000, 30]	150	5	100	Figure 13.6
4	[10 000, 30, 10 000, 30]	50	5	1	Figure 13.7
5	[100 000, 30, 100 000, 30]	150	40	1	Figure 13.8

Table 13.2 Simulation environment 2 parameter settings

No.	$\sigma_{\dot{\rho}}$ (m/s)	γ	Simulation figure
1	0.01	0.9	Figure 13.9
2	0.01	0.1	Figure 13.10
3	10	0.1	Figure 13.11
4	10	0.9	Figure 13.12

The performance of these algorithms can be compared using the RMS position and velocity error:

$$\text{RMS Pos}_k = \sqrt{\frac{1}{M}\sum_{i=1}^{M}\left(x_k^i - \hat{x}_{k|k}^i\right)^2 + \left(y_k^i - \hat{y}_{k|k}^i\right)^2} \tag{13.57}$$

$$\text{RMS Vel}_k = \sqrt{\frac{1}{M}\sum_{i=1}^{M}\left(\dot{x}_k^i - \hat{\dot{x}}_{k|k}^i\right)^2 + \left(\dot{y}_k^i - \hat{\dot{y}}_{k|k}^i\right)^2} \tag{13.58}$$

and the average normalized estimation error square [330]

$$\text{ANEES}_k = \frac{1}{Mn}\sum_{i=1}^{M}\left(x_k^i - \hat{x}_{k|k}^i\right)^T\left(P_{k|k}^i\right)^{-1}\left(x_k^i - \hat{x}_{k|k}^i\right) \tag{13.59}$$

13.4.2 Simulation Results and Analysis

The simulation results of environment 1 are shown in Figures 13.4–13.8, those of environment 2 in Figures 13.9–13.12, and those of environment 3 in Figure 13.14.

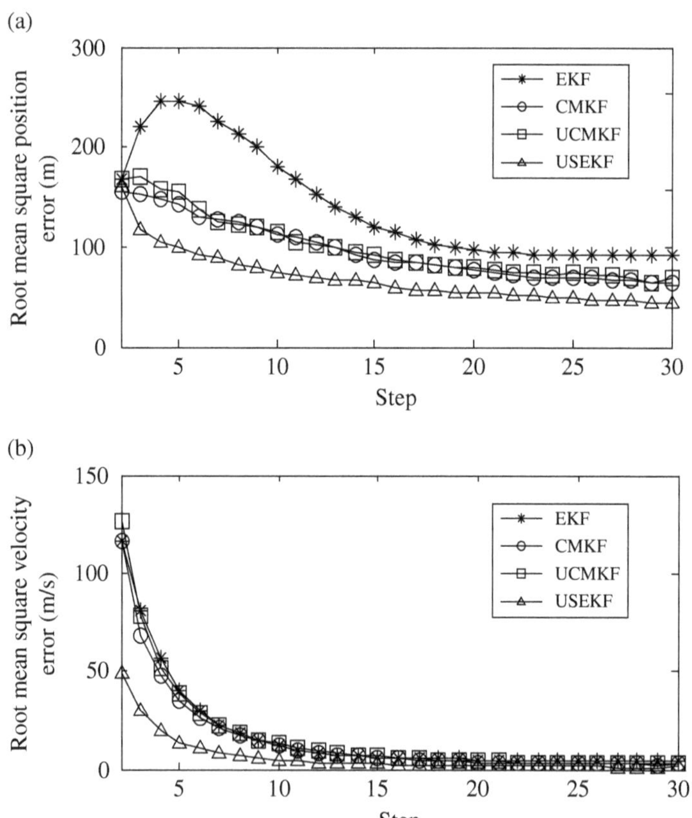

Figure 13.4 (a) RMS position error and (b) RMS velocity error (environment 1, case 1)

As suggested in Figure 13.4, the USEKF algorithm using Doppler measurements has better target tracking performance: both the RMS position error and the RMS velocity error of targets are obviously reduced; the EKF algorithm is largely influenced by the initial states, and shows more position errors in the first few steps, which converge rapidly later on, and has more errors on the whole.

By comparing the simulation condition of Figure 13.5 with that of Figure 13.4, it can be seen that the USEKF algorithm cannot actually improve its tracking accuracy when there is a long distance between the radar and its target. Furthermore, Figure 13.4 illustrates that the tracking accuracy of the USEKF is close to those of other algorithms as a target is going away from the radar.

A comparison of simulation conditions between Figures 13.6 and 13.4 demonstrates that the USEKF algorithm does not show high tracking accuracy either in the presence of low-accuracy Doppler velocity meansurements. This means that the Doppler velocity message, influenced by its own accuracy, cannot necessarily improve the target tracking accuracy.

From the comparison of Figures 13.7 and 13.4, it is seen that the tracking accuracy of the USEKF is not greatly different from those of other algorithms where the accuracy of target ranging is high. This result, combined with the simulation of Figure 13.6, leads to a conclusion that the proportional relationship between velocity measuring accuracy and ranging accuracy has an influence on the behavior of the radial velocity message in improving tracking accuracy.

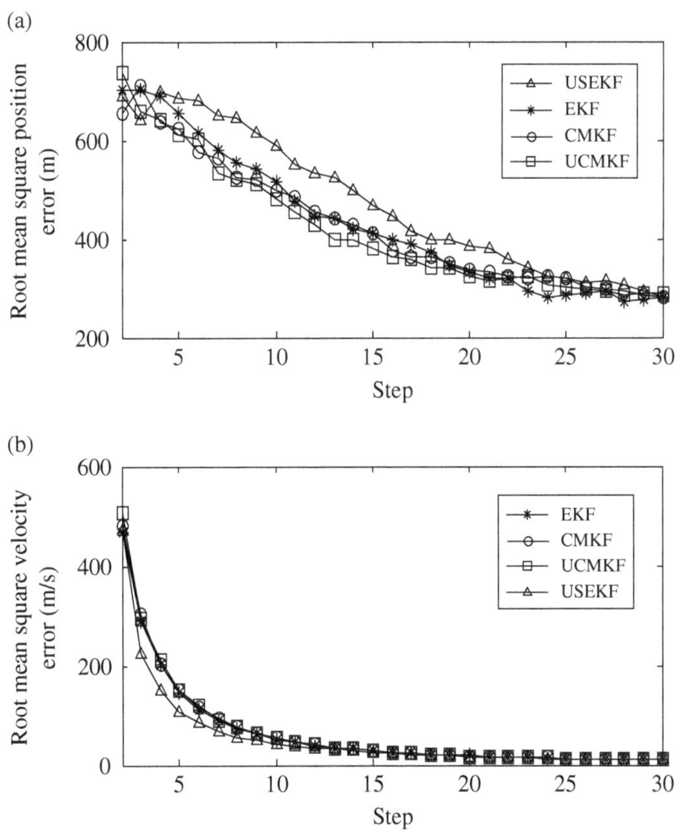

Figure 13.5 (a) RMS position error and (b) RMS velocity error (environment 1, case 2)

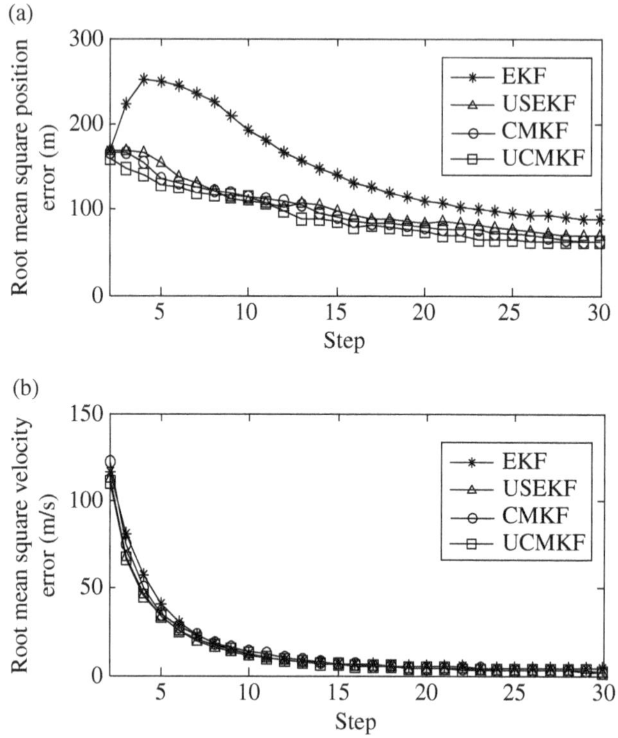

Figure 13.6 (a) RMS position error and (b) RMS velocity error (environment 1, case 3)

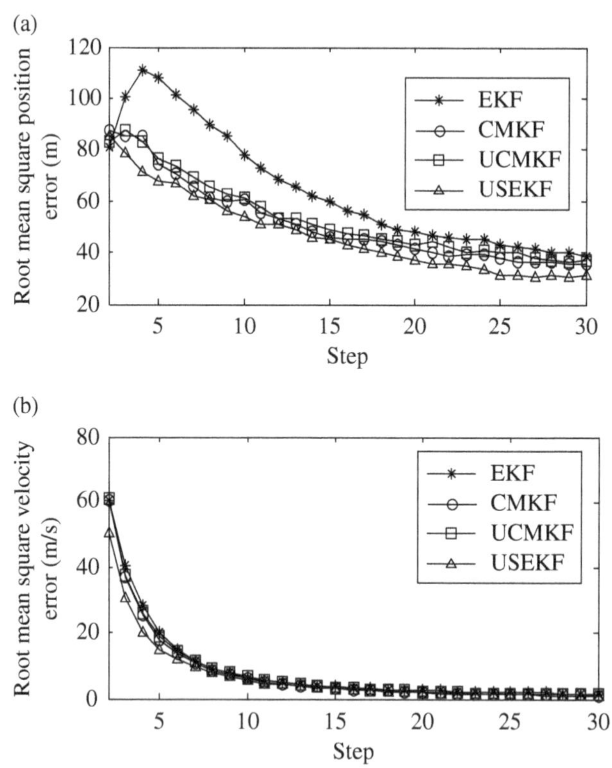

Figure 13.7 (a) RMS position error and (b) RMS velocity error (environment 1, case 4)

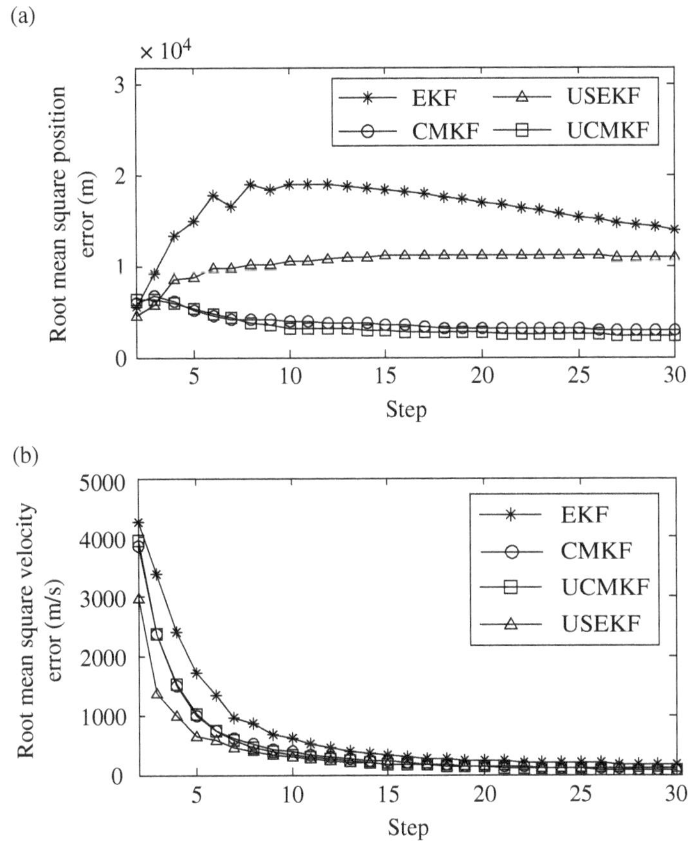

Figure 13.8 (a) RMS position error and (b) RMS velocity error (environment 1, case 5)

As indicated by the comparison of Figures 13.8 and 13.5, in the case of significant long-distance angle-measuring error, the UCMKF shows the highest tracking accuracy, followed by the CMKF, USEKF, and EKF sequentially. Therefore, it can be concluded that under such a simulation condition, the use of a velocity measurement message cannot improve the tracking accuracy, and the UCMKF algorithm displays the best tracking performance while the EKF displays the worst.

Figure 13.9 reveals that the algorithms using a Doppler message can improve their accuracy of position and velocity estimation when the correlative coefficient between ranging measurement and Doppler measurement is quite large and the error of Doppler velocity measuring is small. However, the direct filtering methods of DUKF and UKF show a better estimation consistency than the sequential filtering methods USEKF and SUKF.

By comparing Figure 13.10 with Figure 13.9, it can be seen that the algorithms using a Doppler message can also improve their accuracy of position and velocity estimation when both the error of Doppler velocity measuring and the correlative coefficient between the ranging measurement and the Doppler measurement are quite small. However, the estimation consistency of the sequential filtering methods USEKF and SUKF has been somewhat improved.

It can be discovered from Figure 13.11 that the algorithms using a Doppler message cannot apparently improve their accuracy of position and velocity estimation when the correlative coefficient

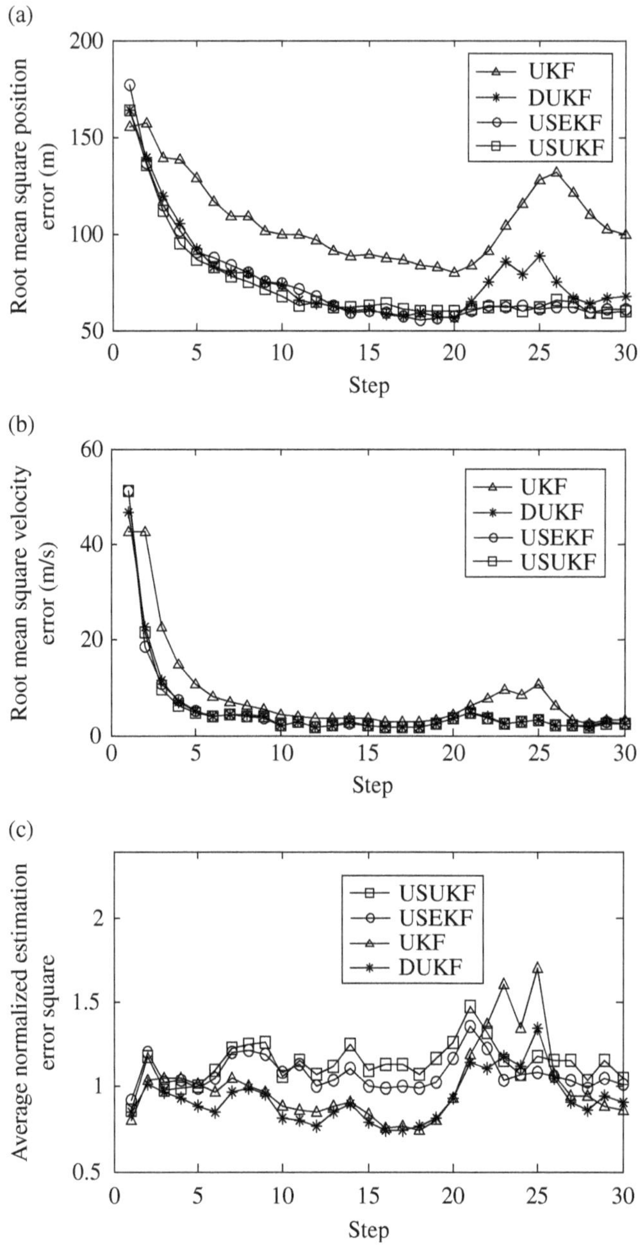

Figure 13.9 (a) RMS position error, (b) RMS velocity error, and (c) average normalized estimation error squared (environment 2, case 1)

between the ranging measurement and the Doppler measurement is quite small and the error of the Doppler velocity measurement is quite large. The estimation consistencies of all the methods are close to one another.

It is seen from Figure 13.12 that when both the error of Doppler velocity measuring and the correlative coefficient between ranging measurement and Doppler measurement are quite large, the

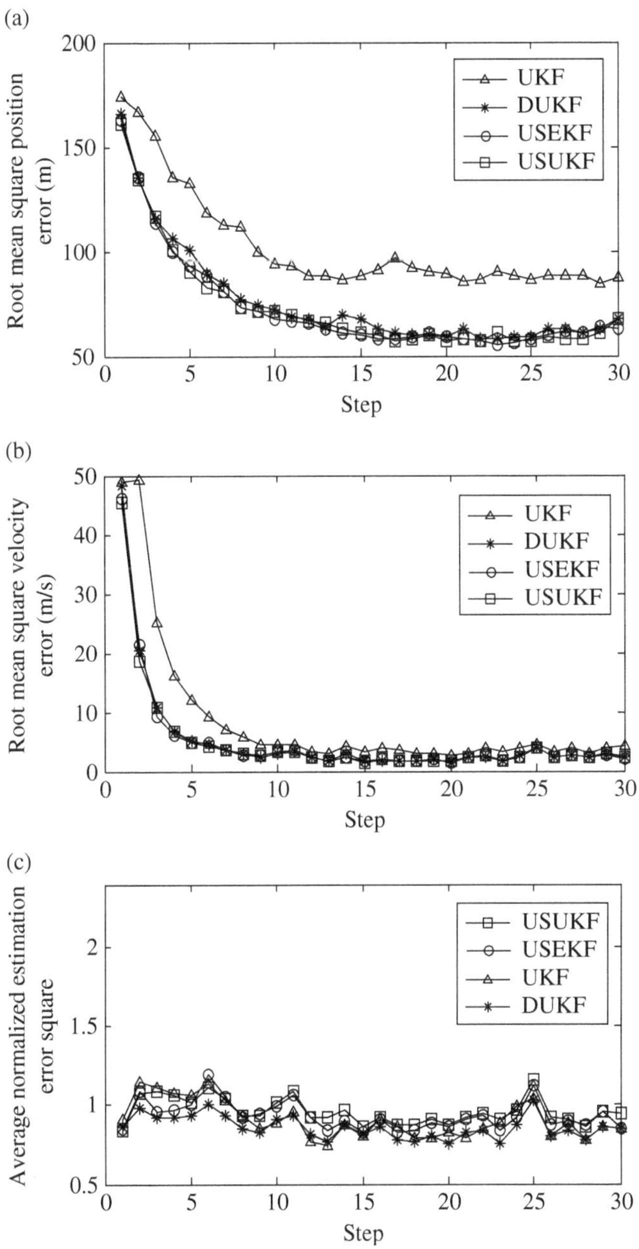

Figure 13.10 (a) RMS position error, (b) RMS velocity error, and (c) average normalized estimation error square (environment 2, case 2)

algorithms using a Doppler message cannot improve their estimation accuracy of the target's state. Instead, they have poorer performance and worse estimation consistency than UKF algorithms without use of Doppler measurements.

It can be deduced from Figures 13.9–13.12 that the accuracy of Doppler measurements should be considered when choosing the algorithm for system design. To be more specific, the accuracy of

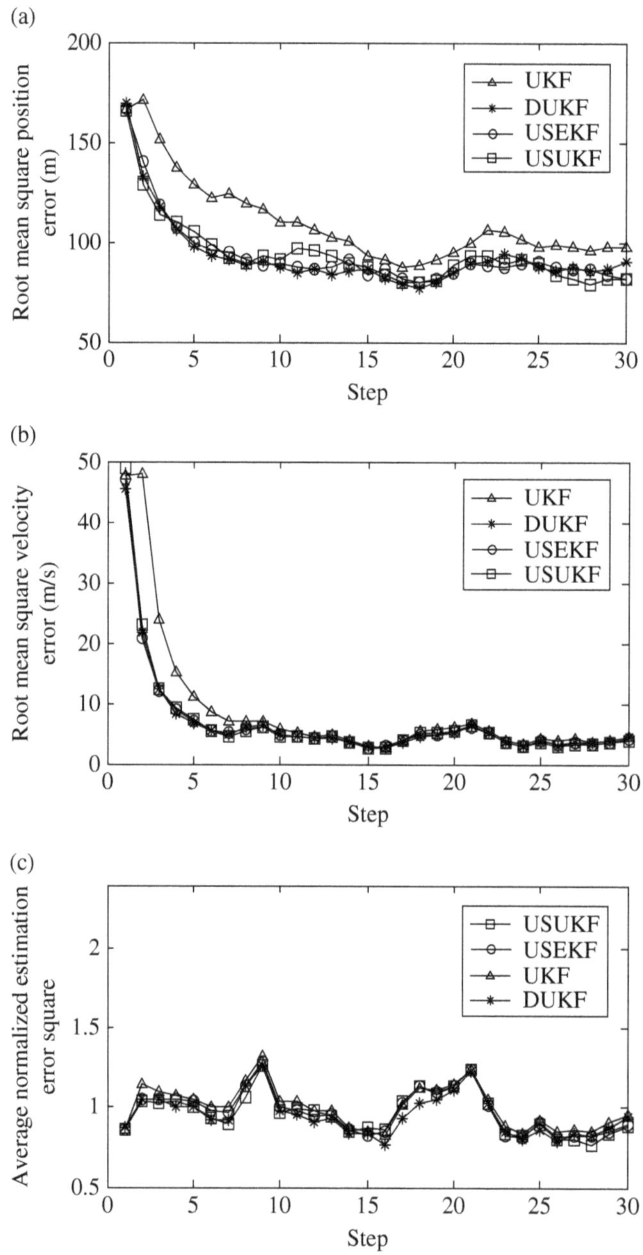

Figure 13.11 (a) RMS position error, (b) RMS velocity error, and (c) average normalized estimation error square (environment 2, case 3)

target tracking can be improved by using Doppler measurements which show high accuracy. On the contrary, the accuracy of target tracking might be reduced by low-accuracy Doppler measurements and by strong correlations between range measurements and Doppler measurements. Doppler measurements are inappropriate in such cases, so further research will focus on how to quantitatively describe the conditions of using Doppler measurements.

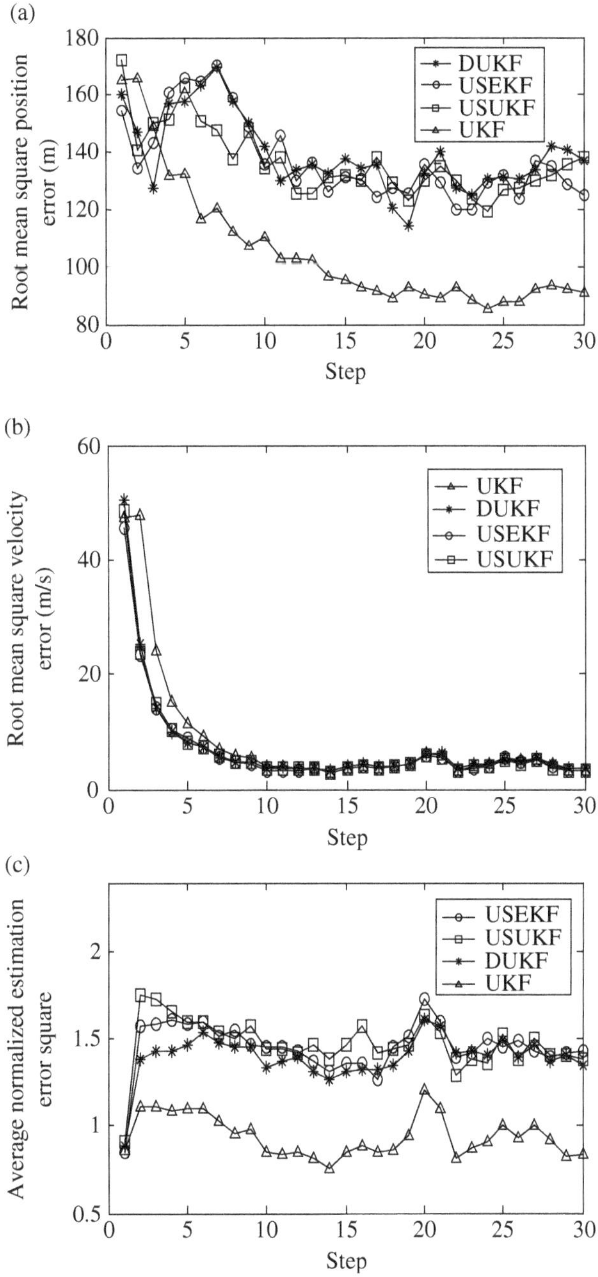

Figure 13.12 (a) RMS position error, (b) RMS velocity error, and (c) average normalized estimation error square (environment 2, case 4)

As shown by Figure 13.14, if the IMM and UKF are combined to track the target under the environment of Figure 13.13, the IMM-DUKF algorithm can apparently improve its performance of target tracking, including position and velocity tracking accuracy, maneuver response time, and maneuver tracking stability.

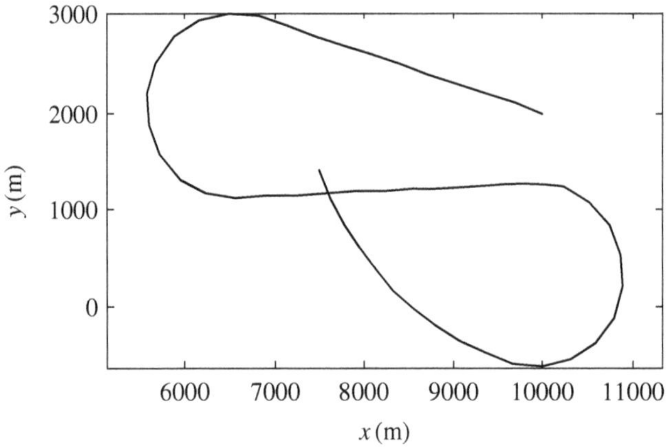

Figure 13.13 True movement track of target (environment 3)

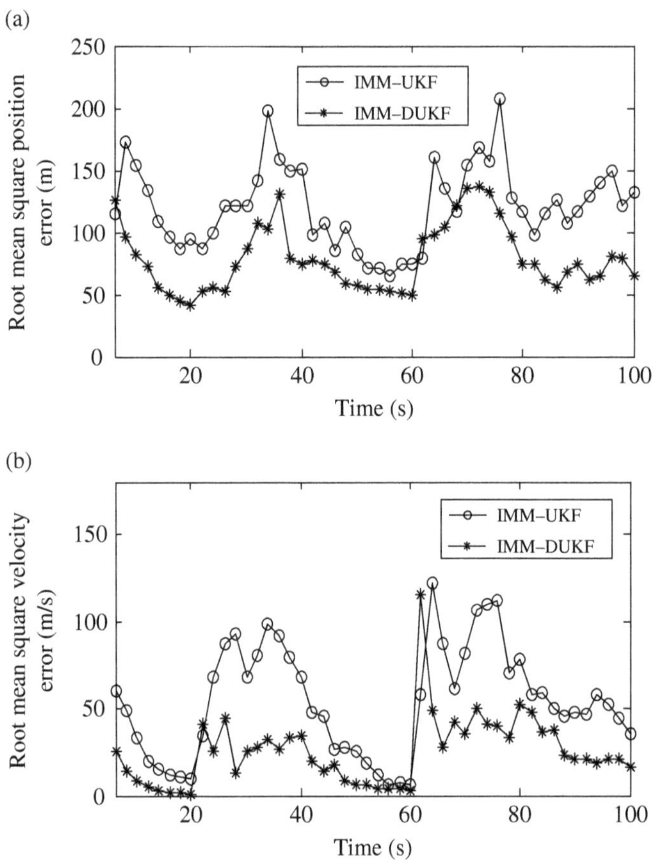

Figure 13.14 (a) RMS position error and (b) RMS velocity error (environment 3)

13.5 Summary

This chapter mainly discusses the data processing technique of PD radar. On the basis of the introduction of PD radar systems, it presents the characteristics of PD radars and the composition of their tracking system. Then, it focuses on certain typical tracking algorithms of PD radar, namely, the USEKF, USUKF, and DUKF algorithms. It also provides a comparative analysis, with respect to two simulation environments, of the requirements of Doppler measurements for improving the accuracy of target tracking and of the performance of several Doppler measurement processing algorithms. Finally, it addresses the problem of tracking maneuvering targets by combining interactive multiple models with the UKF; the simulation results show that the introduction of Doppler measurements can apparently improve the performance of target tracking.

14

Phased Array Radar Data Processing

14.1 Introduction

Phased array radar represents a significant development in radar systems, eliminating the restrictions imposed on conventional radars, such as fixed beam sojourn time, fixed scan modes, fixed emission power, and data rates. With its flexible and swift beam scanning, controllable space power assignment, and time resource assignment, phased array radar shows one important feature – multi-functionality – to conduct scanning and accurate tracking of multiple targets simultaneously.

First, this chapter introduces the main features, system structure, and working process of phased array radars. Second, it puts emphasis on the single/multi-target tracking algorithm in clutter, adaptive sampling period algorithm, and real-time task scheduling strategy in phased array radar data processing.

The first problem that needs to be solved for target tracking in the presence of clutter is data association. Therefore, how to improve the performance of single/multi-target data association using Doppler information obtained from a phased array radar is a hot topic for research. Phased array radars, with their unique variable sampling period ability, are able to adaptively adjust their sampling periods based on the maneuvering state of the target, as a result of which their stability in maneuver tracking increases, accompanied by reductions in their maneuver-free tracking load.

Phased array radars are multifunctional, that is, they are supposed to be capable of executing various tasks such as scanning, tracking, and identification alternately for each of the multiple targets at the same time. Which strategy should be used to schedule these many different tasks for these many targets efficiently in real time? The choice of an appropriate task scheduling strategy (i.e., with flexible, efficient, real-time performance) is another key technique for the data processing of phased array radars.

Finally, this chapter gives a detailed simulation analysis of several adaptive sampling period algorithms of phased array radar.

Radar Data Processing with Applications, First Edition. He You, Xiu Jianjuan, and Guan Xin.

14.2 Characteristics and Major Indexes

14.2.1 Characteristics

With its distinctive capability for beam agility, the phased array radar has many advantages such as multiple functions, multi-target interception and tracking, and adaptive capability. Controlled by computer, it can alter the relevant technical parameters adaptively to fit the changing environment. Specifically, it can select the working modes and technical parameters as needed to accomplish various missions such as target detection, tracking, orbit measuring, cataloging, prediction, and identification.

Compared with conventional radar, the phased array radar has the following technical characteristics [331–334].

1. *Mutual independence between surveillance (scanning) and tracking.* The phased array radar uses the sequential detection method to detect the target. Therefore, it switches to the track mode based on the result of a single detection in the surveillance (scanning) mode, which decided whether the target is in existence (a track has already been initiated) or is a new track. The data rate and beam sojourn time of target tracking are self-adaptive. In the track mode, a phased array radar can conduct the estimation of target track parameters (such as range, angle, and radial velocity) with optimal signal waves, pulse repetition frequencies, and polarization formats, thus enhancing its function of tracking miniature targets in complicated environments.
2. *Function of tracking and processing multiple maneuvering targets.* Compared with ordinary radars, the phased array radar has an incomparable function in tracking and processing multiple maneuvering targets, which relies on the swift agility of its antenna, and its powerful function of filtering tracking and multi-target association processing as well.

   The modern multi-target tracking technique is an integration of data association processing and modern filtering theories. Multi-target tracking includes the formation of association gates, data association, and the initiation, maintenance, and termination of tracks, among which data association is the key and difficult point. The association gate technique, a procedure of association processing essentially, is usually adopted in applications of phased array radar data processing. The space resolution of a radar is decided by its association scope instead of the beam width and bandwidth after the association gate technique is applied.
3. *High data rate.* Since the data rate is only restricted by a beam's sojourn time, the valid data rate of phased array radars can be greatly increased by reducing the time interval of their surveillance (scanning) or increasing the number of parallel channels, on the basis of the flexibility of phased array antennas' beam scanning and mutual independence between surveillance (scanning) and tracking functions.
4. *Power adaptive management.* The phased array radar is capable of exerting optimal control over its transmitting power according to the state/feature and environment of the target, and its own working mode to ensure best possible performance within power and time constraints, and adaptability to the instantaneous requirements of various tasks.
5. *High resolution.* In order to obtain more information from target echoes, the phased array radar can control the beam sojourn time to have better angle and Doppler resolution for the facilitation of target classification and identification.
6. *Space filtering.* The phased array antenna is able to have adaptive zero-point control, especially with the digital beam forming (DBF) technique. The combination of signal processing and antenna techniques leads to the formation of beams of various special shapes and an enhancement of the radars' capability to suppress jamming, clutter, and multi-path effects.

7. *Multiple functions.* The phased array radar can work in the track-while-scan mode based on the time division principle, and implement the scanning, tracking, guiding, and identification of friend or foe (IFF) of multiple batches of targets at the same time by changing parameters such as beam shape, beam sojourn time, signal form, and pulse repetition frequency. In that case, one multiple functional phased array radar could be used in place of several specialized conventional radars.

14.2.2 Major Indexes

The indexes of phased array radars can be divided into two categories: tactical and technical. This section only involves the tactical index, which is closely related to radar data processing.

1. *Radar surveillance airspace.* Includes minimum and maximum detection ranges; azimuth and elevation surveillance scopes; radar tracking, steering, and guidance ranges.
2. *Radar measuring index and measuring accuracy.* The measuring indexes include those which indicate the coordinate position of the target, such as range, azimuth, and elevation, and those which indicate the movement features of the target, for instance, velocity and acceleration. The measuring indexes of special-purpose radars may also contain those indicating such characteristics of the target as the amplitude of fluctuation, and spectral and polarization characteristics of target signals.
3. *Resolution.* Covers range, azimuth, elevation resolution, and velocity resolution.
4. *Capability to process multiple targets.* Mainly includes the capability to track multiple batches of targets in real time and process track associations of multiple batches of targets.
5. *Radar survivability.* Includes such requirements for radar as anti-jamming, anti-ARM, and anti-bombing capabilities.

One advantage that the phased array radar has is the ability to process multiple targets, so the phased array technique is the sole solution to scanning the whole surveillance region while tracking under condition of multiple batches of targets.

14.3 Structure and Working Procedure

14.3.1 Structure

The phased array radar system is similar to the traditional mechanical scanning radar in that it is composed of a transmitter, a receiver, and a data processor. But it also has unique devices such as an array antenna, T/R component, received multi-beams forming network, radar controller, and beam controller. The typical structure of a phased array radar system is shown in Figure 14.1. The antenna may work as a T/R specified antenna or a community antenna, and the latter is shown in Figure 14.1. The received multi-beam forming network can work with overlapping received multiple beams, or sum/difference beams required by monopulse angle measurement. The beam controller is peculiar to the phased array radar, which serves as a replacement for the servo-driven unit in traditional mechanical scanning radars. It receives instruction from the radar controller, and solves the wave control codes for the phase shifter on each antenna unit. With these codes, the phase shifter controls the beam pointing of the array antenna [335].

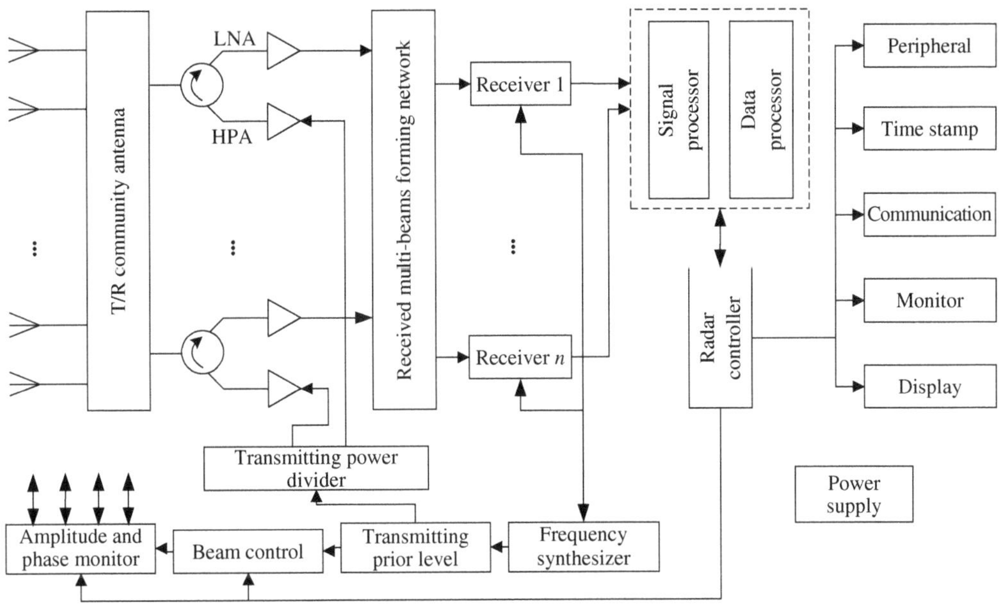

Figure 14.1 Structure of the phased array radar system

The radar controller, also called the "radar scheduler" or "center computer," is the control center of a phased array radar system, which is in control of the operation of other subsystems, and in charge of the management of the whole system's task execution, working modes and parameters, etc. The scan of a specified airspace can be conducted by this controller, which controls the transmitted waveform and the received beam according to pre-written programs. When the target is captured (i.e., under the control of the controller), the track of a new target will be set up by the controller which, at the same time, manages multiple targets being tracked, thus fulfilling the scan-and-track function of the radar. In the case of track loss, the controller should manage and accomplish the complementary illumination of the target in order to maintain its track. Alternatively, it may perform self-adaptive power management, and change the waveform of the transmitted signal, the signal repetition frequency, and the number of transmitted pulses based on the intensity of the target's echo signal.

14.3.2 Working Procedure

The general working procedure of a phased array radar is as follows.

First, the echo signal of the target received by its antenna should be processed by the receiving network and the receiver before being sent into the signal/data processor where the detection, measurement, association, filtering, and prediction of the target are completed.

Then, based on the above results, including the pointing, transmitting time, frequency, working waveform, and sojourn time of the beam, the radar controller generates sojourn commands for the beam, which are sent into the transmitter and beam controller.

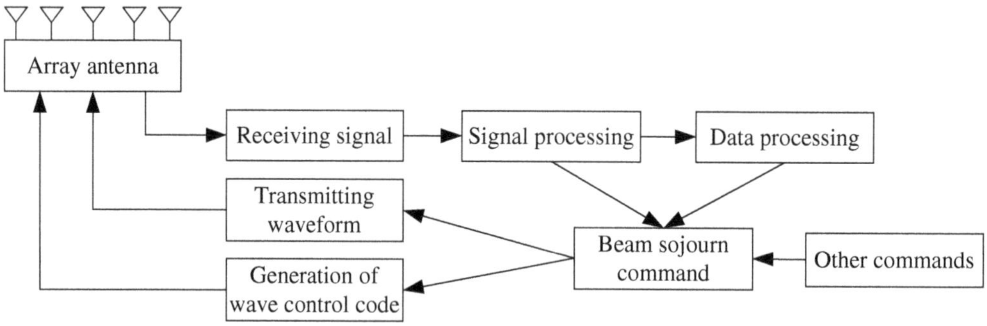

Figure 14.2 Working flowchart of phased array radar

After that, following these commands, the transmitter produces corresponding working waveforms, which will be sent via the transmitting network into the array antenna, where they radiate out.

At the same time, the beam controller calculates, according to the beam angle, the wave control code needed by the phase shifter, which in turn controls the beam direction of the array antenna to execute the task commands given by the radar controller.

Thus, a closed loop of task processing is formed. The specific procedure is shown in Figure 14.2.

After spotting a target by scanning, the phased array radar has to continue searching for other targets while tracking this one. Therefore, it has two typical working modes: track-while-scan (TWS) and track-and-scan (TAS).

14.3.2.1 TWS Working Mode

This working mode is usually adopted by the 1D phased array scanning radar. For radars of this type, the tracking sampling interval is the same as the scanning one (i.e., the tracking data rate is the same as the scanning one). The TWS can easily be controlled while its tracking data rate is not high enough to meet the demands of high-precision tracking or guiding, thus failing to bring the best out of the phased array radar.

14.3.2.2 TAS Working Mode

For radars working in this mode, the data rate of tracking is different from that of scanning. Therefore, a relatively long time interval should be used for scanning to lower the load of the radar, while a relatively low tracking data rate is needed to ensure certain tracking precision and stability. Actually, this problem can be solved effectively by putting tracking tasks into the intervals of scanning ones, which also causes a significant increase in the time utilization ratio of the phased array radar.

14.4 Data Processing

With its technical features such as swift scanning of antenna beams, adjustable beam forms, and synthetic space power, the phased array radar can work in various modes. To be more specific, it can, by enhancing the signal power at some key directions, conduct the "burn through" working mode, widening the scanning or tracking range [336], besides scanning, target interception, and multi-target tracking. This section focuses on the discussion of specific problems in data processing of phased array radars, which mainly covers single/multi-target tracking algorithms of phased array radars in clutter, adaptive sampling period algorithms, and real-time task scheduling strategy.

The application of a multi-target tracking system in phased array radars can accomplish the tracking of single or multiple targets and better realize the capability for multi-target tracking systems, especially with its beam agility and controllability. The basic principle of single-radar multiple-target tracking is shown in Figure 14.3. After a target is detected by the radar, the plot extractor extracts information about the target's position, which forms measurements (known as plots). The new measurements, after pretreatment, perform data associations with the existing track, and the associated ones are used to update the corresponding track information (filtering) while the unassociated ones are used to initiate a new track. A track that fails to be associated with the measurements several successive times should be terminated. Therefore, it can be concluded that the key techniques for multi-target tracking include those of track initiation and termination, data association between measurements and tracks, and tracking filtering [337–339].

The radial velocity information provided by the phased array radar can be used to enhance its tracking capability [41, 340]. Specifically, to accelerate initialization, increase the estimation accuracy of the target's parameters (particularly when the target shows a dramatic change), and decrease the ambiguity about the association between measurements and tracks in multiple-echo cases.

14.4.1 Single-Target-in-Clutter Tracking Algorithms

The IMM is an effective maneuvering target tracking algorithm for targets with complicated movement capability, but it must first handle the data association of measurements and tracks in the presence of clutter. The probability data association filter (PDAF), as one of the typical algorithms to solve data association, can conduct target tracking in clutter [341, 342] by combining maneuvering target tracking algorithms and data association algorithms. The phased array radar is capable of providing Doppler information (i.e., radial velocity measurements) of the target. These measurements, when added to the procedure of phased array radars for data processing, can enhance the target tracking performance. This section focuses on the IMM-PDAF algorithm with Doppler radial velocity.

14.4.1.1 Observation Equation with Doppler Radial Velocity

If it is two-dimensional, then

$$Z(k) = H(k)X(k) + W(k) = [x(k)\ y(k)\ v(k)]' + W(k)$$

$$= \begin{bmatrix} \rho(k)\cos(\theta(k)) \\ \rho(k)\sin(\theta(k)) \\ \dot{\rho}(k) \end{bmatrix} + W(k) = \begin{bmatrix} 1 & 0 & 0 & 0 \\ 0 & 0 & 1 & 0 \\ 0 & \cos(\theta(k)) & 0 & \sin(\theta(k)) \end{bmatrix} \begin{bmatrix} x \\ \dot{x} \\ y \\ \dot{y} \end{bmatrix} + W(k) \tag{14.1}$$

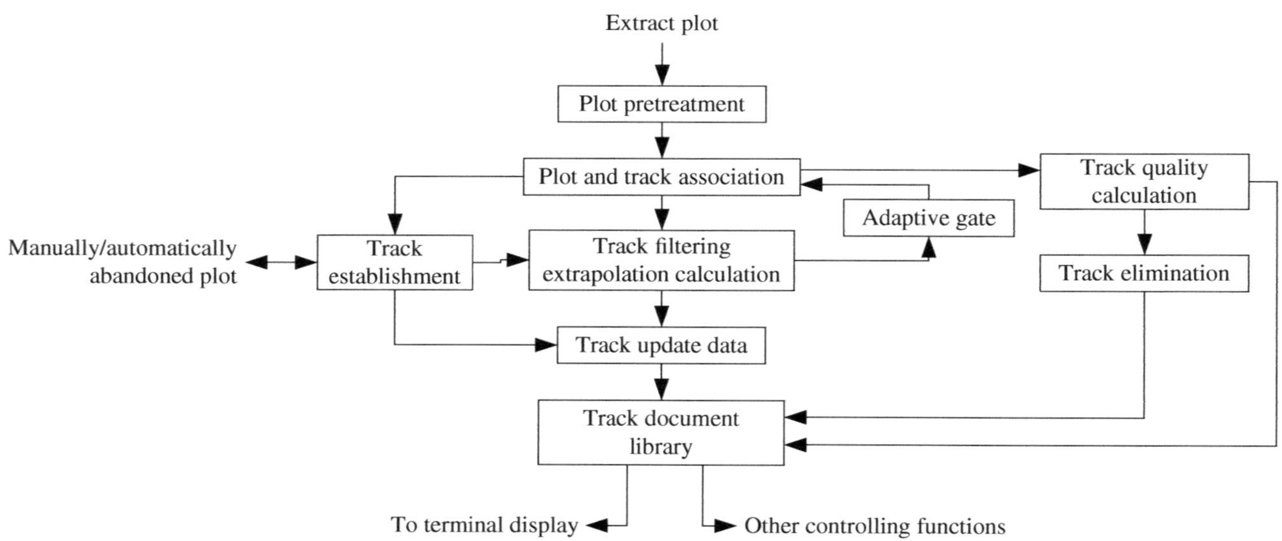

Figure 14.3 Multi-target tracking principle

where the covariance matrix of $\boldsymbol{W}(k)$ is

$$\boldsymbol{R} = \begin{bmatrix} \sigma_x^2 & \sigma_{xy} & 0 \\ \sigma_{xy} & \sigma_y^2 & 0 \\ 0 & 0 & \sigma_v^2 \end{bmatrix} \tag{14.2}$$

which indicates that this measurement equation shows nonlinearity for this state model, therefore, the corresponding optimal filtering algorithm is nonlinear as well. In order to reduce the complexity of nonlinearity, let x, y, and $\varepsilon = \rho(k)\dot{\rho}(k)$ take the place of x, y, and $\dot{\rho}(k)$, then

$$\boldsymbol{Z}^1(k) = \boldsymbol{H}^1(k)\boldsymbol{X}^1(k) + \boldsymbol{W}^1(k) = \begin{bmatrix} \rho(k)\cos(\theta(k)) \\ \rho(k)\sin(\theta(k)) \\ \rho(k)\dot{\rho}(k) \end{bmatrix} + \boldsymbol{W}^1(k) \tag{14.3}$$

Hence, the covariance matrix [128] of $\boldsymbol{W}^1(k)$ is

$$\boldsymbol{R}^1 = \begin{bmatrix} \sigma_x^2 & \sigma_{xy} & \sigma_{x\varepsilon} \\ \sigma_{xy} & \sigma_y^2 & \sigma_{y\varepsilon} \\ \sigma_{x\varepsilon} & \sigma_{y\varepsilon} & \sigma_\varepsilon^2 \end{bmatrix} \tag{14.4}$$

Although the observing error of variable ε is not Gaussian, it can still be supposed to follow the equivalent zero-mean Gaussian distribution with variance

$$\sigma_\varepsilon^2 = \sigma_\rho^2 \sigma_{\dot{\rho}}{}^2 + [\rho(k)]^2 \sigma_{\dot{\rho}}{}^2 + \dot{\rho}(k)\sigma_\rho^2 \tag{14.5}$$

$$\sigma_{x\varepsilon} = \dot{\rho}(k)\sigma_\rho^2 \cos[\theta(k)] \tag{14.6}$$

$$\sigma_{y\varepsilon} = \dot{\rho}(k)\sigma_\rho^2 \sin[\theta(k)] \tag{14.7}$$

Likewise, if it is three-dimensional, then

$$\begin{aligned} \boldsymbol{Z}(k) &= \boldsymbol{H}(k)\boldsymbol{X}(k) + \boldsymbol{W}(k) = [x(k)\ y(k)\ z(k)\ v(k)]' + \boldsymbol{W}(k) \\ &= \begin{bmatrix} \rho(k)\cos[\theta(k)]\cos[\varepsilon(k)] \\ \rho(k)\sin[\theta(k)]\cos[\varepsilon(k)] \\ \rho(k)\sin[\varepsilon(k)] \\ \dot{\rho}(k) \end{bmatrix} + \boldsymbol{W}(k) \\ &= \begin{bmatrix} 1 & 0 & 0 & 0 & 0 & 0 \\ 0 & 0 & 1 & 0 & 0 & 0 \\ 0 & 0 & 0 & 0 & 1 & 0 \\ 0 & \cos[\theta(k)]\cos[\varepsilon(k)] & 0 & \sin[\theta(k)]\cos[\varepsilon(k)] & 0 & \sin[\varepsilon(k)] \end{bmatrix} \begin{bmatrix} x \\ \dot{x} \\ y \\ \dot{y} \\ z \\ \dot{z} \end{bmatrix} + \boldsymbol{W}(k) \end{aligned} \tag{14.8}$$

The analysis of the covariance matrix of $\boldsymbol{W}(k)$ can be done by the reader.

From the above measurement equations, it can be concluded that nonlinear filtering methods should be adopted because of the strong nonlinearity of Doppler radial velocity measurements. The procedure of adding the Doppler radial velocity processing algorithm to the process of filtering can be found in Chapter 13.

14.4.1.2 IMM-PDAF Algorithm Procedure

The procedure of the IMM-PDAF algorithm is shown in Figure 14.4, and the definition for corresponding variables can be found in Table 14.1.

Since the IMM algorithm has been discussed in detail in Chapter 9, this section focuses on the topic of adding Doppler radial velocity information to the process of data association based on specific conditions of phased array radars.

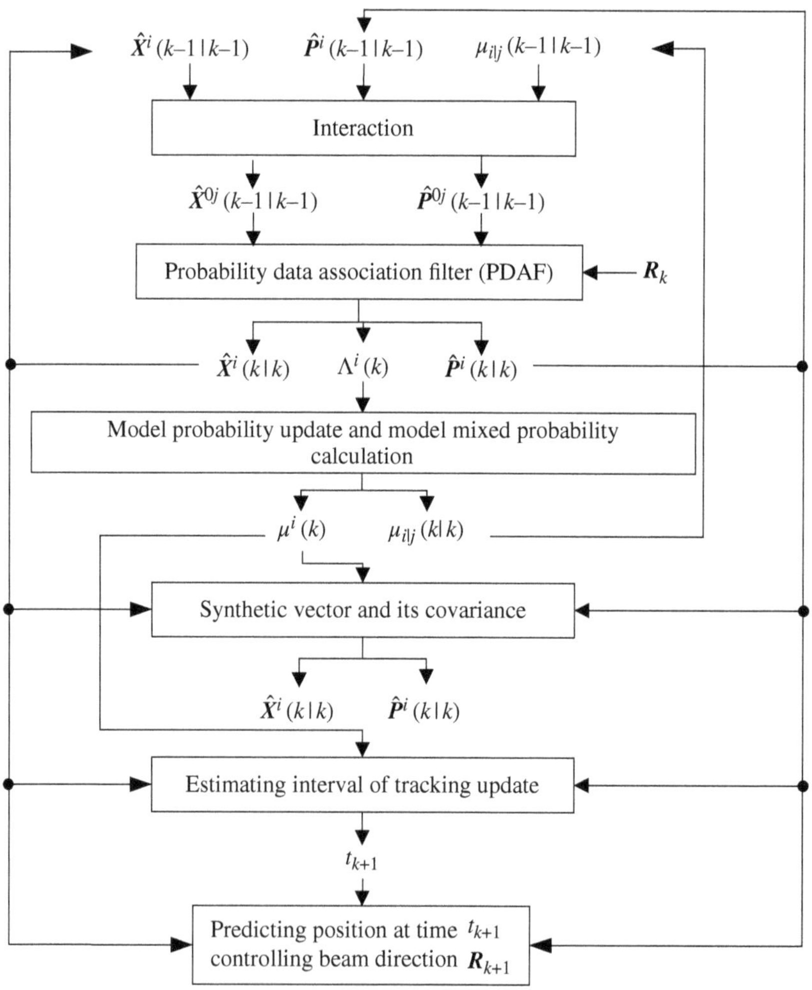

Figure 14.4 IMM-PDAF algorithm flowchart

Table 14.1 Definitions of variables in flowchart

$(i,j=1,\ldots,r)$	Labels for models
$\boldsymbol{Z}^k=[\boldsymbol{R}_1\ \boldsymbol{R}_2\ \ldots\ \boldsymbol{R}_k]$	Cumulative measurement matrix at sampling time k
$\boldsymbol{R}_k=\left[\boldsymbol{r}_{k;1}\ \boldsymbol{r}_{k;2}\ \ldots\ \boldsymbol{r}_{k;N_k}\right]$	Measurement submatrix at sampling time k
$\boldsymbol{r}_{k;s}=\left[x_{k;s}\ y_{k;s}\ z_{k;s}\ v_{k;s}\right]'$	The no. s measurement vector at sampling time k
$\hat{\boldsymbol{X}}^i(k-1\|k-1)$	State vector estimation of model no. i when measurement is $\boldsymbol{Z}^{k-1}$
$\hat{\boldsymbol{P}}^i(k-1\|k-1)$	Estimated covariance matrix of model no. i when measurement is $\boldsymbol{Z}^{k-1}$
$\hat{\boldsymbol{X}}^{0j}(k-1\|k-1)$	Mixed state vector estimation when measurement is $\boldsymbol{Z}^{k-1}$
$\hat{\boldsymbol{P}}^{0j}(k-1\|k-1)$	Mixed state vector estimation covariance matrix when measurement is $\boldsymbol{Z}^{k-1}$
$\mu_{i\|j}(k\|k)$	Mixed probabilities from model i to model j at time k
$\mu^i(k)$	Probability of model i at time k
$\Lambda^i(k)$	Possibility of model j at time k

14.4.1.3 PDAF Algorithm with Doppler Radial Velocity

The state update and covariance update with Doppler radial velocity are almost the same as in the PDAF algorithm of Chapter 8, except for the calculation of association probability.

Decompose $z_i(k)$, the ith measurement at time k, into position measurement $\boldsymbol{e}_i(k)$ and radial velocity measurement $v_i(k)$ (i.e., $z_i(k)=\{\boldsymbol{e}_i(k),v_i(k)\}$), then the PDF $z_i(k)$ is

$$\begin{aligned} p_i(k)\equiv p\left[z_i(k)\middle|\theta_i(k),\boldsymbol{Z}^{k-1}\right]&=p\left[\boldsymbol{e}_i(k),v_i(k)\middle|\theta_i(k),\boldsymbol{Z}^{k-1}\right]\\ &=p\left[\boldsymbol{e}_i(k)\middle|\theta_i(k),\boldsymbol{Z}^{k-1}\right]p\left[v_i(k)\middle|\theta_i(k),\boldsymbol{Z}^{k-1}\right] \end{aligned} \tag{14.9}$$

Denote by $\theta_i(k)$ the event that $z_i(k)$ is target-originated, then the PDF of $\boldsymbol{e}_i(k)$ is

$$p\left[\boldsymbol{e}_i(k)\middle|\theta_i(k),\boldsymbol{Z}^k\right]=P_{\mathrm{G}}^{-1}\mathrm{N}\left[\boldsymbol{e}_i(k);\boldsymbol{H}_e\hat{\boldsymbol{X}}(k|k-1),\boldsymbol{S}(k)\right] \tag{14.10}$$

where P_{G} is the gate probability and $\boldsymbol{H}_e$ the position measurement matrix, then the position predictive error covariance $\boldsymbol{S}(k)$ is

$$\boldsymbol{S}(k)=\boldsymbol{H}_e\boldsymbol{P}(k|k-1)\boldsymbol{H}_e'+\boldsymbol{R} \tag{14.11}$$

where $\boldsymbol{R}$ is the covariance of zero-mean Gaussian noise of the position measurement. $\hat{\boldsymbol{X}}(k|k-1)$ and $\boldsymbol{P}(k|k-1)$ are the state prediction and its covariance at time $k-1$. The standard KF should be used, before the PDF of the radial velocity measurement is determined, to estimate the target state:

$$\left[\hat{\boldsymbol{X}}_{i-}(k|k),\boldsymbol{P}_{i-}(k|k)\right]=\mathrm{KF}\left(\boldsymbol{e}_i(k),\boldsymbol{R},\hat{\boldsymbol{X}}(k|k-1),\boldsymbol{P}(k|k-1),\boldsymbol{H}_e\right) \tag{14.12}$$

If $z_i(k)$ is the target-originated measurement, then the PDF of its $v_i(k)$ is

$$p\left[v_i(k)\middle|\theta_i(k),\boldsymbol{Z}^k\right]=P_{\mathrm{G}}^{-1}\mathrm{N}\left[v_i(k);\boldsymbol{H}_v\left(\hat{\boldsymbol{X}}(k|k-1)\right),\boldsymbol{S}_v\left(\hat{\boldsymbol{X}}_{i-}(k|k),\boldsymbol{P}_{i-}(k|k)\right)\right] \tag{14.13}$$

where $\boldsymbol{H}_v(\cdot)$ is the radial velocity measurement function, and the Doppler measurement prediction covariance $\boldsymbol{S}_v(\boldsymbol{X},\boldsymbol{P})$ is

$$S_v(X,P) = h_v\left(\hat{X}(k|k-1)\right)P(k|k-1)h_v'\left(\hat{X}(k|k-1)\right) + \sigma_v^2 \tag{14.14}$$

where the Jacobian matrix $h_v(\cdot)$ of the radial velocity measurement is defined as

$$h_v(x) \equiv \frac{\partial H_v(x)}{\partial x} \tag{14.15}$$

The position measurement innovation is

$$\nu_{i,e}(k) = e_i(k) - H_e\hat{X}(k|k-1) \tag{14.16}$$

Based on position threshold γ_e, choose measurement

$$\nu_{i,e}(k)'S(k)^{-1}\nu_{i,e}(k) < \gamma_e \tag{14.17}$$

The innovation of the velocity measurement is

$$\nu_{i,v}(k) = v_i(k) - H_v\left(\hat{X}(k|k-1)\right) \tag{14.18}$$

Based on position threshold γ_e, choose measurement

$$\nu_{i,v}(k)'S_v\left(\hat{X}_{i-}(k|k), P_{i-}(k|k)\right)^{-1}\nu_{i,v}(k) < \gamma_v \tag{14.19}$$

Hence, this gives the data association probability [343]. If no measurement is target-originated at time k, then

$$\beta_0(k) = \frac{1 - P_D P_G}{1 - \delta_k}, \quad i = 0 \tag{14.20}$$

If $z_i(k)$ is target-originated, then

$$\beta_i(k) = \frac{P_D P_G}{1 - \delta_k} \cdot \frac{p_i(k)}{\rho_i^*(k)}, \quad i > 0 \tag{14.21}$$

where the factor δ_k can be defined as

$$\delta_k = P_D P_G\left(1 - \sum_{i=1}^{m_k} \frac{p_i(k)}{\rho_i^*(k)}\right) \tag{14.22}$$

with clutter density $\rho_i^*(k) = \rho(e_i(k)) \cdot p^0(v_i(k))$, $\rho(\mathbf{e}_i(k)$ the clutter density of position measurement, $p^0(v_i(k)$ the PDF of Doppler measurements, and m_k the number of measurements at time k.

If none of the measurements is correct, the estimate is

$$\hat{X}_0(k|k) = \hat{X}(k|k-1) \tag{14.23}$$

and the error covariance is

$$P_0(k|k) = P(k|k-1) \tag{14.24}$$

If $z_i(k)$ is target-originated, first the KF should be used to update (14.12) and the state should be updated according to $e_i(k)$. Then, nonlinear filtering methods can be applied to deal with $v_i(k)$ of Doppler measurements to update the state. Here, we take the EKF method as an example:

$$K_v = P_{i-}(k|k)h_v(\hat{X}_{i-}(k|k))S_v(\hat{X}_{i-}(k|k), P_{i-}(k|k))^{-1} \tag{14.25}$$

$$\hat{X}_i(k|k) = \hat{X}_{i-}(k|k) + K_v(v_i(k) - h_v(\hat{X}_{i-}(k|k))) \tag{14.26}$$

$$P_i(k|k) = (I - K_v h_v(\hat{X}_{i-}(k|k)))P_{i-}(k|k) \tag{14.27}$$

The target state estimation and its error covariance are

$$\hat{X}(k|k) = \sum_{i=0}^{m_k} \beta_i(k)\hat{X}_i(k|k) \tag{14.28}$$

$$P(k|k) = \sum_{i=0}^{m_k} \beta_i(k)\left(P_i(k|k) + \hat{X}_i(k|k)\hat{X}_i'(k|k)\right) - \hat{X}(k|k)\hat{X}'(k|k) \tag{14.29}$$

14.4.2 Multi-target-in-Clutter Tracking Algorithm

When dealing with multiple targets, the JPDA method can be used by incorporating procedures for the generation and decomposition of validation matrixes before conducting data associations in the single-target tracking algorithm mentioned in the last section. Since its calculation complexity grows dramatically with the number of measurements and tracks, the JPDA is not applicable in the presence of sparse targets in dense clutter, in which case, sub-optimal multi-target tracking approaches become the best option, say the linear multiple-target (LM) approach whose calculation complexity shows linear growth [344] with the number of tracks and measurements. The LM approach takes all measurements from "other" targets as clutter, so that the clutter density within the tracking gate can be adjusted by these other targets (i.e., it calculates modified clutter density for each tracking threshold, which is used to estimate the data association probability for each track). The specific procedure is described as follows.

$\theta_i^\tau(k)$ denotes the event that the association probability of the ith ($i = 1, 2, \ldots, m_k$) measurement $z_i(k)$ at time k having originated from target τ is approximated as

$$P_i^\tau = \Pr\{\theta_i^\tau(k)|Z^{k-1}\} \approx P_D^\tau P_G^\tau \cdot \frac{p_i^\tau(k)}{\rho_i^\tau(k)} \Big/ \sum_{j=1}^{m_k} \frac{p_j^\tau(k)}{\rho_j^\tau(k)} \tag{14.30}$$

where $p_i^\tau(k)$ is defined by (14.9), $\rho_i^\tau(k)$ is the clutter density in the threshold. As for the non-parametric method, $\rho_i^\tau(k)$ can be considered as a constant for all clutter measurements in the threshold (i.e., $\rho_i^\tau(k) = \rho_i^\tau$). Therefore, (14.30) can be simplified as

$$P_i^\tau = P_D^\tau P_G^\tau \cdot \frac{p_i^\tau(k)}{\sum_{j=1}^{m_k} p_j^\tau(k)} \tag{14.31}$$

When the track of target τ is updated in the single-target environment, (14.21) and (14.22) both rely on the ratio of target measurement density to clutter measurement density:

$$\frac{P_D P_G p_i^\tau(k)}{\rho_i^*(k)} \tag{14.32}$$

When the track of target τ is updated in the multi-target environment, the target measurement density can be modified with the probability of $z_i(k)$ having originated from no target, represented as

$$P_D P_G p_i^\tau(k) \prod_{\eta \neq \tau} (1 - P_i^\eta) \tag{14.33}$$

Likewise, the clutter measurement density is modified. The measurements of other targets $\sigma \neq \tau$ are added to the clutter density because they are unnecessary, in which case the clutter measurement equivalent density at the position of $z_i(k)$ is

$$\rho_i^*(k) \prod_{\eta \neq \tau} (1 - P_i^\eta) + \sum_{\sigma \neq \tau} p_i^\sigma(k) P_i^\sigma \prod_{\eta \neq \tau, \sigma} (1 - P_i^\eta) = \prod_{\eta \neq \tau} (1 - P_i^\eta) \left(\rho_i^*(k) + \sum_{\sigma \neq \tau} p_i^\sigma(k) \frac{P_i^\sigma}{1 - P_i^\sigma} \right) \tag{14.34}$$

where $P_i^\sigma \prod_{\eta \neq \tau, \sigma} (1 - P_i^\eta)$ is the prior probability of $z_i(k)$ having originated from target σ rather than target τ. Therefore, when the track of target τ is updating, the ratio of target measurement density to clutter measurement density becomes

$$\frac{P_D P_G p_i^\tau(k)}{\rho_i^*(k)} \xrightarrow{LM} \frac{P_D P_G p_i^\tau(k) \prod_{\eta \neq \tau} (1 - P_i^\eta)}{\prod_{\eta \neq \tau} (1 - P_i^\eta) \left(\rho_i^*(k) + \sum_{\sigma \neq \tau} p_i^\sigma(k) \frac{P_i^\sigma}{1 - P_i^\sigma} \right)} = \frac{P_D P_G p_i^\tau(k)}{\rho_i^*(k) + \sum_{\sigma \neq \tau} p_i^\sigma(k) \frac{P_i^\sigma}{1 - P_i^\sigma}} \tag{14.35}$$

Hence, the modified clutter density (since there are multi-target measurements) in the threshold of target τ can be expressed as

$$\Omega_i^\tau(k) = \rho(\boldsymbol{e}_i(k)) \cdot p^0(v_i(k)) + \sum_{\sigma \neq \tau} p_i^\sigma(k) \frac{P_i^\sigma}{1 - P_i^\sigma} \tag{14.36}$$

which is used to replace $\rho_i^*(k)$ in (14.21) and (14.22). Therefore, using the LM approach [345] yields

$$\beta_i^\tau(k)=\frac{1}{1-\delta^\tau(k)}\begin{cases}1-P_{\mathrm{D}}^\tau P_{\mathrm{G}}^\tau, & i=0\\ P_{\mathrm{D}}^\tau P_{\mathrm{G}}^\tau \dfrac{p_i^\tau(k)}{\Omega_i^\tau(k)}, & i>0\end{cases} \tag{14.37}$$

where

$$\delta^\tau(k)=P_{\mathrm{D}}^\tau P_{\mathrm{G}}^\tau\left(1-\sum_{i=1}^{m_k}\frac{p_i^\tau(k)}{\Omega_i^\iota(k)}\right) \tag{14.38}$$

14.4.3 Adaptive Sampling Period Algorithm

A specific sojourn time (time when the beam illuminates the target) can be chosen according to certain demands in phased array radar systems, while conventional mechanical scanning radars use the same method to deal with all targets. The probability of finding targets can be controlled by collecting the number of adjustable pulses according to the target characteristics and the battle environment. The phased array radar is capable of arranging beams in any direction within a short time, which enables it to track multiple targets at the same time, and conduct scanning and tracking separately. However, the mechanical scanning radar is different for it has the same data rate in scanning and tracking. While the phased array radar is tracking a certain target, it is not confined to using a fixed data rate, but takes samples from target tracks according to the optimal rule of certain regulations, which means that the sampling rate for maneuvering targets is higher than that for targets in rectilinear motion, thus reducing the tracking errors. It can obtain a new measurement if it suffers from a serious shortage of certain information, for example, the condition for track initiation is not satisfied. Therefore, the time for track initiation can be greatly shortened. More importantly, there is no need to enlarge the association gate during the time interval between measurement loss and repeated observation of the radar, which puts a restriction on the number of false measurements shown in the required region.

However, since the resources of the phased array radar are limited and shared by its multiple functions (scanning, tracking, weapon guiding, etc.), a certain task scheduling method must be adopted to allocate the corresponding beam sojourn time and sampling interval for different targets. So, the tracking beam can be used effectively to minimize the time resources and maximize the tracking quality on the whole. Specifically, the system uses short sampling intervals when the target shows relatively high maneuverability, but long sampling intervals when the maneuverability is low. Meanwhile, the sampling interval should not be too long or too short, because too long intervals may excessively reduce the tracking accuracy, which tends to cause track divergence and target loss, while too short intervals can bring about a greater load to the system but contribute little to the increase in tracking accuracy. Therefore, the adaptive sampling period algorithm should also be capable of balancing tracking accuracy with system load.

Besides, when the filtering model is stably tracking non-maneuvering targets, the improvement in its tracking accuracy by increasing the data rate is insignificant. In the case of track divergence for a maneuvering target, the divergence problem cannot be solved by simply increasing the data rate without adjustment of the filtering model according to the maneuverability of the target. In order to analyze the adaptive sampling period algorithm, therefore, those algorithms for maneuvering target tracking must be employed as well in adjusting the filtering model.

14.4.3.1 Constant Gain Filtering Method

The constant gain α–β filter is typical in phased array radar tracking. It can be described by the following recursive filtering equation set:

$$x_p(n) = x_s(n-1) + v_s(n-1)T(n-1) \tag{14.39}$$

$$x_s(n) = x_p(n) + \alpha\left[x_m(n) - x_p(n)\right] \tag{14.40}$$

$$v_s(n) = v_s(n-1) + \frac{\beta}{T(n-1)}\left[x_m(n) - x_p(n)\right] \tag{14.41}$$

where $x_m(n)$ is the position measurement for sampling time $T(n)$, $x_p(n)$ the corresponding predicted position, $x_s(n)$ the corresponding smoothing position, $v_s(n)$ the smoothing velocity, α and β the filtering parameters of position and velocity, and T the sampling interval.

Assume that the sampling interval is uniform in sections. The parameter relationship for one of the coordinates described in (14.39)–(14.41) is the same with other coordinates. Let

$$e(n) = x_m(n) - x_p(n) \tag{14.42}$$

In order to weaken the noise effect, an α filter processing as follows can be applied to $e(n)$:

$$e_s(n) = e_s(n-1) + \alpha_R\left[e(n) - e_s(n-1)\right] \tag{14.43}$$

The steady lag error of the predicted position caused by inputting a into the constant acceleration can be considered as E for the α–β filter with constant sampling interval T, where

$$E = \frac{aT^2}{\beta} \tag{14.44}$$

Then the predicted position error is proportional to the acceleration and the square of the sampling interval. Since a maneuvering target's acceleration can bring about an increase in the tracking filter's residual, in order to keep the residual constant, the sampling interval should be decreased following the rule of being in reverse ratio to the square root of the acceleration, as shown in (14.44).

The normalized residual $e_0(n)$ is usually adopted in real-time estimations. Here

$$e_0(n) = \frac{e(n)}{\sigma} \tag{14.45}$$

where σ is the standard deviation of measuring noise. Therefore,

$$T(n) = \frac{T(n-1)}{\sqrt{e_0(n)}} \tag{14.46}$$

Some modification of (14.46) is necessary in implementations of the algorithm. For instance, as shown by this equation, $T(n)$ can increase or decrease infinitely, so it should be given the limitation of maximum and minimum according to what takes place in engineering realities. Besides, in practice, discrete values may be required for the change in sampling intervals instead of the continuous value as needed in (14.46), for the sampling interval of discrete jumps may not satisfy (14.46) in that case. Further, if the measurement dimensions of the target are bigger than 1 and the sampling is conducted at the same time, then all dimensions of the measurement residuals should be averaged or consolidated logically. Reference [346] offers an example which uses the residual norm to calculate, and divides sampling intervals into several discrete ranges, so that the adaptive scheduling of sampling intervals can be done by choosing different range values based on the value of the residual norm.

14.4.3.2 Interactive Multiple-Model Algorithm

The IMM-based adaptive data rate algorithm proposed in Ref. [347], based on the Singer maneuvering target model, synthesizes the process noise of the target with IMM calculations, and calculates the sampling intervals with the approximate equation in Ref. [348], which shows the relationship between the sampling update intervals and process noise. The specific procedure is described as follows.

1. *Model target maneuver as correlative noise in the Singer model.* Target acceleration can be modeled as a zero-mean random process with index autocorrelation as follows:

$$R(\tau)=\mathrm{E}[a(t)a(t+\tau)]=\sigma_m^2 e^{-\alpha|\tau|} \tag{14.47}$$

   where σ_m^2 is the variance of target acceleration and α is the maneuvering frequency. The Singer model can be used to predict and estimate the state of the target by Kalman filtering. For details, see Section 9.3.2.
2. *Denote variance of the position prediction error denoted as $\sigma_\rho^2(k+1|k)$.* Obviously, it grows with the extrapolation time. Let the measurement error variance be σ_0^2, then the variance reductive ratio ν_0 is

$$\nu_0^2=\sigma_\rho^2/\sigma_0^2 \tag{14.48}$$

   Given the position error accuracy, the relationship between the next sampling update interval $T\,(T=t_{k+1}-t_k)$ and the prediction accuracy of the steady state is

$$T\approx 0.4\left(\frac{\sigma_0\sqrt{\tau_m}}{\sigma_m}\right)^{0.4}\cdot\frac{\nu_0^{2.4}}{1+0.5\sigma_0^2} \tag{14.49}$$

   where σ_m is the standard deviation of process noise of the target kinematic model, and τ_m is the time constant of target maneuver. It can be concluded that once the standard deviation of position measurement error σ_0 and the target maneuvering parameters σ_m and τ_m are defined, the sampling interval can easily be figured out from (14.49) when the steady prediction accuracy is ν_0.
3. *Estimate maneuvering parameter σ_m with the IMM algorithm.* The IMM algorithm can be used to track the maneuvers of a target by the interactions among N models $\{M_i : i=1,\ldots,N\}$, and the transitions among the models follow the finite Markov chain with known transition probability. Z^k stands for the collection of all the observed values at time t_k. The probability of model M_i within $[t_{k-1},t_k]$ is

$$\mu_i(k) = \Pr\{M_i(k)|Z^k\} \tag{14.50}$$

The detailed updating procedure of the model probability can be found in Section 9.3.6. Suppose that each model is in the form of (14.47) and that the process noise is σ_{mi}^2, then the variance estimation of target acceleration can be synthesized as follows:

$$\sigma_m^2(k|Z^k) = \sum_{i=1}^{N} \mu_i(k)\sigma_{mi}^2 \tag{14.51}$$

Thus, the sampling update interval can be obtained by substituting σ_m^2 in the above equation into (14.48). But this algorithm restricts the type of model collection because it requires that all the models should be isomorphic, namely, Singer models.

14.4.3.3 Predicted Covariance Threshold Algorithm

The predicted covariance of a target's state can be, to a certain extent, a reflection of the maneuvering features of the target. The stronger the maneuverability the target shows, the greater the predicted covariance becomes, and vice versa. Based on this theory, Ref. [349] deduces the predicted covariance threshold algorithm, which compares the position predicted covariance output by the IMM constantly with the given threshold. When it goes beyond the given threshold, the next sampling should be conducted. The sampling time should satisfy

$$\boldsymbol{P}(t_{k+1}) \le \boldsymbol{P}_{\text{th}} \tag{14.52}$$

where $\boldsymbol{P}(t_{k+1})$ is the predicted covariance matrix at time t_{k+1}, $t_{k+1} = t_k + T(t_k)$, and $\boldsymbol{P}_{\text{th}}$ is the given predicted covariance threshold. Likewise, the predicted covariance, using the algorithm based on IMM adaptive sampling, can be calculated as

$$\boldsymbol{P}(t_{k+1}) = \sum_{i=1}^{N} \mu_j(t_{k+1})\left\{\boldsymbol{P}_i(t_{k+1}) + [\hat{\boldsymbol{X}}_i(t_{k+1}) - \hat{\boldsymbol{X}}(t_{k+1})][\hat{\boldsymbol{X}}_i(t_{k+1}) - \hat{\boldsymbol{X}}(t_{k+1})]'\right\} \tag{14.53}$$

$$\hat{\boldsymbol{X}}(t_{k+1}) = \sum_{i=1}^{N} \mu_i(t_{k+1})\hat{\boldsymbol{X}}_i(t_{k+1}) \tag{14.54}$$

where $\hat{\boldsymbol{X}}_i(t_{k+1})$ and $\boldsymbol{P}_i(t_{k+1})$ are the target's predicted state and predicted covariance matrix output from model M_i. Considering that the main diagonals of the covariance matrix denote the error variances of the target's radial range, azimuth, and elevation angle, while the non-diagonals denote their correlation, the traces of the matrix can be used to make a comparison as follows:

$$\text{Tr}[\boldsymbol{P}(t_{k+1})] \le \text{Tr}[\boldsymbol{P}_{\text{th}}] \tag{14.55}$$

The threshold $\boldsymbol{P}_{\text{th}}$ can be a linear function of the measurement noise variance, that is,

$$\text{Tr}[\boldsymbol{P}_{\text{th}}] \le \lambda \text{Tr}[\boldsymbol{R}(t_{k+1})] \tag{14.56}$$

where $\boldsymbol{R}(t_{k+1})$ is the measurement noise variance matrix at time t_{k+1}, and $\lambda > 0$ is the adjustable coefficient. The sampling interval can be controlled flexibly by adjusting this coefficient.

In the predicted covariance threshold algorithm, the definition of each sampling time is done with a number of comparisons by (14.52). Therefore, the amount of calculation can be huge when the sampling interval is quite long. In order to reduce the complexity, a group of sampling intervals can be set in advance before calculating the predicted covariance matrix under each sampling interval. The sampling interval whose predicted covariance is bigger than the threshold can be a selection method. The specific selection procedure is as follows [350]:

1. Pre-set one group of typical sampling interval $\{T_i\}_1^N$, in which $T_1 \le T_i \le T_N$.
2. Calculate the predicted covariance of the target $\boldsymbol{P}^i(t_{k+1})$ in each sampling interval by the tracking filtering algorithm, where $i = 1, 2, \ldots, N$.
3. Identify the next sampling interval based on $\boldsymbol{P}(t_{k+1}) \le \boldsymbol{P}_{\text{th}}$.

The sampling interval can be controlled because the coefficient λ in the predicted covariance control and prior defining sampling interval approaches is controllable. The bigger λ is, the longer the sampling interval becomes and the lower the tracking accuracy of the target shows, and vice versa. In this way, the adaptive scheduling of the time resources of the phased array radar can be realized.

14.4.4 Real-Time Task Scheduling Strategy

The scheduling strategy of the phased array radar is the method by which the computer, conditioned on the given set of task requests of the radar, outputs the execution sequence for each task request in order to implement optimal scheduling in a sense when meeting system limitations. Under the control of computers, the scanning antenna can finish the formation and positioning of radar beams on the microsecond level to enable the radar to conduct scanning, tracking, recognition, and other tasks on multiple targets in an alternate way. Since each task (e.g., scanning and tracking) consumes different resources of the radar, which are limited, the method of allocating and using these limited resources with a flexible and effective task scheduling strategy will exert great influence on the way the radar's advantage of multiple functions comes into full play. This is also the major difference between phased array radars and conventional radars.

14.4.4.1 Influential Factors in Scheduling

The following influential factors must be taken into consideration when effective scheduling policies are planned.

1. *Define relative priority based on task requests.* Since different task requests may compete for the same execution period in the case of multiple tasks, the scheduling algorithm needs to decide which task to schedule at once and which to delay or decline. The relative priority for each task serves as a basis for choice for the scheduling strategy. Because each task of the radar is conducted for a specific target (or airspace), so the definition of its priority is first based on the relative importance of the corresponding target (or airspace). Besides, the degree of urgency for the task should also be considered in defining the relative priority of the task.

2. *Define available working modes and their relative priority.* Generally, the available working modes can be divided into five levels: exclusive, crucial, short-range tracking and scanning, long-range tracking and scanning, and testing and maintaining. Obviously, the more modes are given priority, the higher operating efficiency the system shows, and the more requirements are imposed on the processing and memory of the computer. The number of priority types is usually based on the comprehensive consideration of many factors, such as the scheduling efficiency, time of computation, memory occupancy, and method of sorting the targets.
3. *Define constraints on radar resources and design conditions.* Since each working mode consumes a certain quantity of radar resources, which are limited, the constraints on radar operations should be determined to make the fullest use of these resources. The specific constraints on radar resources and design conditions are as follows.
   i. *On time resources.* The occurrence of each radar event, from the beam location to the completion of the event, calls for corresponding motion time on the part of the radar. Once the scheduling interval is defined, the number of radar events which can be arranged within one scheduling interval is limited even if other constraints are neglected.
   ii. *On energy resources.* As in the case of time resources, each radar event requires one or more pulses in different shapes transmitted from the transmitter. In other words, each such event consumes some energy. In order to ensure adequate data quality, more energy may be consumed – especially for targets which are in a long-range or jamming environment. Different working modes usually require different pulse waveforms (i.e., corresponding duty ratios), therefore, the average duty ratio of one pulse sequence on a fixed time interval (one or more scheduling intervals) – the "comprehensive duty ratio" – should be adopted in the design of the scheduling strategy.
   iii. *On computer resources.* After the completion of each radar event, the radar echo is sent to the computer via the signal processor for data processing, which will occupy certain processing and memory resources of the computer. Typically, the tracking mode needs more computer resources than the scanning mode. However, for convenience, the resources occupied by the former are usually considered 1.5 times those of the latter, and the computer constraints are commonly expressed as the biggest number of tracking beams allowed per unit of time.
   iv. *On the design of the radar.* Constraints in this respect refer to those caused by the design of some hardware. For instance, the material of the phase shifter decides the maximum number of changes in beam locations allowed per time unit for a radar with a closed ferrite phase shifter, and therefore puts constraints on the number of schedulable working modes per scheduling interval.
4. *Choose scheduling intervals.* Scheduling intervals, time intervals for the controlling program of a system to call its scheduling program, are the basis of the structure of the whole program system of a computer, which determines the executive frequency of the major sub-program in the control loop of a radar unit. Since data exchanges are conducted between the radar controller and the antenna front end once every two scheduling intervals, over-long scheduling intervals cannot satisfy the system's requirements for the execution frequency of a certain working mode, while over-short ones lead to excessive consumption of the computer's support and housekeeping programs. Selections are mostly based on the overall compromise between the system's requirements for the frequencies of certain working modes and the principle of minimizing the number of interruptions and management consumptions.

5. *Choose scheduling strategies.* The commonly used scheduling strategies of the phased array radar include the template scheduling strategy and the adaptive scheduling strategy, and the former can be further divided into fixed, multiple, and partial template policies.

14.4.4.2 Template Scheduling Strategy

The template scheduling strategy can be divided into the following three kinds.

1. *Fixed template strategy.* As the simplest one, this strategy can be applied to a fixed group of radar events by allocating the same time intervals in advance in each scheduling interval T. As shown in Figure 14.5, the scheduling program arranges five radar events in sequence in each scheduling interval: verifying–tracking–tracking–scanning–scanning. This strategy is simple: it occupies relatively few resources because it does not need to rank radar events in real time. However, it is suitable for specific target environments rather than multiple dynamic ones since it is short of flexibility and adaptive ability. Therefore, the fixed template strategy is only applicable for single-purpose or single-function radars.
2. *Multiple template strategy.* As a simply modified version of the fixed template strategy, this strategy has better flexibility and adaptive ability. It can design several fixed templates in advance to ensure that each of them matches the specific radar operating environment. The computer, in real-time scheduling, chooses the optimal template according to the current environment of the target and certain principles, as shown in Figure 14.6. This strategy's requirements for the processing ability of the computer tighten with the increase in number of types of template. This means that the flexibility and adaptive ability of the scheduling suffer when there are too many template types. Hence, the multiple template strategy is only applicable for the radar which has prior knowledge of the target environment and the operating environment.

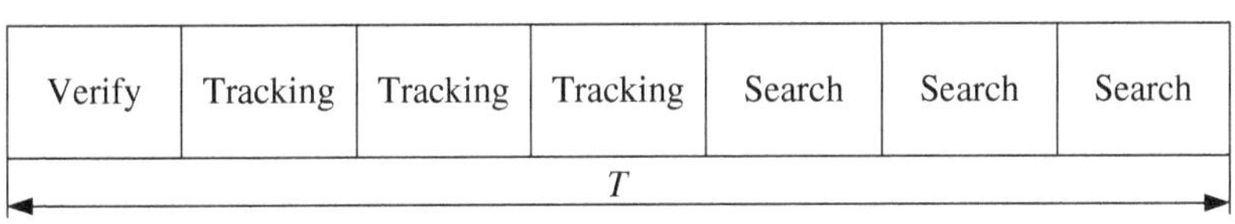

Figure 14.5 Fixed template strategy

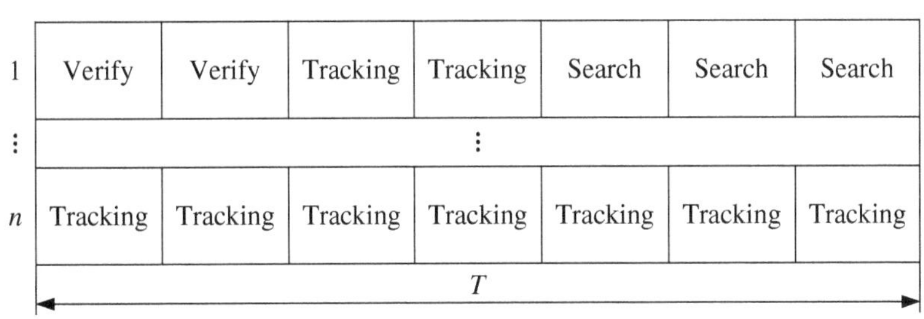

Figure 14.6 Multiple template strategy diagram

3. *Partial template strategy.* This strategy, in effect, is a partial adaptive strategy. It pre-arranges one or several events in a scheduling interval, then deals with other operations of the radar during the remaining time based on the operating priority and various constraints, as shown in Figure 14.7. Compared with the previous two policies, this strategy improves availability of the radar. Because of its flexibility and adaptive ability for target environments, it can be applied to multifunctional and multipurpose radars.

14.4.4.3 Adaptive Scheduling Strategy

The adaptive scheduling strategy is an approach to selecting, subject to the relative priority of various working modes, the optimal sequence of radar events for every scheduling interval by balancing the time, power, and computer resources requested by various radar beams in real time within the scope of the radar's design. Figure 14.8 shows the functional block diagram of this strategy [351]. The task queue inputs the requests of various radar events into the priority filter, which will define the relative priority for each event based on pre-set rules and the external dynamic environment. Then, the constraint filter set decides, on the basis of radar constraints and priority, whether to accept or reject the input events, which will be separately scheduled and sent into the rejection queue. Thus, this strategy satisfies the following adaptive principles:

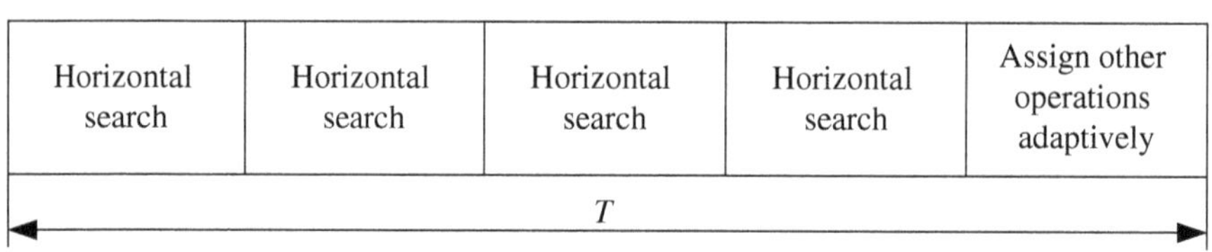

Figure 14.7 Partial template strategy diagram

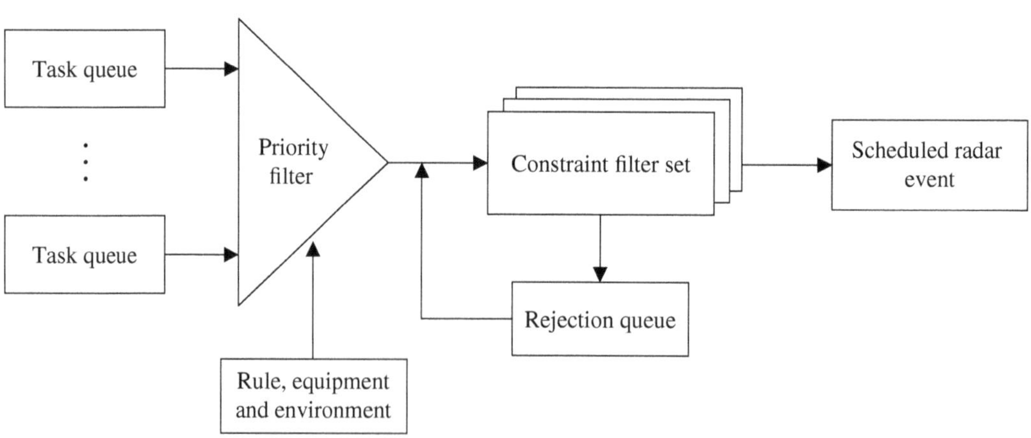

Figure 14.8 Functions of the adaptive scheduling strategy

1. Adaptive to dynamic radar environment.
2. Adaptive to relative priority of different working modes.
3. Able to balance time, power, and computer resources requested by radar's various operations in real time.
4. Able to satisfy the constraints on the design conditions of the radar.

Since the earliest deadline first (EDF) scheduling algorithm of dynamic priority is adaptive to dynamic changes in the phased array radar's tasks, it is a major concern for task scheduling in the phased array radar system. EDF means that the task with the earliest deadline is of uppermost priority in scheduling, and is scheduled prior to other tasks. Since the scheduling algorithm of dynamic priority is adaptive to dynamic changes in the phased array radar's tasks, the classical EDF (earliest deadline first) scheduling algorithm is typical. EDF means that the task with the earliest deadline is of uppermost priority in scheduling, and is scheduled prior to other tasks. This algorithm can be divided, based on the pre-emptive behavior of the task, into two models: pre-emptive and non-pre-emptive EDF scheduling. The other three scheduling models (i.e., researchers' promoted modified EDF scheduling [352], task scheduling based on time windows [353], and task scheduling based on pulse interleaving [354]) are also described.

1. *Pre-emptive.* The pre-emptive model, which was first proposed by Liu and Layland [355], is a scheduling algorithm driven by dynamic priority, in which the priority allocated to each task is decided by its current requirement for the deadline. The task which has an earliest deadline is of uppermost priority, while the one with a latest deadline should be given lowermost priority. This algorithm can guarantee that no idle time is available for the processor before the deadline for a certain task is satisfied. The pre-emptive EDF scheduling algorithm is based on the following assumptions:
   i. Any of the tasks can be pre-empted, and at a negligible cost.
   ii. The task's consumption of processor resources is the only consideration, while the expenses of memory, I/O, and other resources can be ignored.
   iii. The tasks are independent of each other, with no restraint on sequence.
   iv. The relative ultimate deadline for every task equals its period.

   Based on these assumptions, in the pre-emptive EDF scheduling algorithm, the necessary and sufficient condition of schedulability for a given set of periodic tasks is

   $$\sum_{i=1}^{n} e_i/p_i \le 1 \tag{14.57}$$

   where e_i and p_i are, respectively, the execution time and cycle of task i $(1 \le i \le n)$ in the task set. As indicated, its biggest advantage is that for any given periodic task set, the set's schedulability can be guaranteed on the condition that the utilization rate of the processor is not greater than 100%.

   Note: The above-mentioned analysis results are only for the real-time task whose relative deadline equals its period. This algorithm's scheduling conditions for the sets of general tasks (i.e., those whose relative deadlines are not equal to and even earlier than their cycles) can be found in Ref. [356].
2. *Non-pre-emptive.* The non-pre-emptive EDF (NPEDF) scheduling algorithm, proposed by Jeffay *et al.* [357], is applicable to periodic and aperiodic tasks. The execution of a task should be completed without any interruption from other tasks. The scheduling program decides which

task to execute next only when one task is completed, which is different from the pre-emptive model, where the task to be executed is decided at each clock unit. The NPEDF can be described by

$$\forall i, L, 1 < i \leq n, p_1 < L < p_i \quad L \geq e_i + \sum_{j=1}^{i-1} \left| \frac{L-1}{p_j} \right| e_j \tag{14.58}$$

As indicated, the non-pre-emptive algorithm eliminates the scheduling cost of pre-emption, but does not guarantee the execution of the task with high priority. Judging from the features of task scheduling in the phased array radar system, each real-time task corresponds to a radar beam's residence within a period of time in a certain direction, and it cannot be interrupted by other tasks within the illumination period of the beam, which is the reason why the task scheduling in the phased array radar system is considered non-pre-emptive.

3. *Modified EDF scheduling.* The EDF scheduling model is by nature a kind of scheduling strategy driven by priority, in which the system always chooses the task with highest priority to execute at any moment during the process of scheduling analysis. In this scheduling model, the priority of a task is completely decided by its time property (i.e., its deadline). But for a phased array radar system, since each task also has its own property in terms of working mode priority, the task with the earliest deadline is not necessarily scheduled to be executed first in this model. In that case, the ultimate priority of a task is determined by a comprehensive consideration of its various properties. Reference [358] determines a task's comprehensive priority by accounting for its deadline and working mode priority properties, and does the same with other property parameters or with the case of over three property parameters. It can be concluded from the above-mentioned analysis that the determination of priority is the key to the scheduling of tasks with multiple property parameters. Once the priority of a task is determined, a scheduling analysis can be performed along the lines of the EDF model by drawing an analogy between this priority and the deadline of the task in this model. For details, see Ref. [358].
4. *Task scheduling based on time windows.* The concept of a task request time window proposed by Huizing and Bloemen [359] means that the practical execution time requested by a radar task can move in a time window within its expected transmitting time [360]. This concept enables the radar events which conflict in a time arrangement to be scheduled for execution after being adjusted by the time window, improving the scheduling efficiency of radar event requests and the radars' time utilization rate greatly. But this method accounts for the constraints of time resources rather than such resources of the system as energy and computer resources. With the improvement in computer performance, the bottleneck for radar computer resources is gradually disappearing. Hence, emphasis should be put on the overall energy resource constraints on the system. For details, see Ref. [361].
5. *Task scheduling based on pulse interleaving.* In order to enhance radars' time utilization, the pulse interleaving technology [362] is applied to the field of radar, aiming to schedule the transmitting or receiving pulses of other tasks between the receiving and transmitting pulse of one task alternately. Though it enhances the system's efficiency, this technology also extends the radar transmitter's continuous working time accordingly. The constraints of the system's power resources must be thoroughly considered when using pulse interleaving to warrant the effective execution of all radar events, lest the consumption of power goes beyond the physical constraints of the system, which may lead to damage of the transmitter due to overheating. Orman *et al.* [363] made further analyses of this technology, solving the problems of task scheduling for a

radar in practice by the heuristic method, which gives consideration to constraints of both time and power of the radar system. For details, see Ref. [354].

14.5 Performance Analysis of the Adaptive Sampling Period Algorithm

14.5.1 Simulation Environment and Parameter Settings

The data rate of each Monte Carlo simulation is the same for the target tracking simulation with fixed sampling intervals. But when the adaptive sampling period method is adopted, since the number of sample points is different for almost every simulation during a certain simulation period, the traditional sum/average method is not applicable in the performance analysis. Instead, the smoothing process [347] should be applied within a short time. Thus, the average sampling interval $T_{\mathrm{avg}[T_a,T_b]}$ and the RMS position error $X_{\mathrm{rms}[T_a,T_b]}$ during each $[T_a, T_b]$ time are as follows:

$$T_{\mathrm{avg}[T_a,T_b]} = \left(\sum_{j=1}^{M} \sum_{t_k^j \in [T_a,T_b]} 1 \right)^{-1} M(T_b - T_a) \tag{14.59}$$

$$X_{\mathrm{rms}[T_a,T_b]} = \left[\frac{1}{\bar{M}} \sum_{j \in J} \frac{\sum_{t_k^j \in [T_a,T_b]} \left[\left(\hat{x}(t_k^j) - x(t_k^j)\right)^2 + \left(\hat{y}(t_k^j) - y(t_k^j)\right)^2 \right]}{\sum_{t_k^j \in [T_a,T_b]} 1} \right]^{1/2} \tag{14.60}$$

where t_k^j is the kth updating time in the jth Monte Carlo simulation, $\sum_{t_k^j \in [T_a,T_b]} 1$ denotes the total times of $t_k^j \in [T_a, T_b]$, and

$$J = \left\{ j : \sum_{t_k^j \in [T_a,T_b]} 1 \neq 0, j = 1, \ldots, M \right\}, \bar{M} = \sum_{j \in J} 1 \tag{14.61}$$

In the simulation environment, the target sets forth from position (10 000 m, 2000 m) and moves for 100 s with initial velocity (−160 m/s, 50 m/s), as shown in Table 14.2. For the phased array radar, the initial scanning period $T = 2\,\mathrm{s}$, ranging error $\sigma_\rho = 150\,\mathrm{m}$, and angle measurement error $\sigma_\theta = 5\,\mathrm{mrad}$. Assume that the smoothing interval is 2 s, and 500 Monte Carlo runs are performed. The true movement track of the target is shown in Figure 14.9.

Table 14.2 Information on the target's movement

Time (s)	0–12.5	12.5–25	25–50	50–62.5	62.5–100
Moving mode	Constant velocity	Constant acceleration	Constant velocity	Constant acceleration	Constant velocity
Acceleration (m/s^2)	(0, 0)	(10, 50)	(0, 0)	(50, −30)	(0, 0)

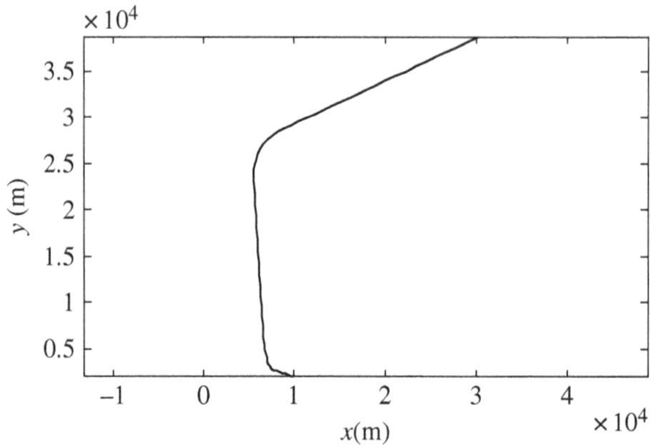

Figure 14.9 Target track

When the constant gain filtering method is adopted, the $\alpha - \beta$ filter is adopted in the constant gain filtering algorithm, with filtering parameters set as $\alpha = 0.5$ and $\beta = 0.167$. Figure 14.10 shows the simulation results by comparing the sampling period of this method with that of the fixed sampling period algorithm, which also uses $\alpha - \beta$ filters.

When the IMM algorithm is adopted, three Singer models with different variances of process noise, whose values are 2.5 m/s^2, 20 m/s^2, and 80 m/s^2, respectively are used. The maneuvering time constant of the target $\tau_m = 10$ s. Suppose that $\nu_0 = 0.8$, the initial probability of the model is [0.8, 0.1, 0.1], and the transfer probability matrix $\boldsymbol{P} = [0.95, 0.025, 0.025; 0.025, 0.95, 0.025; 0.025, 0.025, 0.95]$. Figure 14.11 shows the simulation results by comparing this method with the fixed sampling period IMM algorithm, which also uses three such Singer models.

When the predicted covariance threshold algorithm is adopted, a multiple model set is formed by two constant velocity (CV) models, whose variances of process noise are 0 and 10, and one constant acceleration (CA) model, whose variance of process noise is 20. The initial probability of the model is [0.8, 0.1, 0.1], the transfer probability matrix $\boldsymbol{P} = [0.95, 0.025, 0.025; 0.025, 0.95, 0.025; 0.025, 0.025, 0.95]$, and the sampling period set predefined in this algorithm is $\{0.5\text{ s}, 1\text{ s}, 2\text{ s}\}$. Figure 14.12 shows the simulation results by comparing this method with the fixed sampling period IMM algorithm, which also uses two CV models and one CA model to form a multiple model set. Tables 14.3–14.5 offer a comparison between these three methods.

14.5.2 Simulation Results and Analysis

As can be concluded from Figures 14.10–14.12, the adaptive sampling period algorithm can always yield the expected results of sampling period adaptive control, since the sampling periods are adaptive to changes in the moving states of the targets. Specifically, they become shorter during the maneuvering periods and longer in the absence of maneuvers: all sampling periods in the accelerated maneuver periods of 12.5–25 s and 50–62.5 s are shorter than those in adjacent time periods. The constant gain filtering method can adjust the sampling intervals which are set according to target residuals, the IMM algorithm is able to improve the accuracy of stable prediction, and the predicted covariance threshold algorithm can adjust covariance thresholds, so that the flexibility of adaptive sampling periods can be improved by adjusting the radar's tracking accuracy of the target and the time resources allocated by the system for this tracking task.

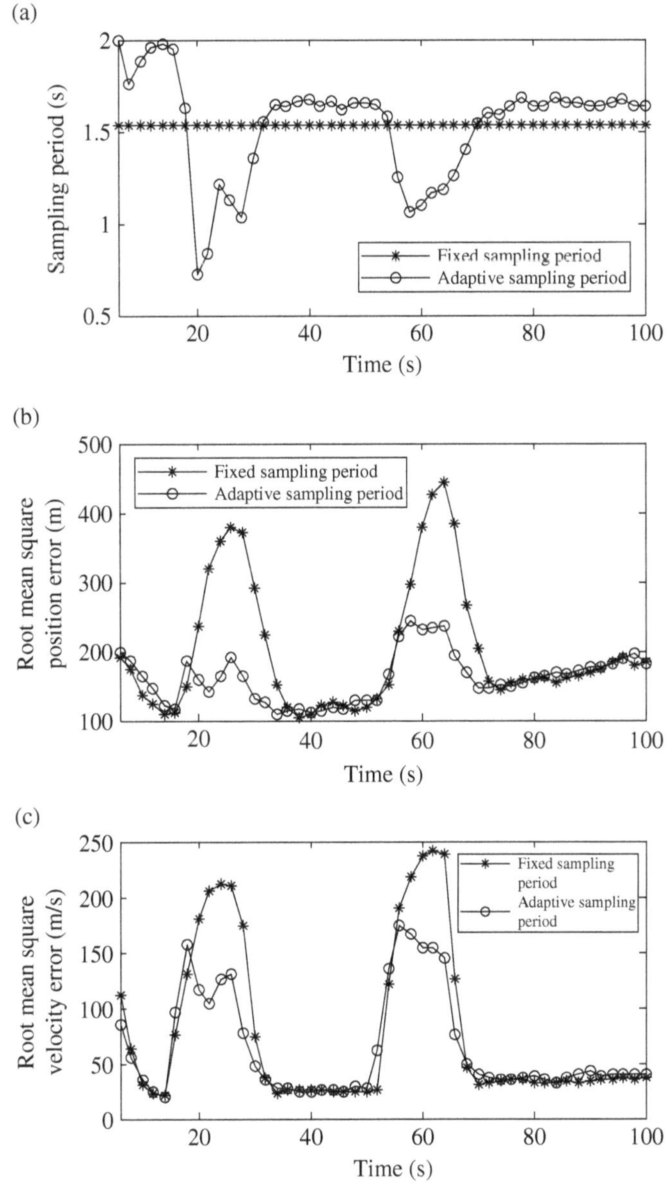

Figure 14.10 (a) Sampling period, (b) RMS position error, and (c) RMS velocity error of the constant gain filtering method

In order to prove the superiority of the adaptive data rate working mode, the adaptive sampling period method is compared with the fixed one. The model sets are assumed the same in both methods, and the latter's sampling period is set as the average one given by the former's modified version, such that they consume the same amount of time resources to conduct the tracking task under consideration. Figures 14.10(b), 14.10(c), 14.11(b), 14.11(c), 14.12(b), and 14.12(c) show

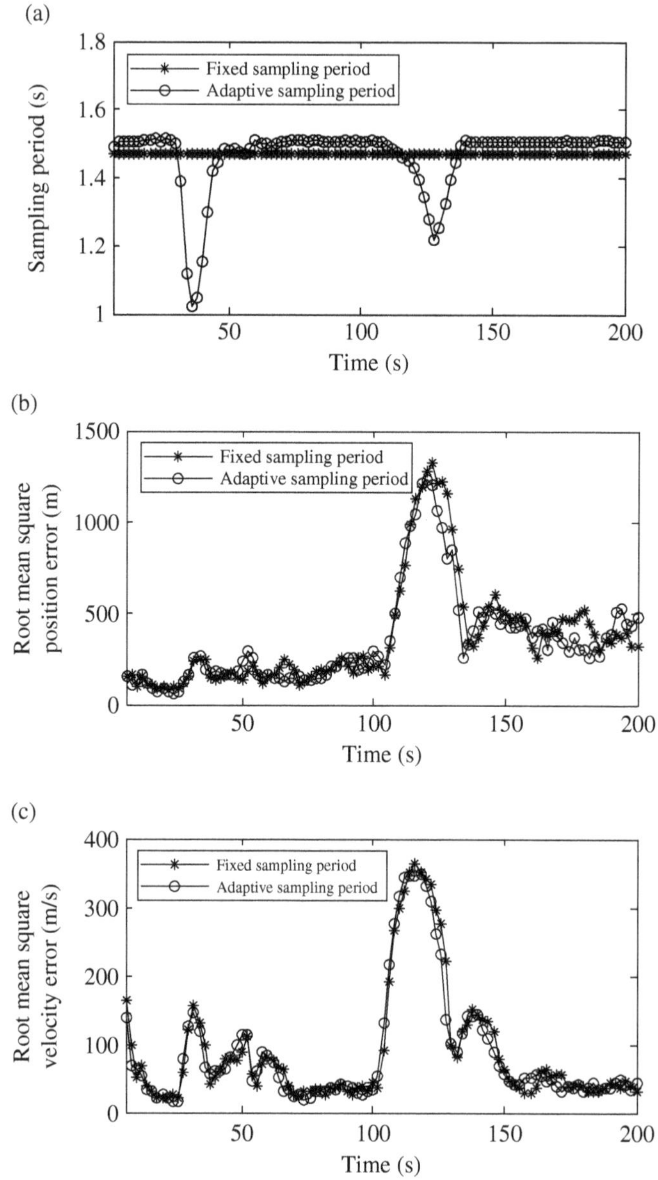

Figure 14.11 (a) Sampling period, (b) RMS position error, and (c) RMS velocity error of the IMM algorithm

a comparison of the tracking performance between the constant gain filtering method, the IMM algorithm, and the predicted covariance threshold algorithm successively, and Tables 14.3–14.5 list the quantitative results of comparison. It can be concluded that both the tracking position and the velocity error generated by the fixed sampling period algorithm are bigger than those of the three adaptive ones.

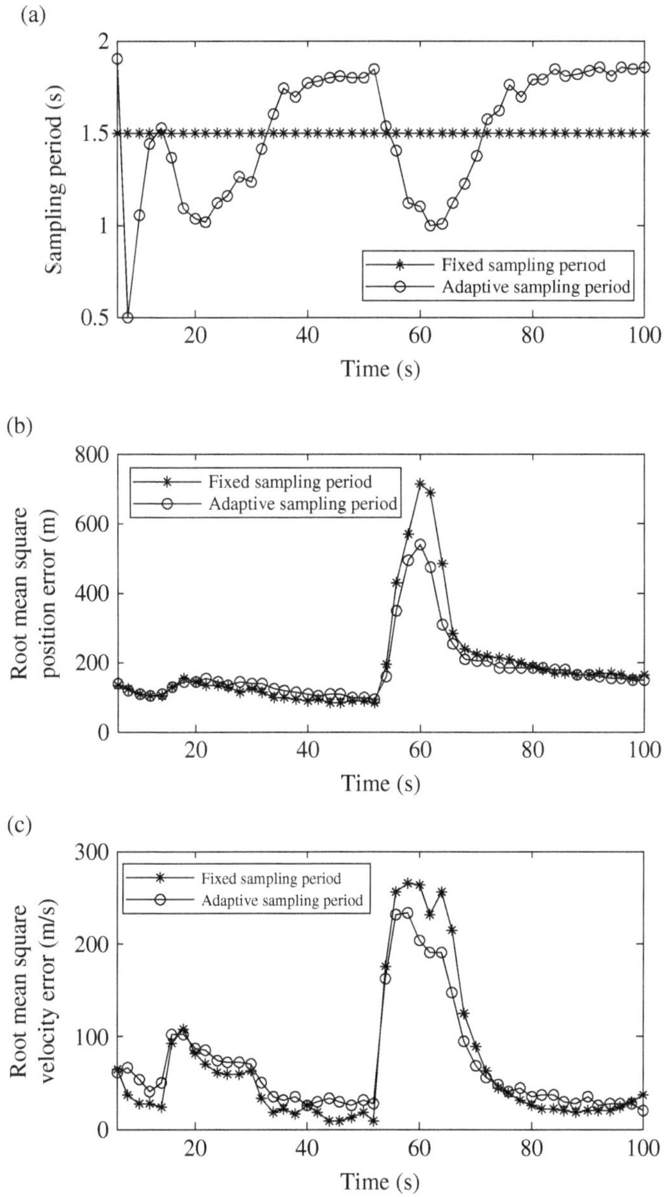

Figure 14.12 (a) Sampling period, (b) RMS position error, and (c) RMS velocity error of the predicted covariance threshold algorithm

Table 14.3 Comparison among position RMS, velocity RMS, and average period of constant gain filtering method

Method	Position RMS (m)	Velocity RMS (m/s)	Average period (s)
Adaptive sampling period	159.70	63.50	1.53
Fixed sampling period	201.42	78.28	

Table 14.4 Comparison among position RMS, velocity RMS, and average period of IMM algorithm

Method	Position RMS (m)	Velocity RMS (m/s)	Average period (s)
Adaptive sampling period	155.70	66.99	1.46
Fixed sampling period	168.04	70.69	

Table 14.5 Comparison among position RMS, velocity RMS, and average period of predicted covariance threshold algorithm

Method	Position RMS (m)	Velocity RMS (m/s)	Average period (s)
Adaptive sampling period	177.12	71.64	1.52
Fixed sampling period	196.64	70.98	

14.5.3 Comparison and Discussion

1. As shown in Figures 14.10(b) and 14.10(c), using the constant gain filtering algorithm, the adaptive sampling period method makes a quicker response to error fluctuations than the fixed method in the presence of target maneuvers. This is mainly because the noise model of the constant gain filter is zero-mean, Gaussian white noise, and is able to rapidly detect the target's maneuver. When the target is maneuvering, the adaptive method can shorten the sampling intervals immediately to detect the position at the next moment with larger position errors, while the fixed method keeps the same intervals and shows even larger position errors when detecting the target at the next moment.
2. As demonstrated by a comparison of Figures 14.11(a) and 14.12(a), using the IMM algorithm, the response of the adaptive sampling period method to changes in sampling periods lags behind the time of target maneuvers remarkably. The reason is that the Singer model applied in the IMM algorithm is a correlative noise model, which brings about decorrelation time in the updating of its probability caused by target maneuvers. It thus follows from Figures 14.11(b) and 14.11(c) that improvements in sampling rates reduce tracking errors during the maneuvering period, while in the maneuver-free period, the adaptive method shows better average tracking performance, which does not suffer from its longer periods than the fixed method.
3. A comparison of Figures 14.12(a) and 14.11(a) suggests that by using the predicted covariance threshold algorithm, the adaptive method makes quicker responses to variations in sampling periods than the fixed method in the presence of target maneuvers, principally because both the CV and CA models applied by this algorithm are white noise models, which are capable of having a quicker response to the model probability changes caused by target maneuvers. Besides, there is a convergent process for the prediction covariance during the tracking phase (i.e., it shows larger values in the initial stage), so this brings about relatively high sampling rates to the adaptive method in the initial phase. But the convergent process has little influence on the whole tracking process, because it is short.
4. By comparing Figures 14.11(c) and 14.12(c), we find that the velocity RMS shows two wave crests during both maneuvering periods. The time of emergence of the crests shows that the target's sampling rate conducted by the phased array radar should be improved when the target is maneuvering, which is, in essence, to speed up the matching process between the filtering and the target motion model. When the filtering model matches the target movement, the tracking accuracy grows gradually, and the sampling rate lowers accordingly when the tracking goes

stable. In addition, the two wave peaks in Figure 14.11(c) are more evident, causing two distinct peaks in Figure 14.11(b), for the Singer model is a correlative noise model whose decorrelation time widens the period of response to changes in the model's probability.

5. Since various maneuvering target tracking algorithms are different in performance, their own adaptive algorithms cannot be compared in a unified way. The above simulations are behavioral comparisons between the fixed and adaptive sampling systems using the same tracking algorithm to demonstrate the superiority of the adaptive method applied by the phased array radar.

14.6 Summary

This chapter mainly introduces the major features, system structure, and working process of phased array radar systems, and then puts emphasis on the discussion of several key typical data processing techniques of the phased array radar. The phased array radar is able to provide Doppler information, and then obtain a target's radial velocity measurements, which can be used to improve the data association and tracking performance for single/multi-targets in clutter. In addition, the tracking performance in the case of high target maneuverability can be promoted and the radar's working load can be better balanced by making full use of the phased array radar's capability to flexibly control the sampling intervals of the target data as needed. Since the phased array radar is capable of simultaneously conducting various tasks – such as scanning, tracking, and identification of multiple targets in an alternate way – the limited radar resources can be fully utilized to have better performance by applying effective real-time task scheduling strategies.

Finally, it lays emphasis on the comparison and simulation of several adaptive sampling period algorithms of the phased array radar, with the results indicating that the radar's load can be effectively reduced and the tracking performance of the target improved at the same time by changing sampling periods adaptively based on the maneuvering behavior of the target.

15

Radar Network Error Registration Algorithm

15.1 Introduction

The radar networking information fusion technology has brought tremendous benefits, recognized across the world, and is still in the process of fast development. However, because of the various radar detection errors existing in practical systems, how to guarantee the results of real-time fusion of a system has become a chronically knotty problem in this technology. As proven by practical implementations, for instance, systematic errors of a multiple-radar network tracking system would lead to bigger target tracking errors than their theoretical values; when the errors are too large, the tracking results of multiple radars are even worse than those of single ones; many tracks of the same target may result in the worst case, which then leads to ambiguities.

Although strict measures can be taken to decrease the systematic errors of networking radars in the entire process regarding them, ranging from design, research, manufacturing, installation, and adjustment to operation and use, there remain severe constraints on these radars due to the systems and methods of measuring systems and equipment, the performance indexes of devices, the residual errors of zero-value calibration, jamming, and noise. Under external influences, the systematic errors, even if eliminated before use, can be regenerated as time goes by. Furthermore, this can be a dynamic changing process.

The grid-lock technology, one of several key technologies advocated currently by the US military in network-centric operations, is an error registration technology by nature. In this chapter, we mainly discuss the registration algorithms of fixed radars and mobile radars.

15.2 The Composition and Influence of Systematic Errors

15.2.1 The Composition of Systematic Errors

In radar observation systems, there are two major kinds of error: stochastic and systematic. The former can be eliminated with various kinds of filtering method. The latter, however, is a kind

Radar Data Processing with Applications, First Edition. He You, Xiu Jianjuan, and Guan Xin.

of deterministic error, which cannot be eliminated with filtering methods. To eliminate systematic errors, estimations should be made in advance, and then the corresponding compensation should be made. This process is known as "radar registration."

Systematic errors of networking radars can be divided, according to their origins, into four categories:

1. radar positioning errors,
2. radar measurement systematic errors,
3. mobile platform attitude angle systematic errors,
4. coordinate transformation systematic errors.

Radar positioning errors are caused by the inaccuracy of position measuring equipment. As GPS devices are widely used, and the Beidou Navigation Satellite System grows mature, the accuracy of radar position measurements will be higher and higher, and the impact of radar positioning errors on the radar networking system is becoming less and less.

Radar measurement systematic errors mainly include range, azimuth, and elevation errors (3D radar). Range error is caused by internal circuit delay in the radar, the zero drift in the system, and velocity incorrectness of the distance clock, and it manifests itself as slow variables of addition errors and gain of errors in proportion to distance. Azimuth error is caused by the deviation which appears when the radar antenna is aligned with due North, and it manifests as slow variables of addition errors. Elevation error is caused by the fixed inclination of the radar antenna and manifests as slow variables of addition errors.

The attitude angle systematic errors of the mobile platform consist chiefly of pitch, yaw, and roll systematic errors. They are caused by factors such as the inaccuracy of attitude measurement equipment (e.g., GPS and gyroscopes) and slow response time, and are related to the maneuverability of platforms. When the platforms make big maneuvers, their attitude angle systematic errors are bigger and have notable influences on the radar network systems, which should be eliminated by registration.

Coordinate transformation systematic errors are caused by errors inherent in transformation formulas between different coordinate systems of the radar. Their impact on radar systems can be decreased by choosing proper public coordinate systems according to the deployment of networking radars.

Because categories 2 (radar measurement systematic errors) and 3 (mobile platform attitude angle systematic errors) have bigger impacts on radar network systems, we will explore, in this chapter, their estimation and elimination methods. In addition, these two types are normally assumed constant unless otherwise specified.

15.2.2 The Influence of Systematic Errors

In the case of a singular radar system, systematic errors are the same for every target except that a fixed rotation and translation appears. This rotation and translation exerts no influence on the estimation of speed and relative position of the target, nor on the tracking performance of singular radars. However, things are quite different in the case of the radar network system: according to Ref. [364], in a multi-radar tracking system, the existence of systematic errors will lead to a bigger tracking mean square root error than its theoretical value. If the systematic errors are too large, it will happen that the effect of multiple radars' fusion tracking is even worse than that of singular radars. In the worst situation, they will cause failure in association of multi-radar measurements from the same trajectory. In this case, many tracks will be generated for the same target, and mistakenly recognized, due to the big difference between them, as belonging to different targets. This wrong recognition will bring difficulty and illegibility to track association and fusion, and a reduction in performance of the system tracks resulting from fusion. Under some complicated circumstances (especially in cases of dense targets and formation flights), it is more likely to create confusion in track association and a decrease in fusion accuracy, and thus render the overall systematic fusion meaningless, which further deprives the radar networking system of the advantages due to it. An example will be cited to explain the influences of systematic errors on target tracks below.

Suppose that 2D fixed radars A and B are deployed at different positions, and a public Cartesian coordinate system is established around radar A, which serves as the system fusion center. In this reference frame, these two radars' coordinates are (0, 0) and (x_{Bs}, 0), respectively. These two radars have some white Gaussian stochastic errors in measurements.

Also suppose that the mean square stochastic errors in range and azimuth measurements are (δr_A, $\delta\theta_A$) and (δr_B, $\delta\theta_B$), respectively, and the range and azimuth systematic errors are (Δr_A, $\Delta\theta_A$) and (Δr_B, $\Delta\theta_B$), respectively. The position state estimates corresponding to the same target among the target tracks reported by the two radars at time k are $(\hat{x}_A(k), \hat{y}_A(k))$ and $(\hat{x}_B(k), \hat{y}_B(k))$, respectively, but the actual position of the target is ($x(k)$, $y(k)$), with its actual polar coordinates being ($r_A(k)$, $\theta_A(k)$) and ($r_B(k)$, $\theta_B(k)$), respectively.

Both radars obtain the track of the target by filtering its measurements, with the influence of filtering error being ignored. This yields

$$\begin{cases} \hat{x}_A(k) = (r_A(k) + \Delta r_A)\sin(\theta_A(k) + \Delta\theta_A) \\ \hat{y}_A(k) = (r_A(k) + \Delta r_A)\cos(\theta_A(k) + \Delta\theta_A) \end{cases} \tag{15.1}$$

$$\begin{cases} \hat{x}_B(k) = (r_B(k) + \Delta r_B)\sin(\theta_B(k) + \Delta\theta_B) + x_{Bs} \\ \hat{y}_B(k) = (r_B(k) + \Delta r_B)\cos(\theta_B(k) + \Delta\theta_B) \end{cases} \tag{15.2}$$

$$\begin{aligned} \hat{x}_A(k) &= (r_A(k) + \Delta r_A)\sin(\theta_A(k) + \Delta\theta_A) \\ &= r_A(k)\sin(\theta_A(k) + \Delta\theta_A) + \Delta r_A \sin(\theta_A(k) + \Delta\theta_A) \\ &= r_A(k)\sin\theta_A(k)\cos\Delta\theta_A + r_A(k)\cos\theta_A(k)\sin\Delta\theta_A + \Delta r_A \sin(\theta_A(k) + \Delta\theta_A) \\ &= x(k)\cos\Delta\theta_A + y(k)\sin\Delta\theta_A + \Delta r_A \sin(\theta_A(k) + \Delta\theta_A) \end{aligned} \tag{15.3}$$

$$\begin{aligned} \hat{y}_A(k) &= (r_A(k) + \Delta r_A)\cos(\theta_A(k) + \Delta\theta_A) \\ &= r_A(k)\cos(\theta_A(k) + \Delta\theta_A) + \Delta r_A \cos(\theta_A(k) + \Delta\theta_A) \\ &= r_A(k)\cos\theta_A(k)\cos\Delta\theta_A - r_A(k)\sin\theta_A(k)\sin\Delta\theta_A + \Delta r_A \cos(\theta_A(k) + \Delta\theta_A) \\ &= -x(k)\sin\Delta\theta_A + y(k)\cos\Delta\theta_A + \Delta r_A \cos(\theta_A(k) + \Delta\theta_A) \end{aligned} \tag{15.4}$$

Thus, we have

$$\begin{cases} \hat{x}_A(k) = x(k)\cos\Delta\theta_A + y(k)\sin\Delta\theta_A + \Delta r_A \sin(\theta_A(k) + \Delta\theta_A) \\ \hat{y}_A(k) = -x(k)\sin\Delta\theta_A + y(k)\cos\Delta\theta_A + \Delta r_A \cos(\theta_A(k) + \Delta\theta_A) \end{cases} \tag{15.5}$$

Likewise, in view of the relative coordinates of radar B, we get

$$\begin{aligned} \hat{x}_B(k) &= (r_B(k) + \Delta r_B)\sin(\theta_B(k) + \Delta\theta_B) + x_{Bs} \\ &= r_B(k)\sin\theta_B(k)\cos\Delta\theta_B + r_B(k)\cos\theta_B(k)\sin\Delta\theta_B + \Delta r_B \sin(\theta_B(k) + \Delta\theta_B) + x_{Bs} \\ &= (x(k) - x_{Bs})\cos\Delta\theta_B + y(k)\sin\Delta\theta_B + \Delta r_B \sin(\theta_B(k) + \Delta\theta_B) + x_{Bs} \\ &= x(k)\cos\Delta\theta_B + y(k)\sin\Delta\theta_B + \Delta r_B \sin(\theta_B(k) + \Delta\theta_B) + x_{Bs}(1 - \cos\Delta\theta_B) \end{aligned} \tag{15.6}$$

Similarly,

$$\hat{y}_B(k) = -x(k)\sin\Delta\theta_B + y(k)\cos\Delta\theta_B + \Delta r_B\cos(\theta_B(k)+\Delta\theta_B) + x_{Bs}\sin\Delta\theta_B \tag{15.7}$$

Hence, it follows that

$$\begin{cases} \hat{x}_B(k) = x(k)\cos\Delta\theta_B + y(k)\sin\Delta\theta_B + \Delta r_B\sin(\theta_B(k)+\Delta\theta_B) + x_{Bs}(1-\cos\Delta\theta_B) \\ \hat{y}_B(k) = -x(k)\sin\Delta\theta_B + y(k)\cos\Delta\theta_B + \Delta r_B\cos(\theta_B(k)+\Delta\theta_B) + x_{Bs}\sin\Delta\theta_B \end{cases} \tag{15.8}$$

Combining (15.5) and (15.8) with the real coordinates $(x(k), y(k))$ cancelled, after derivation, gives

$$\begin{cases} \hat{x}_B(k) = \hat{x}_A(k)\cos(\Delta\theta_B-\Delta\theta_A) + \hat{y}_A(k)\sin(\Delta\theta_B-\Delta\theta_A) \\ \qquad -(-(\Delta r_A\sin(\theta_A(k)+\Delta\theta_B) + \Delta r_B\sin(\theta_B(k)+\Delta\theta_B) + x_{Bs}(1-\cos\Delta\theta_B))) \\ \hat{y}_B(k) = -\hat{x}_A(k)\sin(\Delta\theta_B-\Delta\theta_A) + \hat{y}_A(k)\cos(\Delta\theta_B-\Delta\theta_A) \\ \qquad -(-(\Delta r_A\cos(\theta_A(k)+\Delta\theta_B) + \Delta r_B\cos(\theta_B(k)+\Delta\theta_B) + x_{BS}\sin\Delta\theta_B)) \end{cases} \tag{15.9}$$

Here we define

$$\begin{cases} \theta_0 = \Delta\theta_B - \Delta\theta_A \\ C_x = -(\Delta r_A\sin(\theta_A(k)+\Delta\theta_B) + \Delta r_B\sin(\theta_B(k)+\Delta\theta_B) + x_{Bs}(1-\cos\Delta\theta_B)) \\ C_y = -(\Delta r_A\cos(\theta_A(k)+\Delta\theta_B) + \Delta r_B\cos(\theta_B(k)+\Delta\theta_B) + x_{BS}\sin\Delta\theta_B) \end{cases} \tag{15.10}$$

where, because of the target's relative movement to either radar, only $\theta_A(k)$ and $\theta_B(k)$ change with time. Radar measurement systematic errors are usually relatively small constants or slow variables of addition errors, and the target's position change is very small in a certain period of time. Thus, C_x and C_y can be regarded approximately as constants, and θ_0 is also a constant. Therefore, we have

$$\begin{bmatrix} \hat{x}_B(k) \\ \hat{y}_B(k) \end{bmatrix} = \begin{bmatrix} \cos\theta_0 & \sin\theta_0 \\ -\sin\theta_0 & \cos\theta_0 \end{bmatrix} \begin{bmatrix} \hat{x}_A(k) \\ \hat{y}_A(k) \end{bmatrix} - \begin{bmatrix} C_x \\ C_y \end{bmatrix} \tag{15.11}$$

Thus, it follows from (15.10) and (15.11) that the range systematic errors of networking radars generate relative translations between the target tracks reported by both radars, and azimuth systematic errors mainly produce relative rotations between the target tracks reported by both radars.

Moreover, in some special cases (e.g., when fierce bombardment of the radar position leads to inclination of the antenna), systematic errors in range and azimuth measurements may well grow too large, and the above conclusion should be modified. Then,

$$\begin{aligned} C_x &= -\Delta r_A\sin(\theta_A(k)+\Delta\theta_B) - \Delta r_B\sin(\theta_B(k)+\Delta\theta_B) - x_{Bs}(1-\cos\Delta\theta_B) \\ &= -\Delta r_A\sin(\theta_A(k)+\Delta\theta_A+\Delta\theta_B-\Delta\theta_A) - \Delta r_B\sin(\theta_B(k)+\Delta\theta_B) - x_{Bs}(1-\cos\Delta\theta_B) \\ &= -\Delta r_A\sin(\theta_A(k)+\Delta\theta_A)\cos(\Delta\theta_B-\Delta\theta_A) - \Delta r_A\cos(\theta_A(k)+\Delta\theta_A)\sin(\Delta\theta_B-\Delta\theta_A) \\ &\quad - \Delta r_B\sin(\theta_B(k)+\Delta\theta_B) - x_{Bs}(1-\cos\Delta\theta_B) \\ &= -\frac{\Delta r_A}{\Delta r_A + r_A}\hat{x}_A(k)\cos(\Delta\theta_B-\Delta\theta_A) - \frac{\Delta r_A}{\Delta r_A + r_A}\hat{y}_A(k)\sin(\Delta\theta_B-\Delta\theta_A) - \frac{\Delta r_B}{\Delta r_B + r_B}\hat{x}_B(k) - x_{Bs}(1-\cos\Delta\theta_B) \end{aligned} \tag{15.12}$$

Likewise, we have

$$C_y = -\frac{\Delta r_A}{\Delta r_A + r_A}\hat{x}_A(k)\sin(\Delta\theta_B - \Delta\theta_A) - \frac{\Delta r_B}{\Delta r_B + r_B}\hat{y}_B(k) - \frac{\Delta r_A}{\Delta r_A + r_A}\hat{y}_A(k)\cos(\Delta\theta_B - \Delta\theta_A) - x_{Bs}\sin\Delta\theta_B \tag{15.13}$$

Substituting (15.12) and (15.13) into (15.11), after some simplification, yields

$$\begin{bmatrix}\hat{x}_B(k)\\ \hat{y}_B(k)\end{bmatrix} = \frac{\Delta r_B + r_B}{r_B}\frac{2\Delta r_A + r_A}{\Delta r_A + r_A}\begin{bmatrix}\cos\theta_0 & \sin\theta_0\\ -\sin\theta_0 & \cos\theta_0\end{bmatrix}\begin{bmatrix}\hat{x}_A(k)\\ \hat{y}_A(k)\end{bmatrix} + \frac{\Delta r_B + r_B}{r_B}\begin{bmatrix}x_{Bs}(1-\cos\Delta\theta_B)\\ x_{Bs}\sin\Delta\theta_B\end{bmatrix} \tag{15.14}$$

As can be seen from the above equation, in the case of big systematic errors, range systematic errors will cause a larger influence on tracks. Big range systematic errors will cause translation and rotation, and affine transformation of the tracks, which in turn leads to their overall deformation, namely, the difference in real positions for measurements caused in their scales and translation values. This shows that big range systematic errors will bring forth a departure of the target's course from the real course, and even changes in target speed and heading angle between targets. Azimuth systematic errors, however, will probably only produce a slight increase in translation of the target track and have little influence. Therefore, in order to ensure the overall performance of the radar network system, radar registration has to be done so as to eliminate the systematic errors.

15.3 Fixed Radar Registration Algorithm

According to the difference in radar platforms, radar registration algorithms can be classified into two types: fixed and mobile. The former needs to eliminate only systematic measurement errors by registration, while the latter needs to remove systematic errors in attitude angles in addition because of platform motion. This section primarily approaches registration problems with fixed radars.

With respect to fixed radars, the simplest method is registration conditioned on the given target positions. This condition, however, is hard to satisfy sometimes, hence many radar registration algorithms have been developed with the condition that the target positions are unknown. According to the difference in coordinate systems, fixed radar registration algorithms can be further divided into two subgroups: those based on stereographic projections [365] and those based on earth-centered earth-fixed (ECEF) coordinate systems. The former includes real-time quality control (RTQC) [366], least-squares (LS) [367], generalized least-squares (GLS) [364], and exact maximum likelihood (EML) [368] registration algorithms. These algorithms are discussed below.

15.3.1 Radar Registration Algorithm Based on Cooperative Targets

Under this condition, just align each radar with the targets, whose positions are already known. Suppose that the systematic errors to be compensated do not change with time and space (Figure 15.1).

The measurement error can be written as

$$\begin{cases}\delta_\rho = \rho_M - \rho = \Delta\rho + \varepsilon_\rho\\ \delta_\theta = \theta_M - \theta = \Delta\theta + \varepsilon_\theta\end{cases} \tag{15.15}$$

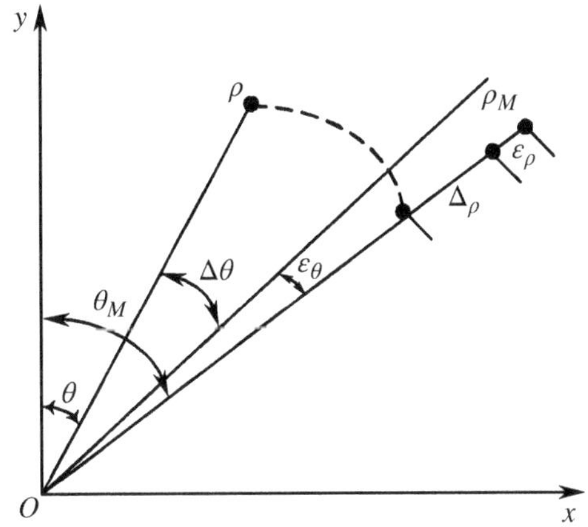

Figure 15.1 Systematic errors and stochastic errors in radar measurement

where (ρ_M, θ_M) is the polar coordinate measurement of the target and (ρ, θ) is the target's real position in polar coordinates. Actually, the above errors are both composed of two terms. One term is $(\Delta\rho, \Delta\theta)$ and, being deterministic error, is unknown. The other term is $(\varepsilon_\rho, \varepsilon_\theta)$ and, being stochastic observation errors, they are not mutually correlated, usually having zero mean and known variance σ_ρ^2 and σ_θ^2. Then, we get the average value of proper measurements and this can decrease the influence of stochastic errors.

A certain radar in the radar network, in a certain period of time, can get n measurement variables $\{(\rho_M(i), \theta_M(i))|i=1,\ldots,n\}$ of a target whose position is already known, the differences between which and the actual positions are n error variables $\{(\delta_\rho(i), \delta_\theta(i))|i=1,\ldots,n\}$. Therefore, the estimates of $(\Delta\rho, \Delta\theta)$ can be obtained from

$$\begin{cases} \Delta\hat{\rho} = \dfrac{1}{n}\sum_{i=1}^{n}\delta_\rho(i) \\ \Delta\hat{\theta} = \dfrac{1}{n}\sum_{i=1}^{n}\delta_\theta(i) \end{cases} \tag{15.16}$$

with variances σ_ρ^2/n and σ_θ^2/n, respectively.

If the range error caused by the distance clock speed deviation is further considered, then its mathematical model can be supposed to be

$$\Delta\rho = a + b\rho \tag{15.17}$$

where a and b are parameters to be estimated. To estimate two unknown variables, we need to know the positions of two different targets P_1 and P_2 whose coordinates are assumed to be (ρ_1, θ_1) and (ρ_2, θ_2). When there is no observation noise $(\varepsilon_\rho = 0)$, the range measurement errors can be expressed as

$$\begin{cases} \delta_{\rho_1} = a + b\rho_1 \\ \delta_{\rho_2} = a + b\rho_2 \end{cases} \tag{15.18}$$

So, the estimates of a and b can be derived as

$$\begin{cases} \hat{a} = \dfrac{-\rho_2\delta_{\rho_1} + \rho_1\delta_{\rho_2}}{\rho_1 - \rho_2} \\ \hat{b} = \dfrac{\delta_{\rho_1} - \delta_{\rho_2}}{\rho_1 - \rho_2} \end{cases} \tag{15.19}$$

When there is observation noise, $\hat{a}$ and $\hat{b}$ are stochastic variables with their means being a and b, respectively, and their variances

$$\begin{cases} \sigma_{\hat{a}}^2 = \dfrac{\rho_1^2 + \rho_2^2}{(\rho_1 - \rho_2)^2}\sigma_\rho^2 \\ \sigma_{\hat{b}}^2 = \dfrac{2}{(\rho_1 - \rho_2)^2}\sigma_\rho^2 \end{cases} \tag{15.20}$$

where ρ_1 and ρ_2 are assumed uncorrelated, with the same variance σ_ρ^2. From (15.20), it can be seen that the longer the radial distance between two observation targets, the more accurate the estimates.

15.3.2 RTQC Algorithm

For the radar registration algorithm with target positions unknown, the systematic errors are usually estimated using many measurements of the same target from two radars (see Figure 15.2) and the radars' measurements are then corrected according to the estimated values.

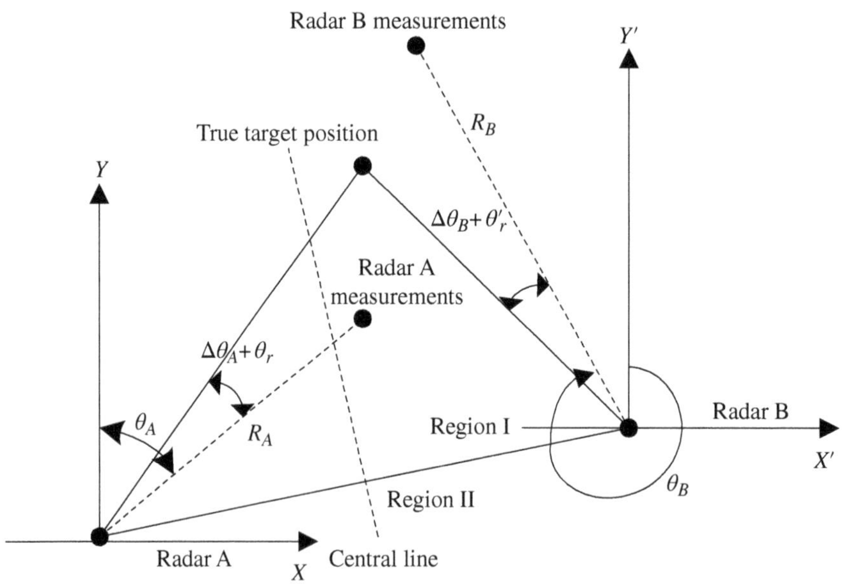

Figure 15.2 Radar registration based on stereographic projection

The RTQC is an algorithm based on stereographic (polar) projection. Before registration, this algorithm projects different radars' measurements of the same target to the public 2D coordinate system and therefore it can estimate the radars' azimuth and range systematic errors only.

Figure 15.2 shows the geometric relationship of two radars located at different places. The systematic errors have two components: range and azimuth errors, namely, $\beta = [\Delta R_A, \Delta R_B, \Delta\theta_A, \Delta\theta_B]'$. For two radar stations S_A and S_B, which are projected on the plane, their coordinates are (x_{S_A}, y_{S_A}) and (x_{S_B}, y_{S_B}), respectively. (x'_A, y'_A) and (x'_B, y'_B) denote the coordinates of target T_k in the local coordinate systems of the radars, respectively, and (R_A, θ_A) and (R_B, θ_B) are the radars' measurements (including systematic and stochastic errors) about target T_k. If the effects of stochastic measurement errors R_r, θ_r, R_r', and θ_r' are neglected, from the geometric relationship in Figure 15.2 it follows that

$$x_A' = (R_A - \Delta R_A)\sin(\theta_A - \Delta\theta_A) \tag{15.21}$$

$$y_A' = (R_A - \Delta R_A)\cos(\theta_A - \Delta\theta_A) \tag{15.22}$$

Because ΔR_A and $\Delta\theta_A$ are minute, and if second-order and above Taylor expansions are neglected, (15.21) and (15.22) can be simplified as

$$x'_A = R_A \sin\theta_A - \Delta R_A \sin\theta_A - R_A \Delta\theta_A \cos\theta_A \tag{15.23}$$

$$y'_A = R_A \cos\theta_A - \Delta R_A \cos\theta_A + R_A \Delta\theta_A \sin\theta_A \tag{15.24}$$

Likewise, we have

$$x_B' = R_B \sin\theta_B - \Delta R_B \sin\theta_B - R_B \Delta\theta_B \cos\theta_B \tag{15.25}$$

$$y_B' = R_B \cos\theta_B - \Delta R_B \cos\theta_B + R_B \Delta\theta_B \sin\theta_B \tag{15.26}$$

For the same target,

$$x_{S_A} + x'_A = x_{S_B} + x'_B \tag{15.27}$$

$$y_{S_A} + y'_A = y_{S_B} + y'_B \tag{15.28}$$

Let

$$x_A = x_{S_A} + R_A \sin\theta_A \tag{15.29}$$

$$y_A = y_{S_A} + R_A \cos\theta_A \tag{15.30}$$

$$x_B = x_{S_B} + R_B \sin\theta_B \tag{15.31}$$

$$y_B = y_{S_B} + R_B \cos\theta_B \tag{15.32}$$

$$A = x_A - x_B = \sin\theta_A \Delta R_A - \sin\theta_B \Delta R_B + R_A \cos\theta_A \Delta\theta_A - R_B \cos\theta_B \Delta\theta_B \tag{15.33}$$

$$B = y_A - y_B = \cos\theta_A \Delta R_A - \cos\theta_B \Delta R_B - R_A \sin\theta_A \Delta\theta_A + R_B \sin\theta_B \Delta\theta_B \tag{15.34}$$

$$PP = A\sin\theta_A + B\cos\theta_A \tag{15.35}$$

$$QQ = -A\sin\theta_B - B\cos\theta_B \tag{15.36}$$

Substituting (15.33) and (15.34) into (15.35) and (15.36) gives

$$PP = \Delta R_A - \cos(\theta_A - \theta_B)\Delta R_B - R_B \sin(\theta_A - \theta_B)\Delta\theta_B \tag{15.37}$$

$$QQ = -\cos(\theta_A - \theta_B)\Delta R_A + \Delta R_B + R_A \sin(\theta_A - \theta_B)\Delta\theta_A \tag{15.38}$$

Averaging all the measurements, it follows from (15.37) and (15.38) that

$$\left[\overline{PP_1}\ \ \overline{QQ_1}\ \ \overline{PP_2}\ \ \overline{QQ_2}\right]' = \boldsymbol{S}[\Delta R_A\ \ \Delta R_B\ \ \Delta\theta_A\ \ \Delta\theta_B]' \tag{15.39}$$

where

$$\boldsymbol{S} = \begin{bmatrix} 1 & \overline{-\cos(\theta_{A1}-\theta_{B1})} & 0 & \overline{-R_{B1}\sin(\theta_{A1}-\theta_{B1})} \\ \overline{-\cos(\theta_{A1}-\theta_{B1})} & 1 & \overline{R_{A1}\sin(\theta_{A1}-\theta_{B1})} & 0 \\ 1 & \overline{-\cos(\theta_{A2}-\theta_{B2})} & 0 & \overline{-R_{B2}\sin(\theta_{A2}-\theta_{B2})} \\ \overline{-\cos(\theta_{A2}-\theta_{B2})} & 1 & \overline{R_{A2}\sin(\theta_{A2}-\theta_{B2})} & 0 \end{bmatrix} \tag{15.40}$$

The bars in the equation denote averaging, and subscript 1 indicates the measurements in area I while subscript 2 indicates the measurements in area II (area I is the area above the line S_AS_B in Figure 15.2 and area II is the area below the line S_AS_B). The error vector $\boldsymbol{\beta} = [\Delta R_A, \Delta R_B, \Delta\theta_A, \Delta\theta_B]'$ can be obtained from (15.40).

In order to keep full rank of matrix $\boldsymbol{S}$, the RTQC algorithm requires that the chosen targets be distributed on both sides of the line S_AS_B and not too close to the line S_AS_B or too far from the central line.

15.3.3 LS Algorithm

The performance of the RTQC algorithm is influenced greatly by the distribution of targets and it requires that the targets be distributed on both sides of the line S_AS_B. Therefore, the radar registration algorithm based on LS has been put forward according to the improvements in RTQC [367].

For N different measurements, from (15.33) and (15.34) we get

$$A(i) = x_A(i) - x_B(i) = \sin\theta_A(i)\Delta R_A - \sin\theta_B(i)\Delta R_B + R_A(i)\cos\theta_A(i)\Delta\theta_A - R_B(i)\cos\theta_B(i)\Delta\theta_B \tag{15.41}$$

$$B(i) = y_A(i) - y_B(i) = \cos\theta_A(i)\Delta R_A - \cos\theta_B(i)\Delta R_B - R_A(i)\sin\theta_A(i)\Delta\theta_A + R_B(i)\sin\theta_B(i)\Delta\theta_B \tag{15.42}$$

where $i = 1, \ldots, N$. This can be expressed in matrix form as

$$\boldsymbol{Z} = \boldsymbol{H\beta} \tag{15.43}$$

where

$$\boldsymbol{Z} = [P(1), Q(1), P(2), Q(2), \ldots, P(N), Q(N)]' \tag{15.44}$$

$$
H = \begin{bmatrix}
\sin\theta_A(1) & -\sin\theta_B(1) & R_A(1)\cos\theta_A(1) & -R_B(1)\cos\theta_B(1) \\
\cos\theta_A(1) & -\cos\theta_B(1) & -R_A(1)\sin\theta_A(1) & R_B(1)\sin\theta_B(1) \\
\sin\theta_A(2) & -\sin\theta_B(2) & R_A(2)\cos\theta_A(2) & -R_B(2)\cos\theta_B(2) \\
\cos\theta_A(2) & -\cos\theta_B(2) & -R_A(2)\sin\theta_A(2) & R_B(2)\sin\theta_B(2) \\
\vdots & \vdots & \vdots & \vdots \\
\sin\theta_A(N) & -\sin\theta_B(N) & R_A(N)\cos\theta_A(N) & -R_B(N)\cos\theta_B(N) \\
\cos\theta_A(N) & -\cos\theta_B(N) & -R_A(N)\sin\theta_A(N) & R_B(N)\sin\theta_B(N)
\end{bmatrix} \tag{15.45}
$$

$\boldsymbol{Z} = \boldsymbol{H\beta}$ is over-determined. Equation (15.46) can be solved by the LS estimation method, as follows:

$$
\boldsymbol{\beta} = (\boldsymbol{H}'\boldsymbol{H})^{-1}\boldsymbol{H}'\boldsymbol{Z} \tag{15.46}
$$

15.3.4 GLS Algorithm

One of the two radars is set as the primary radar, located at the origin of the coordinates, and the other as the secondary radar, located at (u, v). Then, assume that in the absence of stochastic measurement errors and conditioned only on systematic deviations, the measurement vector of the kth target is $\boldsymbol{\Psi}(k) = [r_A(k), \omega_A(k), r_B(k), \omega_B(k)]'$. Let

$$
\Delta x(k) = [r_A(k) + \Delta r_A]\sin[\omega_A(k) + \Delta\theta_A] - u - [r_B(k) + \Delta r_B]\sin[\omega_B(k) + \Delta\theta_B] \tag{15.47}
$$

$$
\Delta y(k) = [r_A(k) + \Delta r_A]\cos[\omega_A(k) + \Delta\theta_A] - v - [r_B(k) + \Delta r_B]\cos[\omega_B(k) + \Delta\theta_B] \tag{15.48}
$$

Let $f[\boldsymbol{\Psi}(k), \boldsymbol{\beta}] = [\Delta x(k), \Delta y(k)]'$ and make a first-order Taylor expansion of it. Then, we have

$$
f[\boldsymbol{\Psi}(k), \boldsymbol{\beta}] \approx f[\boldsymbol{\Psi}'(k), \boldsymbol{\beta}'] + \nabla_{\beta}[f(\boldsymbol{\Psi}'(k), \boldsymbol{\beta}')](\boldsymbol{\beta} - \boldsymbol{\beta}') + \nabla_{\Psi}[f(\boldsymbol{\Psi}'(k), \boldsymbol{\beta}')][\boldsymbol{\Psi}(k) - \boldsymbol{\Psi}'(k)] \tag{15.49}
$$

where $\boldsymbol{\Psi}'(k)$ is the actual target measurements (including systematic and stochastic errors, uncorrected) obtained by radars S_A and S_B at sampling time k, and $\boldsymbol{\beta}'$ is the initial estimate of systematic deviations. Conditioned on the absence of prior information, assume that $\boldsymbol{\beta}' = (0,0,0,0)'$. $\nabla_{\Psi}[f(\boldsymbol{\Psi}'(k), \boldsymbol{\beta}')]$ and $\nabla_{\beta}[f(\boldsymbol{\Psi}'(k), \boldsymbol{\beta}')]$ are as follows:

$$
\nabla_{\Psi}[f(\boldsymbol{\Psi}'(k), \boldsymbol{\beta}')] = \begin{bmatrix}
\dfrac{\partial(\Delta x(k))}{\partial r_A(k)} & \dfrac{\partial(\Delta x(k))}{\partial \omega_A(k)} & \dfrac{\partial(\Delta x(k))}{\partial r_B(k)} & \dfrac{\partial(\Delta x(k))}{\partial \omega_B(k)} \\
\dfrac{\partial(\Delta y(k))}{\partial r_A(k)} & \dfrac{\partial(\Delta y(k))}{\partial \omega_A(k)} & \dfrac{\partial(\Delta y(k))}{\partial r_B(k)} & \dfrac{\partial(\Delta y(k))}{\partial \omega_B(k)}
\end{bmatrix} = \boldsymbol{\kappa}(k) \tag{15.50}
$$

$$
\boldsymbol{\xi} = [\boldsymbol{\kappa}(1)\partial\boldsymbol{\Psi}(1), \boldsymbol{\kappa}(2)\partial\boldsymbol{\Psi}(2), \ldots, \boldsymbol{\kappa}(N)\partial\boldsymbol{\Psi}(N)]' \tag{15.51}
$$

$$
\nabla_{\beta}[f(\boldsymbol{\Psi}'(k), \boldsymbol{\beta}')] = \begin{bmatrix}
\dfrac{\partial(\Delta x(k))}{\partial(\Delta r_A)} & \dfrac{\partial(\Delta x(k))}{\partial(\Delta\theta_A)} & \dfrac{\partial(\Delta x(k))}{\partial(\Delta r_B)} & \dfrac{\partial(\Delta x(k))}{\partial(\Delta\theta_B)} \\
\dfrac{\partial(\Delta y(k))}{\partial(\Delta r_A)} & \dfrac{\partial(\Delta y(k))}{\partial(\Delta\theta_A)} & \dfrac{\partial(\Delta y(k))}{\partial(\Delta r_B)} & \dfrac{\partial(\Delta y(k))}{\partial(\Delta\theta_B)}
\end{bmatrix} = \boldsymbol{\zeta}(k) \tag{15.52}
$$

For the same target, $f[\boldsymbol{\Psi}(k),\boldsymbol{\beta}]=[0,0]'$. Suppose that $[\boldsymbol{\Psi}(k)-\boldsymbol{\Psi}'(k)]$ and $(\boldsymbol{\beta}-\boldsymbol{\beta}')$ are small enough, with negligible high-order components. Then,

$$\boldsymbol{\zeta}(k)\boldsymbol{\beta}+\boldsymbol{\kappa}(k)\partial\boldsymbol{\Psi}(k)=\boldsymbol{\zeta}(k)\boldsymbol{\beta}'-f[\boldsymbol{\Psi}'(k),\boldsymbol{\beta}'] \tag{15.53}$$

where $\partial\boldsymbol{\Psi}(k)=[\boldsymbol{\Psi}(k)-\boldsymbol{\Psi}'(k)]$. For $\boldsymbol{\Psi}(k)$, we only consider systematic deviations instead of stochastic errors. Therefore,

$$\partial\boldsymbol{\Psi}(k)=\left[R_r,\theta_r,R_r',\theta_r'\right] \tag{15.54}$$

In the above, $\boldsymbol{\kappa}(k)\partial\boldsymbol{\Psi}(k)$ is the error caused by measurement noise and $\boldsymbol{\zeta}(k)$ is the matrix of the given parameter. Hence, the right-hand side of (15.53) represents observations and can be expressed as

$$\boldsymbol{X}\boldsymbol{\beta}+\boldsymbol{\xi}=\boldsymbol{Y} \tag{15.55}$$

where

$$\boldsymbol{X}=[\boldsymbol{\zeta}(1),\boldsymbol{\zeta}(2),\ldots,\boldsymbol{\zeta}(N)]' \tag{15.56}$$

$$\boldsymbol{\xi}=[\boldsymbol{k}(1)\partial\boldsymbol{\Psi}(1),\boldsymbol{k}(2)\partial\boldsymbol{\Psi}(2),\ldots,\boldsymbol{k}(N)\partial\boldsymbol{\Psi}(N)]' \tag{15.57}$$

$$\boldsymbol{Y}=[\boldsymbol{\zeta}(1)\boldsymbol{\beta}'-f(\boldsymbol{\Psi}'(1),\boldsymbol{\beta}'),\boldsymbol{\zeta}(2)\boldsymbol{\beta}'-f(\boldsymbol{\Psi}'(2),\boldsymbol{\beta}'),\ldots,\boldsymbol{\zeta}(N)\boldsymbol{\beta}'-f(\boldsymbol{\Psi}'(N),\boldsymbol{\beta}')]' \tag{15.58}$$

Let

$$\boldsymbol{\Sigma}_{\xi}=\mathrm{E}[\boldsymbol{\xi}\boldsymbol{\xi}']=\left\{\boldsymbol{k}(i)\mathrm{E}\left[(\partial\boldsymbol{\Psi}(i))(\partial\boldsymbol{\Psi}(j))'\right]\boldsymbol{k}(j)'|i,j=1,2,\ldots,N\right\} \tag{15.59}$$

If $i\neq j$,

$$\mathrm{E}\left[(\partial\boldsymbol{\Psi}(i))(\partial\boldsymbol{\Psi}(j))'\right]=0 \tag{15.60}$$

If $i=j$,

$$\mathrm{E}\left[(\partial\boldsymbol{\Psi}(i))(\partial\boldsymbol{\Psi}(j))'\right]=\begin{bmatrix}\sigma_r^2(A) & 0 & 0 & 0\\ 0 & \sigma_\theta^2(A) & 0 & 0\\ 0 & 0 & \sigma_r^2(B) & 0\\ 0 & 0 & 0 & \sigma_\theta^2(B)\end{bmatrix} \tag{15.61}$$

Since $\boldsymbol{\kappa}(k)$ is a 2×4 matrix, and $\boldsymbol{\Sigma}_{\Psi}$ is a 4×4 matrix, then $\boldsymbol{\Sigma}_{\xi}$ is a block-diagonal matrix $\{\boldsymbol{\Sigma}_1,\boldsymbol{\Sigma}_2,\boldsymbol{\Sigma}_3,\ldots,\boldsymbol{\Sigma}_N\}$, where

$$\boldsymbol{\Sigma}_k=\boldsymbol{\kappa}(k)\boldsymbol{\Sigma}_{\Psi}\boldsymbol{\kappa}(k)' \tag{15.62}$$

Hence, the resolution of (15.55) can be obtained as follows:

$$\boldsymbol{\beta}^*=\left(\boldsymbol{X}'\boldsymbol{\Sigma}_{\xi}^{-1}\boldsymbol{X}\right)^{-1}\boldsymbol{X}'\boldsymbol{\Sigma}_{\xi}^{-1}\boldsymbol{Y} \tag{15.63}$$

$$\mathrm{cov}(\boldsymbol{\beta}^*)=\left(\boldsymbol{X}'\boldsymbol{\Sigma}_{\xi}^{-1}\boldsymbol{X}\right)^{-1} \tag{15.64}$$

From (15.64), we know that the accuracy of the GLS registration is only correlated with the radar's measurement accuracy and the targets' spatial distribution. And, since $\boldsymbol{\Sigma}_{\xi}$ is a $2N \times 2N$ matrix, (15.63) and (15.64) can be broken down into N small matrixes and then calculated:

$$\boldsymbol{X}'\boldsymbol{\Sigma}_{\xi}^{-1}\boldsymbol{X} = \sum_{k=1}^{N} \boldsymbol{\zeta}(k)'\boldsymbol{\Sigma}_{k}^{-1}\boldsymbol{\zeta}(k) \tag{15.65}$$

$$\boldsymbol{X}'\boldsymbol{\Sigma}_{\xi}^{-1}\boldsymbol{Y} = \sum_{k=1}^{N} \boldsymbol{\zeta}(k)'\boldsymbol{\Sigma}_{k}^{-1}[\boldsymbol{\zeta}(k)\boldsymbol{\beta}' - f(\boldsymbol{\Psi}'(k),\boldsymbol{\beta}')] \tag{15.66}$$

When N is very big, the calculation speed can be improved significantly.

15.3.5 GLS Algorithm in ECEF Coordinate System

The radar registration algorithms introduced in Sections 15.3.2–15.3.4 are all based on stereographic projection. They are widely used in engineering implementations thanks to their advantages, such as simplicity and easy realization. These methods, however, also have disadvantages, as listed below.

1. Although stereographic projection uses high-order approximation to improve accuracy, the earth is an elliptical not a round sphere. Hence, errors are introduced into measurements in projection.
2. Data distortion will arise from stereographic projection. Stereographic conformal projection, for example, can only ensure that the azimuth remains undistorted, but cannot warrant the same for the range. As a result, the systematic errors will not be constants anymore but be correlated to the measurements.
3. In the 2D public coordinate system, they can estimate only azimuth and range errors but not elevation ones.

Therefore, radar registration technologies based on stereographic projection are usually used in the case of short-range radars, while in the case of long-range radars we mainly adopt radar registration technologies based on ECEF coordinates (ECEF-GLS). The ECEF-GLS registration algorithm introduced in this section is derived from the ECEF coordinate system.

15.3.5.1 Coordinate Transformation Relationship

Any point on earth can be denoted by geographical coordinates (L, λ, H), where L stands for altitude, λ longitude, and H the height based on the reference elliptical sphere, namely elevation. Suppose that the radar's geographical coordinates are (L, λ, H) and the ECEF Descartes coordinates are (x_s, y_s, z_s). Then,

$$\begin{cases} x_s = (C + H_s)\cos L_s \cos\lambda_s \\ y_s = (C + H_s)\cos L_s \sin\lambda_s \\ z_s = [C(1 - e^2) + H_s]\sin L_s \end{cases} \tag{15.67}$$

where e is the eccentricity of the earth, and C is defined as

$$C=\frac{E_q}{\left(1-e^2\sin^2 L_s\right)^{1/2}} \tag{15.68}$$

where E_q is the equator radius.

Suppose that the radar measurements are (r_t, θ_t, η_t), with r_t being the range, θ_t the azimuth, and η_t the elevation. The radar measurements can be transformed into the local Descartes coordinate system

$$\begin{cases} x_l = r_t \sin\theta_t \cos\eta_t \\ y_l = r_t \cos\theta_t \cos\eta_t \\ z_l = r_t \sin\eta_t \end{cases} \tag{15.69}$$

Equation (15.69) can be used to transform the target's local Descartes coordinates to the ECEF, namely,

$$\begin{bmatrix} x_t \\ y_t \\ z_t \end{bmatrix} = \begin{bmatrix} x_s \\ y_s \\ z_s \end{bmatrix} + \boldsymbol{T} \times \begin{bmatrix} x_l \\ y_l \\ z_l \end{bmatrix} \tag{15.70}$$

where (x_t, y_t, z_t) is the ECEF coordinates, (x_l, y_l, z_l) denotes local coordinates, and $\boldsymbol{T}$ the rotation matrix, namely,

$$\boldsymbol{T}=\begin{bmatrix} -\sin\lambda_s & -\sin L_s \cos\lambda_s & \cos L_s \cos\lambda_s \\ \cos\lambda_s & -\sin L_s \sin\lambda_s & \cos L_s \sin\lambda_s \\ 0 & \cos L_s & \sin L_s \end{bmatrix} \tag{15.71}$$

15.3.5.2 ECEF-GLS Registration Algorithm

Let (L_A, λ_A, H_A) and (L_B, λ_B, H_B) be the geographical coordinates of radars A and B, respectively, and (x_{As}, y_{As}, z_{As}) and (x_{Bs}, y_{Bs}, z_{Bs}) the ECEF coordinates of the radars. T_k denotes the target at time k. $[r_A(k), \theta_A(k), \eta_A(k)]$ and $[r_B(k), \theta_B(k), \eta_B(k)]$ are their measurements of target T_k, and $\boldsymbol{\beta}=[\Delta r_A, \Delta\theta_A, \Delta\eta_A, \Delta r_B, \Delta\theta_B, \Delta\eta_B]'$ are their systematic errors, $[R_r(k), \theta_r(k), \eta_r(k)]$ and $[R_r'(k), \theta_r'(k), \eta_r'(k)]$ are their stochastic errors. $[r_A''(k), \theta_A''(k), \eta_A''(k)]$ and $[r_B''(k), \theta_B''(k), \eta_B''(k)]$ are their measurements of target T_k when only considering systematic deviations but not stochastic errors, and let $\boldsymbol{\Psi}(k)=[r_A''(k), \theta_A''(k), \eta_A''(k), r_B''(k), \theta_B''(k), \eta_B''(k)]'$. Then, the coordinates of target T_k in radars A and B's local coordinate systems are

$$\begin{cases} x_{Al}'(k) = [r_A''(k)-\Delta r_A]\sin[\theta_A''(k)-\Delta\theta_A]\cos[\eta_A''(k)-\Delta\eta_A] \\ y_{Al}'(k) = [r_A''(k)-\Delta r_A]\cos[\theta_A''(k)-\Delta\theta_A]\cos[\eta_A''(k)-\Delta\eta_A] \\ z_{Al}'(k) = [r_A''(k)-\Delta r_A]\sin[\eta_A''(k)-\Delta\eta_A] \end{cases} \tag{15.72}$$

$$\begin{cases} x'_{Bl}(k) = \left[r''_B(k) - \Delta r_B\right]\sin\left[\theta''_B(k) - \Delta\theta_B\right]\cos\left[\eta''_B(k) - \Delta\eta_B\right] \\ y'_{Bl}(k) = \left[r''_B(k) - \Delta r_B\right]\cos\left[\theta''_B(k) - \Delta\theta_B\right]\cos\left[\eta''_B(k) - \Delta\eta_B\right] \\ z'_{Bl}(k) = \left[r''_B(k) - \Delta r_B\right]\sin\left[\eta''_B(k) - \Delta\eta_B\right] \end{cases} \quad (15.73)$$

The targets' local coordinates can be transformed, using (15.70), into ECEF coordinates:

$$\begin{bmatrix} x_t(k) \\ y_t(k) \\ z_t(k) \end{bmatrix} = \begin{bmatrix} x_{As} \\ y_{As} \\ z_{As} \end{bmatrix} + \boldsymbol{T}_A \times \begin{bmatrix} x'_{Al}(k) \\ y'_{Al}(k) \\ z'_{Al}(k) \end{bmatrix} \quad (15.74)$$

$$\begin{bmatrix} x_t(k) \\ y_t(k) \\ z_t(k) \end{bmatrix} = \begin{bmatrix} x_{Bs} \\ y_{Bs} \\ z_{Bs} \end{bmatrix} + \boldsymbol{T}_B \times \begin{bmatrix} x'_{Bl}(k) \\ y'_{Bl}(k) \\ z'_{Bl}(k) \end{bmatrix} \quad (15.75)$$

Let

$$f(\boldsymbol{\Psi}(k),\boldsymbol{\beta}) = [\Delta x_k, \Delta y_k, \Delta z_k]^T = \begin{bmatrix} x_{As} \\ y_{As} \\ z_{As} \end{bmatrix} + \boldsymbol{T}_A \times \begin{bmatrix} x'_{Al}(k) \\ y'_{Al}(k) \\ z'_{Al}(k) \end{bmatrix} - \begin{bmatrix} x_{Bs} \\ y_{Bs} \\ z_{Bs} \end{bmatrix} - \boldsymbol{T}_B \times \begin{bmatrix} x'_{Bl}(k) \\ y'_{Bl}(k) \\ z'_{Bl}(k) \end{bmatrix} \quad (15.76)$$

Conducting first-order Taylor expansion yields

$$f(\boldsymbol{\Psi}(k),\boldsymbol{\beta}) \approx f(\boldsymbol{\Psi}'(k),\boldsymbol{\beta}') + \nabla_{\boldsymbol{\beta}}[f(\boldsymbol{\Psi}'(k),\boldsymbol{\beta}')](\boldsymbol{\beta}-\boldsymbol{\beta}') + \nabla_{\boldsymbol{\Psi}}[f(\boldsymbol{\Psi}'(k),\boldsymbol{\beta}')][\boldsymbol{\Psi}(k)-\boldsymbol{\Psi}'(k)] \quad (15.77)$$

where $\boldsymbol{\Psi}'(k)$ is radar S_A and S_B's measurements (including systematic and stochastic errors, uncorrected) of target t at the kth sampling time, and $\boldsymbol{\beta}'$ is the initial estimate of systematic errors. In the absence of any prior information, it can be assumed that $\boldsymbol{\beta}' = [0,0,0,0,0,0]'$.

Let $\boldsymbol{X}_A(k) = \left[x'_{Al}(k), y'_{Al}(k), z'_{Al}(k)\right]'$, $\boldsymbol{X}_B(k) = \left[x'_{Bl}(k), y'_{Bl}(k), z'_{Bl}(k)\right]'$. Then, $\nabla_{\boldsymbol{\Psi}}[f(\boldsymbol{\Psi}'(k),\boldsymbol{\beta}')]$ and $\nabla_{\boldsymbol{\beta}}[f(\boldsymbol{\Psi}'(k),\boldsymbol{\beta}')]$ are

$$\nabla_{\boldsymbol{\Psi}}[f(\boldsymbol{\Psi}'(k),\boldsymbol{\beta}')] = [\boldsymbol{T}_A \times \boldsymbol{J}_A(k), -\boldsymbol{T}_B \times \boldsymbol{J}_B(k)] = \boldsymbol{k}(k) \quad (15.78)$$

$$\nabla_{\boldsymbol{\beta}}[f(\boldsymbol{\Psi}'(k),\boldsymbol{\beta}')] = [\boldsymbol{T}_A \times \boldsymbol{L}_A(k), -\boldsymbol{T}_B \times \boldsymbol{L}_B(k)] = \boldsymbol{\zeta}(k) \quad (15.79)$$

where

$$\boldsymbol{J}_A(k) = \begin{bmatrix} \dfrac{\partial(x'_{Al}(k))}{\partial r''_A(k)} & \dfrac{\partial(x'_{Al}(k))}{\partial \theta''_A(k)} & \dfrac{\partial(x'_{Al}(k))}{\partial \eta''_A(k)} \\ \dfrac{\partial(y'_{Al}(k))}{\partial r''_A(k)} & \dfrac{\partial(y'_{Al}(k))}{\partial \theta''_A(k)} & \dfrac{\partial(y'_{Al}(k))}{\partial \eta''_A(k)} \\ \dfrac{\partial(z'_{Al}(k))}{\partial r''_A(k)} & \dfrac{\partial(z'_{Al}(k))}{\partial \theta''_A(k)} & \dfrac{\partial(z'_{Al}(k))}{\partial \eta''_A(k)} \end{bmatrix} \quad (15.80)$$

$$\boldsymbol{J}_B(k)=\begin{bmatrix}\frac{\partial(x'_{Bl}(k))}{\partial r''_B(k)} & \frac{\partial(x'_{Bl}(k))}{\partial \theta''_B(k)} & \frac{\partial(x'_{Bl}(k))}{\partial \eta''_B(k)} \\ \frac{\partial(y'_{Bl}(k))}{\partial r''_B(k)} & \frac{\partial(y'_{Bl}(k))}{\partial \theta''_B(k)} & \frac{\partial(y'_{Bl}(k))}{\partial \eta''_B(k)} \\ \frac{\partial(z'_{Bl}(k))}{\partial r''_B(k)} & \frac{\partial(z'_{Bl}(k))}{\partial \theta''_B(k)} & \frac{\partial(z'_{Bl}(k))}{\partial \eta''_B(k)}\end{bmatrix} \tag{15.81}$$

$$\boldsymbol{L}_A(k)=\begin{bmatrix}\frac{\partial(x'_{Al}(k))}{\partial \Delta r_A} & \frac{\partial(x'_{Al}(k))}{\partial \Delta \theta_A} & \frac{\partial(x'_{Al}(k))}{\partial \Delta \eta_A} \\ \frac{\partial(y'_{Al}(k))}{\partial \Delta r_A} & \frac{\partial(y'_{Al}(k))}{\partial \Delta \theta_A} & \frac{\partial(y'_{Al}(k))}{\partial \Delta \eta_A} \\ \frac{\partial(z'_{Al}(k))}{\partial \Delta r_A} & \frac{\partial(z'_{Al}(k))}{\partial \Delta \theta_A} & \frac{\partial(z'_{Al}(k))}{\partial \Delta \eta_A}\end{bmatrix} \tag{15.82}$$

$$\boldsymbol{L}_B(k)=\begin{bmatrix}\frac{\partial(x'_{Bl}(k))}{\partial \Delta r_B} & \frac{\partial(x'_{Bl}(k))}{\partial \Delta \theta_B} & \frac{\partial(x'_{Bl}(k))}{\partial \Delta \eta_B} \\ \frac{\partial(y'_{Bl}(k))}{\partial \Delta r_B} & \frac{\partial(y'_{Bl}(k))}{\partial \Delta \theta_B} & \frac{\partial(y'_{Bl}(k))}{\partial \Delta \eta_B} \\ \frac{\partial(z'_{Bl}(k))}{\partial \Delta r_B} & \frac{\partial(z'_{Bl}(k))}{\partial \Delta \theta_B} & \frac{\partial(z'_{Bl}(k))}{\partial \Delta \eta_B}\end{bmatrix} \tag{15.83}$$

Because for the same target $f(\boldsymbol{\Psi}(k),\boldsymbol{\beta})=[0,\ 0,\ 0]'$, suppose that $[\boldsymbol{\Psi}(k)-\boldsymbol{\Psi}'(k)]$ and $(\boldsymbol{\beta}-\boldsymbol{\beta}')$ are small enough with negligible high-order components. Then,

$$\boldsymbol{\zeta}(k)\boldsymbol{\beta}+\boldsymbol{\kappa}(k)\partial\boldsymbol{\Psi}(k)=\boldsymbol{\zeta}(k)\boldsymbol{\beta}'-f(\boldsymbol{\Psi}'(k),\boldsymbol{\beta}') \tag{15.84}$$

where $\partial\boldsymbol{\Psi}(k)=(\boldsymbol{\Psi}(k)-\boldsymbol{\Psi}'(k))$. $\boldsymbol{\Psi}(k)$ considers only systematic errors but not stochastic ones, so

$$\partial\boldsymbol{\Psi}(k)=\left[R_r(k),\theta_r(k),\eta_r(k),R'_r(k),\theta'_r(k),\eta'_r(k)\right] \tag{15.85}$$

The right half of (15.84) represents observations. For N times, we can construct a linear relationship similar to that of Section 15.3.4:

$$\boldsymbol{X}\boldsymbol{\beta}+\boldsymbol{\xi}=\boldsymbol{Y} \tag{15.86}$$

It follows from the GLS that

$$\hat{\boldsymbol{\beta}}=\left(\boldsymbol{X}'\boldsymbol{\Sigma}_\xi^{-1}\boldsymbol{X}\right)^{-1}\boldsymbol{X}'\boldsymbol{\Sigma}_\xi^{-1}\boldsymbol{Y} \tag{15.87}$$

$$\operatorname{cov}\left(\hat{\boldsymbol{\beta}}\right)=\left(\boldsymbol{X}'\boldsymbol{\Sigma}_\xi^{-1}\boldsymbol{X}\right)^{-1} \tag{15.88}$$

Similar to Section 15.3.4, in order to improve the calculation speed, (15.87) and (15.88) can be decomposed into N small matrixes for calculation:

$$\boldsymbol{X}'\boldsymbol{\Sigma}_\xi^{-1}\boldsymbol{X}=\sum_{k=1}^{N}\boldsymbol{\zeta}'(k)\boldsymbol{\Sigma}_\xi^{-1}\boldsymbol{\zeta}(k) \tag{15.89}$$

$$X'\Sigma_{\xi}^{-1}Y=\sum_{k=1}^{N}\zeta'(k)\Sigma_{\xi}^{-1}[\zeta(k)\beta'-f(\Psi(k),\beta')] \tag{15.90}$$

15.3.6 Simulation Analysis

This section makes a simulation analysis of ECEF-LS and ECEF-GLS registration algorithms, under the following conditions: radars A and B's measurement accuracy is $\sigma_{\rho A}=\sigma_{\rho B}=50\,\text{m}$, $\sigma_{\theta A}=\sigma_{\theta B}=0.5°$, $\sigma_{\eta A}=\sigma_{\eta B}=0.5°$; systematic errors are $\Delta R_A=\Delta R_B=1842$ m in range,

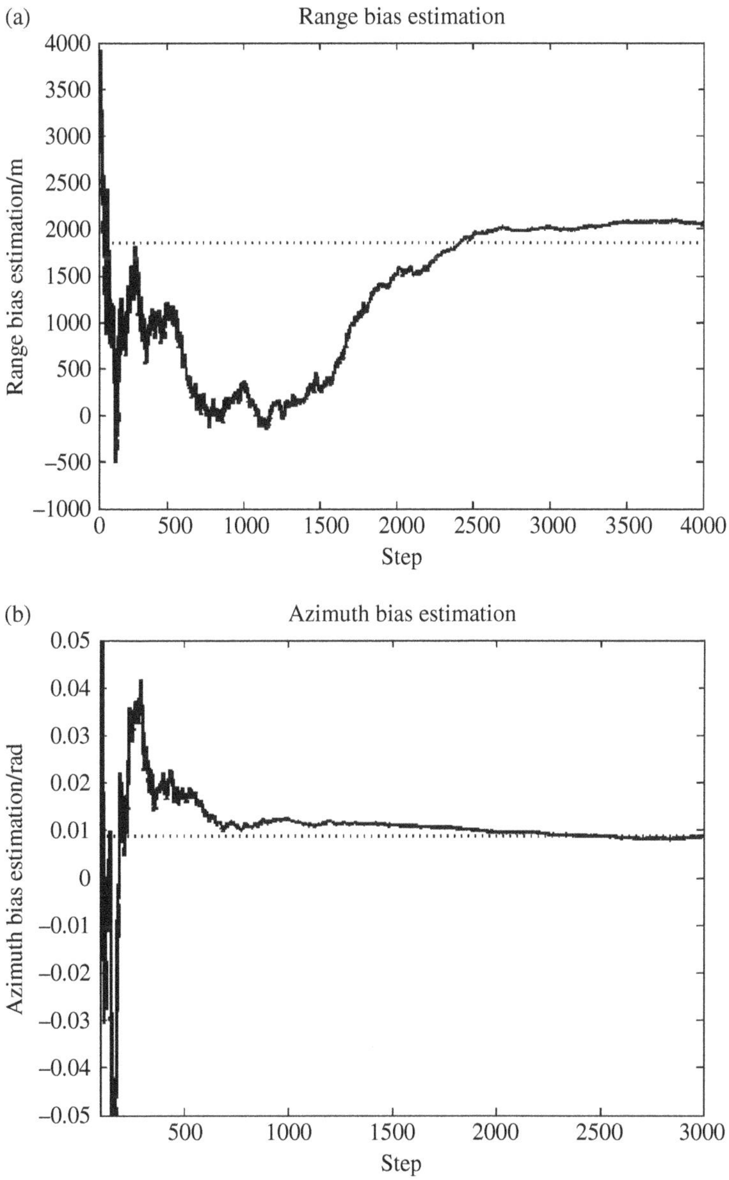

Figure 15.3 Radar A's systematic error estimation curve of ECEF-LS algorithm

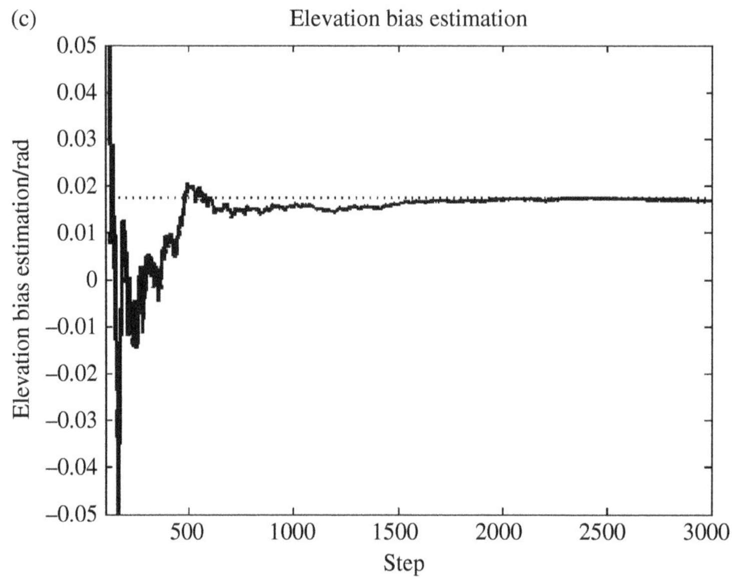

Figure 15.3 (*Continued*)

$\Delta\theta_A(k) = \Delta\theta_B(k) = 0.0087$ rad in azimuth, and $\Delta\eta_A(k) = \Delta\eta_B(k) = 0.0175$ rad in elevation. Radar A and B's geographical coordinates are $(68.923°, -137.2589°, 0.0275\,\text{nmi})$ and $(70.1714°, -124.7250°, 0.1182\,\text{nmi})$, respectively. The earth model is the reference ellipse in the World Geodetic System 1984, whose earth equator radius is $E_q = 3443.9$ nm and eccentricity $e^2 = 0.006694$. The two radar sampling intervals are both 1 s. We use radar A's local coordinate system as the reference coordinate system to generate a registration target track:

$$\begin{cases} x'_A(k) = 2000 + 200k \\ y'_A(k) = -6000 \\ z'_A(k) = 3500 \end{cases} \tag{15.91}$$

Radar A and B's systematic error estimate curve when adopting the ECEF-LS is shown in Figures 15.3 and 15.4. Figures 15.5 and 15.6 are the corresponding curves of the ECEF-GLS for radars A and B, of which the solid line represents the estimation result of systematic errors while the dotted one shows the actual systematic errors.

Figures 15.3 and 15.4 show the following results.

1. When the ECEF-LS algorithm is used, the estimates of range, azimuth, and elevation errors are close to the actual values of systematic errors after certain steps, and hence this algorithm can estimate systematic errors effectively. However, because ECEF-LS calculation is rather complicated, the estimated curves did not converge until 2000 steps, that is, the actual registration speed is very slow.
2. Figures 15.3(a) and 15.4(a) illustrate that the range error estimation results of the ECEF-LS are not ideal.

Two results are suggested in Figures 15.5 and 15.6:

1. When the ECEF-GLS algorithm is employed, the estimates of range, azimuth, and elevation errors are close to the actual values of systematic errors after certain steps, and thus this

algorithm can estimate systematic errors effectively. Moreover, the estimated curves of range, azimuth, and elevation errors converge at around 500 steps. In Figure 15.6, the estimated curve converges after fewer steps, which shows that the registration speed of the ECEF-GLS is much faster.

2. In Figures 15.5(a) and 15.6(a), the range error estimation effects of the ECEF-GLS are better.

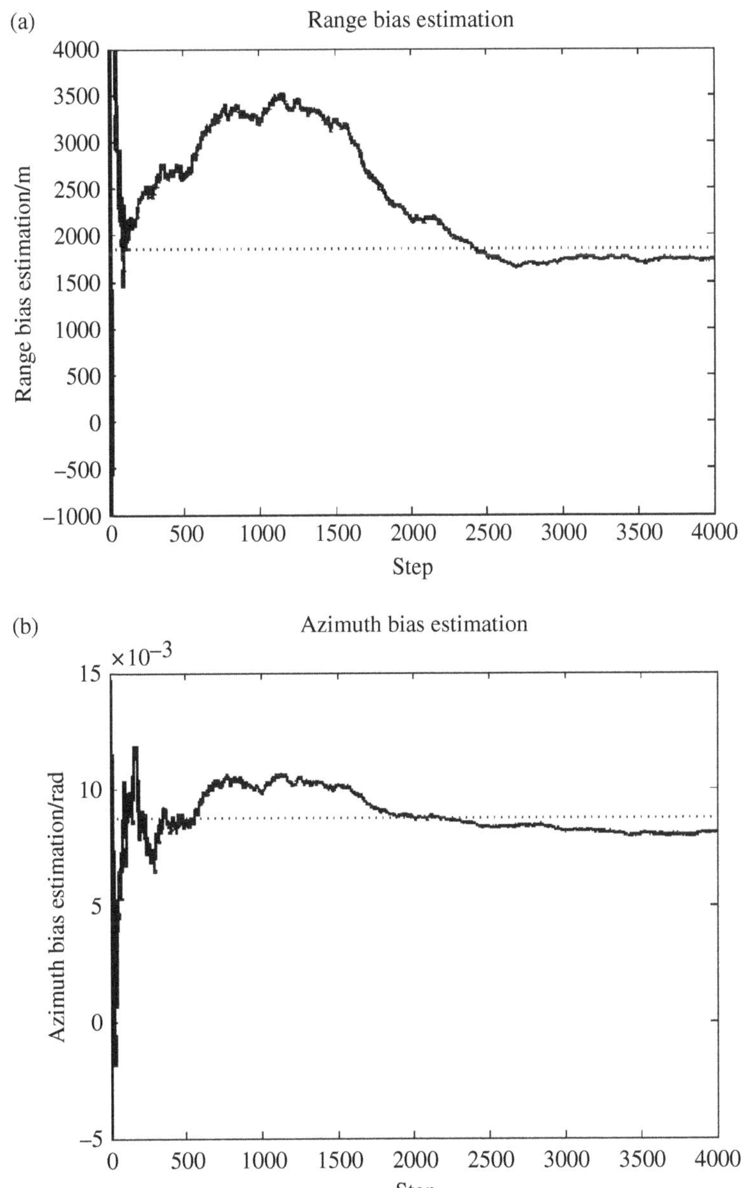

Figure 15.4 Radar B's systematic error estimation curve of ECEF-LS algorithm

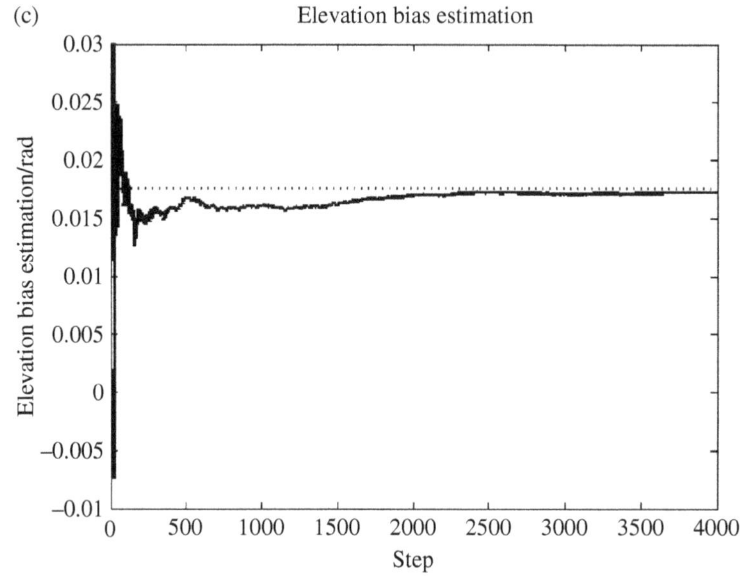

Figure 15.4 (*Continued*)

15.4 Mobile Radar Registration Algorithm

Usually, the fixed radar registration algorithms only consider range, azimuth, and elevation errors, while the mobile ones also consider systematic errors in attitude angle of the mobile platforms and hence registration is more difficult. At present, thorough research has been carried out on fixed radar registration technology, but not yet on mobile radar registration technology. In this section we discuss mobile radar registration technology from four aspects: systematic error modeling, registration based on cooperative targets, off-line batch processing estimation of systematic errors based on non-cooperative targets, and dimension expansion filtering estimation of systematic errors based on non-cooperative targets.

15.4.1 Modeling Method of Mobile Radar Systems

Mobile radar registration often involves measurement transformation between many different coordinate systems with complicated formulas. Therefore, in this section, we discuss the modeling methods of mobile radar systems in the presence of systematic errors.

Suppose that the system is composed of two 3D mobile radars. Suppose also that mobile radar $i\,(i=1,2)$ measures the same target simultaneously in the polar coordinate system, with range error b_i^r, azimuth error b_i^θ, elevation error b_i^η, yaw angle error b_i^ϑ, pitch angle error b_i^ϕ, and roll angle error b_i^α. Then again, assume that they are constant additive errors and can be expressed as

$$\boldsymbol{b}_i=\left[(\boldsymbol{b}_i^l)',(\boldsymbol{b}_i^z)'\right]' \qquad i=1,2 \tag{15.92}$$

where $\boldsymbol{b}_i^l=\left[b_i^r,b_i^\theta,b_i^\eta\right]'$ are measurement systematic errors and $\boldsymbol{b}_i^z=\left[b_i^\vartheta,b_i^\phi,b_i^\alpha\right]'$ attitude angle systematic errors.

Based on the discrete, continuous, white noise acceleration model, the state of targets in uniform motion is modeled as

$$\boldsymbol{X}(k+1)=\boldsymbol{F}(k)\boldsymbol{X}(k)+\boldsymbol{V}(k) \tag{15.93}$$

where the state vector $\boldsymbol{X}(k)$ and state transformation matrix $\boldsymbol{F}(k)$ at time k are defined as

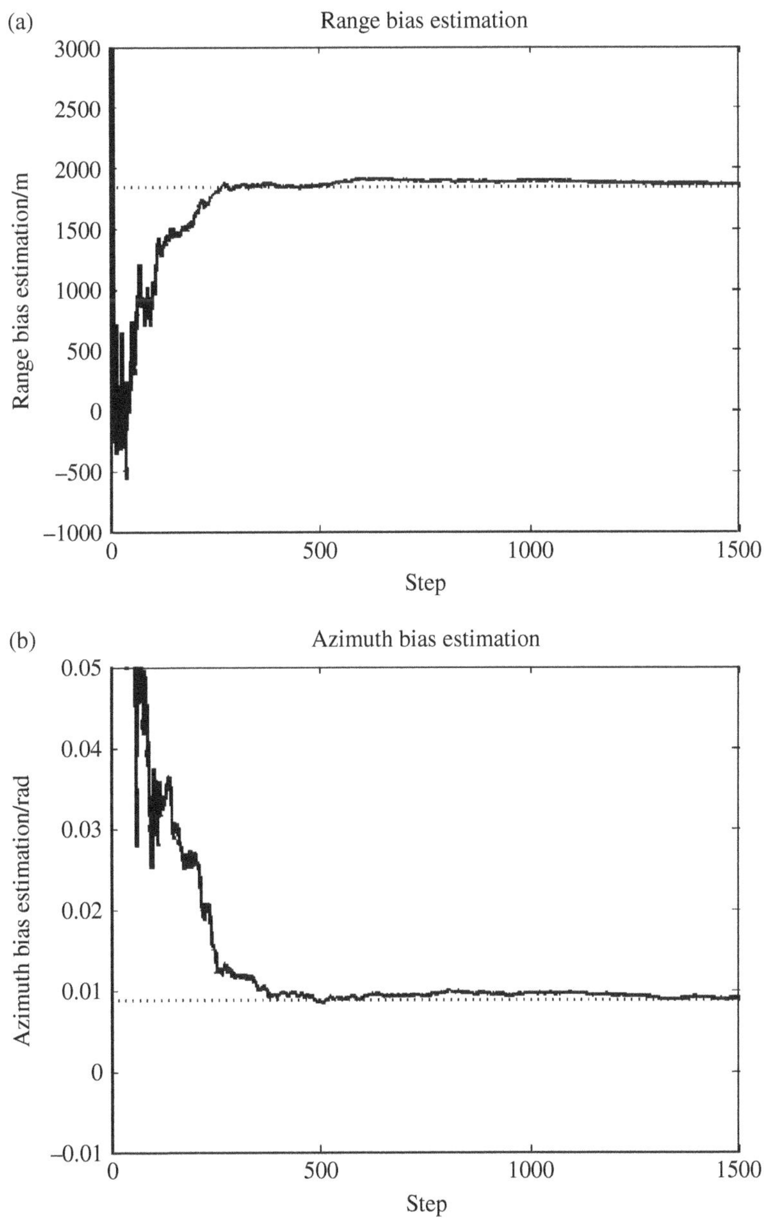

Figure 15.5 Radar A's systematic error estimation curve of ECEF-GLS algorithm

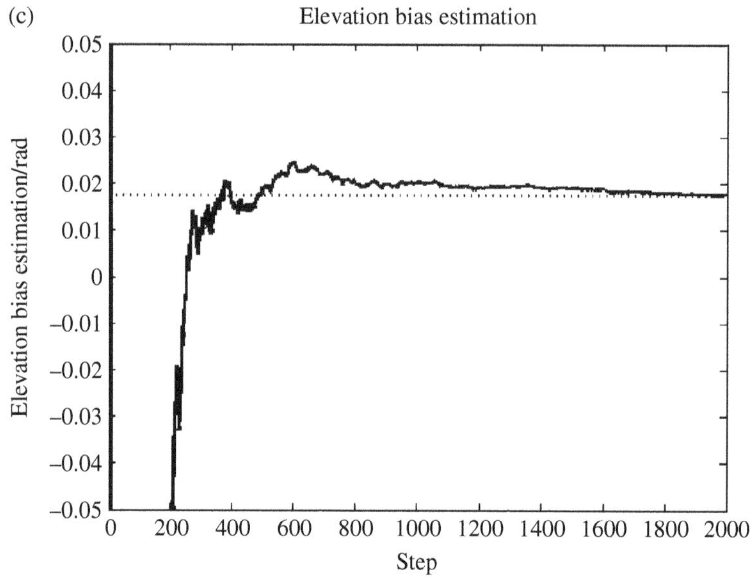

Figure 15.5 (*Continued*)

$$X(k) = [x(k)\ \ \dot{x}(k)\ \ y(k)\ \ \dot{y}(k)\ \ z(k)\ \ \dot{z}(k)]' \tag{15.94}$$

$$F(k) = \begin{bmatrix} 1 & T & 0 & 0 & 0 & 0 \\ 0 & 1 & 0 & 0 & 0 & 0 \\ 0 & 0 & 1 & T & 0 & 0 \\ 0 & 0 & 0 & 1 & 0 & 0 \\ 0 & 0 & 0 & 0 & 1 & T \\ 0 & 0 & 0 & 0 & 0 & 1 \end{bmatrix} \tag{15.95}$$

with T the discrete time interval and $\boldsymbol{V}(k)$ zero-mean, white process noise with variance $\boldsymbol{V}(k)$,

$$\boldsymbol{Q} = \mathrm{diag}\left(\boldsymbol{Q}_x, \boldsymbol{Q}_y, \boldsymbol{Q}_z\right) \tag{15.96}$$

$$\boldsymbol{Q}_x = \begin{bmatrix} \frac{1}{3}T^3 & \frac{1}{2}T^2 \\ \frac{1}{2}T^2 & T \end{bmatrix}\widetilde{q}_x,\ \boldsymbol{Q}_y = \begin{bmatrix} \frac{1}{3}T^3 & \frac{1}{2}T^2 \\ \frac{1}{2}T^2 & T \end{bmatrix}\widetilde{q}_y,\ \boldsymbol{Q}_z = \begin{bmatrix} \frac{1}{3}T^3 & \frac{1}{2}T^2 \\ \frac{1}{2}T^2 & T \end{bmatrix}\widetilde{q}_z \tag{15.97}$$

where $\widetilde{q}_x, \widetilde{q}_y, \widetilde{q}_z$ is the power spectral density of the noise.

At time k, mobile radar i has measured the target range $r_i(k)$, azimuth $\theta_i(k)$, and elevation $\eta_i(k)$, which include measurement systematic errors $\boldsymbol{b}_i$ and measurement noise $\boldsymbol{W}_i(k)$. Suppose that the actual measurements without systematic errors are $r_i'(k), \theta_i'(k), \eta_i'(k)$, where $\boldsymbol{W}_i(k)$ is zero-mean, white measurement noise with variance $\boldsymbol{R}(\boldsymbol{W}_i) = \mathrm{diag}\left(\sigma_{ri}^2, \sigma_{\theta i}^2, \sigma_{\eta i}^2\right)$, with measurement systematic errors and measurement noise being independent of each other. Hence, the measurement equation of mobile radar i is

$$\mathbf{Z}_{idp}(k) = \begin{bmatrix} r_i(k) \\ \theta_i(k) \\ \eta_i(k) \end{bmatrix} = \mathbf{Z}'_{idp}(k) + \boldsymbol{b}_i^l + \mathbf{W}_i(k) = \begin{bmatrix} r_i'(k) + b_i^r + w_i^r \\ \theta_i'(k) + b_i^\theta + w_i^\theta \\ \eta_i'(k) + b_i^\eta + w_i^\eta \end{bmatrix} \tag{15.98}$$

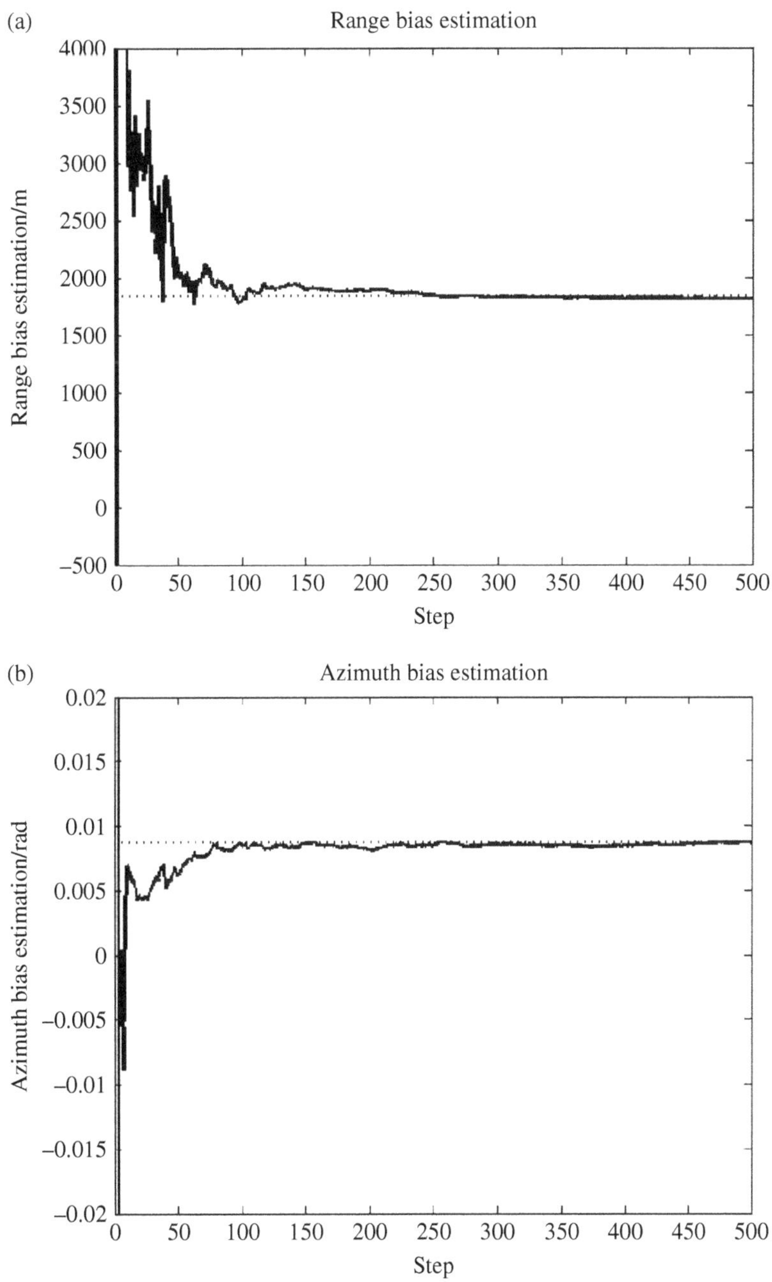

Figure 15.6 Radar B's systematic error estimation curve of ECEF-GLS algorithm

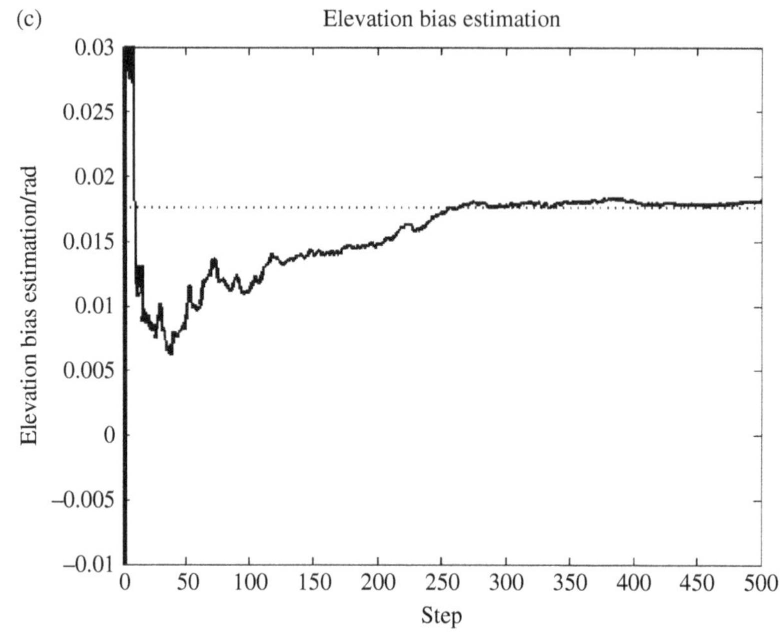

Figure 15.6 (*Continued*)

Transforming polar coordinate measurements into the rectangular coordinate system yields

$$\boldsymbol{Z}_{id}(k)=\begin{bmatrix}x_{id}(k)\\ y_{id}(k)\\ z_{id}(k)\end{bmatrix}=\boldsymbol{h}^{-1}\left(r_i'(k)+b_i^r+w_i^r,\theta_i'(k)+b_i^\theta+w_i^\theta,\eta_i'(k)+b_i^\eta+w_i^\eta\right) \tag{15.99}$$

where

$$\boldsymbol{h}(x,y,z)=\left[\sqrt{x^2+y^2+z^2},\ \arctan\left(\frac{y}{x}\right),\ \arctan\left(\frac{z}{\sqrt{x^2+y^2}}\right)\right]' \tag{15.100}$$

$$\boldsymbol{h}^{-1}(r,\theta,\eta)=[r\cos\theta\cos\eta,\ r\sin\theta\cos\eta,\ r\sin\eta]' \tag{15.101}$$

Suppose that the platform of mobile radar i at time k has attitude angle with systematic errors $\boldsymbol{v}_i(k)=[\vartheta_i(k),\phi_i(k),\alpha_i(k)]'$ and that without systematic errors, $\boldsymbol{v}_i'(k)=[\vartheta_i'(k),\phi_i'(k),\alpha_i'(k)]'$. As shown in Figure 15.7, according to the converting relationship between ship/airborne and END coordinate systems, the measurements obtained by mobile radar i in its carrier's coordinate system can be converted to END as follows:

$$\boldsymbol{Z}_{il}(k)=\begin{bmatrix}x_{il}(k)\\ y_{il}(k)\\ z_{il}(k)\end{bmatrix}=\boldsymbol{A}(\boldsymbol{v}_i(k))\begin{bmatrix}x_{id}(k)\\ y_{id}(k)\\ z_{id}(k)\end{bmatrix}=\boldsymbol{A}\left(\boldsymbol{v}_i'(k)+\boldsymbol{b}_i^z\right)\begin{bmatrix}x_{id}(k)\\ y_{id}(k)\\ z_{id}(k)\end{bmatrix} \tag{15.102}$$

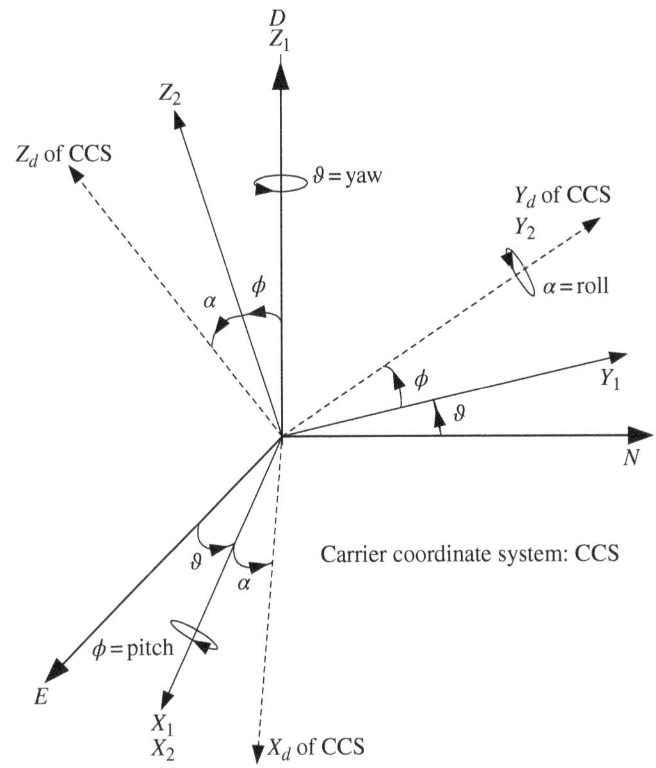

Figure 15.7 The rotation relationship between the END and carrier coordinate systems

where

$$\begin{aligned} \boldsymbol{A}(\vartheta,\phi,\alpha) &= \boldsymbol{A}_{head}\boldsymbol{A}_{pitch}\boldsymbol{A}_{roll} \\ &= \begin{bmatrix} \cos\vartheta\cos\alpha+\sin\vartheta\sin\phi\sin\alpha & \sin\vartheta\cos\phi & -\cos\vartheta\sin\alpha+\sin\vartheta\sin\phi\cos\alpha \\ -\sin\vartheta\cos\alpha+\cos\vartheta\sin\phi\sin\alpha & \cos\vartheta\cos\phi & \sin\vartheta\sin\alpha+\cos\vartheta\sin\phi\cos\alpha \\ \cos\phi\sin\alpha & -\sin\phi & \cos\phi\cos\alpha \end{bmatrix} \end{aligned} \tag{15.103}$$

$$\boldsymbol{A}_{head} = \begin{bmatrix} \cos\vartheta & \sin\vartheta & 0 \\ -\sin\vartheta & \cos\vartheta & 0 \\ 0 & 0 & 1 \end{bmatrix}, \ \boldsymbol{A}_{pitch} = \begin{bmatrix} 1 & 0 & 0 \\ 0 & \cos\phi & \sin\phi \\ 0 & -\sin\phi & \cos\phi \end{bmatrix}, \ \boldsymbol{A}_{roll} = \begin{bmatrix} \cos\alpha & 0 & -\sin\alpha \\ 0 & 1 & 0 \\ \sin\alpha & 0 & \cos\alpha \end{bmatrix} \tag{15.104}$$

Suppose that the geographical coordinates of mobile radar i at time k are $\boldsymbol{X}_{isp}(k) = [L_i(k), B_i(k), H_i(k)]'$ and its coordinates in the earth coordinate system are $\boldsymbol{X}_{is}(k) = [x_{is}(k), y_{is}(k), z_{is}(k)]'$. According to the converting relationship between the END and ECEF coordinate systems, the measurement of mobile radar i in the END coordinate system can be converted to ECEF, which yields

$$Z_{ig}(k)=\begin{bmatrix}x_{ig}(k)\\y_{ig}(k)\\z_{ig}(k)\end{bmatrix}=\begin{bmatrix}x_{is}(k)\\y_{is}(k)\\z_{is}(k)\end{bmatrix}+T\left(X_{isp}(k)\right)\times\begin{bmatrix}x_{il}(k)\\y_{il}(k)\\z_{il}(k)\end{bmatrix} \tag{15.105}$$

where $X_{is}(k)$ can be obtained from (15.67) and T from (15.71).

Combining (15.99), (15.102), and (15.105) yields the measurements of mobile radar i in the ECEF coordinate system:

$$Z_{ig}(k)=[x_{is}(k),\ y_{is}(k),\ z_{is}(k)]'+T\left(X_{isp}(k)\right)A\left(v_i'(k)+b_i^z\right)\ h^{-1}(r_i(k),\theta_i(k),\eta_i(k)) \tag{15.106}$$

Hereafter, according to (15.106), we derive the expression of $Z_{idp}(k)$ about $X(k)$. If mobile radar i has no systematic error and no measurement noise, then $Z_{ig}(k)=X(k)$, namely,

$$X(k)=[x_{is}(k),\ y_{is}(k),\ z_{is}(k)]'+T\left(X_{isp}(k)\right)A\left(v_i'(k)\right)h^{-1}\left(r_i'(k),\theta_i'(k),\eta_i'(k)\right) \tag{15.107}$$

From (15.107) it follows that

$$Z_{idp}(k)=h\left(A\left(v_i(k)-b_i^z\right)^{-1}T\left(X_{isp}(k)\right)^{-1}\left(X(k)-X_{is}(k)\right)\right)+b_i^l+W_i(k) \tag{15.108}$$

From the above, we know that the whole system is composed of the target state transformation equation (15.93) and the radar measurement equation (15.108), and that it involves conversion between three coordinate systems. The target state transformation is modeled in the earth coordinate system and the radar measurement is modeled in its carrier coordinate system.

15.4.2 Mobile Radar Registration Algorithm Based on Cooperative Targets

Given the target's position (e.g., the position of our ships given by the AIS, or the position of our civil planes given by the ADS), a single radar can estimate the systematic errors based on the difference in target position information between its own measurements and the information reported by the data linkage.

Suppose that the real target state is $X(k)=[x(k)\ \ y(k)\ \ z(k)]'$, and the target position gained through data linkages is $\bar{X}(k)=[\bar{x}(k)\ \ \bar{y}(k)\ \ \bar{z}(k)]'$. Then, their relationship can be expressed as

$$X(k)=\bar{X}(k)+W_x(k) \tag{15.109}$$

where $W_x(k)$ is the deviation between the actual movement position of the target and that measured by navigation equipment. Here, we assume zero-mean, Gaussian white noise with covariance $R_x(W_x)=\mathrm{diag}\left(\sigma_{\bar{x}}^2,\sigma_{\bar{y}}^2,\sigma_{\bar{z}}^2\right)$.

Suppose that at time k, the mobile radar has geographical coordinates $X_{sp}(k)=[L(k),B(k),H(k)]'$, coordinates in the earth coordinate system $X_s(k)=[x_s(k),y_s(k),z_s(k)]'$, attitude angle with errors $v(k)=[\vartheta(k),\phi(k),\alpha(k)]'$, and attitude angle without errors $v'(k)=[\vartheta'(k),\phi'(k),\alpha'(k)]'$. Then, from

(15.108) the mobile radar measurement equation in its carrier coordinate system is obtained as follows:

$$\boldsymbol{Z}(k)=[r(k)\ \theta(k)\ \eta(k)]'=\boldsymbol{h}\left(\boldsymbol{A}^{-1}(\boldsymbol{v}(k)-\boldsymbol{b}^z)\boldsymbol{T}^{-1}\left(\boldsymbol{X}_{sp}(k)\right)(\boldsymbol{X}(k)-\boldsymbol{X}_s(k))\right)+\boldsymbol{b}^l+\boldsymbol{W}_z(k) \quad (15.110)$$

where $\boldsymbol{h}(\cdot)$ is given by (15.100), $\boldsymbol{A}$ by (15.103), and $\boldsymbol{T}$ by (15.71). $\boldsymbol{W}_z(k)$ is the mobile radar zero-mean, Gaussian white noise with covariance $\boldsymbol{R}_z(\boldsymbol{W}_z)=\text{diag}\left(\sigma_r^2,\sigma_\theta^2,\sigma_\eta^2\right)$.

From the definition of the ML estimate, we know that the ML estimate of mobile radar systematic errors is

$$\hat{\boldsymbol{b}}_{\text{ML}}=\arg\max_{\boldsymbol{b}} p(\boldsymbol{Z}(1:N)|\boldsymbol{b}) \quad (15.111)$$

where

$$p(\boldsymbol{Z}(1:N)|\boldsymbol{b})=\int\prod_k p(\boldsymbol{Z}(k)|\boldsymbol{X}(k),\boldsymbol{b})p(\boldsymbol{X}(1:N))\text{d}\boldsymbol{X}(1:N) \quad (15.112)$$

Because the airway information of the target is known, by combining (15.109), the target state vectors ($\boldsymbol{X}(k)$) at different times can, given the target's airway position $\bar{\boldsymbol{X}}(k)$, be considered independent of each other. Then,

$$p(\boldsymbol{X}(1:N))=\prod_{k=1}^{N}p(\boldsymbol{X}(k)) \quad (15.113)$$

where $p(\boldsymbol{X}(k))$ is the Gaussian PDF, with mean $\bar{\boldsymbol{X}}(k)$ and variance $\boldsymbol{R}(\boldsymbol{W}_x)$.

Using (15.113) in (15.111) yields

$$\hat{\boldsymbol{b}}_{\text{ML}}=\arg\max_{\boldsymbol{b}}\prod_k p(\boldsymbol{Z}(k)|\boldsymbol{b}) \quad (15.114)$$

where

$$p(\boldsymbol{Z}(k)|\boldsymbol{b})=\int p(\boldsymbol{Z}(k)|\boldsymbol{X}(k),\boldsymbol{b})p(\boldsymbol{X}(k))d\boldsymbol{X}(k) \quad (15.115)$$

From (15.110) it follows that $\boldsymbol{Z}(k)$ is the nonlinear function of $\boldsymbol{X}(k)$ while (15.115) also requires $\boldsymbol{Z}(k)$ integration on $\boldsymbol{X}(k)$. Hence, we need to linearize $\boldsymbol{Z}(k)$.

Let $\boldsymbol{g}(\boldsymbol{X}(k),\boldsymbol{b}^z)=[u(k)\ \ v(k)\ \ m(k)]'=\boldsymbol{A}^{-1}(\boldsymbol{v}(k)-\boldsymbol{b}^z)\boldsymbol{T}^{-1}\left(\boldsymbol{X}_{sp}(k)\right)(\boldsymbol{X}(k)-\boldsymbol{X}_s(k))$ and write $\boldsymbol{h}\left(\boldsymbol{A}^{-1}\right.$ $\left.(\boldsymbol{v}(k)-\boldsymbol{b}^z)\boldsymbol{T}^{-1}\left(\boldsymbol{X}_{sp}(k)\right)(\boldsymbol{X}(k)-\boldsymbol{X}_s(k))\right)$ in the compound function form. Thus,

$$\boldsymbol{h}(\boldsymbol{g}(\boldsymbol{X}(k),\boldsymbol{b}^z))=\boldsymbol{h}\left(\boldsymbol{A}^{-1}(\boldsymbol{v}(k)-\boldsymbol{b}^z)\boldsymbol{T}^{-1}\left(\boldsymbol{X}_{sp}(k)\right)(\boldsymbol{X}(k)-\boldsymbol{X}_s(k))\right) \quad (15.116)$$

For the multi-variable compound function (15.116), solving for the partial derivatives of $\boldsymbol{X}(k)$ and $\boldsymbol{b}^z$ at $\boldsymbol{X}(k)=\boldsymbol{X}_0(k),\boldsymbol{b}^z=\boldsymbol{b}_0^z$, respectively, with the time mark k cancelled for convenience, yields

$$\begin{aligned}F(X_0,b_0^z)&=\left.\frac{\partial h(g(X,b^z))}{\partial X}\right|_{X=X_0,b^z=b_0^z}=\left.\frac{\partial h(g)}{\partial g}\cdot\frac{\partial g(X,b^z)}{\partial X}\right|_{X=X_0,b^z=b_0^z}\\&=\left.\left[\frac{\partial h(g)}{\partial u}\;\frac{\partial h(g)}{\partial v}\;\frac{\partial h(g)}{\partial m}\right]\cdot\left[\frac{\partial g(X,b^z)}{\partial x}\;\frac{\partial g(X,b^z)}{\partial y}\;\frac{\partial g(X,b^z)}{\partial z}\right]\right|_{X=X_0,b^z=b_0^z}\end{aligned}\tag{15.117}$$

$$\begin{aligned}G(X_0,b_0^z)&=\left.\frac{\partial h(g(X,b^z))}{\partial b^z}\right|_{X=X_0,b^z=b_0^z}=\left.\frac{\partial h(g)}{\partial g}\cdot\frac{\partial g(X,b^z)}{\partial b^z}\right|_{X=X_0,b^z=b_0^z}\\&=\left.\left[\frac{\partial h(g)}{\partial u}\;\frac{\partial h(g)}{\partial v}\;\frac{\partial h(g)}{\partial m}\right]\cdot\left[\frac{\partial g(X,b^z)}{\partial b^\vartheta}\;\frac{\partial g(X,b^z)}{\partial b^\phi}\;\frac{\partial g(X,b^z)}{\partial b^\alpha}\right]\right|_{X=X_0,b^z=b_0^z}\end{aligned}\tag{15.118}$$

where

$$\left[\frac{\partial h(g)}{\partial u}\;\frac{\partial h(g)}{\partial v}\;\frac{\partial h(g)}{\partial m}\right]=\begin{bmatrix}\frac{u}{\sqrt{u^2+v^2+m^2}} & \frac{v}{\sqrt{u^2+v^2+m^2}} & \frac{m}{\sqrt{u^2+v^2+m^2}}\\ \frac{-v}{\sqrt{u^2+v^2}} & \frac{u}{\sqrt{u^2+v^2}} & 0\\ \frac{-um}{\sqrt{u^2+v^2}(u^2+v^2+m^2)} & \frac{-vm}{\sqrt{u^2+v^2}(u^2+v^2+m^2)} & \frac{u^2+v^2}{\sqrt{u^2+v^2}(u^2+v^2+m^2)}\end{bmatrix}\tag{15.119}$$

$$\left[\frac{\partial g(X,b^z)}{\partial x}\;\frac{\partial g(X,b^z)}{\partial y}\;\frac{\partial g(X,b^z)}{\partial z}\right]=A^{-1}(v(k)-b^z)T^{-1}(X_{sp}(k))$$

$$\left[\frac{\partial g(X,b^z)}{\partial b^\vartheta}\;\frac{\partial g(X,b^z)}{\partial b^\phi}\;\frac{\partial g(X,b^z)}{\partial b^\alpha}\right]=\frac{\partial A_{roll}(b^\alpha-\alpha)A_{pitch}(b^\phi-\phi)A_{head}(b^\vartheta-\vartheta)}{\partial b^z}\,T(X_{sp})^{-1}(X-X_s)\tag{15.120}$$

$$=\begin{bmatrix}\left(A_{roll}(b^\alpha-\alpha)A_{pitch}(b^\phi-\phi)\frac{\partial A_{head}(b^\vartheta-\vartheta)}{\delta b^\vartheta}T(X_{sp})^{-1}(X-X_s)\right)'\\ \left(A_{roll}(b^\alpha-\alpha)\frac{\partial A_{pitch}(b^\phi-\phi)}{\delta b^\phi}A_{head}(b^\vartheta-\vartheta)T(X_{sp})^{-1}(X-X_s)\right)'\\ \left(\frac{\partial A_{roll}(b^\alpha-\alpha)}{\delta b^\alpha}A_{pitch}(b^\phi-\phi)A_{head}(b^\vartheta-\vartheta)T(X_{sp})^{-1}(X-X_s)\right)'\end{bmatrix}'\tag{15.121}$$

and

$$\frac{\partial A_{head}(b^\vartheta-\vartheta)}{\partial b^\vartheta}=\begin{bmatrix}-\sin(b^\vartheta-\vartheta) & \cos(b^\vartheta-\vartheta) & 0\\ -\cos(b^\vartheta-\vartheta) & \sin(b^\vartheta-\vartheta) & 0\\ 0 & 0 & 0\end{bmatrix}\tag{15.122}$$

$$\frac{\partial A_{pitch}(b^\phi-\phi)}{\partial b^\phi}=\begin{bmatrix}0 & 0 & 0\\ 0 & -\sin(b^\phi-\phi) & \cos(b^\phi-\phi)\\ 0 & -\cos(b^\phi-\phi) & -\sin(b^\phi-\phi)\end{bmatrix}\tag{15.123}$$

$$\frac{\partial A_{roll}(b^\alpha - \alpha)}{\partial b^\alpha} = \begin{bmatrix} -\sin(b^\alpha-\alpha) & 0 & -\cos(b^\alpha-\alpha) \\ 0 & 0 & 0 \\ \cos(b^\alpha-\alpha) & 0 & -\sin(b^\alpha-\alpha) \end{bmatrix} \quad (15.124)$$

Combining (15.117) and (15.118), and carrying out first-order Taylor expansion of (15.110) at the target fairway position $\bar{X}(k)$ and the initial system error estimate $\hat{b}_0$, gives

$$\begin{aligned} Z(k) &\approx h\left(g\left(\bar{X}(k), \hat{b}_0^z\right)\right) + \hat{b}_0^l \\ &+ F\left(\bar{X}(k), \hat{b}_0^z\right)(X(k) - \bar{X}(k)) + G\left(\bar{X}(k), \hat{b}_0^z\right)\left(b^z - \hat{b}_0^z\right) + I_{3\times3}\left(b^l - \hat{b}_0^l\right) + W_z(k) \end{aligned} \quad (15.125)$$

where $I_{3\times3}$ is the 3-rank unit matrix.

Writing (15.125) in compact form yields

$$Z(k) \approx h\left(g\left(\bar{X}(k), \hat{b}_0^z\right)\right) + \hat{b}_0^l + F\left(\bar{X}(k), \hat{b}_0^z\right)(X(k) - \bar{X}(k)) + G_b\left(\bar{X}(k), \hat{b}_0^z\right)\left(b - \hat{b}_0\right) + W_z(k) \quad (15.126)$$

where $G_b\left(\bar{X}(k), \hat{b}_0^z\right) = \left[I_{3\times3} \;\; G\left(\bar{X}(k), \hat{b}_0^z\right)\right]$.

From (15.125), we know that $p(Z(k)|b)$ can be approximated as a Gaussian PDF with mean $\bar{Z}(k)$ and variance $S(k)$:

$$\bar{Z}(k) = h\left(g\left(\bar{X}(k), \hat{b}_0^z\right)\right) + \hat{b}_0^l + G_b\left(\bar{X}(k), \hat{b}_0^z\right)\left(b - \hat{b}_0\right) \quad (15.127)$$

$$S(k) = F\left(\bar{X}(k), \hat{b}_0^z\right) R_x(W_x) F\left(\bar{X}(k), \hat{b}_0^z\right)' + R_z(W_z) \quad (15.128)$$

Hence, JDF $p(Z(1:N)|b)$ is also Gaussian, with mean and variance

$$\begin{bmatrix} \bar{Z}(1) \\ \vdots \\ \bar{Z}(N) \end{bmatrix} = \begin{bmatrix} h\left(g\left(\bar{X}(1), \hat{b}_0^z\right)\right) + \hat{b}_0^l \\ \vdots \\ h\left(g\left(\bar{X}(N), \hat{b}_0^z\right)\right) + \hat{b}_0^l \end{bmatrix} + \begin{bmatrix} G_b\left(\bar{X}(1), \hat{b}_0^z\right) \\ \vdots \\ G_b\left(\bar{X}(N), \hat{b}_0^z\right) \end{bmatrix} \left(b - \hat{b}_0\right) \quad (15.129)$$

$$S = \text{block-diag}(S_1, \ldots, S_N) \quad (15.130)$$

where block-diag$(\cdot)$ is the block-diagonal matrix, and the same below.

According to the standard formula of the Gaussian-distributed ML estimate, we know that the ML estimate of systematic errors is

$$\begin{aligned} \hat{b}_{\text{ML}} &= \left(\begin{bmatrix} G_b\left(\bar{X}(1), \hat{b}_0^z\right) \\ \vdots \\ G_b\left(\bar{X}(N), \hat{b}_0^z\right) \end{bmatrix}' S^{-1} \begin{bmatrix} G_b\left(\bar{X}(1), \hat{b}_0^z\right) \\ \vdots \\ G_b\left(\bar{X}(N), \hat{b}_0^z\right) \end{bmatrix} \right)^{-1} \begin{bmatrix} G_b\left(\bar{X}(1), \hat{b}_0^z\right) \\ \vdots \\ G_b\left(\bar{X}(N), \hat{b}_0^z\right) \end{bmatrix}' S^{-1} \begin{bmatrix} Z(1) - h\left(g\left(\bar{X}(1), \hat{b}_0^z\right)\right) - \hat{b}_0^l + G_b\left(\bar{X}(1), \hat{b}_0^z\right)\hat{b}_0 \\ \vdots \\ Z(N) - h\left(g\left(\bar{X}(N), \hat{b}_0^z\right)\right) - \hat{b}_0^l + G_b\left(\bar{X}(N), \hat{b}_0^z\right)\hat{b}_0 \end{bmatrix} \\ &= \hat{b}_0 + \left[\sum_{j=1}^{N} G_b\left(\bar{X}(j), \hat{b}_0^z\right)' S_j^{-1} G_b\left(\bar{X}(j), \hat{b}_0^z\right)\right]^{-1} \sum_{k=1}^{N} G_b\left(\bar{X}(k), \hat{b}_0^z\right)' S_k^{-1} \left[Z(k) - h\left(g\left(\bar{X}(k), \hat{b}_0^z\right)\right) - \hat{b}_0^l\right] \end{aligned} \quad (15.131)$$

From the above process it follows that (15.127), (15.128), and (15.131) constitute the ML estimate of mobile radar systematic errors, and that a replacement of $\hat{\boldsymbol{b}}_0$ with $\hat{\boldsymbol{b}}_{\mathrm{ML}}$ will have the recursive estimation of systematic errors realized.

15.4.3 Mobile Radar Maximum Likelihood Registration Algorithm

In this section we first introduce the typical batch processing maximum likelihood registration (MLR) algorithm [369]. Then, according to the mobile radar measurement equation, we use the MLR algorithm to solve the mobile radar registration problem and hence get the maximum likelihood registration of mobile radar (MLRM) algorithm.

15.4.3.1 MLR Algorithm

The MLR is a batch processing algorithm. It makes combined estimation of target states and systematic errors, but is only suitable for fixed radar netting systems. Hereafter, we present a detailed introduction to the MLR algorithm.

Consider a system consisting of n fixed radars, each of which can make measurement on all targets in the public area, therefore obtaining subsets of three kinds of measurement: range, azimuth, and elevation. Specifically, the radar can be any type (e.g., 3D, 2D, or passive radar) with accurately known position. For the whole radar system, its measurement equation with systematic errors can be expressed as

$$z(k) = \boldsymbol{h}(\boldsymbol{x}(k)) + \boldsymbol{\beta} + \boldsymbol{w}(k) \tag{15.132}$$

where $k = 1, \ldots, N$ indicates measurements of different targets at the same time, and measurements of the same target at different discrete times. $z(k) = \left[z_1(k)', z_2(k)', \ldots, z_n(k)'\right]'$ represents the vector composed of measurements by n radars at label k. $\boldsymbol{h}(x(k)) = \left[\boldsymbol{h}_1(x(k))', \boldsymbol{h}_2(x(k))', \ldots, \boldsymbol{h}_n(x(k))'\right]'$ denotes the vector composed of the given nonlinear measurement equation of n radars. $\boldsymbol{x}(k)$ is the real position vector of the target. $\boldsymbol{\beta} = [\boldsymbol{\beta}_1', \boldsymbol{\beta}_2', \ldots, \boldsymbol{\beta}_n']'$ is the radar systematic errors vector. $\boldsymbol{w}(k) = \left[\boldsymbol{w}_1(k)', \boldsymbol{w}_2(k)', \ldots, \boldsymbol{w}_n(k)'\right]'$ is the radar random measurement noise vector. Systematic errors $\boldsymbol{\beta}_i$ are fixed and time-invariant, and independent of target state $\boldsymbol{x}(k)$. Measurement noise $\boldsymbol{w}_i(k)$ is zero-mean, Gaussian white noise with covariance $\sum_{z_i}$, and the measurement noise between radars is independent.

Given measurement $\boldsymbol{Z} = \{z(k); k = 1, \ldots, N\}$, the problem to be solved is to get an estimate of the systematic errors $\boldsymbol{\beta}$, and use the estimate $\hat{\boldsymbol{\beta}}$ to modify subsequent measurements with systematic errors. Since the target state $\boldsymbol{x}(k)$ in (15.132) is also unknown, the MLR algorithm needs to make combined estimation of the systematic errors $\boldsymbol{\beta}$ and the target track $\boldsymbol{X} = \{\boldsymbol{x}(k); k = 1, \ldots, N\}$. It does this through maximizing the likelihood function $p(\boldsymbol{Z}|\boldsymbol{X}, \boldsymbol{\beta})$, namely,

$$\left\{\hat{\boldsymbol{X}}, \hat{\boldsymbol{\beta}}\right\} = \arg\max_{\boldsymbol{X},\boldsymbol{\beta}} p(z(1), z(2), \ldots, z(N)|\boldsymbol{X}, \boldsymbol{\beta}) = \arg\max_{\boldsymbol{\beta}} \left\{\prod_{k=1}^{N} \max_{\boldsymbol{x}(k)} p(z(k)|\boldsymbol{x}(k), \boldsymbol{\beta})\right\} \tag{15.133}$$

Because the measurement noise sequence $\boldsymbol{w}_i(k)$ is white noise and the noises at different times are independent of each other, (15.133) can be obtained. Therefore, solving for $\boldsymbol{X}$ which maximizes

$p(\boldsymbol{Z}|\boldsymbol{X},\boldsymbol{\beta})$ is equivalent to solving for $\boldsymbol{x}(k)$, $k=1,\ldots,N$ which maximizes $p(z(k)|\boldsymbol{x}(k),\boldsymbol{\beta})$ and hence its product $\prod_{k=1}^{N} p(z(k)|\boldsymbol{x}(k),\boldsymbol{\beta})$. We first estimate $\hat{\boldsymbol{x}}(k)$ to maximize $p(z(k)|\boldsymbol{x}(k),\boldsymbol{\beta})$, then substitute the estimated value $\hat{\boldsymbol{\beta}}$ in the preceding step for $\boldsymbol{\beta}$.

Under the assumption that the measurement noise between radars is independent, mark k is omitted for simplicity:

$$p(z_1,z_2,\ldots,z_n|\boldsymbol{x},\boldsymbol{\beta}) = \prod_{i=1}^{n} p(z_i|\boldsymbol{x},\boldsymbol{\beta}_i) = K_1 \exp\left\{-\frac{1}{2}\sum_{i=1}^{n}(z_i-\bar{z}_i)'\boldsymbol{\Sigma}_{z_i}^{-1}(z_i-\bar{z}_i)\right\} \tag{15.134}$$

where $\bar{z}_i = \boldsymbol{h}_i(\boldsymbol{x}) + \boldsymbol{\beta}_i$. Now, we use the following equation to project the measurement z_i to the target state space, namely to solve for the target state by z_i:

$$\boldsymbol{x}_i = \boldsymbol{h}_i^{-1}(z_i - \boldsymbol{\beta}_i), \qquad i=1,2,\ldots,N \tag{15.135}$$

Suppose that radar i is passive and then we know that $\boldsymbol{x}_i$ cannot be solved for by z_i. At this time, $\boldsymbol{x}_i$ can be solved for by radar i combined with another passive or active radar. From (15.135) it follows that $\boldsymbol{x}_i$ is a random variable, and we need to solve for the PDF of $\boldsymbol{x}_i$. By first-order Taylor expansion of $\boldsymbol{h}_i$, $\boldsymbol{x}_i$ can also be denoted approximately as a Gaussian random variable and the inverse of its covariance is

$$\boldsymbol{\Sigma}_{\boldsymbol{x}_i}^{-1} = \boldsymbol{H}_i'\boldsymbol{\Sigma}_{z_i}^{-1}\boldsymbol{H}_i \tag{15.136}$$

where

$$\boldsymbol{H}_i = \left[\nabla_{\boldsymbol{x}}\boldsymbol{h}_i(\boldsymbol{x})'\right]' = \begin{bmatrix} \frac{\partial \boldsymbol{h}_{i1}}{\partial x_1} & \frac{\partial \boldsymbol{h}_{i1}}{\partial x_2} & \cdots & \frac{\partial \boldsymbol{h}_{i1}}{\partial x_p} \\ \frac{\partial \boldsymbol{h}_{i2}}{\partial x_1} & \frac{\partial \boldsymbol{h}_{i2}}{\partial x_2} & \cdots & \frac{\partial \boldsymbol{h}_{i2}}{\partial x_p} \\ \vdots & \vdots & \ddots & \vdots \\ \frac{\partial \boldsymbol{h}_{iq_i}}{\partial x_1} & \frac{\partial \boldsymbol{h}_{iq_i}}{\partial x_2} & \cdots & \frac{\partial \boldsymbol{h}_{iq_i}}{\partial x_p} \end{bmatrix} \tag{15.137}$$

$\boldsymbol{H}_i$ is the Jacobian matrix of $\boldsymbol{h}_i(\cdot)$ about $\boldsymbol{x}$, p is the dimension of the target state, and q_i is the dimension of the measurement function $\boldsymbol{h}_i$.

Hence, (15.134) can be expressed approximately in the target state space as

$$\begin{aligned} p(z_1,z_2,\ldots,z_n|\boldsymbol{x},\boldsymbol{\beta}) &\approx K_2 \exp\left\{-\frac{1}{2}\sum_{i=1}^{n}(\boldsymbol{x}-\boldsymbol{x}_i)'\boldsymbol{\Sigma}_{\boldsymbol{x}_i}^{-1}(\boldsymbol{x}-\boldsymbol{x}_i)\right\} \\ &= K_2 \exp\left\{-\frac{1}{2}\left(\boldsymbol{x}'\left[\sum_{i=1}^{n}\boldsymbol{\Sigma}_{\boldsymbol{x}_i}^{-1}\right]\boldsymbol{x} - 2\boldsymbol{x}'\left[\sum_{i=1}^{n}\boldsymbol{\Sigma}_{\boldsymbol{x}_i}^{-1}\boldsymbol{x}_i\right] + \left[\sum_{i=1}^{n}\boldsymbol{x}'\boldsymbol{\Sigma}_{\boldsymbol{x}_i}^{-1}\boldsymbol{x}_i\right]\right)\right\} \end{aligned} \tag{15.138}$$

Using the matrix equation

$$x'Ax-2x'B+B'A^{-1}B=\left(x-A^{-1}B\right)'A\left(x-A^{-1}B\right) \tag{15.139}$$

to break down (15.138) yields

$$\begin{aligned} p(z_1,z_2,\ldots,z_n|\boldsymbol{x},\boldsymbol{\beta}) \approx K_2 \exp\Bigg\{ &-\frac{1}{2}(\boldsymbol{x}-\hat{\boldsymbol{x}})'\left[\sum_{i=1}^{n}\boldsymbol{\Sigma}_{\boldsymbol{x}_i}^{-1}\right](\boldsymbol{x}-\hat{\boldsymbol{x}}) \\ &-\frac{1}{2}\left(\left[\sum_{i=1}^{n}\boldsymbol{x}'\boldsymbol{\Sigma}_{\boldsymbol{x}_i}^{-1}\boldsymbol{x}_i\right]-\left[\sum_{i=1}^{n}\boldsymbol{\Sigma}_{\boldsymbol{x}_i}^{-1}\boldsymbol{x}_i\right]'\left[\sum_{i=1}^{n}\boldsymbol{\Sigma}_{\boldsymbol{x}_i}^{-1}\right]^{-1}\left[\sum_{i=1}^{n}\boldsymbol{\Sigma}_{\boldsymbol{x}_i}^{-1}\boldsymbol{x}_i\right]\right)\Bigg\} \end{aligned} \tag{15.140}$$

where

$$\hat{\boldsymbol{x}}=\left[\sum_{i=1}^{n}\boldsymbol{\Sigma}_{\boldsymbol{x}_i}^{-1}\right]^{-1}\left[\sum_{i=1}^{n}\boldsymbol{\Sigma}_{\boldsymbol{x}_i}^{-1}\boldsymbol{x}_i\right] \tag{15.141}$$

From (15.140) it follows that the likelihood function $p(z_1,z_2,\ldots,z_n|\boldsymbol{x},\boldsymbol{\beta})$ reaches its maximum when $\boldsymbol{x}=\hat{\boldsymbol{x}}$ and hence that $\hat{\boldsymbol{x}}$ is the ML estimate of the target state at time k. It also follows from (15.141) that $\hat{\boldsymbol{x}}$ is the fusion result of measurements from different radars at time k. Hereafter, we estimate the systematic errors $\boldsymbol{\beta}$.

Substitute $\boldsymbol{x}=\hat{\boldsymbol{x}}$ into (15.140) to give

$$\begin{aligned} p(z_1,\ldots,z_n|\hat{\boldsymbol{x}},\boldsymbol{\beta}) &= K\exp\left\{-\frac{1}{2}\left(\left[\sum_{i=1}^{n}\boldsymbol{x}'\boldsymbol{\Sigma}_{\boldsymbol{x}_i}^{-1}\boldsymbol{x}_i\right]-\left[\sum_{i=1}^{n}\boldsymbol{\Sigma}_{\boldsymbol{x}_i}^{-1}\boldsymbol{x}_i\right]'\left[\sum_{i=1}^{n}\boldsymbol{\Sigma}_{\boldsymbol{x}_i}^{-1}\right]^{-1}\left[\sum_{i=1}^{n}\boldsymbol{\Sigma}_{\boldsymbol{x}_i}^{-1}\boldsymbol{x}_i\right]\right)\right\} \\ &= K\exp\left\{-\frac{1}{2}\boldsymbol{X}'(k)\boldsymbol{\Sigma}^{-1}(k)\boldsymbol{X}(k)\right\} \end{aligned} \tag{15.142}$$

where $\boldsymbol{X}(k)=\left[\boldsymbol{x}_1'(k),\ldots,\boldsymbol{x}_n'(k)\right]'$ and $K=1/|2\pi\Sigma(k)|^{1/2}$ is a normalization constant:

$$\Sigma^{-1}(k)=\text{block-diag}\left(\Sigma_{\boldsymbol{x}_1}^{-1},\Sigma_{\boldsymbol{x}_2}^{-1},\ldots,\Sigma_{\boldsymbol{x}_n}^{-1}\right)-\left[\left\{\Sigma_{\boldsymbol{x}_i}^{-1}\left[\sum_{i=1}^{n}\Sigma_{\boldsymbol{x}_i}^{-1}\right]^{-1}\Sigma_{\boldsymbol{x}_j}^{-1}\right\}_{ij}\right] \tag{15.143}$$

Disturb $\boldsymbol{x}_i,\boldsymbol{\beta}_i$ in (15.135) and make linear approximations at $\boldsymbol{x}_{0i},\boldsymbol{\beta}_{0i}$. Then, this becomes

$$\boldsymbol{x}_i-\boldsymbol{x}_{0i}\approx\boldsymbol{H}_i^{-L}(\boldsymbol{\beta}_i-\boldsymbol{\beta}_{0i}) \tag{15.144}$$

where $\boldsymbol{H}_i$ is defined as shown in (15.137) and the superscript $-L$ denotes the left inverse of the matrix.

Using (15.144), $\boldsymbol{X}(k)$ can be expressed as

$$\boldsymbol{X}(k) \approx \begin{bmatrix} \boldsymbol{x}_{01}(k) \\ \boldsymbol{x}_{02}(k) \\ \vdots \\ \boldsymbol{x}_{0n}(k) \end{bmatrix} + \begin{bmatrix} \boldsymbol{H}_1^{-L}(k)\boldsymbol{\beta}_{01} \\ \boldsymbol{H}_2^{-L}(k)\boldsymbol{\beta}_{02} \\ \vdots \\ \boldsymbol{H}_n^{-L}(k)\boldsymbol{\beta}_{0n} \end{bmatrix} - \begin{bmatrix} \boldsymbol{H}_1^{-L}(k)\boldsymbol{\beta}_1 \\ \boldsymbol{H}_2^{-L}(k)\boldsymbol{\beta}_2 \\ \vdots \\ \boldsymbol{H}_n^{-L}(k)\boldsymbol{\beta}_n \end{bmatrix} \tag{15.145}$$

Write (15.145) in compact form as

$$\boldsymbol{X}(k) \approx \bar{\boldsymbol{X}}_0(k) - \boldsymbol{Q}(k)\boldsymbol{\beta} \tag{15.146}$$

where

$$\boldsymbol{Q}(k) = \text{block-diag}\left(\boldsymbol{H}_1^{-L}(k), \boldsymbol{H}_2^{-L}(k), \ldots, \boldsymbol{H}_n^{-L}(k)\right) \tag{15.147}$$

$$\bar{\boldsymbol{X}}_0(k) = \boldsymbol{X}_0(k) + \boldsymbol{Q}(k)\boldsymbol{\beta}_0 \tag{15.148}$$

and $\boldsymbol{X}_0(k) = \left[\boldsymbol{x}'_{01}(k), \boldsymbol{x}'_{02}(k), \ldots, \boldsymbol{x}'_{0n}(k)\right]'$ is the initial state estimate of the target, $\boldsymbol{\beta}_0 = [\boldsymbol{\beta}_{01}{}',$ $\boldsymbol{\beta}'_{02}, \ldots, \boldsymbol{\beta}_{0n}{}']'$ is the initial estimate of systematic errors.

Estimate that $\boldsymbol{\beta}$'s maximizing likelihood function equation (15.140) is equivalent to its maximizing the product of N likelihood functions equation (15.142), namely,

$$\hat{\boldsymbol{\beta}} = \arg\max_{\boldsymbol{\beta}} p\left(z_1, z_2, \ldots, z_n | \hat{\boldsymbol{X}}, \boldsymbol{\beta}\right) = \arg\max_{\boldsymbol{\beta}} \prod_{k=1}^{N} K_k \exp\left\{-\frac{1}{2}\boldsymbol{X}'(k)\boldsymbol{\Sigma}^{-1}(k)\boldsymbol{X}(k)\right\} \tag{15.149}$$

Using (15.139) and (15.146), $p\left(z_1, z_2, \ldots, z_n | \hat{\boldsymbol{X}}, \boldsymbol{\beta}\right)$ can be written as

$$p\left(z_1, \ldots, z_n | \hat{\boldsymbol{X}}, \boldsymbol{\beta}\right) = \bar{K}\exp\left\{-\frac{1}{2}\left(\left(\boldsymbol{\beta} - \hat{\boldsymbol{\beta}}\right)'\left[\sum_{k=1}^{N}\boldsymbol{Q}'(k)\boldsymbol{\Sigma}^{-1}(k)\boldsymbol{Q}(k)\right]\left(\boldsymbol{\beta} - \hat{\boldsymbol{\beta}}\right) + C\right)\right\} \tag{15.150}$$

where C is a constant which is irrelevant to $\boldsymbol{\beta}$ and $\bar{K}$ is a normalization constant:

$$\hat{\boldsymbol{\beta}} = \left[\sum_{k=1}^{N}\boldsymbol{Q}'(k)\boldsymbol{\Sigma}^{-1}(k)\boldsymbol{Q}(k)\right]^{-1}\left[\sum_{k=1}^{N}\boldsymbol{Q}'(k)\boldsymbol{\Sigma}^{-1}(k)\bar{\boldsymbol{X}}_0(k)\right] \tag{15.151}$$

From (15.150) it follows that when $\boldsymbol{\beta} = \hat{\boldsymbol{\beta}}$, the likelihood function $p\left(z_1, z_2, \ldots, z_n | \hat{\boldsymbol{X}}, \boldsymbol{\beta}\right)$ reaches its maximum value. Hence, $\hat{\boldsymbol{\beta}}$ is the maximum likelihood estimate of $\boldsymbol{\beta}$.

15.4.3.2 MLRM Algorithm

From the measurement equation (15.132) of the MLR algorithm assumption we know that $\boldsymbol{h}$ is only the function of target state $\boldsymbol{x}(k)$ and that it is not related to systematic errors. Systematic errors are only constant additive variables. From the mobile radar measurement equation (15.108), we know that $\boldsymbol{h}$ is the function of target state and attitude angle systematic errors, and that systematic errors can be divided into attitude angle systematic errors and measurement systematic errors. Therefore, the measurement model assumed by the MLR algorithm is not suitable for the case of mobile radars. Hence, we first generalize the MLR measurement model. Suppose that

$$\boldsymbol{g}\left(\boldsymbol{X}(k),\boldsymbol{b}_i^z\right)=\boldsymbol{A}\left(\boldsymbol{v}_i(k)-\boldsymbol{b}_i^z\right)^{-1}\boldsymbol{T}\left(\boldsymbol{X}_{isp}(k)\right)^{-1}\left(\boldsymbol{X}(k)-\boldsymbol{X}_{is}(k)\right) \tag{15.152}$$

Then (15.108) can be rewritten as

$$\boldsymbol{Z}_{idp}(k)=\boldsymbol{h}_i\left(\boldsymbol{g}\left(\boldsymbol{X}(k),\boldsymbol{b}_i^z\right)\right)+\boldsymbol{b}_i^l+\boldsymbol{W}_i(k) \tag{15.153}$$

Therefore, the generalized measurement model of the MLR becomes

$$\boldsymbol{z}(k)=\boldsymbol{h}\left(\boldsymbol{g}\left(\boldsymbol{x}(k),\boldsymbol{\beta}^2\right)\right)+\boldsymbol{\beta}^1+\boldsymbol{w}(k) \tag{15.154}$$

After drilling down into the MLR, we have found that using (15.154) as its measurement model has no influence on the derivations from (15.133)–(15.143). That is to say, given the estimates of systematic errors, the estimation of the target state will not be affected but the estimation of the systematic errors will be affected deeply, and deductions need to be done again with (15.144).

Suppose that $\boldsymbol{x}_{0i}(k), \boldsymbol{\beta}_{0i}^1, \boldsymbol{\beta}_{0i}^2$ satisfy (15.154). Now, make a small disturbance of (15.154) around $\boldsymbol{x}_{0i}(k), \boldsymbol{\beta}_{0i}^1, \boldsymbol{\beta}_{0i}^2$ to obtain $\boldsymbol{x}_i(k), \boldsymbol{\beta}_i^1, \boldsymbol{\beta}_i^2$ and $\boldsymbol{x}_i(k), \boldsymbol{\beta}_i^1, \boldsymbol{\beta}_i^2$, which also satisfy (15.154). Then,

$$\boldsymbol{h}_i\left(\boldsymbol{g}\left(\boldsymbol{x}_i(k),\boldsymbol{\beta}_i^2\right)\right)+\boldsymbol{\beta}_i^1=\boldsymbol{h}_i\left(\boldsymbol{g}\left(\boldsymbol{x}_{0i}(k),\boldsymbol{\beta}_{0i}^1\right)\right)+\boldsymbol{\beta}_{0i}^2 \tag{15.155}$$

First-order Taylor expansion of (15.155) at $\boldsymbol{x}_{0i}(k), \boldsymbol{\beta}_{0i}^2$ yields

$$\boldsymbol{h}_i\left(\boldsymbol{g}\left(\boldsymbol{x}_i(k),\boldsymbol{\beta}_i^2\right)\right)+\boldsymbol{\beta}_i^1\approx\boldsymbol{h}_i\left(\boldsymbol{g}\left(\boldsymbol{x}_{0i}(k),\boldsymbol{\beta}_{0i}^1\right)\right)+\boldsymbol{H}_{ix}\left(\boldsymbol{x}_i(k)-\boldsymbol{x}_{0i}(k)\right)+\boldsymbol{H}_{i\beta}\left(\boldsymbol{\beta}_i^2-\boldsymbol{\beta}_{0i}^2\right)+\boldsymbol{\beta}_i^1 \tag{15.156}$$

Combining (15.155) and (15.156) yields

$$\boldsymbol{x}_i(k)-\boldsymbol{x}_{0i}(k)\approx\boldsymbol{H}_{ix}^{-L}\boldsymbol{H}_{i\beta}\left(\boldsymbol{\beta}_{0i}^2-\boldsymbol{\beta}_i^2\right)+\boldsymbol{H}_{ix}^{-L}\left(\boldsymbol{\beta}_{0i}^1-\boldsymbol{\beta}_i^1\right) \tag{15.157}$$

Write (15.157) in compact form as

$$\boldsymbol{x}_i-\boldsymbol{x}_{0i}\approx\boldsymbol{H}_i^{-L}\left(\boldsymbol{\beta}_i-\boldsymbol{\beta}_{0i}\right) \tag{15.158}$$

where $\boldsymbol{\beta}=\left[\boldsymbol{\beta}^{1\prime},\boldsymbol{\beta}^{2\prime}\right]'$, $\boldsymbol{H}_i^{-L}=\left[\boldsymbol{H}_{ix}^{-L},\boldsymbol{H}_{ix}^{-L}\boldsymbol{H}_{i\beta}\right]$.

We can see that (15.158) and (15.144) have identical forms, and that (15.154) has no influence on the derivations of (15.145)–(15.151). Therefore, successive calculations can be done according to the original formula of the MLR algorithm.

Hereafter, we solve for the inverse of (15.152) with $\boldsymbol{A}$, $\boldsymbol{T}$, and the derivative of the compound function.

By the matrix inversion formula, the inverse of $\boldsymbol{A}$, $\boldsymbol{T}$ can be obtained as follows:

$$\boldsymbol{A}^{-1}(\vartheta,\phi,\alpha)=\boldsymbol{A}_{roll}^{-1}(\alpha)\boldsymbol{A}_{pitch}^{-1}(\phi)\boldsymbol{A}_{head}^{-1}(\vartheta)=\boldsymbol{A}_{roll}(-\alpha)\boldsymbol{A}_{pitch}(-\phi)\boldsymbol{A}_{head}(-\vartheta)=\boldsymbol{A}'(\vartheta,\phi,\alpha) \tag{15.159}$$

$$\boldsymbol{T}^{-1}(L_s,B_s)=\boldsymbol{T}'(L_s,B_s) \tag{15.160}$$

Hereafter, we solve for the Jacobian matrix of $\boldsymbol{h}_i$ about $\boldsymbol{x}$ and the Jacobian matrix about the attitude angle systematic errors. Suppose that

$$\boldsymbol{g}\left(\boldsymbol{X}(k),\boldsymbol{b}_i^z\right)=\begin{bmatrix} u_i \\ s_i \\ m_i \end{bmatrix}=\boldsymbol{A}\left(\boldsymbol{v}_i-\boldsymbol{b}_i^z\right)^{-1}\boldsymbol{T}\left(\boldsymbol{X}_{isp}\right)^{-1}\left(\boldsymbol{X}-\boldsymbol{X}_{is}\right) \tag{15.161}$$

From the compound function derivation formula, we have

$$\boldsymbol{H}_{ix}=\left[\nabla_{\boldsymbol{x}}\boldsymbol{h}_i(\boldsymbol{x})'\right]'=\begin{bmatrix} \frac{\partial \boldsymbol{h}_{i1}}{\partial u_i} & \frac{\partial \boldsymbol{h}_{i1}}{\partial s_i} & \frac{\partial \boldsymbol{h}_{i1}}{\partial m_i} \\ \frac{\partial \boldsymbol{h}_{i2}}{\partial u_i} & \frac{\partial \boldsymbol{h}_{i2}}{\partial s_i} & \frac{\partial \boldsymbol{h}_{i2}}{\partial m_i} \\ \frac{\partial \boldsymbol{h}_{i3}}{\partial u_i} & \frac{\partial \boldsymbol{h}_{i3}}{\partial s_i} & \frac{\partial \boldsymbol{h}_{i3}}{\partial m_i} \end{bmatrix}\begin{bmatrix} \frac{\partial u_i}{\partial x} & \frac{\partial u_i}{\partial y} & \frac{\partial u_i}{\partial z} \\ \frac{\partial s_i}{\partial x} & \frac{\partial s_i}{\partial y} & \frac{\partial s_i}{\partial z} \\ \frac{\partial m_i}{\partial x} & \frac{\partial m_i}{\partial y} & \frac{\partial m_i}{\partial z} \end{bmatrix} \tag{15.162}$$

$$\boldsymbol{H}_{ib}=\left[\nabla_{\boldsymbol{b}_i^z}\boldsymbol{h}_i(\boldsymbol{x})'\right]'=\begin{bmatrix} \frac{\partial \boldsymbol{h}_{i1}}{\partial u_i} & \frac{\partial \boldsymbol{h}_{i1}}{\partial s_i} & \frac{\partial \boldsymbol{h}_{i1}}{\partial m_i} \\ \frac{\partial \boldsymbol{h}_{i2}}{\partial u_i} & \frac{\partial \boldsymbol{h}_{i2}}{\partial s_i} & \frac{\partial \boldsymbol{h}_{i2}}{\partial m_i} \\ \frac{\partial \boldsymbol{h}_{i3}}{\partial u_i} & \frac{\partial \boldsymbol{h}_{i3}}{\partial s_i} & \frac{\partial \boldsymbol{h}_{i3}}{\partial m_i} \end{bmatrix}\begin{bmatrix} \frac{\partial u_i}{\partial b_i^\vartheta} & \frac{\partial u_i}{\partial b_i^\phi} & \frac{\partial u_i}{\partial b_i^\alpha} \\ \frac{\partial s_i}{\partial b_i^\vartheta} & \frac{\partial s_i}{\partial b_i^\phi} & \frac{\partial s_i}{\partial b_i^\alpha} \\ \frac{\partial m_i}{\partial b_i^\vartheta} & \frac{\partial m_i}{\partial b_i^\phi} & \frac{\partial m_i}{\partial b_i^\alpha} \end{bmatrix} \tag{15.163}$$

where

$$\begin{bmatrix} \frac{\partial \boldsymbol{h}_{i1}}{\partial u_i} & \frac{\partial \boldsymbol{h}_{i1}}{\partial s_i} & \frac{\partial \boldsymbol{h}_{i1}}{\partial m_i} \\ \frac{\partial \boldsymbol{h}_{i2}}{\partial u_i} & \frac{\partial \boldsymbol{h}_{i2}}{\partial s_i} & \frac{\partial \boldsymbol{h}_{i2}}{\partial m_i} \\ \frac{\partial \boldsymbol{h}_{i3}}{\partial u_i} & \frac{\partial \boldsymbol{h}_{i3}}{\partial s_i} & \frac{\partial \boldsymbol{h}_{i3}}{\partial m_i} \end{bmatrix}=\begin{bmatrix} \frac{u_i}{\sqrt{R}} & \frac{s_i}{\sqrt{R}} & \frac{m_i}{\sqrt{R}} \\ \frac{-s_i}{\sqrt{u_i^2+s_i^2}} & \frac{u_i}{\sqrt{u_i^2+s_i^2}} & 0 \\ \frac{-u_i m_i}{R\sqrt{u_i^2+s_i^2}} & \frac{-s_i m_i}{\sqrt{u_i^2+s_i^2}R} & \frac{u_i^2+s_i^2}{\sqrt{u_i^2+s_i^2}R} \end{bmatrix}, R=\left(u_i^2+s_i^2+m_i^2\right) \tag{15.164}$$

$$\begin{bmatrix} \frac{\partial u_i}{\partial x} & \frac{\partial u_i}{\partial y} & \frac{\partial u_i}{\partial z} \\ \frac{\partial s_i}{\partial x} & \frac{\partial s_i}{\partial y} & \frac{\partial s_i}{\partial z} \\ \frac{\partial m_i}{\partial x} & \frac{\partial m_i}{\partial y} & \frac{\partial m_i}{\partial z} \end{bmatrix}=\boldsymbol{A}\left(\boldsymbol{v}_i-\boldsymbol{b}_i^z\right)^{-1}\boldsymbol{T}\left(\boldsymbol{X}_{isp}\right)^{-1} \tag{15.165}$$

$$\begin{aligned}\begin{bmatrix} \frac{\partial u_i}{\partial b_i^\vartheta} & \frac{\partial u_i}{\partial b_i^\phi} & \frac{\partial u_i}{\partial b_i^\alpha} \\ \frac{\partial s_i}{\partial b_i^\vartheta} & \frac{\partial s_i}{\partial b_i^\phi} & \frac{\partial s_i}{\partial b_i^\alpha} \\ \frac{\partial m_i}{\partial b_i^\vartheta} & \frac{\partial m_i}{\partial b_i^\phi} & \frac{\partial m_i}{\partial b_i^\alpha} \end{bmatrix}&=\frac{\partial \boldsymbol{A}\left(\boldsymbol{v}_i-\boldsymbol{b}_i^z\right)^{-1}\boldsymbol{T}\left(\boldsymbol{X}_{isp}\right)^{-1}}{\partial \boldsymbol{b}_i^z}\left(\boldsymbol{X}-\boldsymbol{X}_{is}\right) \\ &=\frac{\partial \boldsymbol{A}_{roll}\left(b_i^\alpha-\alpha\right)\boldsymbol{A}_{pitch}\left(b_i^\phi-\phi\right)\boldsymbol{A}_{head}\left(b_i^\vartheta-\vartheta\right)\boldsymbol{T}\left(\boldsymbol{X}_{isp}\right)^{-1}}{\partial \boldsymbol{b}_i^z}\left(\boldsymbol{X}-\boldsymbol{X}_{is}\right)\end{aligned} \tag{15.166}$$

$$\frac{\partial \boldsymbol{A}_{roll}(b_i^\alpha-\alpha)\boldsymbol{A}_{pitch}\left(b_i^\phi-\phi\right)\boldsymbol{A}_{head}\left(b_i^\vartheta-\vartheta\right)}{\partial \boldsymbol{b}_i^z}\boldsymbol{T}\left(\boldsymbol{X}_{isp}\right)^{-1}(\boldsymbol{X}-\boldsymbol{X}_{is})$$

$$=\begin{bmatrix}\left(\boldsymbol{A}_{roll}(b_i^\alpha-\alpha)\boldsymbol{A}_{pitch}\left(b_i^\phi-\phi\right)\frac{\partial \boldsymbol{A}_{head}\left(b_i^\vartheta-\vartheta\right)}{\partial b_i^\vartheta}\boldsymbol{T}\left(\boldsymbol{X}_{isp}\right)^{-1}(\boldsymbol{X}-\boldsymbol{X}_{is})\right)' \\ \left(\boldsymbol{A}_{roll}(b_i^\alpha-\alpha)\frac{\partial \boldsymbol{A}_{pitch}\left(b_i^\phi-\phi\right)}{\partial b_i^\phi}\boldsymbol{A}_{head}\left(b_i^\vartheta-\vartheta\right)\boldsymbol{T}\left(\boldsymbol{X}_{isp}\right)^{-1}(\boldsymbol{X}-\boldsymbol{X}_{is})\right)' \\ \left(\frac{\partial \boldsymbol{A}_{roll}(b_i^\alpha-\alpha)}{\partial b_i^\alpha}\boldsymbol{A}_{pitch}\left(b_i^\phi-\phi\right)\boldsymbol{A}_{head}\left(b_i^\vartheta-\vartheta\right)\boldsymbol{T}\left(\boldsymbol{X}_{isp}\right)^{-1}(\boldsymbol{X}-\boldsymbol{X}_{is})\right)'\end{bmatrix}' \tag{15.167}$$

where $\frac{\partial \boldsymbol{A}_{head}\left(b_i^\vartheta-\vartheta\right)}{\partial b_i^\vartheta}$ is defined by (15.122), $\frac{\partial \boldsymbol{A}_{pitch}\left(b_i^\phi-\phi\right)}{\partial b_i^\phi}$ by (15.123), and $\frac{\partial \boldsymbol{A}_{roll}\left(b_i^\alpha-\alpha\right)}{\partial b_i^\alpha}$ by (15.124).

Adopting the formula above and combining the system given in the former section, the detailed process of the MLRM algorithm to solve the mobile radar registration is listed as follows.

1. Set the initial estimate of the systematic errors as $\hat{\boldsymbol{b}}=\left[\hat{\boldsymbol{b}}_1',\hat{\boldsymbol{b}}_2'\right]'=0\boldsymbol{I}$.
2. For radar i $(i=1,2)$, use the current estimate $\hat{\boldsymbol{b}}_0=\left[\hat{\boldsymbol{b}}_{01}',\hat{\boldsymbol{b}}_{02}'\right]'$ of the systematic errors and according to (15.108), project all measurements to the state space. Here, we omit the time tag $k=1,2,\ldots,N$ for convenience and obtain

$$\boldsymbol{x}_i=\boldsymbol{T}\left(\boldsymbol{X}_{isp}\right)\boldsymbol{A}\left(\boldsymbol{v}_i-\hat{\boldsymbol{b}}_{0i}^z\right)h^{-1}\left(\boldsymbol{Z}_{idp}-\hat{\boldsymbol{b}}_{0i}^l\right)+\boldsymbol{X}_{is} \tag{15.168}$$

where $\boldsymbol{X}_{is}$, $\boldsymbol{X}_{isp}$ denote the positions of radar i in the earth coordinate system and geographical coordinate system, respectively, at time k, and they are accurate and known. $\boldsymbol{v}_i$ denotes radar i's attitude angle measurement value with systematic errors at time k and is already known. $\hat{\boldsymbol{b}}_{0i}^l$, $\hat{\boldsymbol{b}}_{0i}^z$ are radar i's systematic error estimates already obtained and are already known. $\boldsymbol{Z}_{idp}$ denotes radar i's measurement of the target at time k and is already known.
3. Substitute $\boldsymbol{x}_i$, $\hat{\boldsymbol{b}}_{0i}$ into (15.162) and solve for $\boldsymbol{H}_{ix}$; and into (15.163) for $\boldsymbol{H}_{ib}$.
4. Substituting $\boldsymbol{H}_{ix}$ and $\boldsymbol{H}_{ib}$ into (15.158) yields $\boldsymbol{H}_i^{-L}$.
5. Substituting $\boldsymbol{H}_i^{-L}$ into (15.147) gives $\boldsymbol{Q}(k)$.
6. Inserting $\boldsymbol{x}_i$, $\hat{\boldsymbol{b}}_{0i}$, $\boldsymbol{Q}(k)$ into (15.148) gives $\bar{\boldsymbol{X}}_0(k)$.
7. Using $\boldsymbol{H}_{ix}$, $\boldsymbol{R}(\boldsymbol{W}_i)$ in (15.136) yields $\boldsymbol{\Sigma}_{\boldsymbol{x}_i}^{-1}$.
8. Inserting $\boldsymbol{\Sigma}_{\boldsymbol{x}_i}^{-1}$ into (15.143) yields $\Sigma^{-1}(k)$.
9. Substituting $\bar{\boldsymbol{X}}_0(k)$, $\boldsymbol{Q}(k)$, $\Sigma^{-1}(k)$ into (15.151), we obtain $\hat{\boldsymbol{b}}$.
10. Judge whether the estimate $\hat{\boldsymbol{b}}$ is convergent or not according to $\left\|\hat{\boldsymbol{b}}-\hat{\boldsymbol{b}}_0\right\|\le\varepsilon$, where $\|\cdot\|$ denotes the number of normal forms of the vector and can be chosen discretionarily. ε denotes the

acceptance threshold. If the estimate $\hat{\boldsymbol{b}}$ is already convergent, continue to go through the process, otherwise, let $\hat{\boldsymbol{b}}_0 = \hat{\boldsymbol{b}}$ and start again from step 2.

11. Substituting $\bar{\boldsymbol{X}}_0(k), \boldsymbol{Q}(k), \hat{\boldsymbol{b}}$ into (15.146), $\boldsymbol{X}(k)$ is obtained.
12. Substitute $\boldsymbol{X}(k), \boldsymbol{\Sigma}_{\boldsymbol{x}_i}^{-1}$ into (15.141) to get the target state estimate $\hat{\boldsymbol{x}}$ after eliminating the error.

The 12 steps covered above form the detailed process of the MLRM algorithm. As can be seen, the MLRM is a batch processing algorithm for measurements, and convergent estimates of systematic errors and target state estimates after registration can finally be obtained through recursive optimization of systematic errors along the maximum gradient of disturbance.

15.4.4 ASR Algorithm

The augmented state registration (ASR) algorithm expands the system state to include systematic errors. It regards systematic errors as the unknown and to-be-estimated state, and conducts joint estimations for target states and systematic errors through conventional state estimation methods. Hereafter, a detailed explanation will be presented.

Unite the target state $\boldsymbol{X}(k)$ and the systematic errors $\boldsymbol{b}_1, \boldsymbol{b}_2$ of two radars, and construct the new system state as

$$\boldsymbol{X}_A = [\boldsymbol{X}' \ \ \boldsymbol{b}_1' \ \ \boldsymbol{b}_2']' = [x \ \ \dot{x} \ \ y \ \ \dot{y} \ \ z \ \ \dot{z} \ \ \boldsymbol{b}_1^l \ \ \boldsymbol{b}_1^z \ \ \boldsymbol{b}_2^l \ \ \boldsymbol{b}_2^z]' \tag{15.169}$$

Since the systematic errors are assumed unchanged constants, we get

$$\boldsymbol{b}_i(k+1) = \boldsymbol{I}_{6\times 6}\boldsymbol{b}_i(k) \tag{15.170}$$

From (15.93) and (15.170), the state transformation equation of the new system state $\boldsymbol{X}_A$ is obtained as follows:

$$\boldsymbol{X}_A(k+1) = \boldsymbol{F}_A(k)\boldsymbol{X}_A(k) + \boldsymbol{V}_A(k) \tag{15.171}$$

where

$$\boldsymbol{F}_A(k) = \mathrm{diag}(\boldsymbol{F}(k), \boldsymbol{I}_{12\times 12}) \tag{15.172}$$

$$\boldsymbol{V}_A(k) = \mathrm{diag}(\boldsymbol{V}(k), 0\boldsymbol{I}_{12\times 12}) \tag{15.173}$$

From (15.173), we know that $\boldsymbol{V}_A(k)$ is still white Gaussian noise, with covariance $\boldsymbol{Q}_A = \mathrm{diag}(\boldsymbol{Q}, 0\boldsymbol{I}_{12\times 12})$.

Combining the measurement equations of mobile radars 1 and 2, and using (15.108), yields

$$\boldsymbol{Z}_{\mathrm{p}}(k) = \boldsymbol{h}_A(\boldsymbol{X}_A(k)) + \boldsymbol{W}(k) \tag{15.174}$$

where

$$\boldsymbol{Z}_{\mathrm{p}}(k) = [\boldsymbol{Z}_{1\mathrm{p}}'(k) \ \ \boldsymbol{Z}_{2\mathrm{p}}'(k)]' \tag{15.175}$$

$$\boldsymbol{h}_A(\boldsymbol{X}_A(k))=\begin{bmatrix} h\left(\boldsymbol{A}\left(\boldsymbol{v}_1(k)-\boldsymbol{b}_1^z\right)^{-1}\boldsymbol{T}(\boldsymbol{X}_{1sp}(k))^{-1}(\boldsymbol{X}(k)-\boldsymbol{X}_{1s}(k))\right)+\boldsymbol{b}_1^l \\ h\left(\boldsymbol{A}\left(\boldsymbol{v}_2(k)-\boldsymbol{b}_2^z\right)^{-1}\boldsymbol{T}(\boldsymbol{X}_{2sp}(k))^{-1}(\boldsymbol{X}(k)-\boldsymbol{X}_{2s}(k))\right)+\boldsymbol{b}_2^l \end{bmatrix} \tag{15.176}$$

$$\boldsymbol{W}(k)=[\boldsymbol{W}_1'(k) \ \ \boldsymbol{W}_2'(k)]' \tag{15.177}$$

From (15.177) it follows that $\boldsymbol{W}(k)$ is still Gaussian, zero-mean white noise with covariance $\boldsymbol{R}(k)=\text{diag}(\boldsymbol{R}_1(k),\boldsymbol{R}_2(k))$.

With the state equation (15.171) and the measurement equation (15.174) after augmentation, adopt the EKF filtering equation set hereafter and make a filtering estimation of the system to constitute the ASR algorithm:

$$\hat{\boldsymbol{P}}_A^-(k+1)=\boldsymbol{F}_A(k)\hat{\boldsymbol{P}}_A(k)\boldsymbol{F}_A'(k)+\boldsymbol{Q}_A \tag{15.178}$$

$$\boldsymbol{S}(k+1)=\boldsymbol{h}_{\boldsymbol{X}_A}(k+1)\hat{\boldsymbol{P}}_A^-(k+1)\boldsymbol{h}_{\boldsymbol{X}_A}'(k+1)+\boldsymbol{R}(k+1) \tag{15.179}$$

$$\boldsymbol{K}(k+1)=\hat{\boldsymbol{P}}_A^-(k+1)\boldsymbol{h}_{\boldsymbol{X}_A}'(k+1)\boldsymbol{S}^{-1}(k+1) \tag{15.180}$$

$$\hat{\boldsymbol{X}}_A(k+1)=\boldsymbol{F}_A(k)\hat{\boldsymbol{X}}_A(k)+\boldsymbol{K}(k+1)\left(\boldsymbol{Z}_\text{p}(k+1)-\boldsymbol{h}\left(\boldsymbol{F}_A(k)\hat{\boldsymbol{X}}_A(k)\right)\right) \tag{15.181}$$

$$\hat{\boldsymbol{P}}_A(k+1)=\hat{\boldsymbol{P}}_A^-(k+1)-\boldsymbol{K}(k+1)\boldsymbol{S}(k+1)\boldsymbol{K}^T(k+1) \tag{15.182}$$

where the Jacobian matrix of $\boldsymbol{h}$ is

$$\boldsymbol{h}_{\boldsymbol{X}_A}(k+1)=\begin{bmatrix} \boldsymbol{H}_{1x}(:,1),0\boldsymbol{I}_{3\times1},\boldsymbol{H}_{1x}(:,2),0\boldsymbol{I}_{3\times1},\boldsymbol{H}_{1x}(:,3),0\boldsymbol{I}_{3\times1},\boldsymbol{I}_{3\times3},0\boldsymbol{I}_{3\times3},\boldsymbol{H}_{1b},0\boldsymbol{I}_{3\times3}, \\ \boldsymbol{H}_{2x}(:,1),0\boldsymbol{I}_{3\times1},\boldsymbol{H}_{1x}(:,2),0\boldsymbol{I}_{3\times1},\boldsymbol{H}_{1x}(:,3),0\boldsymbol{I}_{3\times1},0\boldsymbol{I}_{3\times3},\boldsymbol{I}_{3\times3},0\boldsymbol{I}_{3\times3},\boldsymbol{H}_{2b}, \end{bmatrix} \tag{15.183}$$

with $\boldsymbol{H}_{1x}$ obtained from (15.162) and $\boldsymbol{H}_{1b}$ from (15.163).

15.4.5 Simulation Analysis

This subsection presents the simulative experiments conducted to test the estimation performance of the MLRM and ASR algorithms on mobile radar systematic errors, and a comparison of their estimation performance in the simulations. To guarantee the realness and validity of the target and platform movements, the tracks of the target and platform were all generated through software STK.

The system under consideration in the experiments is composed of two airborne radars (1 and 2) and an aerial target, as shown in Figure 15.8. The measurement noise $\boldsymbol{W}_i(k)$ of both radars is zero-mean, Gaussian white noise, with covariance $\boldsymbol{R}(\boldsymbol{W}_i)=\text{diag}((50\,\text{m})^2,(0.002\,\text{rad})^2,(0.001\,\text{rad})^2)$. The change rule of the attitude angle of radar 1's airborne platform can be expressed as $\boldsymbol{v}_1'(k)=[0.002k,0.01+0.002k,0.01+0.002k]'$, and that of radar 2's airborne platform as $\boldsymbol{v}_2'(k)=[0.002k,0.001k,0.001k]'$. The systematic errors of both radars are $\boldsymbol{b}_i=[1000\,\text{m},\ 0.0087\,\text{rad},\ 0.0047\,\text{rad},\ 0.0087\,\text{rad},\ 0.0047\,\text{rad},\ 0.0037\,\text{rad}]'$.

The radar platform was set as a helicopter platform flying slowly at an elevation of 1 km, and the target as a fighter flying quickly along the meridian at an elevation of 2 km, with a sampling interval of 1 s. The radar measurements with systematic errors are shown in Figure 15.9.

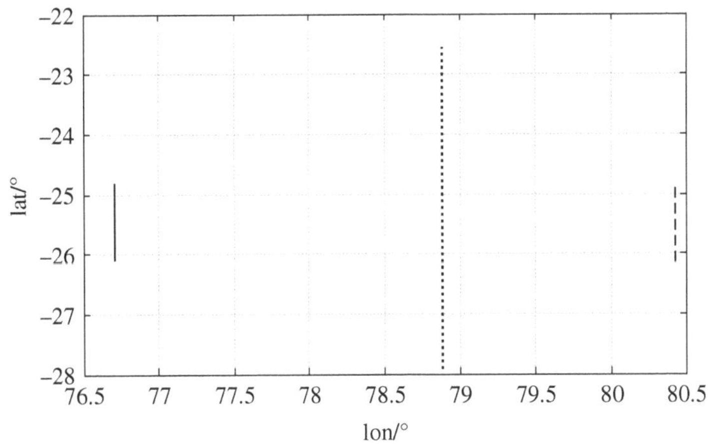

Figure 15.8 The latitude and longitude of the simulative environment

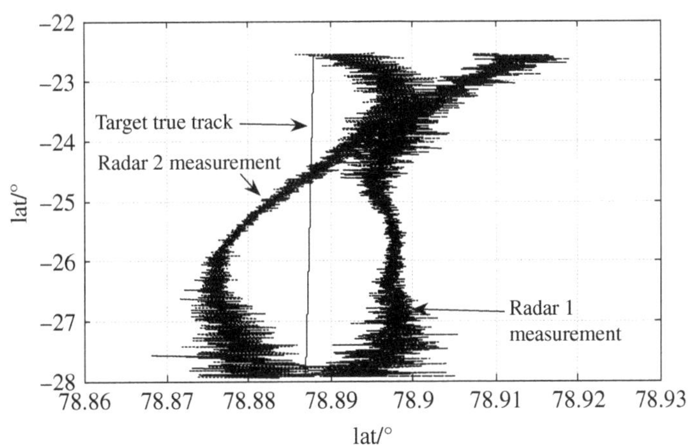

Figure 15.9 The measurements with system errors in mobile radars

1. Since the MLRM is a batch processing algorithm, and the estimate of the target state after registration can be given, the MLRM's estimation performance on systematic errors is explained in Table 15.1.
2. In Table 15.1, K denotes the recursive times of the MLRM. As can be seen, this algorithm has an estimation accuracy of above 95% on every systematic error and converges after three or four recurrence steps. Therefore, it has a good estimation effect on both radars' systematic errors but involves excessive calculations.
3. The results of 50 Monte Carlo simulation experiments on the estimation performance of the ASR algorithm are shown in Figures 15.10–15.17: as the ASR algorithm converges, its estimates of the target state eliminate the influence of systematic errors on the whole.

The comparison of target state estimation results of the ASR and MLRM algorithms is shown in Figure 15.18, and that regarding systematic errors in Table 15.2.

Table 15.1 The systematic error estimation accuracy of the MLRM algorithm

Estimation accuracy of radar 1	$\hat{b}_1^r$	$\hat{b}_1^\theta$	$\hat{b}_1^\eta$	$\hat{b}_1^\vartheta$	$\hat{b}_1^\phi$	$\hat{b}_1^\alpha$
$K=1$	99.8364%	97.8977%	97.9904%	98.1724%	95.9528%	99.6310%
$K=2$	97.5313%	98.9071%	99.1336%	99.2411%	98.0490%	99.3905%
$K=3$	97.5306%	98.9068%	99.1332%	99.2408%	98.0491%	99.3899%
$K=4$	97.5306%	98.9068%	99.1332%	99.2408%	98.0491%	99.3899%
$K=5$	97.5306%	98.9068%	99.1332%	99.2408%	98.0491%	99.3899%
Estimation accuracy of radar 2	$\hat{b}_2^r$	$\hat{b}_2^\theta$	$\hat{b}_2^\eta$	$\hat{b}_2^\vartheta$	$\hat{b}_2^\phi$	$\hat{b}_2^\alpha$
$K=1$	97.0162%	97.8429%	99.1446%	96.4591%	99.5831%	89.9158%
$K=2$	97.5676%	96.8691%	98.5375%	96.9096%	98.0935%	96.3075%
$K=3$	97.5671%	96.8692%	98.5355%	96.9103%	98.0958%	96.3121%
$K=4$	97.5671%	96.8692%	98.5355%	96.9103%	98.0958%	96.3121%
$K=5$	97.5671%	96.8692%	98.5355%	96.9103%	98.0958%	96.3121%

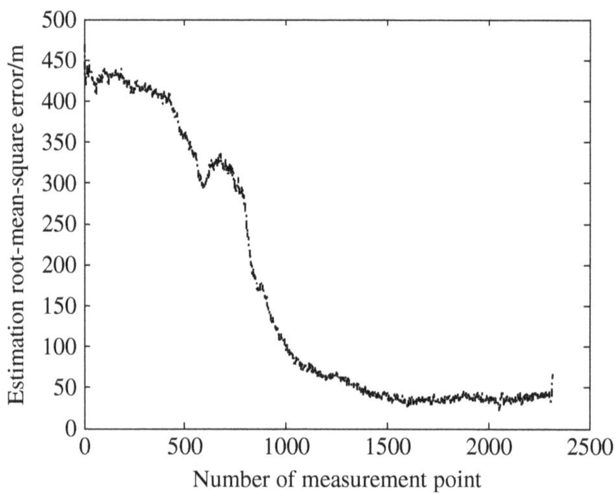

Figure 15.10 Target state estimation effects of ASR algorithm on x axis

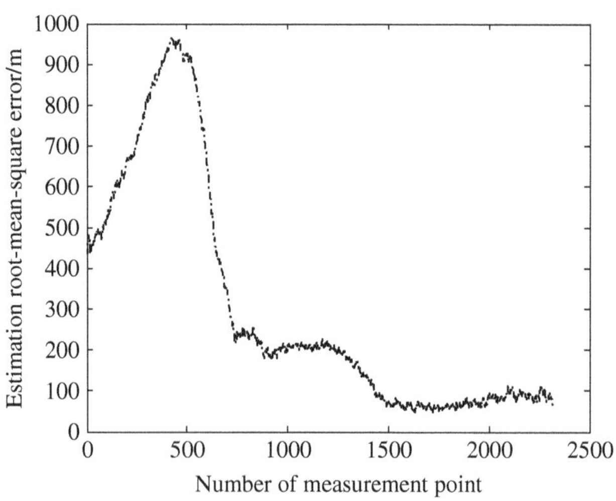

Figure 15.11 Target state estimation effects of ASR algorithm on y axis

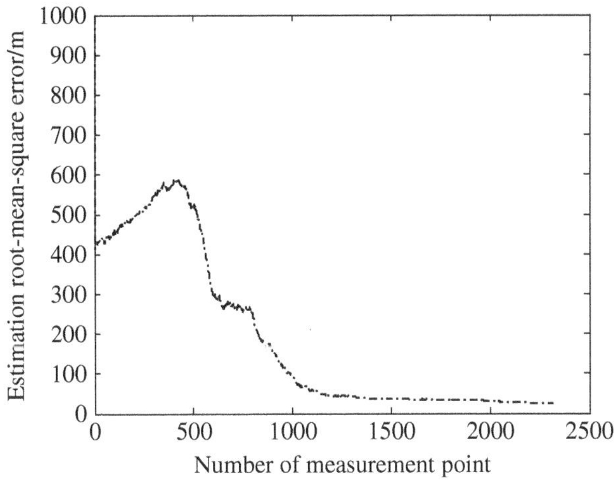

Figure 15.12 Estimation effects of ASR algorithm on radar 1 range systematic error

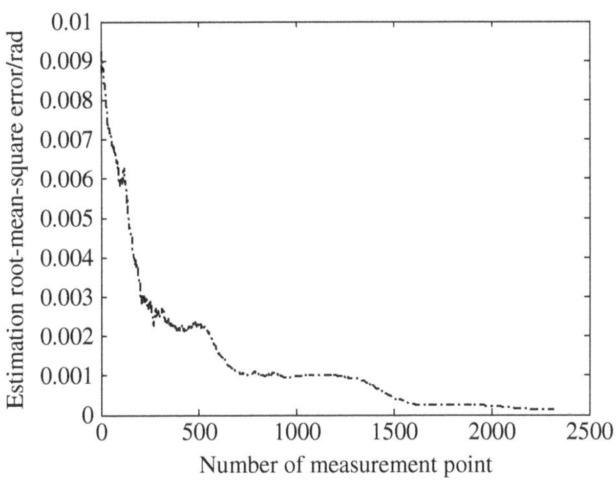

Figure 15.13 Estimation effects of ASR algorithm on radar 1 azimuth systematic error

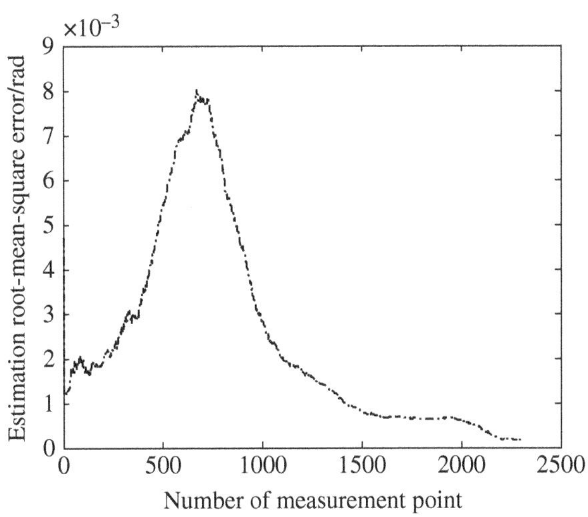

Figure 15.14 Estimation effects of ASR algorithm on radar 2 elevation systematic error

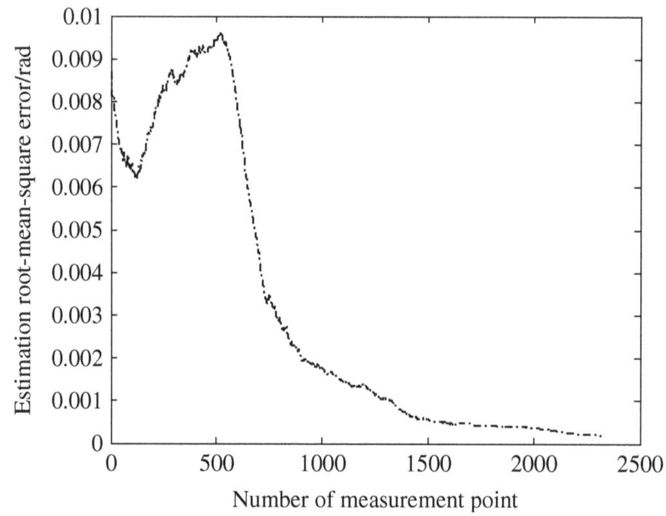

Figure 15.15 Estimation effects of ASR algorithm on radar 2 yaw angle systematic error

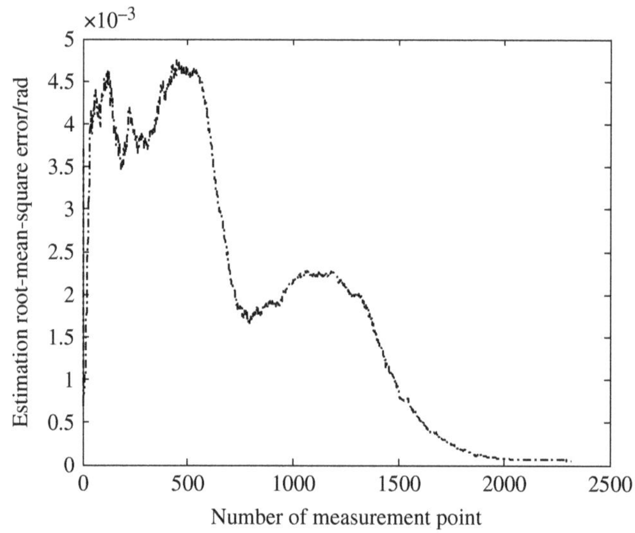

Figure 15.16 Estimation effects of ASR algorithm on radar 1 roll angle systematic error

As illustrated in Figure 15.18 and Table 15.2, both MLRM and ASR yield good estimation results for systematic errors, and can desirably attack problems with mobile radar registration. However, compared solely in terms of estimations of target states and systematic errors, the MLRM performs better than the ASR algorithm, but obviously consumes more time.

15.5 Summary

The radar registration technology for radar network systems is key to the stability of system performance. This chapter has discussed this technology for fixed and mobile radar network systems under assumptions that the target position is known and that it is unknown.

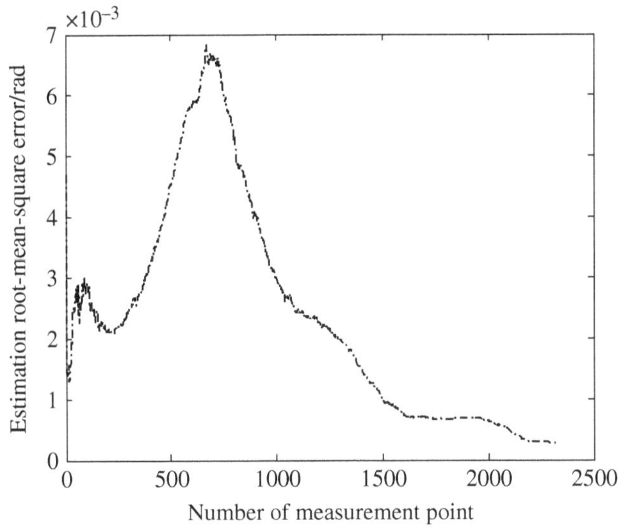

Figure 15.17 Estimation effects of ASR algorithm on radar 2 pitch angle systematic error

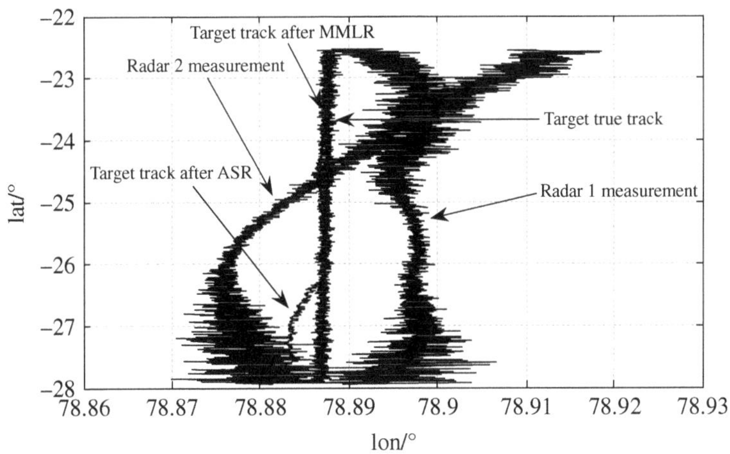

Figure 15.18 Collation map of MLRM and ASR target state estimation

Table 15.2 Collation map of MLRM and ASR systematic error estimation

Radar 1	$\hat{b}_1^r$	$\hat{b}_1^\theta$	$\hat{b}_1^\eta$	$\hat{b}_1^\vartheta$	$\hat{b}_1^\phi$	$\hat{b}_1^\alpha$
ASR	98.1023%	98.8151%	97.9607%	98.6024%	97.5057%	98.9963%
MLRM	98.9969%	99.2715%	98.0792%	99.0052%	97.8646%	98.9880%
Radar 2	$\hat{b}_2^r$	$\hat{b}_2^\theta$	$\hat{b}_2^\eta$	$\hat{b}_2^\vartheta$	$\hat{b}_2^\phi$	$\hat{b}_2^\alpha$
ASR	98.8901%	98.3645%	97.9927%	97.8537%	95.7978%	93.4095%
MLRM	98.9374%	98.5757%	97.9208%	97.9809%	97.1846%	93.5412%

In addition to the registration technology of fixed and mobile radar systems in cases where target positions are given, this chapter has also introduced a real-time radar calibration unit widely adopted for military purposes, which was developed based on this technology.

With respect to the problem of fixed radar registration in the absence of target positions, this chapter has provided a detailed discussion of the RTQC, LS, GLS, and ECEF-GLS algorithms, and a comparative analysis of the performance of the ECEF-LS and ECEF-GLS in a simulative experiment in particular. The results show that the ECEF-GLS has obvious advantages in terms of estimation accuracy and convergence speed.

Regarding the problem of mobile radar registration, we have delved into the MLRM and ASR algorithms, and drawn a simulative comparison of their performance. The results show that the former has obvious advantages in estimation accuracy of target states and systematic errors, but is more time-consuming and thus not suitable for engineering applications.

16

Radar Network Data Processing

16.1 Introduction

The radar network in modern wars is an effective means against major threats. A radar network is a system consisting chiefly of radars with different systems, frequencies, and polarization modes properly distributed such that observations from these radars are trapped and delivered as if through a piece of netting and then synthesized, controlled, and administered in the central station forming an organic whole.

The significance of the radar network lies in the following facts:

1. It has greatly increased the coverage area of the detection zone, providing a clear, full picture of the battlefield; meanwhile, it offers a considerable anti-stealth advantage as a result of the great improvement in probability of target detection in the overlap regions.
2. It integrates data from different radars, improving the precision of target tracking.
3. It detects targets from different angles, which is of great importance for countermeasures against stealth, and low- and ultra-low-altitude targets.
4. It improves the overall ECM performance with its member radars with different systems, frequencies, and polarization modes.

Radar networks can be classified, according to type, into:

1. *Bistatic/multistatic.* This type has desirable ECCM, counter-ARM, and anti-stealth performance and is able to greatly enhance the anti-low-penetration ability with the help of aerial platforms.
2. *Monostatic.* This type is composed of independent radars which form an organic whole by networking with each other. Its constituent radars work independently and flexibly and each of them can finish part of the work even when they lose contact with the central station.
3. *Hybrid.* This type is a combination of monostatic and bistatic/multistatic radar networks and has the merits of both.

Radar Data Processing with Applications, First Edition. He You, Xiu Jianjuan, and Guan Xin.

In terms of spatial deployment, the monostatic type can be divided into common station and distributed radar networks.

Military radar networks may be divided, according to their missions, into regional surveillance (including air surveillance and coastal surveillance) and guidance radar networks.

Section 16.2 introduces the concepts of radar networks including performance, indexes, and optimization of station distribution on the basis of the design and analysis of radar networks. Sections 16.3, 16.4, and 16.5, respectively, address the basic contents of monostatic, bistatic, and multistatic radar network data processing. Section 16.6 deals with the track correlation technique in distributed processing. Section 16.7 is a summary of the chapter.

16.2 Performance Evaluation Indexes of Radar Networks

The principal functions of a radar network are target detection, tracking, and track reporting in its responsible regions. Therefore, its leading evaluation indicator turns out to be its capability to cover the responsible regions, followed by the target track capacity (i.e., the maximum number of targets uploaded, fused, and tracked). In addition, its comprehensive evaluation indicators also include detection probability, anti-stealth technology, anti-jamming ability, counter low-altitude penetration capability, and counter anti-radiation missile capability.

16.2.1 Coverage Performance Indexes

The coverage performance of a radar network consists of coverage continuity and strictness [370, 371]. Generally, the blind zone coefficient, overlap coefficient, average spatial overlap coefficient, and overlap coefficient are used to indicate these.

The blind zone coefficient is defined as

$$C_{\mathrm{BL}} = \frac{\sum A_{\mathrm{BL}}}{A_0} \tag{16.1}$$

where $\sum A_{\mathrm{BL}}$ is the sum of blind zones in a responsible region and A_0 is the total area of the responsible region. In the light of coverage continuity, C_{BL} must approach zero.

The overlap coefficient is defined as

$$C_{\mathrm{OV}} = \frac{A_1 + A_2 + \cdots + A_i}{A_0} \tag{16.2}$$

where $A_1, A_2, \ldots, A_i$ are covering areas of stations in a radar network in its responsible region.

The average spatial overlap coefficient is defined as

$$\bar{C}_{\mathrm{OV}} = \frac{\bar{A}_1 + \bar{A}_2 + \cdots + \bar{A}_i}{A_0} \tag{16.3}$$

where $\bar{A}_1, \bar{A}_2, \ldots, \bar{A}_i$ are the average values of covering areas of radar stations at M flight levels.

The overlap coefficient is the number of radars which simultaneously observe a target in space, represented by

$$K = \frac{\sum s_k}{A_0} = \frac{n\left[\pi R_h^2\right]}{A_0} \tag{16.4}$$

where S_k is the detection area of the kth radar, n the number of radars in the network, and R_h the distance of the target which the network finds at the lower limit of elevation.

16.2.2 Target Capacity

There are three decisive factors in evaluating the target capacity of a radar network: the extraction capacity of the stations, the transmission capacity of the communication network (i.e., the maximum channel transfer rate), and the processing and monitoring capacity of the information center. If the transmission capacity of the communication channels and the monitoring capacity of the information center are not taken into account, then the target capacity N_0 of the network should be the sum of the target capacity of every station in a responsible region, which is written as $N_0 = N_1 + N_2 + \cdots + N_i$ (where N_1, N_2, …, N_i are the target capacities of each of the radar stations, respectively).

16.2.3 Anti-jamming Ability

The anti-jamming ability of a radar network is largely associated with those of the networking monostatic radars, and its bandwidth, spatial signal energy density, and signal types. It is evaluated with three main parameters [371].

1. *Improvement factor of the detection overlap coefficient.* The detection overlap coefficient of a radar network, an index of comprehensive performance evaluation, is closely related to the detection abilities of the member stations, and the anti-jamming abilities of the single stations. The greater the anti-jamming abilities of the member stations, or the better their anti-jamming measures are, the less the detection ability of the network suffers in the presence of jamming, and the higher the overlap coefficient is (and vice versa).

   The improvement factor of the detection overlap coefficient of a radar network refers to the ratio of the overlap coefficient of taking anti-jamming measures to that of taking none of these measures in the presence of jamming. It is used to evaluate the anti-jamming performance of a radar network, that is,

$$C_{\mathrm{OVIF}} = \frac{C'_{\mathrm{OV}}}{C_{\mathrm{OV}}} \tag{16.5}$$

   where C'_{OV} and C_{OV} are the detection overlap coefficients of a radar network, respectively, in the case of taking anti-jamming measures and not taking these measures.
2. *Frequency overlap coefficient.* The frequency domain is the most effective and important anti-jamming method for a radar network: the more bands it occupies and the wider the frequency domain is, the greater its anti-jamming ability. Thus, the frequency domain coverage of a radar network can be considered an important index for evaluating its anti-jamming ability. It consists of band cover and bandwidth coefficients.

If a radar network with N radars occupies M bands with a standard bandwidth for each frequency band F_1, F_2, …, F_M, and the bandwidth of each monostatic radar within each band is $\Delta f_i\ (i=1,\ldots,M)$, then the band overlap coefficient is defined by

$$F_{fd} = \frac{M}{N} \tag{16.6}$$

the bandwidth coefficient is defined by

$$F_{fk} = \frac{\sum_{i=1}^{M} \Delta f_i}{\sum_{i=1}^{M} F_i} \tag{16.7}$$

and the frequency overlap coefficient of the radar network is defined as

$$F_{\mathrm{OV}} = F_{fd} \times F_{fk} \tag{16.8}$$

3. *Signal overlap coefficient.* Multi-radar networks with more signal types and more complicated ones would be more difficult to detect and jam. The ratio of the number and complexity of signal types to the number of radars is therefore used as an important indicator to measure the anti-jamming ability of a radar network, defined as the signal coverage coefficient of a radar network, that is,

$$S_{\mathrm{OV}} = \frac{M_S}{N} \tag{16.9}$$

where N stands for the number of radars and M_s the number of radar signal types.

16.3 Data Processing of Monostatic Radar Networks

According to the type of network, radar networks can be divided into monostatic, bistatic/multistatic, and hybrid. The monostatic radar network can be further divided into co-station and distributed, according to the positions where several radars are deployed in space. In this section we mainly discuss the general procedure of data processing of the monostatic radar network.

16.3.1 The Process of Data Processing of the Monostatic Radar Network

The process of data processing of the monostatic radar network is shown in Figure 16.1. Generally, it can be broken down into the following steps [17]:

1. To correct errors in various radar systems through error registration, as discussed in Chapter 15.
2. To convert observations from different radar stations into the same coordinate system (that of the data processing center) through coordinate transformation, as discussed in Chapter 5.
3. To correlate measurements and tracks from different radars, including correlations between measurements and tracks (see Chapters 7 and 8) and those between tracks (to be discussed in Section 16.6).

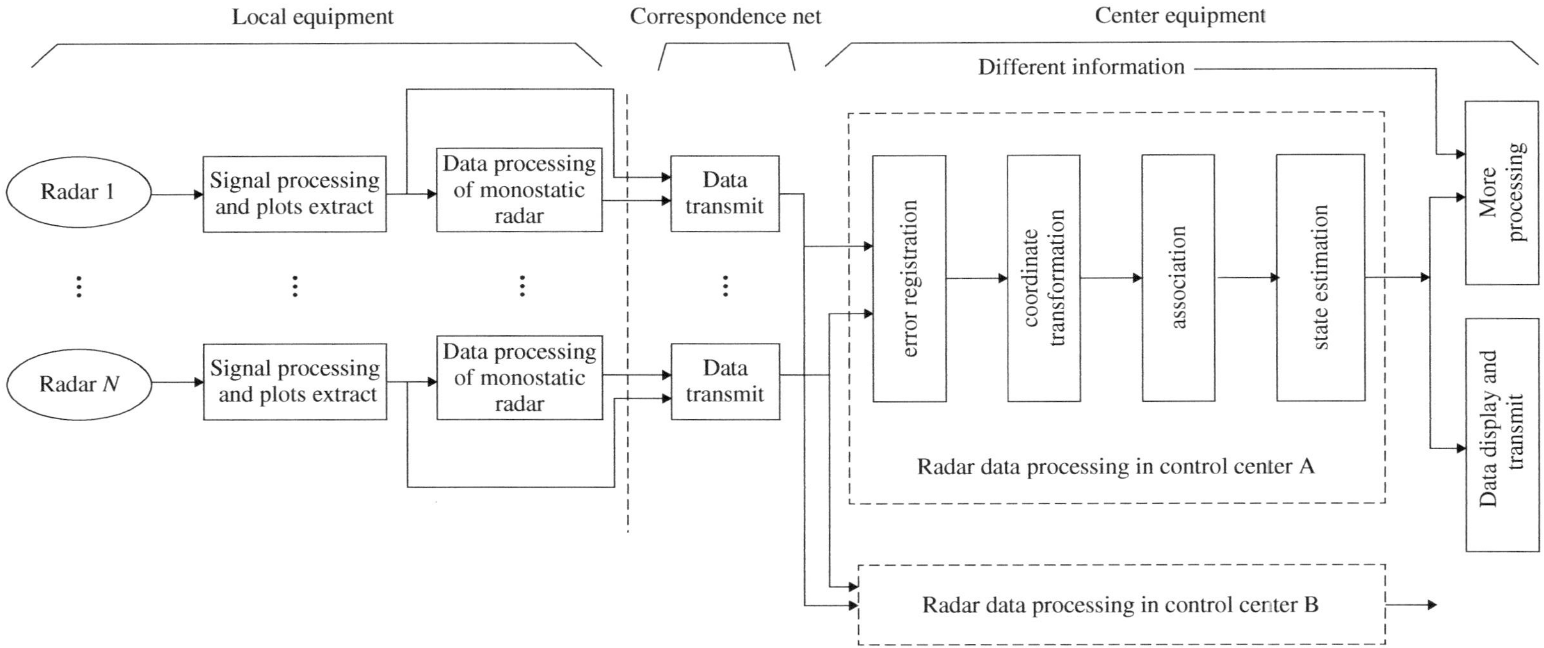

Figure 16.1 The process of data processing of the monostatic radar network

4. Track initiation.
5. To make use of the observed value of radial velocity.
6. To fuse (or integrate) the tracks and measurements from different radar stations (the data derived after system error registration and coordinate transformation), and obtain an estimate of the system's state.

16.3.2 State Estimation of Monostatic Radar Networks

The simplest method of network building is to simplify the data processing work of the local radar which turns with the antenna to that of a common single-radar tracking system. For example, for a primary or secondary surveillance radar that turns with the antenna, the combination of its azimuth data needs attention. The azimuth of the equivalent measurement is

$$\theta = \left(\theta_p \sigma_{\theta_p}^2 + \theta_s \sigma_{\theta_s}^2\right) / \left(\sigma_{\theta_p}^2 + \sigma_{\theta_s}^2\right) \tag{16.10}$$

with the variance of the corresponding error

$$\sigma_\theta^2 = \sigma_{\theta_p}^2 \sigma_{\theta_s}^2 / \left(\sigma_{\theta_p}^2 + \sigma_{\theta_s}^2\right) \tag{16.11}$$

where θ_p and θ_s are the azimuths observed by the primary and secondary radars, with variances $\sigma_{\theta_p}^2$ and $\sigma_{\theta_s}^2$, respectively. The next step is to use the algorithms of the single-radar tracking system to process the measurements, which are to be combined into sequences, and then establish tracks.

The structure of the data processing system has great influence on the performance of the radar network. According to the way in which information flows and is processed, it can be divided into centralized (measurement fusion), distributed (track fusion), and hybrid.

The feature of the centralized structure is that the detection reports recorded by all radars are transmitted to the fusion center where data registration and association, state prediction and updating are completed. The data in the center of the network, after being preprocessed, can be used directly by all single-radar multi-target association algorithms. This structure is characterized chiefly by its small amount of information loss, but it requires large communication capacity and has poor viability.

The feature of the distributed structure is as follows. Before the detection report of a radar enters the fusion stage, its own data processor produces local multi-target tracking tracks, and then this processed information is sent to the fusion center, where track-to-track correlation and track fusion are completed based on the track data from each node, so that an overall estimation is made. This structure has the capability for local independent tracking, and the feature of overall surveillance and evaluation as well. It can be further divided into hierarchy and committee structures, according to the way in which radar information flows. In committee structures, information is also transferred from radar to radar. This structure requires a small system communication capacity and is relatively reliable, but sometimes may incur certain information loss.

The hybrid type is a combination of centralized and distributed structures. It simultaneously transmits detecting reports and track information obtained from local node processing. It has the merits of both types, but incurs high communication and computation expenses.

16.3.2.1 State Estimation of Centralized Structures

Consider a monostatic radar network composed of N radars, where all the measured data are transmitted directly to the fusion center to form a uniform system track. The structure of the centralized radar network decides its specific merit: all the data are processed in the same place, and the resulting target track should be more accurate than that established based on measurements from single radars, which are incomplete.

On the basis of the discrete state equation, the target's motion law can be expressed as

$$\boldsymbol{X}(k+1)=\boldsymbol{F}(k)\boldsymbol{X}(k)+\boldsymbol{V}(k) \tag{16.12}$$

where $\boldsymbol{X}(k)\in\boldsymbol{R}^n$ denotes the target's state vector at time k, $\boldsymbol{V}(k)\in\boldsymbol{R}^n$ denotes the noise vector of a zero-mean, white Gaussian process, and $\boldsymbol{F}(k)\in\boldsymbol{R}^{n,n}$ indicates the state transition matrix. The initial state $\boldsymbol{X}(0)$ is a Gaussian random vector with mean μ and covariance matrix $\boldsymbol{P}_0$, and $\mathrm{Cov}\{\boldsymbol{X}(0),\boldsymbol{V}(k)\}=\boldsymbol{0}$.

The measurement equation of a single radar can be expressed as

$$\boldsymbol{Z}_i(k+1)=\boldsymbol{H}_i(k+1)\boldsymbol{X}(k+1)+\boldsymbol{W}_i(k+1) \tag{16.13}$$

where $\boldsymbol{Z}_i(k+1)\in\boldsymbol{R}^m$, $\boldsymbol{H}_i(k+1)$ is the measurement matrix, $\boldsymbol{W}_i(k+1)$ is the Gaussian sequence with zero mean, and

$$\mathrm{E}\left\{\begin{bmatrix}\boldsymbol{V}(k)\\ \boldsymbol{W}_i(k)\end{bmatrix}[\boldsymbol{V}(k),\ \ \boldsymbol{W}_i(k)]\right\}=\begin{bmatrix}\boldsymbol{Q}(k) & \boldsymbol{0}\\ \boldsymbol{0} & \boldsymbol{R}_i(k)\end{bmatrix} \tag{16.14}$$

Suppose that the three position components of radar i in the Descartes coordinate system of the fusion center are $\tau_i=[a_i b_i c_i]'$. In view of the fact that the fusion center and radar measurement are in different Descartes coordinate systems, the position components of the target (the components on axes x, y, z) are assumed to be subsumed in the measurement vector. Then, let

$$\boldsymbol{\Psi}_i=\begin{bmatrix}\boldsymbol{\tau}_i\\ \boldsymbol{0}\end{bmatrix}_{n\times 1} \tag{16.15}$$

be the augmenting vector of radar i in the Descartes coordinate system of the fusion center. The observation (transformed into the Descartes coordinate system of the fusion center) of radar i at time $k+1$ is

$$\boldsymbol{Y}_i(k+1)=\boldsymbol{Z}_i(k+1)+\boldsymbol{H}_i(k+1)\boldsymbol{\Psi}_i \tag{16.16}$$

So the measurement vectors of N radars are

$$\boldsymbol{Y}(k+1)=[\boldsymbol{Y}_1(k+1),\boldsymbol{Y}_2(k+1),\ldots,\boldsymbol{Y}_N(k+1)]' \tag{16.17}$$

Therefore, the measurement equation can be expressed as

$$\boldsymbol{Y}(k+1)=\boldsymbol{H}(k+1)\boldsymbol{M}(k+1)+\boldsymbol{W}(k+1) \tag{16.18}$$

where

$$\begin{cases} \boldsymbol{H}(k+1) = [\boldsymbol{H}_1(k+1), \boldsymbol{H}_2(k+1), \ldots, \boldsymbol{H}_N(k+1)]' \\ \boldsymbol{W}(k+1) = [\boldsymbol{W}_1(k+1), \boldsymbol{W}_2(k+1), \ldots, \boldsymbol{W}_N(k+1)]' \\ \boldsymbol{M}(k+1) = [\boldsymbol{X}(k+1) + \boldsymbol{\Psi}_1, \boldsymbol{X}(k+1) + \boldsymbol{\Psi}_2, \ldots, \boldsymbol{X}(k+1) + \boldsymbol{\Psi}_N]' \end{cases} \tag{16.19}$$

and

$$\mathrm{E}\left\{ \begin{bmatrix} \boldsymbol{W}(k) \\ \boldsymbol{V}(k) \\ \boldsymbol{X}(k) \end{bmatrix} [\boldsymbol{W}(k) \quad \boldsymbol{V}(k) \quad \boldsymbol{X}(k)] \right\} = \begin{bmatrix} \boldsymbol{R}(k) & \boldsymbol{0} & \boldsymbol{0} \\ \boldsymbol{0} & \boldsymbol{Q}(k) & \boldsymbol{0} \\ \boldsymbol{0} & \boldsymbol{0} & \boldsymbol{P}_0 \end{bmatrix} \tag{16.20}$$

According to the discrete Kalman filtering theory, the state estimation equation of the fusion center of the centralized radar network can be written as [60]

$$\hat{\boldsymbol{X}}(k+1|k) = \boldsymbol{F}(k)\hat{\boldsymbol{X}}(k|k) \tag{16.21}$$

$$\boldsymbol{P}(k+1|k) = \boldsymbol{F}(k)\boldsymbol{P}(k|k)\boldsymbol{F}'(k) + \boldsymbol{Q}(k) \tag{16.22}$$

$$\begin{aligned} \boldsymbol{P}(k+1|k+1)^{-1} &= \boldsymbol{P}(k+1|k)^{-1} + \boldsymbol{H}'(k+1)\boldsymbol{R}(k+1)^{-1}\boldsymbol{H}(k+1) \\ &= \boldsymbol{P}(k+1|k)^{-1} + \sum_{i=1}^{N} \boldsymbol{H}_i'(k+1)\boldsymbol{R}_i(k+1)^{-1}\boldsymbol{H}_i(k+1) \\ &= \boldsymbol{P}(k+1|k)^{-1} + \sum_{i=1}^{N} \left[\boldsymbol{P}_i(k+1|k+1)^{-1} - \boldsymbol{P}_i(k+1|k)^{-1}\right] \end{aligned} \tag{16.23}$$

where $\boldsymbol{P}_i(k+1|k)$ and $\boldsymbol{P}_i(k+1|k+1)$ are the one-step prediction and update of single radars' covariance, which can be obtained using the method discussed in Chapter 3. Since

$$\boldsymbol{K}(k+1) = \boldsymbol{P}(k+1|k+1)\boldsymbol{H}'(k+1)\boldsymbol{R}^{-1}(k+1) \tag{16.24}$$

and

$$\boldsymbol{R}^{-1}(k+1) = diag\left[\boldsymbol{R}_1^{-1}(k+1), \boldsymbol{R}_2^{-1}(k+1), \ldots, \boldsymbol{R}_N^{-1}(k+1)\right] \tag{16.25}$$

then

$$\begin{aligned} \boldsymbol{K}(k+1) &= \boldsymbol{P}(k+1|k+1)\left[\boldsymbol{H}_1(k+1)\boldsymbol{R}_1^{-1}(k+1), \boldsymbol{H}_2(k+1)\boldsymbol{R}_2^{-1}(k+1), \ldots, \boldsymbol{H}_N(k+1)\boldsymbol{R}_N^{-1}(k+1)\right] \\ &= [\boldsymbol{K}_1(k+1), \boldsymbol{K}_2(k+1), \ldots, \boldsymbol{K}_N(k+1)] \end{aligned} \tag{16.26}$$

$$\begin{aligned} \hat{\boldsymbol{X}}(k+1|k+1) &= \hat{\boldsymbol{X}}(k+1|k) + \boldsymbol{K}(k+1)\left[\boldsymbol{Y}(k+1) - \boldsymbol{H}(k+1)\hat{\boldsymbol{X}}(k+1|k)\right] \\ &= \hat{\boldsymbol{X}}(k+1|k) + \sum_{i=1}^{N} \left[\boldsymbol{K}_i(k+1)\left\{\boldsymbol{Z}_i(k+1) + \boldsymbol{H}_i(k+1)\left[\boldsymbol{\Psi}_i - \hat{\boldsymbol{X}}(k+1|k)\right]\right\}\right. \end{aligned} \tag{16.27}$$

16.3.2.2 State Estimation of Distributed Structures

The essence of state estimation of distributed structures is track fusion or correlation. At present there are three main optimal track correlation solution forms [60], whose expressions, similar to those of centralized structures, are all optimal fusion solutions and are equivalent. An optimal track correlation solution of the fusion center is given by

$$\hat{\boldsymbol{X}}(k+1|k+1)=\boldsymbol{P}(k+1|k+1)\Big\{\boldsymbol{P}(k+1|k)^{-1}\hat{\boldsymbol{X}}(k+1|k) \\ +\sum_{i=1}^{N}\Big[\boldsymbol{P}_i(k+1|k+1)^{-1}\big(\hat{\boldsymbol{X}}_i(k+1|k+1)+\boldsymbol{\Psi}_i\big)-\boldsymbol{P}_i(k+1|k)^{-1}\big(\hat{\boldsymbol{X}}_i(k+1|k)+\boldsymbol{\Psi}_i\big)\Big]\Big\} \tag{16.28}$$

where $\boldsymbol{P}(k+1|k+1)$, $\boldsymbol{P}(k+1|k)$, and $\hat{\boldsymbol{X}}(k+1|k)$ are given, respectively, by (16.23), (16.22), and (16.21), and other measurements by the state estimate equation of single radars. However, the optimized performance is usually achieved at the expense of the increase in computation and network communication loads, so a suboptimal fusion algorithm is often adopted in engineering.

When the influence of process noise and initial conditions is ignored, the suboptimal track fusion solution can be expressed as [60]

$$\hat{\boldsymbol{X}}_s(k|k)=\left[\sum_{i=1}^{N}\boldsymbol{P}_i^{-1}(k|k)\right]^{-1}\left\{\sum_{i=1}^{N}\boldsymbol{P}_i^{-1}(k|k)\left[\hat{\boldsymbol{X}}_i(k|k)+\boldsymbol{\Psi}_i\right]\right\} \tag{16.29}$$

16.4 Data Processing of Bistatic Radar Networks

16.4.1 Basic Location Relation

There are two types of bistatic system: T–R and T/R–R. Here, T/R is the transmitting station and R is the receiving station.

The distribution of bistatic radar stations is shown in Figure 16.2. Let the line connecting T/R and R be axis x, the coordinates of R be $(a,0)$, and the coordinates of the target point be (x,y).

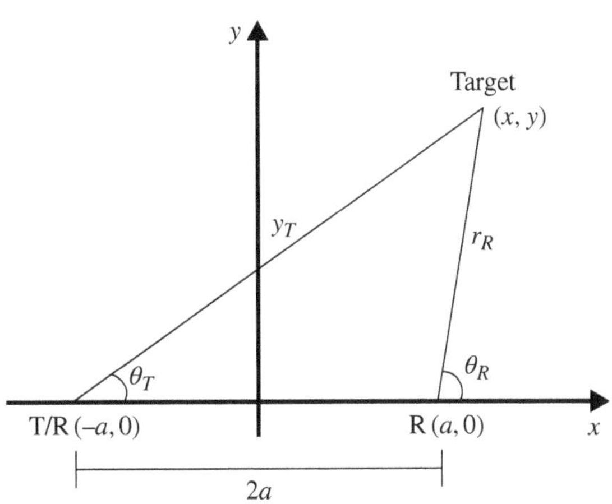

Figure 16.2 Geometric relation of bistatic radar location

The distance between T/R and R is the baseline with length $2a$, and the target is assumed to be not on the baseline.

θ_T and θ_R are, respectively, the direction angles of the transmitted and received beams.
r_T and r_R are, respectively, the distances from the target to stations T/R and R.
$\rho = r_T + r_R$ is defined as the sum of distances.

The four measurements r_T, θ_T, ρ, and θ_R are influenced by the zero-mean, Gaussian noises whose mean square errors are, respectively, σ_{r_T}, σ_{θ_T}, σ_ρ, and σ_{θ_R}, and these measurement noises are independent.

Let η be the correlation coefficient between measurement errors $d\rho$ and dr_T.

In the T–R bistatic system, only station T irradiates, and the target can only be located by the data ρ and θ_R obtained from station R, so there is no information redundancy in the system. In contrast, in the T/R–R bistatic system, station T/R can provide observations r_T and θ_T, while station R can provide observations ρ and θ_R, and therefore information redundancy may arise. To take full advantage of the observations available, combinational estimation of the redundant information needs to be made.

The bistatic radar system under consideration is assumed to be of T/R–R type. The working environment of the bistatic system sometimes makes it impossible to obtain observations from the T/R stations simultaneously. What's worse, it is impossible, in the presence of strong jamming, to observe any information on the target at all, and thus only T/R irradiates. Therefore, according to the possible observations, there are three cases as follows.

1. When the four observations r_T, θ_T, ρ, and θ_R can be obtained simultaneously, they can be combined into six groups of measurement subsets (r_T, θ_T), (ρ, θ_R), (ρ, r_T), (r_T, θ_R), (ρ, θ_T), and (θ_T, θ_R).
2. When the bistatic system can only obtain three observations, three groups of possible measurement subsets (ρ, θ_R), (ρ, θ_T), and (θ_T, θ_R) can be obtained in cases where the three observations are θ_T, ρ, and θ_R; three groups of possible measurement subsets (ρ, θ_R), (ρ, r_T), and (r_T, θ_R) can be obtained in cases where the three observations are r_T, ρ, and θ_R.
3. When only T/R irradiates, only measurements ρ and θ_R can be obtained, in which case the location of the target is decided totally by station R.

If the bistatic system can obtain all the measurements r_T, θ_T, ρ, and θ_R, they can be combined into six groups of measurement subsets, expressed as set $\Omega = \{(\rho,\theta_R),(r_T,\theta_T),(\rho,\theta_T),(\theta_T,\theta_R),(\rho,r_T),(r_T,\theta_R)\} = \{S_1,S_2,\ldots,S_6\}$. If the bistatic system can only obtain part of the measurements, its measurement subset is accordingly part of the six groups of measurement subsets.

Since cases 1 and 2 are special circumstances of case 3, only case 1 is discussed here. The transformation of any measurement subset (a_j, b_j) in set Ω from polar coordinates into rectangular coordinates can result in its transformed measurements in the rectangular coordinate system, illustrated as follows.

When $(a_j, b_j) = (\rho, \theta_R)$, the coordinate transformation equation is

$$\boldsymbol{X}_1 = \begin{bmatrix} x_1 \\ y_1 \end{bmatrix} = \begin{bmatrix} \dfrac{\rho^2\cos\theta_R + 2\rho a}{2\rho + 4a\cos\theta_R} \\ \dfrac{(\rho^2 - 4a^2)\sin\theta_R}{2\rho + 4a\cos\theta_R} \end{bmatrix} \tag{16.30}$$

When $\left(a_j, b_j\right) = (r_T, \theta_T)$, it is

$$\boldsymbol{X}_2 = \begin{bmatrix} x_2 \\ y_2 \end{bmatrix} = \begin{bmatrix} r_T \cos\theta_T - a \\ r_T \sin\theta_T \end{bmatrix} \tag{16.31}$$

When $\left(a_j, b_j\right) = (\rho, \theta_T)$, it is

$$\boldsymbol{X}_3 = \begin{bmatrix} x_3 \\ y_3 \end{bmatrix} = \begin{bmatrix} \dfrac{\rho^2 \cos\theta_T - 2\rho a}{2\rho - 4a\cos\theta_T} \\ \dfrac{\left(\rho^2 - 4a^2\right)\sin\theta_T}{2\rho - 4a\cos\theta_T} \end{bmatrix} \tag{16.32}$$

When $\left(a_j, b_j\right) = (\rho, \theta_R)$, it is

$$\boldsymbol{X}_4 = \begin{bmatrix} x_4 \\ y_4 \end{bmatrix} = \begin{bmatrix} \dfrac{a\sin(\theta_R + \theta_T)}{\sin(\theta_R - \theta_T)} \\ \dfrac{2a\sin\theta_R \sin\theta_T}{\sin(\theta_R - \theta_T)} \end{bmatrix} \tag{16.33}$$

When $\left(a_j, b_j\right) = (\rho, r_T)$, it is

$$\boldsymbol{X}_5 = \begin{bmatrix} x_5 \\ y_5 \end{bmatrix} = \begin{bmatrix} \dfrac{\rho(2r_T - \rho)}{4a} \\ \dfrac{\pm\sqrt{16a^2(r_T^2 - a^2) - \rho(2r_T - \rho)(8a^2 + 2r_T\rho - \rho^2)}}{4a} \end{bmatrix} \tag{16.34}$$

When $\left(a_j, b_j\right) = (r_T, \theta_R)$, it is

$$\boldsymbol{X}_6 = \begin{bmatrix} x_6 \\ y_6 \end{bmatrix} = \begin{bmatrix} a + \left(-2a\cos\theta_R \pm \sqrt{r_T^2 - 4a^2\sin^2\theta_R}\right)\cos\theta_R \\ \left(-2a\cos\theta_R \pm \sqrt{r_T^2 - 4a^2\sin^2\theta_R}\right)\sin\theta_R \end{bmatrix} \tag{16.35}$$

The derivation processes of the above equations can be found in Ref. [47]. The above process can be concluded as

$$\boldsymbol{X}_j = \begin{bmatrix} x_j \\ y_j \end{bmatrix} = \begin{bmatrix} f\left(a_j, b_j\right) \\ g\left(a_j, b_j\right) \end{bmatrix} \tag{16.36}$$

where (a_j, b_j) represents the jth measurement subset in set Ω, and $f(a_j, b_j)$ and $g(a_j, b_j)$ are nonlinear functions related to subset (a_j, b_j).

Therefore, the transformed measurement error in the rectangular coordinate system can be expressed as

$$\mathrm{d}\boldsymbol{X}_j = \begin{bmatrix} \mathrm{d}x_j \\ \mathrm{d}y_j \end{bmatrix} = \begin{bmatrix} \frac{\partial f}{\partial a_j} & \frac{\partial f}{\partial b_j} \\ \frac{\partial g}{\partial a_j} & \frac{\partial g}{\partial b_j} \end{bmatrix} \begin{bmatrix} \mathrm{d}a_j \\ \mathrm{d}b_j \end{bmatrix} \triangleq \boldsymbol{A}_j \begin{bmatrix} \mathrm{d}a_j \\ \mathrm{d}b_j \end{bmatrix} \tag{16.37}$$

16.4.2 Combined Estimation

To improve the location accuracy of bistatic radars, a combinational estimation has to be made on its redundant information. Assume that X_j and $\mathrm{d}X_j$, respectively, represent the estimated location and estimated error of the target in the rectangular coordinate system determined by the jth measurement subset S_j, and satisfy $\boldsymbol{X}_j = \boldsymbol{I}\boldsymbol{X}_0 + \mathrm{d}\boldsymbol{X}_j$, where $\boldsymbol{X}_0 = [x_0, y_0]'$ is the unknown true location of the target, and $\boldsymbol{I}$ is the identity matrix in the same dimension as vector X_0. It follows from any two estimated locations X_k and X_l in the rectangular coordinate system that

$$\boldsymbol{X} = \boldsymbol{H}\boldsymbol{X}_0 + \boldsymbol{V} \tag{16.38}$$

where $\boldsymbol{X} = [\boldsymbol{X}'_k, \boldsymbol{X}'_l]'$, $\boldsymbol{V} = [\mathrm{d}\boldsymbol{X}'_k, \mathrm{d}\boldsymbol{X}'_l]'$, and $\boldsymbol{H} = [\boldsymbol{I}, \boldsymbol{I}]'$. The covariance matrix of the measurement noise vector $\boldsymbol{V}$ is

$$\boldsymbol{R} = \mathrm{E}\left\{ \begin{bmatrix} \mathrm{d}\boldsymbol{X}_k \\ \mathrm{d}\boldsymbol{X}_l \end{bmatrix} [\mathrm{d}\boldsymbol{X}'_k, \mathrm{d}\boldsymbol{X}'_l] \right\} \triangleq \begin{bmatrix} \boldsymbol{R}_{kk} & \boldsymbol{R}_{kl} \\ \boldsymbol{R}_{lk} & \boldsymbol{R}_{ll} \end{bmatrix} \tag{16.39}$$

The element in the covariance matrix of measurement noise vector $\boldsymbol{V}$ obtained from a simple mathematical operation of (16.37) is

$$\boldsymbol{R}_{kl} = \mathrm{E}[\mathrm{d}\boldsymbol{X}_k \mathrm{d}\boldsymbol{X}'_l] = \boldsymbol{A}_k \boldsymbol{B}_{kl} \boldsymbol{A}'_l \tag{16.40}$$

where

$$\boldsymbol{B}_{kl} \triangleq \mathrm{E}\left\{ \begin{bmatrix} \mathrm{d}a_k \\ \mathrm{d}b_k \end{bmatrix} [\mathrm{d}a_l, \mathrm{d}b_l] \right\} \tag{16.41}$$

If covariance matrix $\boldsymbol{R}$ is reversible and the inverse of $\boldsymbol{H}'\boldsymbol{R}^{-1}\boldsymbol{H}$ exists, then the LS estimate of the bistatic radar is

$$\hat{\boldsymbol{X}} = (\boldsymbol{H}'\boldsymbol{R}^{-1}\boldsymbol{H})^{-1}\boldsymbol{H}'\boldsymbol{R}^{-1}\boldsymbol{X} = (\boldsymbol{W}_{ll} + \boldsymbol{W}_{kl} + \boldsymbol{W}_{lk} + \boldsymbol{W}_{kk})^{-1}\boldsymbol{H}'\boldsymbol{W}\boldsymbol{X} \tag{16.42}$$

with the corresponding covariance matrix

$$\boldsymbol{P} = (\boldsymbol{H}'\boldsymbol{R}^{-1}\boldsymbol{H})^{-1} = (\boldsymbol{W}_{ll} + \boldsymbol{W}_{kl} + \boldsymbol{W}_{lk} + \boldsymbol{W}_{kk})^{-1} \tag{16.43}$$

where W_{ll}, W_{kl}, W_{lk}, and W_{kk} are elements [372] in the inverse matrix of covariance matrix $\boldsymbol{R}$, that is,

$$\boldsymbol{R}^{-1} \triangleq \boldsymbol{W} = \begin{bmatrix} \boldsymbol{W}_{kk} & \boldsymbol{W}_{kl} \\ \boldsymbol{W}_{lk} & \boldsymbol{W}_{ll} \end{bmatrix} \tag{16.44}$$

where

$$\boldsymbol{W}_{ll} = \left(\boldsymbol{R}_{ll} - \boldsymbol{R}_{lk}\boldsymbol{R}_{kk}{}^{-1}\boldsymbol{R}_{kl}\right)^{-1} = \left[\boldsymbol{A}_l\left(\boldsymbol{B}_{ll} - \boldsymbol{B}_{lk}\boldsymbol{B}_{kk}^{-1}\boldsymbol{B}_{kl}\right)\boldsymbol{A}'_l\right]^{-1} \tag{16.45}$$

$$\boldsymbol{W}_{kl} = \boldsymbol{W}'_{lk} = -\boldsymbol{R}_{kk}{}^{-1}\boldsymbol{R}_{kl}\boldsymbol{W}_{ll} \tag{16.46}$$

$$\boldsymbol{W}_{kk} = \boldsymbol{R}_{kk}{}^{-1} - \boldsymbol{W}_{kl}\boldsymbol{R}_{lk}\boldsymbol{R}_{kk}{}^{-1} \tag{16.47}$$

Obviously, if R is irreversible, then the LS estimate does not exist. Then, can a combinational estimation be made on the redundant information in the bistatic radar system with use of the LS estimate? Or, is R reversible, and in what condition if so? Theorem 16.1 answers this question [372].

16.4.3 An Analysis of the Feasibility of Combinational Estimation

Theorem 16.1 Suppose that S_1, S_2 are two measurement subsets arbitrarily taken from set Ω, then the necessary and sufficient condition for the LS combinational estimation of S_1, S_2 is that there are no repetitive measurements in S_1, S_2.

Proof: Sufficiency. Let S_k and S_j be any two subsets in set Ω (i.e., $\forall S_k, S_j \in \Omega$, $k \neq j$). When there are no repetitive measurements in the two subsets, that is, when $S_k \cap S_j = \phi$ (ϕ is the empty set), there are three cases:

1. $S_1 = (\rho, \theta_R)$ and $S_2 = (r_T, \theta_T)$;
2. $S_3 = (\rho, \theta_T)$ and $S_6 = (r_T, \theta_R)$;
3. $S_4 = (\theta_T, \theta_R)$ and $S_5 = (\rho, r_T)$.

In the first case (i.e., S_1 and S_2), it can be derived from (16.41) that

$$\boldsymbol{B}_{11} = \mathrm{E}\left\{\begin{bmatrix} \mathrm{d}\rho \\ \mathrm{d}\theta_R \end{bmatrix}[\mathrm{d}\rho, \mathrm{d}\theta_R]\right\} = \begin{bmatrix} \sigma_\rho^2 & 0 \\ 0 & \sigma_{\theta_R}^2 \end{bmatrix} \tag{16.48}$$

$$\boldsymbol{B}_{12} = \mathrm{E}\left\{\begin{bmatrix} \mathrm{d}\rho \\ \mathrm{d}\theta_R \end{bmatrix}[\mathrm{d}r_T, \mathrm{d}\theta_T]\right\} = \begin{bmatrix} \eta\sigma_\rho\sigma_{r_T} & 0 \\ 0 & 0 \end{bmatrix} = \boldsymbol{B}'_{21} \tag{16.49}$$

$$\boldsymbol{B}_{22} = \mathrm{E}\left\{\begin{bmatrix} \mathrm{d}r_T \\ \mathrm{d}\theta_T \end{bmatrix}[\mathrm{d}r_T, \mathrm{d}\theta_T]\right\} = \begin{bmatrix} \sigma_{r_T}^2 & 0 \\ 0 & \sigma_{\theta_T}^2 \end{bmatrix} \tag{16.50}$$

A simple mathematical operation of (16.30), (16.31), and (16.37) yields

$$\boldsymbol{A}_1 = K_1 \begin{bmatrix} K_2\cos\theta_R & 2\rho K_3\sin\theta_R \\ K_2\sin\theta_R & -K_3(2\rho\cos\theta_R + 4a) \end{bmatrix} \tag{16.51}$$

where

$$K_1 = \frac{1}{(2\rho + 4a\cos\theta_R)^2} \tag{16.52}$$

$$K_2 = 2\rho^2 + 8\rho a\cos\theta_R + 8a^2 \tag{16.53}$$

$$K_3 = 4a^2 - \rho^2 \tag{16.54}$$

$$\boldsymbol{A}_2 = \begin{bmatrix} \cos\theta_T & -r_T\sin\theta_T \\ \sin\theta_T & r_T\cos\theta_T \end{bmatrix} \tag{16.55}$$

It follows from (16.40) that

$$|\boldsymbol{R}_{11}| = |\boldsymbol{A}_1\boldsymbol{B}_{11}\boldsymbol{A}_1'| = K_1^3 K_2{}^2 K_3{}^2 \sigma_\rho^2 \sigma_{\theta_R}^2 \tag{16.56}$$

When the target is not on the baseline, $\boldsymbol{R}_{11}$ is reversible, and then it can be derived from (16.45) that

$$|\boldsymbol{R}_{22} - \boldsymbol{R}_{21}\boldsymbol{R}_{11}{}^{-1}\boldsymbol{R}_{12}| = |\boldsymbol{A}_2(\boldsymbol{B}_{22} - \boldsymbol{B}_{21}\boldsymbol{B}_{11}^{-1}\boldsymbol{B}_{12})\boldsymbol{A}_2'| = (1-\eta^2) r_T^2 \sigma_{\theta_T}^2 \sigma_{r_T}^2 \tag{16.57}$$

Since $|\eta| \neq 1$, we have $|\boldsymbol{R}_{22} - \boldsymbol{R}_{21}\boldsymbol{R}_{11}{}^{-1}\boldsymbol{R}_{12}| \neq 0$, which means that matrix $\boldsymbol{R}_{22} - \boldsymbol{R}_{21}\boldsymbol{R}_{11}{}^{-1}\boldsymbol{R}_{12}$ is reversible. $\boldsymbol{W}_{11}$, $\boldsymbol{W}_{13}$, $\boldsymbol{W}_{21}$, and $\boldsymbol{W}_{22}$ follow from (16.45)–(16.47) (i.e., the covariance matrix $\boldsymbol{R}$ is reversible), so the LS estimate of the target can be obtained from (16.42).

Similarly, it can be proved that in cases 2 and 3, covariance matrix $\boldsymbol{R}$ is also reversible, and the LS estimate of the target can also be obtained. That is to say, combinational estimation can be made of the two measurement subsets in all three cases above by means of the LS estimate.

Necessity: It is to be proved that when $S_k \cap S_j \neq \phi (k \neq j)$, matrix $\boldsymbol{R}$ is irreversible, so combinational estimation cannot be conducted on the two measurement subsets by using the LS estimate. When $S_k \cap S_j \neq \phi$, there are two cases.

1. Measurement subsets S_k and S_j ($k \neq j$) have only one element in common, including:

   $S_1 = (\rho, \theta_R)$ and $S_3 = (\rho, \theta_T)$;
   $S_1 = (\rho, \theta_R)$ and $S_4 = (\theta_T, \theta_R)$;
   $S_2 = (r_T, \theta_T)$ and $S_4 = (\theta_T, \theta_R)$;
   $S_2 = (r_T, \theta_T)$ and $S_6 = (r_T, \theta_R)$;
   $S_3 = (\rho, \theta_T)$ and $S_4 = (\theta_T, \theta_R)$;
   $S_4 = (\theta_T, \theta_R)$ and $S_6 = (r_T, \theta_R)$.

   In the first case (S_1 and S_3), using (16.41) yields

$$\boldsymbol{B}_{13} = \mathrm{E}\left\{\begin{bmatrix} \mathrm{d}\rho \\ \mathrm{d}\theta_R \end{bmatrix} [\mathrm{d}\rho, \mathrm{d}\theta_T]\right\} = \begin{bmatrix} \sigma_\rho^2 & 0 \\ 0 & 0 \end{bmatrix} = \boldsymbol{B}_{31}' \tag{16.58}$$

$$\boldsymbol{B}_{33} = \mathrm{E}\left\{\begin{bmatrix} \mathrm{d}\rho \\ \mathrm{d}\theta_T \end{bmatrix} [\mathrm{d}\rho, \mathrm{d}\theta_T]\right\} = \begin{bmatrix} \sigma_\rho^2 & 0 \\ 0 & \sigma_{\theta_T}^2 \end{bmatrix} \tag{16.59}$$

Equation (16.48) can be applied to the mathematical operation of matrix $\boldsymbol{B}_{11}$.

Using (16.32) and (16.37) yields

$$\boldsymbol{A}_3 = K_4 \begin{bmatrix} K_5 \cos\theta_T & 2\rho K_3 \sin\theta_T \\ K_5 \sin\theta_T & 2K_3 K_6 \end{bmatrix} \tag{16.60}$$

where

$$K_4 = \frac{1}{(2\rho - 4a\cos\theta_T)^2} \tag{16.61}$$

$$K_5 = 2\rho^2 - 8\rho a \cos\theta_T + 8a^2 \tag{16.62}$$

$$K_6 = 2a - \rho\cos\theta_T \tag{16.63}$$

Equation (16.51) can be applied to the mathematical operation of matrix $\boldsymbol{A}_1$.

A simple mathematical operation of (16.45) gives

$$\boldsymbol{R}_{33} - \boldsymbol{R}_{31}\boldsymbol{R}_{11}^{-1}\boldsymbol{R}_{13} = \boldsymbol{A}_3\left(\boldsymbol{B}_{33} - \boldsymbol{B}_{31}\boldsymbol{B}_{11}^{-1}\boldsymbol{B}_{13}\right)\boldsymbol{A}_3' = 4K_3^2 K_4^2 \sigma_{\theta_T}^2 \begin{bmatrix} \rho^2 \sin^2\theta_T & \rho\sin\theta_T K_6 \\ \rho\sin\theta_T K_6 & K_6^2 \end{bmatrix} \tag{16.64}$$

Since $\left|\boldsymbol{R}_{33} - \boldsymbol{R}_{31}\boldsymbol{R}_{11}^{-1}\boldsymbol{R}_{13}\right| = 0$ (i.e., matrix $\boldsymbol{R}_{33} - \boldsymbol{R}_{31}\boldsymbol{R}_{11}^{-1}\boldsymbol{R}_{13}$ is irreversible), it is impossible to derive $\boldsymbol{W}_{11}$ from (16.45). In this case, it is impossible to obtain $\boldsymbol{W}_{13}$ and $\boldsymbol{W}_{31}$ from (16.46), or $\boldsymbol{W}_{33}$ from (16.47), which means that $\boldsymbol{W}_{11}$, $\boldsymbol{W}_{13}$, $\boldsymbol{W}_{31}$, and $\boldsymbol{W}_{33}$ do not exist, and matrix $\boldsymbol{R}$ is irreversible.

Likewise, it can be proved that matrix $\boldsymbol{R}$ is irreversible in the other cases, so the LS estimate does not exist in the above cases.

2. Measurement subsets S_k and S_j $(k \neq j)$ share not only an element in common, but also a relevant element, including:

$S_1 = (\rho, \theta_R)$ and $S_5 = (\rho, r_T)$;
$S_1 = (\rho, \theta_R)$ and $S_6 = (r_T, \theta_R)$;
$S_2 = (r_T, \theta_T)$ and $S_3 = (\rho, \theta_T)$;
$S_2 = (r_T, \theta_T)$ and $S_5 = (\rho, r_T)$;
$S_3 = (\rho, \theta_T)$ and $S_5 = (\rho, r_T)$;
$S_5 = (\rho, r_T)$ and $S_6 = (r_T, \theta_R)$.

In the first case (S_1 and S_5), it follows from (16.41) that

$$\boldsymbol{B}_{15} = \mathrm{E}\left\{ \begin{bmatrix} \mathrm{d}\rho \\ \mathrm{d}\theta_R \end{bmatrix} [\mathrm{d}\rho, \mathrm{d}r_T] \right\} = \begin{bmatrix} \sigma_\rho^2 & \eta\sigma_\rho\sigma_{r_T} \\ 0 & 0 \end{bmatrix} = \boldsymbol{B}_{51}' \tag{16.65}$$

$$\boldsymbol{B}_{55} = \mathrm{E}\left\{ \begin{bmatrix} \mathrm{d}\rho \\ \mathrm{d}r_T \end{bmatrix} [\mathrm{d}\rho, \mathrm{d}r_T] \right\} = \begin{bmatrix} \sigma_\rho^2 & \eta\sigma_\rho\sigma_{r_T} \\ \eta\sigma_\rho\sigma_{r_T} & \sigma_{r_T}^2 \end{bmatrix} \tag{16.66}$$

Equation (16.48) can be applied to the mathematical operation of matrix $\boldsymbol{B}_{11}$. It follows from (16.34) and (16.37) that

$$\boldsymbol{A}_5 = \frac{1}{2aK_7}\begin{bmatrix} (r_T-\rho)K_7 & \rho K_7 \\ \pm(r_T-\rho)K_8 & \pm K_9 K_3 \end{bmatrix} \tag{16.67}$$

where

$$K_7 = \sqrt{16a^2(r_T^2-a^2)-\rho K_9(8a^2-2r_T\rho-\rho^2)} \tag{16.68}$$

$$K_8 = \left(\rho^2-2r_T\rho-4a^2\right) \tag{16.69}$$

$$K_9 = (2r_T-\rho) \tag{16.70}$$

Equation (16.51) can be applied to the mathematical operation of matrix $\boldsymbol{A}_1$. A simple mathematical operation of (16.45) yields

$$\boldsymbol{R}_{55}-\boldsymbol{R}_{51}\boldsymbol{R}_{11}^{-1}\boldsymbol{R}_{15} = \boldsymbol{A}_5\left(\boldsymbol{B}_{55}-\boldsymbol{B}_{51}\boldsymbol{B}_{11}^{-1}\boldsymbol{B}_{15}\right)\boldsymbol{A}_5' = \frac{(1-\eta^2)\sigma_{r_T}^2}{4a^2K_7^2}\begin{bmatrix} \rho^2K_7^2 & \rho K_9K_3K_7 \\ \rho K_9K_3K_7 & K_9^2K_3^2 \end{bmatrix} \tag{16.71}$$

and

$$\left|\boldsymbol{R}_{55}-\boldsymbol{R}_{51}\boldsymbol{R}_{11}^{-1}\boldsymbol{R}_{15}\right| = 0 \tag{16.72}$$

Matrix $\boldsymbol{R}_{55}-\boldsymbol{R}_{51}\boldsymbol{R}_{11}^{-1}\boldsymbol{R}_{15}$ is irreversible, so it is impossible to derive $\boldsymbol{W}_{11}$ from (16.45). In this case, it is impossible to obtain $\boldsymbol{W}_{15}$ and $\boldsymbol{W}_{51}$ from (16.46), or $\boldsymbol{W}_{55}$ from (16.47), which means that $\boldsymbol{W}_{11}$, $\boldsymbol{W}_{15}$, $\boldsymbol{W}_{51}$, and $\boldsymbol{W}_{55}$ do not exist, and that matrix $\boldsymbol{R}$ is irreversible.

Similarly, it can be proved that matrix $\boldsymbol{R}$ is irreversible in the other cases, so the LS estimate does not exist in the cases above. Therefore, the theorem is proved.

Under the condition that the measurement data of the bistatic radar satisfies the above theorem, data compression can greatly improve its location accuracy. After the location of the target, the KF method introduced in Chapter 3 can be applied to the tracking filtering of the target. When there is more than one target, it is necessary to make data correlation and multi-target tracking, which are studied in previous chapters and will not be discussed here.

16.5 Data Processing of Multistatic Radar Networks

The multistatic radar network system discussed in this section refers to a system consisting of a transmitting station and many receiving stations located in different places. This section introduces the multistatic radar network system, followed by its general process of data processing.

Since the concept of MIMO (multiple-input, multiple-output) was put forward by Fletcher and Robey in 2003, MIMO radars have attracted worldwide attention in recent years [373,374], and much research has been conducted on target detection. However, the related data processing issues also deserve close attention and extensive research. MIMO is a radar system where multiple

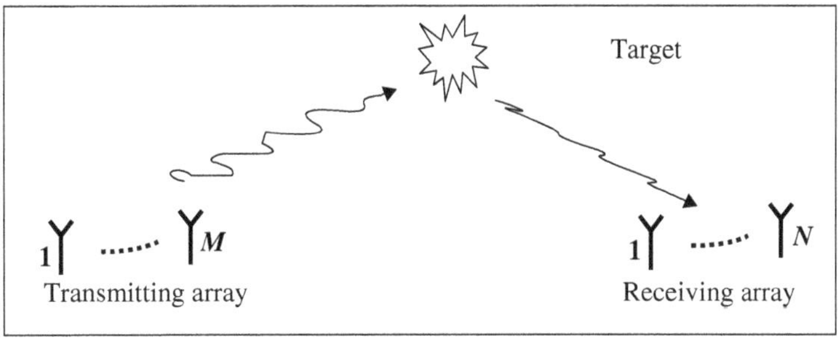

Figure 16.3 Schematic diagram of the MIMO radar

antennas are used to transmit waveforms and receive echoes, and the collected echoes undergo coherent or non-coherent combined treatment (as illustrated in Figure 16.3, in which relative positions of the element antennas are not clearly defined).

MIMO radars fall into two broad categories. The first category is the statistical MIMO radar, where intervals both between the transmitting antenna array elements and between the receiving ones are long (namely, these elements are of the so-called distributed layout, and the MIMO radar can be regarded as a type of multistatic radar in this sense), and each "transmitting–receiving array element pair" corresponds to an independent target scattering response and makes non-coherent combined treatment on target echoes from all the transmitting–receiving array element pairs.

The second category is known as the MIMO array radar or coherent MIMO radar, in which both intervals between the transmitting antenna array elements and those between their receiving equivalents are short, with the same order of magnitude as the wavelength (namely, these elements are of the so-called compact layout), and each "transmitting–receiving array element pair" corresponds to the same target scattering response and conducts coherent combined treatment on target echoes from all the transmitting–receiving array element pairs.

16.5.1 Tracking Principle of Multistatic Radar Systems

Here, we consider only the tracking methods in the 2D case; the principles can be generalized to the 3D case. The mathematical model of target motion adopted by multistatic radars is the same as that of monostatic ones. Figure 16.4 shows the polar observations obtained from the *i*th receiving station in the multistatic radar system:

1. the total path length $\rho_i = \rho_T + \rho_{Ri}$, proportional to the scattering signal transmission time;
2. the arrival angle of scattering signals (θ_i);
3. the transmission beam angle (θ_T).
4. Along the transmitter–target and target–receiver paths, the target's sum of radial velocity components is $\dot{\rho}_i = \dot{\rho}_T + \dot{\rho}_{Ri}$, and this value is proportional to the Doppler frequency shift of scattering signals relative to synchronous chain reference signals.

Because the observation vector composed of the measurements above is nonlinear, so is its optimum filter. In view of the practicality, the quasi-optimal method is generally adopted in combining polar observations, whose rectangular components and covariance matrix are obtained by calculation before going through linear Kalman filtering.

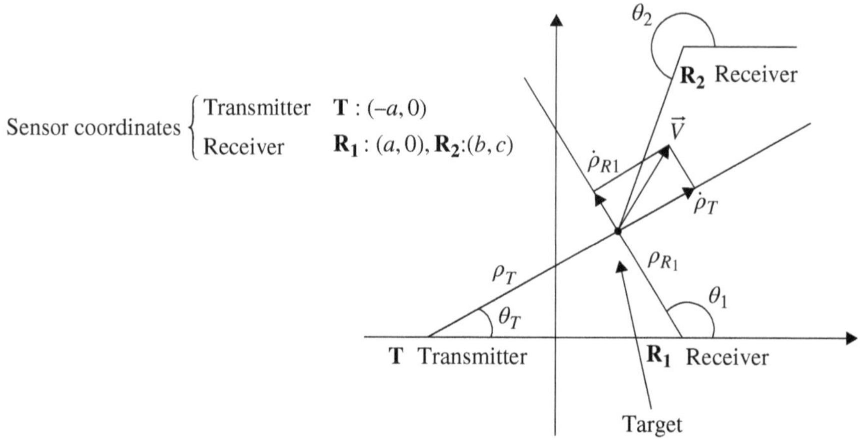

Figure 16.4 The multistatic radar observation under the main reference system

16.5.2 Observation Equation of Multistatic Radar Network Systems

The generic method to define the structure of tracking filters is to establish mathematical models and radar observation equations of target motion. Because the mathematical model of target motion is the same as that of the monostatic network system, its observation equation will be discussed hereafter.

Consider the multistatic system (consisting of a transmitter and two receivers located at different places) shown in Figure 16.4. In order to acquire the rectangular coordinate components of target motion, several different methods can be adopted to combine polar observations. For example, any of the (ρ_1, θ_T), (ρ_2, θ_1), (ρ_2, θ_2), (θ_1, θ_2), and (ρ_1, ρ_2) pairs can be used to solve for the corresponding rectangular coordinate components. Here, the error means of perturbed Cartesian coordinate observations are still zero but interrelated, with covariance matrix

$$\boldsymbol{B} = \begin{bmatrix} \sigma_x^2 & 0 & \sigma_{xy} & 0 \\ 0 & \sigma_{\dot{x}}^2 & 0 & \sigma_{\dot{x}\dot{y}} \\ \sigma_{yx} & 0 & \sigma_y^2 & 0 \\ 0 & \sigma_{\dot{x}\dot{y}} & 0 & \sigma_{\dot{y}}^2 \end{bmatrix} \tag{16.73}$$

Take (ρ_1, θ_T) as an example. The corresponding rectangular coordinate components and observation error covariance are given in Table 16.1.

16.5.3 The Generic Data Processing Process of Multistatic Tracking Systems

The generic data processing process of multistatic radar networks is described as follows.

1. Combine the observation values from all the stations. These observations, after being transmitted to the data processing center, are combined differently according to the method introduced in Section 16.5.2. These different combinations are then estimated with the LS method (as in Section 16.4.2), which converts every group of combined variables to rectangular coordinate

Table 16.1 Cartesian coordinates and observation error covariance corresponding to polar coordinates

$$x = (a - 0.5\rho_1 \cos\theta_T) / \left(\frac{2a}{\rho_1}\cos\theta_T - 1\right)$$

$$y = \left(\frac{2a^2}{\rho_1} - 0.5\rho_1\right)\sin\theta_T / \left(\frac{2a}{\rho_1}\cos\theta_T - 1\right)$$

$$\sigma_x^2 = \left\{\left(2a^2\cos\theta_T - 2a\rho_1\cos^2\theta_T + \frac{\rho_1^2}{2}\cos\theta_T\right)^2 \sigma_{\rho_1}^2 + \left(2a^2\rho_1\sin\theta_T - \frac{\rho_1^3}{2}\sin\theta_T\right)^2 \sigma_{\theta_T}^2\right\}\frac{1}{(2a\cos\theta_T - \rho_1)^4}$$

$$\sigma_y^2 = \left\{\left(\frac{\rho_1^2}{2}\sin\theta_T - a\rho_1\sin 2\theta_T + 2a^2\sin\theta_T\right)^2 \sigma_{\rho_1}^2 + \left[4a^3 - a\rho_1^2 + \left(\frac{\rho_1^3}{2} - 2a^2\rho_1\right)\cos\theta_T\right]\sigma_{\theta T}^2\right\}\frac{1}{(2a\cos\theta_T - \rho_1)^4}$$

$$\sigma_{xy}^2 = \left\{\left(2a^2\cos\theta_T - 2a\rho_1\cos^2\theta_T + \frac{\rho_1^2}{2}\cos\theta_T\right)\left(\frac{\rho_1^2}{2}\sin\theta_T - a\rho_1\sin 2\theta_T + 2a^2\sin\theta_T\right)\sigma_{\rho_1}^2 \right.$$
$$\left. + \left(2a^2\rho_1\sin\theta_T - \frac{\rho_1^3}{2}\sin\theta_T\right)\left[4a^3 - a\rho_1^2 + \left(\frac{\rho_1^3}{2} - 2a^2\rho_1\right)\cos\theta_T\right]\sigma_{\theta_T}^2\right\}\frac{1}{(2a\cos\theta_T - \rho_1)^4}$$

parameters, working out the rectangular coordinate components of target motion and their corresponding covariances. Since many groups of these components are available for target state updating in the multistatic radar network, they may be combined in various ways before track filtering.

2. If each group of rectangular coordinate components is viewed as the measurements from a single radar, then the data processing of the multistatic radar network can also be divided according to the structure into centralized, distributed, and hybrid. The centralized structure involves data registration, track initiation, data association, prediction and comprehensive tracking, in the fusion center, of the data from the combined positioning by the multistatic radar network. In the distributed structure, data registration, track initiation, data association, prediction and tracking are performed separately for each group of data obtained from the combined positioning by the multistatic radar network; then, the generated local multi-target tracks are sent to the fusion center, which completes track association and fusion according to the track data of each node before obtaining a global estimate. The hybrid structure simultaneously transmits detection reports and track information, which is processed at local nodes.

16.6 Track Association

In a distributed multiple-radar environment, each radar has its own information processing system, which contains a large amount of information about target tracks. Then, a very important problem is how to judge if two tracks from different systems represent the same target. This involves track-to-track association (or correlation), "track association" for short. Actually, it is to solve the repeated tracking problem in the coverage region in the radar space. Hence, track association is also called "de-duplication," and it also includes the task to distinguish between different targets. The track association problem is very easy when the track interval reported by radars is large and there is no interference or clutter. But it turns out to be rather complex in the presence of multiple targets, interference, clutter, noise, and many crossing and split tracks, in which case, compounded by other factors including combination mismatches in range or azimuth between radars, and sensor position, target height and coordinate conversion errors, association becomes even more difficult.

Currently, the track association algorithms mainly include approaches based on statistical mathematics, fuzzy mathematics, gray theory, and neural networks. The statistics-based approaches mainly include the weighted [375], independent sequence [376, 377], correction [70, 378], correlation sequential [376, 377], classical assignment [379, 380], generalized classical assignment [376], independent double threshold [376, 381], correlated double threshold [376, 381], nearest neighborhood (NN) [382], K-NN and modified K-NN [51, 383] methods.

When the system contains large errors in navigation, calibration, conversion, and delay, statistical approaches sometimes fail and alternative algorithms are in demand. There is a considerable ambiguity in track association judgment, and this kind of ambiguity can be represented by the membership function of fuzzy mathematics; in other words, the concept of subordinate degree can be used to describe the similarity degree between two tracks. Therefore, a series of fuzzy track correlation algorithms has been put forward [52, 54, 181, 187, 384–391], covering:

- choice of fuzzy factor sets and membership degree functions in sensor track association;
- determination of fuzzy factors and dynamic assignment of fuzzy weight sets;
- fuzzy binary threshold track correlation algorithm;
- track association algorithms based on fuzzy synthetic functions;
- fuzzy comprehensive evaluation track association algorithms;
- fuzzy track correlation algorithms in the case of multiple local nodes; and
- track correlation based on the fuzzy comprehensive analysis in the presence of unequal sample sizes.

In order to solve track association problems, grey theory was introduced [392], where track association is carried out as follows:

- The grey incidence order is obtained by working out the grey association degree between tracks.
- Correlations between tracks are determined according to this sequence.
- Several algorithms are provided, including gray track association, gray track association in the case of multiple local nodes, gray track association in the presence of unequal sample sizes.
- Performance analyses are made of fuzzy track association algorithms and gray track association algorithms.

This section deals only with the sequential track correlation algorithm in the case of multiple local nodes based on statistical mathematics. For other algorithms, see Ref. [8].

For the sake of discussion, suppose that all state estimates $\hat{\boldsymbol{X}}_j^i \ (i=1,2,\ldots,M; j=1,2,\ldots,n_i)$ sent to the fusion center are in the same coordinate system and that all radars take samples synchronously. Here, M is the number of radars (let $M \geq 2$), and n_i is the number of tracks of radar i. Also suppose that the track number set of local nodes 1, 2, …, M (i.e., the corresponding target number set) is

$$U_1 = \{1,2,\ldots,n_1\},\ U_2 = \{1,2,\ldots,n_2\},\ \ldots,\ U_M = \{1,2,\ldots,n_M\} \tag{16.74}$$

Denote

$$\boldsymbol{t}_{ij}^{S_a S_b}(l) = \hat{X}_i^{S_a}(l|l) - \hat{X}_j^{S_b}(l|l) \tag{16.75}$$

as the estimate of

$$\boldsymbol{t}_{ij}^{*S_aS_b}(l) = \hat{X}_i^{S_a}(l) - \hat{X}_j^{S_b}(l) \qquad (i \in U_{S_a}, j \in U_{S_b}) \tag{16.76}$$

where $\boldsymbol{X}_i^{S_a}$ and $\boldsymbol{X}_j^{S_b}$ represent the real states of the *i*th and *j*th target, respectively, and $\hat{\boldsymbol{X}}_i^{S_a}$ and $\hat{\boldsymbol{X}}_j^{S_b}$ the state estimates of node s_a for target i and node s_b for target j, respectively.

Suppose that H_0 and H_1 represent the following events $(i \in \boldsymbol{U}_1, j \in \boldsymbol{U}_2)$:

H_0: $\hat{\boldsymbol{X}}_i^{S_a}(l|l)$ and $\hat{\boldsymbol{X}}_j^{S_b}(l|l)$ are the track estimates of the same target;
H_1: $\hat{\boldsymbol{X}}_i^{S_a}(l|l)$ and $\hat{\boldsymbol{X}}_j^{S_b}(l|l)$ are not the track estimates of the same target.

Thus, the track association problem is converted to the hypothesis testing problem.

For the public surveillance region of M local nodes, the sufficient statistic can be constructed as

$$\rho_{i_{s-1}i_s}(k) = \rho_{i_{s-1}i_s}(k-1) + \left[\hat{\boldsymbol{X}}_{i_{s-1}}(k|k) - \hat{\boldsymbol{X}}_{i_s}(k|k)\right]' \boldsymbol{A}_{i_{s-1}i_s}^{-1}(k)\left[\hat{\boldsymbol{X}}_{i_{s-1}}(k|k) - \hat{\boldsymbol{X}}_{i_s}(k|k)\right] \tag{16.77}$$

where $s = 1, 2, \ldots, M$ is the serial number of local nodes, $i_s = 1, 2, \ldots, n_s$ is the serial number of tracks of local node s, and

$$\boldsymbol{A}_{i_{s-1}i_s}(k) = \boldsymbol{P}_{i_{s-1}}(k/k) + \boldsymbol{P}_{i_s}(k/k) - \boldsymbol{P}_{i_{s-1}i_s}(k/k) - \boldsymbol{P}_{i_{s-1}i_s}(k/k)' \tag{16.78}$$

Now, construct the global statistic

$$a_{i_1 i_2 \cdots i_M}(k) = \sum_{s=2}^{M} \rho_{i_{s-1}i_s}(k) \tag{16.79}$$

Define a binary variable

$$\eta_{i_1 i_2 \cdots i_M}(k) = \begin{cases} 1 & H_0 \quad \textbf{hypothesis} \\ 0 & H_1 \quad \textbf{hypothesis} \end{cases} \tag{16.80}$$

Then, the sequential track correlation problem in the case of multiple local nodes has been converted to that of multi-dimensional assignment,

$$\min_{\eta_{i_1 i_2 \cdots i_M}} \sum_{i_1=1}^{n_1} \sum_{i_2=1}^{n_2} \cdots \sum_{i_M=1}^{n_M} \eta_{i_1 i_2 \cdots i_M} a_{i_1 i_2 \cdots i_M}(k) \tag{16.81}$$

with constraint condition

$$\begin{cases} \displaystyle\sum_{i_2=1}^{n_2} \sum_{i_3=1}^{n_3} \cdots \sum_{i_M=1}^{n_M} \eta_{i_1 i_2 \cdots i_M} = 1, \quad \forall i_1 = 1, 2, \ldots, n_1 \\ \displaystyle\sum_{i_1=1}^{n_1} \sum_{i_3=1}^{n_3} \cdots \sum_{i_M=1}^{n_M} \eta_{i_1 i_2 \cdots i_M} = 1, \quad \forall i_2 = 1, 2, \ldots, n_2 \\ \vdots \\ \displaystyle\sum_{i_1=1}^{n_1} \sum_{i_2=1}^{n_2} \cdots \sum_{i_{M-1}=1}^{n_{M-1}} \eta_{i_1 i_2 \cdots i_M} = 1, \quad \forall i_M = 1, 2, \ldots, n_M \end{cases} \tag{16.82}$$

When $M=2$, (16.81) is degenerated to a 2D assignment problem, namely the sequential track correlation algorithm in Ref. [8]. Then, if $\rho_{i_{s-1}i_s}(k-1)\equiv 0$, it is the modified track association method in Ref. [8].

When the estimation errors of all local nodes are assumed independent, the sufficient statistic of (16.78) becomes

$$\lambda_{i_{s-1}i_s}(k)=\lambda_{i_{s-1}i_s}(k-1)+\left[\hat{\boldsymbol{X}}_{i_{s-1}}(k|k)-\hat{\boldsymbol{X}}_{i_s}(k|k)\right]'\boldsymbol{C}_{i_{s-1}i_s}^{-1}(k)\left[\hat{\boldsymbol{X}}_{i_{s-1}}(k|k)-\hat{\boldsymbol{X}}_{i_s}(k|k)\right] \tag{16.83}$$

where $s=1,2,\ldots,M$ is the serial number of local nodes, $i_s=1,2,\ldots,n_s$ is the track serial number of local node s, and

$$\boldsymbol{C}_{i_{s-1}i_s}(k)=\boldsymbol{P}_{i_{s-1}}(k/k)+\boldsymbol{P}_{i_s}(k/k) \tag{16.84}$$

When $M=2$, (16.84) is degenerated to the independent sequential track association algorithm in Ref. [8]. Then, if $\lambda_{i_{s-1}i_s}(k-1)\equiv 0$, it is the weighted track association method in Refs [387,388].

The track quality design and ambiguity processing are stated in detail in Ref. [8] and will not be discussed further in this book.

16.7 Summary

In the face of increasingly complex and dense electromagnetic spatial environments, radar networking and information fusion have turned out to be an absolute necessity in order to obtain comprehensive, accurate, and timely information maximally. Radar network data processing is a practical implementation of the multi-sensor data fusion theory in engineering, namely an application of this theory in fusing observations from two or more radars to the battlefield situation of the radar network coverage region.

This chapter started with a discussion of performance evaluation indexes and optimal station distribution of radar networks from the perspective of their design and analysis. Then it summarized the basic data processing process of monostatic, bistatic, and multistatic radar networks. Finally, it provided findings on the track association technology in radar network data processing, focusing on the analysis of the sequential track correlation algorithms based on statistical mathematics in the presence of multiple local nodes.

17

Evaluation of Radar Data Processing Performance

17.1 Introduction

Radar data processing technology is applied widely both in military and civilian fields, and is an area of research capturing international attention. The data processing performance of radars depends on many factors: density, quantity and dynamic characteristics of targets, detection performance of sensors, background noise source, and filter performance [40, 60, 393]. Consequently, the evaluation index system for radar data processing involves many diverse contents. As far as the radar data processing technology is concerned, suitable performance evaluation techniques should be used to judge the merits and demerits of its performance, its usability in the applied environment, and various function algorithms (track initiation, data association, and track filtering) [394–399]. For example, Refs [41, 400] propose some indexes of this type including the probabilities of being correct or incorrect for correlations between measurements and tracks, of false tracks and of track loss, but these simple measurement indexes are already inappropriate for the modern multi-target environment with high maneuverability and density.

Therefore, on the basis of analyzing and organizing the related materials [18, 23, 51, 53, 401], this chapter first defines the related terms and then discusses the performance evaluation indexes for radar data processing in several aspects, including average track initiation time, accumulative number of track interruptions, track ambiguity, accumulative number of track switches, track accuracy, maneuvering target tracking capability, false track ratio, divergence, track capacity, radar network detection probability, and response time. Finally, it explores several evaluation methods for radar data processing performance, including Monte Carlo, analytic, semi-physical simulative evaluation, and testing methods.

In addition, it is worth mentioning that the evaluation indexes of different definitions have their own relative rationality and limitations, and that it is very difficult, even unrealistic, to define a radar data processing evaluation system acceptable to all researchers. Instead, there can be different evaluation indexes for users at different levels.

Radar Data Processing with Applications, First Edition. He You, Xiu Jianjuan, and Guan Xin.

17.2 Basic Terms

Before evaluating the data processing performance of radars, we need to design performance test scenarios. This may be done as follows [53, 401].

Scenario 1 A single target in uniform rectilinear motion, used mainly for testing the target tracking accuracy of a radar system.

Scenario 2 A single target in uniform, circular, planar motion, mainly for testing the radar's maneuvering target tracking capability. The main consideration for choosing uniform circular motion is that the maneuvers of a target in track are mostly turning actions in a plane. The acceleration magnitude of the target can be determined according to its speed and radius values.

Scenario 3 Two cross-moving targets, as shown in Figure 17.1, chiefly for testing the radar's false association rate.

Scenario 4 Two targets in "close and off" movement, as shown in Figure 17.2, normally for testing the radar's false association rate.

Scenario 5 Many targets in parallel motion, as shown in Figure 17.3, principally for testing the radar's multi-target tracking capability.

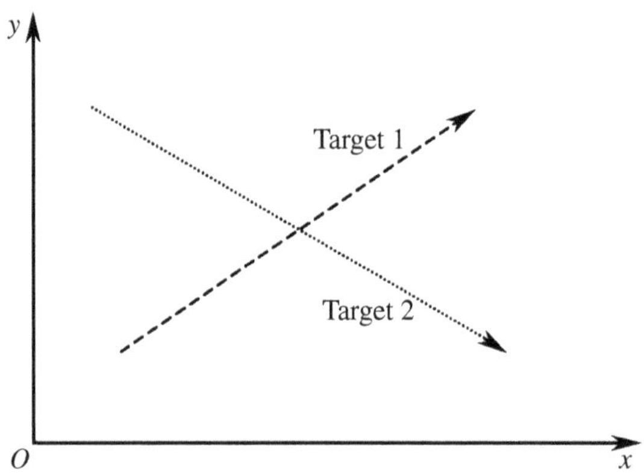

Figure 17.1 Two cross-moving targets

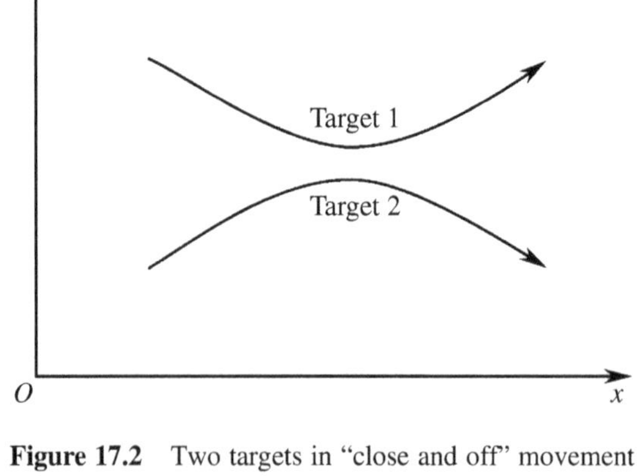

Figure 17.2 Two targets in "close and off" movement

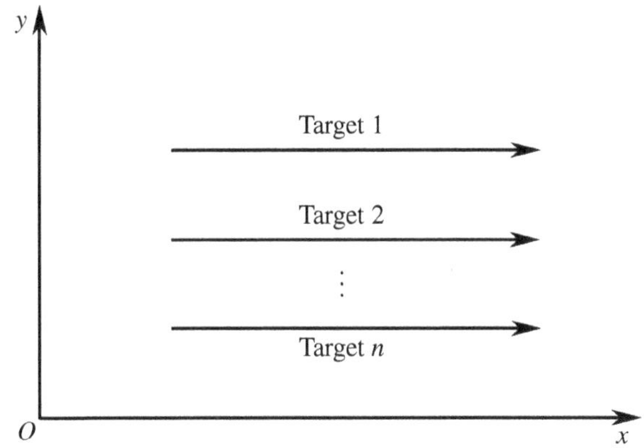

Figure 17.3 Many targets in parallel motion

The tracking capability of a radar is related to target density. The typical definition of target density was given by Farina and Studer [17], who divided target density into three levels: dense, medium, and sparse. The distinction criterion is the ratio of target spacing to radar measurement standard deviation. The target density is "medium" if the ratio is 1; "sparse" if it is bigger than 1.5; and "dense" if it is below 0.5.

The spacing of targets can be considered in terms of bearing, distance, or interval between targets. In the multistatic radar network system, the geographic positions of its member radars are different, as are the target's distances from them. Therefore, even for two radars with the same measurement accuracy, when the target's distances from them are different, so are the target densities calculated according to Farina's definition. In this case, the Euripidean distance between targets can be used directly as the interval between targets.

In the process of evaluating radar data processing performance, time should be specified: chosen randomly or with fixed intervals or designated by the user [53, 401].

17.3 Data Association Performance Evaluation

Radar data processing is equivalent to target state estimation, namely target tracking and data association. As discussed in Chapter 1, data association problems may be divided into three categories according to what is being correlated with what. The first category is the measurement-to-measurement association used in track initiation. Whether the radar can spot targets and initiate tracks for them in the shortest possible time has a significant influence on its survival and the overall outcome of combat, so a smaller mean of the time needed by the data processor to establish target tracks is preferred. Average track initiation time, therefore, is the first topic in the data association performance evaluation indexes below.

17.3.1 Average Track Initiation Time

Suppose that the radar data system discussed here has performed Monte Carlo simulation experiments with M runs and recorded in the mth simulation the time of the confirmed track assigned to target l, $t_{l,\text{first}}^m$. If no track is assigned to target l during the time period of this simulation test (T), then let $t_{l,\text{first}}^m = T$. Hence, the average time of track initiation $t_{l,\text{first}}$ of target l can be defined as

$$t_{l,\text{first}} = \frac{1}{M}\sum_{m=1}^{M} t_{l,\text{first}}^m \tag{17.1}$$

If we take a statistical average of all targets' average time of track initiation measured by radar i, then we can obtain the total average time of track initiation t_{first}^i of radar i, namely

$$t_{\text{first}}^i = \frac{1}{L}\sum_{l=1}^{L} t_{l,\text{first}} \tag{17.2}$$

where L is the total number of real targets.

For a distributed radar data processing system, if there are N radar stations, then its average time of track initiation is

$$t_{\text{first}} = \frac{1}{N}\sum_{i=1}^{N} t_{\text{first}}^i \tag{17.3}$$

17.3.2 Accumulative Number of Track Interruptions

The second category of data association discussed in Chapter 1 is that between measurements and tracks used in track maintenance. If there is no latest measurement assigned to a track, it will get interrupted. Therefore, the data association performance evaluation index to be discussed next is the accumulative number of track interruptions.

The total number of times that the real target is assigned with no track before evaluation time t_{eval} is called the accumulative number of track interruptions at t_{eval} and written

$$NB(t_{\text{eval}}) = \frac{1}{L}\sum_{l=1}^{L} NB_l(t_{\text{eval}}) \tag{17.4}$$

where L is the total number of real targets and $NB_l(t_{\text{eval}})$ is the total number of track interruptions of real target l at t_{eval}, which can be obtained through the Monte Carlo simulation method, represented by

$$NB_l(t_{\text{eval}}) = \frac{1}{M}\sum_{m=1}^{M} NB_l^m(t_{\text{eval}}) \tag{17.5}$$

Here, M is the number of runs of the Monte Carlo simulation and $NB_l^m(t_{\text{eval}})$ is the total number of track interruptions of real target l at t_{eval} in the mth Monte Carlo simulation, which can be obtained through the following method: in the mth Monte Carlo simulation, if a track is assigned to target l at time $(t_{\text{eval}}-1)$ but no track is assigned to it at t_{eval}, then $NB_{l,m}(t_{\text{eval}})$ should be increased by 1.

17.3.3 Track Ambiguity

In the process of data association, when two or more tracks are assigned to the same target, data association ambiguity [52, 187, 376] will occur. Therefore, the performance evaluation index to be discussed below is track ambiguity. In order to describe and measure this phenomenon more clearly, we introduce the following concepts [53, 54, 372, 401].

1. *Possible tracks:* those in the set of validated tracks established by the radar data processor which can be assigned to the real target.
2. *Redundant tracks:* the phenomenon of two or more tracks being assigned to a real target is called track redundancy, and the surplus track is called the redundant track [401].
3. *False tracks:* those which are unrelated to real targets in the set of validated tracks established by the radar data processor.
4. *Number of redundant tracks:* the difference in number between feasible tracks and targets.
5. *Number of false tracks:* the difference in number between the validated tracks established by the radar data processor and feasible targets.
6. *Track capacity:* the maximum batches of radar tracks which the radar data processing center can handle at the same time.
7. *Track interruptions:* if a track is assigned to a target at time t, but no track is assigned to it at time $t+m$, then a track interruption is declared to have occurred at t, where m is a parameter set by the tester (usually $m=1$).
8. *Track switches:* if a track is assigned to a target at time t, but another track is assigned to it at time $t+m$, then a track switch is considered to have occurred at time t, where m is a parameter set by the tester (usually $m=1$).

Based on the concepts given above, define track ambiguity as the ratio of the number of redundant tracks to that of feasible targets [53, 401] t_{eval}, denoted by

$$A(t_{\text{eval}}) = \frac{N_{\text{rt}}(t_{\text{eval}})}{N_{\text{ft}}(t_{\text{eval}})} \tag{17.6}$$

where $N_{\text{rt}}(t_{\text{eval}})$ and $N_{\text{ft}}(t_{\text{eval}})$ denote the numbers of redundant tracks and feasible targets, respectively, at evaluation time t_{eval}. $A(t_{\text{eval}})$ indicates the number of surplus tracks per feasible target, whose value may be bigger than 1 but at least 0. When the validated tracks correspond to the real targets one by one, the value $A(t_{\text{eval}})$ reaches its minimum zero, that is, there is no ambiguity at this time.

Notice that track ambiguity is also related to evaluation time. Its overall evaluation can be made by plotting a curve of it varying with evaluation time, or by calculating the average track ambiguity $\bar{A}$, expressed as

$$\bar{A} = \frac{1}{N_t} \sum_{t_{\text{eval}} \in S} A(t_{\text{eval}}) \tag{17.7}$$

where S is the set of t_{eval} and N_t the number of t_{eval} in set S.

In applications, track ambiguity defined by (17.6) can be obtained normally through the Monte Carlo simulation, as represented by

$$A(t_{\text{eval}}) = \frac{1}{M} \sum_{m=1}^{M} A^m(t_{\text{eval}}) \tag{17.8}$$

where M is the total number of Monte Carlo simulation runs and $A^m(t_{\text{eval}})$ is the ambiguity at t_{eval} in the mth Monte Carlo simulation, calculated by

$$A^m(t_{\text{eval}}) = \frac{N_{\text{rt}}^m(t_{\text{eval}})}{N_{\text{ft}}^m(t_{\text{eval}})} \tag{17.9}$$

where $N_{\text{rt}}^m(t_{\text{eval}})$ and $N_{\text{ft}}^m(t_{\text{eval}})$ denote the numbers of redundant tracks and feasible targets, respectively, at t_{eval} in the Monte Carlo simulation.

17.3.4 Accumulative Number of Track Switches

In the process of data association, changes may occur in the tracks assigned to a target. Therefore, the data association performance evaluation index to be discussed hereafter is the accumulative number of track switches. The total number of track switches in those assigned to a real target before time t_{eval} is called the accumulative number of track switches at t_{eval} and denoted by $NS(t_{\text{eval}})$, expressed as

$$NS(t_{\text{eval}}) = \frac{1}{L} \sum_{l=1}^{L} NS_l(t_{\text{eval}}) \tag{17.10}$$

where L is the total number of real targets and $NS_l(t_{\text{eval}})$ is the total number of track switches of real target l at t_{eval}. In practical applications, $NS_l(t_{\text{eval}})$ is usually obtained through the Monte Carlo simulation, expressed as

$$NS_l(t_{\text{eval}}) = \frac{1}{M} \sum_{m=1}^{M} NS_l^m(t_{\text{eval}}) \tag{17.11}$$

where M is the number of runs of the Monte Carlo simulation, and $NS_l^m(t_{\text{eval}})$ is the total number of track switches of real target l at t_{eval} in the mth Monte Carlo simulation, which can be obtained as follows. In the mth Monte Carlo simulation, if track j is assigned to target l at time $(t_{\text{eval}} - 1)$ and track $(k \neq j)$ is assigned to target l at t_{eval}, then $NS_{l,m}(t_{\text{eval}})$ should be increased by 1.

17.4 Performance Evaluation of Tracking

Data association and tracking are fundamental issues during the process of radar data processing [402]. In Section 17.3, we have analyzed the performance evaluation of data association and given four kinds of evaluation index: average time of track initiation, accumulative number of track interruptions, track ambiguity, and accumulative number of track switches. Hereafter, we discuss the evaluation of track filtering performance and then put forward four kinds of evaluation index: track accuracy, maneuvering target tracking capability, false track ratio, and divergence.

17.4.1 Track Accuracy

Track accuracy is a very important index, used to evaluate the performance of track filtering algorithms, and it embodies the smoothing degrees of different radar data processing algorithms for sensor measurement errors. Broadly, track accuracy includes track position and speed accuracy. The former is defined as the mean square root error of track position estimate error and the latter as the mean square root error of track speed estimate error. During the process of target tracking, the smaller the mean square root of the radar's estimate is, the closer the filtering value of the filter gets to the real value of the target, and the higher the track accuracy. In practice, the Monte Carlo simulation method is usually used to evaluate track accuracy. Suppose that the difference between the filtering and real values of the lth real target at evaluation time t_{eval} is

$$\boldsymbol{E}_l^m(t_{\text{eval}}) = \hat{\boldsymbol{X}}_l^m(t_{\text{eval}}) - \boldsymbol{X}_l(t_{\text{eval}}) \tag{17.12}$$

where $\hat{\mathbf{X}}_l^m(t_{\text{eval}})$ is the estimated state vector of the lth target at time t_{eval} in the mth Monte Carlo simulation, and $\mathbf{X}_l(t_{\text{eval}})$ the real state of the lth target at t_{eval}.

The statistical mean of the squared estimation error, obtained after M runs of Monte Carlo simulation experiments, is defined as

$$\boldsymbol{C}_l(t_{\text{eval}}) = \frac{1}{M}\sum_{m=1}^{M} \boldsymbol{E}_l^m(t_{\text{eval}})\boldsymbol{E}_l^{m'}(t_{\text{eval}}) \tag{17.13}$$

If the target state vector is given as $\boldsymbol{X}(k) = [x \;\; \dot{x} \;\; y \;\; \dot{y} \;\; z \;\; \dot{z}]'$, then $\boldsymbol{C}_l(t_{\text{eval}})$ is a 6×6 matrix, where elements $C_{l,x}(t_{\text{eval}})$, $C_{l,y}(t_{\text{eval}})$, and $C_{l,z}(t_{\text{eval}})$ corresponding, respectively, to (column 1, row 1), (column 3, row 3), and (column 5, row 5) denote the target's variance of position error along axes x, y, and z, respectively. Using $C_{l,x}(t_{\text{eval}})$, $C_{l,y}(t_{\text{eval}})$, and $C_{l,z}(t_{\text{eval}})$ yields the position mean square root error of the tracking to the lth real target at t_{eval}:

$$RMSE_{l,\,\text{position}}(t_{\text{eval}}) = \sqrt{C_{l,x}(t_{\text{eval}}) + C_{l,y}(t_{\text{eval}}) + C_{l,z}(t_{\text{eval}})} \tag{17.14}$$

Likewise, we have the speed mean square root error of the tracking to the lth real target:

$$RMSE_{l,\,\text{velocity}}(t_{\text{eval}}) = \sqrt{C_{l,\dot{x}}(t_{\text{eval}}) + C_{l,\dot{y}}(t_{\text{eval}}) + C_{l,\dot{z}}(t_{\text{eval}})} \tag{17.15}$$

where elements $\boldsymbol{C}_{l,\dot{x}}(t_{\text{eval}})$, $\boldsymbol{C}_{l,\dot{y}}(t_{\text{eval}})$, and $\boldsymbol{C}_{l,\dot{z}}(t_{\text{eval}})$ corresponding, respectively, to (column 2, row 2), (column 4, row 4), and (column 6, row 6) of matrix $\boldsymbol{C}_l(t_{\text{eval}})$ denote the variance of speed error of the target along axes x, y, and z.

At t_{eval}, the total position mean square root error and the total speed mean square root error of all L targets can be defined, respectively, as follows:

$$RMSE_{\text{position}}(t_{\text{eval}}) = \frac{1}{L}\sum_{l=1}^{L} RMSE_{l,\text{ position}}(t_{\text{eval}}) \tag{17.16}$$

$$RMSE_{\text{velocity}}(t_{\text{eval}}) = \frac{1}{L}\sum_{l=1}^{L} RMSE_{l,\text{ velocity}}(t_{\text{eval}}) \tag{17.17}$$

17.4.2 Maneuvering Target Tracking Capability

Maneuver detection delay time can be used as an index to measure the radar network's capability of tracking maneuvering targets. The time interval between a target starting a maneuver and the maneuver being detected by the radar is called the maneuver detection delay time. Note that the value of evaluation time must be taken during the target maneuvering period. The maneuver detection delay time of maneuvering targets can be obtained with the Monte Carlo method. The procedure is described below.

In the centralized data processing system, the target maneuver initiating time is recorded as 0. In the mth Monte Carlo simulation, target l's maneuver initiating time is detected and denoted as $t^{m}_{l,\text{ manouver}}$. If its maneuver has not been detected during this period of simulation T, then let $t^{m}_{l,\text{ manouver}} = T$. Therefore, target l's average maneuver detection time is denoted as

$$t_{l,\text{ manouver}} = \frac{1}{M}\sum_{m=1}^{M} t^{m}_{l,\text{ manouver}} \tag{17.18}$$

where M is the number of runs of the Monte Carlo simulation.

Hence, the total average maneuver detection delay time of L targets is denoted by

$$t_{\text{manouver}} = \frac{1}{L}\sum_{l=1}^{L} t_{l,\text{ manouver}} \tag{17.19}$$

For the distributed data processing system, the average maneuver detection delay time of the ith radar is denoted as t^{i}_{manouver}. Similarly, it can be obtained from (17.18) and (17.19). Assume that there are N radar stations. Then the total average maneuver detection delay time of this system can be denoted as

$$t_{\text{manouver}} = \frac{1}{N}\sum_{i=1}^{N} t^{i}_{\text{manouver}} \tag{17.20}$$

17.4.3 False Track Ratio

In the multi-target case, the target tracks given by the radar may have false ones. Therefore, in the process of evaluating track filtering performance, the ratio of false tracks to total tracks must be given [32]. In the same environment, it is preferable that this ratio is minimized. Hence, define the false track ratio at evaluation time t_{eval} as the ratio of the number of false tracks to that of tracks in total, and denote it by

$$STR(t_{\text{eval}}) = \frac{N_{\text{false track}}(t_{\text{eval}})}{N_{\text{total track}}(t_{\text{eval}})} \tag{17.21}$$

where $N_{\text{false track}}(t_{\text{eval}})$ and $N_{\text{total track}}(t_{\text{eval}})$ denote, respectively, the number of false tracks and that of tracks in total at time t_{eval}.

Note that the ratio of false tracks is related to the evaluation time. Its overall evaluation may be conducted by plotting a curve which shows it varying with the evaluation time, or by calculating the average false track ratio $\overline{STR}$, expressed as

$$\overline{STR} = \frac{1}{N_t} \sum_{t_{\text{eval}} \in S} STR(t_{\text{eval}}) \tag{17.22}$$

where S is the set of evaluation times t_{eval} and N_t is the number of evaluation times in set S.

In practical applications, the average false track ratio $\overline{STR}$ given by (17.22) can be obtained through the Monte Carlo simulation method, hence, so can the false track ratio at time t_{eval}, calculated by

$$STR(t_{\text{eval}}) = \frac{1}{M} \sum_{m=1}^{M} STR^m(t_{\text{eval}}) \tag{17.23}$$

where M is the number of runs of Monte Carlo simulations and $STR^m(t_{\text{eval}})$ the false track ratio at time t_{eval} in the mth Monte Carlo simulation,

$$STR^m(t_{\text{eval}}) = \frac{N^m_{\text{false track}}(t_{\text{eval}})}{N^m_{\text{total track}}(t_{\text{eval}})} \tag{17.24}$$

Here, $N^m_{\text{false track}}(t_{\text{eval}})$ is the number of false tracks at time t_{eval} in the mth Monte Carlo simulation and $N^m_{\text{total track}}(t_{\text{eval}})$ the number of feasible targets at t_{eval} in the mth Monte Carlo simulation.

17.4.4 Divergence

As can be concluded from the discussions in previous chapters, two problems arise in the process of radar data processing:

- uncertainty about the measurements used for filtering (i.e., many plots will arise due to multiple targets and false alarms in the radar environment);
- uncertainty about the parameters of the target model.

Therefore, during the filtering process, if the hypothetical model is compatible with the real model and correct data association has been accomplished in the presence of many targets, the difference between the tracking result and the real value of the targets narrows as the filtering time increases. However, if the hypothetical case does not tally with the real case, or correct data association has failed in the multi-target case, filtering divergence will occur. Once divergence appears in the process of radar data processing, filtering would be rendered meaningless. Thus, the index of divergence will be given here. The decision method of filtering divergence will be given first [403].

Denote by $RMSE_{l,\text{ position}}(t_{\text{eval}})$ the mean square root error of the position tracked by the radar for the lth real target at evaluation time t_{eval}. The calculation of this error is different with two-coordinate radar and three-coordinate radar. Its calculation in the case of three-coordinate radar is shown in (17.14), while in the case of two-coordinate radar it is calculated as

$$RMSE_{l,\text{ position}}(t_{\text{eval}}) = \sqrt{C_{l,x}(t_{\text{eval}}) + C_{l,y}(t_{\text{eval}})} \tag{17.25}$$

where $C_{l,x}(t_{\text{eval}})$ and $C_{l,y}(t_{\text{eval}})$ are defined as in (17.13).

If starting from the stable filtering time k_0, sampling points have been emerging in succession whose position mean square root error $RMSE_{l,\text{ position}}$ is bigger than the divergence test threshold Δ_{P0} [403]; this filtering is regarded as divergent.

Divergence is often evaluated in practice with the Monte Carlo simulation method, that is, it is defined as the ratio of the number of filtering divergences to that of simulation runs in Monte Carlo simulation experiments, denoted by

$$\eta_d = \frac{N_{\text{f d}}}{N_{\text{simulation}}} \tag{17.26}$$

where N_{fd} and $N_{simulation}$ denote, respectively, the number of filtering divergences and the total number of simulation runs ($0 \le \eta_d \le 1$). The bigger the value of divergence, the more likely the filter will lose track of its target during the tracking process.

17.5 Evaluation of the Data Fusion Performance of Radar Networks

In the multi-radar, multi-target case, the evaluation of the system tracking performance is significantly more complicated than in the presence of single radars and targets. Based on the discussion and analysis of performance evaluation methods of single-radar data association and track filtering, in this section we discuss the method of evaluating the data fusion performance of the radar network and lay emphasis on analyzing such indexes as the track capacity, detection probability of radar networks, and response time.

17.5.1 Track Capacity

Radar network track capacity is a basic index in radar networking. It is defined as the maximum number of batches of radar tracks that the radar network information fusion center is capable of processing at the same time.

17.5.2 Detection Probability of Radar Networks

At evaluation time t_{eval}, the ratio of the number of feasible targets to that of real targets is called the detection probability of the radar network at this time. It is denoted by

$$P_D(t_{\text{eval}}) = \frac{\text{number of feasible targets}(t_{\text{eval}})}{\text{number of real targets}(t_{\text{eval}})} \tag{17.27}$$

Note that $P_D(t_{eval})$ is related to t_{eval}. An overall evaluation can be made by plotting a curve of it varying with the evaluation time. At the same time, the average detection probability $\bar{P}_D$ can also be calculated, using

$$\bar{P}_D = \frac{1}{N_t} \sum_{t_{eval} \in S} P_D(t_{eval}) \tag{17.28}$$

where S is the set of evaluation times and $N_t \triangleq \text{Card}\{S\}$ is the number of evaluation times in S.

In practical applications, the radar network average detection probability of (17.28) can be obtained through the Monte Carlo simulation method, using

$$P_D(t_{eval}) = \frac{1}{M} \sum_{m=1}^{M} P_{Dm}(t_{eval}) \tag{17.29}$$

$$P_{Dm}(t_{eval}) = \frac{NV_m(t_{eval})}{NT_m(t_{eval})} \tag{17.30}$$

where M is the total number of Monte Carlo simulation runs, $NV_m(t_{eval})$ the number of feasible targets at t_{eval} in the mth Monte Carlo simulation, and $NT_m(t_{eval})$ the number of real targets in the responsible region of the radar network at t_{eval} in the mth Monte Carlo simulation.

17.5.3 Response Time

Whether a radar has the ability to detect a target as early as possible has significant influence on the survival of a radar network and the outcome of combat. The response time of a radar network is an important index used to measure its target tracking instantaneity.

The response time of the radar network is defined as the statistical mean of the time needed by the system fusion center to establish target tracks.

For the centralized data processing system, denote by $t^m_{l,\,first}$ the time assigned to the validated track of target l recorded in the mth Monte Carlo simulation. If no track is assigned to target l in the period of this simulation (T), then let $t^m_{l,\,first} = T$. Hence, denote the average track initiation time for target l (namely, the response time) by

$$t_{l,\,first} = \frac{1}{M} \sum_{m=1}^{M} t^m_{l,\,first} \tag{17.31}$$

Hence, the total average track initiation time (namely, the response time) is indicated as

$$t_{first} = \frac{1}{L} \sum_{l=1}^{L} t_{l,\,first} \tag{17.32}$$

For the distributed data processing system, denote the average track initiation time of the ith radar by t^i_{first}, and suppose that there are N radar stations, then the total average track initiation time is

$$t_{first} = \frac{1}{N} \sum_{i=1}^{N} t^i_{first} \tag{17.33}$$

17.6 Methods of Evaluating Radar Data Processing Algorithms

The evaluations of data association and track filtering performances discussed above were conducted with the Monte Carlo method. In addition to Monte Carlo, other methods can also be adopted to evaluate the data processing performance of radars [53, 401]: the analytic, semi-physical simulation, and test methods, which will be introduced briefly in this section.

17.6.1 Monte Carlo Method

The Monte Carlo method, also known as the statistical test method, uses statistical sampling theory to approximately solve practical problems. It is a stochastic analysis method using a large quantity of computer simulations to test the system performance and aggregate statistical results. Its theoretical basis is the law of large numbers in probability theory. This method includes the generation of pseudo-random numbers, Monte Carlo simulation design, and result interpretation. Its approach to problems is: first establish a probability model that has similarity to the evaluation of radar data processing performance, then make random simulation or statistical sampling of the model, and finally use the findings to evaluate the data processing performance of radars.

The Monte Carlo method has the following advantages:

1. It is economical, since a large amount of outlay can be saved as a result of large amounts of repeated sampling done on computers.
2. It is feasible for some problems with complex models and hard to solve with numerical solution methods.
3. It is adaptable and hardly subject to the constraints of problems. It is a desirable option especially for some hazardous, hard-to-solve, or high-cost problems.

Its disadvantages include:

1. It tends to converge slowly for some practical problems that require high accuracy, and sometimes cannot satisfy real-time requirements.
2. It only approximates practical problems, but is still different from them. Therefore, its result can be used only for reference or guidance.

The random number generating method, if any, provided by the computer programming language this method adopts can be used directly to generate pseudo-random numbers. For example, when Matlab is used in programming, sentence `rand` can be used directly to generate uniformly distributed random numbers in the interval [0,1], and `randn` to generate standard normally distributed random numbers. If the adopted computer programming language fails to provide the random numbers needed directly, then uniformly distributed random numbers are usually generated first in the interval [0,1], and then the inverse transformation method is used to generate the random numbers distributed as required. For details, see Chapter 18.

17.6.2 Analytic Method

The analytic performance evaluation method involves the establishment of mathematical models for one or more performance evaluation indexes by various methods, and the derivation, through analytic calculation or numerical solution, of the numerical values of performance evaluation indexes of a radar data processing system for its evaluation. In order to adopt the analytic method, we can use one or more theories as the basis for the abstraction of the system, and denote model parameters, initial conditions, and input–output relationships with mathematical expressions to obtain the corresponding mathematical models. In terms of the adopted mathematical theory, the analytic methods for evaluating the data processing performance of radar systems mainly include performance evaluations based on statistical and fuzzy set theory.

When the analytic method is used in performance evaluation, the establishment of mathematical models of the system is a very important step. In the process of establishing mathematical models, the first thing is to decide, based on the analytic result of the problem, the coordinate system and system state variables to be adopted, which are then described in mathematical forms according to the mutual relationship between them and the constraints. Meanwhile, the parameters should be determined, namely: construct a mathematical model for analysis and evaluation. The variables and interactions described by this model must be close to the real system and meanwhile be able to balance the facticity and operating efficiency of the system, such that the models have moderate complexity, neither too simple nor too complex.

17.6.3 Semi-physical Simulation Method

Semi-physical simulation is an indoor simulation test method, which uses hardware and software to simulate the electromagnetic features of signal sources and electronic systems. The typical test procedure is as follows. The computer-controlled test system (simulator) generates real signals in a typical test environment, then the real system is put into in-field semi-physical simulation tools, and computers are employed to simulate the operation or motion of the system, so as to analyze and evaluate its efficiency.

In semi-physical simulation tests, the in-field semi-physical simulation tools need a large amount of data as initial conditions, including the trajectory and RCS characteristics of the target, the radiation characteristics of signal sources, models of radar data processing algorithms, definitions of performance indexes, and data of tactical and position conditions.

This method can provide a flexible and convenient simulation test platform for the test appraisal and evaluation of radar data processing systems. Its advantages mainly include:

1. The test environment is controllable. Various dynamic electromagnetic environments can be simulated as demanded, providing vivid test conditions for the performance evaluation of radar data processing systems.
2. The test process is controllable. The test and evaluation process can be repeated as needed. Alternatively, tests and evaluations can be performed on the pilot processes of interest, providing favorable conditions for evaluating system performance and thus enabling quantitative performance analysis.
3. The test data are easy to record and retrieve, insensitive to environment, high in measurement accuracy, and thus able to facilitate the performance evaluation of the radar data processing system.
4. It has a high cost-to-benefit ratio.
5. It has strict security, generating no electromagnetic radiation and secure from enemy detections.

The major disadvantages of this test are:

1. The credibility of the test result is related to mathematical simulation models and usually different from the real environment.

2. When the test is done in a microwave anechoic chamber, because of restrictions due to the size of the chamber, the operators have to consider how to eliminate the transmission effect of the near-field electromagnetic waves.

17.6.4 Test Validation Method

The test validation method involves putting radar data processing models and/or systems into practical applications, and evaluating their performance through practical tests. The merit of this method is that it can reflect their efficacy and properties objectively, truly, and comprehensively. Its demerits are its high cost and implementation difficulties. This method is typically adopted to analyze the findings of the others discussed in this section, and thus what it produces usually serves as the final evaluation report on the performance of the systems of interest.

17.7 Summary

Rapid advances in radar data processing techniques have led to the emergence of new algorithms in this field over the past few decades. Therefore, how to make desirable evaluations of the algorithms has drawn attention worldwide, and become a hot topic for both theoretical and practical research. In that connection, this chapter opens with definitions of scenarios and target density in Section 17.2.

This is followed by the analysis of key indexes for evaluating radar data processing performance, including average track initiation time, accumulative number of track interruptions, track ambiguity, and accumulative number of track switches for data association in Section 17.3; track accuracy, maneuvering target tracking capability, false track ratio, and divergence for track filtering in Section 17.4.

Then, the evaluation of data fusion performance of the radar network system is described in Section 17.5. The indexes introduced here include track capacity, radar network detection probability, and response time.

Finally, in Section 17.6, we focus on the Monte Carlo method among several others for evaluating radar data processing performance – such as the analytical, semi-physical simulative, and test validation methods.

Note that the evaluation indexes computed through the Monte Carlo method have their limitations, each representing the performance of an algorithm only in one respect. Worthwhile evaluations on the algorithm depend on the overall evaluation of each index on an individual basis, providing a reference for engineers, who usually give preferential consideration to some properties of filtering algorithms in specific applications.

18

Radar Data Processing Simulation Technology

18.1 Introduction

Resulting from a combination of simulation and radar technology, the simulation of radar data processing systems is the modeling of radar data processing with the use of software on digital computers, so that the dynamic process of radar data processing recurs.

Increases in the diversity and complexity of functions of modern radar systems necessitate the use of digital simulation technology to set up a radar data processing model with software so that the dynamic process of the system recurs. The revision and perfection of the model and its parameters as needed can undoubtedly greatly save manpower and resources, and shorten the development cycle. Therefore, it is a wise choice to use digital simulation technology in system design and testing.

This chapter aims to provide some simulation methods which can be used to analyze and design the radar data processing system, briefly introduce the basic knowledge of system simulation, present the method of generating random data, simulate various algorithms, and analyze the estimated result of system performance. Radar data processing algorithms are very suitable, in essence, for simulations on digital computers.

The major topics of this chapter are:

- random data generation methods in the Monte Carlo simulation;
- simulations of movement models, measurement processes, filter prediction algorithms, and multi-target tracking and data association algorithms;
- track initiation and termination;
- statistical evaluation of errors.

Radar Data Processing with Applications, First Edition. He You, Xiu Jianjuan, and Guan Xin.

18.2 Basis of System Simulation Technology

18.2.1 Basic Concept of System Simulation Technology

System simulation, in short, amounts to model testing. It is the process of researching into an existing system or one at the design stage through system model tests. The realization of simulation requires, first of all, a "substitute" for the real system, which is called a model. It is not the reproduction of the prototype, but is the simplification and abstraction of the system according to the emphasis of the research or actual need, so as to help researchers grasp the essence of the problem or the main contradiction. The test technology of this type, which is based on the modeling system, is called the simulation technology.

System simulation is a comprehensive subject arising during the past three decades. It involves theories and technologies of related disciplines, such as system analysis, control theory, computational method, and computer technology. Simulation technology is a crucial and even indispensable tool when it is hard or impossible to do tests on the real system. It plays an important role in modern scientific research, production, and education.

With the advances in digital computer hardware and software, digital computer simulation has been developing rapidly since the 1970s. Digital computer simulation is highly accurate, repeatable, versatile, and cheap. Many computer simulation program packages and simulation languages are available, which is convenient in use. Therefore, simulation technology is widely used in production management, engineering, military research, scientific tests, national economy, important policy decisions, and other areas of social and natural sciences, with impressive results. In the development of ultimate national defense weapon systems and crucial technology, for instance, this technology directly decides the advancement, development cycle, expenditure, and even the outcome of the system being developed.

In the pre-computer days, the physical simulation then in use was a subspecialty of other subjects. With the development and proliferation of computers, however, a large number of common theories, methods, and technologies have been put forward in the simulation field. As a consequence, simulation has emerged as an independent subject.

System simulation is the process of setting up models of the system, process, phenomenon, and environment (such as physical, mathematical, or other logical models), running the established models in a certain period of time, and applying them to system tests and analyses and staff training. The basic idea of system simulation is to set up a test model which is quite similar to the system being studied. Through the model's operation, the information, parameters, and materials necessary to the system under study are obtained to provide a scientific basis for the development of the system. Since the real system is replaced with the model in the test, its characteristics can be studied thoroughly and safely while its environment can remain intact.

18.2.1.1 Classification of System Simulation

System simulation can be divided into three types, according to the models adopted [404].

1. *Object simulation* (also called physical simulation). Object simulation, using all subsystems or components of a real system with artificial factors added, is based on geometrical or physical similarity. The wind tunnel test of an air model, for example, is aimed at accumulating experience for battles and providing a basis for improvement in the system. Physical models have physical properties similar to those of real systems. They can be miniature versions of real objects (e.g., miniature aircraft and ships in wind tunnel tests), or prototypical models with the same function as real systems (e.g., trial manufactured prototypes).

2. *Semi-object simulation* (also called mathematical/physical simulation). In semi-object simulation, some pieces of the hardware of the system being simulated are used while others are simulated by computers. The operational environment (including threats) can be either object- or computer-simulated. The purpose of this type of simulation is to discover problems in the system as early as possible, and add new technologies necessary for the perfection of the system. It is a combination of mathematical models and physical ones (or real subsystems).
3. *Computer simulation*. The characteristic of computer simulation is that the system is expressed with rigid mathematical models and some rules. It is a pure software system, without using any hardware in the real system. Mathematical models refer to those established by describing the relations between physical variables in the system with abstract mathematical equations. Research into the mathematical model of a system can reveal its inner movement and dynamic properties. What interests us most in the computer simulation of a system is its mathematical model. This is the mathematical mode similarity-based simulation, in which mathematical models are tested in place of real systems to simulate changes in their realities, and analyze the whole process of their changes with quantitative methods. The system is expressed in rigid mathematical formulas, figures, or computer programs.

As computers have progressively stronger functions to solve complicated mathematical models, their applications in mathematical simulations have been increasingly emphasized and accepted. Mathematical simulation tests take much less time than physical ones, and have much simpler data processing requirements.

18.2.1.2 Basic Idea of the Monte Carlo Simulation and its Characteristics

Detailed Monte Carlo simulation analyses are often needed in the design and integration of radar data processing systems. The Monte Carlo simulation is a random analysis method which uses large amounts of computer simulation to test the system's dynamic characteristics and summarize statistical findings. It comprises generation of pseudo-random numbers, Monte Carlo simulation design, interpretation of findings, etc. Its function is to simulate the real physical environment with mathematical methods and test the system's reliability and feasibility.

The Monte Carlo method, also known as the statistical test method, is one which approximately solves mathematical, physical, and engineering problems with statistical sampling theory. Its basic approach to problems is: first, a probability model which has similarity to the problem is established and described; then, it undergoes random simulation or statistical sampling; after that, the results are used to find the statistical estimates of the characteristics which serve as the approximate solution to the original problem, whose accuracy is estimated last. This method is primarily based on the law of large numbers in probability theory, with its main means as the sampling analysis of random variables.

The characteristics of the Monte Carlo simulation are as follows:

1. Simple in computational method and program content, because its analysis is realized by many repetitions of simple sampling.
2. The probability and speed of convergence has nothing to do with the dimension of the problem.
3. Highly adaptive, immune to the constraints of the problem's conditions.
4. Slow in convergence, not suitable for practical problems with high accuracy requirements.

A key to the possible practical applications of the Monte Carlo method is whether it can generate random numbers conveniently, economically, and reliably.

18.2.2 Digital Simulation of Stochastic Noise

The simulation research of stochastic systems on digital computers requires the computers to generate the random noise expected. The random noise under study is an ergodic steady stochastic process, so its statistical characteristic can be described by the time average characteristics of a sample function $x(t)$ in the stochastic process $\{x(t)\}$. Theoretically speaking, any random noise can be obtained by the computation of uniformly distributed white noise. Therefore, this section introduces how to use a computer to generate uniformly distributed white noise, and how to generate, from the white noise, the colored noise with other distributions.

18.2.2.1 Production of the Uniformly Distributed Random Number in [0, l] Interval

Mathematically speaking, as long as random numbers with a certain distribution pattern exist, arbitrarily distributed random numbers can be obtained by various mathematical transformations and sampling [405]. Actually, it is always the simplest uniformly distributed random number in interval [0, l] that is generated first, and then it is used to generate random numbers with varied distributions as required. Some languages, like Matlab, have already provided the sentences used to generate the uniformly distributed random number in interval [0, l], and can be applied directly when necessary, while others do not, and users have to program these sentences themselves. Currently, there are many ways to generate uniformly distributed random numbers, and here we only introduce two linear congruential methods: multiplicative and mixed.

Multiplicative Congruential Method

The recurrence formula for using the multiplicative congruential method to generate random numbers is

$$x_i = \lambda x_{i-1} \pmod M \tag{18.1}$$

where λ and M are parameters. Given the initial value x_0, an integer sequence $\{x_i\}$ can be obtained from (18.1). Transform each x_i, that is,

$$u_i = x_i / M \tag{18.2}$$

then u_i is a uniformly distributed random number in interval [0, l], and $\{u_i\},\ i = 1, 2, \ldots$ is a uniformly distributed sequence in interval [0, 1].

For binary system computers, λ and M can be chosen under the following rules:

1. $M = 2^K$, where K is an integer within the number range which can be expressed by the machine;
2. λ is usually the number that is closest to $\lambda \approx 2^{p/2}$ and satisfies $\lambda = 2^3 a \pm 3 = 8a \pm 3$, in which a can be any integer and p is the word length of the machine.

Mixed Congruential Method

The recurrence equation for using the mixed congruential method to generate random numbers is

$$x_{i+1} = \lambda x_i + \varepsilon \pmod M \tag{18.3}$$

where initial value x_0, increment ε, multiplier λ, and mode M are all non-negative integers. Obviously, when $\varepsilon=0$, the mixed congruential method becomes the multiplicative one. When the mixed method is used to generate uniform random numbers, for $M=2^K$, the possible maximum cycle is $T_{\max}=2^K$ (i.e., each integer in interval $\left[0,2^{K-1}\right]$ can appear once in an entire cycle).

18.2.2.2 Production of Normal Distribution Random Numbers

The normal distribution plays an important role in probability statistics, with many statistical phenomena subject to normal distributions. For example, in a radar system, the interior noise of the receiver, radar measurement errors, etc. are mostly assumed normally distributed. Since the closed-form solution to the function of other distributions cannot be obtained, the direct sampling method cannot be used to obtain the random samples with normal distributions, and other measures need to be taken.

Usually, when generating a normally distributed random number, it is necessary to first generate a standard normal distribution, from which the wanted normal distribution is then obtained through transformation. For example, if $X\sim N[0,1]$, let

$$Y=\sigma X+\mu \tag{18.4}$$

so $Y\sim N\left[\mu,\sigma^2\right]$. Therefore, only the method of generating a standard normal distribution random number is introduced here.

Generating a Standard Normal Distribution Random Number with the Central Limit Theorem

The method of generating a standard normal distribution random number with the central limit theorem is also called statistical approximation.

Let $r_1, r_2, \ldots, r_N$ be independent normally distributed random numbers in interval [0, 1]. When N is relatively big, it follows from this theorem that $\sum_{i=1}^{N} r_i$ is a normally distributed random number. Therefore,

$$Y=\frac{1}{N}\sum_{i=1}^{N} r_i \tag{18.5}$$

also follows a normal distribution.

Since r_i is a random number normally distributed in interval [0, 1],

$$\mathrm{E}[Y]=\mathrm{E}\left[\frac{1}{N}\sum_{i=1}^{N} r_i\right]=\frac{1}{N}\sum_{i=1}^{N}\mathrm{E}[r_i]=\frac{1}{2} \tag{18.6}$$

$$\mathrm{D}[Y]=\mathrm{D}\left[\frac{1}{N}\sum_{i=1}^{N} r_i\right]=\frac{1}{N^2}\sum_{i=1}^{N}\mathrm{D}[r_i]=\frac{1}{12N} \tag{18.7}$$

that is, $Y\sim N\left[\frac{1}{2},\frac{1}{12N}\right]$. Then, normalized Y follows the standard normal distribution whose mean is zero and variance one, that is, $\sqrt{12N}\left(\frac{1}{N}\sum_{i=1}^{N} r_i-\frac{1}{2}\right)\sim N[0,1]$. Normally, $N\geq 6$. If $N=12$, then

$\sum_{i=1}^{12} r_i - 6 \sim N[0,1]$. This is a standard normally distributed random number generated according to the central limit theorem.

If $r_1, r_2, \ldots, r_N$ are independent uniformly distributed random numbers in interval $\left[-\frac{1}{2}, \frac{1}{2}\right]$, likewise, when N is relatively big,

$$Y = \frac{1}{N}\sum_{i=1}^{N} r_i \sim N\left[0, \frac{1}{12N}\right] \tag{18.8}$$

Normalized Y follows the standard normal distribution whose mean value is zero and variance one, that is, $\sqrt{12N}\left(\frac{1}{N}\sum_{i=1}^{N} r_i\right) \sim N[0,1]$. If $N=12$, then $\sum_{i=1}^{12} r_i - 6 \sim N[0,1]$. This is a standard normally distributed random number generated according to the central limit theorem.

Transformation Method

The computational equation for generating standard normal distribution random numbers using the transformation method is

$$x_1 = \sqrt{-2\ln r_1}\cos(2\pi r_2), \quad x_2 = \sqrt{-2\ln r_1}\sin(2\pi r_2) \tag{18.9}$$

where r_1 and r_2 are independent random numbers uniformly distributed in interval [0, 1]. Their characteristics are high accuracy, and x_1 and x_2 are independent standard normally distributed random numbers.

Density Approximation Methods

Statistical approximation can yield accurate normally distributed random samples when N is large, but it takes much more time for a computer to do it. To reduce the value of N, the following methods can be adopted to generate standard normal distribution random numbers.

Method 1: Let

$$x = \frac{r}{\sqrt{N}}\left[1 - \frac{a}{bN^2}\left(\frac{r^4}{N^2} - c\frac{r^2}{N} + d\right)\right] \tag{18.10}$$

then x is a standard normally distributed random number, where $a=41$, $b=13\,440$, $c=10$, $d=15$, and r is a uniformly distributed random number in interval [0, 1]. In (18.10), if $N=2$, satisfactory results can be obtained.

Method 2: This is also known as the Hastion rational approximation method. It generates standard normally distributed random samples by approximating the rational fraction to the inverse function of the normally distributed function. Let r be a uniformly distributed random number, and let

$$x = \begin{cases} y - \dfrac{a_0 + a_1 y + a_2 y^2}{1 + b_1 y + b_2 y^2 + b_3 y^3}, & 0 < y \le 0.5 \\ \dfrac{a_0 + a_1 y + a_2 y^2}{1 + b_1 y + b_2 y^2 + b_3 y^3} - y, & 0.5 < y < 1 \end{cases} \tag{18.11}$$

$$y=\begin{cases}\sqrt{-2\ln r}, & 0<r\leq 0.5\\ \sqrt{-2\ln(1-r)}, & 0.5<r<1\end{cases} \tag{18.12}$$

then x is a standard normal distribution random number, where $a_0=13.515517$, $a_1=0.802853$, $a_2=0.010328$, $b_1=1.432788$, $b_2=0.189269$, and $b_3=0.001308$.

18.2.2.3 Production of Arbitrarily Distributed Random Numbers

The exponential, Rayleigh, LS, Weibull, and lognormal distributions are often used in radar and communication systems. How can a computer generate these random numbers? As discussed above, given the random sequence uniformly distributed in interval [0, 1], the simple subsample of the given distribution can be obtained. Actually this process is done by transforming, through some mathematical method, the given uniformly distributed random sequence in interval [0,1] into a random sequence with a given distribution. As long as this given sequence in interval [0,1] passes the test, so can all the simple subsamples with any distribution produced through strict mathematical transformations.

The direct sampling method, also called the method of distribution function characteristics or inverse function, makes use of the characteristics of a distribution function to obtain the random sampling of the given distribution.

Theorem 18.1 Let the distribution function of random variable X be $F_X(x)$, and let $R=F_X(x)$, then R is the variable uniformly distributed in interval [0,1]. The procedure of direct sampling is illustrated next.

Example 18.1
Obtaining uniformly distributed random numbers in interval $[a, b]$ by direct sampling.

Solution
Let $X\sim \mathrm{U}[a,b]$, then when $a\leq x\leq b$,

$$F_X(x)=\int_a^x \frac{1}{b-a}\mathrm{d}x=\frac{x-a}{b-a}$$

Let

$$r=F_X(x)$$

where r is a random number uniformly distributed in interval [0,1], then the sampling equation is

$$x=a+(b-a)r \tag{18.13}$$

or

$$x=a+(b-a)R \tag{18.14}$$

with R a random variable uniformly distributed in interval [0,1]. Equation (18.13) or (18.14) is the sampling equation used to obtain the uniform distribution in interval $[a, b]$.

Example 18.2
Obtaining exponentially distributed random numbers by direct sampling.

Solution
Let $X \sim \mathrm{E}[a]$, that is, $p_X(x) = a\mathrm{e}^{-ax}$, $a > 0$, $x \geq 0$, then when $x \geq 0$,

$$F_X(x) = \int_a^x a\mathrm{e}^{-at}\mathrm{d}t = 1 - \mathrm{e}^{-ax}$$

Let r be a random number uniformly distributed in interval [0,1], and let $r = 1 - \mathrm{e}^{-ax}$, then

$$x = -\frac{1}{a}\ln(1-r) \tag{18.15}$$

so

$$x = -\frac{1}{a}\ln(1-R) \tag{18.16}$$

where $R \sim \mathrm{U}[0,1]$. Since $R \sim \mathrm{U}[0,1]$, $1-R \sim \mathrm{U}[0,1]$. Therefore, (18.15) or (18.16) can be rewritten as

$$x = -\frac{1}{a}\ln r \tag{18.17}$$

or

$$x = -\frac{1}{a}\ln R \tag{18.18}$$

Equation (18.17) or (18.18) is the sampling equation of the exponential distribution.

Example 18.3
Obtaining the random number of Rayleigh distribution by direct sampling.

Solution
Let $X \sim \mathrm{R}(\sigma)$, that is,

$$p_X(x) = \left(x/\sigma^2\right)\exp\left(-x^2/2\sigma^2\right), \quad x \geq 0$$

where σ is the parameter of the Rayleigh distribution. In radar, for example, if X is the output of the linear receiver, then σ is the mean square root of IF noise. When $x > 0$,

$$F_X(x) = \int_0^x p_X(t)\mathrm{d}t = 1 - \exp\left(-x^2/2\sigma^2\right), \quad x \geq 0$$

Let

$$R = F_x(x)$$

then

$$x=\sigma\sqrt{-2\ln(1-R)} \tag{18.19}$$

Similarly, (18.19) can be revised to

$$x=\sigma\sqrt{-2\ln R} \tag{18.20}$$

Equation (18.20) is the sampling equation of the Rayleigh distribution.

18.3 Simulation of Radar Data Processing Algorithms

18.3.1 Simulation of Target Motion Models

18.3.1.1 Motion Equation of Aircraft

In the case of targets like aircraft flying in the atmosphere, it is difficult to give an equation which accurately describes their motion, because of their complicated kinetic properties, atmospheric turbulence, and manned maneuvers. So, we only discuss some common target motion modes that are more complicated than the CV, CA, and CT models of Section 3.2.1.

S Maneuver

The S maneuver is shown in Figure 18.1. In the figure, T is the cycle of an S maneuver, and $\Delta\varphi$ is the deviation angle of the target's flight course from the main course.

Dive Maneuver

The dive is a maneuver that an aircraft takes in dive bombing. The motion model is simplified under some approximate assumptions, with the aircraft viewed as a particle with some rigid characteristics. The typical motion curve is shown in Figure 18.2.

Assumptions:

1. The deflection angle of an aircraft changes uniformly in the descending process, and the changes in angle are the same in size but opposite in direction when the pilot pushes and pulls the lever.
2. The aircraft is in balance when descending; the actions of gravity, lift, etc. are ignored, and what is considered is only the balance relation between the thrust force of the aircraft's engine and the resistance which is in a certain proportion to the aircraft's velocity. The aircraft's thrust forces in horizontal and vertical directions are proportional to its velocity.
3. The speeding up is completed the moment a dive starts, the throttle is fixed in the dive, and the thrust force of the engine is a constant value.
4. There is a fixed proportional coefficient q between the horizontal and vertical velocities created by the same amount of thrust force.

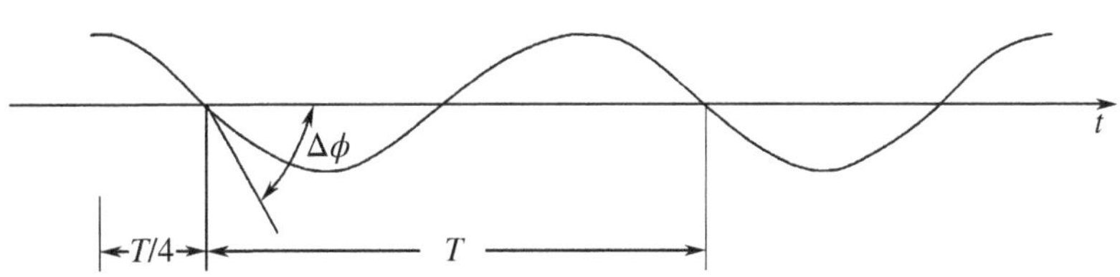

Figure 18.1 S maneuver

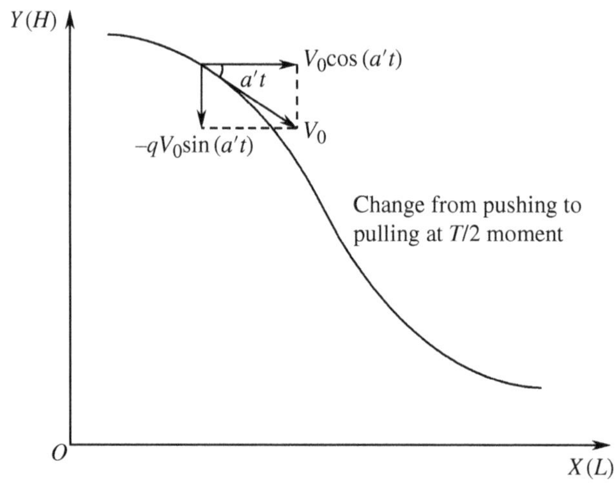

Figure 18.2 Motorized dive

The input of the model is the descending distance ΔL, the descending altitude ΔH, and the descending time T given by the mode editor, and its output is the simulated trajectory of the aircraft in dive and climb motions. The model is built in the 2D (axis L, which is in the same direction as the flight direction and axis H, which is vertical to the ground) longitudinal plane in the flight direction.

If the dive-and-climb process is divided into joystick push and pull modes (the aircraft changes from joystick push to joystick pull mode at moment $T/2$), the following constraints result from the assumptions above.

1. Joystick push process $(t \le T/2)$ $V_H(t) = -V_0 q \sin(a't)$, $V_L(t) = V_0 \cos(a't)$; joystick pull process $(t \ge T/2)$, $V_H(t) = -V_0 q \sin(a'(T-t))$; velocity in direction H $V_H = V_0 \cos[a'(T-t)]$, where a' is the absolute value of angular velocity and V_0 is the initial horizontal velocity in the dive and climb.
2. At moment $t = T/2$, the motion changes from joystick push to joystick pull, the descending altitude is $H/2$, and the descending distance is $L/2$.

Therefore, we get

$$\int_0^{T/2} (\sin a't) V_0 q \mathrm{d}t = \frac{|\Delta H|}{2} \tag{18.21}$$

$$\int_0^{T/2} (\cos a't) V_0 \mathrm{d}t = \frac{\Delta L}{2} \tag{18.22}$$

which are the initiation equations of dives, and a simultaneous solution yields

$$a' = \frac{4}{T} \arctan(q \times |\Delta H| / \Delta L) \tag{18.23}$$

$$v_0 = \Delta L \times a' / 2 \sin(a'T/2) \tag{18.24}$$

From constraint 1 follows the corresponding clock triggering equation

$$x(t) = x(0) + \int_0^t v_0 \cos(a't)\,\mathrm{d}t \tag{18.25}$$

$$y(t) = y(0) - \int_0^t q v_0 \sin(a't)\,\mathrm{d}t \tag{18.26}$$

where $t \le T/2$, $x(0)$ and $y(0)$ are the aircraft's coordinates in axes L and H at the initial moment, then

$$x(t) = x(T/2) + \int_{T/2}^t v_0 \cos[a'(T-t)]\,\mathrm{d}t \tag{18.27}$$

$$y(t) = y(T/2) - \int_{T/2}^t q v_0 \sin[a'(T-t)]\,\mathrm{d}t \tag{18.28}$$

where $t \ge T/2$, $x(T/2)$ and $y(T/2)$ are the aircraft's coordinates in axes L and H at time $T/2$.

Pitch-Up Maneuver

The pitch up is a maneuver which an air raider executes in loft bombing. In the pitch-up process, the flying velocity changes in both direction and magnitude. A typical pitch-up maneuver is the 180° arcing turn made by an aircraft in the vertical plane. As shown in Figure 18.3, the pitch-up maneuver starts at time t_1 and ends at time t_2.

18.3.1.2 Real-Time Track Creation

In system simulation, there may be several targets of interest, and the target fade-in-and-out frequency is high. Therefore, it is necessary to create tracks in real time.

First of all, a track parameter file needs to be set up for the targets in a simulation of the system according to the target environment requirements. These parameters include track initiation, segmentation of target motion types, maneuver types, etc. At the initialization stage in the simulation process, the file is read in, and the coordinates transformed in the meantime.

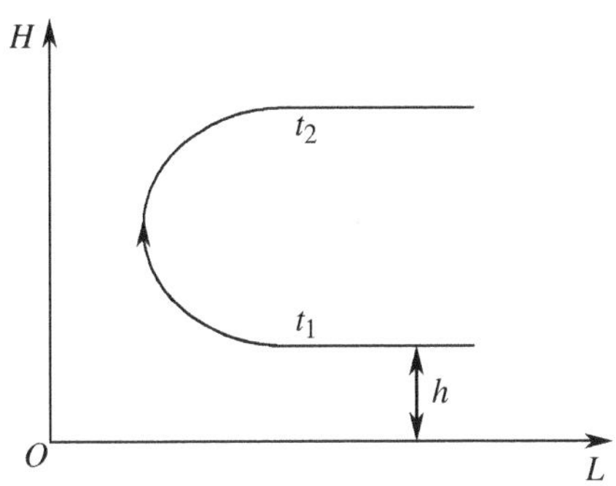

Figure 18.3 Pitch-up maneuver

At the following stages, track data is computed once per sampling interval T according to the system timer. Generally speaking, since several targets are dealt with, the motion type of each target should be determined for track segmentation management.

The creation of tracks in real time can help predetermine the initial target environment in the working airspace when the radar starts up. Then, in the simulation process, parameters are set to control the time sequence in which each flying target fades in and out flexibly and in real time, so as to change the number of targets in the airspace and their characteristics.

18.3.2 *Simulation of the Observation Process*

During each radar scan, if a target is detected successfully, the observations disturbed by the noise have to be simulated. Owing to the interruption from nature and the accuracy of instruments, the signals detected by the radar are not real values, but random variables with noise. The simulative method is to add zero-mean, white Gaussian noise errors to the accurate values, to generate the observation values of range, azimuth, and pitching angle. The assumptions of various noises are given as follows.

18.3.2.1 Range Noise

Suppose that the range ρ detected by the radar satisfies

$$\rho = \bar{\rho} + q_\rho \tag{18.29}$$

where $\bar{\rho}$ is the actual range and q_ρ is the range noise, satisfying

$$\begin{cases} \mathrm{E}\left[q_\rho\right] = 0 \\ \mathrm{E}\left[q_\rho^2\right] = \sigma_\rho^2 \end{cases} \tag{18.30}$$

The standard deviation σ_ρ is related to the pulse width τ, the signal-to-noise ratio S/N, and the velocity of light c. Therefore, the following assumption can be made:

$$\sigma_\rho = \frac{c\tau}{2\sqrt{2}} \cdot \frac{1}{\sqrt{S/N}} = \frac{c\tau}{2\sqrt{2}} \cdot \frac{\rho^2}{\rho_0^2\sqrt{(S/N)_0}} \tag{18.31}$$

where ρ is the detecting range, ρ_0 is the range at the initial tracking moment, S/N is the signal-to-noise ratio, and $(S/N)_0$ is S/N at the initial tracking moment. Thus, by extracting the subsample q_ρ that is subject to the $N[0, \sigma_\rho^2]$ distribution, the radar's simulated detecting range ρ can be obtained from (18.29).

18.3.2.2 Direction Cosine Noise

The measurement of the three-coordinate radar in the direction cosine coordinate system is discussed here. Suppose that the direction cosine $\alpha \neq \beta$ detected by the radar satisfies

$$\begin{cases} \alpha = \bar{\alpha} + q_\alpha \\ \beta = \bar{\beta} + q_\beta \end{cases} \tag{18.32}$$

where $\bar{\alpha}$ and $\bar{\beta}$ are the actual direction cosines, q_α and q_β are the cosine noises, respectively, in the two directions, satisfying

$$\mathrm{E}[q_\alpha] = 0, \quad \mathrm{E}\left[q_\beta\right] = 0 \tag{18.33}$$

$$\mathrm{E}\left[q_\alpha^2\right] = \sigma_\alpha^2, \quad \mathrm{E}\left[q_\beta^2\right] = \sigma_\beta^2 \tag{18.34}$$

The direction cosine noise is influenced by thermal noise and angle glint. Then,

$$\begin{cases} \sigma_\alpha = \sigma_{\text{thermal}} + \sigma_{\text{angle glint}} \\ \sigma_\beta = \sigma_\alpha \end{cases} \tag{18.35}$$

Suppose that

$$\sigma_{\text{thermal}} = \frac{\theta_0}{1.89\sqrt{2}} \cdot \frac{1}{\sqrt{S/N}} = \frac{\theta_0}{1.89\sqrt{2}} \cdot \frac{\rho^2}{\rho_0^2 \sqrt{(S/N)_0}} \tag{18.36}$$

where θ_0 is the half power width of the antenna beam, *S/N* the signal-to-noise ratio, ρ the range, ρ_0 the tracking range at the initial moment, and $(S/N)_0$ the *S/N* at the initial tracking moment. The variance of angle glint is described as

$$\sigma_{\text{angle glint}} = \frac{0.35L}{\rho} \tag{18.37}$$

where ρ is range, L is span.

If transformed into rectangular coordinates (x, y, z), then (ρ, α, β) can be described as $\alpha = \cos\varepsilon\cos\theta = x/\rho$, $\beta = \cos\varepsilon\sin\theta = y/\rho$, and $\rho = \sqrt{x^2 + y^2 + z^2}$, in which ε and θ are, respectively, the pitching angle and azimuth of the measured value. The measured value of azimuth and pitching angle can also be obtained by directly adding the theoretical value to the zero-mean, white Gaussian noise sampling.

Similarly, the simulation of the observation process can also be realized through the measurement equation of the filter. Take the linear system Kalman filter as an example. The measured value is $z(k) = \boldsymbol{H}(k)\boldsymbol{X}(k) + \boldsymbol{W}(k)$, where $\boldsymbol{H}(k)$ is the measurement matrix and $\boldsymbol{W}(k)$ the measurement noise sequence. If the system is nonlinear, the measured value can be obtained from $z(k) = \boldsymbol{h}[k, \boldsymbol{X}(k)] + \boldsymbol{W}(k)$, where the measurement noise can be assumed additive zero-mean white noise and $\boldsymbol{h}(\cdot)$ the nonlinear time-varying function matrix.

18.3.3 Tracking Filtering and Track Management

18.3.3.1 Filtering and Prediction Algorithm

Here we discuss determination of initial values, computation of initial covariance matrixes, and evaluation of statistical indexes. In this filtering algorithm, the target measurement matrix $\boldsymbol{H}$, initial covariance matrix $\boldsymbol{P}(0|0)$, and noise covariance matrix $\boldsymbol{R}$ are unknown. Corresponding equations in rectangular coordinates are given as follows.

Computation of the Target Measurement Matrix

The following equivalent measurement equations give the measurement position of the target:

$$x = \rho\alpha \tag{18.38}$$

$$y = \rho\beta \tag{18.39}$$

$$z = \rho\gamma \tag{18.40}$$

where $\gamma = \sqrt{1-\alpha^2-\beta^2}$. Then, the measurement is linear and the measurement matrix $\boldsymbol{H}$ is

$$\boldsymbol{H} = \begin{bmatrix} 1 & 0 & 0 & 0 & 0 & 0 \\ 0 & 1 & 0 & 0 & 0 & 0 \\ 0 & 0 & 1 & 0 & 0 & 0 \end{bmatrix} \tag{18.41}$$

In the case of nine dimensions, it is only necessary to add three more columns of zero elements to matrix $\boldsymbol{H}$.

***Computation of Measurement Covariance Matrix* R**

When using the equivalent measurement in a rectangular coordinate system, $x \neq y \neq z$ are random variables but are not independent. Now suppose that

$$x = \bar{x} + q_x \tag{18.42}$$

$$y = \bar{y} + q_y \tag{18.43}$$

$$z = \bar{z} + q_z \tag{18.44}$$

where q_x, q_y, and q_z are, respectively, the ranging errors in directions x, y, and z. Therefore, their covariance matrix is

$$\boldsymbol{R} = \begin{bmatrix} R_{11} & R_{12} & R_{13} \\ R_{21} & R_{22} & R_{23} \\ R_{31} & R_{32} & R_{33} \end{bmatrix} \tag{18.45}$$

where coefficient R_{ij} is

$$R_{11} \approx \alpha^2\sigma_\rho^2 + \rho^2\sigma_\alpha^2 \tag{18.46}$$

$$R_{12} = R_{21} \approx \alpha\beta\sigma_\rho^2 \tag{18.47}$$

$$R_{13} = R_{31} \approx \alpha\sigma_\rho^2 - \rho^2\frac{\alpha}{\gamma}\sigma_\alpha^2 \tag{18.48}$$

$$R_{22} \approx \beta^2\sigma_\rho^2 + \rho^2\sigma_\beta^2 \tag{18.49}$$

$$R_{23} = R_{32} \approx \beta\gamma\sigma_\rho^2 - \rho^2 \frac{\beta}{\gamma}\sigma_\beta^2 \tag{18.50}$$

$$R_{33} \approx \gamma^2\sigma_\rho^2 + \rho^2\left(\frac{\alpha}{\gamma}\right)^2\sigma_\alpha^2 + \rho^2\left(\frac{\beta}{\gamma}\right)^2\sigma_\beta^2 \tag{18.51}$$

Besides, the measurement covariance matrix $\boldsymbol{R}$ can also be computed with the method in (3.64) and (3.70).

Computation of the Filtering Initial Value

In the filtering algorithm, if the initial value with a better algorithm is given, there could be great influence on the convergence rate of filtering and better filtering estimates could be obtained. In simulation, the determination of the initial state vector and initial covariance matrix can be found in Section 3.2.3.

18.3.3.2 Multi-target Data Association Methods

There are many multi-target data association methods, and here a brief introduction is given to the system by using the nearest-neighbor algorithm.

In this method, first an association gate has to be set up around the predicted position of each track such that the probability of true observations falling in it is high, without allowing too many extraneous measurements, so as to reduce the computing complexity of the algorithm.

Let

$$\boldsymbol{S}_i = \boldsymbol{HPH}' + \boldsymbol{R} \tag{18.52}$$

denote the innovation covariance matrix of the ith track, and

$$\widetilde{\boldsymbol{Y}}_{ij}(k) = \boldsymbol{Z}_j(k) - \boldsymbol{H}\hat{\boldsymbol{X}}_i(k|k-1) \tag{18.53}$$

the distance between the ith track prediction and the jth observed value in the kth recursion. Then the statistical distance between the ith track and the jth observation can be computed as

$$d_{ij}^2 = \widetilde{\boldsymbol{Y}}_{ij}' \boldsymbol{S}_i^{-1} \widetilde{\boldsymbol{Y}}_{ij} \tag{18.54}$$

If $d_{ij}^2 < G_0$, then the ith track is associated with the jth measurement. Otherwise, they are unrelated. G_0 is an optional parameter.

18.3.3.3 Track Initiation and Termination

The most commonly used method of track initiation is the two-point initiation method, which associates the measurements of two successive frames. Let

$$\boldsymbol{Z}_j(k) = \begin{bmatrix} Z_{jx} \\ Z_{jy} \\ Z_{jz} \end{bmatrix} \tag{18.55}$$

be the jth observed value at time k, and

$$\boldsymbol{Y}_{ij}(k) = \begin{bmatrix} Y_{ijx} \\ Y_{ijy} \\ Y_{ijz} \end{bmatrix} = \boldsymbol{Z}_j(k) - \boldsymbol{Z}_i(k-1) \tag{18.56}$$

the distance between the jth measurement $\boldsymbol{Z}_j(k)$ at time k and the with measurement $\boldsymbol{Z}_i(k-1)$ at time $k-1$.

Thus, one can obtain the statistics as

$$d_{ij}^2 = \boldsymbol{Y}_{ij}{}'\boldsymbol{S}_{ij}^{-1}\boldsymbol{Y}_{ij} \tag{18.57}$$

and

$$\boldsymbol{S}_{ij} = 2\boldsymbol{R}_i + \boldsymbol{\sigma} T_{ij}^2 \tag{18.58}$$

where $\boldsymbol{R}_i$ is the observation covariance matrix computed according to the ith measurement at time $k-1$, $\boldsymbol{\sigma}$ is the covariance matrix of the target velocity vector $\boldsymbol{v}$ (zero mean), and

$$T_{ij} = T_{cj}(k) - T_{ci}(k-1) \tag{18.59}$$

is the time difference in recording between the jth measurement at time k and the ith measurement at time $k-1$.

If $d_{ij}^2 < G_0$, the jth measurement at time k successfully associates with the ith measurement at time $k-1$. By using the nearest-neighbor method, the configuration relation between the measurements at the two moments can be derived. Associated measurements form the initiation of a new track, and those that failed to form tracks are treated as isolated measurements, waiting for the data of the next frame. The isolated measurements, if failing to be initiated in the next frame, will be discarded.

In the multi-target tracking case, besides data association and track maintenance, track termination is another concern. Once a target goes beyond the sensor's detection range, the tracker has to decide to eliminate the files of the redundant tracks accordingly to terminate the tracking process. Current multi-target track termination technologies include the sequential probabilistic ratio test, tracking gate method, and Bayesian track termination method. For details, see Chapter 11.

18.3.3.4 Statistical Evaluation of Errors

In order to compare the performance of various tracking filters, some commonly used evaluation standards have to be given. Here, two main groups of methods are introduced.

The first group is based on "time average." The errors of tracking filter position and velocity are defined as follows.

1. Position error

$$\sigma_x^2 = \frac{1}{N}\sum_{i=k-N+1}^{k} [\hat{x}(i|i) - x(i)]^2 \tag{18.60}$$

where $x(i)$ is the position at the moment of the ith step, $\hat{x}(i|i)$ is the estimated value of position $x(i)$, and N is the number of sampling points.

2. Velocity error

$$\sigma_{\dot{x}}^2 = \frac{1}{N}\sum_{i=k-N+1}^{k} \left[\hat{\dot{x}}(i|i) - \dot{x}(i)\right]^2 \tag{18.61}$$

where $x(i)$ is the velocity at the moment of the ith step, $\hat{x}(i|i)$ is the estimated value of velocity $x(i)$, and N is the number of sampling points.

It follows from the definition above that the smaller σ_x^2 and $\sigma_{\dot{x}}^2$ are, the better the filtering effect will be, and vice versa.

The second group is based on "ensemble average." That is, the estimation of the mean value and variance of position $x(k)$ and velocity $\dot{x}(k)$ at a certain time k.

Denote by $x(k)$ and $\dot{x}(k)$ the position and velocity at time k, by $\hat{x}_i(k|k)$ the estimate of position $x(k)$ in the ith simulation at time k, and by $\hat{\dot{x}}_i(k|k)$ the estimate of velocity $\dot{x}(k)$ in the ith simulation at time k. Then, the mean of position error at time k is estimated as

$$\Delta\bar{x}(k) = \frac{1}{M}\sum_{i=1}^{M} \Delta\hat{x}_i(k) \tag{18.62}$$

where $\Delta\hat{x}_i(k) = x(k) - \hat{x}_i(k|k)$.

The mean of velocity error at time k is estimated as

$$\sigma_{\Delta x(k)}^2 = \frac{1}{M-1}\sum_{i=1}^{M} [\Delta\hat{x}_i(k) - \Delta\bar{x}(k)]^2 \tag{18.63}$$

$$\sigma_{\Delta\dot{x}(k)}^2 = \frac{1}{M-1}\sum_{i=1}^{M} \left[\Delta\hat{\dot{x}}_i(k) - \Delta\bar{\dot{x}}(k)\right]^2 \tag{18.64}$$

where M is the number of Monte Carlo simulation runs, usually $M = 30 - 50$.

As can be seen, error variances $\sigma_{\Delta x(k)}^2$ and $\sigma_{\Delta\dot{x}(k)}^2$ lessen as error means $\Delta\bar{x}(k)$ and $\Delta\bar{\dot{x}}(k)$ approach 0, which indicates a higher accuracy of filtering (and vice versa).

18.4 Simulation Examples of Algorithms

This section deals with the simulation of several radar data association algorithms based on the methods discussed in the previous sections. The performance of these algorithms will be used to evaluate the methods for simulating the radar data processing process, because the association between measurements and tracks is the key to the multi-target tracking technology.

The PDA assumes that all the measurements in the association region of a certain target have originated from the target or random clutter. This algorithm yields the desired results when tracking a single target in dense clutter. In the multi-target environment, however, when the measurements of other targets appear persistently in the association region of a certain target, false associations are likely to arise, which in turn results in track loss.

The JPDA algorithm was thus put forward by Bar-Shalom to solve this problem. Accounting for the probability of several measurements from other targets being in the association region of one target, the JPDA arranges and combines all targets and measurements, from which feasible joint events are selected for joint probability computation. The problem above can thus be solved. However, this algorithm is complicated and has excessive computational requirements. It may give rise to combination explosion as the number of targets increases in the decomposition of the validation matrix. Therefore, it is difficult for the JPDA to be implemented in engineering.

Many simplified versions of this algorithm were proposed for its easy engineering implementations, including an empirical probability computation equation in Ref. [199], a suboptimal JPDA algorithm in Ref. [200], and a DFS approach in its simplified form in Ref. [201]. These models simplify the association probability computation in the JPDA. For their basic filtering equations, see Section 8.5.4.

Next, several typical multi-target scenarios will be set, and a comparison made between the three algorithms above in terms of root mean square position error, time consumption per cycle, and correct association probability. Further, the JPDA and MHT will be compared to highlight the difference between the simplified and original algorithms.

Three typical simulation environments are considered here.

Environment 1: Track two cross-moving targets, whose initial states are, respectively,

$X_1 = [-29\,500\text{ m}, 400\text{ m/s}, 34\,500\text{ m}, -400\text{ m/s}]'$

$X_2 = [-26\,250\text{ m}, 296\text{ m/s}, 34\,500\text{ m}, -400\text{ m/s}]'$.

Environment 2: Track two targets moving in parallel, whose initial states are, respectively,

$X_1 = [-22\,000\text{ m}, 150\text{ m/s}, 23\,000\text{ m}, -150\text{ m/s}]'$

$X_2 = [-22\,600\text{ m}, 150\text{ m/s}, 23\,000\text{ m}, -150\text{ m/s}]'$.

Environment 1

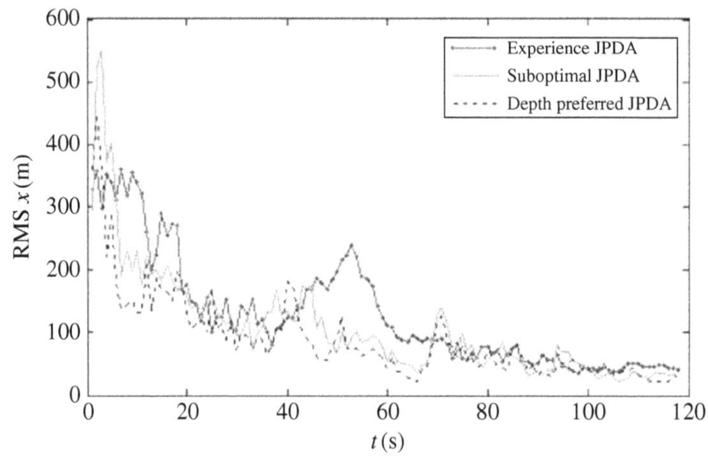

Figure 18.4 A comparison of the position error of target 1's RMS on axis x

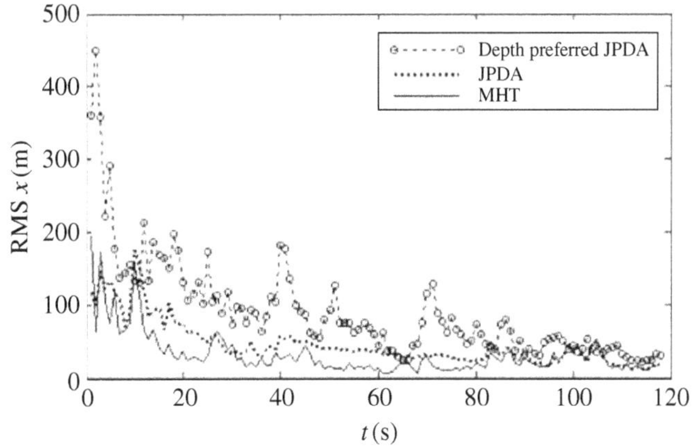

Figure 18.5 A comparison of the position error of target 2's RMS on axis x

Environment 2

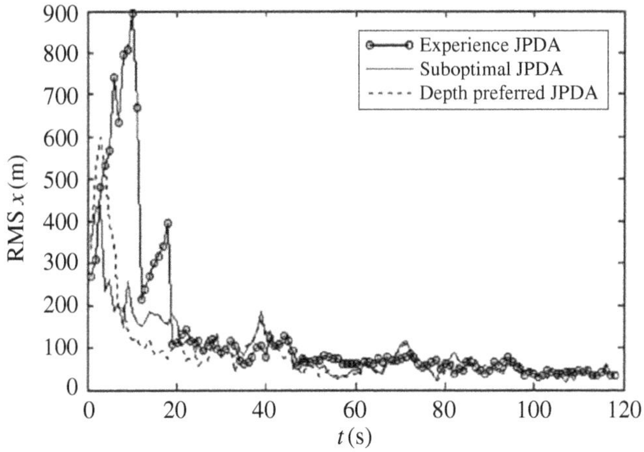

Figure 18.6 A comparison of the position error of target 1's RMS on axis x

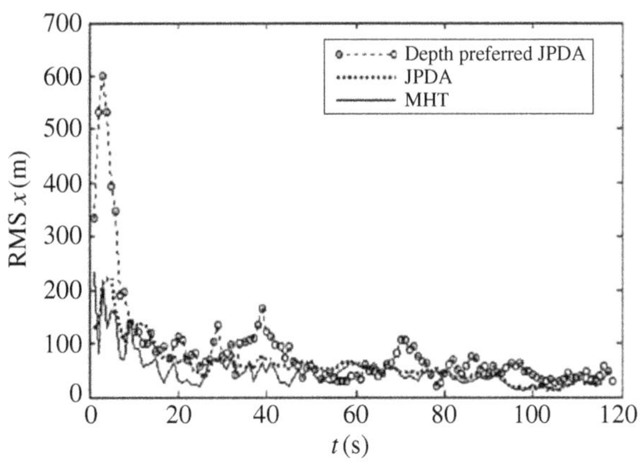

Figure 18.7 A comparison of the position error of target 3's RMS on axis x

Environment 3

Track 10 targets simultaneously, whose initial states are, respectively:

$X_1 = [-15\,600\text{ m}, 500\text{ m/s}, 13\,000\text{ m}, 120\text{ m/s}]'$;
$X_2 = [-7600\text{ m}, 150\text{ m/s}, 30\,000\text{ m}, 250\text{ m/s}]'$;
$X_3 = [-21\,000\text{ m}, 350\text{ m/s}, 23\,000\text{ m}, -50\text{ m/s}]'$;
$X_4 = [-10\,500\text{ m}, -200\text{ m/s}, 24\,500\text{ m}, 200\text{ m/s}]'$;
$X_5 = [-18\,250\text{ m}, 0\text{ m/s}, 26\,500\text{ m}, -300\text{ m/s}]'$;
$X_6 = [-14\,600\text{ m}, -150\text{ m/s}, 9000\text{ m}, 50\text{ m/s}]'$;
$X_7 = [-29\,500\text{ m}, 400\text{ m/s}, 34\,500\text{ m}, -400\text{ m/s}]'$;
$X_8 = [-33\,250\text{ m}, 100\text{ m/s}, 38\,500\text{ m}, -200\text{ m/s}]'$;
$X_9 = [-26\,250\text{ m}, 296\text{ m/s}, 34\,500\text{ m}, -400\text{ m/s}]'$;
$X_{10} = [-23\,250\text{ m}, 296\text{ m/s}, 24\,500\text{ m}, 400\text{ m/s}]'$.

In the process of simulation, suppose that the average number of clutters in the gate is $m = 3$, the range measurement error is $\sigma_r = 100$ m, the angle measurement error is $\sigma_\theta = 0.01$ rad, the sampling interval is $T = 1$ s, the duration of the target's motion is 120 s, and the simulation time is 50 s.

The findings of the simulation, as shown in Figures 18.4–18.10, are analyzed below.

Position Error of Root Mean Square

Figures 18.4–18.7 are, respectively, the curves of RMS position errors of targets 1 and 2's filtering results in the cross and parallel environments with use of the JPDA, MHT, and three simplified JPDA algorithms. As shown, the MHT has the highest tracking accuracy, followed by the JPDA and the DFS sequentially. Among the simplified JPDA algorithms, the DFS converges fastest and functions better in tracking than the suboptimal and the empirical in turn.

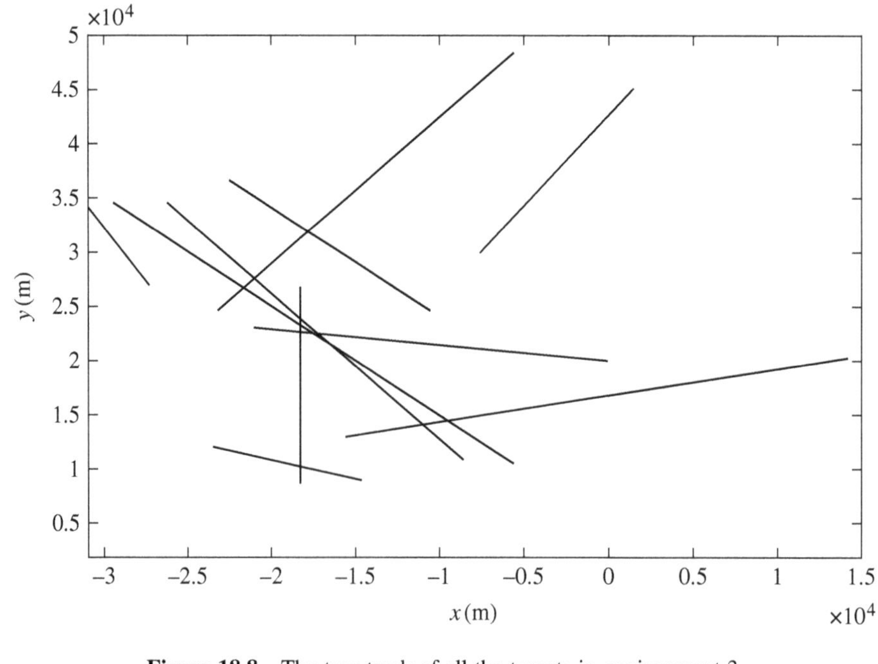

Figure 18.8 The true track of all the targets in environment 3

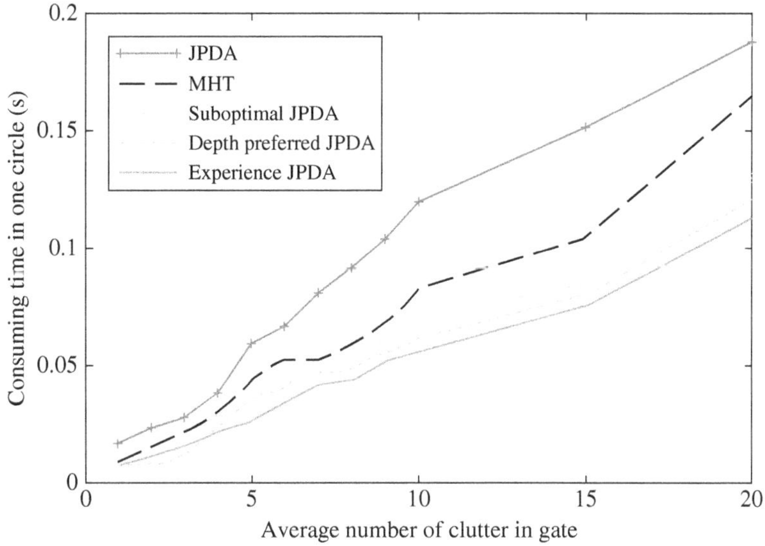

Figure 18.9 A change curve of the time consumed by the algorithm varying with average clutter number

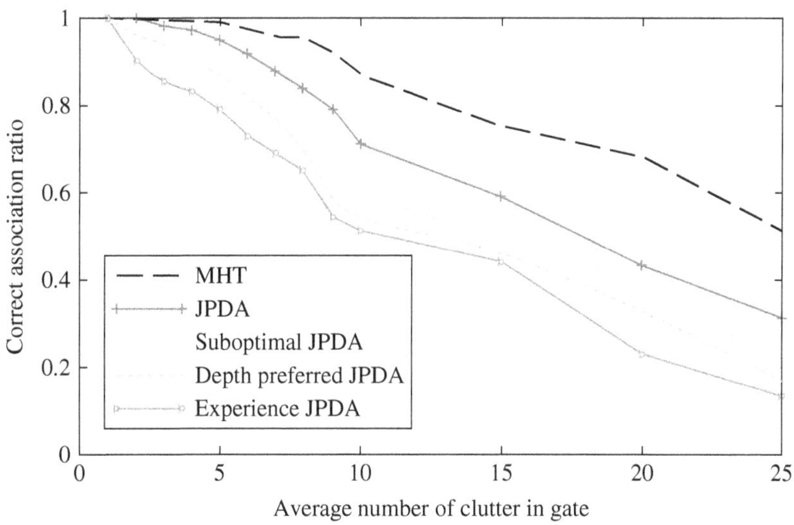

Figure 18.10 A change curve of the correct association probability of the algorithm varying with average clutter number

As indicated in Figure 18.4, the empirical has the largest error when the targets are in cross motion, while the other two have smaller error and are steady in the whole tracking process.

A conclusion can be drawn from the above: the MHT yields the best tracking result, the JPDA ranks second, and the DFS and the suboptimal give similar tracking results, which are better than those of the empirical.

Speed of Algorithm

Figure 18.9 shows a curve of the time consumed by the algorithms varying with the average clutter number. As can be seen, the time consumption of the JPDA grows fastest with this number, followed by that of the MHT, which in turn grows significantly faster than that of the suboptimal. The DFS consumes less time than the empirical when this average number is smaller than 4, but is more time consuming than the latter when it is larger than 4.

Therefore, the JPDA is the most time consuming, succeeded by the MHT, which consumes much more time than the suboptimal. The DFS consumes least time when the number of targets being tracked is small, while the empirical ranks last in time consumption when this number and that of clutters are large.

Correct Association Probability

Figure 18.10 is the curve of the correct association probability of the algorithms varying with the average clutter number, which provides the following results. In dense environments, the correct association probability of the MHT drops slowest with the growth in number of clutters, followed by that of the JPDA, with that of the empirical JPDA ranking last. This probability of the DFS is higher than that of the suboptimal when the number of clutters is smaller than 6, while it is lower than that of the latter but higher than that of the empirical when this number is larger than 6.

As suggested by the results of the various simulations, the MHT is more accurate in tracking, and has a higher correct association probability when targets are dense, but does not have a good real-time performance; the tracking accuracy and correct association probability of the JPDA are lower than the MHT but higher than the simplified JPDA algorithms, and it has a bad real-time performance.

Among simplified JPDA algorithms, the DFS is quick in convergence and has a steady tracking effect. It has a faster computing speed and a higher correct association probability than the other two when the number of clutters is smaller than 4. The convergence speed of the suboptimal is slightly lower than that of the DFS, but the two algorithms are similar in stability and perform better than the empirical. The correct association probability of the suboptimal is slightly lower than that of the DFS when the number of clutters is smaller than 4, but higher than that of the other two algorithms when the number is larger than 4. Its deficiency is that its speed is lower than that of the other two and it is not easy to implement in engineering. The stability, speed of convergence, and correct association probability of the empirical are all lower than those of the other two algorithms, but the speed of the algorithm is higher. When the number of clutters is smaller than 4, its speed is slightly lower than that of the DFS, but much higher than those of the other two when the number is larger than 4.

Several factors may account for this:

1. The empirical uses the essential characteristics of the JPDA to compute association probabilities, so its algorithm is direct and simple. However, the sum of the association probabilities of all the tracks obtained with the algorithm is not 1, so incorrect echoes tend to be over-weighted by the empirical probability.
2. The suboptimal, accounting for partially joint events, has a more reasonable association probability and a higher accuracy. Meanwhile, it is for this reason that the algorithm is comparatively complicated.
3. The simplified form of the DFS, computing the association probability directly, can be as effective as the JPDA when the density of targets is moderate, and it is quick in computation, but the complicated equations of its simplified form have excessive requirements for the memory capacity of the system when targets are dense.

The following conclusion can be drawn from the simulation and theoretical analyses. The tracking accuracy and correct association probability of the MHT are higher than those of the JPDA, and both MHT and JPDA have a bad real-time performance. The tracking accuracy and correct association probability of the JPDA are both higher than those of the simplified JPDA algorithms.

In the three simplified JPDA algorithms, when the number of clutters is not large, the tracking accuracy and correct association probability of the DFS are better than those of the other two. Besides, the DFS has a good real-time performance and is easy to implement in engineering; the suboptimal and the empirical can also satisfy the requirement of engineering implementation, but the tracking accuracy and correct association probability of the suboptimal are higher. When the number of clutters is large, the suboptimal has a higher correct association probability but a poor real-time performance, and it is hard to implement in engineering. The two properties of the DFS are moderate. The empirical has a good real-time performance and is easy to implement in engineering, but its correct association probability is the lowest among them.

This section studies the use of computer simulation technology to analyze and evaluate the problems in radar data processing, and gives examples of the simulation process of several data association algorithms in multi-target tracking.

18.5 Summary

This chapter systematically introduces system simulation technology, so that readers can simulate the algorithms introduced in previous chapters by combining the simulation and the radar data processing technology, and evaluate their advantages and disadvantages in actual engineering applications.

Section 18.2 introduces the basic knowledge of the system simulation technology, with emphasis on the production of random numbers in the Monte Carlo simulation. Section 18.3, the focus of this chapter, discusses the simulation of some radar data processing algorithms. Based on the introduction of the three most basic motion models (CV, CA, and CT models), this section introduces another three maneuver models which occur frequently in aircraft flight: S, dive, and pitch-up maneuvers. Then, it discusses the simulation technology in the observation process. In addition, it makes an introduction to several key issues in track filtering and track management: filtering prediction algorithms, multi-target data association techniques, track initiation and termination, and statistical evaluation of errors. Section 18.4 gives an example of the simulation process.

The main framework of system simulation is presented in this chapter. Still, many problems remain to be further explored. These will be the focus of our further research efforts.

19

Practical Application of Radar Data Processing

19.1 Introduction

Radar data processing is a process of estimating the track of a target and calculating its position with the data provided by radars. In real application systems, the ultimate purpose is not estimating target tracks, but taking advantage of these data to make judgments and take required actions. The radar data processing technology is widely used. Its typical civilian uses include air traffic control (ATC) and maritime surveillance (MS), while its military uses can be found chiefly in air defense, fire control, and missile interception and guidance. Figure 19.1 shows the incorporation of this technology into other systems

Radar data processing technology functions differently in different application systems. For instance, in ATC it is used to carry out functions such as route control, approach control, conflict warning, collision avoidance, and spacing adjustment and measure. In MS, it is used to calculate precisely a ship's course and speed, and to plot the conflict routes quickly and decide proper evasive maneuvers to ensure safe navigation. In defensive applications, the typical functions include early warning detection, threat estimation, weapon assignment, fire control, and so on.

This chapter mainly covers the application of the radar data processing technology in some typical conditions and focuses on the structure of radar data processing and its realization.

19.2 Application in ATC Systems

19.2.1 Application, Components, and Requirement

The basic requirement of the civil aviation air control system is to transport passengers and freight from one place to another safely and in an orderly fashion, which is realized through a service system – ATC. Though emerging as early as in WWII, radar was not developed into a kind of

Radar Data Processing with Applications, First Edition. He You, Xiu Jianjuan, and Guan Xin.

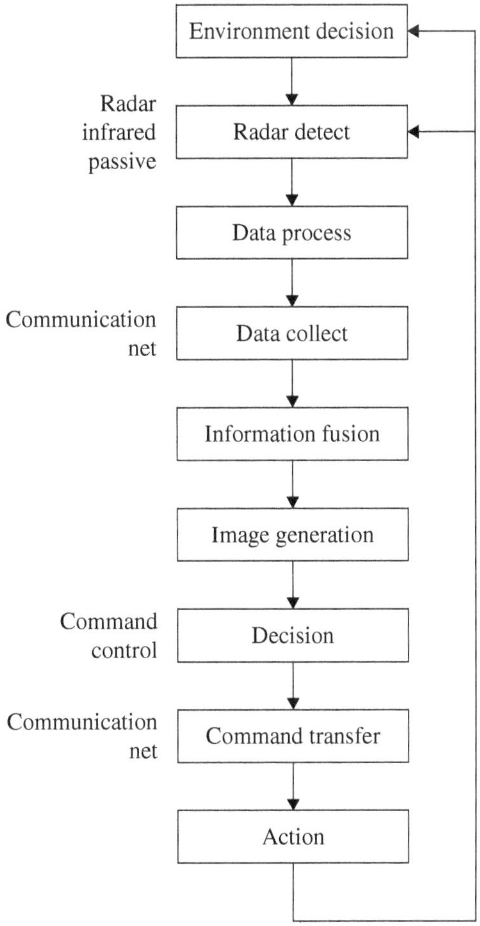

Figure 19.1 Relation between radar data processing and other systems

ATC until the mid- to late 1950s. It helps air traffic controllers to monitor airborne aircraft in almost all weather flight states by precisely positioning and identifying every aircraft in the air.

Common ATC radars fall into two types: primary surveillance radar (PSR) and secondary surveillance radar (SSR). Sometimes SSR is adherent to PSR, forming an integrated ATC radar. An ATC system is mainly composed of four parts: navigation equipment, surveillance and control equipment, communication equipment, and personnel. Its goal is to guarantee safety, efficiency, and orderliness in air traffic. In other words, it aims to reduce service cost and improve service quality through utilizing air traffic resources reasonably, reducing delays and waiting times in schedules, and selecting appropriate, fuel-saving lines.

1. *Navigation equipment.* This enables the aircraft to fly along the specified route, which is realized by locating certain elaborately preset geological positions with radio signals, and by the pilot sending the time and altitude when passing each of these positions to the ground to check whether the flight complies with the plan or not.

2. *Surveillance and control equipment.* This is used to modify an aircraft's deviation from the specified air route, avoiding collision and scheduling aircraft flow. Information about location, course, speed, and properties is mainly provided by PSR and SSR.
3. *Communication equipment.* This consists of air–land–air voice lines, through which controllers insert corrective actions into aircraft controls, while pilots report flight speed and altitude and meteorological conditions to the controllers. The current ATC equipment realizes the complete data fusion through different sensors (multiple-radar structures), computers, and consoles, as a result of which the ATC system becomes a multi-radar data processing system with multiple factors and levels.
4. *Personnel.* These are pilots and air traffic controllers. Pilots are responsible for keeping the aircraft along the planned route and performing controllers' orders. Controllers monitor flight information of the airborne aircraft to identify dangers and decide the most appropriate ways to deal with them. The modern ATC system emphasizes automation and intelligence, while the roles of human beings and human–computer interfaces in this system are still factors to be reckoned with.

19.2.2 Radar Data Processing Structure

First, the coordinates of a target (ρ, θ, φ) are given by radar measurement pretreatment (e.g., the real-time modification of observations' systematic errors, false alarm rate reduction, fixed track measurement deletion, etc.), then they are transmitted and processed in standard formats. The tracks obtained from coordinate transformation and tracking – which are assigned the attributes and altitude of the target as well as unassociated measurements – will be shown to operators on plane position indicators and digital data displays.

To establish a processing structure suitable for an ATC center, the following features should be considered.

1. *Systemic reliability and availability.* When a minor fault occurs, the whole system must be guaranteed to work normally (i.e., the system is able to be in the exception protection state). When a serious fault occurs, the system should also be guaranteed to work in a lowered-performance mode (i.e., the system should be switched to a reliable working state, although its performance has been lowered). When a local fault occurs, the system should be capable of recombining the subsystems.
2. *Systemic automation level.* Caution should be exercised when choosing the functions to be automated, in case the operability degrades and the whole system becomes unnecessarily complicated.
3. *Systemic operational modularization.* The system design should meet the requirements of different ATC systems, which include, for instance, route control conducted in different traffic areas, control in terminal maneuver area, or hybrid control.
4. *Systemic extensibility.*
5. *Systemic feasibility and cost.*

In order to improve an ATC system when the air traffic increases – to enhance its safety and efficiency – it is necessary to integrate data from different radars and control centers. Specifically, radar data processing usually requires many radars' track capability [15, 56, 406, 407], and information about the development of flight plans needs to be exchanged among different centers.

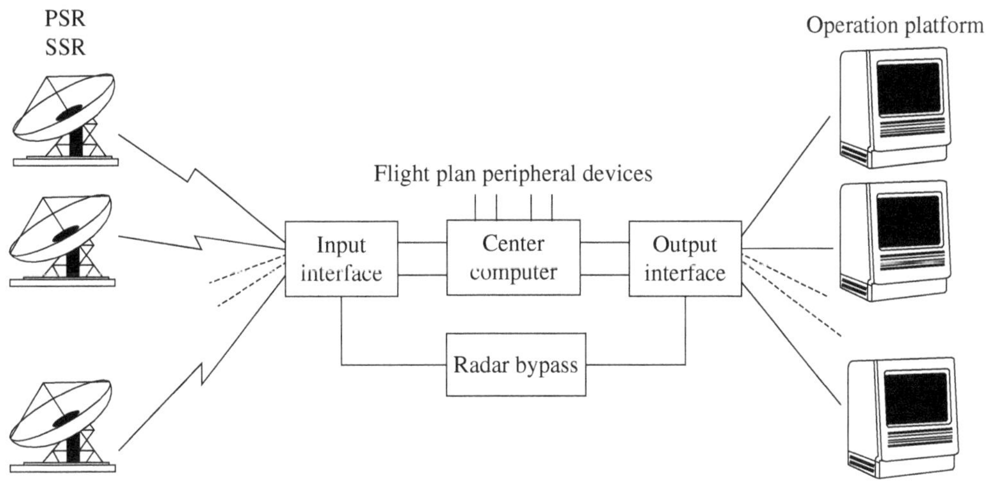

Figure 19.2 Centralized structure of ATC

The structure of a radar data processing system in the centralized ATC system is shown in Figure 19.2. In this system, the general mainframe computer conducts radar data processing and flight plan processing, including coordinate transformation and mono- or multi-radar tracks. The input interface pre-treats radar measurements and also serves as an input data cache. Generally, a specific minicomputer with strong input/output processing performance is needed. In addition, another general minicomputer is used to control the data exchange with the console (i.e., to transmit information – track, flight plan data, etc.), to preprocess data, to make data compatible with the features of indicators and other peripherals, and to implement a human–computer interface.

The measurements recorded by radar stations are sent to the ATC center, where the central computer processes input/output data, radar data, and flight plans to display them in the form of graphs and data. To achieve target tracks, the controlled airspace is divided into several sorted grids, to which the target measurements correspond. In each grid, there is one sensor tagged with "main station," which means that when an aircraft is in this grid, it is right to consider tracking the measurements provided only by this sensor. One standby radar is set in each grid that will function in place of the "main station" out of work, lest the system performance should decline or fail. Similarly, there are also track sorting grids, which overlap with track grids, simplifying the process of measurement-to-track association [126, 199, 408–410]. Classified grid techniques can easily achieve simplified multi-radar tracks: it is unnecessary for the covering regions of all radars to overlap to a large degree, but the overall coverage of the system is much larger than the sum of that of individual radars. As measurements provided by different radars of adjacent grids are used, the system tracks are updated at different rates. The extrapolation of tracks is conducted with the track velocity information.

The main function of radar data processing of the ATC system in the terminal maneuvering area is single-radar tracking, including automatic track initiation [411–413], computation and updating of track quality [54,60], measurement–track association, smoothing and prediction of α–β, and maneuver detection. Both track initiation and termination must comply with the *m*/*n* logic.

19.2.3 ATC Application

As civil aviation booms in China, the increasing number of aircraft sometimes jams an airspace over a certain period of time, which could disable the airspace management center. Thus, ATC is applied to address the problem. ATC systems have already become standard devices for modern civil aircraft, which can effectively warn the pilot of coming flight conflict and provide suggestions about collision avoidance [414–418].

Through collecting information such as radar data, ADS data, meteorological information, and flight plan data, air traffic flow management (ATFM) can be conducted to make full use of ATC capacity and to ensure safe and orderly air traffic with fewer delays and congestions. Radar data, as an important data source for ATFM, is the basis for this. ATFM can be better conducted only on the basis of precise radar data. At present, all management centers in China apply single-radar data processing systems, which are limited in monitoring range and low in accuracy, and wherein data loss may occur during the handover between two management centers. Multiple-radar data processing systems (MRDPSs) are the best option to provide a continuous display of air conditions to ensure seamless, multiple-radar coverage in the entire airspace and flight safety. Multiple-radar networking and data fusion not only extend the radar monitoring range to the entire airspace that all radars cover, but also enhances the radars' target monitoring quality and the reliability of the system itself.

The ATC command monitoring system (ATCCMS), based on information integration, has been proposed to meet the requirements for enhancing safety in ATC. The ATCCMS can monitor ATC operations and provide warnings and quality information about management decisions to ATC and CAA (civil aviation administration) departments at all levels, by integrating all the current ATC information and effectively processing basic data such as voice calls, radar data, and flight plans. This system is divided into four parts: basic data sources, operational subsystems, management decision subsystems, and supporting environments.

19.2.3.1 Basic Data Source

The basic data source includes radar signals, radio signals, flight telegrams, and such auxiliary flight information as meteorological and navigation information.

Format Conversion of Radar Data

Currently, the ATC radars imported by CAAC are of great variety and origin: ranging from Raytheon, Westinghouse, and Telephonics in the USA; Alenia in Italy; Thomson in France; to Toshiba and NEC in Japan. They converge at the information center through specialized digital data network (DDN) lines on the basis of the high-level data link control (HDLC) protocol, with electrical interface following the RS332 or RS422 standard and CCITTV24/V28 criterion. The data are mostly in three kinds of standard datagram format: ASTERIX, MPII, and CD-2. They are greatly different in radar data format, therefore a standard format should be designed first in order to achieve data exchange between different radars and the fusion of these data.

These data are received in real time by several synchronous communication cards configured in the radar data receiver according to corresponding protocols and interface criteria. Then they are converted to IP messages transmitted through the TCP in the Ethernet. The multiple threading method is used in the radar data processing server to set up for each received radar data diagram a single radar processing thread, through which the SOCKET service is established for data reception. For data stocked in the SOCKET buffer in real time and often containing multiple-frame HDLC data blocks, they are first saved and an index is built according to the reception time. The resulting radar data file can be used in data analysis and radar image replay, before the data pretreatment stage.

Data Preprocessing

First, process and check the data block's marker, address, and control fields, then interpret and convert datagrams based on their type. The key point here is considering error handling. Errors are bound to occur in all radar signals in the process of data generation and network transmission, leading to drops in radar data quality. These data errors include discordance between the declared and actual length of data frames, emergence of non-standard datagram heads, and byte loss in datagram tails. Therefore, in terms of hardware, information processing centers commonly adopt two-way signal transmission, real-time monitoring, and auto-switch methods to help the system receive data of higher quality. In terms of system software, the data block with errors may be modified in line with different situations, or this data frame may be discarded directly, and meanwhile a quality report is established for subsequent analyses to guarantee high system stability.

The final step of multi-radar data pretreatment is to form a standard internally unified format in the system. Information in different radar data is expressed in different ways. For example, track position can be shown both in polar coordinates and Cartesian coordinates. Therefore, the received track information is described by the virtual base class in the C++ frame, and the member variables consist of radar ID, receiving time, quality, state, coordinates (x, y, z), speed, acceleration of tracks, Mode-3/A and Mode-C codes, and so on, which are shown in SI units.

The data expression of every radar signal type is the subclass derived from the base class. Then, the special operations are defined according to the characteristics of the radar datagram to initialize and read member variables. For information that does not exist or cannot be provided in a telegram, the corresponding member variable will be set to zero or a special value, which thus guarantees a unified data format in subsequent processing.

19.2.3.2 Functional Subsystem at the Operational Level

The operational part of ATCCMS can be divided into six function subsystems: ATC voice recognition and processing, radar data processing, correlation of flight plans and radar data, correlation of control instructions and flight states, track pre-estimation and crash detection, and graphical user interface (GUI).

Voice Recognition and Processing

VR/LP realizes the semantic conversion of controllers' voice signals to control commands.

Radar Data Processing

The original radar signals processed by the front-end data processor, plus SSR transponder ID code and navigation data (e.g., speed and course), will be sent to the GUI to display radar images after data format interpretation.

1. *Coordinate transformation.* Data from various radars correspond to their own various reference frames, and are expressed in various ways. So, before track association and fusion, all the data should be converted in real time into a unified coordinate system, which is usually the Cartesian coordinate system centered at a certain radar station or another data processing center. Currently, the stereographic projection method is usually adopted to transform coordinates. However, this has some drawbacks. For one thing, errors may be introduced into measurements when projections are conducted, because high-order proximity is used to increase precision, while the earth is ellipsoidal rather than spherical. For another, it distorts data: for example, stereographic conformal projection can guarantee that the azimuth instead of the range remains undistorted, as a result of which system errors are not constant but related to measurements. Another method in common use is the geographical coordinate transformation

method, which adopts geocentric coordinates as the unified coordinate system for high-precision coordinate transformation.

2. *Time registration.* This mainly means unifying tracks measured by various synchronous or asynchronous radars at different moments to the same moments on the time axis.

   First, a unified time benchmark is set up – usually in such a manner that each radar station and information processing center adopts the GPS time system. Then, GPS time stamps are added to the received track information. After these two times are compared, the data communication delay is defined precisely. If there is no GPS time stamp at the radar data output for the moment, the time delay may be defined via average estimation.

   Most radars send track information of a sector after this sector is completely scanned, and some time is needed for data recording and track processing. Consequently, the moment of track output has a delay compared with the moment at which the track was detected. This delay is usually validated as a fixed system parameter through average estimation of measurements.

   Finally, the tracks in one processing period are unified to a systematic period through methods such as linear interpolation/extrapolation and modified Taylor expansion methods. Sometimes, the virtual integration method is applied, which integrates tracks to an "inner track updating time" related to the location of observation area.

3. *Radar head processing delay compensation and due north registration.* Radar head signal processing, data records, and single-radar tracks possess a certain delay, which leads to a track data output time lag compared with the time when the radar antenna beam irradiates the target. Special methods should be applied to modify this delay. Generally, the main part of the delay can be regarded as the system constant of the radar heads and single-radar tracks, and generally a one-time modification is enough – it is unnecessary to make real-time compensations. In addition, the real-time compensation of the processing delay's random shifts also needs to be coped with, but the value of this delay is so minor that it can be regarded as a random error – not to be considered in the process of temporal and spatial registration, but dealt with in the process of fusion.

   Different from geographical north, the nominal due north of ATC radars not only has system error but also random drift. Therefore, when it is transformed to unified coordinates of control centers, the north bias calibration is necessary. In general, the constant part of the difference between the standard north of a radar and geographical north can be regarded as a systemic constant, needing only one-time modification rather than real-time compensation. The fluttering part of the north deviation can be considered as a component of radar angle measurement error, thus special compensation is unnecessary in the process of temporal and spatial registration. After continuous track coordinates' processing and time registration, compensation for the drift in the north deviation amounts to real-time registration with the same target detected by all radars as the ID point.

4. *Error registration.* Random and systemic errors occur in all radar target measurements, so there is deviation between the tracks of the same target from different radars. The process of correcting this error is called "error registration." Random errors arise mainly from random observation noise and unintentional maneuvers of targets, and can be eliminated better in the fusion process. System errors result mostly from the angle and range measurement accuracy of radars, the positions of radar stations and inaccurate due north, as well as the error in the coordinates' transformation approximation algorithm itself. Random errors are eliminated with the SME criterion in both local tracking and data fusion at the center. System errors give rise to fixed deviations of observations from their true values. They are eliminated with the LS method, which is used to obtain the error covariance matrix of each radar serving as a systemic parameter by

getting an estimate of the systemic error of each radar, and making one-time registrations for the measurements and compensations for every single-radar track.

5. *Track association.* In the MRDP system, each single radar station possesses its own information processing system with large amounts of target track information. Then, an important problem is how to determine whether two local tracks from different systems represent the same target or not, which is a problem of correlating tracks to tracks (i.e., track association). Finally, track fusion is conducted. For the processing centers, track association and fusion algorithms are at the core of their work.

   To maintain system tracks, it is necessary to judge the input tracks detected by different radars, and associate tracks of the same target. The main track maintenance algorithms are the weighted statistical distance test, likelihood ratio test, and nearest-neighbor test. With single-radar tracks as input, the ATCCMS system is concerned mainly with information on civil aircraft, and thus usually adopts a method based on the nearest neighbor to associate code Mode-3/A and track batches.

   The association credibility factor FC is established for every new track input by addition or subtraction according to the following level, and the value of FC is used to decide whether they are associated or not.

   i. Identify whether the target is of primary or secondary type, or of joint type.
   ii. Determine whether targets of the same kind have the same track ID.
   iii. Establish position association windows according to the nearest-neighbor rule, and judge whether they share compatible horizontal positions.
   iv. Identify whether they have compatible Mode-3/A codes (i.e., the same SSR codes).
   v. Identify whether they have compatible Mode-C codes or flight altitudes.
   vi. Identify whether they have compatible flight speeds and courses.

   This process and the subsequent process of track fusion guarantee the maintenance of system tracks: tentative system tracks should be established for those new tracks that are not associated; the tentative system tracks correlated to several single-radar tracks in succession are converted into validated system tracks and undergo fusion computing; a tentative or validated system track which fails to be correlated to any single-radar track for several times will be cancelled.

   The classical assignment, maximum likelihood, likelihood ratio, and multiple hypothesis methods all need a great deal of computation, and are hard to implement in engineering.

   The NN, weighted, and modified methods, though simple in algorithm and fast in speed of processing, are suitable only in the case of sparse targets.

   The sequential, double threshold, K-NN, and MK-NN methods are of great efficiency in track association tests, and of relatively high data processing speeds. They are all applicable in the presence of dense targets and/or crossing, split, and maneuvering tracks. The sequential method can be introduced to multiple local nodes to form a multidimensional assignment algorithm, so it is worth wider applications. Compared with the sequential and double-threshold methods, the advantage of the K-NN and MK-NN is that they have no requirement for local systems.

   The fuzzy association algorithm could process fast with lower memory, communication load, and better association results. In terms of speed, it is slower than the K-NN and MK-NN, but quicker than the sequential and double-threshold methods. In the case of highly maneuverable targets, the performance of multiple hypothesis tracking (MHT) is better than that of the IMMJPDA and Kalman filtering.

   Not all methods are appropriate for the data association of multiple local radar tracks in the ATC multi-radar data processing system, therefore, a compromise should be made in practical implementations so as to find, through selection, the most suitable track association method. Currently, the weighted statistical distance test method is fast in processing and easy to realize, so it is the only option in most practical track associations.

6. *Establishment and cancellation of system tracks.* The establishment of multi-radar system tracks is similar to that of single-radar ones. When the single-radar local track of a main radar cannot be associated with any system track in the association gate, a tentative system track starts to be established. When a certain number of single-radar local tracks which serve as system parameters continuously associate with the tentative system track, it is turned into a validated system track. When a tentative or validated system track cannot be associated with any radar track for several successive times (the number of times is a system parameter), it is then cancelled.
8. *Data fusion.* After all the local tracks of single radars are processed, the next issue is track combination, or track fusion, a process of obtaining the new fused track of a target through the fusion of the associated tracks of the target.

   Typical track fusion algorithms include the weighted average, hierarchical fusion, and linear combination algorithms. Because the measurement accuracy of different radars varies, the weighted average method (in order to make the fused target track capable of describing the target better) sets weightings for the fused tracks based on the measurement accuracy of radar – high weightings for those with high accuracy and low weightings for those with low accuracy. The weightings can be flexibly set as required when the system is at work. Here, fusion processing refers to the weighted fusion of normalized position, altitude, and velocity of the associated target track. At last, the track after weighted fusion is applied to update the corresponding system track of the target.

   The linear combination algorithm is deducted from the perspective of optimal combination. The fusion center adopting this algorithm needs only combination instead of filtering to get fusion estimates, but it has to estimate the cross-covariance matrixes between the radars.

   In the hierarchical fusion algorithm, the global state estimate of target = global prediction + correction term. The correction term consists of two parts. The first part refers to the following process: the filtering value of every single-radar track is compared with the global prediction, the resulting differences are then inversely weighted with the corresponding single-radar filtering covariance, and all the results are summed. The second part is carried out as follows: to compare the overall prediction with every single-radar prediction, the resulting differences are then weighted inversely with the corresponding single-radar filtering covariance, before all the results are summed at last. Then, the sum of these two parts is weighted by the covariance of the global estimate. Statistically, the results of the hierarchical fusion and linear combination algorithms are roughly the same.
9. *Reproduction of system track output.* The expected fusion target track is obtained after the multi-radar target tracks have undergone correlative fusion. Then, the remainder of the work is to reproduce the output of the fusion target track on the corresponding monitors so that every controller work station can process them further.

Flight Plan and Radar Data Correlation

Information such as PLN, FPL, and RPL is taken from the flight telegram automatic processing system, and associated with radar identified targets according to the SSR code. This is a process of providing each radar record with a flight call sign.

Control Instructions and Flight Status Correlation
After being associated with flight plans, radar data represents the flight status of the aircraft. Some key data of control instructions – such as flight number, type of control quality, and value of instructions – should undergo correlation matching with flight status data so that radar track records associated with the control quality are picked out for track pre-estimation.

Track Pre-estimation and Collision Detection
Data such as the identified control instructions, current flight conditions, aircraft performance, and meteorological conditions work together in checking whether the relative distance between aircraft has contravened safety interval requirements. Alarm information will be sent out if a possible aircraft collision is detected.

The architecture of the operating function subsystems is shown in Figure 19.3.

It is a necessary option for the current ATC service, and the only approach to ATC automation, to establish multi-radar data processing systems in larger ATC centers and accomplish radar networking. However, the ATC multi-radar data processing system is very complex, involving many aspects that are tough to deal with.

19.2.3.3 Functional Subsystem of Management and Decision Level

The functional subsystems of the management and decision levels are non-real-time; they mainly provide high-level decision support for administrative staff and controllers on duty.

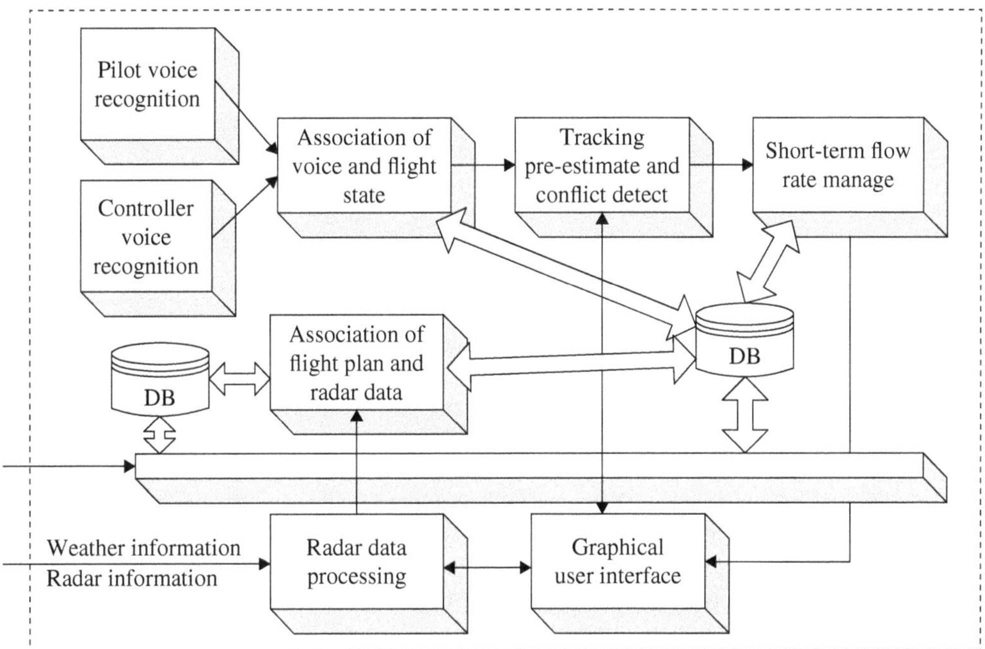

Figure 19.3 Functional architecture of ATCCMS

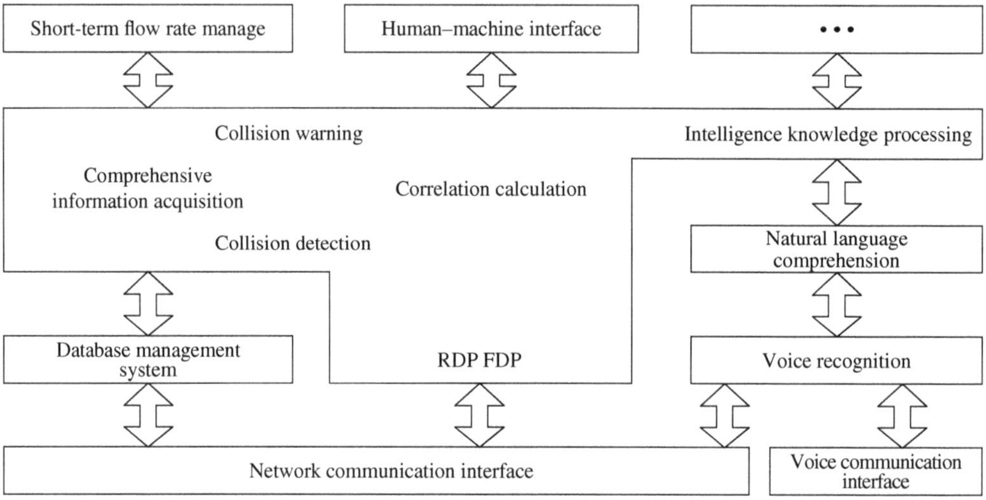

Figure 19.4 Logic architecture of ATCCMS

19.2.3.4 Supporting Environment

The supporting environment of the ATCCMS includes the computer network environment and the database environment to guarantee communication and connection between computers and basic data.

The logic architecture of the ATCCMS is shown in Figure 19.4.

19.3 Application in Shipboard Navigation Radar

Even if visibility is extremely bad, a helmsman must have a clear picture of his ship's surroundings to ensure safe navigation. In addition, he has to spot potential collisions quickly and accurately in advance, and make decisions to prevent collision by performing optimal maneuvers. Navigation radars show the helmsman neighboring ships' relative movements, based on which he will predict any dangerous conditions.

The main function of the radar data processing system in collision avoidance applications is the realization of accurate estimation of target position and speed when the target is moving at a uniform speed in a straight line, which requires accurate positioning of the target and tracking of maneuvering targets at the same time [419–422]. Obviously, the above requirements are contradictory: the more accurate the prediction with linear movement, the worse the performance in tracking maneuvering targets, as is apparent when the α–β filter is used. The α–β filter is widely used because of its simplicity. Usually, tradeoffs are made between the following requirements: for targets with medium speed, a slower response will be applied to achieve higher accuracy; for targets with faster speed, a quick response will be achieved at the cost of lower accuracy.

The anti-collision system with adaptive double-channel tracking filters developed by Roman Selenia, an Italian company (for details, see Ref. [416]), mainly consists of a navigation radar, a digital measurement recorder, a microcomputer, displays, and controllers. The functional block diagram is shown in Figure 19.5.

The original video signal from the radar system, timing data, and antenna position code data are transmitted together to the measurement recorder, which simultaneously provides the computer and

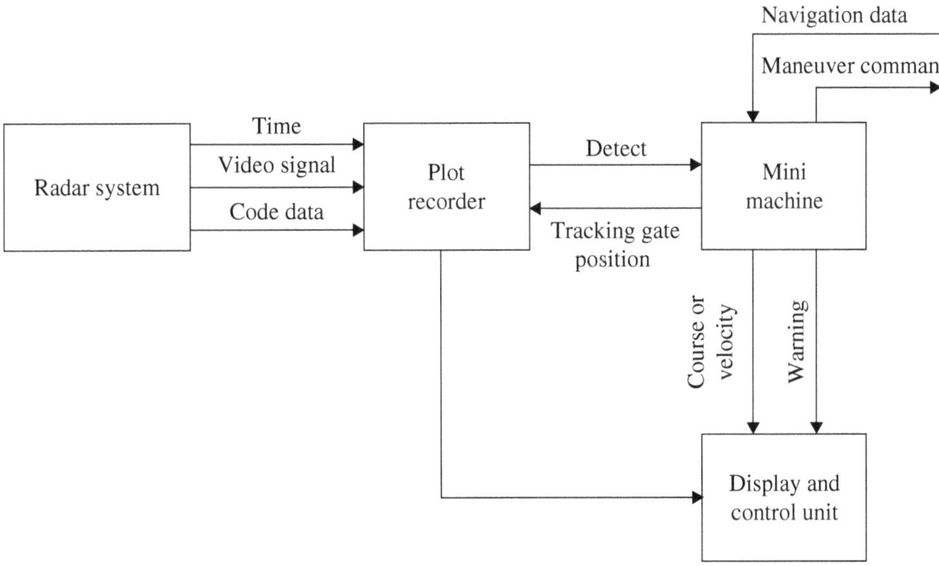

Figure 19.5 Marine collision avoidance system

displays with detection data and coordinate data. The computer executes the track-while-scan algorithm, working out the expected relative course and velocity of all ships being tracked. This system can track more than 40 targets within 25 nautical miles (1 nautical mile = 1.852 km), and the track initiation can be conducted automatically or performed by operators. The track algorithm applied is the adaptive α–β filtering algorithm with time-varying parameters.

Figure 19.6 illustrates the principle of the tracking algorithm of the Selenia collision avoidance system, which is based on two different parallel processing components: the "quick" and the "accurate" filter. The former is used mainly for ship tracking, while the latter is for narrowband filtering so as to reach higher accuracy under stable conditions, and therefore the value of α and β in the former is bigger than in the latter. The two filters both apply constant α–β parameters. But when conducting accurate filtering, three α–β values are allowed and the switch logic circuit is provided. Table 19.1 lists all α–β values and the corresponding response time (calculated by the number of radar scans). In the track initiation phase, only the quick filter channel is working.

After the transient time of N_i scanning, the accurate filter is initialized by target track estimates through gate 1. Then, it processes measurement data in parallel with the quick filter, but here the first permissible pair α–β is used. Under stable conditions, a test will be conducted to determine whether the target speed is low, medium, or high. Its α–β value may be selected appropriately according to the result of the test. The same test should be conducted periodically so as to consider the changes in ship speed. This adaptiveness to target speed aims to make the relative error of speed estimate nearly constant. According to this method, the slower the target speed, the higher the accuracy of the filter, but at the cost of a longer response time. The output of this filter is used to estimate the display, extrapolation and nearest point of proximity (CPA), and time of arriving at the nearest point of proximity (TCA), and it may be used to detect target maneuvers by comparing predicting positions and radar measurements. Once target maneuvers are detected, the position and speed estimates provided by the quick filter are forced via gate 2 to the accurate filter. In this way, adaptiveness to target maneuvers is acquired. Then, the output of the accurate filter will be processed and displayed. As long as the target keeps maneuvering, the periodical data flow from the quick to the accurate filter

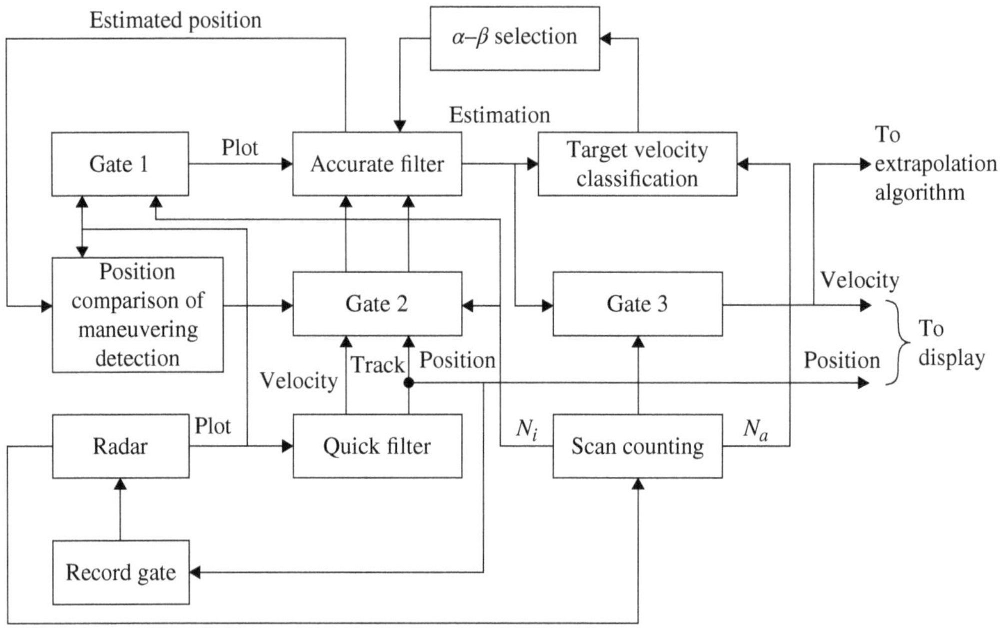

Figure 19.6 Tracking algorithm of Selenia collision avoidance system

Table 19.1 Main parameters of the filtering algorithm

Filter	Target	α–β	Scanning times (s)
Quick	Maneuver	0.54–0.2	6
Accurate	Fast	0.18–0.0178	20
	Medium	0.11–0.0065	40
	Slow	0.07–0.0025	60

will be continuous, hence the interaction between these two channels will minimize the transient time of the whole track filter.

19.4 Application in Shipboard Radar Clutter Suppression

19.4.1 Principle of Clutter Suppression in Data Processing

Because of the special working environment, shipboard radars face great disturbance from clutter, including ground (e.g., the ground, seashores, and islands), sea, and atmosphere (such as cloud, rain, snow, hail, and atmospheric turbulence) clutters. The abnormal atmospheric propagation at low altitude over the sea makes it difficult to detect and track sea-skimming targets. The conventional way to suppress clutter is to take appropriate measures in radar signal processing subsystems. Moving target indication (MTI), moving target detection, digital pulse compression, and constant false alarm rate (CFAR) are the commonly used methods, which have proven effective to a certain extent. As has been proven, whichever technique is used, there are still residual clutters leaked to the radar data processing subsystem, and a great number of false measurements will emerge in severe working conditions. This brings about difficulties in track initiation and filtering of data processors, and even saturates the computer. Therefore, it is necessary to discuss the problem of how to suppress clutter in the radar data processing phase.

After being detected and determined through the signal processor, radar echoes are sent to the data processing subsystem, which includes the echo pretreatment, correlation filter, and radar control modules. The echo pretreatment module has such functions as angle and distance measurement, platform modification, and measurement condensation within beams. The correlative filter has functions such as measurement condensation among beams, track initiation, measurement–track association, track filter prediction, and track management. Radar control is responsible for the scheduling and control of radar coverage, operation missions, jamming countermeasures, signal processing, and radar sources. Connecting with the data processor, the display control subsystem can display radar working states and track information, and radar parameters, area data processing parameters, and interface parameters can be controlled manually through it. The communication module conducts data communications between the serial port, the parallel port, and the network. There are four features in clutter suppression of the data processing subsystem.

1. *Accumulativeness of information.* Radar data processing is based on the software platform, so its real-time requirement is not as strong as that of signal processing. It can accumulate and compare the radar echo information of neighboring periods. Target information can be accumulated sufficiently, and then processed properly when the clutter jamming is serious.
2. *Pretreatment.* A few radar scanning periods will be used to collect clutter information, and store or update the distribution of clutter.
3. *Optionality.* Screening measurements will inevitably cause the loss of some true target measurements, and therefore the methods applied to suppress clutter will be adopted only if necessary (they are optional).
4. *Limitations.* In the data processing phase, the information available to suppress clutter is scarce, and closely coupled with radar work modes and signal processing methods.

19.4.2 Clutter Suppression Method through Shipboard Radar Data Processing

Shipboard radar encounters severe clutter disturbance from the atmosphere and marine environment when in operation. Especially during sea trials, residual clutter affects its overall performance. To solve this problem, a series of specific experiments have been conducted to collect clutter data and observe clutter phenomena in the hope that a solution will be found. The trial data indicates that there are land/sea clutter, angles, and atmospheric duct phenomena in echo data. After the signal processing subsystem adopts corresponding anti-clutter measures, there are still some residual clutters – including cyclic distributed, close low-elevation dense, and noise-like clutters – and those caused by meteorological phenomena. In order to eliminate residual clutters and increase the number of measurements, a series of measures were taken in the data processing phase.

19.4.2.1 Clutter Suppression Method for the Echo Pretreatment Module

As the foremost part of data processing, the echo pretreatment module has the most abundant clutter information. Generally, a complete measurement includes the following information:

1. Radar time and position of a transmitting beam.
2. The number of measurements of echoes of this transmitting beam after sampling and threshold detection.
3. Parameters of the spatial position of each measurement.
4. The radar measurements of some radar systems also containing the Doppler velocity of the target.

Select a set of sea trial data to conduct statistical analyses. The data are all echoes of the radar searching a range of bearing approximately 90°, and the radar working time is 1492 s. Rules to eliminate clutter are as follows.

1. *Echo width judgment.* The amplitude and width of echo signals reflect the energy of target echoes. Echo width refers to the number of continuous distance units which echo analog signals occupy after sampling quantization and amplitude detection. According to statistics, there is an obvious difference in echo width between real targets and clutters. Most clutters may be eliminated effectively according to this feature.
2. *Amplitude judgment.* An amplitude threshold is set for the echo occupying two continuous distance units. The threshold is adjusted according to different distances, and measurements whose amplitude is lower than the threshold will be eliminated.
3. *Doppler velocity judgment.* First, measurements whose radial velocity is about zero are eliminated. Then, in the process of measurement condensation, the most obvious difference is the Doppler velocity, guaranteeing that the adjacent or crossing targets will not be condensed into one measurement. In addition, during measurement–measurement association and measurement–track association, comparing the Doppler radial velocity between measurements or between measurements and track is a criterion for judging whether they are associated or not.
4. *Clutter map judgment.* Specific clutter maps are employed to deal with the severe cyclic clutter phenomenon. The cyclic clutters are concentrated at the scanning layer of low elevation: they are of cyclic distributions in the distance–azimuth figure and of pectination distributions in the distance–elevation figure, and their position is relatively fixed. The method to eliminate cyclic clutters is to count the total number of echoes in a number of adjacent scanning periods within a certain range by the sliding window. When the total number of echoes exceeds the threshold, cyclic clutter phenomena are considered present within the range, and the measurements within the range will be regarded as clutters and eliminated.

19.4.2.2 Clutter Suppression Methods for the Radar Control Module

Radar Control by Manual Intervention

The radar working parameters of the entire airspace should be altered by adjusting the network command of the control subsystem or manual intervention command of the DSP control module, and according to the working environment (e.g., offshore, high sea, sea conditions, and meteorological complexity) of ships. The parameters include STC control, Doppler processing modes (such as number of pulses, notch width, number of sliding windows), movement compensation parameters, clutter thresholds, pulse cancellation, gain, and detection control parameters.

Radar Control with Area Control
Once the distribution of primary clutters is clearly observed on the PPI of the display control subsystem, several clutter areas are established by manual work. Radars' working parameters within clutter areas are then artificially set and revised.

Adaptive Radar Control with Clutter Maps
Specific clutter maps and radar working modes are set to detect land/sea and meteorological clutter areas through applying radar beams with special working parameters, and then land/sea and meteorological clutter maps are established according to the echoes. The residual clutter map is set by processing residual clutters at normal work. The three clutter maps store statistical features of clutters within the airspace, including spatial position, distribution density, amplitude, and Doppler information. When arranging transmitting beams, the radar control module looks up the clutter distribution features of the corresponding area in the clutter maps in accordance with the orientation of the beam center, and determines the working parameters of transmitting beams subject to the corresponding criteria of clutter control. The establishment of clutter control criteria is related to many factors such as battle tasks and radar resources.

19.4.2.3 Methods of Clutter Suppression of the Correlation Filter Module

If the above measures still fail to suppress clutters effectively, the correlation filter module will encounter great difficulties. As dense clutter measurements emerge in many successive periods in a certain area, automatic track initiation and correlation processing consume a large amount of computer resources, and the computer will be out of condition in serious cases. Thus, special track initiation and track filtering rules must be adopted. In areas with many residual clutters, special zones should be established where relatively stricter track initiation rules are adopted for measurements (e.g., the common initiation rule of 3/5 switched to 4/5 or 5/5). Besides the position information of measurements, information such as amplitude and Doppler velocity should also be considered when associating measurements and tracks. Velocity filtering is to be conducted for initiated tracks, and false tracks deleted. Manual zones should be established for areas with excessive residual clutters. The operators should adopt, based on their observations, the track initiation and maintenance rules integrating semi-automatic and manual modes.

19.4.2.4 Radar Control with Manual Intervention

Some special areas where different control and management methods are adopted are set up within the radar coverage range according to clutters' different spatial distributions: silent, shielding, manual, and map control zones. The silent zone is the special area mainly used by the radar control module, where radar beams are only received rather than transmitted in order to counter jamming or suppress clutters. The shielding zone, mostly applied by the echo pretreatment module, is an area used to counter strong clutters where radar waves are only sent. The manual zone is mostly employed by the correlation filter module, and the echoes in it are processed with unconventional track initiation and maintenance algorithms. The map control zone is mainly adopted by the radar control and echo pretreatment modules. The former controls the working parameters of radar beams based on the features of the clutter maps in the map control zone, while the latter filters out clutter measurements according to the characteristics of clutter maps in the map control zone. Based on electrical map background, the radar display and control module can display such geographic information as coastline, islands, and drilling rigs. Such information may serve as the basis for clutter determination, or for the control of radar working modes and regions.

In practical engineering, it is necessary to suppress clutters in the data processing phase. Different radar working modes and environments show different characteristics of clutter distributions, thus different data processing methods should be adopted. Radar signal processing and data processing are two closely related parts, and the clutter suppression system with good performance should be an adaptive one in which the two parts can work in harmony. The rapid development in DSP techniques and computer performance facilitates the integration of the two systems. The method above has proved effective in its application to shipboard radar systems of a certain type.

19.5 Application in Ground-Based Radar

19.5.1 Principle of Data Acquisition

The technology of 3D laser radar imaging scanning is an innovative surveying tool that has emerged with the development of space array scanning technology and reflectorless long-distance quick ranging technology. It is another breakthrough in the field of surveying after GPS.

As a data acquisition technology, it can be divided into airborne and land-based laser radar acquisition systems. The airborne laser radar is the most effective tool to acquire 3D ground digital information massively and promptly. In addition, information about ground, tree height, underwater terrain, water depth, etc. can be acquired with its fore and aft reflection. It is widely used in the surveying of terrain, forest, underwater, and so on.

The ground-based laser radar is small and convenient, precise and efficient, safe and stable, and highly operable. It can set up a detailed and precise 3D image of the area of interest within a few minutes, and provide precise quantitative analyses. Thus, it can be applied in a great variety of fields: local 3D models of cities, ancient building surveying, relics preservation, reverse engineering applications, complex building construction, geological research, building deformation monitoring, to name just a few.

Extensive studies have been carried out on the theory and application of laser radars, but many subjects in this field remain to be explored further. Given its wide application, we focus here on problems relating to ground-based laser radars' working principles, data processing procedures, data precision, and influencing factors.

The ground-based laser radar is a new type of surveying instrument integrating many high and new technologies, in which non-contact fast laser measuring modes are adopted to acquire phased geometric data in the form of point clouds, terrain, and 3D surfaces of complex objects. It consists mainly of the laser ranging and laser scanning system, while integrating CCD digital photography, internal correction, and other systems. Its working principles are as follows.

1. The scanner transmits a laser to the target of interest.
2. The time difference between laser transmission and reception is used to calculate the distance between the plot surveyed and the scanner.
3. The values of horizontal and vertical stepping angles are used to calculate in real time the 3D coordinates of the points being measured.
4. The resulting coordinates are sent to the memory device in order to be saved.
5. The saved coordinates are processed by corresponding software.
6. A 3D geometric model of the target being surveyed is thus obtained.

In fact, the original observations obtained by the laser radar fall into four broad types:

1. The horizontal and vertical values of laser beams acquired according to the two continuously rotating mirror angle values, used to reflect the pulse laser.
2. The distance between the scanner and the scanning spot, calculated based on the time of pulse laser transmission.
3. The reflection strength of the scanning spot.
4. The scene image data acquired by the built-in digital camera.

The first two types are used to calculate 3D coordinates of the scanning spot, and the last two to match colors for the reflecting spot or to carry out texture mapping for models.

19.5.2 Data Processing Procedure

The original data provided by laser radars are 3D coordinate data based on the instrument coordinate system, obtained with the delivery attached commercial software. They are great numbers of attribute-less, discrete point array data suspended in the air, normally called "point clouds." Comprising many outliers and system errors, they cannot be used directly. In addition, they also include a great deal of redundant information, which is of little help in the subsequent data analysis but takes a lot of time to calculate. For these reasons, the pretreatment before data analysis is a necessary step.

The data processing procedure of the ground-based laser radar can be broken down into four steps.

1. *Elimination of outliers.* As mentioned above, the original point cloud data contain a lot of outliers, errors, and irrelevant information, for a variety of reasons: for example, data from the signals reflected from moving targets (e.g., flying birds or some other free targets within the field of view of laser radar), local jumping data (like points below the ground), foreground shielding data, and local echo-less cavities (such as laser penetrating windows or illuminated targets absorbing all laser signals). These outliers, errors, and extraneous information are eliminated, corrected, and/or compensated mostly through man–machine interactive operation, with delivery attached data processing software in the interactive editing environment.
2. *Splice of models.* The ground-based laser radar has a certain penetrating capability, but as with the close-range camera, in its gathered data it still happens that the foreground blocks the background. In addition, to acquire the 3D model of a target, many stations should be set around this target in order to acquire point cloud data from different angles. The direct output of a ground-based radar is based on the local coordinate data of its camera station coordinates. To acquire the overall 3D model of the target under study, the point cloud obtained from different angles should be integrated with the help of overlapping information (i.e., the point cloud of all camera stations should be integrated into the coordinates of one camera station); this is the so-called model splicing process. In the process of integration, there are two basic methods to get the value of two adjacent overlapping areas: one is to select the data of one station as the final data; the other is to resample data on the overlapping area between two camera stations.
3. *Unification of reference coordinates.* The unification of reference coordinates is a prerequisite for subsequent data processing and analysis. There are two methods to unify the reference coordinates: one is to convert to local coordinates, like the splice of models above; the other is to unify in

one set of absolute coordinates, such as projecting to the coordinates of a certain city building or national coordinates. The former can be achieved on the overlapping information, while the latter is actually coordinate projecting conversion. Therefore, the conversion can be accomplished only with enough control points.

4. *Data simplification.* The ground-based laser radar has high measurement accuracy, high efficiency, and fast 3D model establishment, but it also has some demerits: non-continuous coverage, huge amount of data volume, etc. This huge data volume in particular makes the subsequent data processing, transmission, and further processing really difficult. The sampling data of the laser radar comprises lots of redundant data. For example, the edging information is enough to construct the façade of a structure, but the laser radar still provides all the sample information including both the façade edge and the internal part. The redundant information is of little help to model construction or model feature extraction, therefore it is necessary to select or devise the corresponding algorithm to simplify the point cloud data. Currently, data processing methods mainly include two types.
   - *Simple resampling.* Resample the original data set in a certain sampling interval, such as sampling with an interval of one row or one column. This method has an obvious drawback: both redundant and non-redundant information are weakened, and some useful information is lost.
   - *Combination and resampling for homogeneous areas.* Within the original data set, the homogeneous area, such as the façade of a building, has huge redundant data. Through some algorithm, combine this area, replacing the numerous original sampling data with fewer sampling points or combining sampling points, to realize the aim of data simplification. This method has obvious advantages (i.e., data resampling varies adaptively with the actual conditions of targets, which not only simplifies data but also saves useful characteristic information effectively). Data simplification is not only a difficult problem but also a leading edge topic in the field of laser radar data processing. There is no data simplification algorithm applicable to every situation at present. The data simplification tools available can simplify and smooth the original point cloud data only to some extent.
5. *Application analysis.* After the previous data processing, outliers and redundant data are eliminated and the data volume is decreased significantly. These data can be applied in numerous ways. In civil engineering, for example, the edge information of the observed object can be acquired, the fracture information extracted, and the volume calculated through data partitioning. In virtual reality, point cloud data can be converted into the corresponding 3D digital model.

The ground-based laser radar has a bright application prospect for its excellent technical features such as high efficiency, complete coverage and true three-dimensional measurement. From the perspective of technical features, it can be used not only in topographical survey of building construction, highways, bridges, dams and local terrain and deformation monitoring, but also in fields such as industry measurement, cultural relics protection, CAD design and animation production. At present, it has been introduced to many fields, and the relative applied research and theoretical research are developing gradually.

19.6 Applications in Shipboard Monitoring System

19.6.1 Application, Components, and Requirement

With the rapid development of maritime traffic and inland river transport, the tonnage of ships is increasing constantly, and ports and fairways are becoming more and more crowded. In addition, in bad weather conditions, accidents like shipwrecks or crashes often take place in shoals and heavy traffic waters. Nowadays, ships are generally equipped with shipboard navigation radars. Undoubtedly, shipboard radars play a significant role in safe navigation. However, the antenna radius cannot be too big due to the limitation of installation space and therefore the azimuth resolution cannot be too high either.

When it comes to determining the relative movement between ships, the ship being monitored is moving and so is the monitoring radar because it is onboard another ship, so it often happens that the moving state of the monitored target is misjudged in a short period of time, which can result in crash accidents. In a meandering fairway, because of the land block, the blocked ships often cannot be detected, targets with a small echo area (such as wooden ships and buoys) cannot be found easily beyond a certain distance, and some navigation obstacles (such as shoals, shipwrecks, or submerged reef, fishing nets and breakwaters) are difficult to find on screen. Even though the target can be monitored, the radar is unable to determine whether it is sailing into or out of the port or to get related navigation parameters such as speed, course, and collision avoidance warning. Hence, it cannot provide the necessary navigation information for ships. This will inevitably restrain the speed of ships and increase the insecure factors for navigation.

It is to make up the deficiency of shipboard radar that the marine surveillance/vessel traffic control system has been proposed. At present, many countries and regions have established a vessel traffic control system. Typical vessel control systems consist of a radar station, microwave transmission, data processing, terminal display, and operating control subsystems, as well as VHF and UHF communication and hydro-meteorological units, with the following main functions.

1. *Display the chart and topographic/geomorphic map.* Based on the chart and topographic/geomorphic map, the system can display, on a radar synthetic display, a colored radar background situation map, presenting permanent facilities such as ports, anchorage, fairway, and marks, the depth of water and coastline.
2. *Display the radar target and its parameters.* Radar signals sent by radar stations will be processed by the data processing equipment, and the targets being tracked will be nominated and registered. Meanwhile, the speed, course, and collision avoidance information for the targets will be displayed in real time. Usually, the data processing equipment can track and process more than 100 batches of targets and provide navigation information for them all.
3. *Conduct marine supervision and management.* This system can track and monitor ships that infringe regulations or break into a sea area illegally, and give necessary warning. For wrecked ships, this system can report their location and command rescue efforts.
4. *Assist in port affairs management.* The data collected by this system, such as inward and outward time, tonnage, load, destinations, registration/cancellation, and anchorages can be input to the vessel data management network and presented in data forms. These data can also be recorded for replay afterwards.

19.6.2 Structure of the Marine Control System

Marine traffic control depends mainly on data acquisition, processing, and transmission systems.

The data acquisition system consists of primary or secondary radars, meteorological and marine sensors, telegrams, telephones, and other data. The data processing system includes several computers capable of a variety of processing tasks (general-purpose, mini-, and micro-computers). The types of data to be processed include [57, 423, 424]:

1. Radar echoes, aiming to survey the location and speed of all ships, and if possible, the distance between ships.
2. Shipping information on the berth and goods of ships.
3. Technical information on monitoring all equipment performance.
4. Information provided by additional radio positioning systems to complement information provided by radars.

The data transmission system is made up of conventional or special inner communication and display units. Transmission modes depend on users, such as traffic control personnel and port staff. In the daytime, the TV coordinates displaying method can be adopted, which can gather information provided by radars in ports or along the fairway.

Normally, it takes more than one radar to realize the entire coverage of a port or fairway. To avoid the advent of blind areas, radars must be carefully deployed at fixed points and usually should be positioned high. Another issue related to radar positioning and choice of beam pitching is to reduce the influence of sea clutters. In the surveillance phase, radar range is restricted to dozens of thousand meters, but when in navigation, it is 5–10 km. The latter situation requires high range resolution and range accuracy of 7–10 m. Because they normally work on band X or K, radars can have perfect bearing resolution with limited antenna sizes. As to the scanning periods of radars, if the target speed is 5 m/s, a 3 s scanning period is sufficient. Actually, the target moves 15 m during this period, which is nearly twice as much as the position error. To realize the track-while-scan function of all radars, both $\alpha-\beta$ filter and KF algorithms need to be applied. Meanwhile, the multiple-radar tracking system is also widely used, monitoring and controlling the overall traffic in the central control room.

19.7 Application in the Fleet Air Defense System

Data processing is widely used in fleet air defense systems, like the Aegis system. Aegis is an all-weather, all-airspace, shipboard air defense missile weapon system of the US Navy, safeguarding aircraft carrier or destroyer and frigate formation, undertaking area air defense combat tasks. It has been the main air defense system for the US Navy since the 1980s, and is regarded as the third generation of ship-to-air missile weapon system [425]. The multi-functional phased array radar is the core of the Aegis combat system. The data processing technique relevant to the phased array radar can be found in Chapter 14, so the following subsection briefly discusses the application of the data processing technique in Aegis.

19.7.1 Components and Function of the Aegis Fleet Air Defense System

The Aegis fleet air defense system consists mainly of the multi-function phased array radar, weapon control system, missile fire control system, missile and launching system, and maintenance and test equipment.

19.7.1.1 AN/SPY-1A Multi-function Phased Array Radar

The AN/SPY-1A multi-function phased array radar is an S-band electronically scanned phased array radar, with functions such as the detection, search, and tracking of multiple targets and guided missiles; its maximum detecting range for aircraft is 370 km. Through the work of computers and beam controllers, the AN/SPY-1A can control time and energy resources flexibly to enable parameters such as beam pointing, beam energy, beam width, pulse repetition frequency, and radiant power changeable in real time; it can produce complex waveforms required by transmitter excitation, choose and process proper echo signals, and adjust strategies on the requirement of detection, tracking, and electronic countermeasures; it can simultaneously process information in 11 modes, such as search, track, display search and track moving targets, missile guidance, passive search and track, barrage jamming detection, warning and recognition. In the track-while-search mode, it always scans the whole airspace to find new targets. Considering that it is difficult to detect anti-ship missiles because of their low-altitude penetration, the multi-beam high-data-rate search is used in low-altitude areas, while in other airspaces fewer beams and lower data rates are applied in search; if necessary, beams can be concentrated on an estimated fan threaten area to conduct search. To tackle anti-radiation missiles, radars can be temporarily shut in a flash (a fraction of a second), and revert to the track mode after being powered on in 1 s.

An important function of AN/SPY-1A is to send guidance command to flying missiles and control illuminating radars to illuminate targets for the semi-active radar terminal guidance system. The midcourse guidance commands sent from the weapon control system are transmitted to the flying missiles and reported to the command decision-making system via AN/SPY-1A.

19.7.1.2 MK1 Weapon Control System

This system undertakes some command and fire control functions.

19.7.1.3 MK99 Missile Fire Control System

This system receives the command from the MK1 weapon control system, cooperating with the AN/SPY-A1 in providing target illumination in the terminal guidance phase for the SM-2. In addition, it can control the Phalanx close-in weapon system and harpoon missiles as well.

19.7.1.4 MK1 Operational Readiness and Test System

This system is chiefly responsible for the real-time and pre-work tests of the AN/SPY-1A multifunction phased array radar, MK1 command decision-making system, and MK1 weapon control system in the Aegis system, as well as their centralized coordination, working state diagnosis, fault isolation, and maintenance.

19.7.2 Main Performance Indexes

1. *Target type:* aircraft with high performance, anti-ship missiles, and supersonic sea-skimming anti-ship missiles.
2. *Operation airspace:* far boundary 74 km (median altitude), 20–25 km (super-low altitude), 150 km (high altitude); high boundary 24 km; low boundary 25–30 km.
3. *Guidance system:* inertial navigation/instruction + semi-active radar target homing.
4. *Guidance law:* proportional guidance.

5. *Missile fusillade interval:* 5 s, 1 s.
6. *Multi-target tackling capability:* intercepting 12–16 targets simultaneously in 360° and whole airspace.
7. *System anti-jamming capability:* capable of resisting various types of electronic jamming and sea clutter/meteorological disturbances.
8. *Radar detection performance:*
   - azimuth scope of electronic scanning, ±50°;
   - elevation scope of electronic scanning, ±60°;
   - number of surveillance targets, 400 batches;
   - number of track targets, 200 batches.

19.8 Applications in AEW Radar

19.8.1 Features, Components, and Tasks

For ground-based and sea-based radars, the capability for surveillance in larger areas is restricted by the earth's curvature, and natural or artificial obstacles. Visual restrictions make the defense system more vulnerable to attack from low-altitude aircraft invading at high speed or long-distance terrain-following cruise missiles. It is natural to mount the surveillance radar higher to overcome the limitations of the coverage range, but for marine systems the benefits brought about by this method are limited. Setting radar sensors in the air platform can elevate radar stations by hundreds of meters. Compared with ground-based and sea-based radars, if the coverage capability of airborne radars is strong enough, the extension of their low-altitude coverage is considerable.

Suppose that a target is moving at an altitude of 100 m at a speed of 1000 km/h. If the altitude of the ground-based radar is 300 m, then the target may be detected 120 km away, and the interception time is 7 minutes. But if the flight altitude of the airborne radar is 9000 m, then it can detect the target hundreds of kilometers away, and the time it takes is 20 minutes earlier before the target's entry into the effective range of the weapons.

The airborne early warning (AEW) system detects targets and conducts command missions by means of aircraft, and surveillance radar is a main part of this. Integrating functions such as early warning, command, control, and communication, the AEW acts as a mobile radar station and airborne command center, with features such as good low-altitude performance, large surveillance range, powerful survivability, strong command and control capability, high flexibility and maneuverability. In general, the AEW consists of the following subsystems:

1. high-performance long-range surveillance radar;
2. friend-or-foe recognition device;
3. one or more computers for radar data processing, navigation, and control;
4. consoles and displays used by all kinds of operators;
5. data channels for communication with the ground base or friendly aircraft;
6. electronic support devices for foe radiant source identification and orientation.

In wartime, the system can undertake the following typical tasks:

1. *Early warning.* The main task of an AEW aircraft is to warn about air targets, and command and control of air operations. Warning against air targets is the main and basic task of AEW aircraft. The AEW aircraft can detect, recognize, and track various kinds of target far away from its own airspace, providing sufficient early warning time for its own forces. It can conduct air defense early warning tasks independently, and construct an air defense early warning network with the long-range and close-range radar.
2. *Command and control.* AEW aircraft can conduct search, detection, and command tasks within battlefields, perform long-range information associations with commanding centers in the wide front and many operation directions, and undertake relay communication tasks. Advanced AEW aircraft can conduct effective guidance, command, and control of massive air surprise attacks and multiple aircraft in air combat. Combined with air surprise attack forces, they can greatly enhance the capability for quick response and efficiency of air surprise attacks. Coordinating with the installations and devices of the interception and attack system and the C^3I system, they can increase the integrated combat capability of the air defense system. When executing interception tasks, with the guidance of ground-based radars, they can guide their own fighters to conduct long-range interceptions.
3. *Passive detection or secret early warning.* When an AEW aircraft conducts this task, the detection radar should be kept silent and only the electronic support measure (ESM) and communication surveillance measure (CSM) subsystems are used to find the positions of enemy forces and weapons. The passive detection system can be employed to detect electronic radiant sources beyond the detection range of the early warning radar, with capabilities for fast response and highly accurate frequency measurement. In dense electromagnetic environments, it has a high interception probability. Through the cross location of direction finding with other surveillance stations or different points on its own route, it can quickly detect the bearing and electronic parameters of enemy electronic signals. The attributes of targets can be identified through the analysis of received radiant signals. The CSM subsystem can also acquire the enemy's other tactical information.
4. *Air communication center.* The AEW aircraft is usually equipped with a super-powered UHF radio station, UHF transceiver, VHF and UHF AM/FM radio station, HF data link, and joint tactical information distribution system (JTIDS). Acting as an air tactical information center, it can receive and relay joint tactical communications.
5. *Enhancing efficiency of air defense combats.* In the air defense system of a nation, the use of AEW aircraft can effect a reduction in number of CAP operational aircraft, and even their abolishment, as well as greatly decreasing the size of land-based air defense forces on duty. The working efficiency of a high-performance AEW aircraft equals that of 8 to 10 high-performance, high-power, ground-based air defense radar stations.

19.8.2 Data Processing Technology

There is no essential difference in radar data processing methods between AEW systems and surface-based radars, but differences do exist. The most obvious difference is that the movement of the platform should be taken into consideration in the AEW radar system, which requires continuous modification of the detected measurements according to time-variable data provided by the airborne navigation system. The double parallel line patrol route is the typical movement mode of the AEW aircraft, as shown in Figure 19.7.

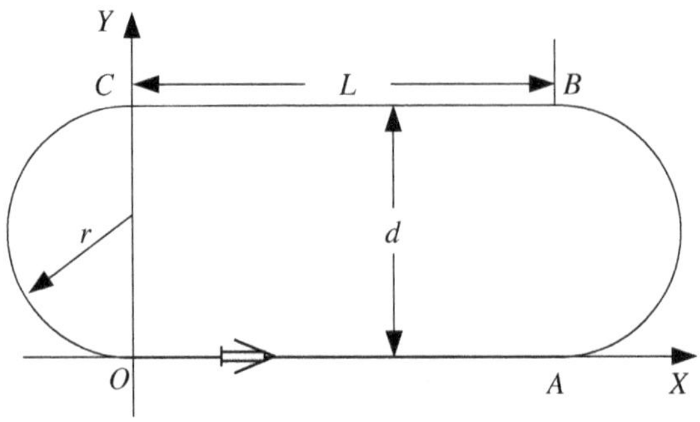

Figure 19.7 Double parallel line patrol route

Without loss of generality, suppose that an AEW aircraft is moving at altitude H with constant velocity V, from point O along the double parallel patrol route (i.e., the AEW aircraft is moving cyclically along $O \to A \to B \to C \to O$), in which sections OA and BC are straight lines, sections AB and OC are two semi-circles, the distance between the two parallel airways is d, the patrol distance of straight flight is L, and the turning radius is $r = d/2$.

19.8.2.1 Active Phased Array Radar Technology

As the phased array radar system has great flexibility in antenna beam formation and scanning control, the electronically scanning phased array radar consists of thousands of independent transmitting and receiving array elements, and it can realize beam scanning with the computer-controlled array. Compared with the ordinary rotating antenna radar, it has such merits as flexibility of scanning, high reliability, strong anti-jamming capability, minor aerodynamic influence on the aerial carrier, and excellence in stealth performance. Capable of detecting and processing multiple targets, it is a new radar system appropriate to future information warfare.

In the phased array system, the advanced space-time adaptive processing (STAP) technology can be applied to endow AEW aircraft radar with superior performance. The features are as follows.

1. High reliability. As proven by tests, with 10% of the units failing, there is no significant influence on system performance and instant repair is unnecessary. When the unit failure rate rises to 30%, the system gain lessens by 3 db, but the basic working performance can still be maintained.
2. Fast scanning beam, reaching the millisecond or even microsecond level.
3. Strong anti-jamming capability.
4. Strong capability to distinguish short-range targets.

19.8.2.2 AEW Aircraft Data Fusion Technology

Various sensors of AEW aircraft platforms provide target information, which can be shown on the controllers' display or displayed in the form of signals generated by computers. Automatic fusion of these data from sensors will greatly relieve the burden on the controllers of assessing or recognizing a specific target and deciding on ways to deal with this target.

In the electronic task system of AEW aircraft, there are three pairs of sensor systems: inertial navigator and GPS receiver; PSR & SSR/IFF interrogator; PSR & ESM/CSM receiver. Through data fusion, they can provide ample and more accurate information, which enhances the system performance of target detection and recognition [426].

To provide target track data to the air defense information system, the early warning system must depend on the coordination of radar and inertial navigation systems. Because of the influence of drift error on inertial navigators and the Schuler error resulting from the rotation of the earth, as well as minor random errors, the position data from the inertial navigator for the aerial carrier is not accurate, and therefore it is necessary to correct the inertial navigator. At present, the inertial navigator of the early warning system is combined with the GPS receiver to obtain highly accurate position data. It can determine its position by receiving code signals from four GPS satellites at the same time, with the application of universal C/A code (standard positioning service), whose position accuracy 95% probability is less than 100 m (equal to 40 CEPm) – far greater than that of the Tacan or Loran system.

If an air target is detected by the AEW radar, the target echo signals will form measurements. If there is an SSR transponder, after being received and decoded by it, the response signals will also form SSR measurements, which may form tracks after being processed by the data processor. Friend-or-foe identification responding signals work by associating with their corresponding target signals. They are often acquired simultaneously, and the association between them is usually realized by a gate set at the position of the measurements, the size of which is used to balance the reliability of these associations and the decrease in probability of false associations.

There are many difficulties in data fusion between the radar and ECM/CSM. The radar provides complete position information on targets and their motion condition information, while the ESM/CSM only provides their angle information, whose accuracy is lower than that of the radar. In addition, radar data and ESM/CSM data are asynchronous and cannot be associated simply. First of all, the position data of the radar and ESM/CSM should be adjusted to the same time, and then some criteria should be taken to determine if the two data sources originate from the same target.

There have been many studies on association determination algorithms in recent years, mainly on the tracking and filtering of targets' position data and radiant sources' angle data, angle value comparison between each object and each radiant source after being filtered, and association value calculation of any two angle values. The definition of an association value makes algorithms different from each other. This value is often defined with the angle mean square error over a period of time, but sometimes by means of the fuzzy mathematical method. In addition, the tracking algorithm with the association of angle information is a new research field as well. It has a higher accuracy of determination by converting the association of position data acquired by the radar and ESM/CSM to tracks into that of the two sensors to multiple targets.

19.8.3 Typical Working Mode

Take E-3A for example. The working modes of its airborne radars include AEW, over-the-horizon, marine, passive, and so on. The AEW mode can be divided into look-up and look-down modes. The former is mainly for the detection of flying targets in the airspace above the aircraft's horizontal line of sight, while the latter is mainly for the detection of flying targets in the airspace below, especially so with flying targets at low altitude.

19.8.3.1 AEW Mode

The AEW mode is mainly for the detection of high or low-altitude moving targets. The radar usually adopts pulse compression technology to ensure maximum detection range and range resolution. The look-up AEW mode is mainly used to detect moving targets above the aircraft's horizontal line of sight, and its antenna elevation should be above 0°. As there is no clutter jamming in the detection region, it can adopt PD and PE detecting methods. The look-down AEW mode is generally applied to detect moving targets in strong ground and sea-wave clutters in the area below this line of sight. Its antenna should be below 0°. Because there is strong clutter jamming in the detection region, it must adopt the PD detecting method. Applications of the AEW mode are as follows.

1. *Air defense and detection of moving targets at high and low altitudes.* The AEW mode of large AEW aircraft is mainly for air defense. In this mode, the radar's detection range for small aerial targets is about 300 km. For instance, the detection range of E-3A and E-2C is 400 km and 270 km, respectively. For large targets, the maximum detection range should be about 600 km, and the maximum detection range of E-3A and E-2C is 667 km and 560 km, respectively. Though capable of early air warning, the small AEW aircraft's radar detection range toward air targets is limited, and its detection range is only 50–70% of that of big AEW aircraft. In addition, as the beam width of radar antenna is limited, it has difficulty considering both high and low altitudes, and therefore a certain search mode will be employed by the antenna. The airborne PD radar usually changes its pulse repetition frequency in the search process to achieve optical detection. In the look-up AEW mode, the flight altitude of AEW aircraft depends on the air defense warning task, warning area, and performance of the AEW aircraft itself, etc.

   Because of the influence of ground clutter, the look-down detection range of the radar in AEW mode is smaller than that in look-up mode (only about 80% that of the latter). While in look-down mode, the radar can detect the target in the frequency domain and its velocity with application of the PD method. However, detecting the moving targets at close range makes no sense, so the look-down mode is usually applicable to intermediate and long-range warning. In look-down mode, the radar's detection capability relies on the flight altitude, because of the influence of its line of sight, and if the flight altitude is relatively lower, the AEW function may be affected. Besides, the direction of the antenna beam also affects the warning airspace of an AEW aircraft: if it is horizontal, targets beyond the line of sight may be detected; if it lowers one level, the middle range targets and surface targets could be detected. Therefore, flight altitude and antenna direction are two important factors affecting the detection capability of a radar.
2. *Guiding and intercepting air targets.* The radar of the giant AEW aircraft can provide continuous intercept vectors for multiple targets at the same time, which means that it can provide "one-to-one" or "several-to-several" guidance modes. When intercepting air targets, guidance can be divided into rough guidance and accurate guidance according to the time interval between information provisions and information amounts.

As modern war fighters are equipped with good-performance fire control radars, when intercepting air targets, AEW aircraft usually adopt "several-to-several" rough guidance. Which guidance mode should be used depends on the type of operations. Generally, AEW aircraft should undertake warning tasks within secure regions in which their own security can be guaranteed. Accurate guidance is necessary in cases where targets are judged to possibly be very dangerous, so continuous tracking is needed. For faraway targets whose threat degree cannot be determined, rough guidance can be used. For the same batch of targets, rough guidance should be used first, and when conditions are right, accurate guidance can be applied.

As the detection range of the radar in the AEW mode of giant AEW aircraft is relatively far, the same guidance and interception method can be adopted in the look-up and look-down modes. For small AEW aircraft, the detection range in AEW mode is relatively near. If adopting continuous accurate guidance, the AEW aircraft itself will be in danger, and therefore rough guidance should be the first priority.

The radar of the E-3A AEW aircraft has all-elevation target surveillance capability. Only when the radar measures the altitude of targets can it realize accurate guidance. If the radar provides no target altitude, then it can only work in the plane's accurate guidance mode, in which case the AEW aircraft must acquire altitude information on targets through other methods.

19.8.3.2 Over-the-Horizon Mode

When in over-the-horizon mode, the AEW aircraft is mainly responsible for long-range warning tasks and detecting enemy aircraft carriers or warship formations. In this case, the radar antenna beam should be directed toward the aircraft's horizontal line of sight. The application of the radar in this mode is as follows.

1. *Long-range warning.* To guard against enemy aircraft carriers and warship formations, it is required that the maximum detection range of the AEW aircraft radar be fully applied. In this mode, the aircraft should adopt a maximum cruise altitude of 9000–12,000 m, and the radar should also work in its maximum range, so that it has about 10 hours to warn against these enemy targets.
2. *Command and navigation.* AEW aircraft attacking marine targets can be used to command aircraft and warship formations to attack enemy surface targets. For aircraft carrying air-to-ship missiles and with good-performance airborne fire control radars, rough guidance can be adopted to brief the target position, properties, and kinematic factors. For those carrying bombs but without airborne fire control radars, accurate guidance should be used to brief their own flight elements and enemy targets' positions and kinematic factors. When commanding the operation of a fleet, combat command can be handed over to the commanding center of the fleet because the fleet moves slowly, and what the AEW aircraft should do is to brief position, kinematic factors, formation, target properties, and quantity of enemy surface targets.

19.8.3.3 Marine Mode

When in the marine mode, the AEW aircraft is used mainly to search for small marine targets and submarine targets in cluttered environments such as coastlines and islands. The applications of this mode are listed below.

1. *Detecting and attacking small marine targets.* The AEW aircraft is moving along its coastline or within a responsible region within its native land, undertaking the offshore defense air command task, or providing the ground commanding center with information about the position, kinematic factors, formation, target properties, and quantity of enemy marine targets.
2. *Searching for submarines and commanding anti-submarine operations.* When used to search for submarines, the application of AEW aircraft is essentially the same as that of detecting and attacking small marine targets. In the surveillance region, the flight altitude can be lowered to about 1000 m. The aircraft can practice submarine search directly, or command and coordinate anti-submarine helicopters and ships in searching for and attacking submarines.

19.8.3.4 Passive Working Mode

1. *Applicable in electronic warfare.* The passive working mode of the E-3A AEW aircraft is applicable to electronic warfare. Being a passive detector, it can cooperate with other ESMs to detect the position of enemy radar signal sources and detect enemy jam transmitters. Especially when enemy forces conduct detection and jamming toward the aircraft, the passive working mode enables the aircraft to take counter-reconnaissance and anti-jamming measures. In this mode, the data fusion performance of the radar with other passive detection systems will influence the engagement competency directly.
2. *Applicable to self-defense survivability.* Being an effective means of low-altitude air defense, the AEW aircraft poses a great threat to enemy air defense, and thus it is also a prime target for enemy air defense and air forces. In this mode, the AEW aircraft can effectively avoid enemy anti-radiant missile strikes, so the passive mode is appropriate for self-defense survival. In actual combat, the passive mode of the radar should be used together with other self-defense equipment of the aircraft.

19.9 Application in Air Warning Radar Network

Air warning missions cannot be satisfactorily completed by a single radar or many radars working independently. Compared with single radars, radar networks have some remarkable advantages such as high detection probability, broad detection region, high tracking accuracy, good anti-stealth performance, and strong electronic countermeasures. We discuss in brief the method of data processing in the air warning radar network.

19.9.1 Structure of Radar Network Data Processing

A radar network is an organic whole integrating radars with various systems, frequencies, and polarization modes deployed in such a manner that information about air situations from these radars is trapped and delivered as if through a piece of netting and then integrated and controlled in the central station, forming an organic whole. It is mainly made up of three parts: terminal radar, data transmission, and central processing systems. Data calibration, correlation, and fusion are the core of the last system [427]. Figure 19.8 shows the data processing structure of a radar network.

In this figure, a radar station in the network consists of radars, measurement recorders, a series of receivers, track data computers, a transmitter, and a display. Tracking takes place at every station at the same time, and includes track management of different levels [375, 428–432]. Tracks can be categorized into two groups: local and systemic. The former is produced and updated according to the measurements transmitted by local radars, while the latter is a unique clear track acquired from the association and fusion of several local tracks. Both local and system tracks are stored at the memory device of each station. The target report conveys the most practical and latest information to a radar station, but the not-so-reliable data are not filtered yet. It may also receive from many other stations a great deal of track information, which may be obsolete but relatively reliable: track ID and quality, latest update time, target coordinates, and speed and type.

Measurement data is primarily used to associate with local tracks or to produce new tracks. Track information should first be associated with ultimate tracks to update system tracks, or should be fused with local tracks to produce new system tracks [13, 54, 60]. The initiation of a new track,

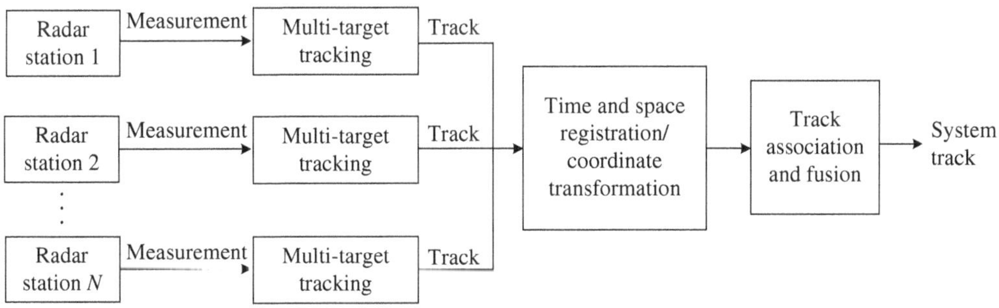

Figure 19.8 Data processing structure of the radar network

or the updating of local or system tracks, or the production of new system track, is always sent out by system tracks. As long as at least one radar detects a target, tracks of all the stations will be monitored. The track will be cancelled when it fails to be associated with any measurements within two scanning periods of the slowest antenna. The same track ID for the same target of different stations is provided through track ID management. This means that when a false track ID switch occurs, track ID management serves to calibrate the different track IDs of the same initial track, and re-change the final track ID of all stations.

19.9.2 Key Technologies of Radar Network Data Processing

19.9.2.1 Time Registration

The scanning of the antennas within a radar network is usually asynchronous, and it would be difficult to realize data fusion if there was no unified time standard. Hence, the same time benchmark is required for the information reported by the radars of a network. In other words, these observations must be synchronized before being fused, which is also called "time registration." Time registration refers mainly to the process of synchronizing tracks detected by various synchronous or asynchronous radars at different times to the same time on the time axis. First, a unified time benchmark should be set. This can be done by means of the high-stability clock of satellite positioning systems, like the double satellite positioning system of China, which meets accuracy requirements but is susceptible to interference. Alternatively, it can be completed by utilizing high-stability rubidium atomic clocks in every station as synchronous signals, but time registration should be made before utilization. Refer to Section 5.2 for more details on time registration.

19.9.2.2 Space Registration and Coordinate Transformation

Accurate Position of Single Station Radar

The accurate positioning and spatial geometric calibration of monostatic radars in the radar network is the basis for stations to share data through coordinate transformation, and to conduct accurate positioning for targets and data fusion for measurements. The positioning and spatial calibration of radars can be realized with the accurate positioning service provided by China's double satellite positioning system.

Coordinate Transformation
In the radar network system, the target measurements reported by each radar are located in its local coordinates. Therefore, before information fusion for multiple stations, the measurement values of all other radars should be transformed into a unified coordinate system. Common unified coordinate systems include the following.

1. *Earth-centered Cartesian coordinates:* inertial coordinates, appropriate for strategic early warning and detection systems with large surveillance ranges.
2. *Cartesian coordinates whose fusion center is the coordinate origin:* inertial coordinates. When these coordinates are adopted, the altitude of targets out of the fusion center's line of sight but detected by other radar is negative.
3. *Geographical coordinates:* non-inertial coordinates. The fusion result can be displayed directly on electronic maps.
4. *Stereographic coordinates whose fusion center is the coordinate origin:* inertial coordinates. Conformal mapping is the most important merit here, which keeps the angle values between tracks unchanged after coordinate transformation. However, it involves complicated calculations during coordinate transformation.

To sum up, the selection of a unified coordinate system depends on the mission of the radar network and the performance of the radars. For coordinate transformation, refer to Section 5.3.

19.9.2.3 System Error Registration

As the target observations of the radar network always have random errors and system errors, errors may occur between different radar tracks of the same target. The modification process of these errors is called "error registration." The random errors mainly derive from random observation noise and targets' random maneuvers, which can be eliminated through filtering methods. But system errors (such as inaccuracy in radars' angle and range measurement, radar stations' position and due north, and the error of coordinate transformation approximation algorithms) cannot be eliminated through filtering, so it is necessary to estimate errors in advance and make compensation afterwards. The most desirable method of error registration requires that the target position is known, which is difficult to meet in practical circumstances. At present, most studies focus on the error matching algorithms in unknown target circumstances, such as the RTQC algorithm based on spherical/polar coordinates and the GLS algorithm based on geodetic coordinates.

19.9.2.4 Multi-station Track Association and Fusion

The radars in the air warning radar network have their own information processing systems, which collect a great number of target tracks, make comprehensive judgments on the large number of tracks from different systems, and determine target tracks. This process is multi-station track association. Multi-station track association algorithms normally include statistics-based and fuzzy mathematics-based methods. The former mainly consist of the weighted, modified, sequential, classical assignment, double threshold [433] algorithms. The latter mainly consist of the fuzzy double threshold, fuzzy comprehensive function, and multi-factor fuzzy comprehensive decision algorithms. Following track association, the fusion center fuses the associated tracks to obtain more accurate system tracks; this process is known as "track fusion." Track fusion algorithms include the simple convex combination, Bar-Shalom–Campo, and maximum posterior probability state estimation fusion algorithms [433].

19.10 Application in Phased Array Radar

The past decade has witnessed the increasing importance of phased array radars in warfare. They have replaced traditional mechanical scanning radars on a large scale, and are used widely in ground-based, shipboard, and airborne radars. The application of varied new and high technologies has greatly enhanced the performance of phased array radars, and also compounded the complexity of the process of radar data processing. The example of the MESAR will be cited below to illustrate the application of radar data processing techniques.

19.10.1 Functional Features

Developed by Britain in the late 1980s and early 1990s, the MESAR is an advanced experimental active phased array radar exploiting the state-of-the-art digital and solid-state microwave technologies then available [434]. As an experimental platform, it is used to test new technologies to be adopted in the phased array radar system. It adopts an octagon array face with a diameter of 1.8 m and possesses 918 transmitting elements, in which only 156 elements are installed with transceiver modules to reduce the cost. The system frequency range is 2.7–3.3 GHz, the peak power of each module is 2 W, the average power is 100 W, the beam width is approximately 3°, and it transmits horizontally polarized waves.

19.10.2 Data Processing Procedure

19.10.2.1 Measurements Processing

When the radar updates a known track, the pencil beams can only illuminate the track's adjacent space, which may contain the association thresholds of other targets, so the updating of this track cannot be accomplished at once. Associating a false measurement with the track will decrease the track accuracy, and therefore the MESAR adopts the simple nearest-neighbor algorithm in most cases when conflicts occur in the association and updating of tracks.

19.10.2.2 Track Filter

The MESAR uses a constant-velocity Kalman filter with a maneuver detector strategy for track smoothing. It provides a filter with adaptive memory space according to the mobility of the target. When the target is judged to be in motion at constant speed, the filter adopts low process noises; when the target is judged to be maneuvering, process noises are enhanced to adapt the filter to the maneuvers of the target. This is the adjustable white noise maneuvering target tracking method discussed in Chapter 9.

19.10.2.3 Adaptive Track Update Rate

The next update time of the MESAR is calculated according to two factors.

1. Guaranteeing a high probability that the target will fall within the 3 dB beam on the next update. This value depends on the filtering precision of the azimuth and the elevation of the track, and on the value of azimuth and elevation of the track's possible maneuver. It implies that an update of the track must be completed before the target falls outside the 3 dB beam width (i.e., before the target is lost).
2. Obtaining the expected location and speed tracking accuracy. This means that if the track is not updated in time, the expected tracking accuracy cannot be achieved even if the target is not missing.

If the updating time is too long, this means that one of the two above conditions is not satisfied. If the updating time is too short, this means that the time the tracking function takes has exceeded the optimal value.

19.10.3 Test Examples

19.10.3.1 Multiple-Function Operation

Figure 19.9 shows a typical display output screen of the MESAR. This PPI screen displays the result of search and tracking of the targets within its scanning airspace, where the range scale is 55 km, bearing 45°, and elevation 50°. The screen information includes the following.

1. *Search for measurements:* these measurements are obtained mainly thanks to the search function and they include false measurements due to false alarms.
2. *Track update validated:* these measurements are validated as true targets.
3. *Track update lost:* an updated track fails to be detected.
4. *Slow tracks:* these tracks fall below the specified velocity threshold.

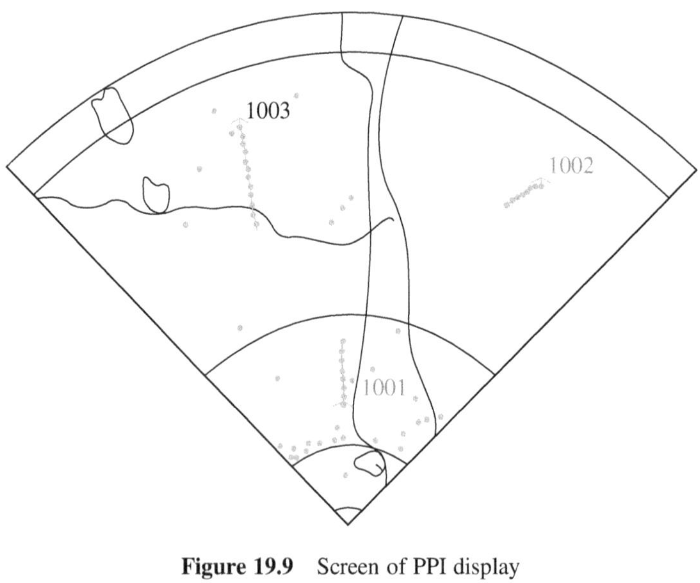

Figure 19.9 Screen of PPI display

19.10.3.2 Adaptive Tracking

The MESAR's track update rate and waveform are adaptive to target environments. Figure 19.10 shows the tracks and measurements of two aircraft taking highly maneuverable turns, flying from the top left, zigzagging and taking detours, making turns close to the radar and flying straight to the top right. The change in measurement density reflects the process that the track update rate varies with the prediction confidence. In other words, the update rate is the lowest in straight flight, increases when the target is approaching the radar, and once maneuvers are detected, reaches a maximum to maintain tracking accuracy.

19.10.3.3 Split Track Tracking

When the radar detects a track splitting process, it must make the two split tracks sufficiently accurate as soon as possible. The MESAR can initiate new tracks for "extra" measurements, while updating tracks. Figure 19.11 illustrates the radar's function when an aircraft flying at 10 km altitude projects a bomb, which produces split tracks.

The MESAR is a multi-functional phased array radar system capable of adaptive tracking and control in real time. But it needs to be improved with respect to the tracker in two ways. For one thing, more advanced track filters and measurement association methods can be adopted, such as the IMM, JPDA, and MHT techniques. For another, it can be improved by developing new functions, such as variable beam width, simultaneous multiple receiving beam processing, and super-resolution techniques, as well as optimizing the performance for the rotating system of the antenna. These problems have been tackled in succeeding research on MESAR II.

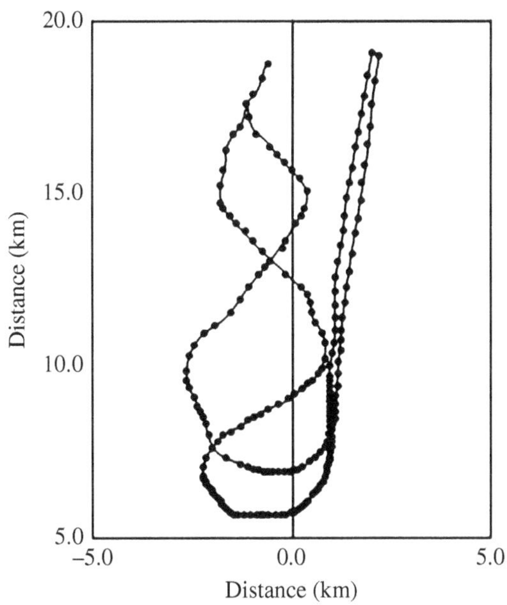

Figure 19.10 Adaptive update rate

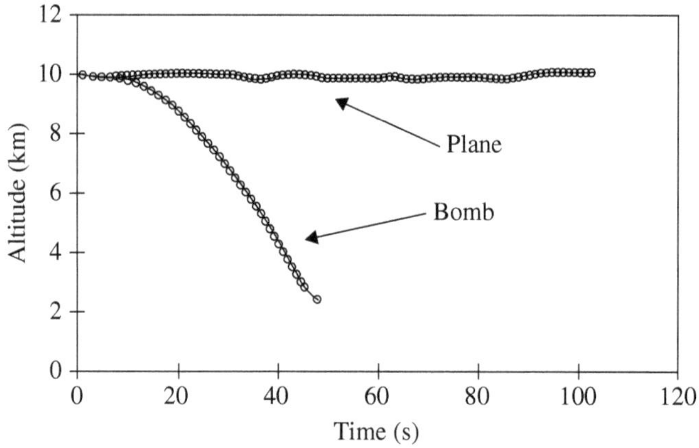

Figure 19.11 Split track tracking

19.11 Summary

Aiming to provide readers with a clearer picture of how the radar data processing technology is applied in engineering implementations, this chapter presents a comprehensive introduction to some of its typical applications in the following systems: air traffic control, shipboard navigation, shipboard radar clutter suppression, ground-based laser radar, marine surveillance, fleet air defense, shipboard artillery control radar, AEW radar, air warning radar network, and phased array radar.

These applications, although in different fields, have some similarities. Firstly, all the algorithms suitable for radar data processing have common mathematical bases: they are based on dynamic system simulation, filtering, statistical decision, optimal control, and management theory. Secondly, the common purpose of incorporating radar data processing techniques is to selectively reduce the data quantity from sensors to users, to select information related most closely to judgment and control, and to display the information in a way most satisfying to operators. Lastly, the demand for radar data processing varies with the application system, which requires different solutions.

20

Review, Suggestions, and Outlook

20.1 Introduction

This book features topical discussions. Each special topic starts with an introduction to the basic concepts and filtering methods, followed by theoretical analyses and studies on the topic, and concludes with examples illustrating the practical implementations of the theories, technologies, and techniques with some dos and don'ts. With target tracking and data association as the main thread, it presents comprehensive, systematic, in-depth discussions of the latest developments in the radar data processing technology, delves further into some major subjects, and provides a multitude of references.

This chapter offers a review of the main theoretical achievements covered in this book, suggestions on several issues, and a forecast of future trends in radar data processing techniques.

20.2 Review of Research Achievements

20.2.1 The Basis of State Estimation

Chapters 1 and 2 focus on some basic concepts of radar data processing and several time constant parameter estimation methods commonly used in basic linear systems: ML, MAP, LS, and MMSE.

This leads to parameter estimation in dynamic situations in Chapter 3, which starts with an introduction to state estimation methods in linear systems, including the establishment of a system model, corresponding filtering models, filter initialization methods, definition and judging method of filter stabilization, controllability and observability of random linear systems, and steady-state KF.

State estimation of nonlinear systems is dealt with in Chapter 4, in which nonlinear filtering methods including EKF, UKF, and PF are discussed. A comparative analysis is made of the tracking accuracy and computational load of some linear and nonlinear filtering algorithms with the same target in the same simulation environment.

Radar Data Processing with Applications, First Edition. He You, Xiu Jianjuan, and Guan Xin.

20.2.2 Measurement Preprocessing Technology

The preprocessing technology of measurement data is essential in the correct processing of radar data. Effective techniques can reduce the computational load, with tracking accuracy improved. Chapter 5, therefore, introduces some techniques for preprocessing measurements, including spatial and temporal synchronization of data from different sensors and data compression. Temporal synchronization serves mainly to resolve the inconsistency between data rates from different sensors, while spatial synchronization chiefly guarantees that all data formats can be unified into the same coordinate system. Data compression techniques are used to reduce the calculation load of subsequent data processing and improve tracking efficiency.

20.2.3 Track Initiation in Multi-target Tracking

In terms of multi-target track initiation in Chapter 6, this book describes initial wave gates and correlation wave gates and then introduces techniques for multi-target track initiation.

Track initiation techniques can be conveniently divided into two general categories. One is target-oriented sequential processing techniques, including the logic-based and modified logic-based methods. These are mostly used in track initiation of targets in sparse clutter, for their fast initiation speed but poor initiation results in the presence of dense clutter.

The other category is batch processing techniques, including the HT and MHT methods. These are very effective for initiating tracks of targets in dense clutter, but need many rounds of scanning to initiate a track. Finally, a comparison is made in performance between the two categories in the same simulation environment.

20.2.4 Multi-target Data Association Method

ML-based and Bayesian data association approaches are compared and analyzed in Chapters 7 and 8, respectively. The former is based on the likelihood ratio of observation sequences; in other words, it does not yield the probability that the sequence is correct. It includes track splitting, UML, 0-1-IP, and GC algorithms. All these algorithms adopt batch processing tools, which involve extra calculation.

Bayesian approaches to data association are based on Bayesian rules. They perform recursive calculations and can be realized with computers. Two Bayesian approaches are discussed: suboptimal and optimal. The first deals only with the latest set of validated measurements, mainly including NN, PDA, IPDA, and JPDA. The second computes association probabilities for each sequence of measurements, mainly including the OBF and MHT. The OBF can have excessive computing requirements.

20.2.5 Maneuvering Target and Group Tracking

Maneuvering target tracking approaches in Chapter 9 mainly include maneuver detection integrated and adaptive tracking algorithms. The former can be further divided into two types: filter-gain adjustment and filter-structure adjustment. Adaptive ones include the MIE, multiple, Singer, current, IMM, and jerk models. The performances of these algorithms were compared through simulation analyses.

In the group tracking part (i.e., Chapter 10), several typical algorithms are compared with respect to group initiation, separation and association, and group speed estimation. A detailed discussion is made of two typical (centroid and formation) group tracking algorithms in terms of track update, merging and splitting of group targets.

20.2.6 *Multi-target Tracking Termination Theory and Track Management*

Chapter 11 starts with a discussion of several multi-target tracking termination techniques based on NNA algorithms: SPRT, TG, CF, Bayesian, and AN Bayesian. It then deals with track ID and track quality management techniques, including methods of single and double-track ID management, selection of initiation criteria, and track cancellation by means of track quality. In this part, the track quality management is also expounded in single and multi-station cases.

20.2.7 *System Error Registration Issue*

Radar observation systems have two types of error: random and system. Random errors can be eliminated through various filtering methods but system errors are deterministic, and cannot be handled in this way. They need pre-estimation and compensation, and this process is called error registration. As proven in practical applications, target tracking errors may become larger than their theoretical values due to system errors in multi-radar network tracking systems. When they are too large, the tracking results of multiple radars are even worse than those of single radars. In the worst situation, many tracks may be generated for the same target, which leads to ambiguities.

Therefore, Chapter 15 is devoted to system error registration, covering modeling methods and registration algorithms of mobile radar systems, with a comparative analysis of MLRM and ASR algorithms.

20.2.8 *Performance Evaluation of Radar Data Processors*

The performance of radar data processors involves myriad factors, so does its evaluation index systems. In Chapter 17 of this book, the performance evaluation index systems of radar data processors are discussed in terms of average track initiation time, accumulated number of track interruptions, track ambiguities, accumulated number of track switches, track accuracy, maneuver tracking capability, ratio of false track, divergence, track capacities, detection probability of radar networks, response time, etc.

20.2.9 *Simulation Technology of Radar Data Processing*

This book gives an overview of basic knowledge on system simulation in Chapter 18, and presents methods of generating uniformly and normally distributed random numbers in Monte Carlo simulative tests. Simulation examples of radar data processing algorithms are given to illustrate the integration of system simulation and radar data processing technologies with the aim of tackling practical problems in radar data processing.

20.2.10 Applications of Radar Data Processing Techniques

This book addresses itself to some data processing methods developed according to the intrinsic features of different radars, including passive, phased array, and PD radars in Chapters 12, 13, and 14.

Passive radars are dealt with in terms of spatial association of measurements, optimal deployment of passive sensors under the rule for the minimum area of position concentration ellipses, and passive location using time difference of arrival.

The discussion of PD radars focuses on such typical algorithms as optimal distance–velocity coupled tracking and target tracking of Doppler measurement integrated radars. For algorithms of target tracking by Doppler measurement integrated radars, the discussion focuses on USEKF, USUKF, DUKF, and IMM-DUKF algorithms, with the performance of CMKF, UCMKF, EKF, and USEKF in two simulation environments compared and analyzed.

When it comes to phased array radars, tracking filtering methods are discussed in terms of multi-target processing, variable sampling interval filtering, and resource scheduling strategies, with stress laid on three adaptive sampling approaches: SSGF, IMM, and PCT, whose performance in simulations is compared and analyzed.

With regard to radar data processing applications, research on the data processing technology of network radars is presented in Chapter 16 in terms of performance, indexes, and optimal deployment from the perspective of the design and analysis of radar networks. Data processing procedures of mono-, bi-, and multistatic radar networks are also introduced, as are track association techniques – the core of the data processing process of radar networks.

Radar data processing is the process of estimating the tracks of targets and forecasting their future positions with the data provided by radars. In practical implementations, track estimation is not the ultimate aim of radar systems, but users need to use the estimations to make judgments and then take action as required. Therefore, the last section of this part (i.e., Chapter 19) is devoted to the applications of radar data processing techniques in the ATC, shipborne navigation radar, shipborne radar clutter suppression, ground laser radar, marine surveillance, fleet air defense, AEW radar, aircraft warning radar network, and phased array radar systems.

20.3 Issues and Suggestions

20.3.1 The Application of Data Processing Technology in Other Sensors

The extension of the radar data processing approaches discussed in this book to other sensors is a topic to be further explored, because modern combat systems cannot depend only on radars to achieve optimum combat results. It is essential for them to incorporate multi-sensor assemblies integrating various active and passive (TV, sonar, infrared ray, laser, ESM, and ELINT) detectors covering a broad band of frequencies to provide various observation data, from the optimization and integration of which combat information (including target detection, state estimation, target properties, behavioral intention, situation evaluation, threat analysis, fire control, ECM, combat modes, and decision aids) can be acquired in real time.

20.3.2 Track Initiation in Passive Sensor Tracking

Passive sensors have difficulty determining the initial states of single targets. If the errors in the assumed values of the initial state and covariance are large, the filters tend to diverge. In the multi-target case, passive sensors face more difficulties with the effective initiation of tracking filters, and dynamic choice and allocation of association gates.

Track initiation techniques of passive sensors, therefore, are worthy of extensive studies. Besides, research efforts in the area of multi-radar integrated tracking should be directed at track initiation techniques, and the determination of the number of radars needed in track initiation based on the compromise between performance and computing load according to tactical requirements.

20.3.3 Non-Gaussian Noise

The radar data processing techniques discussed in this book are based on the assumption that both measurement noise and process noise are Gaussian. This is an approximation to some extent, but in reality neither of them is Gaussian. Therefore, techniques of radar data processing in the presence of non-Gaussian noise are a topic for further practical research.

In addition, colored noise-related subjects need to be explored, since both the measurement noise and process noise in this book are assumed white noise.

20.3.4 Data Processing in Non-standard and Nonlinear Systems

Except in Chapters 4 and 12, all data processing models are based on discrete standard system models. Although tremendous achievements have been made in the research on data processing of non-standard and nonlinear systems in their long research histories, numerous problems have not been solved yet, and many need further study. An example of these in non-standard system models is system state equations, including control terms, process noise, and measurement noise.

20.3.5 Data Processing in Multi-radar Networks

Up to now, enormous research efforts have been channeled into the data processing problems of multi-radar networks, but the theoretical findings have not yet produced the desired practical results. This is because many of them are based on ideal assumptions (e.g., Gaussian distribution, measurement synchronization, and no system deviations), which cannot be satisfied in all engineering environments. Therefore, modifications and improvements in the present theories and methods according to the actual engineering requirements have become an urgent problem.

Further, while the possibly great benefits brought about by radar network information fusion have been widely recognized, the guarantee of real-time fusion efficiency of practical radar systems (where various detection errors tend to occur) has long become knotty in this field and further research still needs to be undertaken.

20.3.6 Joint Optimization of Multi-target Tracking and Track Association

In practical applications, the problem to be tackled concerning data processing of distributed radars is not the respective optimization of each processor's performance, but the joint optimization of multi-target tracking and track association in the whole processing system. So far, little research has been done on this topic and no definite conclusions can be drawn in the unclassified literature.

20.3.7 Comprehensive Utilization of Target Features and Attributes in Multi-radar Tracking

Section 12.7 touches briefly on the attribute data association issue of passive radars, instead of delving deeply into how to make comprehensive use of position, kinetic, feature, and attribute parameters, and subjective knowledge.

In the military field where electronic warfare is becoming increasingly complex, system errors are liable to arise in sensor systems due to serious clutter and false alarms, and especially false associations are apt to result from the tracking and association based only on statistical distances. In this case, it is vital that various features and attributes of targets be fully tapped in track processing so as to improve the reliability of the fusion system for effective, accurate tracking.

Whereas some approaches to using such information have been introduced in the references, they are still far from enough for practical applications.

20.3.8 Comprehensive Optimization of Multi-radar Information Fusion Systems

Chapters 3 to 10 are concerned with target state estimation, multi-target tracking, maneuvering targets, and group tracking. However, each of these problems is approached separately. When it comes to multi-target tracking, track association, and state estimation, the comprehensive optimization of information fusion performance is not considered against the overall performance of the system.

In practical applications, the front ends of distributed multi-radar information fusion systems usually cascade multi-target tracking data processors. What needs to be considered is not the separate optimization of every processor's performance, but the overall optimization of multi-target processors and distributed information fusion systems in cascade connections. This necessitates adjusting every local performance index, adding feedback layers, constructing unitary performance indexes (which are then embedded in each processing process at every stage), and thus completing the optimization of unitary fusion performance.

Introducing feedback information at distributed fusion nodes to sensor stages can improve the tracking accuracy of these stages significantly, certainly on condition that both data and track associations are correct. Little research has been done on how to obtain joint optimized results in the presence of false and missed associations in both stages, and no definite practical conclusions are available in open-source references up to now.

20.3.9 Tracking Multi-targets in Complex Electromagnetic Waves and Dense Clutter

The four groups of approaches to radar data association (ML and Bayesian, maneuvering and group target tracking) described in Chapters 7 to 10 are incapable of responding to the fresh challenges posed by the complex electromagnetic environment in modern warfare with multi-radar and multi-target tracking. Topics for future research in this area of practical significance to engineering include:

1. multi-radar, multi-target tracking in the presence of blanket jamming;
2. multi-radar, multi-target tracking in the presence of massive deceptive jamming;
3. multi-radar, multi-target tracking in the presence of compound jamming (i.e., various types of man-made interference simultaneously);
4. tracking multi-targets in complex electromagnetic waves and dense clutter;
5. multi-target tracking with low resolution and multiple paths.

20.4 Outlook for Research Direction

Despite continuous advances in the radar data processing technology, there are still many areas needing further research and exploration. Some main research directions are expounded below.

20.4.1 Information Fusion and Control Integration Technology of Multi-radar Networks

In modern warfare, multi-radar networks are needed in the joint detection of the same area and the integration of time, space, and detection information. So, the integration and coordination of the information from multiple radars in networks is a new topic of study.

Some approaches to single-radar, multi-target tracking (including JPDA and MH) are covered in Chapter 8, but further research is needed to investigate whether and how they can be extended to multi-radar cases. For example, there are no plans as yet for the OBF, 0–1 integer programming, NNPDA, IMM-PDA, and DPDA techniques to be used directly in multi-radar, multi-target integrated tracking. The research on both how to use them in multi-radar systems and their performance analysis is of great real significance.

Furthermore, in order to perform effective management and centralized resource allocation, multi-radar information fusion centers need (according to the realistic requirement of information fusion and situation awareness) to control each radar's work and mode, and hence provide complete, correct, universal, continuous, and timely air situation information about the entire combat zone.

20.4.2 Joint Optimization of Target Tracking and Identification

This book is concerned mostly with target tracking. However, in the complex battlefield nowadays, it is essential that the sensor system be able to track targets in hundreds of batches, and identify these various and numerous targets, with a timely and quick response.

20.4.3 Integration Technology of Search, Tracking, Guidance, and Command

The integrated use of multiple radars and multi-functional radars requires that the data processing system have functions such as search, tracking, guidance, and command. In reality, it is natural that integrated data processing systems with more comprehensive functions and stronger performances will be needed with the continuous advancement in technology.

20.4.4 Multi-radar Resource Allocation and Management Technology

Multiple radars constitute the complementary system of multi-sensor systems, and hence these sensors should be properly managed according to some working criteria so as to obtain optimal data collection performance. Sensor management usually involves space, mode, and time management. Primary topics of study in this field are performance prediction, target assignment of sensors, control standards for spatial and temporal ranges, configuration and control strategies, interface techniques, target assignment priority techniques, and indication and transfer techniques.

20.4.5 *Database and Knowledge Base Technology in Radar Data Processing*

According to specific tactical backgrounds, the establishment of radar data processing databases and knowledge bases, and the adoption of high-speed parallel inference mechanisms, are key issues in the engineering and practical applications of radar data processing techniques, which require special attention in the research to come.

20.4.6 *Engineering Realization of Advanced Radar Data Processing Algorithms*

The radar data processing technology has been widely applied in both military and civil fields, but data processing algorithms are usually very simple and need approximation to different degrees. The growing complexity in engineering applications, and the increasingly demanding requirements of data processing in effect, call for the application of advanced radar data processing algorithms, enabled by the rapid development in computer technology. How to apply the theoretically advanced data processing algorithms to engineering reality has become an important and arduous task in the radar data processing field.

20.4.7 *High-Speed Calculation and Parallel Processing Technology*

In order to satisfy real-time requirements and realize continuous tracking of high-speed and highly maneuverable targets in complex environments, it is necessary that data processing systems be able to tap more detection information, advanced algorithms, and hardware and software platforms. Besides, they are also required to be capable of breaking down the data processing algorithms into parallel ones suitable for parallel machines, and developing corresponding software and hardware for parallel calculations, so as to adapt themselves to the requirements of these algorithms in their processing capability.

Research topics in this respect include the following issues for engineering implementations: real-time data transmission with high speed and large volume, systematic calculations with high speed and efficiency, and task assignment and synchronous control in parallel processing.

20.4.8 *Establishment of System Performance Evaluation Methods and Test Platforms*

The explanation of radar data processing performance prediction and evaluation methods given in Chapter 17 is confined to theoretical analysis. Moreover, some other problems need to be solved urgently. One of them is how to establish evaluation systems and test platforms, so as to measure the performance of data processing algorithms and analyze and evaluate these algorithms, and the system performance, in a comprehensive and objective manner.

20.4.9 *Common Theoretical Models for Variable Structure State Estimation*

Because of the rapid development in sensor network systems and the mobile ad hoc network (MANET) technology, the structure and parameters of multi-source information systems can change with sensor nodes. Therefore, state estimation models adaptable to changes in the structure and parameters of these systems are an important subject of research in state estimation fusion in the future trend toward networking.

20.4.10 Automatic Tracking of Targets in Complex Environments

An urgent problem in application is the automatic tracking of targets with low detection probabilities, especially of small targets with high speed in the presence of massive residual ground and meteorological clutters and acute electromagnetic interference. This problem can be resolved through inter-frame filtering techniques, tracking before detection approaches, and advanced algorithms with the radar front end being unchanged.

20.4.11 Tracking and Invulnerability of Multi-radar Network Systems

In the era of information warfare, the network is the basic platform for multi-radar, multi-target tracking systems. When one or several radars are subject to observation ranges, enemy jamming, atmospheric conditions, and malfunctions, they can observe the target only over a part of the time period, which leads to discrete tracks and even track loss. However, through the real-time data exchange between radars in the net, different data can be shared in the shortest possible period of time, hence filling in the observation blanks, which gives rise to improvements in observation accuracy and minimization of the blind zones and instability of target tracks in the interference region.

Therefore, invulnerability to destruction is a guarantee of performance for any radar network system. That is to say, after some nodes of a radar system are destroyed, its target tracking system can maintain normal operation or resume it effectively in due time.

References

[1] Wang, X., Zhang, G., He, R., *et al.* 2000. *Radar and Detection – Modern war piercing eye*. Beijing: National Defense Industry Press.

[2] Huang, P., Yin, H., and Xu, X. 2005. *Radar Target Characteristics*. Beijing: Publishing House of Electronics Industry.

[3] Tong, Z. 2008. *Integrated Electronic Information System*. Beijing: National Defense Industry Press.

[4] Elias-Fuste, A.R., Broquetas-Ibars, A., Antequera, J.P., and Yuste, J.C.M. 1992. CFAR data fusion center with inhomogeneous receivers. *IEEE Transactions on Aerospace and Electronic Systems*, 28(1): 276–285.

[5] He, Y., Guan, J., Meng, X., *et al.* 2011. *Radar Target Detection and CFAR Processing*, 2nd edn. Beijing: Tsinghua University Press.

[6] Wu, S. and Mei, X. 2008. *Radar Signal Processing and Data Processing Techniques*. Beijing: Publishing House of Electronics Industry.

[7] Yang, J. 1994. *Battlefield Data Fusion Technology*. Beijing: Publishing House of Ordnance Industry.

[8] Chair, Z. and Varshney, P.K. 1986. Optimal data fusion in multiple sensor detection system. *IEEE Transactions on Aerospace and Electronic Systems*, 22: 98–101.

[9] He, Y., Wang, G., and Guan, X. 2010. *Information Fusion Theory with Application*. Beijing: Publishing House of Electronics Industry.

[10] Quan, T. 2009. *Target Tracking Advanced Theory and Techniques*. Beijing: National Defense Industry Press.

[11] He, Y. and Zhang, J. 2006. New track correlation algorithms in a multisensor data fusion system. *IEEE Transactions on Aerospace and Electronic Systems*, 42(4): 1359–1371.

[12] Mao, E., Long, T., and Han, Y. 2001. Digital signal processing of stepped frequency radar. *Acta Aeronautica et Astronautica Sinica*, 22(6): 16–20.

[13] Liu, X. 2010. *Digital Signal Processing*. Beijing: Publishing House of Electronics Industry.

[14] Maroulas, V. and Panos, S. 2012. Improved particle filters for multi-target tracking. *Journal of Computational Physics*, 231(2): 602–611.

[15] Llinas, J. and Waltz, E. 1990. *Mutisensor Data Fusion*. Boston, MA: Artech House.

[16] Bar-Shalom, Y. and Fortmann, T.E. 1988. *Tracking and Data Association*. New York: Academic Press.

[17] Farina, A. and Studer, F.A. 1985. *Radar Data Processing*, Vols 1, 2. Letchworth, UK: Research Studies Press.

[18] He, Y., Tang, J., *et al.* 1991. *Integrated Multi-radar Tracking Algorithm*. Yantai: Naval Aeronautical and Astronautical University Research Reports.

[19] He, Y., Wang, G., Lu, D., and Peng, Y. 2000. *Multisensor Information Fusion with Applications*, 1st edn. Beijing: Publishing House of Electronics Industry.

[20] Fu, M., Deng, Z., and Zhang, J. 2003. *Kalman Filtering Theory and Application in Navigation System*. Beijing: Science Press.

Radar Data Processing with Applications, First Edition. He You, Xiu Jianjuan, and Guan Xin.

[21] Kalman, R.E. and Bucy, R.S. 1961. New results in linear filtering and prediction theory. *ASME Journal of Basic Engineering*, 83D: 95–108.

[22] Benedict, T.R. and Bordner, O.W. 1963. Synthesis of an optimal set of radar track-while-scan smoothing equations. *IEEE Transactions on Automatic Control*, AC-8(2): 27–32.

[23] He, Y., Tang, J., *et al.* 1991. *Research on the Algorithm of Short Range Weapon System Multi-target Track Processing*. Yantai: Naval Aeronautical and Astronautical University Research Reports.

[24] Cantrell, B.K. 1973. *Description of an α, β Filter in Cartesian Coordinates*. Naval Research Laboratory Report, Distributed by NTIS.

[25] David, L.H. and James, L. 2008. *Handbook of Multisensor Data Fusion*. New York: CRC Press.

[26] Singer, R.A. 1970. Estimating optimal tracking filter performance for manned maneuvering targets. *IEEE Transactions on Aerospace and Electronic Systems*, 6(4): 473–483.

[27] Pearson, J.B. and Stear, E.B. 1972. Kalman filter applications in airborne radar tracking. *IEEE Transactions on Aerospace and Electronic Systems*, 10: 319–329.

[28] Peterson, I.R. and Savkin, A.V. 1999. *Robust Kalman Filtering for Signal and System with Large Uncertainties*. Boston: Birkhauser.

[29] Wax, N. 1955. Signal-to-noise improvement and the statistics of track populations. *Applied Physics*, 26(5): 586–595.

[30] Sittler, R.W. 1964. An optimal data association problem in surveillance theory. *IEEE Transactions on Military Electronics*, MIL-8: 125–139.

[31] Sea, R.G. 1971. An efficient suboptimal decision procedure for associating sensor data with stored tracks in real-time surveillance systems. Proceedings of the 10th IEEE Conference on Decision & Control, Miami Beach, FL, December, pp. 33–37.

[32] Singer, R.A. and Stein, J.J. 1971. An optimal tracking filter for processing sensor data of imprecisely determined origin in surveillance systems. Proceedings of the 10th IEEE Conference on Decision & Control, Miami Beach, FL, December, pp. 171–175.

[33] Bar-Shalom, Y. and Tse, E. 1975. Tracking in cluttered environment with probabilistic data association. *Automatica*, 11(5): 451–460.

[34] Fortmann, T.E., Bar-Shalom, Y. and Scheffe, M. 1983. Sonar tracking of multiple targets using joint probabilistic data association. *IEEE Journal of Oceanic Engineering*, 8(3): 173–183.

[35] Reid, D.B. 1979. An algorithm for tracking multiple targets. *IEEE Transactions on Automatic Control*, AC-24: 843–854.

[36] Singer, R.A. and Sea, R.D. 1971. A new filter for optimal tracking in dense multitarget environments. Proceedings of the 9th Allerton Conference on Circuit and System Theory, Urbana, IL, pp. 210–211.

[37] Bar-Shalom, Y. and Birmiwal, K. 1982. Variable dimension filter for maneuvering target tracking. *IEEE Transactions on Aerospace and Electronic Systems*, 18(5): 611–619.

[38] Kirubarajan, T. and Bar-Shalom, Y. 1998. IMMPDAF for radar management and tracking benchmark with ECM. *IEEE Transactions on Aerospace and Electronic Systems*, 34(4): 1115–1134.

[39] Zhou, H., Jing, Z., and Wang, P. 1991. *Tracking of Maneuvering Targets*. Beijing: National Defense Industry Press.

[40] Blackman, S.S. 1986. *Multiple-Target Tracking with Radar Applications*. Boston, MA: Artech House.

[41] Carlson, N.A. 1990. Federated square filtering for decentralized parallel processes. *IEEE Transactions on Aerospace and Electronic Systems*, 26(3): 517–525.

[42] Julier, S.J. and Uhlmann, J.K. 1997. A new extension of the Kalman filter to nonlinear systems. *SPIE*, 3068: 182–193.

[43] Gordon, N.J. 1997. A hybrid particle filter for target tracking in clutter. *IEEE Transactions on Aerospace and Electronic Systems*, 33(1): 353–358.

[44] Deng, Z. 2007. *Information Fusion Filtering Theory with Applications*. Harbin: Harbin Institute of Technology Press.

[45] Blackman, S.S. and Popoli, R. 1999. *Design and Analysis of Modern Tracking Systems*. Boston, MA: Artech House.

[46] Dong, Z. 1995. *Warships Command and Control System Theoretical Basis*. Beijing: National Defense Industry Press.

[47] Sun, Z., Zhou, Y., and He, L. 1996. *Active and Passive Positioning Technology by Single or Multiple Observers*. Beijing: National Defense Industry Press.

[48] Jing, Z. 1995. *Neural Networks-Based Tracking Theory with Applications*. Beijing: National Defense Industry Press.

[49] Sun, Z., Guo, F., Feng, D., *et al.* 2008. *Passive Location and Tracking Technology by Single Observer*. Beijing: National Defense Industry Press.

[50] Peng, D., Wen, C., and Xu, A. 2010. *Multi-sensor Multi-source Information Fusion Theory with Applications*. Beijing: Science Press.

[51] Xia, P. 2010. *Target Tracking and Information Fusion*. Beijing: National Defense Industry Press.

[52] He, Y. and Tang, J. 1996. Multiradar integrated tracking. *Journal of Electronics (in Chinese)*, 018(003).

[53] Wang, B., He, Y., Wang, G., and Xiu, J. 2010. Optimal allocation of multi-sensor passive localization. *Science China Information Sciences*, 53: 2514–2526.

[54] Wang, B., He, Y., Wang, G., and Xiu, J. 2011. Optimal allocation of multi-sensor passive localization. *Science China Information Sciences*, 41: 1251–1267.

[55] He, Y., Tan, Q., and Jiang, R. 1989. Multi-sensor integrated systems track correlation algorithm. *Fire Control & Command Control*, 1989(1): 1–12.

[56] He, Y., Peng, Y., Lu, D., and Wang, G. 1999. Track correlation algorithm based on fuzzy synthetic function. *Journal of Electronics* (in Chinese), 1999(1): 91–96.

[57] Wang, G. 2004. *Radar Network Key Technology Research* (Postdoctoral Research Report). Nanjing: Fourteenth Institute, CETC.

[58] He, Y. 1996. Distributed multi-target multi-sensor data fusion algorithm. Ph.D. thesis, Tsinghua University, Beijing.

[59] Yang, W. 2004. *Multi-sensor Data Fusion with Applications*. Xian: Xidian University Press.

[60] He, Y., Wang, G., Lu, D., and Peng, Y. 2007. *Multi-sensor Information Fusion with Application*, 2nd edn. Beijing: Publishing House of Electronics Industry.

[61] Hall, D.L. 1992. *Mathematical Techniques in Multisensor Data Fusion*. Boston, MA: Artech House.

[62] Bar-Shalom, Y. and Li, X.R. 1995. *Multitarget–Multisenser Tracking: Principles and Techniques*. Storrs, CT: YBS Publishing.

[63] Kang, Y. 1997. *Data Fusion Theory with Application*. Xian: Xidian University Press.

[64] Han, C., Zhu, H., and Duan, Z. 2006. *Multi-source Information Fusion*. Beijing: Tsinghua University Press.

[65] Ashraf, M.A. 2013. A new nearest-neighbor association approach based on fuzzy clustering. *Aerospace Science and Technology*, 26(1): 87–97.

[66] Guo, G., Zhuang, Z., and Chen, C. 1995. *Electromagnetic Feature Extraction and Object Recognition*. Changsha: National University of Defense Technology Press.

[67] Aidala, V.J. 1976. *Behavior of the Kalman Filter Applied to Bearings-Only Target Motion Analysis*. Naval Underwater Systems Center (NUSC), Technical Report 4984, November.

[68] Sorensor, W. 1985. *Kalman Filtering: Theory and Application*. New York: IEEE Press.

[69] Carlson, E.A. and Berarducci, M.P. 1994. Federated Kalman filter simulation results. *Navigation*, 41(3): 297–321.

[70] Hall, D.L., Linn, R.J., and Lins, J. 1991. A survey of data fusion systems. Proceedings of SPIE Conference on Data Structure and Target Classification, Orlando, FL, April, Vol. 1470, pp. 13–36.

[71] Gong, Y. 2003. *Adaptive Filtering*. Beijing: Publishing House of Electronics Industry.

[72] Zhang, J. 2008. The algorithm study of multi-sensor multi-target tracking. Ph.D. thesis, Naval Aeronautical and Astronautical University, Yantai.

[73] He, Y., Wang, B., Wang, G., and Xiu, J. 2010. A clustering localization algorithm with adaptive threshold in passive sensor network. *Journal of Astronautics*, 31(4): 1125–1130.

[74] Su, W. 2009. Single or multiple station passive direction finding crossover positioning technology research. Master's thesis, Naval Aeronautical and Astronautical University, Yantai.

[75] Wang, B., He, Y., Wang, G., and Xiu, J. 2010. Dual station passive: average location algorithm precision analysis. *Journal of Sichuan Ordnance*, 31(4): 78–81.

[76] Hu, L. 2004. *Passive Locating*. Beijing: National Defense Industry Press.

[77] Zhou, Y., Jia, Y., and Wang, H. 1999. Passive radar. Proceedings of 7th Annual National Radar Conference, pp. 60–63.

[78] Xiu, J., He, Y., and Wang, G. 2000. The realization of passive detector for multiple target tracking. *Ship Electronic Engineering*, 2000(1): 25–28.

[79] Cheng, Y., Pan, Q., Zhang, H., *et al.* 2003. Multistation passive fusion tracking based on extended Kalman filter. *Journal of System Simulation*, 15(4): 548–550.

[80] Li, S., Zeng, T., Long, T., *et al.* 2002. Improvement on passive location algorithm based on extended Kalman filter. *Journal of Beijing Institute of Technology*, 22(4): 521–524.

[81] Zhou, D., Hu, Z., and Wu, H. 1997. Adaptive extended Kalman filtering for passive missile guidance problem. *Journal of Astronautics*, 18(4): 31–36.

[82] Zhang, H., Zhang, Y., and He, Z. 1992. A robust adaptive extended Kalman filter and its application to flight state estimation. *Information and Control*, 21(6): 343–348.

[83] Hu, X., Wang, C., and Guo, Z. 1995. Extended Kalman filter in flight target high angle and low angle estimation. IEEE International Radar Conference, pp. 803–807.

[84] Song, T.L. and Speyer, J.L. 1985. A stochastic analysis of a modified gain extended Kalman filter with application to estimation with bearing only measurements. *IEEE Transactions on Automatic Control*, AC-30: 940–949.

[85] Wang, G., Mao, S., and He, Y. 2002. Optimal unbiased converted measurement Kalman filtering in the mean-square sense. *Journal of System Simulation*, 14(1): 119–122.

[86] Wang, G. and He, Y. 1999. Unbiased converted measurement covariance minimum mean square error estimation. Beijing: Aviation Institute of Electrical Branch Annual Meeting.

[87] Phanenf, R.J. 1968. Approximate nonlinear estimation. Ph.D. thesis, MIT, Cambridge, MA.

[88] Julier, S.J. and Uhlmann, J.K. 2000. A new method for the nonlinear transformation of means and covariances in filters and estimators. *IEEE Transactions on Aerospace and Electronic Systems*, 45(3): 477–482.
[89] Xiong, W., Chen, L., He, Y., and Zhang, J. 2007. Unscented Kalman filter with colored noise. *Journal of Electronics & Information Technology*, 29(3): 598–600.
[90] Wang, G., Xiu, J., and He, Y. 2004. An unbiased transform based Kalman filter for 3D radar. *Chinese Journal of Electronics*, 2004(4): 697–700.
[91] Xiong, W., Zhang, J., and He, Y. 2004. A debiased unscented transform based Kalman filter. Proceedings of International Radar Conference, France.
[92] Merwe, R. and Wan, E.A. 2001. Efficient derivative-free Kalman filters for online learning. European Symposium on Artificial Neural Networks, pp. 205–210.
[93] VanDyke, M.C., Schwartz, J.L., and Hall, C.D. 2004. Unscented Kalman filtering for spacecraft attitude state and parameter estimation. AAS/AIAA Space Flight Mechanics Conference.
[94] Wang, S., Cheng, Y., Yang, D., and Cui, H.T. 2003. UKF and its application to bearings-only tracking problem. *Flight Dynamics*, 2(2): 59–62.
[95] Gordon, N. and Salmond, J. 1993. Novel approach to nonlinear/non-Gaussian Bayesian state estimation. *IEE Proceedings on Radar and Signal Processing*, 140(2): 107–113.
[96] Wan, E.A. and van der Merwe, R. 2000. The unscented Kalman filter for nonlinear estimation. Proceedings of IEEE Symposium 2000 (AS-SPCC), Lake Louise, Alberta, Canada, October, pp. 153–158.
[97] Carvalho, H. and Del Moral, P. 1997. Optimal nonlinear filtering in GPS/INS integration. *IEEE Transactions on Aerospace and Electronic Systems*, 33(3): 835–849.
[98] Pitt, M. and Shephard, N. 1999. Filtering via simulation: auxiliary particle filters. *Journal of the American Statistical Association*, 94(446): 590–599.
[99] Carpenter, J., Clifford, P., and Fearnhead, P. 1999. An improved particle filter for non-linear problems. *IEE Proceedings of Radar, Sonar and Navigation*, 146(1): 2–7.
[100] van der Merwe, R. and Doucet, A. 2000. The Unscented Particle Filter. Technical Report CUED/F- INFENG/TR 380, Cambridge University, Cambridge, 2000.
[101] Karlsson, R. and Bergman, N. 2000. Auxiliary particle filters for tracking a maneuvering target. Proceedings of the 39th IEEE Conference on Decision & Control, 4: 3891–3895.
[102] Doucet, A., Gordon, N., and Krishnamurthy, V. 2001. Particle filters for state estimation of jump Markov linear systems. *IEEE Transactions on Signal Processing*, 49(3): 613–624.
[103] Iba, Y. 2001. Population Monte Carlo algorithms. *Transactions of the Japanese Society for Artificial Intelligence*, 16(2): 279–286.
[104] Doucet, A., Freitas, N.D., and Gordon, N. 2001. *Sequential Monte Carlo Methods in Practice*. Berlin: Springer-Verlag.
[105] Orton, M. and Marrs, A. 2001. Incorporation of Out-of-Sequence Measurements in Non-Linear Dynamic Systems Using Particle Filters. Technical Report.
[106] Herman, S.M. 2002. *A particle filtering approach to joint passive radar tracking and target classification*. Ph.D. thesis, University of Illinois, Urbana, IL.
[107] Sanjeev Arulampalam, M., Maskell, S., and Gordon, N. 2002. A tutorial on particle filters for online nonlinear/non-Gaussian Bayesian tracking. *IEEE Transactions on Aerospace and Electronic Systems*, 55(2): 174–188.
[108] Farina, A. and Ristic, B. 2002. Tracking a ballistic target: comparison of several nonlinear filters. *IEEE Transactions on Aerospace and Electronic Systems*, 38(3): 477–482.
[109] Gustafsson, F., Gunnarsson, F., Bergman, N., and Forssell, U. 2002. Particle filters for positioning, navigation and tracking. *IEEE Transactions on Signal Processing*, 50(2): 425–437.
[110] Karlsson, R. 2002. *Various Topics on Angle-Only Tracking Using Particle Filters*. Technical Report LiTH-ISY-R-2473.
[111] Kim, S.J. and Iltis, R.A. 2002. Performance comparison of particle and extended Kalman filter algorithms for GPS C/A code tracking and interference rejection. Conference on Information Sciences and Systems, Princeton University, Princeton, NJ.
[112] de Freitas, N. 2002. Rao-Blackwellised particle filtering for fault diagnosis. IEEE Aerospace Conference Proceedings, Vol. 4, pp. 1767–1772.
[113] Doucet, A., Godsill, S., and Andrieu, C. 2003. On sequential Monte Carlo sampling methods for Bayesian filtering. *Statistics and Computing*, 10(3): 197–208.
[114] Karlsson, R., Gusfafsson, F., and Karlsson, T. 2003. Particle filtering and Cramer–Rao lower bound for underwater navigation. IEEE International Conference on Acoustics, Speech and Signal Processing, Vol. 6, pp. 65–68.
[115] Bruno, M.G.S. 2003. Sequential importance sampling filtering for target tracking in image sequences. *IEEE Transactions on Signal Processing*, 10(8): 246–249.

[116] Kwok, C., Fox, D., and Meila, M. 2004. Real-time particle filters. Proceedings of the IEEE, 92(3): 469–484.

[117] Bolic, M. 2004. Architectures for efficient implementation of particle filters. Ph.D. thesis, State University of New York at Stony Brook, Stony Brook, NY.

[118] Wei, X., Zhang, J., and He, Y. 2005. Multisensor multitarget tracking methods based on particle filter. Proceedings of International Symposium on Autonomous Decentralized Systems, pp. 306–309.

[119] Wei, X., Zhang, J., and He, Y. 2005. A new multisensor particle filter method. Proceedings of 5th International Conference on Machine Learning and Computing, Vol. 8, pp. 614–617.

[120] Wei, X., He, Y., and Zhang, J. 2005. Multisensor sequential particle filter. *Acta Electronica Sinica*, 33(6): 1116–1119.

[121] Zhao, R. and Gu, Q. 2000. New filtering algorithms with applications in navigation systems. *Tsinghua University Journal*, 40(5): 24–27.

[122] Yuan, Z., Zheng, N., and Jia, X. 2003. The Gauss–Hermite particle filter. *Acta Electronica Sinica*, 31(7): 970–973.

[123] Liu, J.S. and Chen, R. 1998. Sequential Monte Carlo methods for dynamical systems. *Journal of the American Statistical Association*, 93: 1032–1044.

[124] Kitagawa, G. 1996. Monte Carlo filter and smoother for non-Gaussian nonlinear state space models. *Journal of Computational and Graphical Statistics*, 5(1): 1–25.

[125] Yi, X., He, Y., and Guan, X. 2006. Study on a novel coordinate transform method. *Geomatics and Information Science of Wuhan University*, 31(3): 237–239.

[126] Deng, Z. and Sun, Z. 1994. Robust Kalman filtering insensitive to continuous outlier. *Journal of Tsinghua University*, 1994(1).

[127] Hu, S. and Sun, G. 1992. Statistical diagnosis method for outliers from spacecraft tracking data. *Journal of Astronautics*, 1999(2).

[128] Hu, F. and Sun, G. 1995. Fault-tolerant improvement on Kalman filter. *Acta Automatica Sinica*, 1999(5).

[129] Liu, C., Sun, F., Chen, X., and Cao, J. 2002. Fault tolerant on integrated navigation system when existing outliers. *Journal of Chinese Inertial Technology*, 2002(6).

[130] Bar-Shalom, Y. and Li, X.R. 1995. *Multitarget–Multisensor Tracking: Principles and Techniques*. Storrs, CT: YBS Publishing.

[131] Bai, J., Wang, G., Kong, M., *et al.* 2009. Study on data association methods for distributed passive sensors with long baseline. *Chinese Journal of Electronics*, 18(2): 270–274.

[132] Bai, J., Wang, G., Wang, N., *et al.* 2009. Study on optimum cut angles in bearing-only location systems. *Acta Aeronautica et Astronautica Sinica*, 30(2): 298–304.

[133] Taek, L.S. and Darko, M. 2012. Smoothing innovations and data association with IPDA. *Automatica*, 48(7): 1324–1329.

[134] Zhang, N. 2002. Application of coordinate conversion in radar real time display software. *Modern Radar*, 2002(5): 30–32.

[135] Wei, X. 2001. Surface warship formation combat system information fusion technology research. Master's thesis, Naval Aeronautical and Astronautical University, Yantai.

[136] Johson, G.W. 1974. Choice of coordinates and computational difficulty. *IEEE Transactions on Aerospace and Electronic Systems*, 19(2): 77–80.

[137] Cantrell, B.H., Grindlay, A., and Dodage, C.H. 1976. *Formulation of a Platform-to-Platform Radar Integration System*. NRL Memorandum Report 3404.

[138] Broida, T.J. 1991. Choice of coordinate systems for multiple sensor fusion. *SPIE*, 1611.

[139] Cheng, H. and Sun, Z. 1997. On the influence of coordinate transform upon measurement error of long-baseline distributed sensors system. *SPIE*, 3067: 136–145.

[140] He, Y., Wang, G., and Xiu, J. 2000. Redundant data compression and location accuracy analysis in T/R-R bistatical radar system. International Conference on Signal Processing, August, China, pp. 1951–1955.

[141] Xiu, J., He, Y., and Wang, G. 1999. Analysis on the feasibility and the positioning accuracy of bistatic radar redundant data compression. *Naval Aeronautical and Astronautical University*, 1999(3).

[142] Xiu, J., He, Y., Wang, G., and Yan, H. 1999. A theorem on bistatic radar redundant data compression. Proceedings of 7th Annual National Radar Conference, Nanjin, pp. 542–545.

[143] Wei, X. and He, Y. 2003. Research of different dimension process in one centralized multi-radar system. *Journal of System Simulation*, 15(6): 845–848.

[144] He, Y. and Wei, X. 2003. State estimation techniques for radars with different observation dimension in one distributed data fusion system with feedback information. *Journal of Astronautics*, 24(2): 156–161.

[145] Dong, Z. 1999. The track initiation method. *Information Command Control System and Simulation Technology*, 1999 (2): 1–6.

[146] Xing, F., Wei, X., and Wang, H. 2010. Dense multi-formation track initiation algorithm based on K-means clustering and Hough transform. *Journal of Naval Aeronautical and Astronautical University*, 25(6): 624–628.

[147] Farina, A. and Pardini, S. 1979. Multi-radar tracking system using radial velocity measurement. *IEEE Transactions on Aerospace and Electronic Systems*, 15(3): 555–562.

[148] He, Y. 1988. *The comprehensive algorithm study of multi-sensor multi-target*. Master's thesis, Naval University of Engineering, Wuhan.

[149] Su, F., Wang, G., and He, Y. 2004. Track initiation algorithm based on the modified logic. *Modern Defence Technology*, 32(5): 66–68.

[150] Su, F., Wang, G., and He, Y. 2001. A new fast track initiation method in clutter environments. Radar System Simulation and Aided Design Algorithm in Radar Technology Application Seminar, Chinese Institute of Electronics – Radar Branch, Yantai, Vol. 8, pp. 132–137.

[151] Duda, R.O. and Hart, P.E. 1972. Use of the Hough transformation to detect lines and curves in pictures. *Communications of the ACM*, 135: 11–15.

[152] Sklansky, J. 1978. On the Hough technique for curve detection. *IEEE Transactions on Computing*, 27(10): 923–926.

[153] Wang, G., Kong, M., and He, Y. 2005. Hough transform and its application in information processing. Beijing: Publishing House of Ordnance Industry.

[154] Carlson, B.D., Evans, E.D., and Wilson, S.L. 1995. Search radar detection and track with the Hough transform. Part I: System concept. *IEEE Transactions on Aerospace and Electronic Systems*, 30(1): 102–108.

[155] Carlson, B.D., Evans, E.D., and Wilson, S.L. 1995. Search radar detection and track with the Hough transform. Part II: Detection statistics. *IEEE Transactions on Aerospace and Electronic Systems*, 30(1): 109–115.

[156] Carlson, B.D., Evans, E.D., and Wilson, S.L. 1995. Search radar detection and track with the Hough transform. Part III: Detection performance with binary integration. *IEEE Transactions on Aerospace and Electronic Systems*, 30(1): 116–124.

[157] Casasent, D.P. and Slaski, J. 1988. Optical track initiator for multitarget tracking. *Applied Optics*, 22: 4546–4553.

[158] Chen, J., Leung, H., Lo, T., *et al.* 1996. A modified probabilistic data association filter in real clutter environment. *IEEE Transactions on Aerospace and Electronic Systems*, 32: 300–314.

[159] Leung, H., Hu, Z., and Blanchette, M. 1996. Evaluation of multiple target track initiation techniques in real radar tracking environments. *IEE Radar Proceedings, Sonar Navigation*, 143(4): 246–254.

[160] Cai, Q., Xue, Y., and Zhang, B. 1997. *Phased Array Radar Data Processing and Its Simulation Technologies*. Beijing: National Defense Industry Press.

[161] Xiu, J., He, Y., and Jiu, J. 2012. Study on passive tracking algorithm of targets in clutter. *Systems Engineering and Electronics*, 34(2): 227–230.

[162] Zhang, J. 2004. Multi-sensor multi-target tracking algorithm performance comparison analysis and research. Master's thesis, Naval Aeronautical and Astronautical University, Yantai.

[163] Smith, P. and Buechler, G. 1975. A branching algorithm for discriminating and tracking multiple objects. *IEEE Transactions on Automatic Control*, AC-20: 101–104.

[164] Morefield, C.L. 1974. Solution of multiple choice estimation problems via 0–1 integer programming. Proceedings of IEEE Conference on Decision & Control, November, pp. 753–754.

[165] Morefield, C.L. 1975. Efficient computational forms for Bayesian multitarget tracking. Proceedings of 6th Symposium on Nonlinear Estimation Theory and its Applications, September, pp. 208–216.

[166] Morefield, C.L. 1975. Application of integer programming to track assembly problems. Proceedings of IEEE Conference on Decision & Control, December, pp. 428–433.

[167] Morefield, C.L. 1976. Application of Bayesian decision theory to multitarget surveillance problems. Proceedings of National Aerospace Electronic Conference.

[168] Morefield, C.L. 1977. Application of 0–1 integer programming to multitarget tracking problems. *IEEE Transactions on Automatic Control*, AC-22: 302–312.

[169] Morefield, C.L. 1979. Decision directed multitarget tracking. Proceedings of IEEE Conference on Decision & Control, pp. 1197–1201.

[170] Mao, S., Zhang, R., Xu, W., *et al.* 1990. *Pulse Doppler Radar*. Beijing: National Defense Industry Press.

[171] Stein, J.J. and Blackman, S.S. 1975. Generalized correlation of multitarget track data. *IEEE Transactions on Aerospace and Electronic Systems*, 11(6): 1207–1217.

[172] He, Y., Song, Q., and Wei, X. 2010. Track alignment–correlation technique based on phase correlation. *Acta Electronica Sinica*, 38(12): 2718–2723.

[173] He, Y., Song, Q., and Wei, X. 2010. A track registration–correlation algorithm based on Fourier transform. *Acta Aeronautica et Astronautica Sinica*, 31(2): 356–362.

[174] Bar-Shalom, Y. 1992. *Multitarget–Multisensor Tracking: Advanced Application*, Vol. II. Decham, MA: Artech House.

[175] Yi, X., Guan, X., and He, Y. 2005. Gray track correlation model for distributed multitarget tracking system. *Signal Processing*, 21(6): 653–655.

[176] Gennari, G. and Hager, G.D. 2004. Probabilistic data association methods in visual tracking of groups. Proceedings of the 2004 IEEE Computer Society Conference on Computer Vision and Pattern Recognition, Vol. 2, pp. 876–881.
[177] van Keuk, G. 2002. MHT extraction and track maintenance of a target formation. *IEEE Transactions on Aerospace and Electronic Systems*, 38(1): 288–294.
[178] Gning, A. and Mihaylova, L. 2010. Ground target group structure and state estimation with particle filtering. IEEE International Conference on Information Fusion, pp. 1–8.
[179] Singer, R.A. and Sea, R.G. 1973. New results in optimizing surveillance system tracking and data correlation performance in dense multitarget environments. *IEEE Transactions on Automatic Control*, AC-18: 571–582.
[180] Zhang, J., Xiu, J., He, Y., and Wei, X. 2006. Based on the theory of D–S distributed interactive multi-sensor data association algorithm. *Science China*, 36(2): 182–190.
[181] Jin, G., Zhao, D., and Zhang, Y. 1991. Relevant target track. *Radar & Ecm*, 1991(2).
[182] Park, S.-T. and Lee, J.G. 2001. Improved Kalman filter design for three-dimensional radar tracking. *IEEE Transactions on Aerospace and Electronic Systems*, 37(2): 727–739.
[183] Papadimitriou, C.H. and Steiglitz, K. 1982. *Combinatorial Optimization: Algorithms and Complexity*. Englewood Cliffs, NJ: Prentice-Hall.
[184] Yang, C. and Liu, G. 1999. Multi-sensors for a long range target tracking. *Systems Engineering and Electronics*, 21(3): 42–47.
[185] Li, X.R. 1998. Tracking in clutter with strongest neighbor measurements. I. Theoretical analysis. *IEEE Transactions on Automatic Control*, 43(11): 1560–1578.
[186] Song, T.L., Lee, D.G., and Ryu, J. 2005. A probabilistic nearest neighbor filter algorithm for tracking in a clutter environment. *Signal Processing*, 85(10): 2044–2053.
[187] Wang, H.P., Wei, X., He, Y., and Wang, Y. 2010. Parallel centralized multisensor general association algorithm. *Chinese Journal of Scientific Instruments*, 31(11): 2500–2507.
[188] Suo, J. 1999. *Research on target tracking of radar*. Ph.D. thesis, Dalian Maritime University.
[189] Wang, G. 2002. The key technology research of multi-sensors information fusion. Ph.D. thesis, Aerospace University, Beijing.
[190] Yi, X., He, Y., and Guan, X. 2004. Cooperative location model under the nearest neighbor criterion. *IEEE Position Location and Navigation Symposium, Monterey, CA*, Vol. 4, pp. 658–661.
[191] Zhu, Z. 1993. Modern radar data processing technology – the development of multi-target tracking technology. Multiple Target Detection Technology to Track and Attack Album, China Leihua Electronic Technology Research Institute.
[192] Brookner, E. 1998. *Tracking and Kalman Filtering Made Easy*. New York: John Wiley & Sons.
[193] Ding, Z., Leung, H., and Hong, L. 1999. Decoupling joint probabilistic data association algorithm for multiple target tracking. *IEE Radar Proceedings, Sonar Navigation*, 146(5): 251–254.
[194] Song, X. and Sun, Z. 1999. Data association in multi-target tracking with multi-sensor. *Systems Engineering and Electronics*, 1999(1): 27–33.
[195] Wei, X., Zhang, J.W., and He, Y. 2004. Filtering algorithms for passive location. *Journal of Naval Aeronautical and Astronautical University*, 19(3): 309–311.
[196] Musicki, D. and Evans, R. 1994. Integrated probabilistic data association. *IEEE Transactions on Automatic Control*, 39(6): 1237–1241.
[197] Musicki, D. and Evans, R. 2004. Joint integrated probabilistic data association: JIPDA. *IEEE Transactions on Aerospace and Electronic Systems*, 40(3): 1093–1099.
[198] Chang, K.C., Chong, C.Y., and Bar-Shalom, Y. 1986. Joint probabilistic data association in distributed sensor networks. *IEEE Transactions on Automatic Control*, AC-31: 889–897.
[199] Bar-Shalom, Y. 1992. *Multitarget–Multisensor Tracking: Applications and Advances*. Norwood, MA: Artech House.
[200] Roecher, J.A. and Phillis, G.L. 1993. Suboptimal joint probabilistic data association. *IEEE Transactions on Aerospace Electronic Systems*, 29(2): 510–517.
[201] Zhou, B. and Bose, N.K. 1993. Multitarget tracking in clutter: fast algorithms for data association. *IEEE Transactions on Aerospace and Electronic Systems*, 29(2): 352–363.
[202] Xu, Y., Gao, X., Liu, Q., *et al.* 1993. Multi-target tracking and attack detection technology. China Lei Hua Institute of Technology.
[203] Roecker, J.A. 1999. A class of near optimal JPDA algorithms. *IEEE Transactions on Aerospace and Electronic Systems*, 30(2): 504–510.
[204] Bar-Shalom, Y. 1978. Tracking methods in multitarget environment. IEEE-AC, August.
[205] Zhang, J., Wei, X., and He, Y. 2004. Extended adjustable white noise model for maneuvering target. *Fire Control & Command Control*, 29(5): 28–30.

[206] Rago, C., Willett, P., and Streit, R. 1995. A comparison of the JPDA and PMHT algorithm. Proceedings of the International Conference on Acoustics, Speech, and Signal Processing, pp. 3571–3574.

[207] Houles, A. and Bar-Shalom, Y. 1989. Multisensor tracking of a maneuvering target in clutter. *IEEE Transactions on Aerospace and Electronic Systems*, 25: 176–188.

[208] Li, S. and Mao, S. 1993. Some problems in multi-target tracking. China Leihua Electronic Technology Research Institute.

[209] Chang, C.B., Whiting, R.H., and Athans, M. 1977. On the state and parameter estimation for maneuvering reentry vehicles. *IEEE Transactions on Automatic Control*, AC-22: 99–105.

[210] Khaloozadeh, H. and Karsaz, A. 2009. Modified input estimation technique for tracking maneuvering targets. *IET Proceedings on Radar, Sonar and Navigation*, 3(1): 30–41.

[211] Mazor, E., Dayan, J., and Bar-Shalom, Y. 1998. Interacting multiple model in target tracking: a survey. *IEEE Transactions on Aerospace and Electronic Systems*, 34(1): 103–124.

[212] Guu, J.A. and Wei, C.H. 1991. Tracking technique for maneuvering target with correlated measurement noises and unknown parameters. *IEE Proceedings-F*, 138(3): 278–288.

[213] Wei, X., Zhang, J., and He, Y. 2004. Extended current statistical model for maneuvering target. *Electronics, Optics & Control*, 11(2): 15–17.

[214] Birmiwal, K. and Bar-Shalom, Y. 1984. Maneuver target tracking a cluttered environment with a variable dimension filter. *IEEE Transactions on Aerospace and Electronics*, 20(9): 635–645.

[215] Hong, L. 1994. Multiresolutional multiple-modal target tracking. *IEEE Transactions on Aerospace and Electronics*, 30(2): 518–524.

[216] Li, X.R. and Bar-Shalom, Y. 1992. Model-set adaptation in multiple-model estimators for hybrid systems. American Control Conference, pp. 1794–1799.

[217] Li, X.R. and Bar-Shalom, Y. 1996. Multiple-model estimation with variable structure. *IEEE Transactions on Automatic Control*, AC-41: 478–439.

[218] Li, X.C. and Bar-Shalom, Y. 1993. Performance prediction of interacting multiple model algorithm. *IEEE Transactions on Automatic Control*, 29(3): 755–771.

[219] Rice, T.R. and Alouani, A.T. 1998. Multiple model filtering. SPIE Conference on Acquisition, Tracking and Pointing, Orlando, FL, Vol. 3365, pp. 100–112.

[220] Zhang, J., He, Y., and Wei, X. 2002. Modified current model algorithm for tracking maneuvering targets. Proceedings of 8th National Radar Conference, pp. 764–768.

[221] Fortmann, T.E., Bar-Shalom, Y., Scheffe, M., and Gelfand, S. 1981. Detection thresholds for multitarget tracking in clutter. Proceedings of 20th IEEE Conference on Decision & Control, December.

[222] Fortmann, T.E., Bar-Shalom, Y., Scheffe, M., and Gelfand, S. 1985. Detection thresholds for tracking in clutter – a connection between estimation and signal processing. *IEEE Transactions on Automatic Control*, AC-30: 221–228.

[223] Jazwinski, A. 1970. *Stochastic Processes and Filtering Theory*. New York: Academic Press.

[224] Chan, Y.T., Hu, A.G.C., and Plant, J.B. 1979. A Kalman filter based tracking scheme with input estimation. *IEEE Transactions on Aerospace and Electronics*, 15(2): 237–244.

[225] Mehrotra, K. and Mahapatra, P.R. 1997. A jerk model for tracking highly maneuvering targets. *IEEE Transactions on Aerospace and Electronics*, 33(4): 1094–1105.

[226] Munir, A., *et al.* 1995. Adaptive interacting multiple model algorithm for tracking a maneuvering target. *IEEE Proc-F*, 142(1): 11–16.

[227] Chen, L. and Tong, M. 2001. Interacting multiple model algorithm with neural networks. *Acta Aeronautica et Astronautica Sinica*, 22(1): 54–56.

[228] Gao, S. and Pan, Q. 1996. Efficient square-root covariance factorization algorithms for interacting multi-model filters. *Journal of Xi'an Institute of Technology*, 16(1): 20–25.

[229] Pan, Q., Liu, G., and Dai, G. 1999. Combined interacting multiple models probabilistic data association algorithm. *Acta Aeronautica et Astronautica Sinica*, 20(3): 234–238.

[230] Qiao, X. 2003. Study of target tracking techniques in information fusion system. Ph.D. thesis, Xidian University, Xian.

[231] Campo, L., Mookerjee, P., and Bar-Shalom, Y. 1991. State estimation for systems with a sojourn-time-dependent Markov switching model. *IEEE Transactions on Automatic Control*, 36(2): 238–243.

[232] Zhang, J., Wei, X., and He, Y. 2004. Analysis of algorithms for estimating a maneuvering target in clutter. *Fire Control & Command Control*, 29(4): 71–74.

[233] Wang, H. 2009. *The algorithm study of multi-sensor multi-target tracking*. Master's thesis, Naval Aeronautical and Astronautical University, Yantai.

[234] Jia, P., Wu, J., Wang, L., *et al.* 1991. Dense multi-target tracking methods. *Modern Defence Technology*, 1991(2): 49–55.

[235] Sun, C. and Yuan, T. 1995. 2D Sea search radar target tracking group. *Radar & Ecm*, 1995(4).

[236] Geng, W. 2008. Summarizing of group-target tracking. Proceedings of 10th National Radar Conference, Vol. 10, pp. 367–371.

[237] Taenzer, E. 1980. Tracking multiple targets simultaneously with a phased array radar. *IEEE Transactions on Aerospace and Electronic Systems*, 16(9): 604–614.

[238] Yang, C.-Y., Qu, J.-M., Mao, S.-Y., *et al.* 1995. An initialization method for group tracking. IEEE Electronics Conference, Dayton, OH, pp. 303–308.

[239] Clark, D. and Godsill, S. 2007. Group target tracking with the Gaussian mixture probability hypothesis density filter. Proceedings of the International Conference on Intelligent Sensors, Sensor Networks and Information Processing, Melbourne, AU, pp. 149–154.

[240] Zhou, D., Geng, W., and Ni, C. 2010. Study of track initiation method based on barycenter of formation target. *Radio Engineering*, 40(2): 32–34.

[241] Tang, Q., Huang, J., and Yang, X. 2007. Algorithm of track initiation and performance evaluation. *Journal of System Simulation*, 19(1): 149–152.

[242] Zhu, H.Y., Han, C., Han, H., *et al.* 2004. Study on approaches for track initiation. *Acta Aeronautica et Astronautica Sinica*, 25(3): 284–288.

[243] Tang, Q., Huang, J., and Yang, X. 2007. Algorithm of track initiation and performance evaluation. *Journal of System Simulation*, 19(1): 149–152 (in Chinese).

[244] Zhao, Z., Chao, R., Wang, X., *et al.* 2010. Multi-radar track initiation algorithm based on probabilistic grid Hough transform. *Acta Aeronautica et Astronautica Sinica*, 31(11): 2209–2215.

[245] Jin, S., Liang, Y., and Wang, Z. 2008. A two-hierarchy Hough transform based track initiation algorithm. *Acta Electronica Sinica*, 36(3): 590–593.

[246] Wang, H.-L., Wang, D.-S., Tian, L.-S., *et al.* 2006. A new algorithm for group tracking. ICR2001: 1159–1163.

[247] Geng, W.-D., Liu, H.-Y., *et al.* 2006. A study of Kalman-based algorithm for the maneuvering group-target tracking. ICR2001: 1211–1214.

[248] Frazier, A.P. and Scott, J.A. 1976. ATOMS-1: An Algorithm for Tracking of Moving Sets. Report No. ECOM-0510-4, AD-B015080L, System Planning Corporation, Arlington, VA, August.

[249] Binias, G. 1978. The formation tracking procedure for tracking in dense target environment. AGARD Conference Proceedings No. 252, Strategies for Automatic Track Initiation, Monterey, CA, Vol. 8, pp. 1–11.

[250] Flad, E.H. 1977. Tracking of formation flying aircraft. Proceedings of the IEE International Radar Conference, London, October, pp. 160–163.

[251] Taenzer, E. 1977. Tracking multiple targets simultaneously with a phased array radar. Proceedings of EASCON '77, Washington, D.C., September, pp. 10-6A–10-6R.

[252] Ramachandra, K.V. 1979. Multitarget Kalman tracking filter. *Electro-Technology (India)*, 23: 1–8.

[253] Binias, G. 1977. Computer controlled tracking in dense target environment using a phased array antenna. Proceedings of IEE International Radar Conference, London, pp. 155–159.

[254] Tou, J.T. and Gonzalez, R.C. 1974. *Pattern Recognition Principles*. London: Addison-Wesley.

[255] Song, Q. 2010. Study on target track alignment correlation and sensor system error estimation technology. Ph.D. thesis, Naval Aeronautical and Astronautical University, Yantai.

[256] Wei, X., He, Y., and Zhang, J. 2006. Particle filter method for a centralized multisensor system. In Yeung, D.S., Liu, Z.-Q., Wang, X.-Z., and Yan, H. (eds), *Advances in Machine Learning and Cybernetics*. Berlin: Springer Lecture Notes in Computer Science No. 39, pp. 64–69.

[257] Wei, X., Zhang, J., and He, Y. 2005. Multisensor joint probabilistic data association algorithm based on S–D assignments. *Journal of Tsinghua University*, 45(4): 452–455.

[258] Zhang, J., He, Y., and Wei, X. 2007. Centralized multisensor fuzzy joint probabilistic data association algorithm. *Journal of Tsinghua University*, 47(7): 1188–1192.

[259] Koch, W. and van Keuk, G. 1997. Multiple hypothesis track maintenance with possibly unresolved measurements. *IEEE Transactions on Aerospace and Electronic Systems*, 33(3): 883–892.

[260] Shea, P.J., Alexander, K., and Peterson, J. 2003. Group tracking using genetic algorithms. Proceedings of 6th International Conference on Information Fusion, pp. 680–687.

[261] Ferry, J.P. 2009. Group tracking on dynamic networks. Proceedings of 12th International Conference on Information Fusion, Seattle, WA, Vol. 7, pp. 930–937.

[262] Mori, S. and Chong, C.-Y. 2009. Tracking of groups of targets using generalized Janossy measure density function. IEEE International Conference on Radar, pp. 1–7.

[263] Mihaylova, L. 2009. Group object structure and state estimation in the presence of measurement origin uncertainty. Proceedings of 15th IEEE Workshop on Statistical Signal Processing, pp. 473–476.

[264] Feldmann, M. and Franken, D. 2009. Advances in tracking of extended objects and group targets using random matrices. Proceedings of 12th International Conference on Information Fusion, Seattle, WA, pp. 1029–1036.

[265] Lian, F., Han, C.Z., and Liu, W.F. 2010. Sequential Monte Carlo implementation and state extraction of the group probability hypothesis density filter for partly unresolvable group targets-tracking problem. *IET Radar, Sonar and Navigation*, 4(5): 685–702.

[266] Wei, X., Pan, X., and Peng, Y. 2010. Unscented bias estimation technique for maneuvering sensor. *Acta Aeronautica et Astronautica Sinica*, 31(4): 819–824.

[267] Wei, X., Xing, F., and Peng, Y. 2011. Bias estimation for moving sensors network using cooperation targets. *Systems Engineering and Electronics*, 33(3): 544–547.

[268] Dong, Y., He, Y., Wang, G., *et al.* 2006. Generalized least squares registration algorithm with earth-centered earth-fixed (ECEF) coordinate system. *Acta Aeronautica et Astronautica Sinica*, 27(3): 463–467.

[269] Cui, Y., Wei, X., and He, Y. 2012. Mobile platform sensor registration algorithm based on MLR. *Acta Aeronautica et Astronautica Sinica*, 33(1): 118–129.

[270] Xiong, W., Pan, X., Peng, Y., and He, Y. 2012. Height estimation in distributed 2-D radar network. *Sensor Letters*, 10: 1–5.

[271] Geng, W. 2007. A study on group target merging and splitting method based on PDA. *Radio Engineering*, 37(2): 24–26.

[272] Liu, H. and Geng, W. 2010. A study of merging and splitting methods for group targets based on pattern space. *Radio Engineering*, 40(2): 53–56.

[273] Septier, F. and Pang, S.K. 2009. Tracking of coordinated groups using marginalized MCMC-based particle algorithm. *IEEE Transactions on Automatic Control*, AC-31: 1–11.

[274] Tao, R., Deng, B., and Wang, Y. 2009. *Fractional Fourier Transform with Applications*. Beijing: Tsinghua University Press.

[275] Holmes, J.E. 1977. Development of algorithms for the formation and updating of tracks. Proceedings of IEE International Radar Conference, London, pp. 81–85.

[276] Fieskes, W. and Van Keuk, G. 1980. Adaptive control and tracking with the ELRA phased array radar experimental system. Proceedings of IEE International Radar Conference, Arlington, VA, pp. 8–13.

[277] Sea, R.G. 1973. Optimal correlation of sensor data with tracks in surveillance systems. Proceedings of the 6th International Conference on Systems Sciences, Honolulu, HI, pp. 424–426.

[278] Maged, Y.A. 1980. Critical probabilities for optimum tracking system. Proceedings of IEE International Radar Conference, Arlington, VA, pp. 330–335.

[279] Alspach, D.L. and Lobbia, R.N. 1979. A score for correct data association in multi-target tracking. Proceedings of IEEE Conference on Decision & Control, Fort Lauderdale, TX, pp. 389–393.

[280] Casner, P.G. and Prengaman, R.J. 1977. Integration and automation of multihole collocated radars. Proceedings of IEE International Radar Conference, London, pp. 145–149.

[281] Bath, W.G. 1982. False alarm control in automated radar surveillance systems. Proceedings of IEE International Radar Conference, London, pp. 71–75.

[282] Zhang, J., Xiu, J., He, Y., and Wei, X. 2006. Distributed interacted multisensor joint probabilistic data association algorithm based on D–S theory. *Science in China, Series F – Information Sciences*, 49(2): 219–227.

[283] Castella, F.R. 1976. Sliding window detection. *IEEE Transactions on Aerospace and Electronic Systems*, 12(6): 815–819.

[284] Hammer, D.E. 1976. Techniques for automatic target detection in scanning 3D radar. Proceedings of AGARD No. 197, The Hague, June.

[285] He, Y., Lu, D., and Peng, Y. 1999. Track file management techniques in distributed multisensor data fusion. *Modern Radar*, 1999(6): 65–70.

[286] He, Y., Huang, X., and Tang, X. 2000. Research object model information fusion system. *Ship Electronic Engineering*, 2000(4): 23–26.

[287] Tang, X., He, Y., and Huang, X. 2002. Management techniques of track ID in multi-sensor fusion system. *Fire Control & Command Control*, 27(4): 18–22.

[288] He, Y., Tang, J., and Wang, G. 1995. Optimal tracking-quality-management in multiradar tracking systems. *Modern Radar*, 1995(1): 14–19.

[289] Wan, Z. and Ping, H. 2001. Track file management in multi sensor data fusion. *Ship Electronic Engineering*, 125(5): 24–28.

[290] Zhai, W. and Dai, G. 2003. The analysis of passive radar in brief. *Electronic Warfare*, 2003(2): 41–46.

[291] He, Y., Xiu, J., Tang, X., and Dong, S. 2003. Location and tracking of the over-the-horizon target on TDOA. *Acta Electronica Sinica*, 31(12): 1917–1920.

[292] Liu, C., Yan, L., and Zhou, D. 2003. A survey of passive target tracking. *Modern Radar*, 25(9): 5–7.

[293] Dong, Z. 2003. Pure azimuth multi-station multi-target information fusion set theory described method. *Information Command Control System and Simulation Technology*, 2003(11): 20–32.

[294] Nardone, S.C., Lindgren, A.G., and Gong, K.F. 1984. Fundamental properties and performance of conventional bearing only target motion analysis. *IEEE Transactions on Automatic Control*, AC-29: 775–787.

[295] Passericux, J.M., Piuon, D., Blanc-Benon, P., *et al.* 1989. Target motion analysis with bearings and frequencies measurements via instrumental variable estimator. Proceedings of International Conference on Acoustics, Speech, and Signal Processing, pp. 2645–2652.

[296] Rajagopal, R., *et al.* 1997. Adaptive bearing estimation and tracking of multiple targets in a realistic passive sonar scenario. *SPIE*, 3086: 139–150.

[297] Hammel, S.E. and Aidaia, V.J. 1985. Observability requirements for three-dimensional tracking via angle measurements. *IEEE Transactions on Aerospace and Electronic Systems*, 21(2): 200–207.

[298] Song, T.L. 1996. Observability of target tracking with bearings-only measurements. *IEEE Transactions on Aerospace and Electronic Systems*, 32(4): 1468–1471.

[299] Koteswara Rao, S. 2001. Pseudo-linear estimator for bearings-only passive target tracking. *IEE Radar Proceedings, Sonar Navigation*, 148(1): 16–22.

[300] Kirubarajan, T., Bar-Shalom, Y., and Lerro, D. 2001. Bearings-only tracking of maneuvering targets using a batch-recursive estimator. *IEEE Transactions on Aerospace and Electronic Systems*, 37(3): 770–779.

[301] Wei, X. and Sun, Z. 1999. Passive location of fixed emitter using phase rate of change. *Systems Engineering and Electronics*, 21(3): 34–37.

[302] Wei, X., Sun, Z., and Zhou, Y. 1999. Passive location and observability analysis of moving target with phase rate of change measurements. *Systems Engineering and Electronics*, 21(8): 34–37.

[303] Becker, K. 1999. Passive localization of frequency-agile radars from angle and frequency measurements. *IEEE Transactions on Aerospace and Electronic Systems*, 35(4): 1129–1143.

[304] Guo, F., Gong, X., Feng, D., and Sun, Z. 2004. A location method in single platform passive radar using Doppler-shift changing rate. Proceedings of 9th Annual National Radar Conference, Vol. 8, pp. 717–720.

[305] Leung, D.S.P. and Williams, D.S. 1991. A multiple hypothesis based multiple sensor spatial data fusion algorithm. *SPIE Automatic Object Recognition*, 1471: 314–325.

[306] Yuan, G., Wu, Y., and Zhang, J. 2005. Fast track initiation algorithm for passive location system. *Journal of System Simulation*, 17(6): 1484–1486.

[307] Dunham, D.T. and Hutchins, R.G. 1997. Tracking multiple targets in cluttered environments with a probabilistic multi-hypothesis tracker. *SPIE*, 3086: 284–295.

[308] Xiu, J., He, Y., Wang, G., and Xiu, J. 2002. Study on multitarget tracking algorithm in two direction-finding location systems. *Acta Electronica Sinica*, 30(12): 1763–1767.

[309] Peach, N. 1995. Bearings-only tracking using a set of range-parameterised extended Kalman filters. *IEE Proceedings on Control Theory Application*, 142(1): 73–80.

[310] Kronhamn, T.R. 1998. Bearings-only target motion analysis based on a multihypothesis Kalman filter and adaptive ownership motion control. *IEE Radar Proceedings, Sonar Navigation*, 145(4): 247–252.

[311] Wang, G.H., Bai, J., He, Y., *et al.* 2009. Optimal deployment of multiple passive sensors in the sense of minimum concentration ellipse. *IET Radar, Sonar & Navigation*, 3(1): 8–17.

[312] Bai, J., Wang, G., Xiu, J., *et al.* 2009. New deghosting method based on generalized triangulation. *Journal of System Engineering and Electronics*, 20(3): 504–511.

[313] Wang, B., Wang, G., and He, Y. 2010. Clustering algorithm of passive sensor network for locating interference source. *Electronics, Optics & Control*, 17(3): 15–18.

[314] Xiu, J., He, Y., Wang, G., *et al.* 2005. Constellation of multisensors in bearing-only location system. *IEE Proceedings on Radar, Sonar and Navigation*, 152(3): 215–218.

[315] Weisstein, E.W. 1999. *Concise Encyclopedia of Mathematics*. Chapman & Hall/CRCnetBASE.Fan, L., Zhang, P., and Ji, H. 1999. The techniques of small signal detection and velocity tracking for pulse Doppler radar. *Journal of Electronics (in Chinese)*, 17(6): 607–612.

[316] Mellen, G., Pachter, M., and Raquet, J. 2003. Closed-form solution for determining emitter location using time difference of arrival measurements. *IEEE Transactions on Aerospace and Electronic Systems*, 39(3): 1056–1058.

[317] Chen, G. 1997. Analysis of fighting effectiveness for airborne PD fire-control radar. *Modern Radar*, 4: 21–24.

[318] Ben, D. 2000. The property of airborne pulsed Doppler radar and the methods of counter countermeasures. *Modern Radar*, 8: 1–6.

[319] Weber, P., Hayk, S., and Cray, R. 1985. Simultaneous resolution of unambiguous range and Doppler in a pulse-Doppler radar using multiple PRFs. *Proceedings of IEEE*, 73(6): 1144–1145.

[320] Trunk, G. and Brockett, S. 1993. Range and velocity ambiguity resolution. *Proceedings of IEEE National Rader Conference*, pp. 146–149.

[321] Skillman, W.A. and Mooney, D.H. 1986. Multiple high-PRF ranging. *CW and Doppler Radar*: 205–214.

[322] Redy, N.S. and Swamy, M.N. 1986. Resolution of range and Doppler ambiguities in medium PRF radar in multiple target environment. *Signal Processing*, 11(3): 223–236.
[323] Doviak, R.J. and Zrnic, D.S. 1993. *Doppler Radar and Weather Observations*. New York: Academic Press.
[324] Trungk, G. and Brockett, S. 1993. Range and velocity ambiguity resolution. IEEE National Radar Conference.
[325] Shnitkin, H.A. 1991. Unique joint STARS phased-array antenna. *Microwave Journal*, Jan: 131–141.
[326] Bar-Shalom, Y. and Negative, Y. 2001. Correlation and optical tracking with Doppler measurements. *IEEE Transactions on Aerospace and Electronic Systems*, 37(3): 1117–1120.
[327] Lei, M. and Han, C. 2007. Sequential nonlinear tracking using UKF and raw range-rate measurements. *IEEE Transactions on Aerospace and Electronic Systems*, 43(1): 239–250.
[328] Duan, Z., Li, X.R., Han, C., and Zhu, H. 2005. Sequential unscented Kalman filter for radar target tracking with range rate measurements. 8th International Conference on Information Fusion, pp. 130–137.
[329] Smith, M.A. 2008. On Doppler measurements for tracking. Proceedings of International Conference on Radar, pp. 513–518.
[330] Li, X.R., Zhao, Z., and Jilkov, V.P. 2002. Estimator's credibility and its measures. Proceedings of IFAC 15th World Congress, July.
[331] Zhang, G. 1994. *Phased Array Radar System*. Beijing: National Defense Industry Press.
[332] Wang, Z. 2000. The prospect of the development of phased array radar. *Lingbayi Technology*, 2000(1): 5–11.
[333] Zhang, Z. 1995. Development and prospects of phased array radar. *Modern Electronic*, 1995(3): 1–6.
[334] Zhang, G. 2000. Comparison between passive PARs and active PARs. *Modern Radar*, 22(4): 7–13.
[335] Hu, W., Yu, W., Lu, J., *et al.* 2010. *Theory and Method of Resource and Management for Phased Array Radars*. Beijing: National Defense Industry Press.
[336] Zhang, G. 2003. Effect of operation modes on phased array radar range. *Information and Electronic Engineering*, 2003 (1): 1–6.
[337] Navarro, A.M. 1977. *General Properties of Alpha–Beta and Gamma Tracking Filters*. Technical Report 77-24347, US Department of Commerce, National Technical Information Services, The Hague, January.
[338] Bar-Shalom, Y. 1997. Multitarget–multisensor tracking and fusion. Proceedings of National Radar Conference, May.
[339] Bar-Shalom, Y. 1978. Tracking methods in a multitarget environment. *IEEE Transactions on Automatic Control*, 24 (8): 618–626.
[340] Castella, F.R. 1981. Tracking accuracies with position and rate measurements. *IEEE Transactions on Aerospace and Electronic Systems*, 17(3): 433–438.
[341] Kural, F. 2010. Performance evaluation of track association and maintenance for a MFPAR with Doppler velocity measurements. *Progress in Electromagnetics Research*, 108: 249–275.
[342] Bar-Shalom, Y., Chang, K., and Blom, H. 1990. Automatic track formation in clutter with a recursive algorithm. In Bar-Shalom, Y. (ed.), *Multitarget–Multisensor Tracking*, Vol. 1. Norwood, MA: Artech House, pp. 25–42.
[343] Musicki, D. 2010. Doppler-aided target tracking in heavy clutter. Paper presented at the 13th International Conference on Information Fusion, Edinburgh, UK, July 26–29.
[344] Musicki, D. and La Scala, B. 2008. Multi-target tracking in clutter without measurement assignment. *IEEE Transactions on Aerospace and Electronic Systems*, 44(3), 877–896.
[345] Wang, X., Musicki, D., Ellem, R., and Fletcher, F. 2009. Efficient and enhanced multi-target tracking with Doppler measurements. *IEEE Transactions on Aerospace and Electronic Systems*, 45(4), 1400–1417.
[346] Shi, L., Wang, X., and Xiao, S. 2005. Adaptive data rate tracking of phased-array radar based on residue norm. *Shipboard Electronic Warfare*, 28(5): 45–47.
[347] Shin, H.J., Hong, S.M., and Hong, D.H. 1995. Adaptive-update-rate target tracking for phased-array radar. *IEE Proceedings on Radar, Sonar and Navigation*, 142(3): 137–143.
[348] van Keuk, G. 1987. Software structure and sampling strategy for automatic tracking with a phased array radar. Proceedings of AGARD Conference, Monterey, CA, Vol. 11, pp. 1–13.
[349] Watson, G.A. and Blair, W.D. 1994. Tracking performance of a phased array radar with revisit time controlled using the IMM algorithm. Proceedings of IEEE National Radar Conference, Atlanta, GA, pp. 160–165.
[350] Wang, F. 2002. Study on phased array radar adaptive resource scheduling. Master's thesis, Northwestern Polytechnical University, Xian.
[351] Baugh, R.A. 1973. *Computer Control of Modern Radars*. New York: RCA Corporation, pp. 37–54.
[352] Lu, J. 2007. Theory and method of resource optimization and management for phased array radars. Ph.D. thesis, National University of Defense Technology, Changsha.
[353] Mao, Y. 2011. Resource management under tracking mode and task scheduling for phased array. Master's thesis, Xidian University, Xian.
[354] Cheng, T. 2008. Research on adaptive resource management technology for phased array radar. Ph.D. thesis, University of Electronic Science and Technology of China, Chengdu.

[355] Liu, C.L. and Layland, J.W. 1973. Scheduling algorithms for multiprogramming in a hard real time environment. *Journal of the ACM*, 20(1): 44–61.
[356] Goossens, J. and Devillers, R. 1999. Feasibility intervals for the deadline driven scheduler with arbitrary deadlines. IEEE Conference on Real-Time Systems.
[357] Jeffay, K., Stanat, D.F., and Martel, C.U. 1991. On non-preemptive scheduling of periodic and sporadic tasks. Proceedings of the 12th IEEE Symposium on Real-Time Systems.
[358] Lu, J. 2007. Theory and method of resource optimization and management for phased array radars. Ph.D. thesis, National University of Defense Technology, Changsha.
[359] Huizing, A.G. and Bloemen, A.A.F. 1996. An efficient scheduling algorithm for a multifunction radar. IEEE International Symposium on Phased Array Systems and Technology, Boston, MA, October 15–18, pp. 359–364.
[360] Huizing, A.G. and Bosse, E. 1998. A high-level multifunction radar simulation for studying the performance of multi-sensor data fusion systems. Proceedings of SPIE on Signal Processing Sensor Fusion Target Recognition, Orlando, FL, Vol. 3374, pp. 129–138.
[361] Mao, Y. 2011. Resource management under tracking mode and task scheduling for phased array. Master's thesis, Xidian University, Xian.
[362] Farina, A. and Neri, P. 1980. Multitarget interleaved tracking for phased array radar. *IEE Proceedings, Part F: Communication, Radar, and Signal Processing*, 127(4): 312–318.
[363] Orman, A.J., Potts, C.N., Shahani, A.K., *et al.* 1996. Scheduling for a multifunction phased array radar system. *European Journal of Operational Research*, 90(1): 13–25.
[364] Dana, A.M.P. 1990. Registration: a prerequisite for multiple sensor tracking. In Bar-Shalom, Y. (ed.), *Multitarget–Multisensor Tracking: Advanced Applications*. Norwood, MA: Artech House, pp. 155–185.
[365] Zhou, Y., Henry, L., and Martin, B. 1993. Sensor alignment with earth-centered earth-fixed (ECEF) coordinate system. *IEEE Transactions on Aerospace and Electronic Systems*, 35(2): 410–417.
[366] Burke, J.J. 1996. *The SAGE Real Time Quality Control Function and its Interface with BUIC II/BUIC III*. MITRE Corporation Technical Report No. 308, November.
[367] Leung, H. and Blanchett, M.A. 1994. Least square fusion of multiple radar data. Proceedings of IEEE International Radar Conference, Paris.
[368] Zhou, Y. and Henry, L. 1997. An exact maximum likelihood registration algorithm for data fusion. *IEEE Transactions on Signal Processing*, 45(6): 1560–1572.
[369] Okello, N.N. and Ristic, B. 2003. Maximum likelihood registration for multiple dissimilar sensors. *IEEE Transactions on Aerospace and Electronic Systems*, 39(3), 1074–1083.
[370] Li, N. 2000. Study on the assessment of efficacy and cost-effectiveness ratio of surveillance radar net. *Modern Radar*, 2000(2).
[371] Xie, H., Yang, Z., and He, F. 2001. Study on the modeling of anti-jamming ability of ground surveillance radar net. *Electronic Warfare*, 2001(18).
[372] He, Y., Xiu, J., Wang, G., and Xiu, J. 2003. Theorem for the combination of bistatic radar measurements using least squares. *IEEE Transactions on Aerospace and Electronic Systems*, 39(4): 1441–1445.
[373] Zhang, X. 2010. Conception of MIMO radar countermeasures. *Modern Radar*, 32(4): 1–4.
[374] Li, S., Zhang, L., Chen, J., *et al.* 2010. Two-dimensional DOA estimation for orthogonal MIMO radar based on complete complementary sequence. *Chinese Journal of Radio Science*, 25(4): 617–624.
[375] Ditzler, W.R. 1987. A demonstration of multisensor tracking. Proceedings of the Tri-Service Data Fusion Symposium, June, pp. 303–311.
[376] He, Y., Peng, Y., and Lu, D. 1996. Survey of multisensor data fusion models. *Journal of Tsinghua University*, 1996(9): 14–20.
[377] He, Y., Lu, D., Peng, Y., and Gao, Z. 1997. Two new track correlation algorithms in a multisensor data fusion system. *Acta Electronica Sinica*, 1997(9): 10–14.
[378] Bar-Shalom, Y. 1981. On the track-to-track correlation problem. *IEEE Transactions on Automatic Control*, AC-26: 571–572.
[379] Wei, X., Zhang, J., and He, Y. 2010. Track correlation algorithm based on multi-dimension assignment and gray theory. *Journal of Electronics & Information Technology*, 32(4): 898–901.
[380] Chang, C.B. and Youens, L.C. 1982. Measurement correlation for multiple sensor tracking in a dense target environment. *IEEE Transactions on Automatic Control*, AC-27: 1250–1252.
[381] He, Y., Peng, Y., Lu, D., and Gao, Z. 1997. Binary track correlation algorithms in a distributed multisensor data fusion system. *Journal of Electronics* (in Chinese), 1997(6): 721–728.
[382] Kosaka, M., Miyamoto, S., and Ihara, H. 1983. A track correlation algorithm for multisensor integration. Proceedings of the IEEE/AIAA 5th Digital Avionics Systems Conference, Vol. 10, pp. 1–8.

[383] Wang, G. and He, Y. 1997. Radar-to-ESM correlation based on fuzzy synthetic function and statistical hypothesis testing. *Systems Engineering and Electronics*, (4): 13–16.
[384] Liu, G., Wang, G., and He, Y. 1994. Multiple radar track is fuzzy in the relevant calculation model and simulation comparison. *Fire Control Radar Technology*, 1994(3): 13–16.
[385] He, Y., Huang, X. 1999. Track correlation algorithms based on fuzzy synthetic decision. *Journal of Naval University of Engineering*, 1999(4).
[386] Wilson, J.F. 1997. *A fuzzy logic multisensor association algorithm. SPIE*, 3068: 76–87.
[387] Tummala, M., Glem, I., and Midwood, S. 1996. *Multisensor Data Fusion for the Vessel Traffic System*. NPS EC-96-055.
[388] Tummala, M. and Midwood, S. 1998. *A Fuzzy Associative Data Fusion Algorithm for Vessel Traffic System*. NPS EC-98-004.
[389] Kim, K.H. 1994. Development of track to track fusion algorithms. Proceedings of the American Control Conference, Baltimore, MD, June, pp. 1037–1041.
[390] He, Y., Lu, D., and Peng, Y. 1998. Fuzzy track correlation algorithms for multitarget and multisensor tracking. *Acta Electronica Sinica*, 26(3): 15–19.
[391] He, Y. and Huang, X. 1999. Distributed multi factor fuzzy comprehensive decision track correlation algorithm. Proceedings of 7th Annual National Radar Conference, Nanjing, pp. 417–420.
[392] Guan, X., He, Y., and Yi, X. 2006. Gray track-to-track association algorithm for distributed multitarget tracking system. *Signal Processing*, 86(11).
[393] Kang, Y. 2006. *Data Fusion Theory with Application*, 2nd edn. Xidian University Press, Xian.
[394] Gong, Y., Yang, H., Hu, W., *et al.* 2006. Performance evaluation of tracking fusion systems. The performance of fire and command control. *Fire Control & Command Control*, 31(9): 4–7.
[395] Meng, Z. and Zhang, N. 2002. Evaluation of six tracker actual radar tracking environment. *Information Command Control System and Simulation Technology*, 12: 27–32.
[396] Lian, X. 2003. Radar multi-target tracker and its evaluation criteria. *Radar & Ecm*, 3: 23–25.
[397] Chen, H. 2002. *Performance Evaluation of Multitarget Tracking Algorithms*. University of Connecticut, Storrs, CT.
[398] Alouani, A.T., Gray, J.E., and McCabe, D.H. 2003. Performance evaluation of an asynchronous multisensor track fusion filter. *Proceedings of SPIE*, 2003: 1–12.
[399] Lian, X. 2003. Radar multi-target tracker and its evaluation criteria. *Radar & Ecm*, 3: 23–25.
[400] Manson, K., *et al.* 1992. Taxonomic performance evaluation for multitarget tracking systems. *IEEE Transactions on Aerospace and Electronic Systems*, 28(3): 775–787.
[401] Wang, G. 2006. *Study on Distributed Detection, Tracking, and Heterogeneous Sensor Association and Cueing*. Beijing: Higher Education Press.
[402] He, Y., Peng, Y., and Lu, D. 1997. New track correlation algorithm for multitarget and multisensor tracking. *Journal of Tsinghua University*, (9): 108–113.
[403] Yan, H., Wang, X., and Wang, G. 2008. Study on performance evaluation for tracking filter algorithms. *Modern Radar*, 30(4): 33–36.
[404] Wang, H. 1994. *The Principle and Application of Computer Simulation*. Changsha: National University of Defense Technology Press.
[405] Gu, Q. 1995. *Application of Simulation Technology*. Beijing: National Defense Industry Press.
[406] Hall, D.L. and Llinas, J. 1997. An introduction to multisensor data fusion. *IEEE Transactions on Aerospace and Electronic Systems*, 85(1): 6–23.
[407] Kokar, K., Bedworth, M., and Frankel, C. 2000. A reference model for data fusion systems. In *Sensor Fusion: Architectures, Algorithms and Applications IV*, Proceedings of SPIE No. 4051.
[408] Wang, J. and Luo, J. 2004. Data association algorithm based on fuzzy synthetic evaluation of multiple features in multi-target passive tracking. *Acta Aeronautica et Astronautica Sinica*, 25(2): 172–176.
[409] Liu, J., Zhu, W., and Li, D. 2001. Tracks correlation decision method based on bearing measures for heterogeneous sensors. *Journal of Qingdao University*, 16(3): 8–11.
[410] Lin, L., Kirubarajan, T., and Bar-Shalom, Y. 2002. New assignment-based data association for tracking move–stop–move targets. Proceedings of 5th International Conference on Information Fusion.
[411] Su, F. 2001. Detection and tracking based on Hough transform. Master's thesis, Naval Aeronautical and Astronautical University, Yantai.
[412] Pawlak, R.J. and Beex, A.A. 1995. Fusion technique for multisensor track initiation. *IEE Radar Proceedings, Sonar Navigation*, 142(5).
[413] Tuveny, O., Artaud, M., Tomasimi, B., and Alengrin, G. 1993. Track initiation within a multi-sensor environment radar/IR. Part 2: Evaluation and multisensor approach. In *Signal Process, Sensor Fusion and Target Recognition*, Proceedings of SPIE No. 1956.

[414] Johnston, S.L. 1997. An efficient decentralized multiradar multitarget tracker for air surveillance. *IEEE Transactions on Aerospace and Electronic Systems*, 33(4): 1357–1363.
[415] Furcolo, B., Spatola, A., and Tarantino, M. 1983. SATCAS-80: a new generation of air traffic control systems. *Alta Frequenza*, Vol. LII, No. 5.
[416] Grasso, G., Paoli, L., and Pardini, S. 1975. Comparison between a manual and an automatic collision avoidance system. *Alta Frequenza*, Vol. XLIV, No. 2.
[417] Liu, W. and Hu, M. 2005. ATC multiradar data processing system. *Jiangsu Aviation*, 2005(3): 17–19.
[418] Shu, X., Cui, D., and Jiang, W. 2007. Multi-radar data processing in air traffic control command monitoring system. *Journal of Computer Applications*, 28(4): 337–342.
[419] He, M. 2002. Data fusion of multi-sensor target tracking technology research. Ph.D. thesis, National University of Defense Technology, Changsha.
[420] Mazor, E., Averbuch, A., and Bar-Shalom, Y. 1998. Interacting multiple model methods in target tracking: a survey. *IEEE Transactions on Aerospace and Electronic Systems*, 34(1): 103–123.
[421] Desai, U.B. and Das, B. 1985. Parallel algorithms for Kalman filtering. Proceedings of the American Control Conference, pp. 920–921.
[422] Meyer, G.G. and Weinert, H.W. 1984. Parallel algorithms and computational structures for linear estimation problems. *Statistical Signal Processing*: 507–516.
[423] Bar-Shalom, Y., Li, X.R., and Kirubarajan, T. 2001. *Estimation with Applications to Tracking and Navigation: Theory, Algorithms and Software.* New York: John Wiley & Sons.
[424] Lobbia, R. and Kent, M. 1994. Data fusion of decentralized local tracker outputs. *IEEE Transactions on Aerospace and Electronic Systems*, 30(3).
[425] You, X. 2008. *The Typical C4I and Weapon Systems of Foreign Navy.* Beijing: National Defense Industry Press.
[426] Zhu, L., Feng, C., and Zhang, Y. 2005. Key technologies analysis of airborne early warning system. *Information Command Control System and Simulation Technology*, 27(5): 67–70.
[427] Zhang, J., Dang, L., and Diao, H. 2010. Study on the key problem of air defense fire control radar network data fusion. *Science and Technology Information*, 1: 31.
[428] Deb, S., Pattipati, K.R., and Bar-Shalom, Y. 1993. A multisensor multitarget data association algorithm for heterogeneous sensors. *IEEE Transactions on Aerospace and Electronic Systems*, 29(2).
[429] Bath, W.G. 1982. Association of multisite radar data in the presence of large navigation and sensor alignment errors. IEE International Conference on Radar.
[430] Keneic, R.J. 1993. Local and remote track file registration using minimum description length. *IEEE Transactions on Aerospace and Electronic Systems*, 29(3): 245–249.
[431] Castella, F.R. 1995. Theoretical performance of a multisensor track-to-track correlation technique. *IEE Radar Proceedings, Sonar Navigation*, 142(6): 281-285.
[432] Lin, Y. 2003. Study on multiple moving targets passive tracking and data association algorithms. Ph.D. thesis, Zhejiang University, Hangzhou.
[433] Han, C., Zhu, H., Duan, Z., *et al.* 2010. *Multisource Information Fusion*, 2nd edn. Beijing: Tsinghua University Press.
[434] Butler, J.M. 1998. Tracking and control in multi-function radar. Ph.D. thesis, University of London.

Index

www.ingramcontent.com/pod-product-compliance
Ingram Content Group UK Ltd.
Pitfield, Milton Keynes, MK11 3LW, UK
UKHW050942290926
15470UKWH00005B/1591

9 781118 956861